Smart Innovation, Systems and Technologies

Volume 174

The Smart Innovation, Systems and Technologies book series encompasses the topics of knowledge, intelligence, innovation and sustainability. The aim of the series is to make available a platform for the publication of books on all aspects of single and multi-disciplinary research on these themes in order to make the latest results available in a readily-accessible form. Volumes on interdisciplinary research combining two or more of these areas is particularly sought.

The series covers systems and paradigms that employ knowledge and intelligence in a broad sense. Its scope is systems having embedded knowledge and intelligence, which may be applied to the solution of world problems in industry, the environment and the community. It also focusses on the knowledge-transfer methodologies and innovation strategies employed to make this happen effectively. The combination of intelligent systems tools and a broad range of applications introduces a need for a synergy of disciplines from science, technology, business and the humanities. The series will include conference proceedings, edited collections, monographs, handbooks, reference books, and other relevant types of book in areas of science and technology where smart systems and technologies can offer innovative solutions.

High quality content is an essential feature for all book proposals accepted for the series. It is expected that editors of all accepted volumes will ensure that contributions are subjected to an appropriate level of reviewing process and adhere to KES quality principles.

**** Indexing: The books of this series are submitted to ISI Proceedings, EI-Compendex, SCOPUS, Google Scholar and Springerlink ****

More information about this series at http://www.springer.com/series/8767

Sanjay Yadav · D. B. Singh ·
P. K. Arora · Harish Kumar
Editors

Proceedings of International Conference in Mechanical and Energy Technology

ICMET 2019, India

Springer

Editors
Sanjay Yadav
Physico-Mechanical Metrology
CSIR–National Physical Laboratory
New Delhi, India

D. B. Singh
Department of Mechanical Engineering
Graphic Era Deemed to be University
Dehradun, Uttarakhand, India

P. K. Arora
Galgotias College of Engineering
and Technology
Greater Noida, Uttar Pradesh, India

Harish Kumar
Department of Mechanical Engineering
National Institute of Technology
Delhi, India

ISSN 2190-3018 ISSN 2190-3026 (electronic)
Smart Innovation, Systems and Technologies
ISBN 978-981-15-2646-6 ISBN 978-981-15-2647-3 (eBook)
https://doi.org/10.1007/978-981-15-2647-3

This Springer imprint is published by the registered company Springer Nature Singapore Pte Ltd.
The registered company address is: 152 Beach Road, #21-01/04 Gateway East, Singapore 189721, Singapore

Preface

International Conference on Mechanical and Energy Technologies (**ICMET-2019**) has been the first conference of its series organized by the Department of Mechanical Engineering of Galgotias College of Engineering and Technology, Greater Noida, Uttar Pradesh, India. It was considered as a forum to bring together scientists, university professors, graduate students, and mechanical engineers, presenting new science, technology, and engineering ideas and achievements.

The motive of this conference was of great importance. The time and location of the conference were quite excellent. The month of November in Greater Noida is very pleasant with the advantage of proximity with Delhi as the capital of the country with excellent connectivity. The conference attracted many participants working in various fields of engineering: design, mechanics, materials, energy, etc. The success of the conference inspired the organizers to turn the conference into a regular event.

More than 270 manuscripts were submitted to the conference, topics ranging from the latest research in the field of aerodynamics and fluid mechanics, artificial intelligence, rapid manufacturing technologies, remanufacturing, refrigeration and air conditioning, renewable energies technology, I.C engines, turbo machinery metrology, and computer-aided design and inspection, etc.

Furthermore, we thank the management and director of Galgotias College of Engineering and Technology, Greater Noida, India, for their cooperation and support. We are thankful to all the members of the organizing committee for their contribution in organizing the conference. Last but not least, we thank Springer for its professional assistance and particularly Mr. Aninda Bose who supported this publication.

New Delhi, India — Sanjay Yadav
Dehradun, India — D. B. Singh
Greater Noida, India — P. K. Arora
Delhi, India — Harish Kumar

Contents

About the Editors

Sanjay Yadav is a Senior Principal Scientist & Head of the Pressure, Vacuum and Ultrasonic Metrology Section and Prof. Academy of Scientific and Innovative Research (AcSIR), National Physical Laboratory, New Delhi, India. He received his Ph.D. in Physics from Meerut University in 1990. He has published more than 270 papers in international journals and conference proceedings and is the Editor-in-Chief of the journal "MAPAN/Journal of Metrology Society of India," published by Springer.

D. B. Singh is a Professor at Mechanical Engineering Department of Graphic Era Deemed to be University, Dehradun, India. Earlier he was working as Professor at Mechanical Engineering Department of Galgotias College of Engineering and Technology, Greater Noida, India. He has 17 years of teaching experience and received his Ph.D. in Solar Thermal Energy from the IIT Delhi in 2017. Dr. Singh has published more than 40 papers in international journals and conference proceedings. He has also published one book chapter in the "CRC Handbook of Thermal Engineering, Second Edition", published by Taylor and Francis.

P. K. Arora is a Professor at the Mechanical Engineering Department of Galgotias College of Engineering and Technology, Greater Noida, India. He received his Ph.D. in Mechanical Engineering from Jamia Millia Islamia, New Delhi, in 2014. With 16 years of academic experience, his main research areas are industrial engineering and optimization. Dr. Arora has published more than 30 papers in international journals and conference proceedings, as well as two book chapters.

Harish Kumar is currently working as an Assistant Professor at the National Institute of Technology Delhi. With more than 15 years of research and academic experience, he is an active researcher in the area of mechanical measurement and metrology. He served as a scientist at the National Physical Laboratory, India, for more than a decade, and as a Guest Researcher at the National Institute of Standards and Technology, USA, in 2016. He has authored more than 50 publications in peer-reviewed journals and conference proceedings; is an active reviewer for many prominent journals on measurement, metrology and related areas; and is an editor of the upcoming book "Recent Advances in Mechanical Engineering" in the Springer book series Lecture Notes in Mechanical Engineering.

Chapter 1
Comprehensive Study of Additives and Corrosion in Biodiesel

Khushbu Yadav, Naveen Kumar and Rajiv Chaudhary

Abstract India is a diesel-driven country with the extensive use of diesel engines in transportation, farming, and industrial sector. This is primarily due to high thermal efficiency and better fuel economy. However, diesel engines produce higher harmful emissions such as NOx, particulate matter (PM) as compared to diesel fuel, which is harmful to human health. Biodiesel is one of the attractive alternative fuels made from renewable resources and can be used in the existing CI engine without any major modification. Biodiesel has several drawbacks like high density, low heating value, and high fuel consumption. Additive plays an important role in alleviating problems of the use of biodiesel in diesel engine, and additives in biodiesel can result in improved engine performance, engine combustion, emissions, and physicochemical properties. Metal-based additives can result in improved emissions. The oxidation stability is also a major issue in the use of biodiesel. Oxidation stability leads to storage problem for a longer time period due to slush formation, and it also had their impact on stored metal and metal parts of the engine. Antioxidant additives provide protection against corrosion and enhance combustion process, and engine performance properties with the cleanliness of fuel lead to less wear. Antioxidants help to improve storage and stability characteristics. In the present study, a comprehensive review is done to study the different additives and their impact on engine performance, combustion, emission, and metal parts.

Keywords Biodiesel · Additives · Performance · Emission · Corrosion

K. Yadav (✉)
Amity University, Noida 201301, India

K. Yadav · N. Kumar
Centre for Advanced Studies and Research in Automobile Engineering, Delhi Technological University, Delhi 110042, India

R. Chaudhary
Department of Mechanical Engineering, Delhi Technological University, Delhi 110042, India

S. Yadav et al. (eds.), *Proceedings of International Conference in Mechanical and Energy Technology*, Smart Innovation, Systems and Technologies 174,
https://doi.org/10.1007/978-981-15-2647-3_1

1.1 Introduction

The development of a country is directly affected by the available energy. Increasing demand of energy by different sectors of Indian economy like commercial and domestic needs, agriculture, and the industrial sector plays an important role toward the fast depletion of fossil fuel along with an increase in environmental issues that affect the growth of the nation. Worldwide consumption of energy increases by 20-fold in the last century and sources which are finite in nature going to be exhausted except hydropower and nuclear energy in the near future [1]. The utilization of energy worldwide of crude oil is about 12.2×10^9 tons. Energy utilization rises approximately by 1.75×10^9 tons of crude oil by 2035 [2, 3]. The ever-increasing demand for petroleum fuel increases the consumption of fossil fuel indicating an increase in global mean temperature by emission to an atmosphere having an impact on climate change and harmful to human health. Therefore, scientists and researchers are concern to develop eco-friendly and infinite renewable source energy [3].

Among all available renewable fuels, biodiesel is one of the promising non-toxic and eco-friendly fuels having properties comparable with petroleum diesel. Biodiesel used in diesel engine produces oxides of nitrogen (NO_X) because of more quantity of oxygen (O_2) content and reduces hydrocarbon (HC) and carbon mono-oxides (CO). As compared to diesel higher molecular weight, higher viscosity, higher pour point, and lower volatility are considered as drawbacks of biodiesel which leads to incomplete combustion [4, 5]. Many investigations have been done with different feedstock without or with less modification in a diesel engine. Shahiret al. [6] compared pure diesel with different blends of biodiesel produced from animal fats on CRDI engine, and they found increase in brake-specific fuel consumption (BSFC), brake thermal efficiency (BTE), exhaust gas temperature and lower value of CO, O_2 with increasing amount of CO_2, NO_X, and HC for 30% blend of biodiesel compared to diesel. Bueno et al. [7], Conclude from experimental work on caster and soybean biodiesel, shows not having any impact on thermal efficiency as compared to diesel but increases NO_X. Caster biodiesel blend produces a low level of PM (Partculate Matter) emission than soybean biodiesel. Dutta et al.'s [8] results show brake thermal efficiency slightly increases by adding ethanol and methanol alcohols to palm stearin biodiesel. Maximum in-cylinder temperature lowers with lower the value of peak pressure rise by adding alcohol blend with biodiesel, decreases in NO_X emission, and increases PM and smoke. Nabi et al. [9], Experiment performed on biodiesel obtained from waste cooking oil to check emission and engine performance under the 13-Mode European Stationary Cycle. The result shows reduction in emission by 13%, 11–15% reduction in CO, 47–70% reduction in unburnt hydrocarbon emission, 84% reduction in PM and slightly increases in NO_X emission with reference fuel. Nair et al. [10] performed experiment on diesel and neem biodiesel blending. By using neem biodiesel, brake thermal efficiency found to higher than diesel and brake power is approximately same. CO, HC, NO_X emission decreased and O_2 emission increases as compared to reference fuel. At B10 blend of diesel with neem biodiesel gives lower emission and higher performance than other blends. Oxygen-based emission caused by biodiesel reduced by using additives.

1.2 Additives and Their Effects

Additives are chemicals used to enhance the properties of fuel by mixing it in diesel, biodiesel, and gasoline. Additives play a major role to improve quality of biodiesel fuel, improve engine performance, and environmental emission. Additives provide benefit in regard to the performance of the engine, oxidation stability, deposit formation, emission and corrosion while storing fuel for long time, cold flow properties, and contamination. Selection of additives based on several parameters like blending of fuel, viscosity, density, toxicity, autoignition, flash point, additive solubility, and cetane number. Additives can resolve problem associated with renewable fuel and reduce dependency on conventional fuel [11, 12].

1.2.1 Metal-Based Additives

These additives are added in the diesel, biodiesel, and gasoline to reduce the emission of exhaust gases to environment and for more complete combustion. This is done by the reaction of metal-based additives in two ways: first enhancing soot oxidation by reaction of metal with water vapor to produce hydroxyl and, second, lowering oxidation temperature by reaction directly with carbon atoms. These metal-based additives consist of copper (II) chloride ($CuCl_2$), copper (II) oxide (CuO-nanostructured), copper (II) sulfate ($CuSO_4$), iron (III) chloride ($FeCl_3$), cobalt (II) chloride ($COCl_2$), cerium (Ce), platinum (Pt), cerium–iron (Ce–Fe), platinum–cerium (Pt–Ce), manganese (Mn), barium, calcium, nickel (Ni), zinc (Zn), titanium (Ti), and zirconium (Zr) [5, 13, 14]. Nano-sized particles getting more attention due to the behavior of superparamagnetic, good plasticity nature, lower freezing point, high ignition value, lower sintering temperature, and high density. Most effective nano-additives used to reduce emission are iron (Fe), manganese (Mn), and cerium (Ce) [14].

1.2.2 Oxygenated Additives

These are the most commonly used additives to fulfill the requirement of complete combustion of fuel. Oxygenated additives are fuel having oxygen content and at least one atom of oxygen side by hydrogen and carbon atoms in blending component. Emission from the diesel engine is more because of the lack of oxygen content in the combustion chamber and incomplete combustion. Oxygenated additives having oxygen content in extent lead to complete combustion and minimum ignition temperature. These additives are alcohols (methanol, ethanol, propanol, and butanol). Widely used additives are group of ether and group of ester. Some more combinations used as oxygenated additives are polyethoxy ester, methyl butanoate, ethylene glycol monoacetate (EGM), diethyleneglycol diethyl ether (DGE), dimethyl carbonate (DMC), triacetin, acetone, di-n-pentyl ether (DNPE), and glycerinetri-acetate [13, 14].

1.2.3 Cetane Number Improver Additives

They have a direct effect on the ignition delay period. Minimum ignition delay refers to complete combustion of the fuel in the engine which reduces vibrations, noise, and harmful emission produced to the environment. Cetane number of the fuel varies with properties of the fuel. Different additives are used to enhance the cetane number to meet the fuel specification which is set at the highest cetane number. Additives used as peroxides (tertiary-butyl peroxide), alkyl nitrates (mixed octyl-nitrates, hexyl-nitrates, and amyl-nitrate), nitrites, aldehydes, and tetraazole. 2-ethyl hexyl nitrate (2EHN) is a commonly used additive to increase cetane number. Di-tertiary butyl peroxide is promising additives having potential to reduce NO_X emission. Insufficient cetane number leads to increased consumption of fuel, poor starting characteristics in cold weather, increase level of harmful emission in diesel and biodiesel gases [5, 13].

1.2.4 Antioxidant Additives

These additives are used to enhance oxidation stability. The stability of biodiesel is less in accordance with other fuel because of unsaturated fatty acid derived from long-chain fatty acid esters present in the fuel which leads to oxidation. Antioxidant additives used in biodiesel fuels to overcome these drawbacks like deposit and formation of slush, change in color of fuel, corrosion, effects on engine performance, injector fouling, fuel filter plugging while storing fuel for long time. This is done by interrupting the long-chain formation of unsaturated fatty acid and by peroxide decomposition. Most commonly used antioxidants additives are tert-Butylhydroquinone (TBHQ), butylated hydroxylanisole (BHA), diphenylamine (DPA), butylated hydroxyl toluene (BHT), propyl gallate (PG), and pyrogallol (PL). Many researchers are working in that field, and they found that TBHQ, PG, and PL give the best results [5, 15, 16] (Table 1.1).

Table 1.1 List of additives commonly used in specific biodiesel and comparative analysis of performance and emission effects of additives on different biodiesels

S. no	Fuel	Additives	Brake power	BTE	BSFC	CO	HC	NOx	References
1.	Palm biodiesel–diesel blend	Diethyl ether	Decrease	Decrease	Decrease	–	–	–	[17]
2.	Palm biodiesel	Partially stabilized zirconia (PSZ) Di-tertiary butyl peroxide (DTBP),	–	Increase	–	Decrease	Decrease	Decrease	[18]
3.	Jatropha and palm biodiesel blend	Ethanol, n-butanol and diethyl ether	Increase	Increase	–	Decrease	Increase	Decrease	[19]
4.	Emulsified bio-solution/palm biodiesel/diesel blends	Bio-solution additive	–	–	Decrease	–	Decrease	–	[20]
5.	Palm olein oil	Irgalube 349, anti-corrosion additive	Increase		Increase	Decrease	Decrease	Decrease	[21]
6.	Mahua methyl ester	Dimethyl carbonate	Increase	Increase	Decrease	Decrease	Decrease	Decrease	[22]
7.	Mahua biodiesel	n-Octanol	–	–	–	Decrease	Decrease	Decrease	[23]
8.	Mahua methyl ester	Copper oxide (nano-additives)	–	Increase	Decrease	Decrease	Decrease	Increase	[24]
9.	Mahua methyl ester	Aluminum oxide nanoparticles (ANPs)	–	Increase	Decrease	Increase	Decrease	Increase	[25]
10.	Mahua biodiesel	Dimethyl carbonate	Increase	Increase	Decrease	Decrease	Decrease	Decrease	[26]

(continued)

Table 1.1 (continued)

S. no	Fuel	Additives	Brake power	BTE	BSFC	CO	HC	NOx	References
11.	H2 biodiesel dual-fueled jatropha methyl ester (JME) biodiesel	Zinc oxide nanoparticle fuel additive	–	–	–	Increase	Decrease	Increase	[27]
12.	Jatropha methyl ester	Antioxidant additives, ethylenediamine, butylated hydroxytoluene, α tocopherol acetate, p-phenylenediamine, and L-ascorbic acid	–	–	Decrease	Increase	Increase	Decrease	[28]
13.	Jatropha biodiesel	N′-diphenyl-1,4-phenylenediamine antioxidants	–	Same	Same	Decrease	Decrease	Decrease	[29]
14.	Jatropha biodiesel	Nano-aluminum oxide (n-Al_2O_3)	Increase	Increase	–	–	–	–	[30]
15.	Jatropha–palm blended biodiesel	Blended biodiesel	–	–	–	Increase	–	Increase	[31]
16.	Jatropha biodiesel	Pentanol, di-tertiary butyl peroxide (DTBP), and 2-ethylhexyl nitrate (EHN)	Decrease	Decrease	Decrease	Increase	Increase	Decrease	[32]
17.	Water–biodiesel emulsion fuel (Jatropha)	Carbon nanotubes and diethyl ether	–	Increase	–	–	–	Decrease	[33]
18.	Calophyllum inophyllum methyl ester (CIME) biodiesel	Zinc oxide (ZnO) and ETHANOX	–	Increase	–	Increase	Increase	Decrease	[34]

(continued)

Table 1.1 (continued)

S. no	Fuel	Additives	Brake power	BTE	BSFC	CO	HC	NOx	References
19.	Pongamia biodiesel	Ferrofluid (nanofuel additives)	–	Increase	Decrease	Decrease	Decrease	Decrease	[35]
20.	Pongamia biodiesel	Zinc oxide nanoparticles	–	Increase	–	Decrease	Decrease	Decrease	[36]
21.	Pongamia biodiesel	Ethanol	–	Increase	Decrease	–	–	Increase	[37]
22.	Neem biodiesel	Diethyl ether	–	Increase	Decrease	Decrease	Increase	Increase	[38]
23.	Waste cooking oil	n-pentanol	–	Increase	Decrease	Decrease	Decrease	Decrease	[39]
24.	Waste cooking oil	n-butanol	–	Increase	Same	Decrease	–	Decrease	[40]
25.	Waste cooking oil	Carbon nanotubes	Increase	Increase	Decrease	Decrease	Decrease	Decrease	[41]
26.	Waste cooking oil	Iso-butanol	–	–	Increase	Decrease		Decrease	[42]
27.	Sunflower biodiesel	Polystyrene	–	–	Decrease	Decrease	Decrease	Decrease	[43]
28.	Sunflower biodiesel	Bio-oil extraction	Same	Increase	Increase	Increase	Increase	Increase	[44]

1.3 Sustainability of Biodiesel Toward Metal Corrosion

Biodiesel is found to be the most promising fuel with several advantages, but biodiesel also moving with snag like unsaturated fatty acid which causes problems of deposition of impurities leads to corrosion. Most of the engine parts made of metals like aluminum, mild steel, stainless steel, and cast iron. These metals are receptive to corrosion [45]. Alves et al. [46] show an experimental analysis of corrosion by stainless steel on oxidation stability of biodiesel. They have performed test on AISI 316 stainless steel at room temperature and in dark place in the presence of soybean biodiesel. Fuel properties and oxidation stability measured by gas chromatography and Rancimat. Corrosion on stainless steel is evaluated by gravimetric. SEM/EDS microscopy techniques and XRF analysis found micro-pitting covered by oxide film. Metal ion released from stainless steel, when comes in contact with biodiesel, causes oxidation process and degradation of fuel reduces induction period. Fazal et al. [47], Effects of inhibitors like butylate-hydroxytoluene (BHT), benzotriazole (BTA), pyrogallol (PY) and tert-Butylamine on corrosion of low carbon steel and cast iron through immersion test done by using palm biodiesel. TBA acts as good inhibitors of corrosion in comparison with others for selected materials. It also shows good efficiency, and SEM images shows smooth surface with less corrosion. Hu et al. [48] performed assessment of corrosion on mild carbon steel, copper, stainless steel, and aluminum in biodiesel and diesel. They found 0.01819, 0.002334, 0.00087, and 0.00324 nm/year corrosion rate at 43 °C for 60 days considered a condition for biodiesel. Copper and carbon steel found to be more susceptible to corrosion than aluminum and stainless steel. Fazal et al. [49] investigated the effects of additives on material exposed to biodiesel. Author used palm biodiesel with and without additives like propyl gallate, butylated hydroluene, benzotriazole, tert-Butylamine, and pyrogallol for 2160 h at 25–27 °C. Metal surface degradation notably reduced by adding additives like benzotriazole and tert-Butylamine measured by energy-dispersive spectroscopy, X-ray diffraction, and scanning electron microscopy. Fernandes et al. [50] performed experiment on CS 1015 carbon steel with and without additive, i.e., tert-Butylhydroquinone (TBHQ) in the absence of light for 84 h with Moringa oleifera Lam biodiesel. Induction period was increased by 1.3 h after adding TBHQ additive to biodiesel in amount of 500 mg per kg. Till 21 days IP value decreased. Iron value was found when CS 1015 exposed to biodiesel in the presence of TBHQ. On the other hand, CS 4140 did not show any iron value and looks more compatible with biodiesel. They also show iron does not affect the oxidation stability of biodiesel. Deyab et al. [51] investigated cardanol as natural corrosion inhibitor made from nutshell liquid on 5052 H32 aluminum in the presence of soybean biodiesel. The efficiency of inhibitor enhances with the addition of concentration of cardanol performed. They found 93.2% maximum efficiency by taking 220 ppm of cardanol. They concluded cardanol works as effective corrosive inhibitor for aluminum AA 5052 H32.

1.4 Conclusion

The literature review shows that performance and emission can be altered by adding additives in biodiesel. Most commonly used biodiesels are jatropha, mahua, palm, pongamia, neem, waste cooking oil biodiesel etc. shows improvement in their performance and harmful emission when some specific quantity of additives used in it. Brake thermal efficiency and brake power of most of these biodiesels are increased, but brake-specific fuel consumption decreases as positive sign. HC, CO, and NOx emission also decrease in many cases; in some cases, it may increase. Due to oxidation, reaction on the surface of metal due to biodiesel leads to corrosion. Corrosion rate can be measured by SEM, EDS, and X-ray diffraction methods. This corrosion problem can also be reduced by using antioxidants additives as inhibitors of corrosion toward metal parts.

References

1. Money, A., Gerpen, J.H.V.: Biomass Bioenerg **20**, 317–325 (2001)
2. Shell Energy Scenarios to 2050, Shell International BV (2011)
3. Brower, D.: An alarming global energy outlook, petrol. Econ **78**(10), 4–7 (2011)
4. Dwivedi, G., Jain, S., Sharma, M.P.: Diesel engine performance and emission analysis using biodiesel from various oil sources—review. J. Mater. Environ. Sci. **4**, 434–447 (2013)
5. Kumar, M.V., Babu, A.V., Kumar, P.R.: The impacts on combustion, performance and emissions of biodiesel by using additives in direct injection diesel engine. Alexandria Eng. J. **57**, 509–516 (2018)
6. Shahir, V.K., Jawahar, C.P., Suresh, P.R., Vinod, V.: Experimental investigation on performance and emission characteristics of a common rail direct injection engine using animal fat biodiesel blends. Energy Procedia **117**, 283–290 (2017)
7. Bueno, A.V., Pereira, M.P.B., Pontes, J.V.O., Luna, F.M.T., Cavalcante, C.L.: Performance and emission characteristics of castor oil biodiesel fuel blends. Appl. Therm. Eng. **125**, 559–566 (2017)
8. Dutta, A., Mandal, B.K.: Engine performance, combustion and emission characteristics of a compression ignition engine operating on different biodiesel-alcohol blends. Energy **125**, 470–483 (2017)
9. Nabi, M.N., Hossain, F.M., Ristovski, Z.D., Brown, R.J.: Reduction in diesel emission including PM and PN emission with diesel-biodiesel blends. J. Clean. Prod. **166**, 860–868 (2017)
10. Nair, N.J., Kaviti, A.K., Daram, A.K.: Analysis of performance and emission on compression ignition engine fuelled with blends of neem biodiesel. Egypt. J. Petrol **26**, 927–931 (2017)
11. Alleman, T.L., Mckormick, R.L.: Biodiesel handling and use guide. Energy Efficiency and Renewable Energy. DOE/GO-102016-4875 (2016)
12. Biodiesel handling and use guide. National Renewable Energy Laboratory (NREL) (2009)
13. Reshedul, H.K., Masjuki, H.H., Ashraful, A.M., Rehamn, S.M.A., Shahir, S.A.: The effect of additives on properties, performance, and emission of biodiesel fuelled compression ignition engine. Energy Convers. Manag. **88**, 348–364 (2014)
14. Khlife, E., Tabatabaei, M., Demirbas, A., Aghbashlo, M.: Impact of additives on performance and emission characteristics of diesel engine during steady state operation. Prog. Energy Combust. Sci. **59**, 32–78 (2017)
15. Yaakob, Z., Narayanan, B.N., Padikkaparambil, S., Unni, S., Akbar, M.: A review on the oxidation stability of biodiesel. Renew. Sustain. Energy Rev. **35**, 136–153 (2014)

16. Dwivedi, G., Sharma, M.P.: Impact of antioxidant and metals on biodiesel stability. J. Mater. Sci. **5**, 1412–1425 (2014)
17. Ali, O.M., Mamat, R., Najafi, G., Yusaf, T., Ardebili, S.M.S.: Optimization of biodiesel-diesel blended fuel properties and engine performance with ether additives using statistical analysis and response surface methods. Energies **8**, 14136–14150 (2015)
18. Musthafa, M.M.: Development of performance and emission characteristics on coated diesel engine fuelled by biodiesel with cetane number enhancing additives. Energy **134**, 234–239 (2017)
19. Imtenan, S., Masjuki, H.H., Verman, M., Kalam, M.A., Arbab, M.I., Sajjad, H., Rahman, S.M.A.: Impact of oxygenated additives to palm and jatropha biodiesel blends in the context of performance and emission characteristics of a light-duty diesel engine. Enegy Convers. Manag. **83**, 149–158 (2014)
20. Chena, K.S., Lina, Y.C., Tehsiehb, L., Linc, L.F., ChiehW, C.: Saving energy and reducing pollution by use of emulsified palm-biodiesel blends with bio-solution additives. Enegy **5**, 2043–2048 (2010)
21. Ashraful, A.M., Masjuki, H.H., Kalam, M.A., Rashedul, H.K., Sajjad, H., Abedin, M.J.: Influence of anti- corrosion additive on the performance, emission and engine component wear characteristics of an IDI diesel engine fueled with palm biodiesel. Energy Convers. Manag. **87**, 48–57 (2014)
22. Nayak, S.K., Pattanaik, B.P.: Experimental investigation on performance and emission characteristics of a diesel engine fuelled with mahua biodiesel using additives. Energy Procedia **54**, 569–579 (2014)
23. Mahalingam, A., Devarajan, Y., Radhakrishnan, S., Vellaiyana, S., Nagappan, B.: Emission analysis on mahua oil biodiesel and higher alcohol blends in diesel engine. Alexandria Eng. J. **57**, 2627–2631 (2018)
24. Chandrasekharan, V.K., Arthanarisamy, M., Nachiappan, P., Dhanakotti, S., Moorthy, B.: The role of nano additives for biodiesel and diesel blended transportation fuels. Transp. Res. Part: D Transp. Environ. **46**,145–156 (2016)
25. Aalam, C.S., Saravanan, C.G.: Effect of nano metal oxide blended mahua biodiesel on CRDI diesel engine. Ain shams Eng. J. **8**, 689–696 (2017)
26. Kumar, S., Bhabani, N., Pattanaik, P.: Experimental investigation on performance and emission characteristics of a diesel engine fuelled with mahua biodiesel with additive. **54**, 569–579 (2014)
27. Javeda, S., Murthya, S., Satyanarayana, S., Reddy, R., Rajagopala, K.: Effect of a zinc oxide nanoparticle fuel additive on the emission reduction of a hydrogen dual-fuelled engine with jatropha methyl ester biodiesel blends. J. Clean. Prod. **20**, 490–506 (2016)
28. Varatharajan, K., Cheralathan, M., Velraj, R.: Mitigation of NO_X emission from a jatropha biodiesel fuelled DI diesel engine using antioxidant additives. **8**, 2721–2725 (2011)
29. Palash, S.M., Kalam, M.A., Masjuki, H.H., Arbab, M.I., Masum, B.M., Sanjid, A.: Impact of NO_X reducing antioxidant additives on performance and emission of a multi-cylinder diesel engine fueled with jatropha biodiesel blends. Energy Convers. Manag. **77**, 577–585 (2014)
30. Kumar, H., Kumar, P.S.: Experimental analysis on performance of diesel engine using mixture of diesel and biodiesel as a working fuel with aluminum oxide nanoparticle additives. Thermal Sci. Eng. Prog. **4**, 252–2258 (2017)
31. Amirul, S.Y., Rashid, A., Abdullah, N.R., Mamat, R., Abdul Latip, S.: Emission of transesterification Jatropha-Palm blended biodiesel. Procedia Eng. **68**, 265–270 (2013)
32. Imdadula, H.K., Rashedab, M.M., Shahinb, M.M., Masjukia, H.H., Kalama, M.A., Kamruzzamanab, M., Rashedula, H.K.: Quality improvement of biodiesel blends using different promising fuel additives to reduce fuel consumption and NO emission from CI engine. Energy Convers. Manag. **138**, 327–337 (2017)
33. Basha, J.S.: Impact of carbon nanotubes and Di-Ethyl Ether as additive with biodiesel emulsion fuels in a diesel engine—An experiment investigation. J. Energy Inst. **91**, 289–303 (2018)
34. Ashok, B., Nanthagopal, K., Mohan, A., Johny, A., Tamilarasu, A.: Comparative analysis on the effect of Zinc oxide and ethanox as additives with biodiesel in CI engine. Energy **140**, 353–364 (2017)

35. Kumar, S., Dinesha, P., Bran, I.: Influence of nanoparticles on the performance and emission characteristics of a biodiesel fuelled engine: an experimental analysis. Energy **140**, 98–105 (2017)
36. Seela, C.R., Ravisankar, B., Raju, B.M.V.A.: A GRNN based frame work to test the influence of nano zinc additive biodiesel blends on CI engine performance and emission. Egypt. J. Petrol. **27**, 641–647 (2018)
37. Paul, A., Panua, R., Debory, D.: An experimental study of combustion, energy and emission characteristics of a CI engine fueled by Diesel-Ethanol-Biodiesel blends. Energy **141**, 839–852 (2017)
38. Sivalakshmi, S., Balusamy, T.: Effect of biodiesel and its blends with diethyl ether on the combustion, Performance and emission from a diesel engine. Fuel **106**, 106–110 (2013)
39. Babu, D., Anand, R.: Effect of biodiesel-diesel-n-pentanol and biodiesel-diesel-n-hexanol blends on diesel engine emission and combustion characteristics. Energy **133**, 761–776 (2017)
40. Isik, M.Z., Bayindir, H., Iscan, B., Aydin, H.: The effect of n-butanol additive on low load combustion, performance and emission of biodiesel-diesel blend in a heavy duty diesel power generator. J. Energy Inst. **90**, 174–184 (2017)
41. Hosseini, S.H., Alisaraei, A.T., Ghobadian, B., Mayvan, A.A.: Performance and emission characteristics of a CI engine fuelled with carbon nanotubes and diesel-biodiesel blends. Renew. Energy **111**, 201–213 (2017)
42. Kumar, K.S., Raj, R.T.K.: Effect of Di-Teriary Butyl peroxide on the performance, combustion and emission characteristics of ethanol blended cotton seed methyl ester fuelled automotive diesel engine. Energy Convers. Manag. **127**, 1–10 (2016)
43. Hazrat, M.A., Rasul, M.G., Khan, M.M.K., Ashwath, N., Rufford, T.E.: Emission characteristics of polymer additives mixed diesel-sunflower biodiesel fuel. Energy Procedia **156**, 59–64 (2019)
44. Dueso, C., Munoz, M., Moreno, F., Arroyo, J., Noemi, G., Bautista, A., Gonzalo, A., Sanchez, J.L.: Performance and emission of a diesel engine using sunflower biodiesel with a renewable antioxidant additives from bio-oil. Fuel **234**, 276–285 (2018)
45. Kovacs, A., Toth, J., Isaak, GY., Keresztenyi, I.: Aspect of storage and corrosion characteristics of biodiesel. Fuel Process. Technol. **134**, 59–64 (2015)
46. Alves, S.M., Dutra-pereira, F.K., Bicudo, T.C.: Influence of stainless steel corrosion on biodiesel oxidative stability during storage. Fuel **249**, 73–79 (2019)
47. Fazal, M.A., Sazzad, B.S., Haseeb, A.S.M.A., Masjuki, H.H.: Inhibition study of additives towards the corrosion of ferrous metal in palm biodiesel. Energy Convers. Manag. **122**, 290–297 (2016)
48. Hu, E., Xu, X., Hu, X., Pan, L., Jiang, S.: Corrosion behaviors of metals in biodiesel from rapeseed oil and methanol. Renew. Energy **37**, 371–378 (2012)
49. Fazal, M.A., Suhaila, N.R., Haseeb, A.S.M.A., Rubaiee, S.: Sustainability of additives-doped biodiesel: analysis of its aggressiveness toward metal corrosion. J. Clean. Prod. **181**, 509–516 (2018)
50. Fernandes, D.M., Squissato, A.L., Lima, A.F., Richter, E.M., Munoz, R.A.A.: Corrosive character of Moringa oleifera Lam biodiesel exposed to carbon steel under simulated storage conditions. Renew Energy **139**, 1263–1271 (2019)
51. Deyab, M.A., Correa, R.G.C., Mazzetto, S.E., Dhmees, A.S., Mele, G.: Improving the sustainability of biodiesel by controlling the corrosive effects of soybean biodiesel on aluminum alloy 5052 H32 via cardanol. Ind. Crops Prod. **130**, 146–150 (2019)

Chapter 2
Modified Twisted Tape Inserts: Significance of Geometric Non-dimensional Parameters

Kalpana Gupta, Raj Kumar Singh and M. V. Tendolkar

Abstract Augmentation of heat transfer is a basic process essential in variety of industrial applications. Different techniques are in use to raise the turbulence and heat transfer coefficient. Twisted tape is one of the very popular techniques. In the present paper, varieties of twisted tapes are discussed and novelty of this paper is the collection, description and optimized values of various geometric non-dimensional parameters (GNDPs) which has rarely been discussed collectively. New terms which are independently been described in references are together discussed in details. Their physical significance is emphasized towards efficient performance of the heat exchangers. This includes various geometric non-dimensional parameters defined in the literature, which are very important factors to decide performance of twisted tape heat exchangers. The present work is very useful for researchers to understand concept of such non-dimensional parameters from the point of view of performance evaluation. Accordingly, the present work is highly supportive towards proposing a new design with efficient geometries of twisted tapes.

Keywords Heat transfer enhancement · Augmentation · Modified twisted tape · Friction factor · Thermal performance factor

Nomenclature

Nu	Nusselt number
f	Friction factor
y	Pitch length
w	Width of tape
d_e	Depth of cut

K. Gupta (✉) · R. K. Singh
Department of Mechanical Engineering, Delhi Technological University (DTU), Delhi, India
e-mail: Kalpana_phd2k15@dtu.ac.in

M. V. Tendolkar
Department of Mechanical Engineering, Veermata Jijabai Technological Institute (VJTI), Mumbai, India

S. Yadav et al. (eds.), *Proceedings of International Conference in Mechanical and Energy Technology*, Smart Innovation, Systems and Technologies 174,
https://doi.org/10.1007/978-981-15-2647-3_2

w_i	Width of cut
d_h	Diameter of hole
s_h	Distance between holes
θ_t	Taper angle
GNDP	Geometric non-dimensional parameter
η	Enhancement efficiency
TT	Twisted tape
θ	Twist angle
Re	Reynolds number
A_p	Total pores' area
A_t	Total tape area
C-CC	Clockwise counterclockwise

2.1 Introduction

Effective use of energy has gained highest attention in the present world. For that purpose, one needs to improve performance of various equipments. Heat exchangers are widely and popularly used at many places which comprise thermal, nuclear, solar power plants, refrigeration and air conditioning, petroleum refining, aviation, spacecraft, automobile, food processing, chemical, cosmetics industries, waste heat recovery systems and many more. Improvement of effectiveness of any heat exchanger ultimately improves the performance of various systems. Various techniques are proposed by researchers for enhancing performance of heat exchangers, which basically includes passive, active, and compound techniques. In active techniques, power from external source is utilized to increase heat transfer, e.g. jet impingement, surface vibration, and use of electrostatic field. On the other hand, passive techniques do not require power from any external source. Few passive techniques are surface modification, use of swirl flow devices, use of different types of roughness elements, use of nano-fluids, etc. In compound techniques, two or more active/passive techniques are employed. Different researchers have contributed significantly in improving performance of heat exchanger using various heat transfer enhancement techniques.

Bergles and Webb [1, 2] reviewed different techniques for heat transfer enhancement in heat exchangers. Heat transfer enhancement is brought by modifications on heat transfer surface by using different types of ribs or use of vortex or swirl flow devices or by varying shape of cross section of tube. These techniques are called passive heat transfer augmentation techniques as no external power source used. Different researchers proposed different heat transfer enhancement techniques over the years. Out of them, employment of twisted tapes inside heat exchanger tubes is significant passive heat transfer augmentation technique. As documented by Manglik and Bergles [3], TT inserts improve heat transfer by the generation of swirl flow, increasing flow velocity by flow blockage and tube portioning and by providing effectively longer flow path. Depending upon contact between tube wall and tape edge, minor fin effect may also be a supporting factor for enhanced heat transfer. Twisted tape inserts are extensively used to generate swirl. These inserts yield performance improvement based on type of ducts employed. Ray and Date [4] predicted numerically heat transfer in square and circular tube inserted with twisted tape. It is reported that the thermohydraulic performance of an empty square or circular duct is much lower when compared with same duct comprising twisted tapes. It is further stated that the increase in heat transfer in case of twisted tapes in square tubes is significantly higher than those in circular tubes. Naphon [5] reported investigations on f and Nu in the horizontal double pipe (plain, i.e. without any insert) and same double pipe with twisted tape inserts. Twisted tapes give better results combining with other enhancement techniques. Zimparao [6, 7] experimentally obtained enhancement in heat transfer by combining spirally corrugated tubes (single start or three starts) with a twisted tape. Zimparao [8] developed mathematical model for the prediction of friction factor. He combined spirally corrugated tube and twisted tape insert for his studies. Their model also revealed impact of wall roughness on the axial, secondary and swirl flow velocities. Bhardwaj et al. [9] found heat transfer and pressure drop in a spirally grooved tube (75 starts) with TT inserts taking water as working fluid. They observed further augmentation in heat transfer enhancement for spirally grooved tube with twisted tape inserts, compared to smooth tube. It is reported that the direction of twist either clockwise or anticlockwise impacts the thermohydraulic performance. In most of earlier works, researchers have used plain TT [1–8]. However, many researchers contributed further towards designing different geometries of twisted tapes. Rahimi et al. [10] compared performance of plain tube and tube with three kinds of inserts, namely jagged, perforated and classic. He carried out both experimental and computational fluid dynamics investigations for comparison. Smith Eimsa-ard et al. [11, 12] carried out comparative investigations on enhanced heat transfer and pressure loss. They used dual twisted tapes and serrated twisted tapes, respectively. Murugesan et al. [13] examined effect of TT inserts with V-cut on edges of tape. Saha et al. [14] used combined enhancement technique, i.e. circular tube with axial rib roughness inserted with centre-cleared twisted tape. Bhuiya et al. [15, 16] studied the effects of triple TT and double counter TT, respectively, on heat transfer rate, friction factor and thermal enhancement efficiency. Saha and Dutta [17] have reported the effect of regularly spaced twisted tape inserts in circular tubes with respect to the heat transfer coefficient and friction factor.

Before entering into the discussion and interpretation of various modified tapes, the fundamental non-dimensional parameters used towards performance enhancement are discussed in depth in the next section. Discussion of effect of these GNDPs is emphasized in upcoming section as it provides valuable information to designers for making effective modifications in TT geometry.

2.2 Various Geometric Non-dimensional Parameters

The number of GNDP is reported by different researchers in the literature. The subsequent description provides the basic definition, impact on performance and physical interpretation of each of them in detail. The expressions reported are valid for prescribed range of Reynolds number. Mostly cases discussed here are for turbulent flow region.

2.2.1 Twist Ratio, y/w *[18]*

Most of the researchers used this GNDP in their study. It may be defined as the ratio of pitch length (y) to width of tape (w). It plays a crucial role in determining the performance of every kind of twisted tape. Generally, range of twist ratio used in experiments lies between 1 and 15. For plain tape, twist ratio is infinite. Decrease in twist ratio results in appreciable performance enhancement of TT. It is because of the rise in swirl intensity passed on to tube wall with the decrease of twist ratio. Empirical correlations developed for finding effect of various non-dimensional parameters on friction factor, enhancement efficiency and heat transfer coefficient generally include twist ratio of twisted tape, which could be perceived in correlations given in subsequent segments.

2.2.2 Twist Angle, θ *[18]*

It is the angle between two edges of the tape which, after some depth of cut, twisted opposite to previous direction, as shown in Fig. 2.1b. Fundamentally, it is an arrangement for the production of swirl flow from anticlockwise direction to clockwise or vice versa, with respect to former twisted length. As twisted angle increases, heat transfer rate and thereby Nusselt number also increase. Larger the twisted angle, more is the fluid directed to flow in opposite direction, resulting more disturbance in fluid for adjacent twisted length. In future researches, if some other modified design of twisted tape comes into account which may have twist angle as one of GNDP, then similar kind of effect may be observed under same working conditions. Correlations which communicate the effect of twist angle on Nu and f are reported as:

$$\mathrm{Nu} = 0.31\mathrm{Re}^{0.6}\mathrm{Pr}^{0.4}(y|w)^{-0.36}(\sin\theta + 1)^{0.44} \tag{2.1}$$

$$f = 46.39\mathrm{Re}^{-0.544}(y|w)^{-0.77}(\sin\theta + 1)^{0.45} \tag{2.2}$$

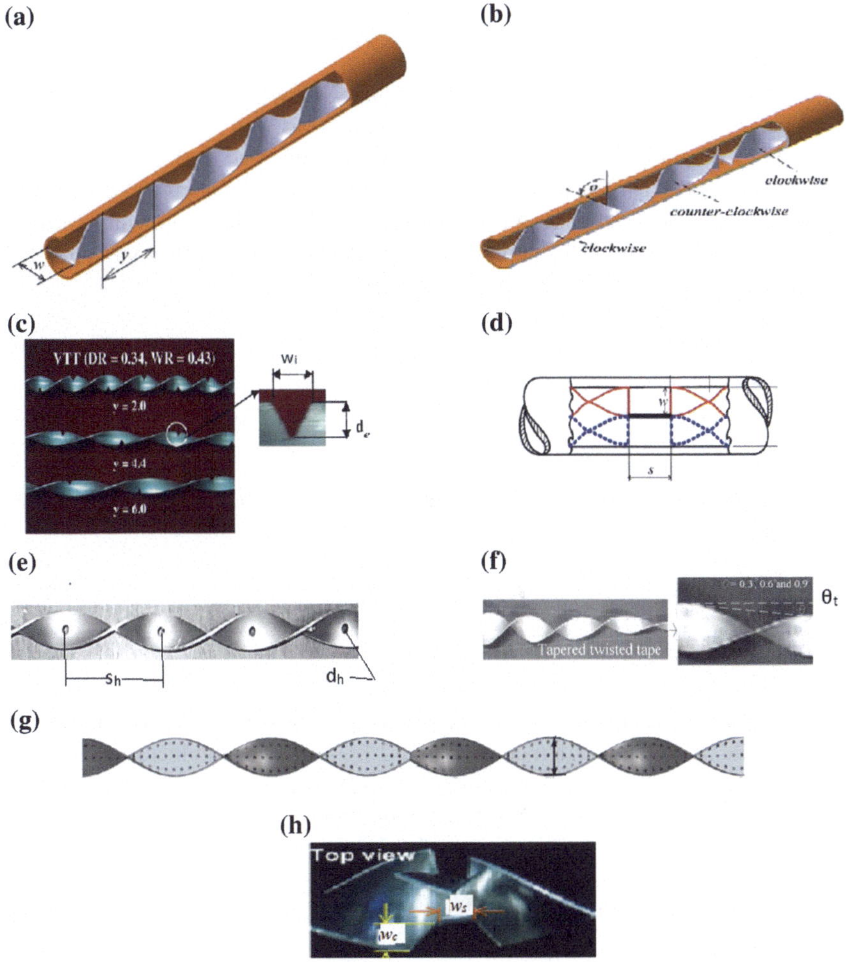

Fig. 2.1 Diagram showing various non-dimensional parameters [11, 13, 18, 20–22]

2.2.3 Depth Ratio, d_e/w *[13]*

It is defined as the ratio of depth of V/serrated/square/rectangular cut to width of tape, as shown in Fig. 2.1c. It has been observed for different shapes of cut that with increase in depth ratio, there is also increment in friction factor, heat transfer as well as overall enhancement efficiency. There is increase in fluid mixing near tape edge because of cut at edges of tape. This attributes to increase in velocity near wall which gives rise to thinner boundary layer hence produces greater temperature gradient and heat transfer rates. Numerical values of depth ratio and percentage change in Nu are described in subsequent section for V-cut and serrated twisted tape.

2.2.4 *Width Ratio*, w_i/w *[13]*

It can be defined as the ratio of width of V/square/serrated cut to width of tape, as shown in Fig. 2.1c. There is increment in heat transfer as well as friction factor with the reduction of width ratio. It is attributed to the fact that as width decreases, there is increase in turbulence intensity hence effective mixing of fluid at core and near wall region. The influence of the depth ratio is more prevailing compared width ratio for investigated range of Reynolds number. Nu, f and enhancement efficiency for V-cut TT can be calculated by correlations as reported:

$$\mathrm{Nu} = 0.0296\mathrm{Re}^{0.853}\mathrm{Pr}^{0.33}(y|w)^{-0.222}((d_e|w)+1)^{1.148}((w_i|w)+1)^{-0.751} \tag{2.3}$$

$$f = 8.632\mathrm{Re}^{-0.615}(y|w)^{-0.269}((d_e|w)+1)^{1.148}((w_i|w)+1)^{-1.148} \tag{2.4}$$

$$\eta = 1.392\mathrm{Re}^{-0.01}(y|w)^{-0.124}((d_e|w)+1)^{0.252}((w_i|w)+1)^{-0.058} \tag{2.5}$$

2.2.5 *Space Ratio*, s/d *[11]*

It is the ratio of length (free spacing between two consecutive separated parts of tape) to inside tube diameter, as shown in Fig. 2.1d. As space ratio increases, friction factor and heat transfer decrease because of weaker swirl generation and lesser fluid mixing. Effect of space ratio on Nu, f and enhancement efficiency of dual TT with regular spacing can be analysed using the following correlations:

$$\mathrm{Nu} = 0.069\mathrm{Re}^{0.74}\mathrm{Pr}^{0.4}(y|w)^{-0.26}(1.5(s|d)+1)^{-0.1} \tag{2.6}$$

$$f = 30.5\mathrm{Re}^{-0.56}(y|w)^{-0.54}(1.5(s|d)+1)^{-0.2} \tag{2.7}$$

$$\eta = 1.9\mathrm{Re}^{-0.05}(y|w)^{-0.08}(1.5(s|d)+1)^{-0.034} \tag{2.8}$$

2.2.6 *Perforation Diameter Ratio*, d_h/w *[19]*

It is the ratio of hole diameter to the width of tape, as shown in Fig. 2.1e. As the diameter ratio increases, heat transfer and Nusselt number decrease. It is because of the fact that larger the perforation diameter more freely fluid flow in axial direction hence lesser tangential contact between tube wall and core fluid. On the other hand, as perforation diameter ratio increases, there is reduction in friction factor because of diminishing fluid flow blockage.

2.2.7 *Perforation Pitch Ratio,* s_h/w *[19]*

It is the ratio of distance between two consecutive holes on twisted tape and width of twisted tape, as indicated in Fig. 2.1e. Friction factor decreases with diminishing perforation pitch ratio (*s*/*w*), and this is attributed to reduction in flow blockage because of increased perforated area or decreased perforation pitch ratio. Variation of Nu, f and η with diameter ratio and perforation pitch ratio is given by Nanan et al. [19]. The empirical correlations reported show the effect of diameter ratio and perforation pitch ratio on performance of helical perforated twisted tapes.

$$\mathrm{Nu} = 0.035\mathrm{Re}^{0.795}\mathrm{Pr}^{0.4}(d_h|w)^{-0.068}((s_h|w))^{0.086} \tag{2.9}$$

$$f = 1.91\mathrm{Re}^{-0.299}(d_h|w)^{-0.068}((s_h|w))^{0.094} \tag{2.10}$$

$$\eta = 4.058\mathrm{Re}^{-0.145}(d_h|w)^{-0.045}((s_h|w))^{0.054} \tag{2.11}$$

2.2.8 *Taper Angle* θ_t *[20]*

The regular TT, initially tapered as per the required geometry, and further modified with desired level of twist, serves the basic idea about this revised version of twisted tape inserts, as shown in Fig. 2.1f. Heat transfer enhancement along with increase in frictional loss is observed with reducing taper angle. Similarly, thermal performance factor is reported to be increased with rising taper angle for turbulent flow (Re range 6,000–20,000). This could be attributed to the increase in taper angle which results in reduction of swirl strength induced via tapered twisted tape, yielding poor fluid mixing between wall and core regions. Variation in Nu, f and η with respect to the taper angle is analysed with the following correlations:

$$\mathrm{Nu} = 0.076\mathrm{Re}^{0.75}\mathrm{Pr}^{0.4}(y/w)^{-0.39}(1+\theta)^{-0.1} \tag{2.12}$$

$$f = 16.559\mathrm{Re}^{-0.49}\mathrm{Pr}^{0.4}(y/w)^{-0.51}(1+\theta)^{-0.53} \tag{2.13}$$

$$\eta = 1.871\mathrm{Re}^{-0.04}\mathrm{Pr}^{0.4}(y/w)^{-0.22}(1+\theta)^{0.08} \tag{2.14}$$

2.2.9 *Porosity* R_p, A_P/A_t *[21]*

It is defined as the ratio of total pores' area to the tape area, as shown in Fig. 2.1g. It is another way to define dimensionless parameter of perforated tape suggested. It has been found out that tube with perforated twisted tape gives higher friction factor compared to plain tube for same value of Reynolds number as a result of perturbation and flow blockage. It has been determined that there is an optimum value of porosity which provides highest heat transfer. It is attributed to the fact that at higher perforation, there is lesser swirl intensity and at lower perforation, lesser agitation produced. More efficient disruption of boundary layer occurs because of stronger swirl intensity at optimal value of porosity. Effect of porosity on Nu, f and η with dual TT can be analysed with the following correlations:

$$\mathrm{Nu} = \left(0.0002R_p^3 - 0.046R_p^2 + 0.033R_p + 0.6569\right) \times \mathrm{Re}^{\left(0.00005R_p^3 - 0.0013R_p^2 + 0.0073R_p + 0.5501\right)} \cdot \mathrm{Pr}^{0.33} \tag{2.15}$$

$$f = \left(-0.0027R_p^3 + 0.0583R_p^2 + 0.0455R_p + 24.536\right) \times \mathrm{Re}^{\left(0.00009R_p^3 - 0.0022R_p^2 - 0.012R_p - 0.6006\right)} \tag{2.16}$$

$$\eta = 36.995 \cdot C \cdot C_1 \times \mathrm{Re}^{\left(-0.000011R_p^3 + 0.000187R_p^2 - 0.0000808R_p - 0.07168\right)} \tag{2.17}$$

where $C = \left(0.0002R_p^3 - 0.046R_p^2 + 0.033R_p + 0.6569\right)$ and $C_1 = \left(-0.0027R_p^3 + 0.0583R_p^2 + 0.0455R_p + 24.536\right)$

2.2.10 *Wing Chord Ratio,* w_c *[22]*

This parameter is defined for TT with different shapes of wings. It is defined as the ratio of wing chord to width of TT, as shown in Fig. 2.1h. As wing chord ratio increases, there is an increase in heat transfer as well as friction factor. This is attributed to stronger interruption of fluid stream. The correlations reported represent the effect of wing chord ratio on friction factor, Nusselt number and enhancement efficiency. These are applicable for TT with trapezoid shape wing and for Re between 5,500 and 20,200. Shapes of different wings can be grasped from Fig. 10 in the subsequent section.

$$\mathrm{Nu} = 0.625\mathrm{Re}^{0.547}\mathrm{Pr}^{0.4}(w_c|w)^{0.113} \tag{2.18}$$

$$f = 64\mathrm{Re}^{-0.587}(w_c|w)^{0.189} \tag{2.19}$$

$$\eta = 6.56\text{Re}^{-0.175}(w_c|w)^{0.049} \quad (2.20)$$

Equations (2.1)–(2.20) are vital to find the effect of various GNDPs. These correlations may be used for specified range of Re. In next section, some modified TT has been discussed.

2.3 Description of Some Modified Twisted Tapes Stating Range of Non-dimensional Parameters Used and Their Effects

Eimsa and Promvonge [18] used nine different combinations of C–CC TT. They used three twist ratios (3, 4 and 5) and three twist angles (30°, 60°, 90°). Uniform heat flux is maintained during experimentation with water as working fluid. Range of Re selected is 3,000–27,000. Figure 2.1 (a) and (b) demonstrates the typical geometry of such TT. It has been found that the mean Nu for the C–CC twisted tapes is greater than that with the typical in the range of 12.8–41.9% and with the plain tube between 27.3 and 90.5% depending upon values of non- dimensional parameter, i.e. twist ratio, twist angle and Re. Experimental and numerical (CFD) investigations were carried out for Nu and f characteristics in a tube equipped with the classic and three modified TT inserts as shown in Fig. 2.2a by Rahimi et al. [9]. Re is varied between 2,950 and 11,800. It is found from the experimental study that performance with jagged tape inserts is superior to others. The results show maximum increase of 31% in Nusselt number and 22% in overall performance of the jagged inserts, compared with the classic one. The main reason for improved performance of jagged twisted tape is

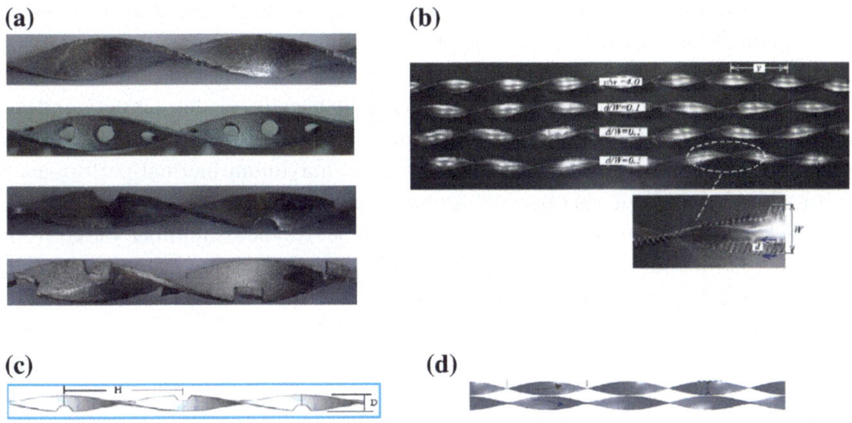

Fig. 2.2 **a** Classic, perforated, notched, jagged TT **b** Serrated TT **c** Trapezoidal cut TT **d** Double counter TT [9, 11, 12, 20]

found to be greater turbulence intensity of the fluid adjacent to the tube wall. Eiamsa-ard and Promvonge [12] carried out experimental analysis for finding heat transfer and pressure drop behaviour of circular tube with serrated twisted TT as shown in Fig. 2.2b. The measurements are conducted for Re in the range 4,000–20,000. Twist ratio kept constant at 4 and other geometric non-dimensional parameters; i.e., serration depth and width ratio both have three values 0.1, 0.2 and 0.30. Thermal performance factor for serrated TT is evaluated and is found to be more than unity. Durga Prasad et al. [20] examined the effect of trapezoidal cut TT insert as shown in Fig. 2.2c on performance of U-tube heat exchanger. They performed experiments with Re (3,000–30,000) and twist ratio (5, 10, 15, 20). Decreasing twist ratio is favourable for increasing heat transfer. For dual TT, Eiamsa-ard et al. [11] used three different space ratios ($s/D = 0.75$, 1.5 and 2.25). The efficiency value is found to be increased from 1.03 to 1.11 for space ratio of 0.0; from 1.02 to 1.09 for space ratio of 0.75; from 1.01 to 1.08 for space ratio of 1.5, and from 1.0 to 1.07 for space ratios of 2.25. They found smaller space ratio more attractive in heat transfer applications. Bhuiya et al. [16] explored the influence of double counter TT inserts as shown in Fig. 2.2d. They used double counter TT with four different twist ratios (1.95, 3.85, 5.92 and 7.75). Re varied in the range 6,950–50,050.

Bhuiya et al. [15] analysed the performance of triple twisted tape as shown in Fig. 2.3a. They used four different twist ratios (1.92, 2.88, 4.81 and 6.79). Re is varied from 7,200 to 50,200. The experimental results show that the Nu, f and thermal enhancement efficiency increase with decreasing twist ratio. For these inserts, Nu increased up to 3.85 times and f increased up to 4.2 times. Maximum enhancement efficiency achieved found to be 1.44. Wongchareea and Eiamsa-ard [22] performed experimental investigations to calculate the effect of TT with alternate axes and wings as shown in Fig. 2.3b on thermal performance characteristics. In experiments, Re range is varied (5,500–20,200), twist ratio kept constant at 4 and wing chord ratio varied with three different values (0.1, 0.2 and 0.3). Increment in wing chord ratio is favourable for enhancement in thermal performance because of stronger interruption of main stream. TT with alternate axes and wings found to give superior performance than TT with only alternate axes. This is due to the collective effects of the strong collision of fluid behind the alternate point because of alternate axis and the additional fluid disturbance near tube wall made by wings. Alternate axis TT with trapezoidal wings and wing chord ratio 0.3 gives maximum thermal performance factor of 1.42. Similar trend observed for varying width and depth ratio as observed for other configurations of TT is shown in Fig. 2.3c. Reynolds number varied from 4,000 to 20,000 with air as working fluid. Twist ratio has kept constant at 4 while length ratio (LR) has four different values (0.29, 0.43, 0.57 and 1). Here, length ratio (LR) is defined as the ratio of short-length TT (ls) to full-length (lf) TT. Introduction of short-length TT reduces the swirling effect while full-length tape (LR = 1) gives stronger swirling effect. Short-length tapes with LR 0.29, 0.43, 0.57 produce lower heat transfer by about 14%, 9.5% and 6.7%, respectively, than the full-length tape. On the other hand, the friction factor values also get reduced. Short-length tape insert is found to be inferior to that with the full-length tape because of lower η. At highest value of blockage ratio (0.19), maximum value of heat transfer and pressure drop and

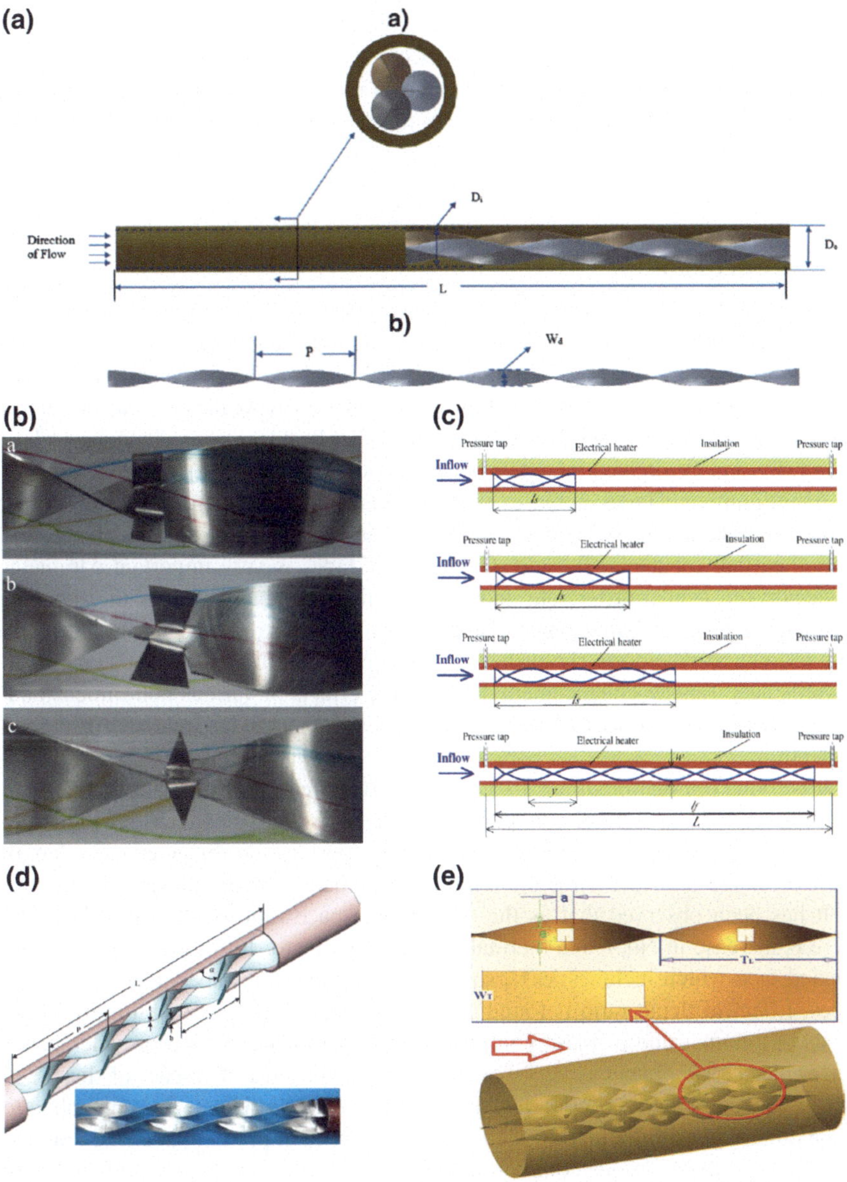

Fig. 2.3 **a** Triple twisted tape inserts in circular tube **b** Geometries trapezoidal, rectangular, triangular wings twisted tapes **c** Short-length twisted tape inserts in tube at different tape-length ratios: LR = 0.29, LR = 0.43, LR = 0.57 and LR = 1.0 **d** Test section with V-ribbed twisted tapes **e** Multiple square perforated twisted tape insert in tube [15, 22]

thermal performance factor has been observed. Optimum value of width ratio and twist ratio is found to be 0.25 and 2.5, respectively, comparing thermal performance factor at different width ratios and twist ratios. From the above discussion of various TT respective GNDPs studied and optimum values, we can conclude effectively in the subsequent section.

2.4 Conclusion

In commercial heat exchangers, twisted tapes are extensively employed to enhance the performance as it is an effective method of heat transfer augmentation. Thinner boundary layer and proper mixing of fluid result in improving the heat transfer. It has been found out that TTs integrated with some other heat transfer techniques such as use of ribs and corrugated tube give better results.

As the number of tapes increases, thermal enhancement efficiency increases. Value of thermal enhancement efficiency for triple TT insert is found to be higher by around 10% than that for dual TT insert. Geometry of tapes is very important parameter to decide the heat transfer rate and pressure drop behaviour inside the tubes of heat exchangers. Different types of tape geometries are reported in the literature. The present work reports about the various geometries of twisted tapes employed towards heat transfer enhancement. Further, the effects of various non-dimensional parameters in terms of Nu, f and heat transfer enhancement index have been discussed in depth.

It is found out that modified TT generally offers better performance than typical TT, wherein the non-dimensional parameters like twist ratio, depth ratio, width ratio, space ratio, twist angle, perforation pitch ratio, perforation diameter ratio, length ratio, blockage ratio and wing chord ratio play critical role for efficient design.

It has been observed that as the twist ratio, space ratio, width ratio decrease, there is increase in Nu, f and thermal performance factor for most of the modified twisted tape which includes V, U, square, rectangular and trapezoidal cut twisted tapes. Effect of depth ratio is found to be more prevailing than width ratio. There exists optimum value of these non-dimensional parameters which provides better performance. This paper is very beneficial to get the idea of impact of discussed GNDP on heat transfer, friction factor and enhancement efficiency, when applied to some other types of modified twisted tape. Exclusivity of present paper is combined with discussion of various common GNDPs for different kinds of modified TT. In future, more detailed practical and CFD work may be very helpful for researchers to determine the effect of same non-dimensional parameter on the performance of different modified twisted tapes.

Acknowledgements First author thanks "DST (WOS-A) KIRAN" division for financial support under grant number SR/WOS-A/ET-176/2016.

References

1. Bergles, A.E.: Techniques to augment heat transfer. In: Hartnett, J.P., Rohsenow, W.M., Ganic, E.N. (eds.) Handbook of Heat Transfer Applications, 2nd edn, Chapter 1. McGraw-Hill, New York, NY, USA (1985)
2. Webb, R.L.: Principle of Enhanced Heat Transfer. Wiley, New York, NY, USA (1994)
3. Manglik, R.M., Bergles, A.E.: Swirl flow heat transfer and pressure drop with twisted tape inserts. Adv Heat Transf. **36**, 183–266 (2003)
4. Ray, S., Date, A.W.: Laminar flow and heat transfer through square duct with twisted tape insert. Int. J. Heat Fluid Flow **22**, 460–472 (2001)
5. Naphon, P.: Heat transfer and pressure drop in horizontal double pipe with and without twisted tape insert. Int. Commun. Heat Mass Transf. **33**, 166–175 (2006)
6. Zimparov, V.: Enhancement of heat transfer by a combination of three- start spirally corrugated tubes with a twisted tape. Int. J. Heat Mass Transf. **44**, 551–574 (2001)
7. Zimparov, V.: Enhancement of heat transfer by a combination of single start spirally corrugated tubes with a twisted tape. Exp. Thermal Fluid Sci. **35**, 535–546 (2002)
8. Zimparov, V.: Prediction of friction factors and heat transfer coefficients for turbulent flow in corrugated tubes combined with twisted tape inserts. Int. J. Heat Fluid Flow **47**, 385–393 (2004)
9. Bharadwaj, P., Khondge, A.D., Date, A.W.: Heat transfer and pressure drop in a spirally grooved tube with twisted tape insert. Int. J. Heat Mass Transf. **52**, 1938–1944 (2009)
10. Rahimia, M., Shabaniana, S.R., Alsairafi, A.A.: Experimental and CFD studies on heat transfer and friction factor characteristics of a tube equipped with modified twisted tape inserts. Chem. Eng. Process. **48**, 762–770 (2009)
11. Eiamsa-ard, S., Thianpong, C., Eimsa-ard, P., Promvonge, P.: Thermal characteristics in a heat exchanger tube fitted with dual twisted tape elements in tandem. Int. Commun. Heat Mass Transf. **37**, 39–46 (2010)
12. Eiamsa-ard, S., Promvonge, P.: Thermal characteristics in round tube fitted with serrated twisted tape. Appl. Therm. Eng. **30**, 1673–1682 (2010)
13. Murugesan, P., Mayilsamy, K., Suresh, S., Srinivasan, P.S.S.: Heat transfer and pressure drop characteristics in a circular tube fitted with and without V-cut twisted tape insert. Int. Commun. Heat Mass Transf. **38**(3), 329–334 (2011)
14. Saha, S.K., Bhattacharyya, S., Pal, P.K.: Thermohydraulics of laminar flow of viscous oil through circular tube having integral axial rib roughness and centre cleared twisted tape. Exp. Thermal Fluid Sci. **41**, 121–129 (2012)
15. Bhuiya, M.M.K., Chowdhury, M.S.U., Shahabuddin, M., Saha, M., Memon, L.A.: Thermal characteristics in a heat exchanger tube fitted with triple twisted tape insert. Int. Commun. Heat Mass Transf. **48**, 124–132 (2013)
16. Bhuiya, M.M.K., Sayem, A.S.M., Islam, M., Chaudhary, M.S.U., Shahabuddin, M.: Performance assessment in a heat exchanger tube fitted with double counter twisted tape inserts. Int. Commun. Heat Mass Transf. **50**, 25–33 (2014)
17. Saha, S.K., Dutta, A., Dhal, S.K.: Friction and heat transfer characteristic of laminar swirl through a circular tube fitted with regularly spaced twisted-tape elements. Int J. Heat Mass Transf. **44**, 4211–4223 (2001)
18. Eiamsa-ard, S., Promvonge, P.: Performance assessment in heat exchanger tube with alternate clockwise and counter clockwise twisted tape. Int. J. Heat Mass Transf. **53**, 1364–1372 (2010)
19. Nanan, K., Thianpong, C., Promvonge, P., Eiamsa-ard, S.: Investigation of heat transfer enhancement by perforated helical twisted-tapes. Int. Commun. Heat Mass Transfer **52**, 106–112 (2014)
20. Durga Prasad, P.V., Gupta, A.V.S.S.K.S., Deepak, K.: Investigation of Trapezoidal-cut twisted tape insert in a double pipe U-tube heat exchanger using Al_2O_3/water nanofluid. Procedia Mater. Sci. **10**, 50–63 (2015)

21. Bhuiya, M.M.K., Chowdhury, M.S.U., Saha, M., Islam, M.T.: Heat transfer and friction factor characteristics in turbulent flow through a tube fitted with perforated twisted tape inserts. Int. Commun. Heat Mass Transfer **46**, 49–57 (2013)
22. Wongchareea, K., Eiamsa-ard, S.: Heat transfer enhancement by twisted tapes with alternate-axes and triangular, rectangular and trapezoidal wings. Chem. Eng. Process. **50**, 211–219 (2011)

Chapter 3
In Situ Non-destructive Testing of Man Riding Chair Lift System

Mohd Ahtesham Hussain Siddiqui, Anil Kumar Agrawal and Somnath Chattopadhyaya

Abstract Man riding chair lift (MRCL) system is a transportation media for the people working in underground mines. Underground mines are opened in two ways: shaft sinking vertically from the surface to the coal seam and through an incline drifted horizontally following some gradient pattern normally 1:4.5 up to the coal seam. People have to travel a long distance to reach up to the working faces in underground mines through the incline. This long travelling distance consumes more time and energy to reach working places. Due to that, the miners get tired. Production losses are incurred due to less working hours and inefficiency of workers. MRCL is installed to facilitate the people working in underground mines. It reduces man effort to reach the face and also saves time. Non-destructive testing of MRCL is mandatory to ensure the safety of the people. In this paper, ultrasonic flaw detection technique and liquid dye penetration test were conducted for NDT of MRCL.

Keywords Man riding chair lift · Non-destructive testing · Ultrasonic flaw detection technique · Liquid dye penetration testing

3.1 Introduction

Coal production from underground mines is a tedious job. In bord-and-pillar method of mining, we have to go advanced day to day during the development of coal seams. Miners have to walk long, about 3–4 km, on the uneven, gradient surface, maximum depth (RL-132) of 200 m from the surface (RL-332) for reaching working places. This travelling distance enhances with progress in coal faces. The productivity of people goes down as we move ahead. Their time and energy are wasted in this unproductive work of walking [1]. Coal Mines Regulations, 2017, Regulation 93(6) stated that "In

M. A. H. Siddiqui (✉) · A. K. Agrawal · S. Chattopadhyaya
Indian Institute of Technology (Indian School of Mines), Dhanbad, India

M. A. H. Siddiqui
Coal India Limited, Western Coal Fields Limited, Saoner, Nagpur, India

M. A. H. Siddiqui · A. K. Agrawal · S. Chattopadhyaya
Springer Heidelberg, Tiergartenstr. 17, 69121 Heidelberg, Germany

S. Yadav et al. (eds.), *Proceedings of International Conference in Mechanical and Energy Technology*, Smart Innovation, Systems and Technologies 174,
https://doi.org/10.1007/978-981-15-2647-3_3

case the travelling distance from the incline or adit mouth or pit bottom exceeds one kilometre or the travelling is arduous, the owner, agent and manager shall provide suitable man-riding arrangement as approved by the Chief Inspector, within one year from the date of coming into force of these regulations" [2]. Therefore, a system was designed which increases the effectiveness of utilisation of workmen with reducing the drudgery of walking. MRCL is a device utilising for transporting of the workforce to mines and back. MRCL was manufactured by M/s Indicon Westfalia Limited and installed in incline No. 6 of Saoner Mine No. 3, Nagpur Area, Western Coalfield Limited. The total length of man rider is 1003 m from the surface with a vertical lift of 105.04 m. Its drive head is at the surface (RL-332.90) and tail end at 21 Rise/0 Level, Seam V (RL-227.86). The chair lift system for man riding is an electrohydraulically operated and an endless haulage-type machine, equipped with all statuary safety devices. Induction motor (110 KW, 550 V, 1450 rpm) used as a prime mover to give torque to the axial piston pump with swash plate. A pump delivers 298 L/min of hydraulic fluids up to maximum 400 bars to the high-torque hydraulic motor, which transmits torque to the drive sheave. MRCL consists of two main pulleys, one driving station and one at the return station on which the rope is roaming for endless operation. These pulleys can bear the maximum loading capacity of the rope. While testing of hydraulic brakes' effectiveness, maximum torque is applied by the hydraulic pump via high-torque hydraulic motor on the driving pulley. It can stand against that maximum torque and load, without shear or weld failure.

In situ investigation via NDT technique can be done to ensure feasibility, reliability and integrity of all vital parts of MRCL. A vital component of MRCL can bear and transmit load, without impairing their originality, structural changes, physical and chemical properties and usefulness. NDT is the best engineering tool for detecting a flaw or defect developing and inherently present in the system. It is an economical and time-saving procedure. On that basis, the testing, safety and reliability of machines can be ensured [3].

An objective behind this research work of NDT is to ensure the intactness of MRCL vital component which bears the maximum load [4]. Examine the component for any discontinuity due to inhomogeneity. The discontinuity can be developed in any stage, during the life cycle of the component. This may be present inherently or developed in processing or during providing services. NDT can ensure the actual condition of the component; it provides information on defects whether present or developed internally. NDT evaluates the levels of reliability, materials' behaviour and internal characteristics of a test piece without impairing their operational condition and structural dimensions [5]. NDT suggests us to go for further use or needs some repairing or replacing of the tested components. MRCL is installed to facilitate the people working in underground mines. It reduces man effort to reach the working places and also saves time. Non-destructive testing of MRCL is mandatory to ensure the safety of the people and machines itself. Figure 3.1 indicates the line diagram of MRCL system used in transportation media for the people working in underground mines.

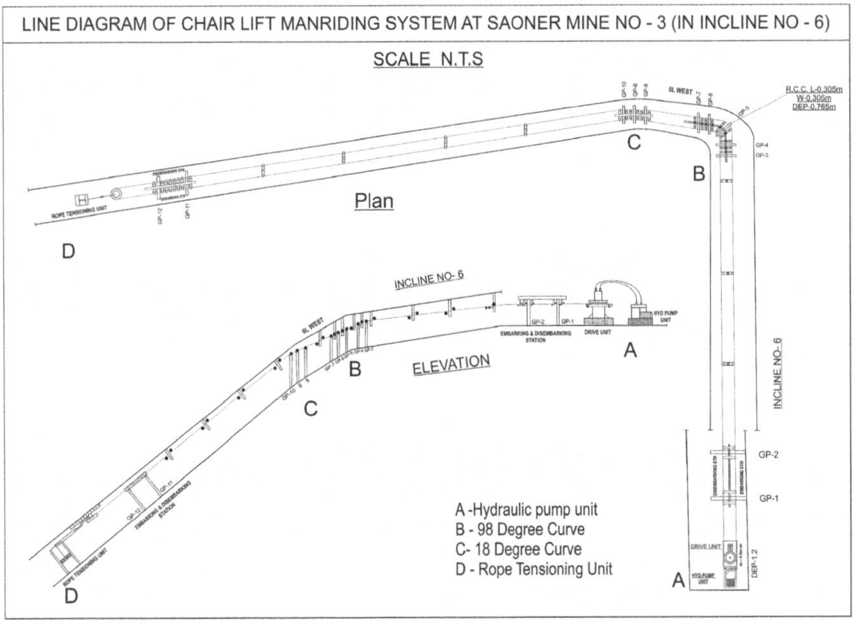

Fig. 3.1 Line diagram of MRCL system

3.2 Literature Review

Wahed and Farhan [6] performed an investigation of non-destructive testing of the pressure vessel. They utilised ultrasonic flow detection and magnetic particle testing techniques for checking long seam weld and circumferential seam weld, undercuts, porosity and hairline cracks. No defect was found in the pressure vessel. Some limitations are there for checking the varying cross-sectional parts and high-current magnetising effects in magnetic particle testing. Various cross-sectional jobs could not be tested by ultrasonic flow detection techniques. Pohl et al. [7] investigated NDT techniques for railroad wheel and gauge corner inspection. They performed in situ inspection of rail load wheel and gauge corner especially of high-speed trains to ensure safety norms of wheels with mechanised or non-destructive inspection systems. Ultrasonic wheel inspection technique and eddy current techniques were used to inspect the wheel and railway gauge corners, respectively. Gholizadeh [8] reviewed non-destructive testing methods of composite materials. In this review study, various techniques are analysed based on their capabilities and account their advantages and disadvantages. NDT techniques, such as visual testing, thermography, radiographic testing, ultrasonic testing, acoustic emission, electromagnetic testing and thermography tests, were engaged for performing these methods. These techniques are characterised based on their intrinsic characteristics and applications. Singh et al.

[9] did a review of MRCL system effectiveness in underground coal mines with reference to Ballarpur Colliery 3 and 4 pits using rank–weight method. They suggest the basic safety features to be included with the system for riskless operation of the man riders.

3.3 Test Techniques

NDT of MRCL system was done by two methods: (i) ultrasonic flaw detection technique and (ii) liquid dye penetration technique.

3.3.1 Ultrasonic Flaw Detection Technique (UFDT)

Ultrasonic flaw detection/Ultrasonic testing of a component consists of sending a beam of energy into the component. High-frequency sound energy (between 0.5 and 10 MHz) is propagating from the transducer into the testing specimen. The speed of propagation is proportional to the density and elastic behaviour of the medium. Ultrasonic waves are reflected from a dissimilar surface or interface. This is the working principle of ultrasonic waves used in UFDT. The amount of reflection of ultrasonic waves from interfaces depends on the acoustic properties of the material. Laws of light are followed by the ultrasonic waves. These waves are generated from the transducer, which transforms electrical signals to mechanical vibration and vice versa. These waves collect information and displayed on a cathode ray tube. The cathode ray tube is calibrated to enable accurate thickness measurement of locating the defect. Height, amplitude and percentage of the reflected waves are given the exact orientation and size of the defects. Figure 3.2 shows the line diagram of UFDT set-up. The drive sheave pin and return pulley shafts were tested by UFDT.

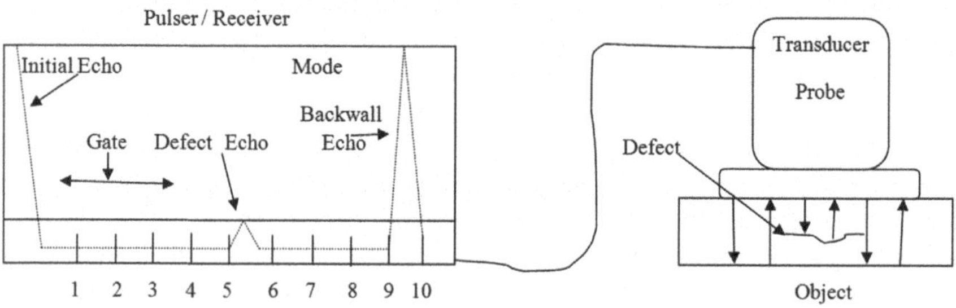

Fig. 3.2 Line diagram of UFDT set-up

3.3.2 Liquid Dye Penetration Testing Technique (LDPTT)

It is the most widely used non-destructive testing technique for the detection of surface flaws by a capillary phenomenon. In this NDT technique, the liquid form of penetrant is spread over the cleaned surface of the test specimen which was cleaned by cleaner or remover and waits for the next ten minutes. This liquid can be penetrated effectively in the surface discontinuities vicinity. The capillary phenomenon helps in finding the surface discontinuities, cracks and blowholes or pinhole.

After ten minutes, clean the surface with cleaner and then spray the contrast colour developer on the surface. If any discontinuity is there, penetrant comes out from the vicinity of defects. Shape and size of contrast colour penetrant show the intensity of defects or cracks.

3.4 Testing

3.4.1 Ultrasonic Testing Process of NDT

Ultrasonic testing process of NDT follows the reference standard ASME Sec. V, Article 4, ASTM A609 for test procedure and ASME Sec. VIII, Div.1 Apx-12 (welding checking with angle probe) for acceptance standard. Equipment used was Digiscan DS-322 with the pulse-echo method with a light intensity of 1080 lx. The different probe is shown in Fig. 3.3. Table 3.1 presents a different parameter of UFDT. Figure 3.4 shows the different test of UFDT, and Fig. 3.5 presents the procedure of testing.

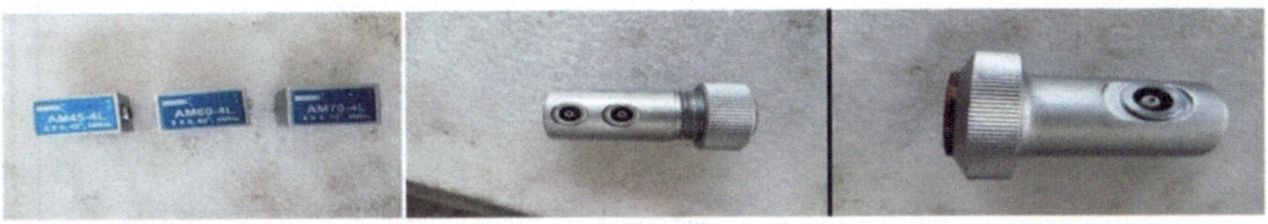

Fig. 3.3 Image of types of probes

Table 3.1 Specification of MRCL

Range	Speed	Capacity	Vertical gradient	Horizontal curve
2000 m	0–2 m/s	200 person/h	1:4.5 maximum	15–120°

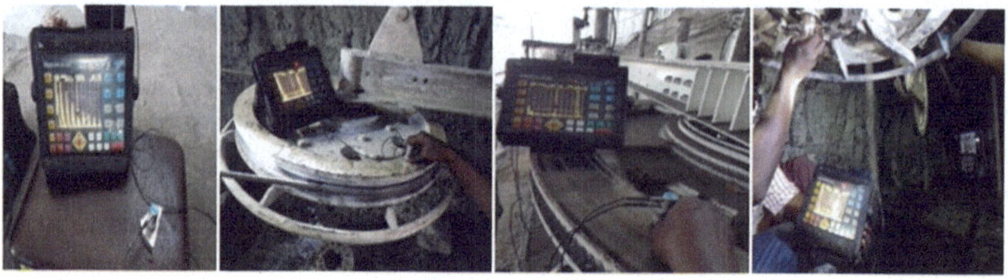

Fig. 3.4 Photographs of conducting ultrasonic flaw detection of MRCL

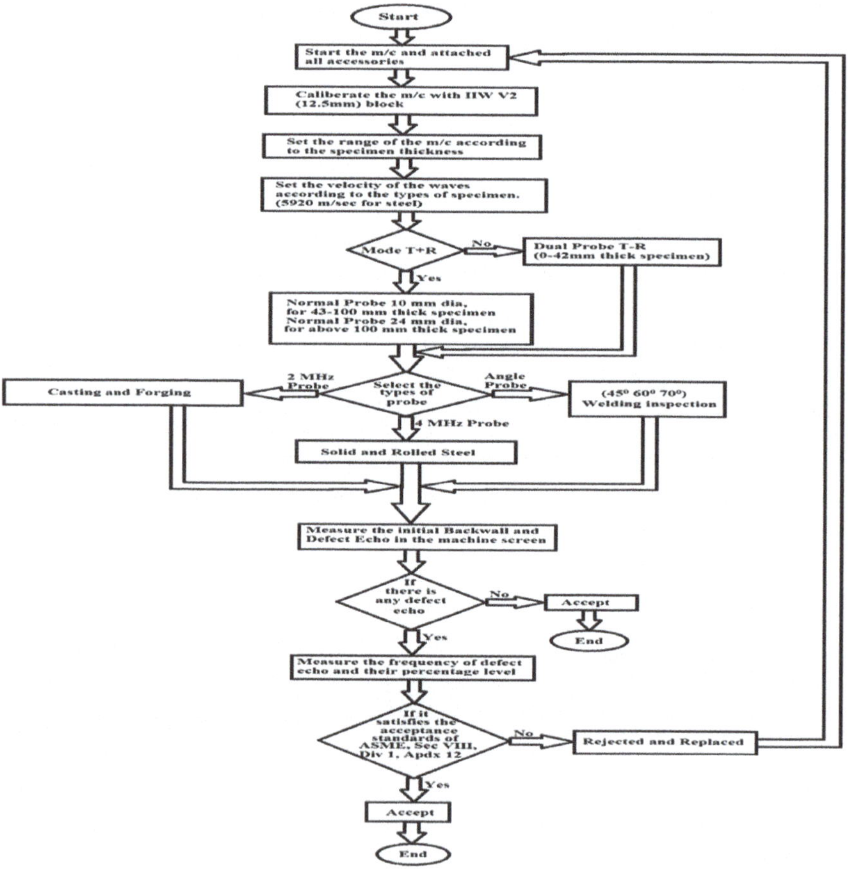

Fig. 3.5 Flowchart of UFDT testing procedure

Fig. 3.6 Images of sprays used in dye penetration test

3.4.2 Liquid Dye Penetrant Test Process of NDT

Liquid dye penetrant test process of NDT follows the reference standard ASME Sec. V, Article-24, SE-165 (IS3658-1999) for test procedure and ASME Sec. VIII, Div.1 Apx.8 for acceptance standard. The light intensity is of 1080 lx. **Cleaner/Remover** used was SKC-1, Batch No. 19A02, Brand Name: Spotcheck (Magnaflux), Manufacturing Date: January 2019, Due Date: December 2021, Cleaning Time: 05 min, done with lint-free cloth and cleaner. **Penetrant** used was Solvent Removable-type SKL-SP1, Batch No. 18F15, Brand Name: Spotcheck (Magnaflux), Manufacturing Date: June 2018, Due Date: May 2020, Dwell time: 10 min. **The Developer** used was SKD-S2, Batch No. 18J03, Brand Name: Spotcheck (Magnaflux), Manufacturing Date: September 2018, Due Date: July 2020, Developing Time: 1–2 min. Figure 3.6 shows liquid dye penetrant bottles.

Figures 3.7 and 3.8 show the use of liquid dye penetration on MRCL. Figure 3.9 shows the flowchart of liquid penetration test procedure. Different parts of MRCL were tested by NDT. Figures 3.10, 3.11 and 3.12 show the different components of MRCL. Table 3.2 presents the different components of MRCL on which NDT performed.

3.5 Result

All the above components (Table 3.3) of MRCL gone under NDT with UFDT and LDPTT, whichever applicable mentioned in front of those components. The NDT was conducted as per ASME and ASTM references, as mentioned in Tables 3.4 and 3.5. The ultrasonic test reported that no recordable indication was observed. Hence, the result was satisfactory as per the above-mentioned acceptance standards. Liquid penetrant test report stated that no surface discontinuities were observed. Hence, the results are satisfactory as per the above-mentioned standards. Except for Chair No. 66 Bracket with handle—Identification No. WCL/SAON-143 and Chair No. 77

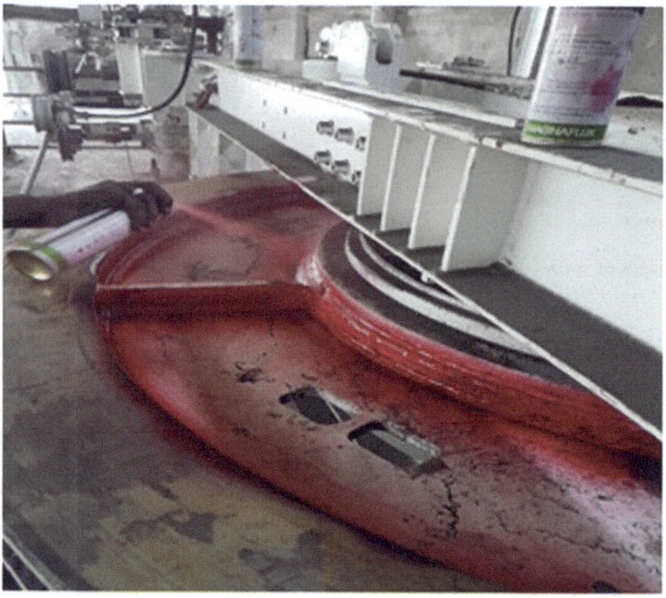

Fig. 3.7 Spraying penetrant on drive sheave

Fig. 3.8 Spraying developer on drive sheave

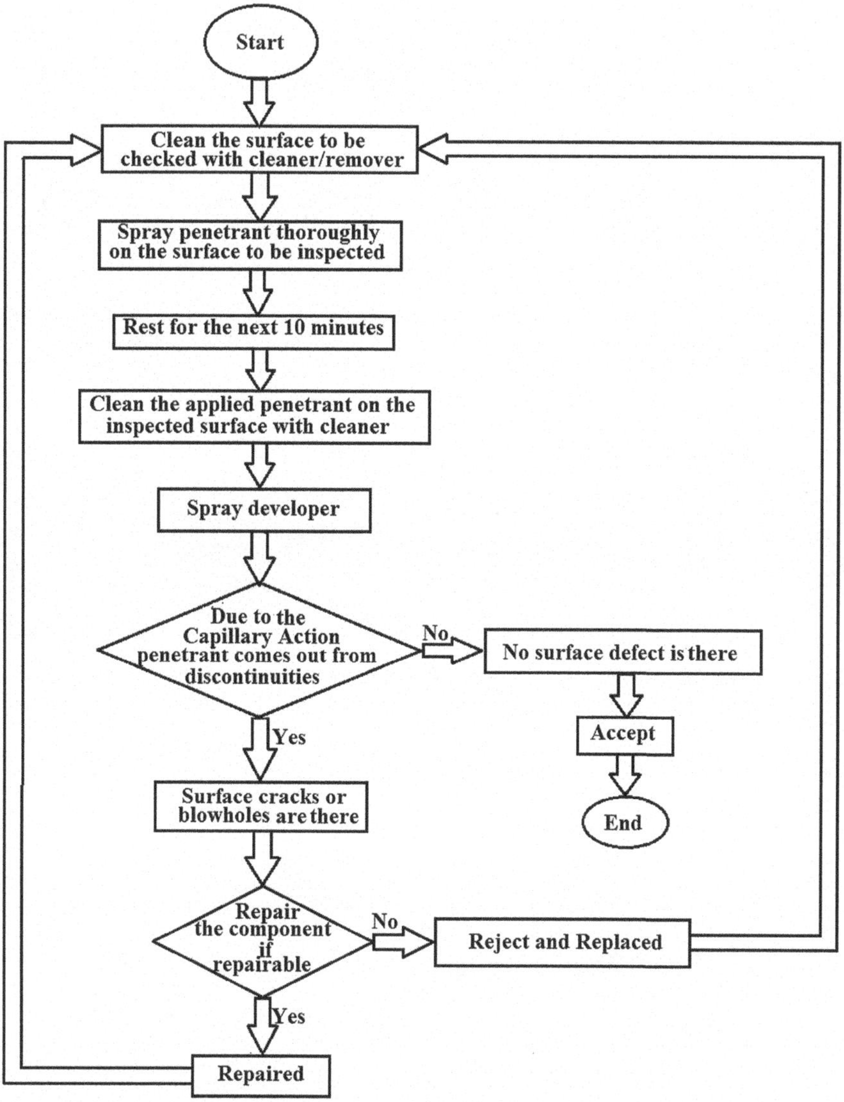

Fig. 3.9 Flowchart of liquid penetration test procedure

Bracket with handle—Identification No. WCL/SAON-16 were faulted. A 50- and 70-mm welding surface cracks were found (Fig. 3.13), respectively. These cracks were welded with suitable welding electrodes, and the Chair Nos. 66 and 77 can be utilised for further use.

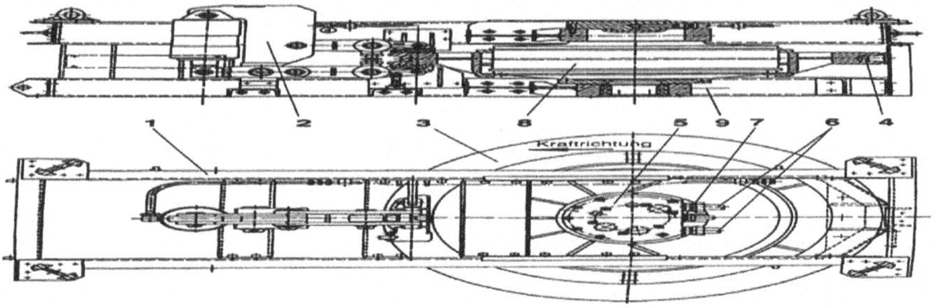

Fig. 3.10 Drive unit of MRCL. 1. Frame, 2. disc brake, 3. drive sheave, 4. drive sheave lining, 5. hydraulic motor connection set, 6. high-pressure line connection, 7. leakage line connection, 8. hydraulic motor

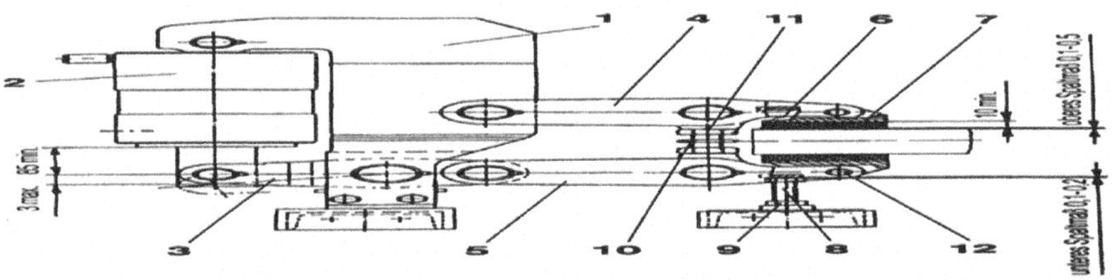

Fig. 3.11 Brake assembly of MRCL 1. Mounting, 2. hydraulic cylinder, 3. intermediate lever, 4. upper brake lever, 5. lower brake lever, 6. brake shoe, 7. brake pad, 8. adjusting screw, 9. lock nut

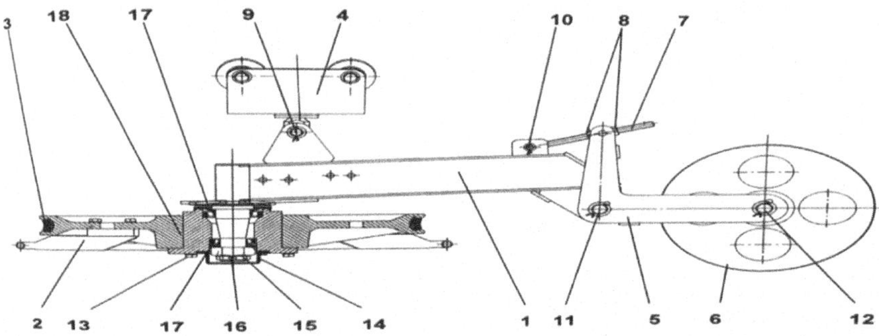

Fig. 3.12 Return station of MRCL 1. Beam, 2. return pulley, 3. lining, 4. trolley, 5. wing, 6. tensioning pulley, 7. adjusting bolt, 8. counter nut, 9. pin, split pin, washer, 10. pin, split pin, washer, 11. pin, split pin, washer, 12. pin, split pin, washer, 13. fastening bolts, 14. cover, 15. split pin, 16. adjustable nut, 17. bearing

3.6 Conclusion

Underground coal mines are governed by Coal Mines Regulation 2017. Director-General of Mines Safety (DGMS), Directorate General of Mines Safety, Dhanbad, India, is the inspecting and permitting authority for installation and use of MRCL.

Table 3.2 UFDT testing specification

Probe	Frequency	Input pulse and waves	Couplant	Pulse energy (dB)	Calibration block	Sensitivity
Normal probe	4 MHz	Ø10 mm Longitudinal	Oil	42 dB (36 + 6)	IIW V1 and V2 block	100% FSH +06 dB
Normal probe	2 MHz	Ø24 mm Longitudinal	Oil	40 dB (34 + 6)	IIW V1 and V2 block	100% FSH +06 dB

Table 3.3 Components of MRCL on which NDT performed

S. No.	Component description	Qty in nos.	UFDT	LDPTT	Identification number
1	Drive sheave casting	01	Done	Done	WCL/SAON-01
2	Return wheel casting	01	Done	Done	WCL/SAON-13
3	Bottom pin of drive sheave	01	Done	–	WCL/SAON-02
4	Brake assembly pins	05	Done	–	WCL/SAON-03 to 07
5	Turn buckle assembly	02	Done	–	WCL/SAON-08 to 09
6	Pin of brake shoe	02	Done	–	WCL/SAON- 10 to 11
7	Eye bolt	01	Done	–	WCL/SAON- 12
8	Return wheel shaft	01	Done		WCL/SAON-14
9	Man rider chair with bracket pin	150	Done	Done	WCL/SAON-15 to 311
10	Carrying pulley shaft	90	Done	–	WCL/SAON-T01 to T90

Table 3.4 UFDT test report

Surface condition of the part	Material description	Procedure ref standard	Acceptance standard	Equipment used	Method used	Light intensity
Smooth	Steel	ASME Sec. V Article-4 ASTM A609	ASME Sec VIII Div. 1 Apx-12	DS-322	Pulse-echo contact type	1080 lx

DGMS permitted for the use of MRCL in underground mines under Regulation 208(4) of CMR 17 after conducting NDT on its vital parts. Non-destructive testing conducted on MRCL to ensure whether load bearing components are in intact condition and safe for the operation of transportation of people in the underground, both level and gradient of mines. UFDT and LDPT are conducted as an NDT of MRCL.

Table 3.5 Liquid penetrant test report

Surface condition of the part	Procedure ref standard	Acceptance standard	Cleaning application	Penetrant application	Developer application
Smooth	ASME Sec. V Article-24 SE-165 (IS3658-1999)	ASME Sec. VIII Div. 1 Apx-08	Lint-free cloth and cleaner (SKC-1, Batch No. 19A02, Brand Name Spotcheck, Magnaflux)	Removable solvent (SKL-SP1, Batch No. 18F15, Brand Name Spotcheck, Magnaflux)	SKD-S2, Batch No. 18j03, Brand Name Spotcheck, Magnaflux

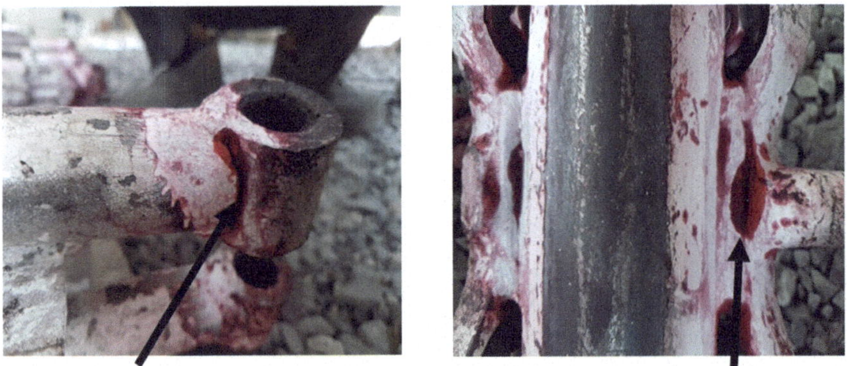

Fig. 3.13 70- and 50-mm welding surface cracks on chairs

Result of these tests ensures for continuing the use of MRCL for transporting miners in underground mines for the further next six months.

Few limitations are there in ultrasonic testing procedure of NDT. We cannot test the material which unable to transmit ultrasound waves. The velocity of sound waves propagation is fixed for every material and depends upon the elastic properties and density of parent material. Thickness, permeability, grained structure and shape of the material to be tested can create difficulty in ultrasonic testing. Thin size, course-grained structure and irregular shapes' components are difficult to check by ultrasonic method of NDT testing. The orientation of the defects concerning the sound beam, size, nature and distribution of the defects within the test material can affect the ability to detect the flaws. Large surface area test piece can consume more time for testing, and therefore, this technique is expensive. Highly skilled and experienced manpower is required to conduct the UFDT as compared with other methods of NDT.

Acknowledgements Authors are thankful to Shubham Enterprises, Nagpur, for helping in conducting NDT on MRCL.

References

1. Clemente, J.: The Expanding International Coal Market. Earth Sciences. IntechOpen (2012)
2. Coal Mines Regulations (2017)
3. Basak, D.: Performance evaluation of chairlift ropes in two man-riding systems in an incline coal mine using a non-destructive testing instrument. Insight-Non-Destructive Test. Cond. Monit. **58**(4), 206–209 (2016)
4. Saleem M.: Evaluating the pull-out load capacity of steel bolt using Schmidt hammer and ultrasonic pulse velocity test. Struct. Eng Mech. **65**(5), 601–609 (2018)
5. Kumar, S., Mahto, D.G.: Recent trends in industrial and other engineering applications of non destructive testing: a review. Int. J. Sci. Eng. Res **4**(9) (2013)
6. Wahed, M.A., Farhan, M.: An investigation of non destructive testing of pressure vessel. ISO 9001: 2008 Certified J. 3.1 (2013)
7. Pohl, R., et al.: NDT techniques for railroad wheel and gauge corner inspection. NDT & e Int. **37**(2), 89–94 (2004)
8. Gholizadeh, S.: A review of non-destructive testing methods of composite materials. Procedia Struct. Integrity **1**, 50–57 (2016)
9. Singh, U.K., Shahare, A.S., Awachat, P., Kant, R., Singh, V.A.: A review of chair lift man riding system effectiveness in underground coal mines with reference to Ballarpur Colliery 3 & 4 Pits using rank- weight method. Imperial J. Interdisc. Res. (IJIR), **2**(9) (2016)

Chapter 4
Video Processing Using Data Mining

Kanika Singhal and Abhineet Anand

Abstract Data mining is the extraction of useful information from complex data sets. The data can be complex in terms of its variety, volume and values. Video mining is becoming quite popular nowadays. Video is a combination of several types of data such as audio, video and text data. Mining video requires tremendous efforts. Starting from shooting the video, editing it and posting it in optimized form is itself a big process. And mining data from these videos involves a lot of processing. This paper provides review of various data mining techniques used in the mining of a video. In this analysis, we also describe some methods to improve the performance of video mining by using data mining techniques.

Keywords Video processing · Segmentation · Data mining

4.1 Introduction

This is the age of videos. They are increasing at a very fast pace. Extracting useful information from such a fast growing database becomes a critical process. Tremendous amount of videos is getting released on the Internet every day. Starting from YouTube channels to producing short movies, videos are the main dominating factor. The platforms like Hotstar, Netflix and Amazon prime are the by-product of emerging video technology. Videos are not only the combination of pictures; it also includes audio and text. Video processing not only involves the content of the video but also involves the color, texture and other properties involved in the video.

Video processing involves a number of steps. After shooting the videos, the foremost and most important step is to divide the video into smaller units. It is not possible to apply any operation on the complete video; dividing the video into the smaller units is necessary. The operations such as trimming the video, adding text, adding transitions, adding background music and muting the video can be easily

K. Singhal
Galgotias College of Engineering and Technology, Greater Noida, India

A. Anand (✉)
Chitkara University Institute of Engineering and Technology, Rajpura, Punjab, India
e-mail: abhineet.mnnit@gmail.com

S. Yadav et al. (eds.), *Proceedings of International Conference in Mechanical and Energy Technology*, Smart Innovation, Systems and Technologies 174,
https://doi.org/10.1007/978-981-15-2647-3_4

done after segmenting the video into smaller units. The unwanted data is removed, and the video is merged and concatenated into one video [1]. This process can be done by relevance analysis where the irrelevant attributes are removed. Every key frame can be processed according to the requirement. The patterns are observed, and the information is extracted from these videos.

The video that is concise and depicts maximum information in short time is the need of the hour. Video processing not only involves segmentation of video into smaller units but also involves the analysis of the text embedded in the video, background music added to the video and still images used in the video. Different operations can be applied by using data mining techniques.

After segmentation of the video, clustering can help group videos in the same cluster according to the content of the video. Video is not a single entity; it is the collection of audio, video and text. Audio plays an important role in video processing [2]. Time domain features are the key component. The audio can be extracted by using frequent item sets. For example, in a movie, if there is a scene of bomb explosion, it must be accompanied by high audio. The audio and the sound are inseparable, so they will frequently occur together.

Videos can be analyzed using the evolution analysis technique of data mining. Videos change with time; it involves the behavior that is not constant; adapting with the changes requires mining of associations and patterns [3]. After the careful observation of pattern, the next step is to compress the video before distributing it. Figure 4.1 depicts the segmentation of original video and its concatenation.

Text is an important information source for the video. The text can be added manually or it can automatically occur in the video. The text data can be characterized using the outlier analysis. The values outside the boundary are rejected. The relevant values can be used for further analysis. The text properties such as range of values and numeric data can be classified into different groups. The background noise can be eliminated by data cleaning process [4].

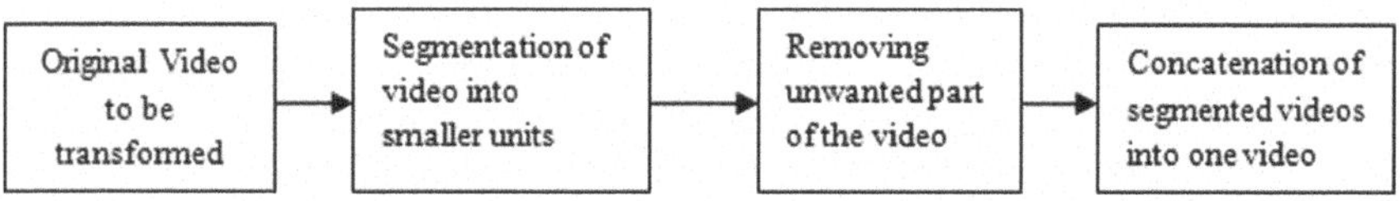

Fig. 4.1 Video processing

4.2 Video Data Mining Approaches

4.2.1 Video Clustering Mining

Whenever a video is processed, it has to be segmented into smaller units for better processing. The segments are divided into clusters according to their similar properties. It is easy to process the smaller clusters than the complete video. There is a high probability of pattern detection in clusters rather than the complete stream of video.

4.2.2 Video Rule-Based Classification

After splitting the video into different segments, the next step is to remove irrelevant data and concatenate the videos to form meaningful pattern [5]. After merging the videos, a categorical label can be assigned to them. A classifier can be built with some predefined rules, and the videos can be processed according to the predefined rules.

4.2.3 Video Decision Tree Mining

A decision tree depicting a hierarchal structure can be formed. The root node consists of processing the video. The first and foremost step is the segmentation of the video which can be done by video dividing the key frames into clusters and shot boundary detection. The feature extraction is also an important part of video processing [6]. Adding keywords to the video adds to its value which can be done by video annotation mechanism. Video summarization can be done by rule-based classification. And the video can be posted and distributed. Figure 4.2 shows the hierarchal structure of video processing.

4.2.4 Video Association Rule Mining

The association rules from the videos can be generated by using the association rule mining. This can be done by generation of frequent patterns. Every video is processed according to some fixed pattern. The associations between audio and text are also extracted, and classification rules can be generated using a priori algorithm [7]. The patterns are observed, and the video is mined according to that pattern.

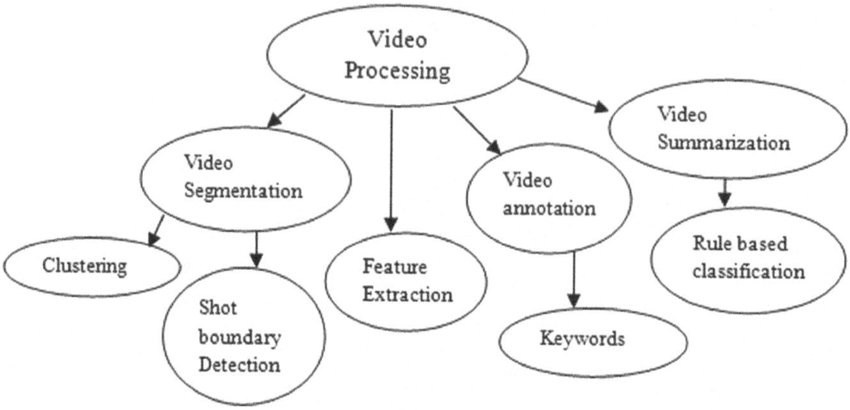

Fig. 4.2 Hierarchal structure of video processing

4.2.5 Video Prediction Mining

Prediction is the phenomenon of predicting the data values which are unknown. For example, after the video is live and posted, many factors come into prediction category. For example, how many people will watch the video, the views on the videos, subscriber increase after watching the video and the watch time of the video. These all factors need to be predicted.

4.2.6 Data Mining Applications in Video Processing

Video processing involves a number of steps. The main concern is to optimize the video before distribution by using data mining techniques. This can be easily possible if we follow some steps in proper manner and in sequence.

Shooting the video—The foremost thing about a video is shooting it in a proper manner. This can be avoided by using the shot boundary detection technique of data mining where the video is segmented in between the shots of the camera to detect the boundaries and to produce high-quality videos.

Video processing—It is the most important step after shooting the video. The long-frame video must be segmented, and the data must be divided into small clusters and then subclusters. The processing on the data will become easy if the long-frame videos are divided into small time frames for editing using clustering approach of data mining [8]. Clustering helps to divide the small intervals into clusters, and processing becomes quite simple.

Video information retrieval—After processing the video, the relevant information is to be extracted. Suppose, if we want to omit certain portion of the video that is not in relevance, the relevant information needs to be retrieved. This can be done by dividing the video into smaller units and forming a hierarchy using hierarchical clustering method of data mining. Forming a hierarchy and then retrieving the necessary information will be a simple task.

Video summarization—This refers to the concise content that we create for our target audience. The viewers should understand the content in the proper manner, and it should seem useful to them [9]. Here, rule-based classification can be used to extract information from the videos that draw the attention of the viewer viewing it.

Video annotation—Before posting the video, the keywords are to be determined. The video will be more famous and get more views if the keywords are used correctly. Selecting proper keywords for the video is an important task. This can be done by frequent pattern association where we can observe the pattern associated with the video and select the keywords according to that association [10]. For example, if we are processing a travel video, one of the keywords that can be used is itinerary.

Figure 4.3 shows the process of video processing by using different data mining techniques.

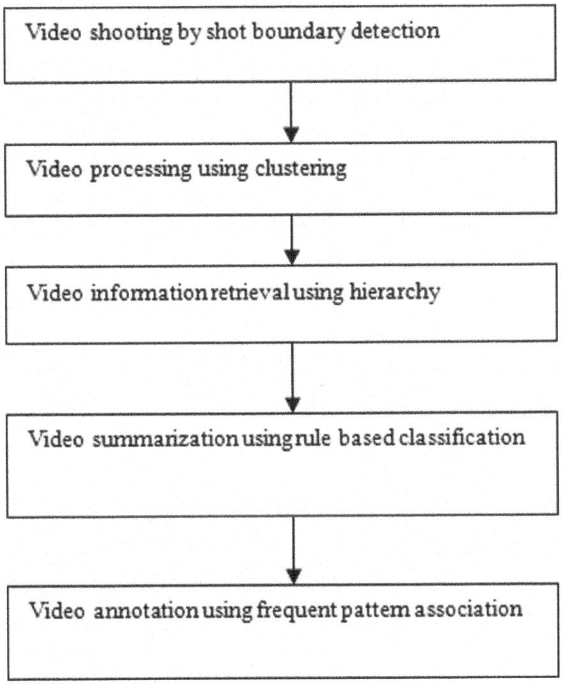

Fig. 4.3 Data mining applications in video processing

4.3 Work Analysis

Video processing consists of many steps that are used to produce a quality video. Table 4.1 shows the video mining process along with data mining techniques used to help in the video processing process.

The foremost step is shooting the video. Frequent pattern can be generated by the shot boundary detection technique. It mainly uses the technique association rule mining and prediction of the parameters of shooting a video. The association between parameters is of utmost importance.

Video segmentation consists of dividing the video into smaller interval by removing irrelevant data. This step can be done by clustering which divides the video into small clusters, classification which assign the categorical labels and hierarchy of data.

Video processing makes use of clustering, classification, frequent pattern association, hierarchal method, association rule mining and prediction to process a video [11]. Video information retrieval generates rules to retrieve the relevant information from the large database which can be done by frequent pattern association and association rule mining.

Video summarization uses frequent pattern association to generate a frequent item set, hierarchal method to generate a hierarchy among the key frames of the video and associate rule mining to generate a pattern among the key frames of the video.

Video annotation consists of adding keywords which can make the search of video easier. This can be done by hierarchal method which generates the hierarchy among keywords on the priority basis, the associations among the keywords and prediction to predict the appropriate keyword.

Table 4.1 Video mining process with data mining techniques

Video mining process	Clustering	Classification	Frequent pattern association	Hierarchical method	Association rule mining	Prediction
Video shooting	No	No	Yes	No	Yes	Yes
Video segmentation	Yes	Yes	No	Yes	No	No
Video processing	Yes	Yes	Yes	Yes	Yes	Yes
Video information retrieval	No	No	Yes	No	Yes	No
Video summarization	No	No	Yes	Yes	Yes	No
Video annotation	No	No	No	Yes	Yes	Yes

4.4 Research Issues in Video Mining

The main research issue that comes under video processing is the way the segmentation of the video is done. Video segmentation is the most important step of video processing. But the way it is done is not concise and clear. Next problem is in the selection of data mining algorithm. The implementation of data mining techniques along with video processing steps is an arduous task. Thirdly, the extraction of useful information from the key frame of the video is a complex process.

4.5 Conclusion

Video mining has grown at a great pace in the coming years. Lot of work has been done, but the scope of improvement still remains. The analytics of the video, proper optimization of the key parameters and selection of appropriate data mining algorithm still remain a concern. Removal of irrelevant data from the main stream of video requires careful observation of video key frames. The manner in which the operations are performed on the video sometimes becomes a great challenge. Video processing is a growing field and the challenges will be detected and solved in the coming time.

References

1. Anwar, F., Petrounias, I., Morris, T., Kodogiannis, V.: Mining anomalous events against frequent sequences in surveillance videos from commercial environments. Exp. Syst. Appl. **39**, 4511–4531 (2012)
2. Anjulan, A., Canagarajah, N.: A unified framework for object retrieval and mining. IEEE Trans. Circ. Syst. Video Technol. **19**(1), 63–76 (2009)
3. Ahmed, A.: Video representation and processing for multimedia data mining. Semantic Mining Technologies for Multimedia Databases, pp. 1–31. IGI Press (2009)
4. Anjulan, A., Canagarajah, N.: A novel video mining system. In: Proceedings of 14th IEEE International Conference on Image Processing, pp. 185–189. San Antonio, Texas (2007)
5. Aradhye, H., Toderici, G., Yagnik, J.: Video2Text: learning to annotate video content. In: Proceedings of IEEE International Conference on Data Mining Workshops, pp. 144–152 (2009)
6. Bhatt, C.A., Kankanhalli, M.S.: Multimedia data mining: state of the art and challenges. Multimedia Tools Appl. **51**, 35–76 (2011)
7. Brezeale, D., Cook, D.J.: Automatic video classification: a survey of the literature. IEEE Trans. Syst. Man Cybern. Part C Appl. Rev. **38**(3), 416–430 (2008)
8. Burl, M.C.: Mining patterns of activity from video data. In: Proceedings of the SIAM International Conference on Discrete Mathematics, pp. 532–536 (2004)
9. Cui, P., Liu, Z.-Q., Sun, L.-F., Yang, S.-Q.: Hierarchical visual event pattern mining and its applications. J. Data Mining Knowl. Disc. **22**(3), 467–492 (2011)

10. Chen, S.-C., Shyu, M.-L., Zhang, C., Strickrott, J.: Multimedia data mining for traffic video sequences. In: Proceedings Second International Workshop on Multimedia Data Mining MDM/KDD'2001 in Conjunction with ACM SIGKDD Seventh International Conference on Knowledge Discovery and Data Mining, pp. 78–86 (2001)
11. Choudhary, A., Chaudhury, S., Basnerjee, S.: A framework for analysis of surveillance videos. In: Proceedings of Sixth Indian Conference on Computer Vision, Graphics & Image Processing, pp. 344–350 (2008)

Chapter 5
Impact of Nanoparticles on the Performance and Emissions of Diesel Engine Using Mahua Biodiesel

Aprajit Jasrotia, Anoop Kumar Shukla and Niraj Kumar

Abstract This experiment confronts the performance as well as the exhaust gas emission highlights (characteristics) of the four-stroke diesel engine (single-cylinder direct injection) by introducing three fuel samples: biodiesel (B100), diesel–biodiesel (B40) and diesel–biodiesel–nanoparticles (B40T105MWCNT45). Titanium dioxide (TiO_2) and multi-walled carbon nanotubes (MWCNT) are taken in a quantity of 105 and 45 mg/l and are mixed with diesel–biodiesel fuel sample by using the ultrasonicator. N-butyl alcohol (1% by volume) is also added in nanoparticles dispersed fuel sample to increase the nanoparticles' stability into diesel–biodiesel fuel blend. The experiment is performed on the diesel engine at 1500 rpm which is kept constant throughout the experiment. The results obtained from the experiment show clearly that by adding the nanoparticles, namely TiO_2 and MWCNT into the diesel–biodiesel fuel mixture (i.e. B40T105MWCNT45), there is a significant improvement of 27.6 and 10.3% in the brake thermal efficiency (BTE) as compared to B100 and diesel. Also, the reduction of 22.8 and 20.4% in NO_x; 22.4 and 27.9% in CO; 35.8 and 52.7% in HC is observed when compared to B100 and diesel.

Keywords Mahua biodiesel · Titanium dioxide · Multi-walled carbon nanotubes · N-butyl alcohol · Engine performance · Engine emissions

Nomenclature

BTE	Brake thermal efficiency
TiO_2	Titanium oxide
MWCNT	Multi-walled carbon nanotubes
BSFC	Brake-specific fuel consumption
NO_x	Nitrogen oxide
CO	Carbon monoxide
bTDC	Before top dead centre
HC	Hydrocarbon

A. Jasrotia (✉) · A. K. Shukla · N. Kumar
Amity University, Uttar Pradesh Noida, India

S. Yadav et al. (eds.), *Proceedings of International Conference in Mechanical and Energy Technology*, Smart Innovation, Systems and Technologies 174,
https://doi.org/10.1007/978-981-15-2647-3_5

B100	Mahua biodiesel
B40	40%B100 + 60% diesel
B40T105MWCNT45	40%B100 + 60% diesel + 105 mg TiO_2 + 45 mg MWCNT

5.1 Introduction

Today, there is a great demand of diesel engines in the automobile industry, construction work and marine industries as they are considered more efficient and economical [1]. But the future of diesel engines has become enigmatic as the oil reserves are falling rapidly due to increase in world population and demand for more energy [2]. Astonished threats, like ozone layer depletion and change in climate due to global warming, have prompted the governments of many countries to issue stringent norms on harmful emissions released by such engines, hence developing an urgent need to use renewable fuels [3]. Biodiesel is being used all over the world today and is considered as renewable and eco-friendly fuel over petrol and diesel [4, 5]. It has various advantages like lower smoke level, unburned hydrocarbons and carbon monoxide than petrol and diesel fuel [6]. Biodiesel is found as the most suitable source available to meet the rising energy demand throughout the world [7] and also its use would limit the reliance on fossil fuels in the situation of world energy oil crisis [8].

The growing research feels the account of biofuels in lowering down the engine emissions with lower ramifications on the performance of the engine [9].

In past some years, various researchers have revealed that by introducing the nanoparticles as nano-additives in diesel–biodiesel fuel blends, the performance of engine can be raised and the harmful exhaust gas emissions can be lowered down [9]. The most frequently used nanoparticles in the study are alumina and cerium oxide [10], carbon-coated aluminium nanoparticles [9], multi-walled carbon nanotubes [11], titanium dioxide [12] and zinc oxide [13, 14].

However, very few studies have focused on the combined effect of titanium dioxide (TiO_2) and multi-walled carbon nanotubes (MWCNT) that can be used as additives to check the performance as well as the emission parameters of diesel engine by introducing mahua biodiesel. Thus, the present paper aims to find out the combined effect of two nano-additives, namely MWCNT and TiO_2 on the performance and emission characteristics of the diesel engine.

5.2 Experimental Procedure

In this experiment, mahua biodiesel (B100) (supplied by: M/s. Indo-Bio Energy Industries, Nagpur, India) and diesel are used. Nanoparticles, such as titanium oxide and multi-walled carbon nanotubes, are purchased from Platonic Nanotech Private Ltd., Jharkhand, India, and their specifications are mentioned in Tables 5.1 and 5.2.

Table 5.1 TiO_2 nanoparticles specifications

Item	Specification
Manufacturing company	Platonic Nanotech Private Ltd.
Chemical name	Titanium oxide (TiO_2)
CAS No.	13463-67-7
Atomic weight	79.8658 g/mol
Average size of particle	30–50 nm
Surface area per unit mass	200–230 m^2/g
Appearance	White

Table 5.2 Specification of MWCNT nanoparticles

Item	Specification
Manufacturer	Platonic Nanotech Private Ltd.
Chemical name	Multi-walled carbon nanotubes (MWCNT)
CAS No.	308068-56-6
Diameter	10–15 nm
Length	2–10 μm
Surface area per unit mass	250–270 m^2/g
Appearance	Black

Also, the characteristics' study of both the nanoparticles is given in Figs. 5.2, 5.3, 5.4 and 5.5 (Fig. 5.1).

During the experiment, three test fuel samples, namely B100 (containing 100% biodiesel), B40 (containing 40% biodiesel and 60% diesel) and B40T105MWCNT45 (containing 40% biodiesel, 105 mg titanium dioxide and 45 mg multi-walled carbon

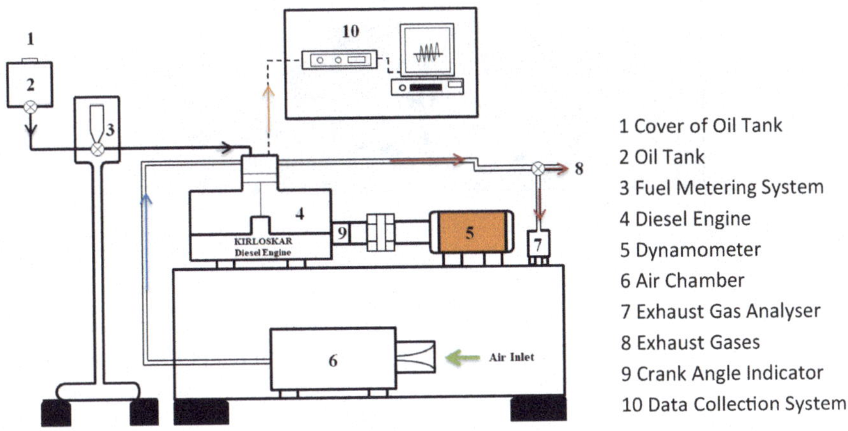

Fig. 5.1 Block diagram of experimental set-up

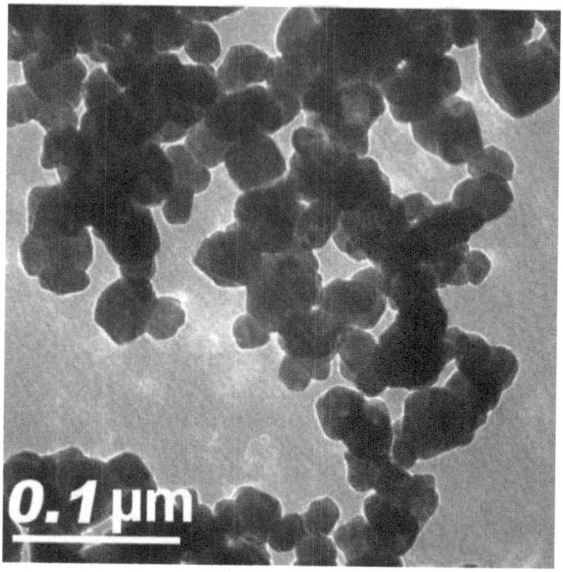

Fig. 5.2 SEM image of TiO_2

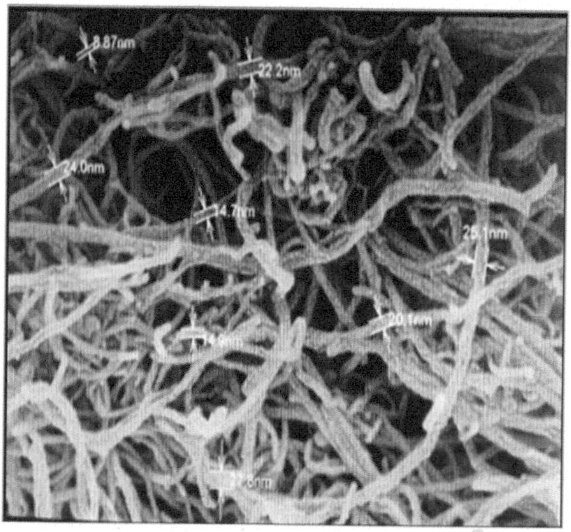

Fig. 5.3 SEM image of MWCNT

nanotubes) are prepared. The nanoparticles are incorporated into the diesel–biodiesel fuel blend by using the ultrasonicator (Fig. 5.6) for 90 min to get the homogeneous

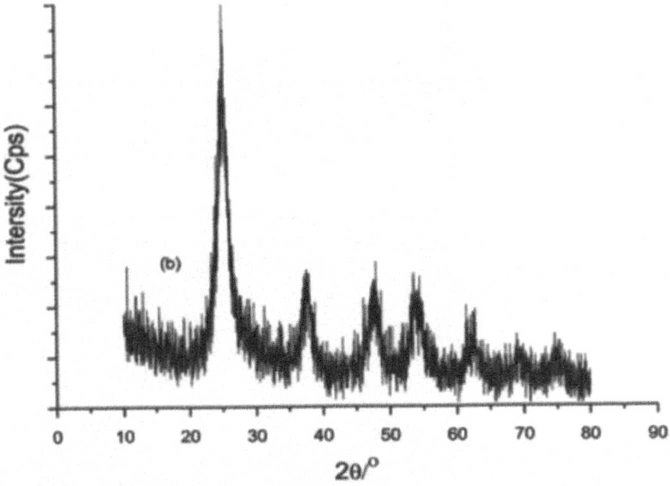

Fig. 5.4 XRD image of TiO_2

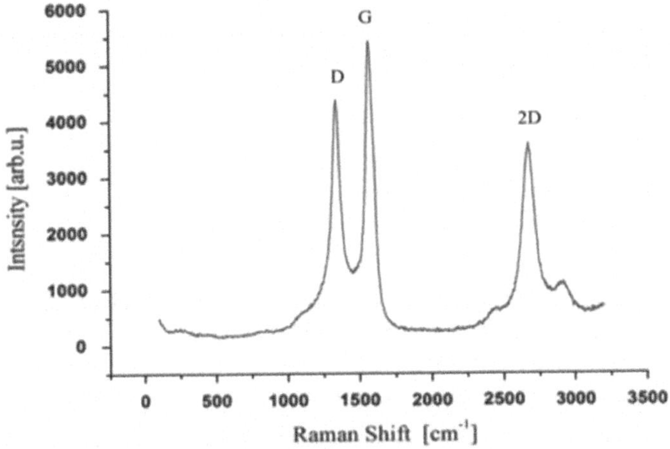

Fig. 5.5 Raman spectra of MWCNT

solution. N-butyl alcohol is added (1% by volume) in the test fuel sample containing the nanoparticles to enhance the uniformity of the nanoparticles in the diesel–biodiesel test fuel samples. The physiochemical properties of diesel, B100, B40 and B40T105MWCNT45, are listed in Table 5.3.

In the present work, all experiments are carried out with Kirloskar, single-cylinder, four-stroke diesel engine having a rated power of 5.2 kW. The engine runs with the injection timing of 26° bTDC at the same speed of 1500 rpm. Firstly, the engine is fuelled with diesel to obtain initial data, and after that it is fuelled with B100,

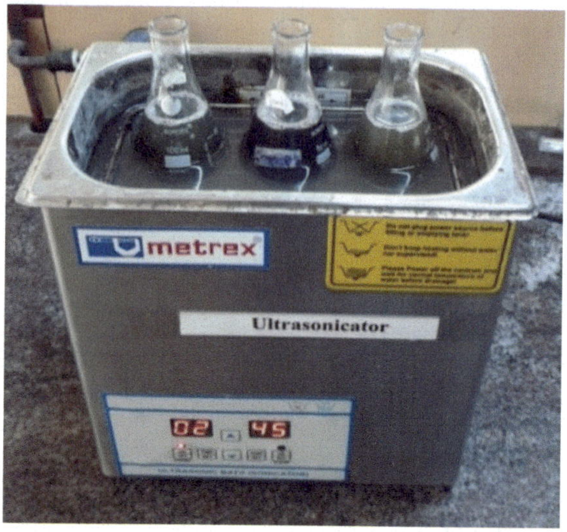

Fig. 5.6 Ultrasonicator used for homogeneous mixing

Table 5.3 Properties of test fuels

Properties	Diesel	B100 (mahua biodiesel)	B40	B40T105MWCNT45
Viscosity cST	2.72	5.93	3.76	3.89
Density at 40 °C (g/cc)	0.8	0.89	0.84	0.83
Calorific value (MJ/kg)	45.25	41.81	42.45	42.98
Flash point (°C)	63	154	76	72
Cetane number	46	52.4	51.7	53.6

B40 and B40T105MWCNT45. The specification of the diesel engine is cited in Table 5.4. Eddy current dynamometer is being used to add load on the engine. The i3SYS exhaust gas analyser (EGA) is employed to check the engine emissions like CO, NO_x and UBHC. The schematic diagram is shown in Fig. 5.1.

Table 5.4 Specifications of diesel engine

Engine details	Specification
Make	Kirloskar
Type	Single-cylinder, four-stroke, water-cooled DI engine
Compression ratio	17.5:1
Bore × stroke	85.7 × 110 mm
Swept volume	661.45 cm^3
Rated output	5.2 kW at 1500 rpm
Injection timing	26° bTDC
Connecting rod length	234 mm

5.3 Results and Discussions

Impact of nano-additives (TiO_2 and MWCNT) in mahua biodiesel–diesel blend is investigated, in which the performance and emission parameters of diesel engine are calculated for B100, B40, B40T105MWCNT45 and diesel test fuels. The results of which are discussed in below section.

5.3.1 Performance Characteristics

It is perceived from Fig. 5.7 that the brake thermal efficiency for all the fuel samples is showing the increasing trend with the increase in loading conditions. Figure 5.7 also shows that the B40T105MWCNT45 shows the highest brake thermal efficiency (BTE) of 27.6, 14.6 and 10.3% in comparison with B100, B40 and neat diesel. This could be attributed to the combined effect of TiO_2 and MWCNT nano-additives, which enhanced the combustion, atomized the fuel and fast evaporation of nanoparticles dispersed test fuel. B100 is showing the minimum brake thermal efficiency over all other fuel samples because of the high viscosity as well as high density of neat mahua biodiesel (B100) which leads to the poor combustion process.

Figure 5.8 depicts that the BSFC for all fuel samples has a decreasing trend from no load to full load condition, and it also shows that the B40T105MWCNT45 has the minimum fuel consumption when compared to B100, B40 and diesel probably because of fine atomization properties of both TiO_2 and MWCNT nanoparticles leading to the better combustion process.

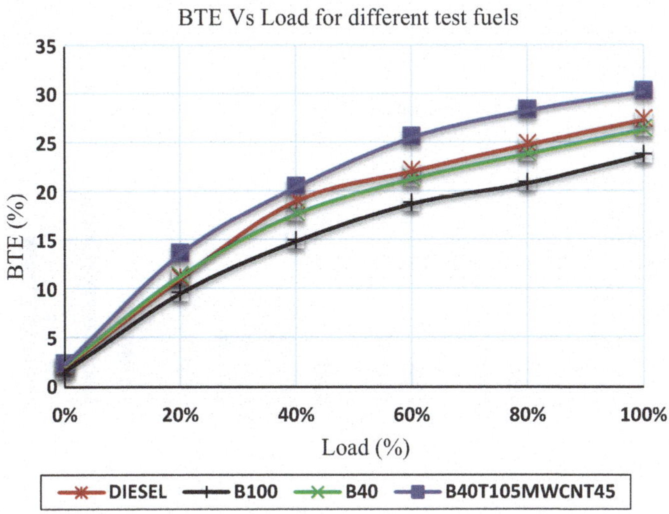

Fig. 5.7 Variation of BTE with sample fuels at different load

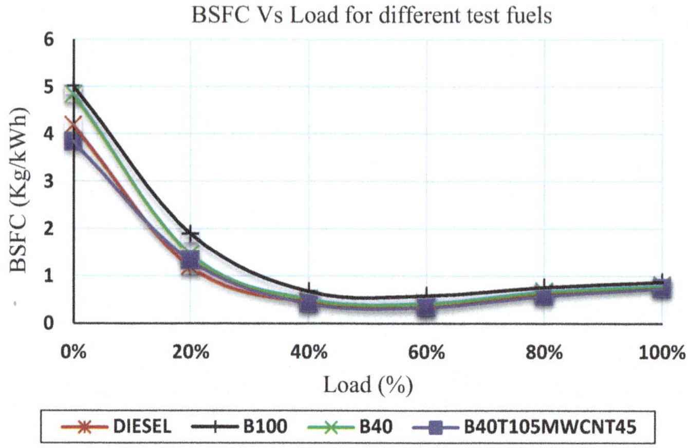

Fig. 5.8 Variation of BSFC with sample fuels at different load

5.3.2 *Emission Characteristics*

Figure 5.9 shows the change in NOx emissions w.r.t different loading conditions for all three test fuels. NO_x emissions of all the three fuels increase with the increase in loading condition. As load increases, the fuel supply also increases which causes poor vaporization, incomplete combustion, long delay period and high NO_x emissions. It is reported that NO_x emissions for B100 are maximum as compared to rest of the test fuel samples (i.e. B40, B40T105MWCNT45 and diesel) which could be possible due to the more availability of oxygen (O_2) in mahua biodiesel (B100) when compared with diesel which improves the combustion process and rises the combustion temperature which yields higher NO_x emissions. However, the mixing of TiO_2 and

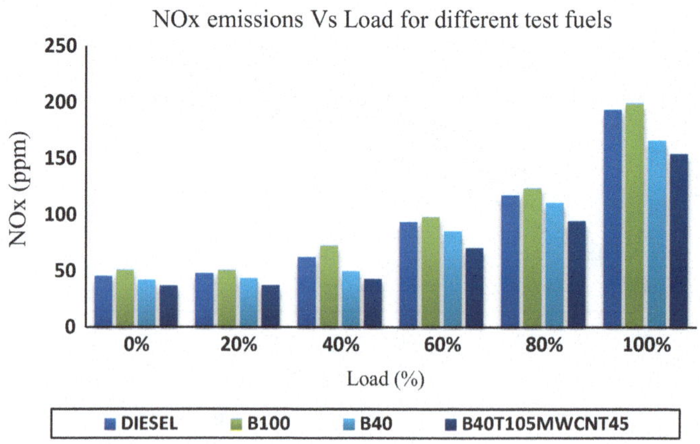

Fig. 5.9 NO_x variations at different loads for various test fuels

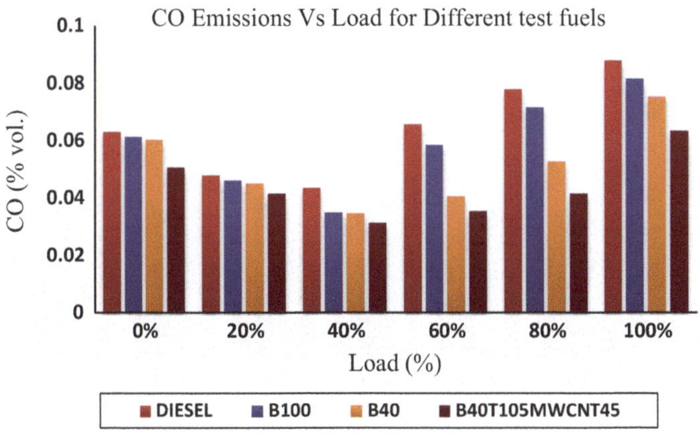

Fig. 5.10 Change in CO emissions for various test fuels at different loads

MWCNT nanoparticles into biodiesel–diesel fuel mixture causes a reduction in NO_x emissions at all loads. B40T105MWCNT45 shows the minimum NO_x emissions of 22.8, 7.2 and 20.4% when compared to B100, B40 and neat diesel fuel.

Figure 5.10 shows the CO emissions change for various fuels at different loading conditions. At full load, CO emission of 0.081% is reported for B100 as compared to the diesel of 0.088%. This is because of ingrained oxygen molecule that changes carbon monoxide (CO) to carbon dioxide (CO_2). Adding TiO_2 and MWCNT nanoparticles into biodiesel–diesel fuel mixture (i.e. B40T105MWCNT45) improved the atomization and evaporation of fuel which pilot the perfect combustion that yields lower CO emissions. B40T105MWCNT45 shows the lowest CO emissions of 22.4, 16 and 27.9% when compared to B100, B40 and neat diesel fuel.

Figure 5.11 shows that the HC emission for diesel is maximum when it is compared with other test fuels. However, the TiO_2 and MWCNT nanoparticle fuel sample (i.e. B40T105MWCNT45) have minimum HC emissions of 35.8, 18.5 and 52.7% when compared to B100, B40 and neat diesel. The remarkable reduction in HC emissions is noticed for B40T105MWCNT45. This is because of the excellent air–fuel mixing ability of nanoparticle dispersed test fuel over all the other fuel samples which changes HC to H_2O.

5.4 Conclusions

Based on the present study on the performance and emission parameters of diesel engine by using titanium dioxide (TiO_2) and MWCNT into mahua diesel–biodiesel fuel, the following conclusions have been drawn:

1. The performance of the diesel engine got enhanced, and the emissions were reduced when the nanoparticles were mixed into the diesel–biodiesel fuel sample.

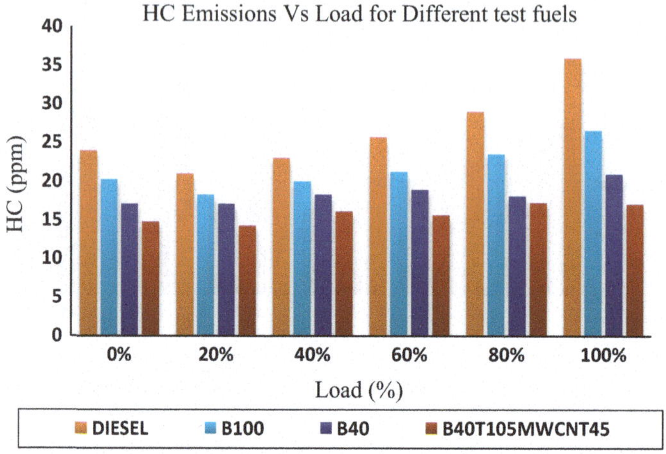

Fig. 5.11 Variation of HC emissions at different loads for various test fuels

2. The remarkable improvement of 27.6, 14.6 and 10.3% is observed in brake thermal efficiency (BTE) for B40T105MWCNT45 in comparison with B100, B40 and diesel.
3. There was the reduction in ignition delay period as well as the boost in combustion process when nanoparticles were added.
4. NO_x emissions reduced drastically for B40T105MWCNT45 with percentage of 22.8% against B100 test fuel sample due to the combined effect of titanium dioxide and multi-walled carbon nanotubes.
5. Lowest CO emissions of 22.4, 16 and 27.9% are observed for B40T105MWCNT45 when compared to B100, B40 and diesel.
6. HC emissions are lower for B100, B40 and B40T105MWCNT45 at all loads as compared to the diesel.
7. The fuel B40T105MWCNT45 showed the best engine performance along with reduced exhaust emissions as compared with neat diesel, B40 and B100 test fuels.

References

1. Selvan, V.A., Anand, R.B., Udayakumar, M.: Effects of cerium oxide nanoparticle addition in diesel and diesel-biodiesel-ethanol blends on the performance and emission characteristics of a CI engine. J. Eng. Appl. Sci. **4**(7), 819–6608 (2009)
2. Al-Maamary, H.M.S., Kazem, H.A., Chaichan, M.T.: Climate change: the game changer in the Gulf Cooperation Council region. Renew. Sustain. Energy Rev. **76**, 555–576 (2017)
3. Al-Maamary, H.M.S., Kazem, H.A., Chaichan, M.T.: The impact of oil price fluctuations on common renewable energies in GCC countries. Renew. Sustain. Energy Rev. **75**, 989–1007 (2017)

4. Meng, Y.-L., et al.: Transesterification of rapeseed oil for biodiesel production in trickle-bed reactors packed with heterogeneous Ca/Al composite oxide-based alkaline catalyst. Bioresour. Technol. **136**, 730–734 (2013)
5. Dwivedi, G., Pillai, S., Shukla, A.K.: Study of performance and emissions of engines fueled by biofuels and its blends. In: Agarwal, A., Gautam, A., Sharma, N., Singh, A. (eds.) Methanol and the Alternate Fuel Economy. Energy, Environment, and Sustainability. Springer, Singapore (2019). https://doi.org/10.1007/978-981-13-3287-6_5
6. Salinas, D., Araya, P., Guerrero, S.: Study of potassium-supported TiO_2 catalysts for the production of biodiesel. Appl. Catal. B **117**, 260–267 (2012)
7. Ghadge, S.V., Raheman, H.: Biodiesel production from mahua (*Madhuca indica*) oil having high free fatty acids. Biomass and Bioenergy **28**(6), 601–605 (2005)
8. Sadhik Basha, J., Anand, R.B.: Role of nanoadditive blended biodiesel emulsion fuel on the working characteristics of a diesel engine. J. Renew. Sustain. Energy **3**(2), 023106 (2011)
9. Wu, Q., et al.: Experimental investigations on diesel engine performance and emissions using biodiesel adding with carbon coated aluminum nanoparticles. Energy Procedia **142**, 3603–3608 (2017)
10. Prabu, A.: Nanoparticles as additive in biodiesel on the working characteristics of a DI diesel engine. Ain Shams Eng. J. **9**(4), 2343–2349 (2018)
11. EL-Seesy, A.I., et al.: The influence of multi-walled carbon nanotubes additives into non-edible biodiesel-diesel fuel blend on diesel engine performance and emissions. Energy Procedia **100**, 166–172 (2016)
12. Praveen, A., Lakshmi Narayana Rao, G., Balakrishna, B.: Performance and emission characteristics of a diesel engine using *Calophyllum inophyllum* biodiesel blends with TiO_2 nanoadditives and EGR. Egypt. J. Petrol. **27**(4), 731–738 (2018)
13. Selvaganapthy, A., et al.: An experimental investigation to study the effects of various nano particles with diesel on DI diesel engine. ARPN J. Sci. Technol. **3**(1), 112–115 (2013)
14. Javed, S., et al.: Effect of a zinc oxide nanoparticle fuel additive on the emission reduction of a hydrogen dual-fuelled engine with jatropha methyl ester biodiesel blends. J. Cleaner Prod. **137**, 490–506 (2016)

Chapter 6
Performance Analysis of Vapor Compression Refrigeration System Using R-600a and R-134a

Sudhanshu Sharma, Amit Kumar Yadav, Akash Verma, Aditya Shankar and Manvendra Singh

Abstract The objective of this project is to conduct a comparative examination of two refrigerants in the vapor compression refrigeration system. One of the refrigerants is R-134a (tetrafluoroethane) which is commonly utilized as a refrigerant in the vapor compression refrigeration system, and the other is R-600a (isobutene) which is an isomer of butane. R-134a has zero ozone depletion potential (ODP), and its global warming potential (GWP) is tall, though R-600a has zero ODP and has almost nil GWP. Subsequently, R-600a may be distant better refrigerant as compared to the R-134a. Execution of both the refrigerants has been watched on the same test model. Coefficient of performance has been calculated to compare the two refrigerants. Results demonstrate that COP of R-600a is more than that of R-134a in this way R-600a may be a superior refrigerant.

Keywords Thermoelectric refrigerator · Laws of thermodynamics · COP · Rate of refrigeration

6.1 Introduction

Refrigeration is the science of reducing and keeping up the temperature underneath the surrounding temperature. The household refrigerator works on the vapor compression refrigeration cycle. Vapor compression refrigeration system employs a circulating fluid (refrigerant) as the medium which abstracts and discharges heat from the space to be cooled and hence rejects that heat within the environment. Coefficient of performance could be a prominent performance measuring parameter of a refrigeration system. A review of various refrigerants, their applications, and utility has been done. In this, the history of refrigerants based on their selection criteria has been examined beside the rising refrigerants basically within the zone of normal refrigerants [1]. The Emissions Gap Report 2018 reports the most recent logical studies on current and evaluated future greenhouse gas outflows and compares these with the outflow levels reasonable for the world [2]. There have continuously been

S. Sharma (✉) · A. K. Yadav · A. Verma · A. Shankar · M. Singh
Department of Mechanical Engineering, Galgotias College of Engineering and Technology, Greater Noida, UP 201306, India

S. Yadav et al. (eds.), *Proceedings of International Conference in Mechanical and Energy Technology*, Smart Innovation, Systems and Technologies 174,
https://doi.org/10.1007/978-981-15-2647-3_6

endeavors to diminish the greenhouse gas emissions and assurance of the ozone layer by staging out the destructive substances causing these issues [3, 4]. Performance of the domestic refrigerator containing R600a as a refrigerant was investigated. TiO_2-R600a nano-refrigerant was charged into the system and compared with performance pure R600a [5]. An experimental study was done for R152a and R32, environment-friendly refrigerants with zero ODP, and very small global warming potential (GWP) to replace R134a used in a household refrigerator. A refrigerator designed and developed to work with R134a was tried, and its performance using R152a and R32 was assessed and compared [6]. Many other researchers have also compared the performance of different refrigerants used in vapor compression cycle [7, 8].

6.2 Experimental Procedure

- In both the experiments, 3 kg of water has been used for a system to be cooled.
- Plug the compressor to AC supply and switch on the board.
- Take the reading for 40 min.
- Take the reading of pressure gauge at inlet and outlet.
- Take the initial temperature of water.
- Take the blink per minute reading from electric meter.
- Take the temperature of water after cooling.
- Calculate the coefficient of performance of the refrigerator.
- Repeat the process for second refrigerant.

Experimental setup developed to perform the experiment and vapor compression refrigeration cycles is shown in Fig. 6.1.

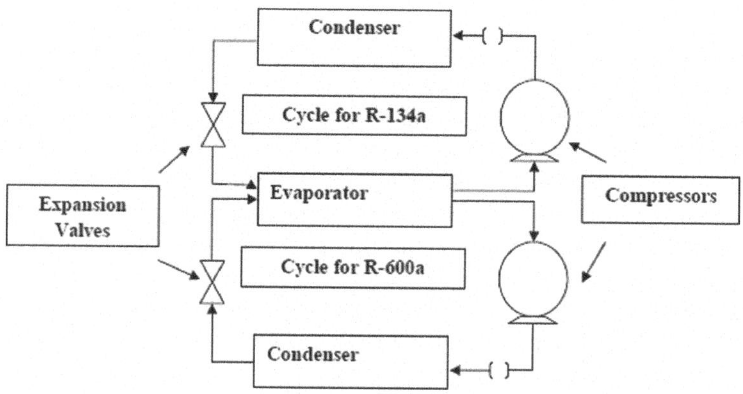

Fig. 6.1 Line diagram of experimental setup

6.3 Observation

See Table 6.1.

$$\text{Refrigeration Effect} = m * Cp * \Delta t$$

m Mass of water (kg)
Cp Specific heat of water (kJ/kg K)
Δt Change in temperature of water (°C).

$$\text{Power input} = 270\,\text{kJ}$$

$$\text{Actual COP} = \frac{\text{Refrigeration Effect}}{\text{Power Input}}$$

$$\text{Theoritical COP} = \frac{h1' - h4'}{h2 - h1}$$

Table 6.1 Observation table for the two refrigerants

Parameter	R-134a	R-600a
(T_1) Inlet temp to the compressor (°C)	8.3	8.3
(T_2) Outlet temp from the compressor (°C)	83.3	63.4
(P_1) Pressure at compressor suction (bar)	1.1	1.103
(P_2) Pressure at compressor discharge (bar)	11.37	8.75
Condenser outlet temp (°C)	44.2	60.5
Initial temp of the water (°C)	39.4	39.8
Final temp of water (°C)	6.8	4.2
Duration of experiment (min)	40	40
Electric meter blink per minute	3	3
(T_1') Saturation temp at 1.10 bar (°C)	−25	−9.5
(T_2') Saturation temp at 11.37 bar (°C)	44.2	60.5
h_1' (kJ/kg)	383.55	542.33
h_2' (kJ/kg)	421.3	634.77
h_3 (kJ/kg)	262.4	351.05

Table 6.2 Theoretical COP

Refrigerant	Theoretical COP
R-134a	3.54
R-600a	4.10

Table 6.3 Actual COP

Refrigerant	Actual COP
R-134a	1.514
R-600a	1.653

Table 6.4 Refrigeration effect obtained

Refrigerant	Refrigeration effect (kJ)
R-134a	408.802
R-600a	446.42

6.4 Result and Discussion

Tables 6.2, 6.3 and 6.4 show the comparative examination of actual COP and theoretical COP of the two refrigerants. It can be noted that theoretical as well as actual COP of R600a is higher than that of R134a. Refrigeration effect of R-134a is less than that of R-600a with the same compressor work.

6.5 Conclusions

An experimental investigation has been done to compare the performance of a vapor compression refrigeration system on the basis of coefficient of performance utilizing R-600a (isobutene) and R-134a (tetrafluoroethane) as refrigerants at steady state conditions. The ODP of R-134a is zero, GWP is tall, the ODP of R-600a is zero, and GWP is insignificant. Coefficient of performance of R-600a is higher than that of R-134a at steady compressor work, i.e., 270 kJ. Refrigerating capacity of R-600a is higher than that of R-134a. As the coefficient of performance and refrigeration effect both are the alluring performance parameters of a refrigeration system, subsequently it can be concluded with this examination that R 600a has a critical potential to replace R134a as a refrigerant in vapor compression refrigeration systems.

References

1. Calm, J.M.: The next generation of refrigerants–Historical review, considerations, and outlook. Int. J. Refrig. **31**(7), 1123–1133 (2008)
2. Olhoff, A., Christensen, J.M., Burgon, P., Bakkegaard, R.K., Larsen, C., Schletz, M.C.: The Emissions Gap Report 2015: A UNEP Synthesis Report (2015)
3. Brack, D.: International Trade and the Montreal Protocol. Routledge (2017)
4. Breidenich, C., Magraw, D., Rowley, A., Rubin, J.W.: The Kyoto protocol to the United Nations framework convention on climate change. Am. J. Int. Law **92**(2), 315–331 (1998)
5. Ohunakin, O.S., Adelekan, D.S., Babarinde, T.O., Leramo, R.O., Abam, F.I., Diarra, C.D.: Experimental investigation of TiO_2-, SiO_2-and Al_2O_3-lubricants for a domestic refrigerator system using LPG as working fluid. Appl. Therm. Eng. **127**, 1469–1477 (2017)
6. Bolaji, B.O.: Experimental study of R152a and R32 to replace R134a in a domestic refrigerator. Energy **35**(9), 3793–3798 (2010)
7. Mohanraj, M.: Energy performance assessment of R430A as a possible alternative refrigerant to R134a in domestic refrigerators. Energy Sustain. Dev. **17**(5), 471–476 (2013)
8. Mohanraj, M., Jayaraj, S., Muraleedharan, C.: Environment friendly alternatives to halogenated refrigerants—a review. Int. J. Greenhouse Gas Control **3**(1), 108–119 (2009)

Chapter 7
Experimental Study to Evaluate the Impact of Insulation with Radiative Shield on the Performance of Panel-Based Radiant Cooling System

Prateek Srivastava and Vivek Kumar

Abstract HVAC systems consume significant amount of energy in buildings to maintain thermal comfort. To make buildings energy efficient, it is desirable to develop alternative low-energy means to achieve cooling without compromising thermal comfort in the buildings. Radiant cooling system has proved to consume less energy with enhanced indoor air quality as compared to conventional air system. The main thrust of this research is to enhance the performance of radiant panels by restricting the heat transfer of panels from the exposed roof. In this study, we have experimentally estimated the impact of insulation with radiative shield on the cooling performance of radiant cooling panel system in terms of peak power and energy consumption. Effect of variation of internal load is analyzed by varying the zone sensible load. We have considered two cases: (i) Conventional panel-based radiant cooling system is used and (ii) radiant panels were insulated from the top using nitrile rubber with radiative shield. Roof is directly exposed to ambient condition, roof absorbs high heat gains, and this heat is transferred to room interior zone by re-radiation, which increases the thermal load on cooling system from the top. Radiant cooling panel system receives maximum amount of heat by re-radiation from the roof. Applying thermal insulation with radiation reflective shield increases the thermal performance of radiant cooling panel system, and the insulated panels show 30% reduction in thermal load and 20% electrical energy saving compared to conventional radiant panel.

Keywords Radiant panel cooling · Energy savings · Insulation

P. Srivastava (✉)
Chitkara College of Applied Engineering, Chitkara University, Punjab, India
e-mail: prateek.srivastava@chitkara.edu.in

V. Kumar
Centre for Energy and Environment, MNIT, Jaipur, India

S. Yadav et al. (eds.), *Proceedings of International Conference in Mechanical and Energy Technology*, Smart Innovation, Systems and Technologies 174,
https://doi.org/10.1007/978-981-15-2647-3_7

7.1 Introduction

Air-conditioning system consumes more than 30% energy of the total building energy consumption [1]. Improvising performance of HVAC system can give major boost for the businesses and facility owner, and it can also reduce carbon footprint. Majority of HVAC systems work on the basis of pulsated air, cold air is supplied in the room to cool the hot air in the room, and this air is continuously recycled. Using the same air in recirculation mode impacts on people working in the zone [2]. In conventional HVAC system, the majority of the heat transfer takes place in convection mode; cold draft of air leads to formation of hot air pockets in the zone and reduces the thermal comfort [3]. In radiant cooling system, more than 50% heat transfer takes place through radiation [4]. Development and interest in radiant cooling systems have increased in recent years because they have been shown to be high energy efficient in comparison to all air HVAC systems [5]. Radiant cooling system has saved 30% energy compared to conventional HVAC system [6]. Water-based radiant cooling systems can achieve high energy efficiency because of the possibility of high temperature cooling and reduced distribution losses due to which the COP of the heat extracting equipment can increase [7]. The surface temperature of the radiant cooling ceiling panels are commonly kept 18–22 °C [8]. Radiant cooling system can further improve its performance if supply water temperature can be achieved through evaporative cooling system [9]. In dry climatic conditions, if WBT is lower, appropriate water temperature can be achieved through cooling tower. Heat transfer in radiant cooling systems are complex because they involve different mechanisms of heat transfer, convection heat exchange between the heating panel and the indoor air, heat conduction through the walls, heat radiation exchange between the cooling panel and the surrounding surfaces, and the heat conduction between the ground and floor. Radiant cooling system provides better thermal comfort as compared to all other HVAC systems by proper control of the surface temperature and less air movement. Different types of radiant cooling systems are available, i.e., radiant panel cooling, embedded surface cooling system, and thermal activated building cooling system [10]. In the retrofit building projects, only radiant panel cooling system can be used, while the embedded surface cooling system and thermal activated cooling system can only be used in new building construction. Cooling capacity of radiant panel cooling system is highest among the three types of radiant cooling system. Performance of radiant cooling system is influenced by the climatic conditions [11]. Controlled outdoor air for cooling coupled with radiant cooling operation can provide more than 20% energy savings for cooling compared to existing radiant cooling systems in the hot and humid climate (Korea) [12]. Experimental study was conducted to compare theoretical heat flux and the measured data; the deviation is less than 4% for the heating mode and 7% for the cooling mode [13]. Radiant cooling surface gets condensation if the surface temperature falls below dew point temperature of the space; this generally happens in humid condition or due to sudden increment of latent load in the zone [14]. In this paper, a full-scale experimental study was conducted to investigate the performance improvement by insulating the panel to avoid heat loss for radiant cooling system. Jeong and Mumma [15] show that the radiant panels are top side insulated to reduce heat gain from plenum, but the actual performance and the energy savings due to insulation are not quantified and clarified. In this paper, an experimental study is conducted to share system running experience, and energy-saving potential of top side insulated radiant cooling panel is presented.

7.2 Methodology

A typical day-time office use building (old administrative building) of Malaviya National Institute of Technology, Jaipur, India, located in composite climate of India is considered in this study. The building is a two story, having floor to ceiling height of 3.5 m and total floor area of 1500 m^2. Experimental study was conducted in zone located at the second floor of building. A total of 24 sensors are placed according to guidelines of ASHRAE 55. Sensors and instrumentation are placed in the experimental room to measure different parameters, i.e., room air dry bulb temperature (DBT), mean radiant temperature (MRT), relative humidity (RH), dew point temperature (DPT), and panel surface temperature of room. Turbine-type flow meters and temperature sensors (thermocouples) were installed at the supply and return of chilled water pipe supplying water to the zone from chiller to measure the flow rate and water inlet and outlet temperature, flow rate and temperature difference along the supply and return water will be used in calculating the instantaneous thermal energy and cooling capacity of the system. Chiller set point is fixed at 16.5 °C and varies from 16.5 to 18.5 °C (due to differential setting of 3 °C), and the internal load is kept fixed for both experiments, 2000 W of sensible load is continuously being added while conducting experiments. Internal load is added in the system by lighting ten numbers of filament bulbs of 200 W each. Figure 7.1 shows the flowchart showing the methodology of the study. Experiments were conducted in two different phases. The methodology adopted for the experiments is described below in detail.

Phase 1: Experiments were conducted in a room of 36 m^2 (390 ft^2), and the cooling performance and the thermal load for the panel-based radiant cooling system are calculated from experimental results.
Phase 2: Nitrile rubber insulation of 20 millimeter thickness with reflective radiation shield was applied at the top of the panels to minimize the heat exchange between the roof and the panel; experiments were conducted to evaluate thermal performance and heat load of panel-based radiant cooling system after modification.

Experimental results from both phases were analyzed and compared in terms of thermal load and energy-saving potential.

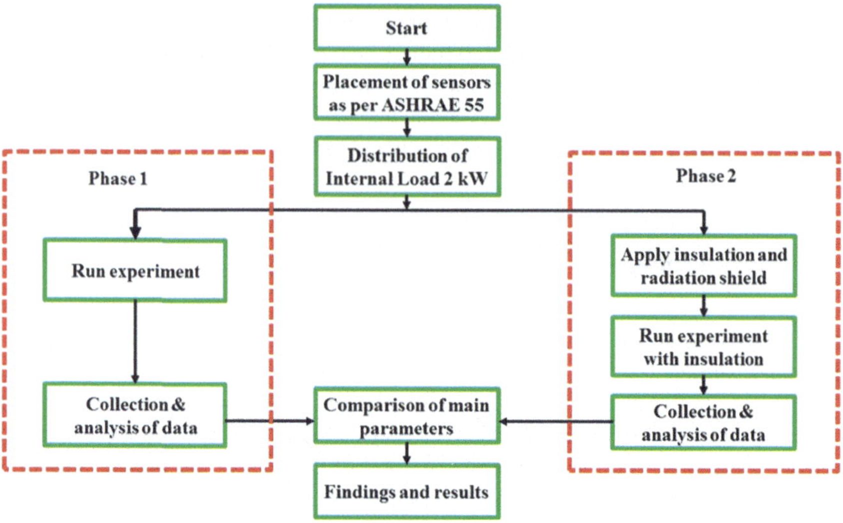

Fig. 7.1 Flowchart of methodology

7.3 Building Description and Sensor Placement

Figure 7.2a shows the old administrative building, and experimental set-up is installed in the second floor of the building. HVAC plant is installed on the roof. The floor plan of second floor of the building is shown in Fig. 7.2b. Floor to ceiling height of the second floor is 3.5 m. Panel-based radiant cooling system is installed in the zone (highlighted room colored with blue of second floor was termed as Radiant lab). The total floor area of the building is (31.6 × 16) 505.6 m^2. Experiments were conducted in room of 36 m^2 (6 × 6) area. Experimental room is located at the south

Fig. 7.2 **a** Old administrative building and **b** floor plan of second floor

side of the building, and there are four windows on the south wall of the experimental room. Window-to-wall ratio is 29%. Sensors were placed according to ASHRAE 55 for measuring different parameters. For measuring dry bulb temperature of indoor air, eight thermocouples were placed inside the room. Thermocouples were installed at 1 m away from the center of each wall and at 1.1 m from the floor, and four thermocouples were placed at the center of experimental room at the heights of 0.1, 0.6, 1.1, and 1.7 m as mentioned in ASHRAE 55. No mechanical ventilation system is used, so the air velocity is measured at the center of room at 1.1 m from the floor and 0.6 m from the floor, and relative humidity is measured by RH sensors at five different places inside the room. Water inlet temperature is measured in supply and return pipe after 1 feet inside in the room, and two turbine-type flow meters were used (one for each pipe) to measure the flow rate of chilled water. Testo 435 thermal comfort equipment kit was used to measure the air velocity and mean radiant temperature inside the room. One covered thermocouple is used to measure the DBT of air outside the experimental room (ambient temperature). Two thermocouples were used to measure the temperature of air between the ceiling panel and the roof. Two surface-temperature-measuring sensors were placed at panel to measure the temperature of the panel surfaces. A Fluke 435 power analyzer is used to measure the total energy consumption of the system, and pressure gauges were used in the header pipe to measure the pressure in the pipe. Figure 7.3 contains seven different pictures which show (1) the position temperature sensors, (2) the radiant panel with the tubes impression (water flows inside these tubes), (3) stand placed in the center containing different temperature at a heights of 0.1, 0.6, 1.1, and 1.7 m, (4) the filament bulbs used to add sensible load in the zone, (5) the loggers used to log temperature and flow readings, (6) the globe for measuring MRT and Testo 435 thermal comfort kit, and (7) the insulation which is placed on the back side of the panels shown in first picture.

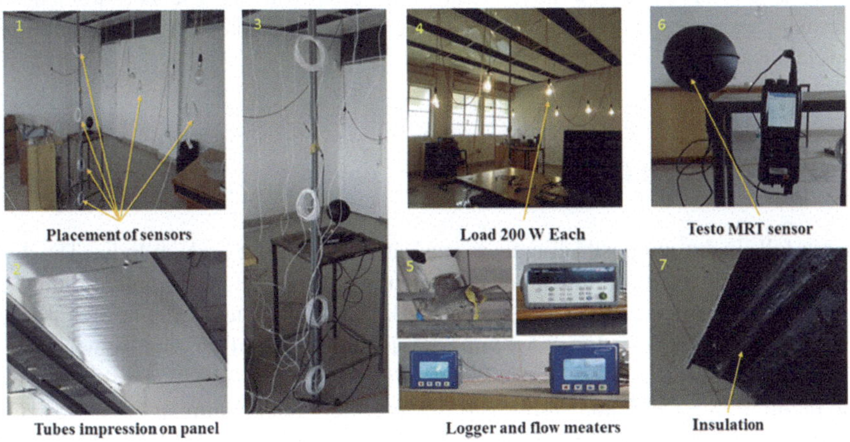

Fig. 7.3 Positions of sensors and distribution of sensible thermal load in the zone

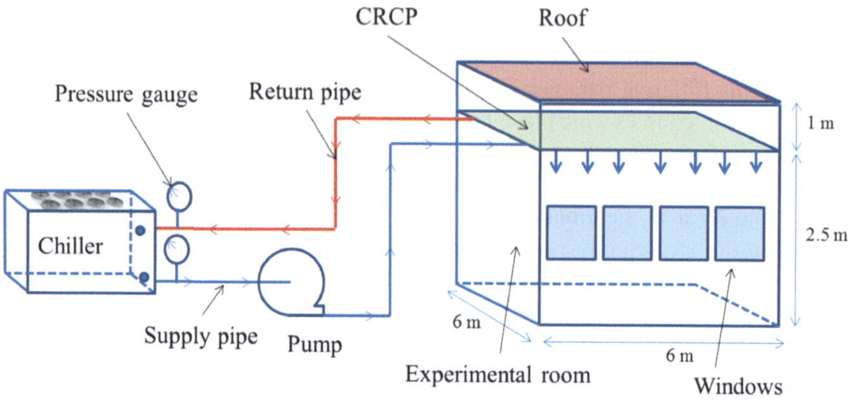

Fig. 7.4 System description

7.4 System Description

Panel-based radiant cooling system is installed in the zone, and experiments were conducted in the room of floor area (6 × 6) 36 m^2. The total projected area of panel is 54% of the total roof area of experimental room. Panels were installed 1 m below the ceiling of the room. Panels were made from two aluminum sheets seam welded at periphery and have tubes' impression for water flow. The inside diameter of the aluminum tubes was 6 mm. A constant COP chiller with capacity of 4.5 TR is used for continuous cooling of water, and a variable speed pump is used to flow the water. Figure 7.4 shows the schematic of the system description installed panel-based radiant cooling system. Experiments were conducted from 10 a.m. to 5 p.m. in both the phases. The water flow rate is 25.3 LPM, and the total internal sensible load is 2000 W is constant in Phase 1 and Phase 2 experiments. The measurement of thermal parameters inside the room and other parameters like energy consumption was taken at a time interval of 15 min.

7.5 Results and Discussion

Jaipur falls in the composite climate as per National Building Code of India. The dry bulb temperature of the ambient air goes high in the summer, so air conditioning is essential for achieving thermal comfort. Measured dry bulb temperature of ambient air in summer is shown in Fig. 7.5. Figure 7.6 shows the monthly dry bulb temperature and wet bulb temperature of Jaipur (ISHRAE weather handbook). Water flow rate of the pump is 25.3 LPM, and chiller is providing water temperature at 16.5–18.5 °C. It was found from Phase 1 experiment that the total thermal load on panel is 5.44–6.90 kW, while 2000 W internal load (constant) and variable external load according to outdoor conditions. The measured room air dry bulb temperature is varied from

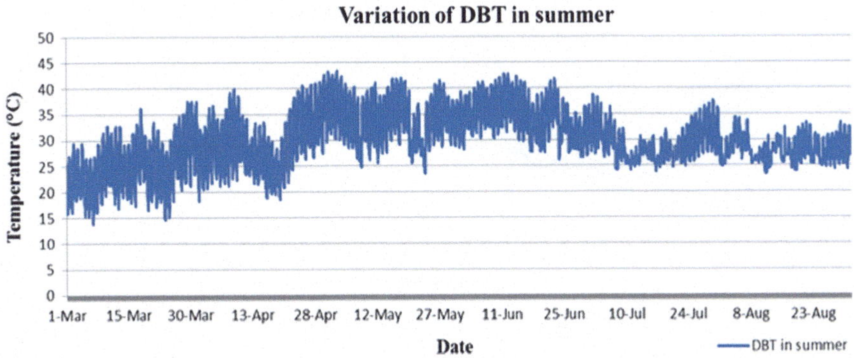

Fig. 7.5 Variation of dry bulb temperature in summer measured during experiments

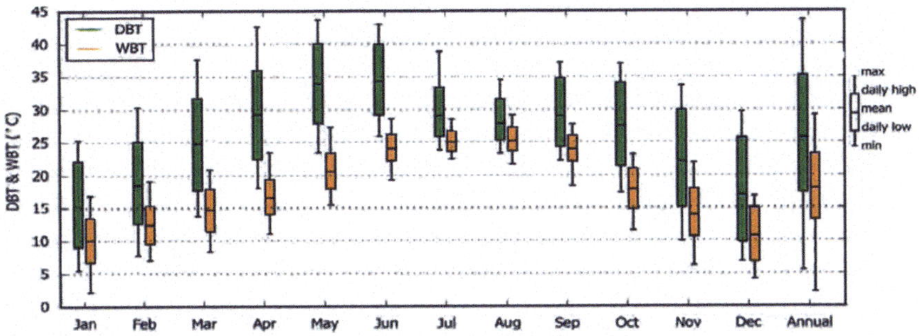

Fig. 7.6 Monthly dry bulb temperature and wet bulb temperature

27.1 to 29.3 °C, and the mean radiant temperature of the room is 23.9–26.4 °C which is lower than the room air dry bulb temperature. Figure 7.7 shows the variation of different measured parameters, i.e., room air dry bulb temperature, panel surface

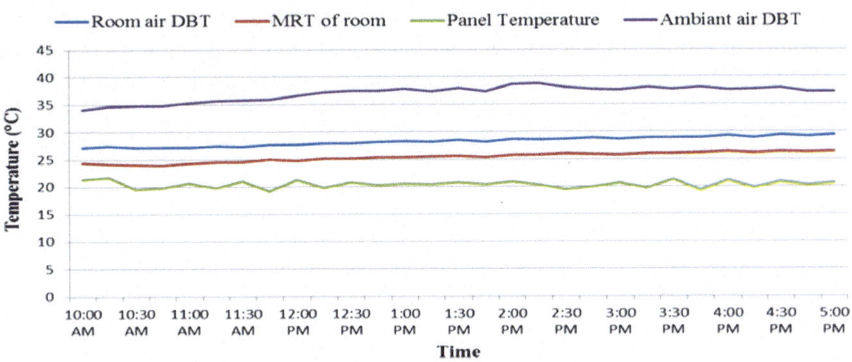

Fig. 7.7 Variations of thermal parameters in experimental room

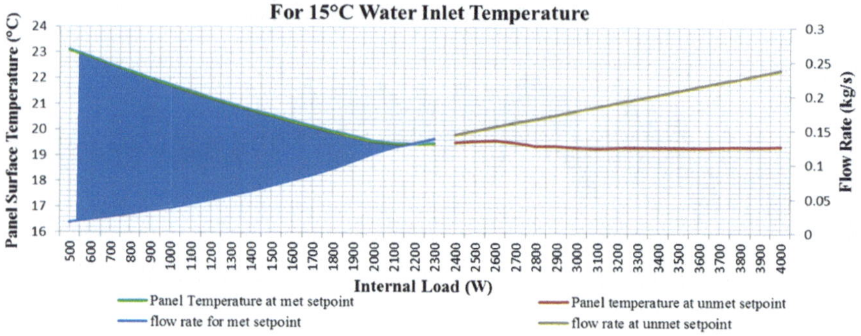

Fig. 7.8 Variation of flow rate and panel temperature with different loading conditions

temperature, mean radiant temperature of room, and ambient temperature. The panel surface temperature is 20 ± 1 °C. As the time increases, the outdoor thermal load increases due to which the room air temperature and ambient air temperature increase more rapidly as compared to mean radiant temperature. In between 2:30 and 3:30 p.m., the ambient temperature is maximum. From these conditions, thermal comfort can easily be achieved. For analyzing effect of internal load, internal load was varied from 500 to 4000 W. Figure 7.8 shows the variation of flow rate and panel temperature with different loading conditions. We can see that as the internal load increases, the panel temperature decreases due to increase in flow rate of water, but after a certain value of internal load (2300 W), the panel temperature is almost constant but the flow rate is continuously increasing; we can conclude that for any value if the water flow rate and panel temperature lie inside the highlighted area, then the operative temperature of room will not increase from set point temperature. Table 7.1 shows the comparison of measured values with the recommended ISO 7730 values; measured values show good result with the recommended value for radiant cooling system. Figure 7.9 shows variation of system capacity and flow rate at different internal load.

Table 7.2 shows the variation of total thermal load on panel at different outdoor air temperature. It was found from experiment that as the outdoor air temperature increases, the thermal load on panel also increases. When the maximum outdoor air temperature is 38 °C, the average thermal load on radiant cooling panel system is 6.90 kW. The average total thermal load on panel system is reduced to 5.44 kW when outdoor air maximum temperature is 35 °C. Figure 7.10a shows the comparison of total thermal load in Phase 1 and Phase 2 experiments with different outdoor air

Table 7.1 Recommended factors of ISO 7730

	Operative temperature (°C)	Vertical temperature difference (°C)	Vertical temperature asymmetry (°C)	Surface temperature of floor (°C)
ISO 7730	23–26	<3	<10	19–26
Measured	26–29	<3	<6	19–24

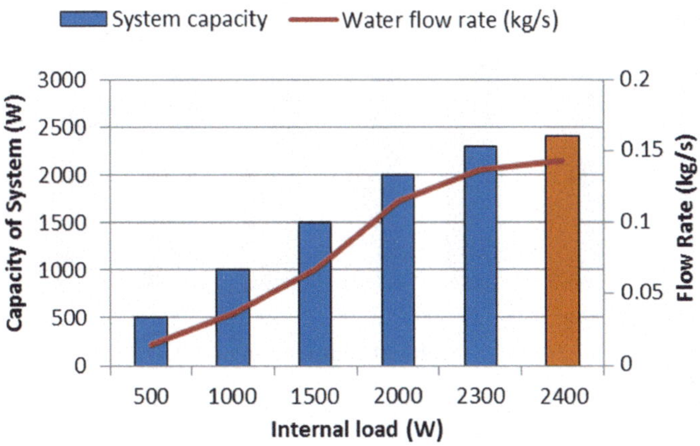

Fig. 7.9 Variation of system capacity and flow rate at different internal load

Table 7.2 Thermal load on Phase 2

Ambient temperature (max.)	Thermal load on panel (kW)
38	6.90
38	6.79
37	6.70
36	6.64
36	6.56
35	5.44

temperature. From experimental results, it was found that by applying insulation with reflective radiation shield, the total thermal load on panel in Phase 2 experiments is 30% less as compared to Phase 1 experiments.

The total electrical energy consumption in Phase 2 reduces from 19.8 to 14.1 kWh, while outdoor air temperature is 38 °C, and when outdoor conditions are at 35°C, the cooling energy reduced from 17.12 to 13.31 kWh because of reduction of total thermal load on panel. The total electricity consumption is shown in Fig. 7.10b. It was found from experiment that the temperature above panel (between panel and the roof) is higher from the room temperature. The roof of the room is directly exposed to ambient so the panel top surface receives maximum heat load from the roof by radiation load. Figure 7.9 shows the variation of temperature above the panel surface

$$\text{Electricity consumption}_{\text{Average}} \text{by radiant cooling panel system in phase1} = 18.12\,\text{kWh}$$

$$\text{Electricity consumption}_{\text{Average}} \text{by radiant cooling panel system in phase2} = 13.85\,\text{kWh}$$

$$\text{Thermal load reduction} = 18.12 - 13.85 = 4.27\,\text{kWh}$$

$$\begin{aligned}\text{Percentage saving} &= \frac{(18.12 - 13.85)}{18.12} \times 100\\ &= 23.50\%\end{aligned}$$

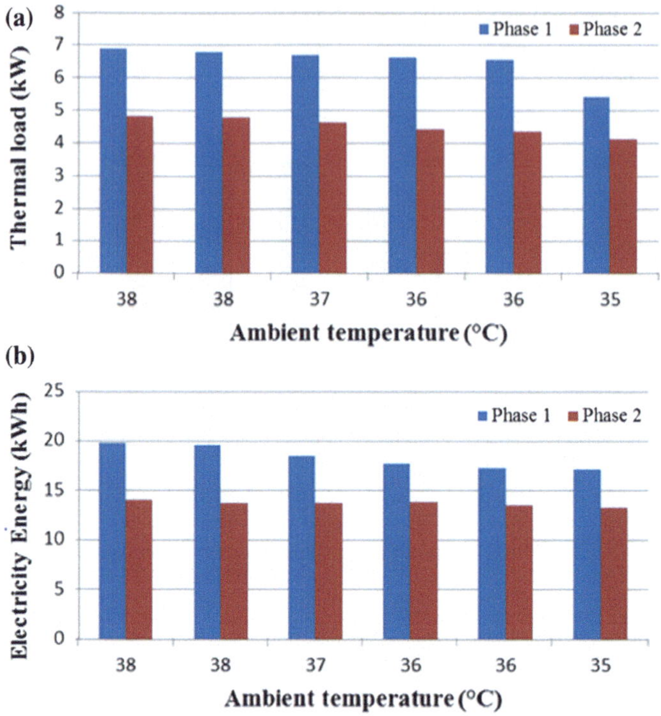

Fig. 7.10 **a** Comparison of thermal load in Phase 1 and Phase 2; **b** comparison of electricity consumption

By using the thermal insulation and reflective shield, we can reduce the 23.5% of electrical energy consumption achieved by reduction in thermal load received from the roof internal surface. In Phase 2 experiments, when the thermal insulation with reflective radiation shield is applied at the top surface of the panel, heat load from the roof has been minimized to a great extent, and temperature difference in supply and return water is reduced which has reduced the load on the plant. Figure 7.11 shows the temperature variation above panel in both the phases.

7.6 Conclusion

In this study, the thermal performance and effect of insulation with reflective shield on the performance of radiant panel were analyzed in terms of energy-saving potential. Experiments have been conducted in two phases: Phase 1 conventional panel-based radiant cooling system is compared with one-side insulated panel-based radiant cooling system in Phase 2. Impact of variation of internal load on the performance of radiant cooling system was also analyzed. Thermal performance of radiant cooling

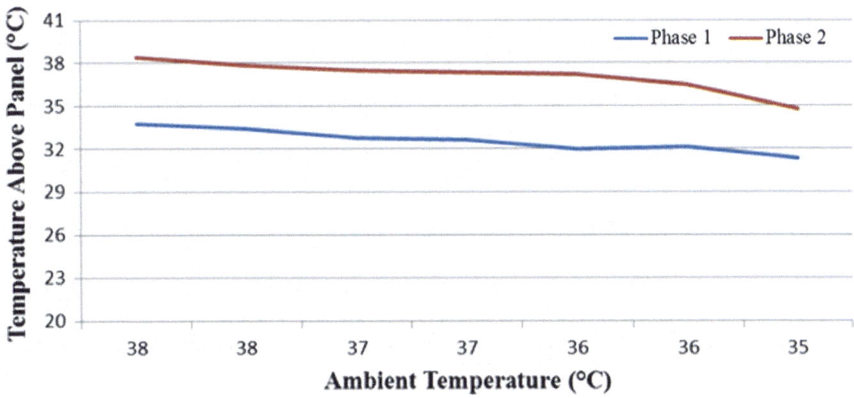

Fig. 7.11 Comparison of temperature above panel in Phase 1 and Phase 2

system has been improved with the application of insulation. It was found from experimental results that the thermal load on panel is reduced from an average value of 6.5–4.52 kW. So by applying insulation thermal load reduced to 30%, the total electrical energy consumption is reduced from 18.32 to 13.69 kWh. Based on the experimental analysis, it is recommended to use top side insulated panel-based radiant cooling system for top floor exposed to sun.

References

1. Chiang, W.H., Wang, C.Y., Huang, J.S.: Evaluation of cooling ceiling and mechanical ventilation systems on thermal comfort using CFD study in an office for subtropical region. Build. Environ. **48**(1), 113–127 (2012)
2. Catalina, T., Virgone, J., Kuznik, F.: Evaluation of thermal comfort using combined CFD and experimentation study in a test room equipped with a cooling ceiling. Build. Environ. **44**(8), 1740–1750 (2009)
3. Corgnati, S.P., Perino, M., Fracastoro, G.V., Nielsen, P.V.: Experimental and numerical analysis of air and radiant cooling systems in offices. Build. Environ. **44**(4), 801–806 (2009)
4. Tian, Z., Love, J.A.: A field study of occupant thermal comfort and thermal environments with radiant slab cooling. Build. Environ. **43**(10), 1658–1670 (2008)
5. Niu, J.L., Zhang, L.Z., Zuo, H.G.: Energy savings potential of chilled-ceiling combined with desiccant cooling in hot and humid climates. Energy Build. **34**, 487–495 (2002)
6. Khan, Y., Rai, V., Mathur, J., Bhandari, M.: Performance evaluation of radiant cooling system integrated with air system under different operational strategies. Energy Build. **97**, 118–128 (2015)
7. Liu, W., Lian, Z., Radermacher, R., Yao, Y.: Energy consumption analysis on a dedicated outdoor air system with rotary desiccant wheel. Energy **32**(9), 1749–1760 (2007)
8. Chiang, W.-H., Wang, C.-Y., Huang, J.-S.: Evaluation of cooling ceiling and mechanical ventilation systems on thermal comfort using CFD study in an office for subtropical region. Build. Environ. **48**, 113–127 (2012)

9. Srivastava, P., Khan, Y., Bhandari, M., Mathur, J., Pratap, R.: Calibrated simulation analysis for integration of evaporative cooling and radiant cooling system for different Indian climatic zones. J. Build. Eng. **19**
10. Ning, B., Schiavon, S., Bauman, F.S.: A novel classification scheme for design and control of radiant system based on thermal response time. Energy Build. **137**, 38–45(2017)
11. Saber, E.M., Iyengar, R., Mast, M., Meggers, F., Tham, K.W., Leibundgut, H.: Thermal comfort and IAQ analysis of a decentralized DOAS system coupled with radiant cooling for the tropics. Build. Environ. **82**, 361–370
12. Seo, J., Song, D., Ho, K.: Possibility of coupling outdoor air cooling and radiant floor cooling under hot and humid climate conditions. Energy Build. **81**, 219–226 (2014)
13. Li, R., Yoshidomi, T., Ooka, R., Olesen, B.W.: Field evaluation of performance of radiant heating/cooling ceiling panel system. Energy Build. **86**, 58–65 (2015)
14. Yin, Y.L., Wang, R.Z., Zhai, X.Q., Ishugah, T.F.: Experimental investigation on the heat transfer performance and water condensation phenomenon of radiant cooling panels. Build. Environ. **71**, 15–23 (2014)
15. Jeong, J.-W., Mumma, S.A.: Simplified cooling capacity estimation model for top insulated metal ceiling radiant cooling panels. Appl. Therm. Eng. **24**(14–15), 2055–2072 (2004)

Chapter 8
Important Parameters Influencing Total Quality Management: A Comparative Study

Nidhi Gupta, Anil Kumar Shrivastava and Prashant Rawat

Abstract In today's competitive environment (static and dynamic), for achieving the business excellence, companies want to implement new and/or modified marketing strategies continuously. Improved features with good-quality products or services result in improving satisfaction in the customer's mind. The guidelines provided in ISO 9000 are ultimate tool to achieve the above-mentioned objective by any organization. Although several multi-disciplinary approaches have been proposed to offer quality products as per customer's expectations or demands, total quality management (TQM) is a systematic management instrument that synchronizes the resources and staff in order to accomplish higher product standards, maintaining companies and customer's prospects. This paper reviews and relates the important parameters like ISO 9000, product quality and customers' satisfaction in TQM.

Keywords Total quality management (TQM) · Product quality · Customer satisfaction · ISO 9000

8.1 Introduction

It is certainly true that today's world is full of competition; business excellence has become very difficult to manage and maintain. Therefore, in order to achieve this goal, the total quality management has become one of the predominant factors in any organization. The market is led by different alternatives of quality management tools and ISO 9000 from raw material to final or finished product. Looking at the present scenario, Japan's quality improvement programme has changed the view for the present trade situation of the rest of the world. Japanese approach of quality management is very perceptible. By process control, the Japanese companies have developed many stages of quality improvement, i.e. from inspection to new product development (NPD).

N. Gupta · P. Rawat (✉)
School of Mechanical Engineering, Lingaya's Vidyapeeth, Faridabad, India

A. K. Shrivastava
Mechanical Engineering Department, Galgotias College of Engineering & Technology, Greater Noida, India

S. Yadav et al. (eds.), *Proceedings of International Conference in Mechanical and Energy Technology*, Smart Innovation, Systems and Technologies 174,
https://doi.org/10.1007/978-981-15-2647-3_8

Total quality control is not just methodology but also a procedure to combine all the effective basic elements to maintain the product standards, hence customer satisfaction. The five basic elements (majorly) are (i) customer focus, (ii) planning, (iii) management, (iv) improvement and (v) total participation [1]. Ishiwaka is a well-known person for his cause-and-effect diagram (Fig. 8.1) also called 'fishbone diagram'. Quality improvement is a continuous aspect of human living and needs training at all level.

Deming's cycle, shown in Fig. 8.2, defines the necessity of a continuous quality improvement model. The first element of Deming's cycle is dealing with process planning which is based on customer/client perspective; second element 'do' is related to implementation/production process. The third step 'check' is for evaluating or checking the quality of goods or services. The final or fourth step is to 'act' in addition to the main focus to analyse/understand the profit or gain in the organization [1]. Figure 8.2 is a continuous process of growing any organization, i.e. production unit or service agencies.

This paper focuses on the four essential parameters: International Organization for Standardization (ISO), business excellence, customer satisfaction and product quality (Fig. 8.3), and their importance in any organization. ISO documents are the systematic philosophy that states 'what to do?' [2]. For many organizations, the ISO standards are the key guidelines to maintain the 'product quality'. ISO 9000 is the family of standards or strategies to maintain the quality of products or services. Both (ISO and product quality) are the necessary parameters deal with the internal performance of the business. Furthermore, 'business excellence' capabilities enable any industry to achieve international benchmarks. In today's competitive environment, 'customer satisfaction' is the ultimate goal to be accomplished; without this key parameter, any business/services cannot run. Therefore, these four parameters deal with the internal as well as external capabilities of any organization.

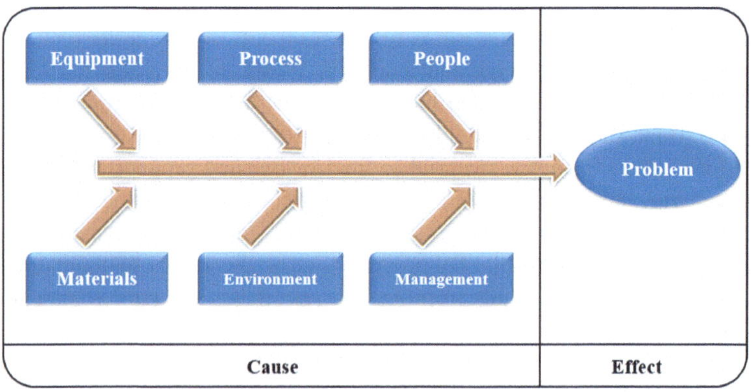

Fig. 8.1 Cause-and-effect diagram of Ishiwaka [1]

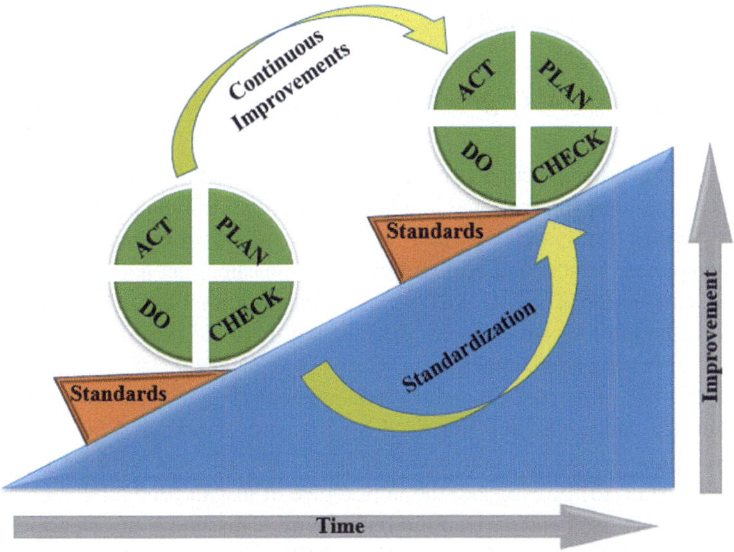

Fig. 8.2 Deming's cycle

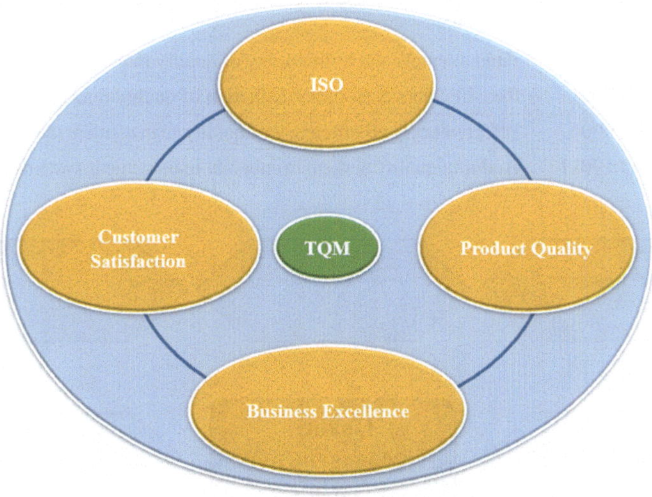

Fig. 8.3 Four essential parameters of TQM

8.2 Parameters of TQM

8.2.1 International Organization for Standardization (ISO)

ISO is the global governance (non-governmental institution listed under united nation agencies) to maintain international standards. Currently, ISO has 17,000 international standards, with a history of sixty years [3]. These standards abridged the gap between public and private sectors and deal with the size, measurement and weight to ensure customer satisfaction. The ISO 9000 series that deals with the 'quality management systems' is the most successful drive [4]. The family of ISO 9000 is summarized in Table 8.1 for better understanding.

However, the principles of quality management (Fig. 8.4) are well known to every industry and commonly practised by almost everywhere. Nevertheless, due to the continuously changing environment, these are still very vital for the growth [5]. Dreyfus et al. [6] investigated the importance of just-in-time implementation in QM efforts. They also categorized the responding firms into four categories based on certification and implementation of ISO 9000 standards. The conclusion of this

Table 8.1 ISO family

S. No.	Name	Deals with
1	ISO 9000	Fundamentals and terminology of quality management systems
2	ISO 9001	Requirements/customer satisfaction of quality management systems
3	ISO 9004	Effectiveness and efficiency of quality management systems
4	ISO 19011	Auditing quality and environmental management systems

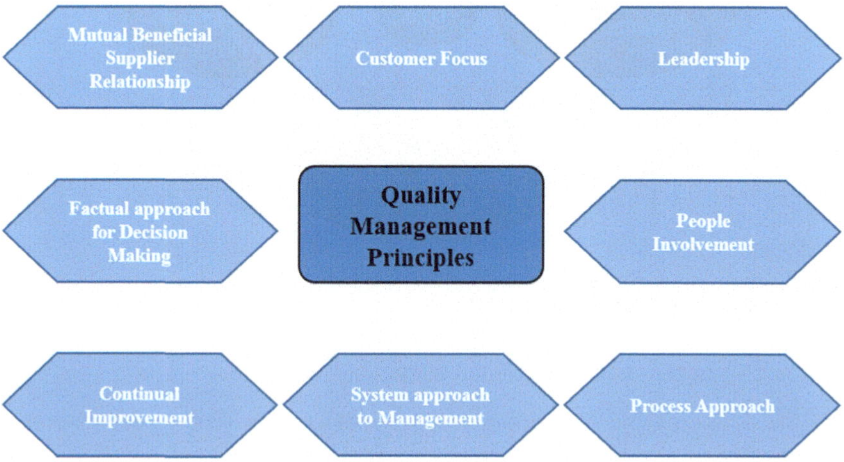

Fig. 8.4 Principles of quality management

study strongly supported the transformation of firms by implementing ISO standards during the mid-90s. The quality of the product also highly influenced by ISO 9000 standards [7], and it directly improves customer satisfaction.

8.2.2 Customer Satisfaction and Business Excellence

Modern industries are having clear objectives for attracting their customers by providing them maximum facilities related to discounts, offers or free services, moreover exchange options too. No company wants its customers to go anywhere for maximum possible time. Although it is not happing in nowadays several researchers are focusing on this important aspect and for a long time, this parameter is one of the central parameters of any business policy. Tse and Wilton [8] have investigated this parameter in 1988 and interrelated this parameter with the expectation of the customer. They highlighted one more factor that improves customer satisfaction is performance. Product services and excellence [9] are the key features of business excellence which undoubtedly improves customer satisfaction in so many aspects. Nevertheless, attaining excellence in any business is not one-night task although it takes long time and continuous efforts. Kim et al. [10] focused on long-term investment in quality as a prime feature which repays as customer satisfaction. However, it requires long-term planning and continuous efforts which should be planned properly. Deng et al. [11] focused on the bonding between production industry, supplier and customer which can be developed by the good-quality product and service providers when customer is in need. This mutual bonding is the key to attaining customer satisfaction. As a result of this bonding, a better relationship between the company and customer may lead a long-term business [12, 13].

Nowadays, the market is completely based on Internet, and many analysis tools are in existence for calculating customer satisfaction and other factors. Blessing and Natter [14] investigated the mystery shoppers' behaviours, which directly influences customer satisfaction. As a conclusion, this investigation justified that the parameters sales and customer satisfaction are highly correlated and must be analysed seriously. Kim et al. [15] studied the feature of promotion of wines in South Korea and influence of positive promotion on customer satisfaction over customers.

This research concluded that the positive promotion highly influences the customer perspective; moreover, the age of customer is another important parameter. Nevertheless, positive promotion is very helpful to improve customer satisfaction. Figure 8.5 represents the different factors that affect customer satisfaction.

8.2.3 Product Quality

This factor is completely dependent on the production industry and process of manufacturing. However, with the improvement in the technologies of production, it is

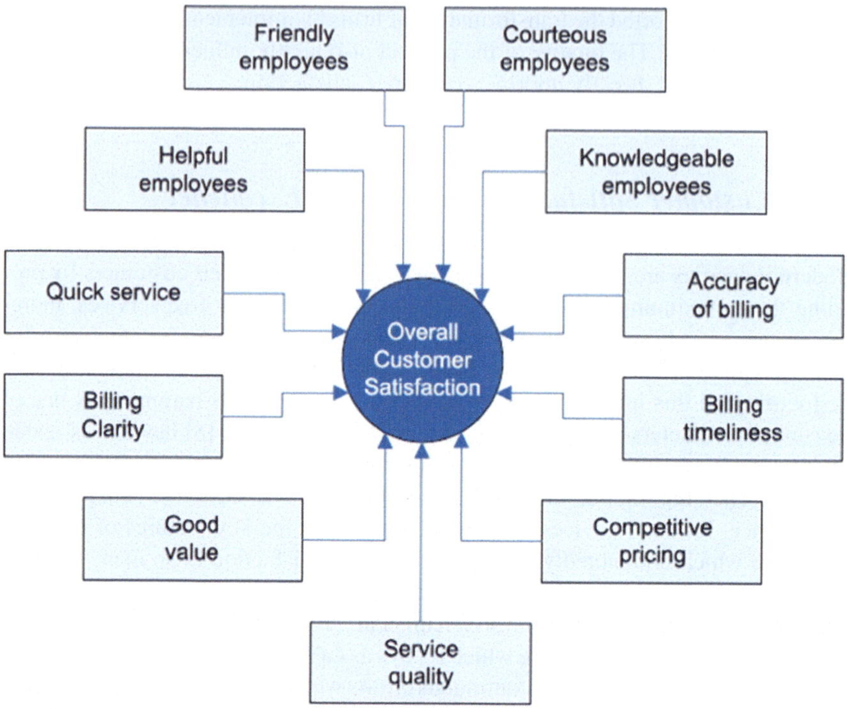

Fig. 8.5 Factors affecting customer satisfaction [16]

the responsibility of any industry to improve and/or update the quality of available product after certain time. From the perspective of small business, once the quality is a serious concern, poor quality may lead to a major loss in business [17]. Moreover, if any industry follows the international guidelines and/or ISO standards for maintaining the quality of product, it becomes easy to deal with quality production. Berry [18] focused on the positive effect of continuous implementation of ISO 9000 as a result of investment cost with improved quality of the product. This is best suitable for making long-term policies in the continuous production business.

Implementation of ISO 9000 technique at small- to large-scale industries has been investigated, and their benefits are highly influential for any industry for long terms [18]. Singels et al. [19] investigated the influence of using ISO certification in any organizations. This work focused on both aspects, i.e. internal and external benefits of any industry which follows the proper certification techniques. The performance is considered with five important parameters (process of production, production results, customer satisfaction, self-motivation and cost of investment). Lipovatz et al. [20] interrelated the producer and customer by stating them 'the two important factors'. They have focused on the points that may improve the customer satisfaction with customer prospective, quality and branding of the product. Therefore, if the organization follows all the standards of production and produces a quality product, this will undoubtedly improve customer satisfaction resulting in a good benefit for the industry.

In present time, the market is very sensitive for new upcoming products with better features. Therefore, the linkage between product variety and quality choice cannot be ignored [21]. This is the key factor for better bonding with customer for the long term; the only characteristic required from the organization side is improving 'customer satisfaction' by providing the best quality product as per market expectations.

8.3 Discussion and Conclusion

The term 'total quality management' is not particularly bound with any specific feature of organization, while there are several different parameters which influence the production, quality of production, customer satisfaction, etc. The expectations of all these unite, when fulfilled in a well-organized manner, the outcome is very satisfactory from the perspective of people involved in producing the products/services or the consumers. This paper reviews the major involved approaches for attaining high standards. The conclusion of the review can be summarized as follows:

- The market environment changes with the time and society requirements, and this is most important to judge the needs before production.
- The organization benefits are long termed if proper standards (like ISO 9000) are implemented properly as per guidelines.
- Size of the production organization does not matter for implementing the production standards. In addition, the benefits of implementing proper standards are internal as well as external in nature.
- The standards are needed to be modified/updated with time to time for better possibilities of benefit.
- Quality of the product is necessary from the customer's point of view. This is also responsible for making better bonding and abridges the gap between consumer and manufacturing organization.
- The well-organized and balanced application of all the factors provides benefit for all.

References

1. Al-ibrahim, A.: Quality management and its role in improving service quality in public sector. J. Bus. Manag. Sci. **2**, 123–147 (2014). https://doi.org/10.12691/jbms-2-6-1
2. Levene, T.: Life Cycle Management for Dependability. Springer-Verlag London Limited (1997). https://doi.org/10.1007/978-1-4471-0939-6
3. Murphy, C.N., Yates, J. (ed.): The International Organization for Standardization (ISO). Taylor & Francis Group, Routledge (2009)

4. Boli, J., Thomas, G.M. (eds.): Standardization in the World Polity. Technical Rationality over Power. Stanford University Press, Stanford, CA (1999)
5. Priede, J.: Implementation of quality management system ISO 9001 in the world and its strategic necessity. Procedia—Soc. Behav. Sci. **58**, 1466–1475 (2012). https://doi.org/10.1016/j.sbspro.2012.09.1133
6. Dreyfus, L.P., Ahire, S.L., Ebrahimpour, M.: The impact of just-in-time implementation and ISO-9000 certification on total quality management. IEEE Trans. Eng. Manag. **51**, 125–141 (2004). https://doi.org/10.1109/tem.2004.826024
7. Korane, F.: Making sense of ISO-9000. Mach. Des. **65**, 74–77 (1993)
8. Tse, D.K., Wilton, P.C.: Models of consumer satisfaction formation. An extension. J. Mark. Res. **25**, 204 (1988). https://doi.org/10.2307/3172652
9. Hennig-Thurau, T., Klee, A.: The impact of customer satisfaction and relationship quality on customer retention. A critical reassessment and model development. Psychol. Mark. **14**, 737–764 (1997). https://doi.org/10.1002/(sici)1520-6793(199712)14.8%3c737..aid-mar2%3e3.0.co;2-f
10. Kim, M.-K., Park, M.-C., Jeong, D.-H.: The effects of customer satisfaction and switching barrier on customer loyalty in Korean mobile telecommunication services. Telecommun. Policy **28**, 145–159 (2004). https://doi.org/10.1016/j.telpol.2003.12.003
11. Deng, Z., Lu, Y., Wei, K.K., Zhang, J.: Understanding customer satisfaction and loyalty. An empirical study of mobile instant messages in China. Int. J. Inf. Manag. **30**, 289–300 (2010). https://doi.org/10.1016/j.ijinfomgt.2009.10.001
12. Eshghi, A., Haughton, D., Topi, H.: Determinants of customer loyalty in the wireless telecommunications industry. Telecommun. Policy **31**, 93–106 (2007). https://doi.org/10.1016/j.telpol.2006.12.005
13. Amin, S.M., Ahmad, U.N.U., Hui, L.S.: Factors contributing to customer loyalty towards telecommunication service provider. Procedia—Soc. Behav. Sci. **40**, 282–286 (2012). https://doi.org/10.1016/j.sbspro.2012.03.192
14. Blessing, G., Natter, M.: Do mystery shoppers really predict customer satisfaction and sales performance? J. Retail. (2019). https://doi.org/10.1016/j.jretai.2019.04.001
15. Kim, W.-H., Cho, J.-L., Kim, K.-S.: The relationships of wine promotion, customer satisfaction, and behavioral intention. The moderating roles of customers' gender and age. J. Hospitality Tourism Manag. **39**, 212–218 (2019). https://doi.org/10.1016/j.jhtm.2019.03.001
16. https://www.van-haaften.nl/branding/corporate-branding/84-brand-equity (2019)
17. Dale, B.G., Duncalf, A.J.: A study of quality assurance in small businesses. Proc. Inst. Mech. Eng., Part B. Manag. Eng. Manuf. **198**, 135–139 (1984). https://doi.org/10.1243/pime_proc_1984_198_054_02
18. Berry, T.H.: Managing the total Quality Transformation. McGraw-Hill (1991)
19. Singels, J., Ruël, G., Water, H-V-D.: ISO-9000 series—Certification and performance. Int. J. Qual. Reliab. Manag. **18**, 62–75 (2001). https://doi.org/10.1108/02656710110364477
20. Lipovatz, D., Stenos, F., Vaka, A.: Implementation of ISO-9000 quality systems in Greek enterprises. Int. J. Qual. Reliab. Manag. **16**, 534–551 (1999). https://doi.org/10.1108/02656719910278575
21. Granero, L.M.: Strategic product variety and quality choice. Econ. Lett. (2019). https://doi.org/10.1016/j.econlet.2019.05.033

Chapter 9
Numerical Studies on Thermally Induced Flow of Nanofluid in a Vertical Annulus

Suhail Ahmad Khan, M. Altamush Siddiqui and Shahid Husain

Abstract A numerical investigation of the thermally induced flow of nanofluid (Al_2O_3 + water) inside a vertical annulus has presently been carried out. The simulations are carried for various nanoparticle concentrations at a constant heat flux of 3 kW/m^2 using computational fluid dynamics (CFD) solver ANSYS Fluent. Two viscosity models namely Brinkman and Buongiorno and thermal conductivity model of Maxwell's are adopted to estimate heat transfer rates. Uncertainty in the results of heat transfer due to the adoption of two different formulae of viscosity is discussed in detail. The average heat transfer coefficient estimated on the basis of both viscosity models increases with nanoparticle concentration. Nanoparticle concentration has a much more pronounced effect on heat transfer coefficient when estimated on the basis of the Brinkman model. The heat transfer coefficient is found to enhance by 10.86% when estimated using the Brinkman viscosity model while it is only just 3.91% in the case of Buongiorno viscosity model-based nanofluids.

Keywords Natural convection · Vertical annulus · Nanofluids · Thermophysical properties

Nomenclature

A_f	Flow area
c	Specific heat
d	Diameter of tube (m)
h	Heat transfer coefficient (Wm^{-2} °C^{-1})
k	Thermal conductivity (Wm^{-2} °C^{-1})
L	Length (m)
g	Acceleration due to gravity (m s^{-2})
p	Pressure (Pa)
q_w	Heat flux (W/m^2)

S. A. Khan (✉) · M. Altamush Siddiqui · S. Husain
Department of Mechanical Engineering, Zakir Husain College of Engineering & Technology, AMU, Aligarh, UP 202002, India

S. Yadav et al. (eds.), *Proceedings of International Conference in Mechanical and Energy Technology*, Smart Innovation, Systems and Technologies 174,
https://doi.org/10.1007/978-981-15-2647-3_9

r Radial distance (m)
R Non-dimensional radial distance
T Temperature (°C)
u Radial velocity (m s^{-1})
U Non-dimensional radial velocity
v Axial velocity (m s^{-1})
V Non-dimensional axial velocity
y Axial distance (m)
Y Non-dimensional axial distance

Greek

δ Annular gap (m)
λ Thermal diffusivity
β Expansion coefficient (K^{-1})
ρ Density (kg m^{-3})
μ Dynamic viscosity (N s/m^2)
θ Non-dimensional temperature
ν Kinematic viscosity (m^2 s^{-1})

Dimensionless Parameters

A Aspect ratio (L/δ)
Nu Nusselt number $\left(\frac{h\delta}{k}\right)$
Pr Prandtl number $\left(\frac{\nu}{\lambda}\right)$
Ra Rayleigh number $\left(\frac{g\beta|q_w|\delta^4}{k\nu\lambda}\right)$
RR Radius ratio $\left(\frac{r_o}{r_i}\right)$

Subscripts and Superscripts

a Ambient
h Heated
i Inner
o Outer
nf Nanofluid
w Wall

9.1 Introduction

Researchers from all over the world proposed many methods to enhance the heat transfer rates of fluid flowing inside the heat exchange equipment. Addition of solid ultrafine particles as an additive in the base fluid is one of the methods adopted to increase thermal conductivity was emerged in 1873 [1]. But due to large particles (size of the order of millimetres or micrometres), problems such as pressure loss, clogging and abrasion make these mixtures inappropriate for heat transfer application. Consequently, Choi [2] from Argonne National Laboratory [USA] dispersed the nanosized particles in the conventional fluid and developed the innovative nanofluid. Experimental studies [3–5] have revealed that adding the small amounts of solid nanoparticles to the base fluids increases its thermal conductivity. The higher thermal conductivity and smaller size of nanoparticles make these nanofluids attractive for heat transfer applications. In the literature, several studies are reported concerning thermal conductivity [6, 7] and viscosity [8, 9] of nanofluids. From the available literature, it is found that inconsistencies exist between experimental data and theoretical models regarding the viscosity and thermal conductivity of nanofluids.

Various researchers study the heat transfer characteristics of nanofluids experimentally and numerically in the last previous years. Pak and Cho [8] carried out experiments by using water-based Al_2O_3 and TiO_2 nanofluids inside the cylinder and found enhancement of Nusselt number with nanoparticle concentration. Using the Buongiorno [10] model of viscosity, Yang et al. [11] theoretically examined the convective heat transfer of nanofluids inside annulus and observed improvement in the heat transfer rates. The buoyancy induced flow of water-based alumina nanofluid in a 2-D rectangular cavity was first studied by Khanafer [12] and his co-workers. They revealed that the heat transfer rate increases proportionally with nanoparticle concentration. On the other hand, Putra and co-workers [13] and Wen and Ding [14] carried out experiments on natural convection heat transfer and reported a reduction in heat transfer with volume concentration of nanoparticle. Ho et al. [15] numerically investigate the laminar natural convection heat transfer of (Al_2O_3 + water) nanofluid in a square enclosure and discussed suspicions in the calculation of viscosity and thermal conductivity of nanofluids. They ascertain that Brinkman [16] model of viscosity underrates the viscosity of nanofluid mainly at higher nanoparticle concentration and subsequently gives improper calculation for natural convection heat transfer in different enclosures.

In recent years, a lot of researchers studied the effect of magnetic and electric field on heat transfer of nanofluids in various enclosures. Sheikholeslami and Chamkha [17] carried out a numerical analysis on thermally induced flow of Fe_3O_4-ethylene glycol nanofluid in semi-annulus enclosure. They studied the effect of electric field, supplied voltage and nanoparticle concentration on flow and heat transfer characteristics. They revealed that electric field strength has favourable effects on heat transfer rates and effects are much more prominent at small Rayleigh number. The natural convection of magnetic nanofluid inside porous media was numerically studied by Sheikholeslami [18] using inventive numerical technique of control volume finite element method (CVFEM). He states that Darcy number, Rayleigh number and nanoparticle concentration have beneficial effects on Nusselt number. More recently, Dogonchi et al. [19] numerically investigate the magnetohydrodynamic flow of nanofluid (Cu + water) using the Corcione [20] correlation for evaluating the viscosity and thermal conductivity of nanofluid by considering Brownian motion. They ascertain that Rayleigh number increases the natural convective flows while Hartman number and contraction ratio have detrimental effects.

The literature review reveals that the experimental and numerical approaches do not report any detailed inspection of nanofluids inside different geometrical enclosures. Further, inconsistencies seem to appear between experimental and numerical approaches concerning natural convection heat transfer of nanofluids. These contradictory results arise due to the adoption of different formulae of thermophysical properties. Therefore, a numerical study of natural convective heat transfer of water-based alumina nanofluids inside a vertical annulus of high aspect ratio (300) has presently been carried out. The analysis is carried out for various volume fractions of nanoparticles by considering Brinkman [16] model and Buongiorno [10] correlation of viscosity. Uncertainty in the results of heat transfer due to the adoption of these two different formulae is discussed in detail.

9.2 Mathematical Modelling

The computational geometry of vertical annulus is shown in Fig. 9.1. The annulus comprised of two coaxial cylinders, the inner wall of which is partly heated and the outer wall is kept adiabatic. The annulus is divided into a lower adiabatic and upper heated zone. The annulus understudied has inner and outer diameters of 45 mm and 38 mm that gives gap of 3.5 mm. The total height of the annulus is 1048 mm that makes an aspect ratio of 300. Alumina–water nanofluid enters from the bottom and

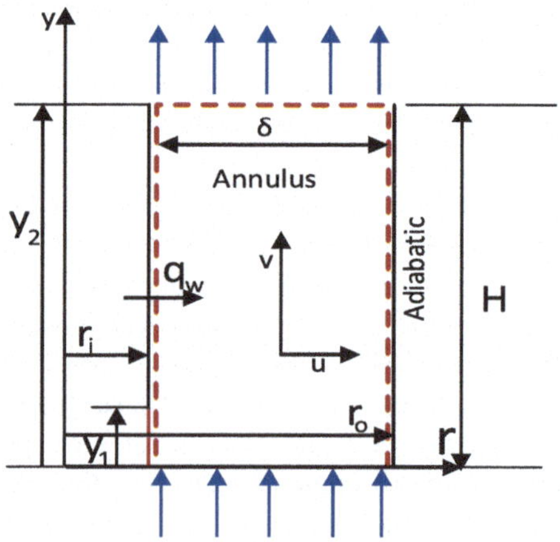

Fig. 9.1 Computational geometry

flows upward inside the vertical annulus. The heat flux of 3 kW/m^2 is applied in the upper heated region from $y = y_1$ to y_2.

Following assumptions are made during the analysis:

1. The problem will be considered as two-dimensional and axis-symmetric.
2. The test fluid flow inside the annulus is assumed to be Newtonian and incompressible.
3. The thermophysical properties of nanofluid are assumed to be independent on temperature except density (*Boussinesq Approximation*).
4. Single-phase approach is used to model the flow of nanofluids.

The energy transfer and flow of nanofluid inside the vertical annulus are governed by continuity equation, Navier–Stokes equation and the energy equation which in their elliptic forms represent as follows:

$$\frac{1}{r}\frac{\partial(ru)}{\partial r} + \frac{\partial v}{\partial y} = 0 \tag{9.1}$$

$$\left(u\frac{\partial u}{\partial r} + v\frac{\partial u}{\partial y}\right) = -\frac{1}{\rho_{\text{nf}}}\frac{\partial p}{\partial r} + \nu_{\text{nf}}\left[\frac{1}{r}\frac{\partial}{\partial r}\left(r\frac{\partial u}{\partial r}\right) - \frac{u}{r^2} + \frac{\partial^2 u}{\partial y^2}\right] \tag{9.2}$$

$$\left(u\frac{\partial v}{\partial r} + v\frac{\partial v}{\partial r}\right) = -\frac{1}{\rho_{\text{nf}}}\frac{\partial p}{\partial y} + g\beta_{\text{nf}}(T - T_a) + \nu_{\text{nf}}\left[\frac{1}{r}\frac{\partial}{\partial r}\left(r\frac{\partial v}{\partial r}\right) + \frac{\partial^2 v}{\partial y^2}\right] \tag{9.3}$$

$$\left(u\frac{\partial T}{\partial r} + v\frac{\partial T}{\partial y}\right) = \lambda_{\text{nf}}\left(\frac{1}{r}\frac{\partial}{\partial r}\left(r\frac{\partial T}{\partial r}\right) + \frac{\partial^2 T}{\partial y^2}\right) \tag{9.4}$$

The non-dimensionalization is done by using the following parameters:

$$R = \frac{r}{\delta}, Y = \frac{y}{\delta}, U = u\left(\frac{\delta}{\lambda_{\text{nf}}}\right), V = v\left(\frac{\delta}{\lambda_{\text{nf}}}\right), P = p\left(\frac{1}{\rho_{\text{nf}}}\right)\left(\frac{\delta}{\lambda_{\text{nf}}}\right)^2, \theta = \frac{T - T_a}{\frac{|q_w|\delta}{k_{\text{nf}}}},$$

$$\text{Aspect ratio, } A = \frac{L}{\delta}, \text{ Radius ratio, RR} = \frac{r_o}{r_i}, \text{ Annular gap, } \delta = r_o - r_i$$

Thus, governing equations in non-dimensional form are represented as follows:

$$\frac{1}{R}\frac{\partial(RU)}{\partial R} + \frac{\partial V}{\partial Y} = 0 \tag{9.5}$$

$$U\frac{\partial U}{\partial R} + V\frac{\partial U}{\partial Y} = -\frac{\partial P}{\partial R} + \Pr\left[\frac{1}{R}\frac{\partial}{\partial R}\left(R\frac{\partial U}{\partial R}\right) - \frac{U}{R^2} + \frac{\partial^2 U}{\partial Y^2}\right] \tag{9.6}$$

$$U\frac{\partial V}{\partial R}+V\frac{\partial V}{\partial R}=-\frac{\partial P}{\partial Y}+\text{Ra Pr }\theta+\Pr\left[\frac{1}{R}\frac{\partial}{\partial R}\left(R\frac{\partial V}{\partial R}\right)+\frac{\partial^2 V}{\partial Y^2}\right] \tag{9.7}$$

$$U\frac{\partial\theta}{\partial R}+V\frac{\partial\theta}{\partial Y}=\frac{1}{R}\frac{\partial}{\partial R}\left(R\frac{\partial\theta}{\partial R}\right)+\frac{\partial}{\partial Y}\left(\frac{\partial\theta}{\partial Y}\right) \tag{9.8}$$

The set of governing equations for natural convection of alumina–water inside vertical annulus must be solved simultaneously. The non-dimensional form of length in axial direction is represented as: lower adiabatic section is $Y = 0$ to $Y = Y_1 = 21$ and the heating zone is $Y = Y_1 = 21$ to $Y = \mathrm{Y}_2 = 300$. For radial direction $R_i = \frac{r_i}{\delta}$, and $R_o = \frac{r_o}{\delta}$.

Boundary Conditions

(a) Inner wall ($R = R_i$):

(i) Lower unheated zone:

$$U = V = 0 \text{ and } \frac{\partial\theta}{\partial R} = 0 \quad 0 \le Y < Y_1$$

(ii) Upper heated zone:

$$U = V = 0 \text{ and } \frac{\partial\theta}{\partial R} = -1 \quad Y_1 < Y \le Y_2$$

(b) Outer wall ($R = R_o$):

$$U = V = 0 \text{ and } \frac{\partial\theta}{\partial R} = 0 \quad 0 < Y \le A$$

(c) Inflow (inlet) condition ($Y = 0$):

$$\frac{\partial U}{\partial Y} = \frac{\partial V}{\partial Y} = 0 \text{ and } \theta = 0 \quad R_i < R < R_o$$

(d) Outflow (outlet) condition ($Y = A$):

$$\frac{\partial U}{\partial Y} = \frac{\partial V}{\partial Y} = 0 \text{ and } \frac{\partial^2\theta}{\partial Y^2} = 0 \quad R_i < R < R_o$$

The heat transfer coefficient at any location on the heated wall can be evaluated as:

$$h = q_w/\left(T_w - T_l\right) \tag{9.9}$$

The bulk liquid temperature is given by:

$$T_l = \frac{\int \rho C_{pl} \vec{V} T \, \mathrm{d}A}{\int \rho C_{pl} \vec{V} \, \mathrm{d}A} \tag{9.10}$$

where $\vec{V}$ is the resultant velocity.

9.3 Solution Methodology

The CFD solver ANSYS Fluent is used for simulation and ANSYS ICEM CFD is used in the generation of mesh. The grid distribution in the complete computational field is shown in Fig. 9.2. The generated grid has a uniform and non-uniform distribution, respectively, in the axial and radial directions. The third-order MUSCL and second-order upwind scheme are selected to discretized convective terms of the momentum and energy equations, respectively. The least-square gradient scheme is adopted to solve diffusion terms and PRESTO! formulation is employed for the discretization of pressure terms. The pressure–velocity decoupling is performed by means of a semi-implicit method for the pressure-linked equations (SIMPLE).

9.3.1 Thermophysical Properties

The density, specific heat and volumetric expansion coefficient of nanofluids are evaluated using the mixing theory. For the thermophysical properties of nanofluids, the following correlations are used:

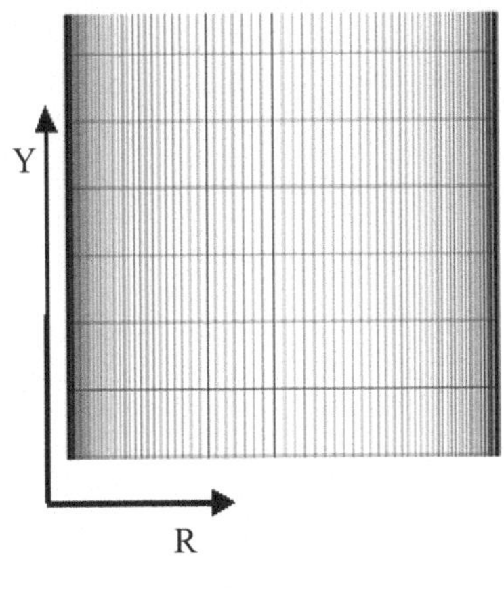

Fig. 9.2 Representative numerical grid

Density

$$\rho_{\text{nf}} = (1-\alpha)\rho_f + \alpha\rho_p \tag{9.11}$$

α is a volume fraction of nanoparticles.

Specific heat

$$Cp_{\text{nf}} = (1-\alpha)Cp_f + \alpha Cp_p \tag{9.12}$$

Volumetric expansion coefficient

$$\beta_{\text{nf}} = (1-\alpha)\beta_f + \alpha\beta_p \tag{9.13}$$

Figure 9.3 shows the variation of ratio of nanofluids properties to the base fluid properties with the concentration of nanoparticles. The density increases while specific heat and volume expansion coefficient decrease linearly with nanoparticle concentration as shown in Fig. 9.3.

Viscosity

(a) Brinkman [16] model:

$$\mu_{\text{nf}} = \left(1 + 2.5\alpha + 4.375\alpha^2 + \cdots\right)\mu_f \tag{9.14}$$

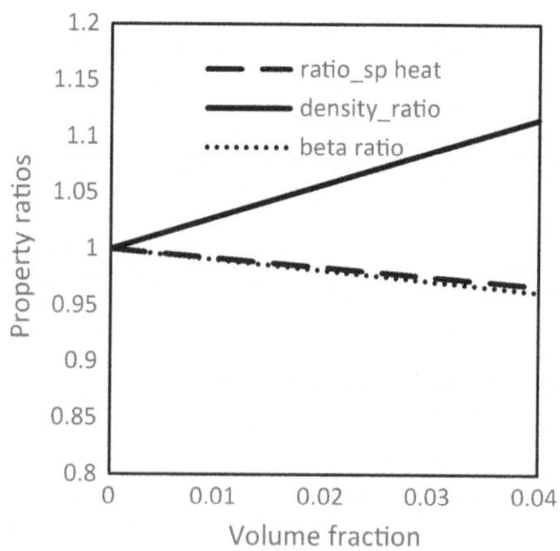

Fig. 9.3 Variation of property ratios (nanofluid/base fluid) with volume fraction of nanoparticles

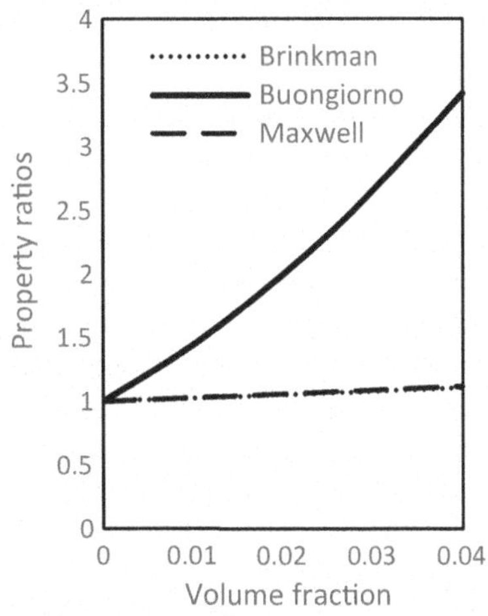

Fig. 9.4 Variation of thermal conductivity and viscosity with volume fraction of nanoparticles

(b) Buongiorno [10] correlation:

$$\mu_{\mathrm{nf}} = \left(1 + 39.11\alpha + 533.9\alpha^2\right)\mu_f \tag{9.15}$$

The viscosity of nanofluids based on Brinkman model increases linearly with nanoparticle concentration while Buongiorno correlation-based viscosity increases drastically as shown in Fig. 9.4.

Thermal conductivity

Maxwell [1] model:

$$\frac{k_{\mathrm{nf}}}{k_f} = \frac{k_p + 2k_f + 2\alpha(k_p - k_f)}{k_p + 2k_f - \alpha(k_p - k_f)} \tag{9.16}$$

9.4 Results and Discussion

The thermally induced flow of nanofluid (Al_2O_3 + water) inside a vertical annulus of high aspect ratio (300) is numerically investigated for various volume fractions of nanoparticle and heat flux of 3 kW/m^2. Numerical results are presented for the base fluid (water) and nanofluids, and uncertainty in the results concerning heat transfer due to the adoption of two formulae of viscosity is discussed in detail.

9.4.1 Wall and Liquid Temperatures

The numerical results for steady-state wall and liquid temperature have been presented. The variation of the wall and liquid temperature along the axial direction of annulus for nanofluids based on Brinkman and Buongiorno viscosity models are, respectively, shown in Figs. 9.5 and 9.6. From the figures, it is realized that the wall temperature increases rapidly in lower heated region and then rises linearly in the entire remaining heated length of the vertical annulus while the bulk liquid temperature follows a linear variation along heated axial length. Thus, the region where the wall temperature increases rapidly and liquid temperature increases linearly signifies the developing boundary layer, followed by the fully developed region where the wall and liquid temperature profiles become parallel to each other. The nanoparticle concentration lowers down the wall temperature while no significant effect seems on liquid temperature when predicted using Brinkman viscosity model as shown in Fig. 9.5. On the other hand, in the case of nanofluids based on Buongiorno viscosity model, the wall and liquid temperatures increase significantly with nanoparticle concentration as depicted in Fig. 9.6. Since the viscosity estimated from Buongiorno model is high that results in lower mass flow rates of the nanofluids. Therefore, wall and liquid temperatures may probably increase with nanoparticle concentration due to lower mass flow rates.

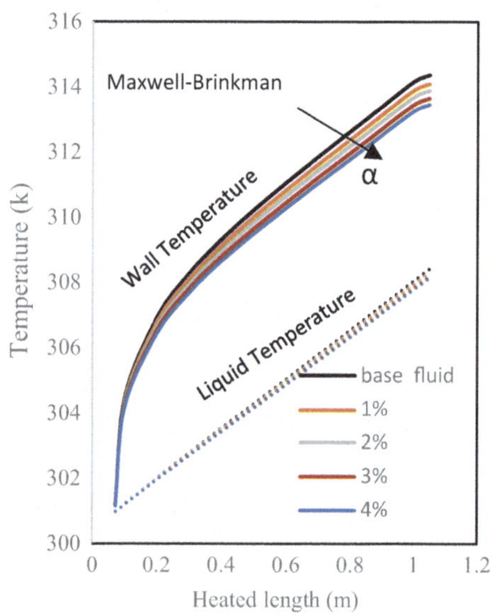

Fig. 9.5 Variation of wall and bulk liquid temperatures (nanofluid based on Brinkman viscosity model) along heated axial length of annulus

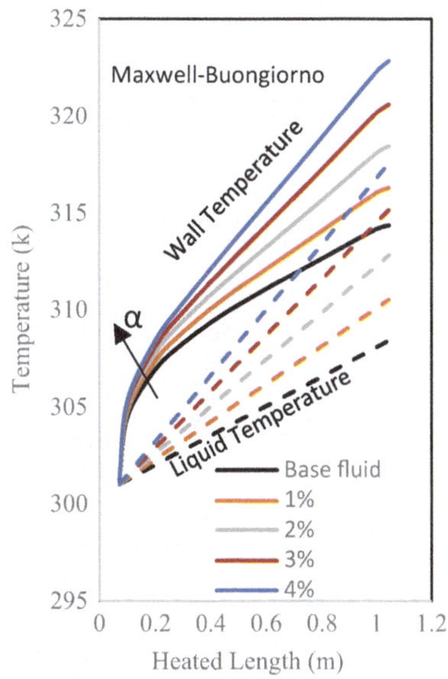

Fig. 9.6 Variation of wall and bulk liquid temperatures (nanofluid based on Buongiorno viscosity model) along heated axial length of annulus

9.4.2 *Heat Transfer Coefficient*

The thermal performance of nanofluids inside a computational geometry is judged by heat transfer coefficient that is estimated on the basis of the wall and liquid temperatures as given in Eq. (9.9). The variation of heat transfer coefficient along the heated axial length for the nanofluids based on Brinkman and Buongiorno viscosity models are, respectively, shown in Figs. 9.7 and 9.8. In the entrance region, the local heat transfer coefficient seems to drops abruptly represents the developing flow followed by a fully developed flow where the heat transfer coefficient becomes almost constant. Figures 9.7 and 9.8 reveal that heat transfer coefficient increases with nanoparticle concentration when predicted using both Brinkman and Buongiorno viscosity models. Figure 9.9 shows the variation of average heat transfer coefficient of the nanofluid based on Brinkman and Buongiorno viscosity models with nanoparticle concentration. The average heat transfer coefficient increases linearly with nanoparticle concentration based on Brinkman viscosity model, whereas, in the case of Buongiorno viscosity model-based nanofluid, it follows nonlinear variation. At heat flux of 3 kW/m^2 and volume concentration of 4%, the heat transfer coefficient is found to be enhanced by 10.86% when estimated using Brinkman viscosity model while it is only just 3.91% in the case of Buongiorno viscosity model-based nanofluid.

There seems to be the disparity in results concerning thermally induced flow of water-based alumina nanofluid inside the vertical annulus. These contradictions arise due to adoption of two different formulae for evaluating the viscosity of nanofluid available in the literature.

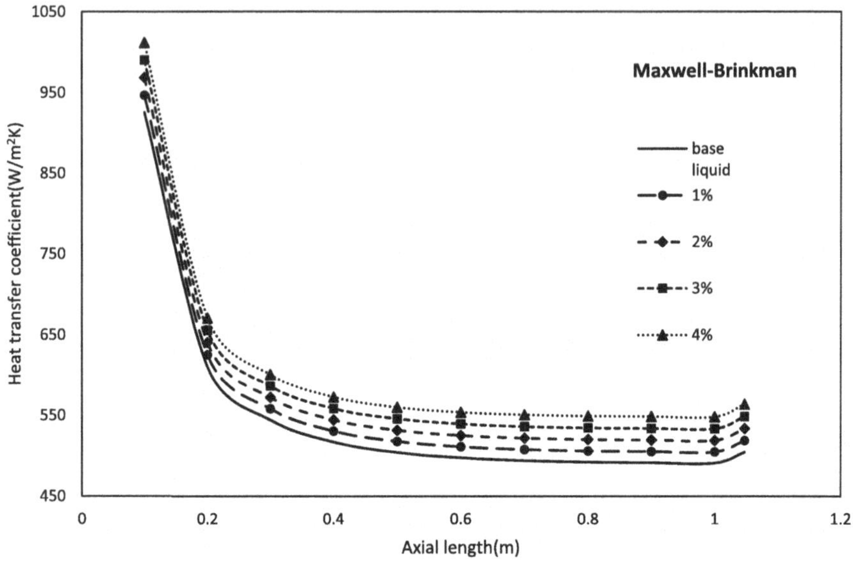

Fig. 9.7 Variation of local heat transfer coefficient (nanofluid based on Brinkman viscosity model) along heated axial length of annulus

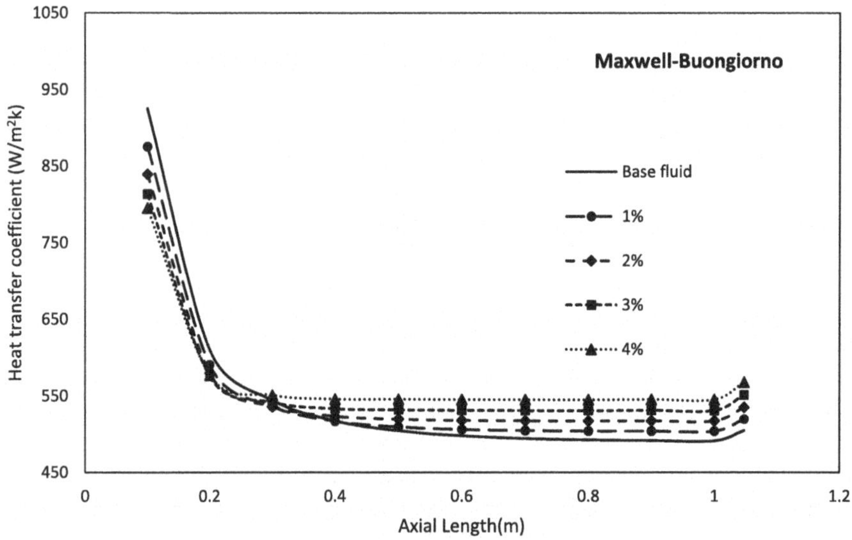

Fig. 9.8 Variation of local heat transfer coefficient (nanofluid based on Buongiorno viscosity model) along heated axial length of annulus

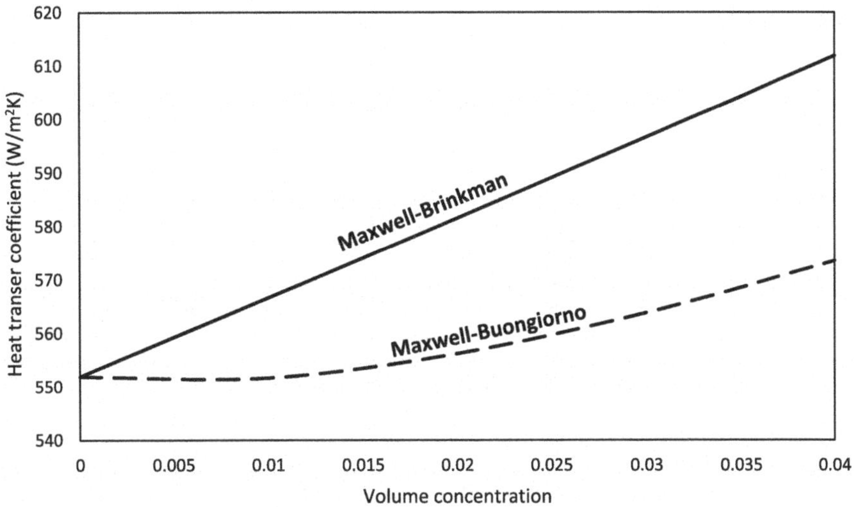

Fig. 9.9 Variation of average heat transfer coefficient with concentration of nanoparticles

9.5 Conclusion

The thermally induced flow of the nanofluids (Al_2O_3 + water) inside a vertical annulus has been numerically investigated. Following conclusions are drawn from the present study:

(i) For a given heat flux, the lower heated region where the wall temperature rises suddenly and the liquid temperature rises linearly represents the developing boundary layer, followed by the fully developed region where the wall and liquid temperature profiles become parallel to each other.
(ii) The wall and liquid temperatures decrease with nanoparticle concentration when estimated using the Brinkman viscosity model.
(iii) The wall and liquid temperatures increase with nanoparticle concentration when estimated using the Buongiorno viscosity model.
(iv) The average heat transfer coefficient increases linearly with nanoparticle concentration based on Brinkman viscosity model, whereas, in the case of Buongiorno viscosity model-based nanofluid, it follows nonlinear variation.
(v) When the heat transfer coefficient is estimated using a Brinkman viscosity model, maximum enhancement is found to be 10.86% while it is only just 3.91% in the case of Buongiorno viscosity model-based nanofluid.

Acknowledgements The authors are indebted to the Council of Scientific and Industrial Research (CSIR) for providing financial support under SRF Direct award No. 09/112(579)2K18.

References

1. Maxwell, J.C.: A Treatise on Electricity and Magnetism, 2nd edn. Clarendon Press, Oxford, UK (1873)
2. Choi, S.U.S.: Enhancing thermal conductivity of fluids with nanoparticles. ASME FED **231**, 99–105 (1995)
3. Lee, S., Choi, S.U.S., Eastman, J.A.: Measuring thermal conductivity of fluids containing oxide nanoparticles. J. Heat Transfer **121**, 280–289 (1999)
4. Wang, X., Xu, X., Choi. S.U.S.: Thermal conductivity of nanoparticle–fluid mixture. J. Thermophys. Heat Transfer **13**(4), 474–480 (1999)
5. Xie, H., Wang, J., Xi, T., Liu, Y., Ai, F., Wu, Q.: Thermal conductivity enhancement of suspensions containing nanosized alumina particles. J. Appl. Phys. **91**(7), 4568–4572 (2002)
6. Das, S.K., Putra, N., Thiesen, P., Roetzel, W.: Temperature dependence of thermal conductivity enhancement for nanofluids. ASME Trans. J. Heat Transfer **125**, 567–574 (2003)
7. Jang, S.P., Choi, S.U.S.: Role of Brownian motion in the enhanced thermal conductivity of nanofluids. Appl. Phys. Lett. **84**, 4316–4318 (2004)
8. Pak, B.C., Cho, Y.I.: Hydrodynamic and heat transfer study of dispersed fluids with submicron metallic oxide particles. Exp. Heat Transf. **11**, 151–170 (1998)
9. Lee, J.H., Hwang, K.S., Jang, S.P., Lee, B.H., Kim, J.H., Choi, S.U.S., Choi, C.J.: Effective viscosities and thermal conductivities of aqueous nanofluids containing low volume concentrations of Al_2O_3 nanoparticles. Int. J. Heat Mass Transfer **51**, 2651–2656 (2008)
10. Buongiorno, J.: Convective transport in nanofluids. ASME J. Heat Transfer **128**, 240–250 (2006)
11. Yang, C., Li, W., Nakayama, A.: Convective heat transfer of nanofluids in a concentric annulus. Int. J. Thermal Sci. **71**, 249–257 (2013)
12. Khanafer, K., Vafai, K., Lightstone, M.: Buoyancy-driven heat transfer enhancement in a two-dimensional enclosure utilizing nanofluids. Int. J. Heat Mass Transfer **46**, 3639–3653 (2003)
13. Putra, N., Roetzel, W., Das, S.K.: Natural convection of nanofluids. Heat Mass Transfer **39**(8/9), 775–784 (2003)
14. Wen, D., Ding, Y.: Formulation of nanofluids for natural convective heat transfer applications. Int. J. Heat Fluid Flow **26**(6), 855–864 (2005)
15. Ho, C.J., Chen, M.W., Li, Z.W.: Numerical simulation of natural convection of nanofluid in a square enclosure: effects due to uncertainties of viscosity and thermal conductivity. Int. J. Heat Mass Transfer **51**, 4506–4516 (2008)
16. Brinkman, H.C.: The viscosity of concentrated suspensions and solutions. J. Chem. Phys. **20**, 571–581 (1952)
17. Sheikholeslami, M., Chamkha, A.: Electro hydrodynamic free convection heat transfer of a nanofluid in a semi-annulus enclosure with a sinusoidal wall. Numer. Heat Transfer **69**, 781–793 (2016)
18. Sheikholeslami, M.: Numerical simulation of magnetic nanofluid natural convection in porous media. Phys. Lett. A **381**, 494–503 (2017)
19. Dogonchi, A.S., Chamkha, A., Ganji, D.D.: A numerical investigation of magneto-hydrodynamic natural convection of Cu-water nanofluid in a wavy cavity using CVFEM. J. Thermal Anal. Calorim. **135**, 2599–2611 (2019)
20. Corcione, M.: Empirical correlating equations for predicting the effective thermal conductivity and dynamic viscosity of nanofluids. Energy Convers. Manag. **52**, 789–793 (2011)

Chapter 10
Abrasive Flow Machining Process Hybridization with Other Non-Traditional Machining Processes: A Review

Sachin Dhull, Qasim Murtaza, R. S. Walia, M. S. Niranjan and Saloni Vats

Abstract Abrasive flow machining is a process in which two piston cylinder arrangements are utilized to provide extrusion pressure to the abrasive laden polymer media. This media helps to finish the internal surfaces of the hollow cylindrical workpiece. In this paper, different variants of AFM processes are explained. These include electric discharge machining (EDM)/electrochemical machining lapping, grinding, rotary EDM ball-burnishing process, vibration assisted AFM, etc. The paper has been broadly classified into three main headings of advanced hybrid AFM processes viz electrochemical, discharge and magnetic AFM process. The hybridization has successfully resulted in more material removal and better surface finish of the job. Various modelling and simulation models utilizing different types of in-house developed media and hybrid electrochemo magneto rotato fixtures used have been shown and explained in detail. The use of hybridization of the conventional AFM process is the demand of industries since complex geometries of workpiece with difficult to reach areas need to be finished in less time.

Keywords AFM variants · Magnetic field · Polymer media · Modelling · Optimization

10.1 Introduction to Abrasive Flow Machining and Its Variants

Abrasive flow machining is generally a nano-finishing operation in which the internal surface of the workpiece is finished by the abrading action of the abrasive particles held by polymer media. The semi-paste media is extruded using the machine piston cylinder arrangement. In one-way [1] AFM process the media is extruded inside the work piece only in one direction, Two-way AFM machine [2] has two hydraulic cylinders and two medium cylinders and in orbital process, good surface finishing

S. Dhull (✉) · Q. Murtaza · R. S. Walia · M. S. Niranjan
Delhi Technological University, Delhi, India

S. Vats
MSIT, Delhi, India

S. Yadav et al. (eds.), *Proceedings of International Conference in Mechanical and Energy Technology*, Smart Innovation, Systems and Technologies 174,
https://doi.org/10.1007/978-981-15-2647-3_10

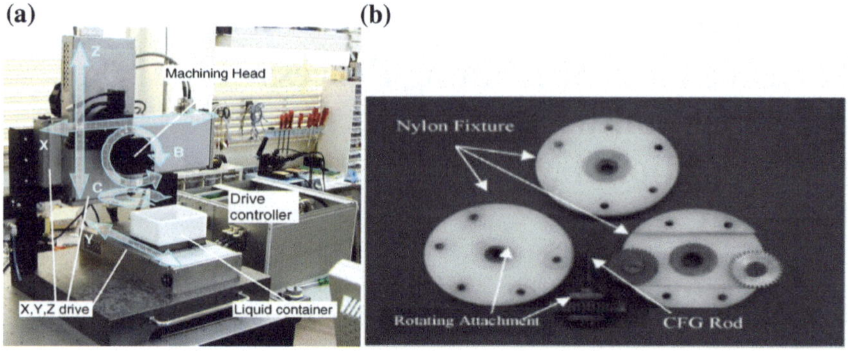

Fig. 10.1 **a** Machine tool system, **b** CFG rod and nylon fixture arrangement [6]

is obtained by producing low-amplitude oscillations of the work piece [3]. The various elements are fixture or tooling, the machine and abrasive laden media. Fixture plays a vital role to increase productivity of the process, and it holds the workpiece to be machined. This technique uses a non-Newtonian liquid polymer containing abrasive particles of Al_2O_3, SiC, boron carbide or diamond as the grinding medium and additives [4]. All AFM machines regardless of size are positive displacement hydraulic systems, where work piece is clamped between two vertically opposed media cylinder.

10.1.1 EDM and ECM/ECM-Lapping

Same machine tool, same electrode, i.e. copper and same machining liquid water were used, and the experiment was performed using formed electrode and simple shaped electrode. The EDM surface of Ra, i.e. 1 mm was reduced to 0.2 mm by further ECM application. If EDM and ECM-lapping were used, then surface roughness was improved up to 0.07 mm in 2 min [5]. The machine tool system of ECM-lapping process is shown in Fig. 10.1a. ECM is applied after EDM because it produces smooth and stress-free surface.

10.1.2 Abrasive Electrochemical Grinding (AECG)

The electrochemical grinding of sintered carbides G20, titanium alloys WT31 was done and it was concluded that wear in conventional grinding was 6 times more as compared to AECG. The grinding of tool was 15 times lesser in AECG as compared to mechanical grinding [7].

10.1.3 Rotary EDM with Ball-Burnishing of Al_2O_3/6061 Al Composites

The parameters were peak current, dielectric flushing pressure, electrode rotational speed, and non-load voltage [8].

10.1.4 Modelling of Hybrid Electrochemical Turning Magnetic Abrasive Finishing

DC motor was used for rotating workpiece. Magnetic flux of 0–0.24 T, working gap 1.5 mm, copper wire diameter 1 mm, soft iron was used as magnetic pole and core since it has high magnetic permeability [9]. The solenoid coils were connected in series., feed 0.02–0.1 mm/min, voltage 8–24 V, electrolyte pressure 0.4 MPa, speed 125–750 RPM, flow rate 2 lit/min, machining time 1.5 min. The electrolytic ions path changed from linear to curved in the presence of magnetic field.

10.1.5 Vibration Force-Assisted AFM Process

In the hybrid vibration abrasive flow machining process, the media is extruded to and fro through two-media piston cylinder arrangement and the process is assisted with the help of vibration force. In the paper, the workpiece was in vibration motion orthogonal in direction to media flow. The frequency is in the range 5–20 kHz. Responses were velocity, pressure of fluid, temperature and the pressure–velocity diagrams well explained the advantage of ultrasonic assisted abrasive flow machining over conventional AFM process. The temperature had no effect on the media stability [10]. The flexible fixture was used to hold hollow cylindrical workpiece as shown in Fig. 10.2. ANSYS FLUENT software was used to simulate the velocity, pressure and temperature distribution and CFD was used to study the path of flowing media.

10.1.6 Abrasive-Assisted Electrochemical Process

Reinforcement particles have a significant effect in abrasive-assisted ECM of Al–B–carbide graphite composite. In the paper, ECM was used and surfaces obtained were smoother as compared to that obtained by laser and electric discharge machining that produced heat-affected zone [13]. The magnetic field-assisted AFM process and its variants as well as magnetorheological finishing processes are explained with details of tool, workpiece and dielectric media used.

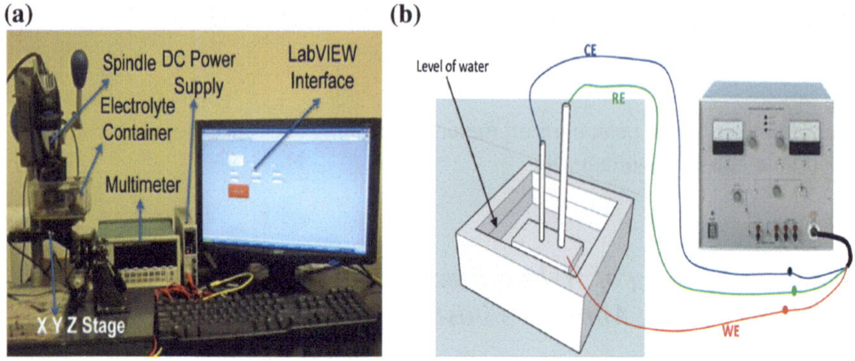

Fig. 10.2 **a** ECDM apparatus set-up [11], **b** Electrochemical cell and polarization systems [12]

10.1.7 Electrochemical and Centrifugal Force-Assisted AFM

ANOVA was applied for significant parameters identification so that maximum material could be removed during machining process. The various input variable parameters selected were RPM, salt molal concentration, operating pressure and the pressure was kept at 6 MPa [6]. The process set-up is shown in Fig. 10.1b.

10.1.8 AFM of Slit Fabricated by EDM

Analysis of variance (ANOVA) and signal-to-noise ratio (S/N) were used to identify significant combination of parameters so that optimum material removal and surface quality are obtained. After eliminating the recast layer produced by EDM, the optimum parameters were 150 μm, concentration of 50%, pressure 6.7 MPa and time of machining was 30 min [14]. If the concentration of media was high, the viscosity increased, thereby improving the surface finish of workpiece. The precision of surface was reduced due to high extrusion pressure.

The different types of electric discharge machining process and the hybrid processes are explained with proper details of workpiece, electrolyte and tool used. The electrochemical processes, both conventional and hybrid ECM, are explained in detail with all the parameters used.

10.2 Electrochemical Machining-Assisted AFM Process and Its Variants

In electrochemical abrasive flow machining process, the abrasive media is mixed in electrolyte solution and extruded through the inner surface of workpiece assisted by electrolytic force. The workpiece is connected to positive, while tool is connected to negative terminal and the electrolysis occurs through electrolyte between anode and cathode. In a paper, the glass was micromachined by ECDM process used in various applications like micro reactors, micro fuel cells, etc., as shown in Fig. 10.2a. It reduced overcut of hole and aspect ratio was improved [11]. A microstage for feed with step size 0.1 μm, multimeter of resolution 0.01 mA, to control current supply was used. The cross-innovation and modelling of ECM were done. ECF/ECP and its variants helped to achieve micro- to nano-range of surface finish, the parameters affecting the output results included electrolyte properties, position, motion of electrode, etc., [15]. A model was developed in which electrode gap was simulated and modelled and final micro-tool diameter was predicted within the accuracy of 94% or 6% variation range to develop tungsten tools [16]. Grey relational analysis was applied to study and analyse the output results obtained in ECM process [17].

10.2.1 ECM Using Mineral Water Electrolyte

Green machining, low cost, high-efficiency electrochemical process was explained, polarization cure was measured in potentiostat of 300 V, working electrode (WE) was workpiece, i.e. SUS304, reference electrode (RE) is Ag/AgCl, counter electrode (CE) is tool, i.e. WC rod of diameter 0.3 mm, as shown in Fig. 10.2b.

The mineral water was used in the process. It was concluded that short-pulsed supply was rapid than electrolyte $NaNO_3$ [12].

10.2.2 Electrochemical Micro-Machining Process (ECMM)

Various types of profiles generated by sinking and milling methods using straight, conical and reverse taper tool were investigated. It was found that lesser the micro-tool taper angle, lesser would be taper angle of profile generated during ECMM process. 14.5° taper angle of profile was generated by conical micro tool of 10.5° angle, while 5.14° angle was obtained by 4.34° angle tool [18]. If molarity value of electrolyte is less than 0.05 M, then electrolytic micromachining (EMM) does not give result, rather micro sparking removes material from the work surface.

10.2.3 Electrolytic Magnetic Abrasive Finishing Process and Lorentz Force Study

The effect of magnetic field on abrasive particles was studied. In case of both magnetic and electric fields, the oppositely charged ions move towards electrode in cycloidal path due to the action of Lorentz force. If approach angle of ions towards electrode is not proper, it resulted in poor electrolytic effect. Zinc stearate helped to stabilize steel grit movement, hence used for rough machining. High current and speed of rotation helped in more material removal and better surface finish [19].

10.2.4 Electrochemical Discharge Machining (ECDM)

Response surface methodology, i.e. RSM was used in ECDM process since material removal was non-linear. Aluminium frame was used for machine tool, worm and worm gear speed reduction of 25:1 was obtained, worm was attached to stepper motor shaft [20]. Taguchi method of optimization was used to obtain best result.

10.3 EDM and Its Hybrid Processes

As compared to conventional Cu electrodes, Cu–ZrB_2 electrodes showed more MRR, less tool wear. Since the difference between melting points of Cu and ZrB_2 is there, the spark mainly passes through Cu, hence the erosion of ZrB_2 decreases [21].

10.3.1 Conventional and Micro-EDM of WC

It is used where high-dimensional accuracy is required in hard materials [22].

10.3.2 Gas Film Study in ECDM

Gas film plays a vital role in machining quality and efficiency. The electric supply affects the produced gas film, which further affects the electrolyte properties, hence the output material removal and surface finish of the workpiece varies [23]. The development of bubble occurs and film thickness was in the range 17–29 mm. The advantage of gas bubble is that critical current as well as voltage gets decreased, hence overcutting problem in ECM is prevented.

10.3.3 Wire EDM

Addition of SiC abrasive particles in electrolyte improved wire EDM efficiency. The shape accuracy of workpiece was found to be significantly affected by wire vibration and introduction of bubbles in the electrolyte. The advantage of addition of silicon

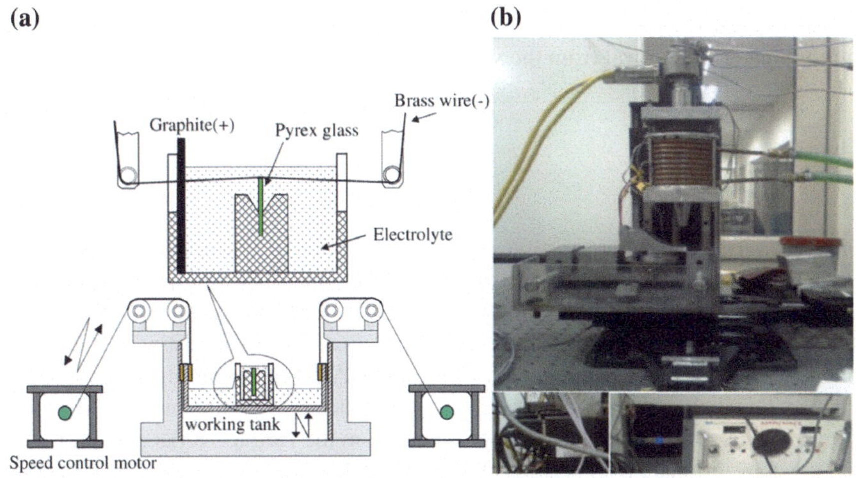

Fig. 10.3 **a** Wire EDM machine set-up [25], **b** MR finishing set-up [26]

carbide abrasive particles in the electrolyte is that it prevents the accumulation of bubbles, thereby decreasing the energy used in discharge. The expansion of slit and roughness value was 0.024 and 0.84 μm, respectively [24]. The brass wire diameter 0.25 mm was put in contact with cathode and graphite was used as anode, hydrogen bubbles were produced when applied voltage exceeded critical voltage, as shown in Fig. 10.3a. The temperature controller (401C) was used to control electrolyte properties, while the workpiece was driven by speed controlled motor.

10.3.4 Magnetic EDM

Grey relational analysis optimization technique applied in magnetic-assisted EDM of SKD 61 workpiece [27]. Taguchi L18 was applied, parameters taken were machining polarity, high voltage auxiliary current (IH), servo reference voltage (SV). The machine set-up comprised digital oscilloscope, current probe and passive voltage probe were used for monitoring discharge voltage and current waveforms.

10.4 Magnetic Field Assisted AFM Process and Its Variants

In this paper, applied magnetic field was used to calculate the normal force on abrasive grain, and microstructure of abrasive and magnetic particle was studied. The MR finishing set-up is shown in Fig. 10.3b. The abrasive fluidized bed machining process

[28] incorporated set of high pressure fluid piston cylinder arrangements, that resulted in more material removal from the work surface.

10.5 Conclusion

Different variants of AFM processes have been explained. Mineral water electrolyte usage in ECM process resulted in eco-friendly, economical and efficient EC micromachining process, further Lorentz force study and ECDM process are also explained. In EDM process, conventional micro EDM process, gas film study, wire EDM and magnetic EDM have been explained. In magnetic AFM, ball-end MR process, double disc, flow behaviour and force study in magnetic abrasive finishing (MAF) have been summarized. Hence, it can be concluded from the various works of researchers that the conventional AFM process efficiency can be increased by making it hybrid, i.e. different forces like magnetic, rotational, electrochemical, etc., resulted in higher material removal and better roughness of the surface. The different mathematical and analytical model of the fixtures used and of the forces incorporated in the hybrid processes can be further utilized for the process improvement. Further the newly developed media utilizing CNT particles had better results as compared to the conventional media used.

References

1. Rhoades, L.J.: Abrasive flow machining. Technical paper of the society of manufacturing engineers (SME), MR, pp. 89–145 (1989)
2. Tzeng, H.J., Yan, B.H., Hsu, R.T., Lin, Y.C.: Self-modulating abrasive medium and its application to abrasive flow machining for finishing micro channel surfaces. Int. J. Adv. Manuf. Technol. **32**, 1163–1169 (2007)
3. http://dc225.4shared.com/doc/LfcclGh0/preview.html
4. www.physorg.com/news609.html
5. Kurita, T., Hattori, M.: A study of EDM and ECM/ECM-lapping complex machining technology. Int. J. Mach. Tools Manuf. **46**, 1804–1810 (2006)
6. Vaishya, R., Walia, R.S., Kalra, P.: Design and development of hybrid electrochemical and centrifugal force assisted abrasive flow machining. In: 4th International Conference on Materials Processing and Characterization, Materials Today, Proceedings, vol. 2, pp. 3327–3341 (2015)
7. Zaborski, S., Łupak, M., Poro, D.: Wear of cathode in abrasive electrochemical grinding of hardly machined materials. J. Mater. Process. Technol. **149**, 414–418 (2004)
8. Yan, B.H., Wang, C.C., Lin, Y.C.: Feasibility study of rotary electrical discharge machining with ball burnishing for Al_2O_3/6061Al composite. Int. J. Mach. Tools Manuf. **40**, 1403–1421 (2000)
9. Yan, B.H., Chang, G.W., Cheng, T.J., Hsu, R,T.: Electrolytic magnetic abrasive finishing. Int. J. Mach. Tools Manuf. **43**, 1355–1366 (2003)
10. Venkatesh, G., Sharma, A.K., Singh, N.: Simulation of media behaviour in vibration assisted abrasive flow machining. Simul. Model. Pract. Theory **51**, 1–13 (2015)
11. Jui, S.K., Kamaraj, A.B., Sundaram, M.M.: High aspect ratio micromachining of glass by electrochemical discharge machining. J. Manuf. Process. **15**, 460–466 (2013)

12. Yanga, Y., Natsub, W., Zhaoa, W.S.: Realization of eco-friendly electrochemical micromachining using mineral water as an electrolyte. Precis. Eng. **35**, 204–213 (2011)
13. Sankara, M., Babu, A.G.: Effect of reinforcement particles on the abrasive assisted electrochemical machining of Aluminium–Boron carbide–Graphite composite. In: 12th Global Congress on Manufacturing and Management, Procedia Engineering, vol. 97, pp. 381–389 (2014)
14. Tzeng, H.Z., Yan, B.H., Hsu, R.T., Chow, H.M.: Finishing effect of abrasive flow machining on micro slit fabricated by wire-EDM. Int. J. Adv. Manuf. Technol. **34**, 649–656 (2007)
15. Kumar, P., Tailor, B., Agrawal, A., Joshi, S.S.: Evolution of electrochemical finishing processes through cross innovations and modeling. Int. J. Mach. Tools Manuf. **66**, 15–36 (2013)
16. Mathew, R., Sundaram, M.M.: Modeling and fabrication of micro tools by pulsed electrochemical machining. J. Mater. Process. Technol. **212**, 1567–1572 (2012)
17. Liua, J., Zhua, D., Zhaoa, L., Yang, Z.: Experimental investigation on electrochemical machining of TiAl intermetallic. In: 15th Machining Innovations Conference for Aerospace Industry (2012)
18. Ghoshal, B., Bhattacharyya, B.: Investigation on profile of microchannel generated by electrochemical micromachining. J. Mater. Process. Technol. **222**, 410–421 (2015)
19. Taweel, T.A.E.: Modelling and analysis of hybrid electrochemical turning magnetic abrasive finishing of 6061 Al/Al_2O_3 composite. Int. J. Adv. Manuf. Technol. **37**, 705–714 (2008)
20. Paul, L., Shekhar, S., Hiremath, S.: Response surface modelling of micro holes in electrochemical discharge machining process. In: International conference on design and manufacturing (2013)
21. Zhang, M., Jin, D.G.: EDM performance of electroformed Cu–ZrB_2 shell electrodes. Rapid Prototyp. J. **15**(2), 150–156 (2009)
22. Jahan, M.P., Rahman, M., Wong, Y.S.: A review on the conventional and micro-electrodischarge machining of tungsten carbide. Int. J. Mach. Tools Manuf. **51**, 837–858 (2011)
23. Jiang, B.Y., Lan, S.H., Ni, J.: Modeling and experimental investigation of gas film in micro-electrochemical discharge machining process. Int. J. Mach. Tools Manuf. **90**, 8–15 (2015)
24. Yang, C.T., Song, S.L., Yan, B.H., Huang, F.Y.: Improving machining performance of wire electrochemical discharge machining by adding SiC abrasive to electrolyte. Int. J. Mach. Tools Manuf. **46**, 2044–2050 (2006)
25. Das, M., Jain, V.K., Ghoshdastidar, P.S.: Analysis of magnetorheological abrasive flow finishing (MRAFF) process. Int. J. Adv. Manuf. Technol. **38**, 613–621 (2008)
26. Singh, A.K., Jha, S., Pandey, P.M.: Nanofinishing of a typical 3D ferromagnetic workpiece using ball end magnetorheological finishing process. Int. J. Mach. Tools Manuf. **63**, 21–31 (2012)
27. Lin, Y.C., Lee, H.S.: Optimization of machining parameters using magnetic-force-assisted EDM based on gray relational analysis. Int. J. Adv. Manuf. Technol. **42**, 1052–1064 (2009)
28. Barletta: Progress in abrasive fluidized bed machining. J. Mater. Process. Technol. **209**, 6087–6102 (2009)

Chapter 11
Microstructural and Microhardness Analysis of Nickel-Based Ceramic Composite Coating on AISI 304 Stainless Steel by TIG Coating Method

Anjani Kumar, Neeraj Singh, Sonu Nagar and Anil Kumar Das

Abstract In present work, nickel-based metal matrix composite (MMC) coating was produced on AISI 304 stainless steel by tungsten inert gas (TIG) coating technique. TIG heat source used to develop intermetallic bonding between coating and substrate. The metal ceramics composition was employed viz. Ni + SiC and Ni + Si + B_4C. Two different proportions of SiC and B_4C in coating mixture were used and assess their effect on microstructure and microhardness. Field emission scanning electron microscopy (FESEM) analysis revealed that fine martensitic microstructure evolved, for S4 (85%Ni + 7.5%Si + 7.5%B_4C) sample. For this, maximum microhardness and average hardness of coated layer were found 676 $HV_{0.1}$ and 523 $HV_{0.1,}$ respectively, whereas average hardness of substrate was 200 $HV_{0.1}$. The overall result concludes that Si + B_4C is better choice in comparison to direct inclusion of SiC in nickel-based MMC fabrication.

Keywords TIG coating · FESEM · Microhardness · Silicon carbide

11.1 Introduction

AISI 304 stainless steel is extensively used in various engineering application like automobile, agricultural equipment, manufacturing industries, milk and food processing unit [1–3]. Although, stainless steel has high stiffness, corrosion resistance and strength, but their surface hardness and wear resistance are not enough to survive sliding motion against high load [4]. This leads to hamper the multifaceted application of AISI 304 steel in the practice [5]. The surface coating process is recently evolving as attractive coating technique, which has potential to improve weakness of steel and make it versatile and cost effective. In recent decades, incorporation of metal matrix composites layer on ordinary metal through coating has been gained attention to researcher. The metal matrix composite consists of two phases: one is matrix (ductile phase) and other one is reinforcement (hard phase). The matrix phase imparts ductility and toughness in composite, whereas reinforcements impart

A. Kumar (✉) · N. Singh · S. Nagar · A. K. Das
Department of Mechanical Engineering, National Institute of Technology Patna, Bihar 800005, India

S. Yadav et al. (eds.), *Proceedings of International Conference in Mechanical and Energy Technology*, Smart Innovation, Systems and Technologies 174,
https://doi.org/10.1007/978-981-15-2647-3_11

hardness and modulus of elasticity [6–8]. The ceramic material like, metal carbides, hard oxides and nitrides such as SiC, TiC, TiB_2 WC, B_4C and Al_2O_3 are very hard and wear resistance [9–14]. Moreover, alone ceramic material is incoherent for surface coating on steel due to large melting-point difference, poor adhesion and low fracture toughness. Typically, a transition metal is introduced in ceramic due to the metal's good ductility and toughness. Therefore, ceramics reinforced metal matrix composite coating is good choice for surface improvement [4, 11]. Silicon carbide is a promising ceramic material due to its excellent strength and wear resistance at room and elevated temperature [15]. Silicon carbide is compatible with nickel (Ni) because of high melting point and thermal conductivity almost similar.

In recent decades, several research groups are attempted to explore the TIG coating characteristics and mechanical performance on steel with several compatible combination of ceramic and transition metal. Bouyaz and Ulutan [5] reported in situ SiC coating by TIG on 304 steel. Saroj et al. [16] fabricated TiC reinforced inconel825 metal matrix composite and achieved seven times more wear resistance than 304 steel substrates. Çelik et al. [17] investigated SiC coating by TIG with different percentage of graphite powder and result found that improvement in wear characteristics. Mridha [18] fabricated TiN in nitrogen environment by TIG on the surface of pure titanium (Ti) plate and achieved hardness 2000 HV. Tosun [19] studied nickel-based WC coating and observed microstructural changes and microhardness variation of steel. All reported results evident that TIG coating is a viable technique for modification in stainless steel. There are still very few records available about SiC and B_4C reinforced nickel-based composite, developed through TIG method. Through this literature, authors try to add some more information in the field of TIG coating by using some specific material.

In this investigation, the surface of AISI 304 stainless steel was coated with preplaced layer of Ni + SiC and Ni + Si + B_4C, using TIG arc as heat source for the melting of preplaced layer. The influence of SiC and B_4C proportion on microstructure and microhardness is reported. It is hoped that this information could give useful support to fabricate hard and wear resistant composite surface.

11.2 Experimental Procedure

First of all, AISI 304 stainless steel was taken as substrate material of size $100 \times 50 \times 10\ mm^3$ and surface rubbed with 220 grades SiC emery paper to remove oxide layer followed by cleaning with acetone to remove other oily contamination. The different powders: Silicon carbide (Size 400 mesh, purity 99%, make: Sigma Aldrich), Boron carbide (Size 200 mesh, purity 98%, make: Sigma Aldrich), Silicon (Size 200 mesh, purity 98.5%) and nickel (Size 200 mesh, purity 98%) were used as coating material; make: Sigma Aldrich. Four samples were prepared with different material composition (shown in Table 11.1). The mixture of powders and 5% polyvinyl alcohol were blended with the help of magnetic stirrer to form homogeneous semisolid paste. The polyvinyl alcohol works as binder and imparts strength in preplaced layer

Table 11.1 Sample identification and corresponding composition

Sample identification	Composition (by weight%)
S1	Ni (95) + SiC (5)
S2	Ni (85) + SiC (15)
S3	Ni (95) + [Si + B_4C = 1:1] (5)
S4	Ni (85) + [Si + B_4C = 1:1] (15)

to withstand arc pressure; higher concentration of polyvinyl alcohol causes formation of void and porosity. The paste was uniformly dispersed over the clean surface and dried up to 72 h to remove moisture.

A semi-automatic TIG set-up, make: Electra TIG 200 DC, was used to generate arc. The heat input was controlled by regulating current and scan speed of arc; the heat input was calculated with the assistance of mention Eq. 11.1. The preplaced dried sample was processed with TIG arc at optimum parameter (current 120 A, voltage 25 V, speed 2.297 mm/s and heat input 626.905 J/mm); the preplaced powder got melted and form metallurgical bond with substrate upon solidification of melt pool. The coated sample were sectioned and cold mounted for microhardness test. The mounted sample was polished with standard metallurgical process; initially, sample was polished on SiC emery paper from 220 to 3000 grades followed by diamond paste on polishing cloth for mirror-like finishing. The morphology of coated sample was carried out on FESEM (Make: ZIESS) to examine evolution of microstructure.

$$Q = \eta \times V \times I/v \tag{11.1}$$

where, Q = heat input in (J/mm), η = efficiency of heat transfer (0.48), V = voltage in (Volt), I = current in (A) and v = scan speed in (mm/s).

11.3 Results and Discussion

11.3.1 Microstructural Analysis

The mechanical properties of coated layer depend on the morphology and distribution of ceramic particle in the region. The details of composition are listed in Table 11.1. Figure 11.1a shows different zone of coating; the coated layer, the interface layer and the substrate. Figure 11.1a–c shows the microstructure of S1; which shows small size SiC particles distribution in phases and irregular dendritic solidification occurs for sample S1. Figure 11.2a–d shows micrograph of sample S2, in which Fig. 11.2b shows a thick interface layer; the grain near to the substrate surface is equiaxed and near to coated layer is irregular due to different solidification rate. As the percentage of SiC increased the interface layer thicker and coated zone retain a fine dendritic structure reinforced with grey SiC particle. This results due to rapid solidification and

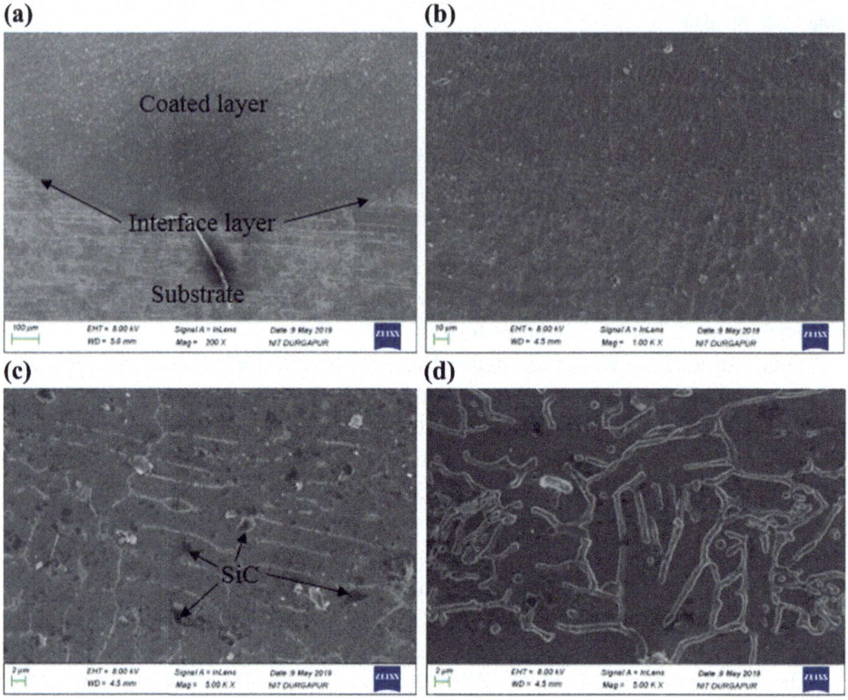

Fig. 11.1 **a** FESEM micrograph of sample S1; **b** the microstructure of coated layer; **c** and **d** enlarged view of coated layer at different magnification

other reason also dominated due to larger different in melting point. In addition, the dissociation of SiC will also form hard carbide phases in the coating. Furthermore, the Marangoni effect in the melt pool increased the rate of heat dissipation, along with the mixing of the hard phase, leading to the formation of finer dendritic structures.

Figure 11.3a–d shows morphology of sample S3 in which SiC formed by reaction between Si and B_4C. A mixed grain was observed; the equiaxed grain, and irregular larger grains. Moreover, the eutectic phases also observed in coated layer in Fig. 11.3c, d. Figure 11.4a–d shows micrograph of sample S4, in which 15% (Si + B_4C = 1:1) was used in powder mixture composition (details of composition in Table 11.1). A thin layer of interface layer was observed in Fig. 11.4c, also micrograph revealed that solidification front starts from the interface due to high cooling rate facilitated near the interface. At the bottom, heat dissipation rate is higher than that of middle and top layer, the undercool region also more at the bottom. Moreover, if the rate of nucleation is high, it impedes the growth of grains, leading to a finer martensitic microstructure and disordered eutectic phases, which would be able to improve mechanical performance of coated layer.

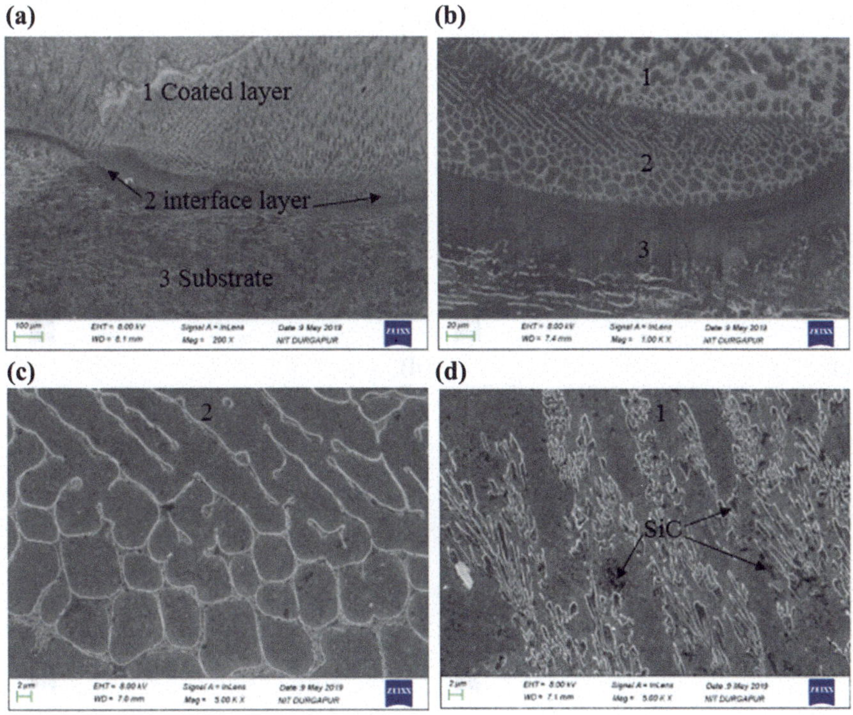

Fig. 11.2 **a** FESEM micrograph of sample S2; **b** the enlarged view of the coated layer and interface; **c** the microstructure of interface layer **d** the microstructure of the coated layer

11.3.2 Microhardness Analysis

The microhardness test was performed (on Vickers hardness tester, make: INNOVATEST, FALCON-503) to investigate the mechanical performance of coating. The microhardness is mainly depending upon the microstructure and phases present in coating. The microhardness graph shows that as the percentage of ceramic increased microhardness value increases, the variation in hardness shown in Fig. 11.5. More interestingly, microhardness value observed better for introducing Si + B_4C in nickel matrix rather than direct inclusion of SiC. The higher hardness is possible due to evolution of fine homogenous microstructure, also Vickers hardness of B_4C is slightly higher than that of SiC. The maximum hardness was observed 676 $HV_{0.1}$ and average hardness was 523 $HV_{0.1}$ for S4 as compared to average microhardness of steel was 200 $HV_{0.1}$. In sample S3, maximum hardness observed 318 $HV_{0.1}$, whereas maximum hardness for S1 and S2 were observed 240 $HV_{0.1}$ and 353 $HV_{0.1}$, respectively.

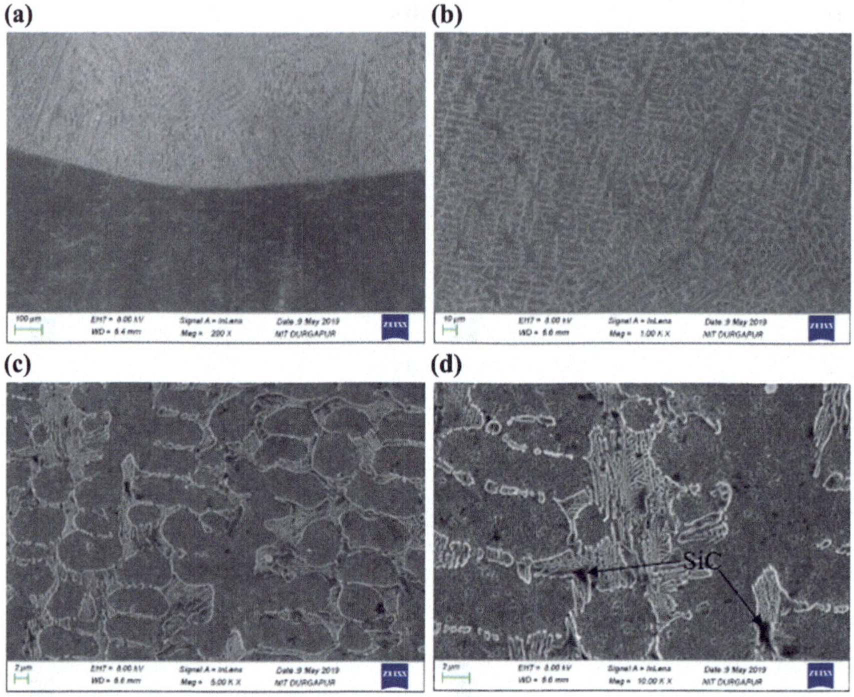

Fig. 11.3 **a** FESEM micrograph of sample S3; **b** the microstructure of coated layer; **c** and **d** enlarged view of coated layer at different magnification

11.4 Conclusions

The nickel-based MMC was developed on 304 steel through TIG coating method. In coating composition, directly employed SiC and Si + B_4C in nickel-based matrix and their results were investigated.

- The MMC coating on 304 steel was successfully developed and good interfacial bonding was observed.
- A eutectic phases and homogenous growth of microstructure were observed for S3 and S4.
- The higher hardness was observed for Si + B_4C than SiC for same proportion (shown in Table 11.1) taken in nickel-based coating.
- For sample S4, Maximum microhardness and average microhardness were observed 676 $HV_{0.1}$ and 523 $HV_{0.1,}$ respectively, due to formation of fine martensitic structure, whereas average hardness of the substrate was 200 $HV_{0.1}$.
- A thin and homogeneous interface layer was observed for S4, whereas thick interface layer for S2 with varying grain from equiaxed at bottom to irregular large grain at top.

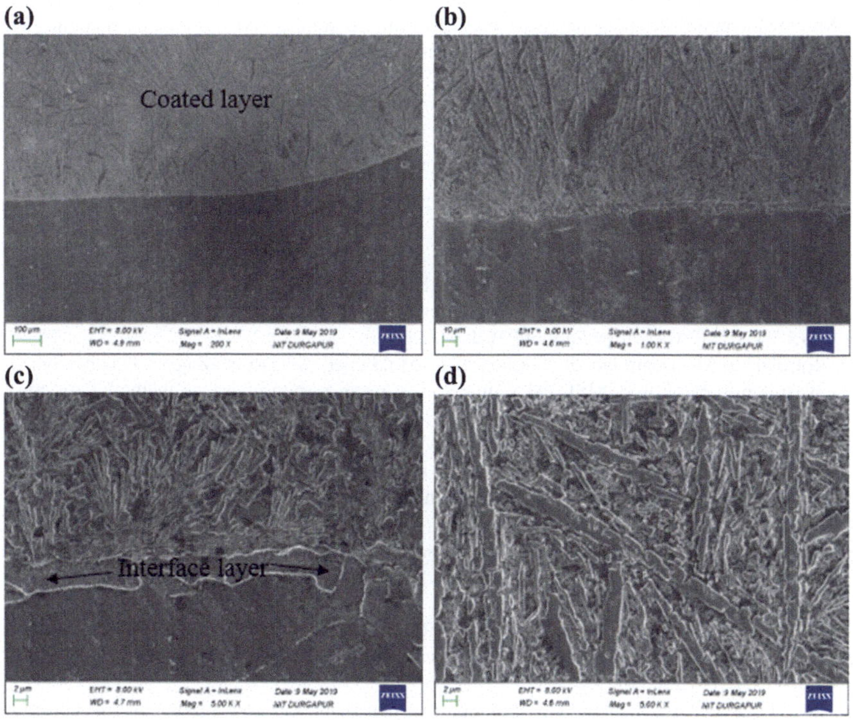

Fig. 11.4 **a** FESEM micrograph of sample S4; **b** the enlarged view of sample; **c** the microstructure of interface layer and coated layer; **d** the enlarge microstructure of the coated layer

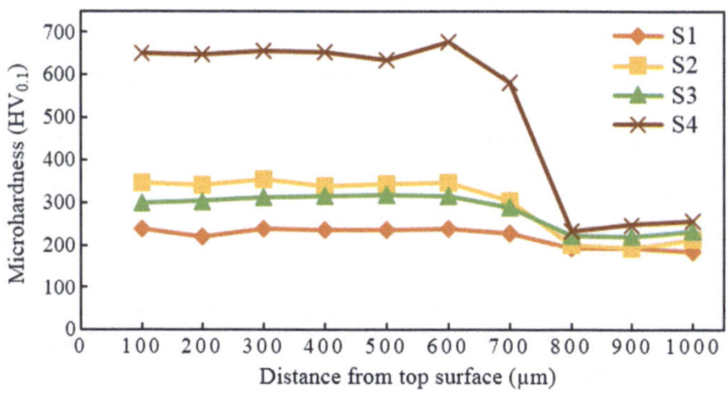

Fig. 11.5 Microhardness profile of coated sample across the crosssection from the top surface of sample

Acknowledgements Authors are extremely thankful to staff of NIT Durgapur metallurgical department and CMERI Durgapur for conducting microstructure and microhardness analysis, respectively.

References

1. Sun, G.F., Zhang, Y.K., Zhang, M.K., Zhou, R., Wang, K., Liu, C.S., Luo, K.Y.: Microstructure and corrosion characteristics of 304 stainless steel laser-alloyed with Cr–CrB_2. Appl. Surf. Sci. **295**, 94–107 (2014). https://doi.org/10.1016/j.apsusc
2. Kumari, M.M., Natarajan, S., Alphonsa, J., Mukherjee, S.: Dry sliding wear behaviour of plasma nitrocarburised AISI 304 stainless steel using response surface methodology. Surf. Eng. **26**, 191–198 (2010). https://doi.org/10.1179/174329409X439041
3. Saklakoğlu, N., Saklakoğlu, I.E., Ceyhun, V., Ağaçhan, N., Short, K.T., Collins, G.A.: Tribological behaviour of plasma immersion ion implanted AISI 304 stainless steel against polymer and ceramic counterfaces. Surf. Eng. **23**, 257–260 (2007). https://doi.org/10.1179/174329407X215177
4. Majumdar, J.D., Ramesh C.B., Manna, I.: Laser composite surfacing of AISI 304 stainless steel with titanium boride for improved wear resistance. Tribol. Int. **40**, 146–152 (2007). https://doi.org/10.1016/j.triboint.2006.04.006
5. Buytoz, S., Ulutan, M.: In situ synthesis of SiC reinforced MMC surface on AISI 304 stainless steel by TIG surface alloying. Surf. Coat. Technol. **200**, 3698–3704 (2006). https://doi.org/10.1016/j.surfcoat.2005.02.178
6. Jha, P., Gupta, P., Kumar, D., Parkash, O.: Synthesis and characterization of Fe–ZrO_2 metal matrix composites. J. Compos. Mater. **48**, 2107–2115 (2014). https://doi.org/10.1177/0021998313494915
7. Gupta, P., Kumar, D., Parkash, O., Jha, A.K.: Effect of sintering on wear characteristics of Fe–Al_2O_3 metal matrix composites. Proc. Inst. Mech. Eng. Part J J. Eng. Tribol. **228**, 362–368 (2014). https://doi.org/10.1177/1350650113508934
8. Gupta, P., Kumar, D., Parkash, O., Jha, A.K.: Structural and mechanical behaviour of 5% Al_2O_3-reinforced Fe metal matrix composites (MMCs) produced by powder metallurgy (P/M) route. Bull. Mater. Sci. **36**, 859–868 (2013). https://doi.org/10.1007/s12034-013-0545-1
9. Rasool, G., Stack, M.M.: Wear maps for TiC composite based coatings deposited on 303 stainless steel. Tribol. Int. **74**, 93–102 (2014). https://doi.org/10.1016/j.triboint.2014.02.002
10. Choudhury, A.R., Ezz, T., Chatterjee, S., Li, L.: Microstructure and tribological behaviour of nano-structured metal matrix composite boride coatings synthesized by combined laser and sol-gel technology. Surf. Coat. Technol. **202**, 2817–2829 (2008). https://doi.org/10.1016/j.surfcoat.2007.10.013
11. Feng, J., Ferreira, M.G.S., Vilar, R.: Laser cladding of Ni–Cr/Al_2O_3 composite coatings on AISI 304 stainless steel. Surf. Coat. Technol. **88**, 212–218 (1997). https://doi.org/10.1016/S0257-8972(96)02909-X
12. Majumda, J.D., Kumar, A., Li, L.: Direct laser cladding of SiC dispersed AISI 316L stainless steel. Tribol. Int. **42**, 750–753 (2009). https://doi.org/10.1016/j.triboint.2008.10.016
13. Tijo, D., Masanta, M., Das, A.K.: In-situ TiC-TiB_2 coating on Ti-6Al-4 V alloy by tungsten inert gas (TIG) cladding method: part-I. Microstructure evolution. Surf. Coat. Technol. **344**, 541–552 (2018). https://doi.org/10.1016/j.surfcoat.2018.03.082
14. Chatterjee, S., Shariff, S.M., Padmanabham, G., Majumdar, J.D., Choudhury, A.R.: Study on the effect of laser post-treatment on the properties of nanostructured Al_2O_3-TiB_2-TiN based coatings developed by combined SHS and laser surface alloying. Surf. Coat. Technol. **205**, 131–138 (2010). https://doi.org/10.1016/j.surfcoat

15. Lusquiños, F., Pou, J., Quintero, F., Pérez-Amor, M.: Laser cladding of SiC/Si composite coating on Si–SiC ceramic substrates. Surf. Coat. Technol. **202**, 1588–1593 (2008). https://doi.org/10.1016/j.surfcoat
16. Saroj, S., Sahoo, C.K., Masanta, M.: Microstructure and mechanical performance of TiC-Inconel825 composite coating deposited on AISI 304 steel by TIG cladding process. J. Mater. Process Technol. **249**, 490–501 (2017). https://doi.org/10.1016/j.jmatprotec.2017.06.042
17. Çelik, O.N., Ulutan, M., Gaşan, H., Er, Ü., Buytoz, S.: Effects of graphite content on the microstructure and wear properties of an AISI 8620 steel surface modified by tungsten inert gas (TIG). Surf. Coat. Technol. **206**, 1423–1429 (2011). https://doi.org/10.1016/j.surfcoat.2011.09.009
18. Mridha, S.: Titanium nitride layer formation by TIG surface melting in a reactive environment. J. Mater. Process. Technol. **168**, 471–477 (2005). https://doi.org/10.1016/j.jmatprotec.2005.02.247
19. Tosun, G.: Ni–WC coating on AISI 1010 steel using TIG: microstructure and microhardness. Arab. J. Sci. Eng. **39**, 2097 (2014). https://doi.org/10.1007/s13369-013-0754-3

Chapter 12
Analyzing Overall Equipment Effectiveness of HEMM Using LTB and CTB Approaches for Open-Pit Mines: A Case Study

Kartick Bhushan, Anil Kumar Agrawal and Somnath Chattopadhyaya

Abstract Open-pit or surface mining is considered to be one of the most capital extensive industries in the world. Massive costing and condensed productivity are enforcing organizations for automation of mining machinery especially heavy earth moving machineries (HEMMs) like dumpers, shovel and dozers. Employment of advanced and efficient machinery is a paramount need. The recorded losses such as breakdowns and idling during mining operations are serious challenges. This research article presents an effective methodology to evaluate the efficacy of HEMMs which is commonly termed as "Overall Equipment Effectiveness (OEE)." OEE is a quite familiar name in the manufacturing sector but found a very rare application in the mining field. OEE comprises availability, performance and quality to determine overall efficiency. A practical case study on HEMM (shovel) for open-surface mining has been performed in one of the biggest mining industry of India, using loading time-based (LTB) and standardized calendar time-based approach (CTB) for validation of OEE technique.

Keywords OEE · LTB and CTB · Mining · Time losses · TPM · HEMM

12.1 Introduction

Mining is a very capital-exhaustive industry, and it is an undeniable fact that the equipment utilization and correct assessment of these setups are very significant. Mining sectors need to utilize their equipment as efficiently as possible to get the primary yield on their investments and reducing production cost as well [1]. The mining sector has improved gradually but its complexities like costing, competition; size and safety are still the biggest concern [2]. Failure of capital-intensive HEMM and its concerns have the major influence on coal production, productivity and production costs of the pit surface mining operation. Failure of equipments (HEMM) such as shovels, dumpers and dozers during the operational shift will result in an unavoidable stoppage of coal production and revenue losses [3]. In the present scenario of stiff

K. Bhushan (✉) · A. K. Agrawal · S. Chattopadhyaya
Indian Institute of Technology (Indian School of Mines), Dhanbad, India

S. Yadav et al. (eds.), *Proceedings of International Conference in Mechanical and Energy Technology*, Smart Innovation, Systems and Technologies 174,
https://doi.org/10.1007/978-981-15-2647-3_12

competition, any organization can only be exceedingly productive, if there exists a high operational efficiency and effectiveness. Merely maintenance strategies have been considered and employed in order to escalate the availability of equipment [4]. Therefore, the intention of this study is to present a technique, to increase the equipment efficiency (HEMMs) and productivity in open-pit mines by utilizing overall equipment effectiveness (OEE) approach.

OEE method was developed by Nakajima for assessment of total productive maintenance (TPM) during the late 1980s. TPM advocates equipment maintenance through productive involvement of all workforces from production and maintenance department [5]. The main motto of TPM is to obtain almost no defects, zero accidents and negligible breakdowns to enhance the quality, cost and productivity. To evaluate the effectiveness of the equipment, OEE is generally used as a parameter of TPM [6]. OEE is inferred as the combination of availability, performance and quality. By thoroughly analyzing the literature, it is evident that OEE methodology provides eloquent explanations to real-time problems of industries [7].

In this paper, an effort has been made to determine the equipment effectiveness of HEMMs using OEE methodology. For analyzing OEE, LTB and CTB approaches are used. LTB approach advocates the significance of proper planning and scheduling for estimation of time losses; whereas, CTB approach computes the effectiveness based on scheduled and non-scheduled time of operation.

12.2 Overall Equipment Effectiveness (OEE)

OEE is an extensively used performance indicator for accessing manufacturing operations such as time loses and breakdowns[8]. In order to increase productivity and reduce loses, OEE plays a significant role. It accounts for the utmost and significant causes of productivity loss, which are shown in Table 12.1 [9].

OEE can be accessed in terms of these loss factors, i.e., availability, performance and quality. It is preferred to have availability greater than 90%; whereas, performance and quality should be 90% and 95%, respectively. If these standards are fulfilled, then it is expected to have overall effectiveness greater than 70%, i.e., benchmark value [10, 11].

Table 12.1 Losses category

Category	Losses	Factors
Breakdown	Downtime	Availability (Av)
Setup and adjustment		
Idling and minor stoppages	Speed	Performance (Pr)
Reduced speed		
Reduced yield	Defects	Quality (Qt)
Quality defects		

$$\textbf{OEE} = \textbf{Availability}(\textbf{Av}) \times \textbf{Performance}(\textbf{Pr}) \times \textbf{Quality}(\textbf{Qt}) \tag{12.1}$$

Availability refers to the elapsed time during which the production schedule halts. It is the probability that a system will perform its required function under instant time [12]. It can be considered in terms of total time duration for which the equipment is in operation. This is generally due to equipment breakdowns, waiting time, etc. [13]. Then, the availability can be determined using formulae as follows:

$$\begin{aligned}\textbf{Availability}\ (\textbf{Av}) = \{&(\text{Net Available Time} - \text{Downtime Losses})\\ &\div (\text{Net Available time})\} \times 100\end{aligned} \tag{12.2}$$

The performance consists of various factors due to which equipment operates at lesser speed, i.e., below optimum values. Probable causes may be inferior quality products, machinist incompetence and working conditions, etc. The performance can be calculated as follows:

$$\begin{aligned}\textbf{Performance}\ (\textbf{Pr}) = \{&(\text{Operating Time}-\text{Speed Losses})\\ &\div (\text{Operating Time})\} \times 100\end{aligned} \tag{12.3}$$

The quality rating can be defined as the losses incurred due to technical snag, i.e., breakdowns, malfunctioning of equipment and reworks.

$$\begin{aligned}\textbf{Quality}\ (\textbf{Qt}) = \{&(\text{Net Operating Time} - \text{Defect Losses})\\ &\div (\text{Net Operating Time})\} \times 100\end{aligned} \tag{12.4}$$

For equipment to function efficiently, it needs a higher level of performance against all three of these constraints. The parameters given in the OEE equation are generally considered as shown in Table 12.2.

Once the OEE estimation is performed, the subsequent step will be in comparison with the benchmarked value which is considered to be 85% approximately. If the probable OEE is lesser than the standard value, the method should be evaluated for

Table 12.2 OEE parameters

Overall time		
Net available time		Scheduled downtime
Operating time	Downtime losses	Av = (Operating time) ÷ (Net available time)
Net operating time	Speed losses	Pr = Net operating time ÷ Operating time = Actual output ÷ Target output
Valuable productive time	Defect losses	$\text{Qt} = \dfrac{\text{Fully productive time}}{\text{Net operating time}}$ = Good output ÷ Actual Output

Table 12.3 OEE evaluation

Total time (24 h × 7 days)	Recorded losses		OEE parameters
Actual available time (AAT)	Availability	Non-scheduled time Scheduled maintenance Unscheduled maintenance Setup and adjustment Idling Dumper waiting time	Availability = (Actual available time) ÷ (Total time)
Net production time (NPT)	Performance	Propelling duration Operational losses Speed loss	Performance = (Net production time) ÷ (Actual available time)
Valuable production time (VPT)	Quality	Quality loss Below standard material Defective product	Quality = (Valuable production time) ÷ (Net production time) = (Volume of average load ÷ Actual bucket capacity)

enhancement [14]. It is apparent that the effectual computation equipment efficiency relies on proper data compilation capability. Erratic or insufficient data may not replicate authentic equipment utilization [15]. It is also very vital to distinguish that each loss is categorized corresponding to an equipment state. Good output refers to scrap/rework and setup losses.

12.3 LTB and CTB Approaches

In the evaluation of OEE, total time plays a vital role. Total time (Table 12.2) is the time accessible for the production over a certain time period. This process of total time evaluation is termed as loading time-based approach **(LTB)** and was quite common earlier [16]. However, LTB methodology results in an overestimation of OEE. Another methodology, i.e., calendar time-based approach is also known as standardized calendar time approach **(CTB)** and is more preferable as it measures actual equipment utilization [17]. Calendar time is a theoretical approach during which a firm possesses the equipment. CTB approach computes the effectiveness based on scheduled and non-scheduled time of operation. The LTB approach advocates the significance of proper planning and scheduling for estimation of time losses [18]. The focal variance amid the two approaches relies on the assessment to operational availability at the expense of accountable time. It has been observed that the calendar time approach provides lower availability than loading time analysis [19]; hence, the choice and selection of these approaches should be done accordingly during mining operations.

12.4 Equipment Effectiveness (OEE) of HEMM

OEE implementation in mining sector varies from the manufacturing industry. Unfortunately, the implementation of OEE application is rarely observed in the mining industry as the complexities in this sector are quite high. That is why very few literatures are available in this context. The massive size and expenses of HEMMs involved in operation make the situation critical to apply OEE methodology. Hence, proper classification of equipment for a recorded loss like downtime and stoppages should be done accordingly [20]. Improper and fabricated data lead to poor effectiveness of mining types of machinery; therefore, it is essential to have authentic and factual data but it is quite challenging due to various reasons such as

- Surface mining involves a lot of serialized processes such as loading, dumping and excavating. The operating and environmental conditions are also not ideal for equipment.
- The dimensions of mining equipment are massive. Therefore, the outcome of utilization on overall production is also very high.

12.5 Case Study 4.1: OEE Estimation for Shovel

In open-pit mining, shovels are considered to be one of the most important equipments for loading purposes. It is bucket-equipped machinery, generally electrically power-driven. It should function efficiently without any faults throughout the week, utilizing every hour, i.e., 24 × 7 a day. It is expected to have efficiency between 80 and 90%. Any fault with shovel operation leads to an increment in total production cost. It is essential to record all-time losses that occurred during the operation. Scheduled as well as non-scheduled time will not be considered while estimating OEE through LTB approach. Remaining parameters will be the same. Here CTB has been employed, and all losses which may occur should be recorded as such.

No.	Recorded losses	Descriptions
1	Non-scheduled losses	Overall losses as machineries have no fixed timetable for operation
2	Schedule maintenance time	Interval during which shovel is maintained or repaired

(continued)

(continued)

No.	Recorded losses	Descriptions
3	Unscheduled maintenance time	Losses during machinery failure
4	Setup and adjustment	Time interval during shovel is all set for operation
5	Idling	Unproductive time in the absence of operators
6	Dumpers waiting time	Shovel waiting time for loading dumpers
7	Propel duration	Duration lost due to shovel movement from one load spot to other
8	Time losses due to job conditions	Climatic conditions, management delay, etc.
9	Speed loss	Time delay due to swing time, loading time, etc.
10	Quality loss	Losses due to filling factor, i.e., shovel bucket cannot be over loaded

12.5.1 *Data Collection and OEE Calculation*

For evaluating equipment effectiveness (OEE), shovel data have been collected from one of the biggest coal mine companies of North India. Data of total working hours (WH), maintenance hours (MH), propel hours (PH), idle hours (IH), etc., were collected from the mentioned mine during (2017–18). Totally, nine shifts are operable during a month. Percentage availability and percentage utilization have been calculated from the below-collected data.

12.5.1.1 Filling Factor

Fill factor designates the actual capacity of shovel bucket. It explains the percentage capacity of the material, which is to be filled. There are various aspects on which filling factor depends like bucket size, shape and dimension of material, digging ability of bucket, operator skill, repose angle of material which is to be loaded, etc. [21, 22]. In actual practice, the bucket fill factor should lie in the range of 0.4–1.2 [23].

12.5.2 *OEE Calculation for Shovel EX300-LC*

Using recorded data from Table 12.4 and co-relating with Table 12.3, for calculation of equipment effectiveness, the calculations for OEE using CTB and LTB approaches are as follows:

Table 12.4 Time delays for shovel operation

SHOVEL MODEL NO.: EX300-LC (TATA HITACHI)		
Recorded loss	Description	Duration (h. per month)
Total time(TT)	24 h. per day × 30 days per month	720
Non-scheduled time (NT)	2 days in a month	48
Schedule maintenance (SM)	3 of days in a month	72
Unscheduled maintenance (UM)	Breakdowns	95
Setup and adjustment (ST)	06 h/shift	54
Idle time (IT)	05 h/shift	45
Truck waiting time (TW)	06 h/shift	54
Job conditions (JC)	Equipment not operating (rain, fog, etc.)	27
Speed loss (SL)	05 h/shift	45
Propel time (PL)	4 moves/month, 2 h per move	8
Quality loss (QL)	Bucket filling factor (87%)	–
Shovel bucket capacity (BC)	15 m^3	–
Ideal production (IP)	15 bucket per minute	–

(A) **Calendar Time-Based Approach (CTB)**

Total time = 720 h.

Now, next step is to check for availability (A) using formulae from Table 12.3.

$$
\begin{aligned}
\text{Availability (Av)} &= \text{(Actual Available Time)} \div \text{(Total Time)} \\
&= \{720 - \text{(operational time losses)}\} \div 720 \text{ hrs.} \\
&= \{720 - (NT + SM + UM + ST + IT + TW)\} \div 720 \\
&= \{720 - (48 + 72 + 95 + 54 + 45 + 54)\} \div 720 \\
&= (720 - 368) \div 720 \\
&= 0.4888 \text{ hrs. i.e. } 49\% \text{ approx.}
\end{aligned}
$$

Availability (Av) = 49% approx.

After availability calculation, the next step is to determine performance and quality rating as well.

$$
\begin{aligned}
\textbf{Performance (Pr)} &= \text{(Net Production Time)} \div \text{(Actual Available Time)} \\
&= \{\text{Operational time losses} - (JC + SL + PL)\} \div \{AAT\} \\
&= \{352 - (27 + 45 + 8)\} \div \{352\} \\
&= 0.7727 \text{ hrs. i.e. } 77\% \text{ approx.}
\end{aligned}
$$

Hence, the performance rating is found to be **77%** approx.
Quality (Qt) = 87% (from Table 12.3).
Now, calculate the equipment effectiveness using Eq. (12.1)

$$\begin{aligned}\mathbf{OEE} &= \mathbf{Av} \times \mathbf{Pr} \times \mathbf{Qt}\\ &= 0.4880 \times 0.7727 \times 0.8777\\ &= 0.3315 \text{ i.e. } 33\% \text{ approx.}\\ \therefore \text{OEE} &= 33\%\end{aligned}$$

Therefore, the overall equipment effectiveness (OEE) for shovel EX300-LC is found to be 33% approx., and overall production or total production will be calculated as

$$\begin{aligned}\text{Total Production} &= \text{TT(mins)} \times \text{BC} \times \text{OEE}\\ &= (720 \times 60) \times 15 \times 15 \times 0.33\\ &= 3,207,600 \text{ m}^3.\\ \therefore \text{Total Production} &= 3,207,600 \text{ m}^3.\end{aligned}$$

(B) **Loading Time-Based Approach (LTB)**

Loading time approach is also based on standardized calendar time. The loading metric is a dimension of schedule effectiveness, and it is intended to eliminate loses and running operations smoothly. On scheduled time (NT), scheduled maintenance(SM) is not considered here.

Therefore,

$$\begin{aligned}\textbf{Total Time} &= \{\text{Total calendar hrs.} - (0\text{NT} + \text{SM})\}\\ &= \{720 - (48 + 72)\}\\ &= 600 \text{ hrs.}\end{aligned}$$

Now,

$$\begin{aligned}\text{Availability (Av)} &= (\text{Actual Available Time}) \div (\text{Total Time})\\ &= \{600 - (\text{operational time losses})\} \div 600 \text{ hrs.}\\ &= \{600 - (95 + 54 + 45 + 54)\} \div 600\\ &= 0.586 \text{ hrs.}\\ \textbf{Availability (Av)} &= \mathbf{58\%} \textbf{ approx.}\end{aligned}$$

Next step is the estimation of performance and quality rating as follows:

$$\begin{aligned}\text{Performance (Pr)} &= (\text{Net Production Time}) \div (\text{Actual Available Time})\\ &= \{\text{Operational time losses} - (\text{JC} + \text{SL} + \text{PL})\} \div \{\text{AAT}\}\end{aligned}$$

Table 12.5 Comparison of OEE between LTB and CTB approaches for EX300-LC

Items	CTB	LTB
Total time hours	720 h	600 h
Availability (Av) %	49	58
Performance (pr) %	77	77
Quality (Qt) %	87	87
OEE %	33	40
Total production	3,207,600 m^3	3,207,600 m^3

$$= \{352 - (27 + 45 + 8)\} \div \{352\}$$
$$= 0.772 \text{ hrs.}$$
$$\text{Performance (Pr)} = 77\% \text{ approx.}$$

Quality (Qt) = 87% (from Table 12.3).

$$\mathbf{OEE} = \mathbf{Av} \times \mathbf{Pr} \times \mathbf{Qt}$$
$$= 0.586 \times 0.772 \times 0.877$$
$$= 0.3967 \text{ approx.}$$
$$\therefore \text{OEE} = 40\% \text{ approx.}$$

Therefore, the overall equipment effectiveness (OEE) for shovel EX300-LC is found to be 40% approx., and overall production or total production will be calculated as

$$\text{Total Production} = \text{TT mins.} \times \text{BC} \times \text{IP} \times \text{OEE}$$
$$= (600 \times 60) \times 15 \times 15 \times 0.396$$
$$= 3,207,600 \text{ m}^3.$$
$$\therefore \mathbf{Total\ Production} = \mathbf{3,207,600\ m^3}.$$

From Table 12.5, it is clearly evident that equipment efficacy is 33% and 40%, respectively, using CTB and LTB approach, which is far below the benchmarked values. But, overall total production is almost the same.

12.6 Conclusion

Equipment effectiveness can be enhanced and measured effectively if an appropriate performance measurement technique is adopted. OEE is a well-recognized technique to access equipment efficiency in manufacturing industries, and this methodology has also been adapted for the coal mining industry for evaluating the effectiveness of

HEMMs. The purpose of this paper is to categorize infertile time losses during opencast mining operations which affect the availability, performance as well as quality. This paper has also discussed the possible time losses for HEMM, more specifically shovel and procedures to estimate OEE. CTB and LTB approaches have also been discussed to set a benchmark for improvement. Loading time-based methodology delivers greater OEE standards than the calendar time approach. LTB approach provides 40% OEE whereas CTB 33%, it means LTB is better than CTB. This article also reflects the selection of proper analyzing techniques for determining the effectiveness of HEMMs.

References

1. Krellis, O., Singleton, T.: Mine maintenance-the cost of operation (1998)
2. Unger, R.L., Conway, K.: Impact of maintainability design on injury rates and maintenance costs for underground mining equipment. In: Improving Safety at Small Underground Mines, pp. 18–94. Bureau of Mines SP (1994)
3. O'Connor, P., Kleyner, A.: Practical reliability engineering. Wiley (2012)
4. Başçetin, A., Kesimal, A.: A new approach in selection of loading-hauling systems in surface mining. In: 16th Mining Congress of Turkey (1999)
5. Nakajima, S.: Introduction to TPM: total productive maintenance (Translation), vol. 129. Productivity Press, Inc. (1988)
6. Shirose, K.: Equipment effectiveness, chronic losses, and other TPM improvement concepts in TPM development program: implementing total productive maintenance, pp. 27–84 (1989)
7. Jain, A., Bhatti, R., Singh, H.: Total productive maintenance (TPM) implementation practice: a literature review and directions. Int. J. Lean Six Sigma **5**(3), 293–323 (2014)
8. Singh, K., Singh, A.I.: An evaluation of transfusion of TQM-TPM implementation initiative in an Indian manufacturing industry. J. Qual. Maint. Eng. **21.2**, 134–153 (2015)
9. Tsarouhas, P.H.: Evaluation of overall equipment effectiveness in the beverage industry: a case study. Int. J. Product. Res. **51.2**, 515–523 (2013)
10. Méndez, J.D.M., Rodriguez, R.S.: Total productive maintenance (TPM) as a tool for improving productivity: a case study of application in the bottleneck of an auto-parts machining line. Int. J. Adv. Manuf. Technol. **92**(1–4), 1013–1026 (2017)
11. Zandieh, S., Tabatabaei, S.A., Ghandehary, M.: Evaluation of overall equipment effectiveness in a continuous process production system of condensate stabilization plant in Assalooyeh. Interdiscip. J. Contemp. Res. Bus **3**(10), 590 (2012)
12. Agrawal, A.K., Murthy, V.M.S.R., Chattopadhyaya, S.: Investigations into reliability, maintainability and availability of tunnel boring machine operating in mixed ground condition using Markov chains. Eng. Fail. Anal **105**, 477–489 (2019)
13. Chiarini, A.: Improvement of OEE performance using a lean six sigma approach: an Italian manufacturing case study. Int. J. Product. Qual. Manag. **16**(4), 416–433 (2015)
14. Fore, S., Zuze, L.: Improvement of overall equipment effectiveness through total productive maintenance. World Acad. Sci. Eng. Technol. **61** (2010)
15. Ireland, F., Dale, B.G.: A study of total productive maintenance implementation. J. Qual. Maint. Eng. **7**(3), 183–192 (2001)
16. Muchiri, P., Pintelon, L.: Performance measurement using overall equipment effectiveness (OEE): literature review and practical application discussion. Int. J. Product. Res. **46**(13), 3517–3535 (2008)
17. Relkar, A.S., Nandurkar, K.N.: Optimizing and analysing overall equipment effectiveness (OEE) through design of experiments (DOE). Proc. Eng. **38**, 2973–2980 (2012)

18. Mohammadi, M., et al.: Analysis of availability and utilization of dragline for enhancement of productivity in surface mines-a case study. In: Proceedings of the 23rd World Mining Congress, Montreal (2013)
19. Ericsson, J.: Disruption analysis-an important tool in lean production (1997)
20. Jonsson, P., Lesshammar, M.: Evaluation and improvement of manufacturing performance measurement systems-the role of OEE. Int. J. Oper. Product. Manag. **19**(1), 55–78 (1999)
21. Dhillon, B.S.: Mining equipment reliability. Springer, London (2008)
22. Osanloo, M., Hekmat, A.: Prediction of shovel productivity in the Gol-e-Gohar iron mine. J. Min. Sci. **41**(2), 177–184 (2005)
23. Doktan, M.: Impact of blast fragmentation on truck shovel fleet performance. In: 17th International Mining Congress and Exhibition of Turkey (2001)

Chapter 13
Smart Strain Approximation Surface-Mounted Optical Fiber Strain Sensor

A. S. Wali and Amit Tyagi

Abstract This study is aimed to develop a smart neural network perceptron model for strain prediction using fiber optic sensor signals. Optical parameters corresponding to surface-mounted optical fiber are obtained experimentally under static loading conditions. Four variations are used by creating external damages to study the strain variations on healthy, single damage and multiple damage beam structures. The strain values are obtained by using phase difference and change in intensities as input data for the feed-forward backpropagation neural network model. A comparative study of pre-existing analytical solutions, conventional strain gauge measurement, and finite element analysis is performed. The neural network model proposed in this work provides more close results to the results obtained by strain gauge and FEA analysis as compared to analytical analysis carried out by Haslach.

Keywords Optical parameters · Neural network · Finite element analysis (FEA) · Strain

Nomenclature

FOS	Fiber optic sensor
FBG	Fiber Bragg grating
R	Resistance of the wire
L	Length of the wire
A/C	Cross-section area of the wire
G.F	Strain gauge factor
LM	Levenberg–Marquardt backpropagation algorithm
BR	Bayesian regularization backpropagation algorithm
RMSE	Root mean square error
MSE	Mean square error

A. S. Wali (✉) · A. Tyagi
Department of Mechanical Engineering, Indian Institute of Technology (B.H.U), Varanasi 221005, India
e-mail: ashwary.rs.mec14@itbhu.ac.in

S. Yadav et al. (eds.), *Proceedings of International Conference in Mechanical and Energy Technology*, Smart Innovation, Systems and Technologies 174,
https://doi.org/10.1007/978-981-15-2647-3_13

N	Number of samples
$\Delta\emptyset$	Change in phase
ΔI	Change in intensity
η_0	Refractive index of unstrained state
p_{11}, p_{12}	Pockel strain optic constants
ν_f	Poisson's ratio
ε	Strain
ρ	Resistivity of the material
V_r	Ratio of voltage values under strained to unstrained condition
Y_i	Predicted neural network output values
Y_i'	Actual neural network values
λ	Wavelength of light used

13.1 Introduction

Smart structures are structures having inbuilt intelligence that can assess and respond to an external change in the environment. In the last few decades, some researchers have used fiber optic sensors to analyze and measure the mechanical properties of structures. SHM is a method of integrity analysis of different parameters, which characterize the integrity of the object. Presently, there are several ongoing research initiatives toward the development and deployment of fiber optic sensors for sensing applications in structural health monitoring. Fiber optic sensors carry some advantages over traditional sensors. Fiber optic sensors are: immune to electromagnetic interference, lightweight, small, easy to install, corrosion-resistant, durable, and can be multiplexed.

Chan et al. studied the FBG sensors' performance, comparing results with the conventional structural health monitoring system [1]. Her and Yang investigation shows that the Mach–Zehnder interferometric optical fiber sensor can be successfully used for measuring the dynamic strain of a vibrating cantilever beam [2]. Regez et al. showed that the optic fiber strain measurement sensing mechanism gives extremely low strains predictions that are in the order of 0–25 microstrains [3]. Kadhim used the phase modulated technique for optical fiber strain analysis [4]. A comparison of the results is made with analytical solutions Lunia et al. performed a smart wing analysis using composite material [5]. Gholamzadeh and Nabovati demonstrated the use of the fiber optical sensor in different civil engineering structures [6]. James and Tatam discussed recent developments in fiber optics with a focus on the fabrication of optical long period gratings and the level of their sensitivity related to different mechanical parameters [7]. Heredia et al. introduced an alternative non-contact method to measure the stress–strain behavior of ductile materials under compression testing using a fiber optic sensor [8]. Present experimental work explores the area of smart strain sensing, using an optical signal with a trained neural network under static loading. This work deals with designing a neural network with a noisy signal. For validation of the proposed work analytical, finite element and conventional strain gauge is considered. The strain is a structural parameter that is analyzed for a cantilever beam structure under static loading. The developed ANN model gives positive predictions under the acceptable error range.

13.2 Methodologies

A strain is the primary mechanical parameter directly related to the displacement, and accurate measurement of strain developed in a structure is always questionable as the magnitude of displacement is very small. A recent development in artificial intelligence opens new areas for strain measurement. Different techniques have their disadvantages like some test is not possible over large structures, high-quality skill is required for strain gauge measurement, it is impossible to measure strain on active structures. The advantage of using the FOS technique, is that they can be multiplexed, has high flexibility with low propagation loss, accurate measurement, and immunity to electromagnetic interference. Different methods are applied using FOS to measure strain values of various structures. The present work includes experimental investigation based on the FOS signal as input to a neural network for strain approximation under static loading. A neural network is a major tool in artificial intelligence. A neural network can be single or multilayer as per the problem and function definition. In the present work, a single layer neural network is developed. Method of selection of network training function and selection of the minimum number for the neuron in the hidden layer is discussed using analytical solutions and the finite element method.

13.2.1 Fiber Optic Sensors

Butter and Hocker [9] introduced an equation for the straight fiber which relates the change in phase to the change in the axial fiber strain which is used in the present study as

$$\Delta\emptyset = \frac{2\pi n_0 L}{\lambda}\left[1 - \frac{n_0^2}{2}\{p_{12} - \nu_f(p_{11} + p_{12})\}\right]\epsilon_f \tag{13.1}$$

Based on the Helmholtz equation, it has been shown analytically that the absolute value, as well as the direction of the phase gradient, can be expressed using the three-dimensional intensity distribution of the light. Haslach and Sirkis [10] have presented the theory of surface-mounted fiber optic sensor defining the path selection for structures under different loading conditions. Butter and Hocker [9] strain equation is obtained for the strained and unstrained condition, as shown in Eq. 13.2.

$$\Delta\emptyset = \beta(1-c)\int_0^L \epsilon_n \mathrm{d}s = \beta(1-c)\epsilon_{\mathrm{avg.}} \tag{13.2}$$

where $\epsilon_{\mathrm{avg.}}$ the average strain is developed as an optical fiber. To develop a sensor design using this phase change property of optical fiber, we have to express the sensing and reference path as a function of optical parameters. As derived by Haslach [10], sensing and reference path can be expressed in the x-y coordinate system, and surface strain variation can be obtained in the x-y coordinate system is shown in Eq. 13.3. On applying coordinate condition for one dimension Eq. 13.3 get reduce to Eq. 13.4.

$$\begin{aligned}\Delta\emptyset &= \beta(1-c)\int_0^L \epsilon_n \mathrm{d}s \\ &= \beta(1-c)\int_0^L \frac{\left(\epsilon_{xx}x'^2 + \epsilon_{yy}y'^2 + \gamma_{xy}x'y'\right)}{\left(x'^2 + y'^2\right)} \cdot \sqrt{x'^2 + y'^2}\mathrm{d}t \end{aligned} \tag{13.3}$$

$$\Delta\emptyset = \beta(1-c)[\epsilon_{xx}k_1] \tag{13.4}$$

As change in phase is directly related to the change in the axial strain for a surface-mounted fiber optic sensor. The paper presents strain measurement in a cantilever beam, under static loading, using transmission type fiber optic sensor. A comparative study is presented relative to the initial analytical solution by Haslach [10].

13.2.2 Numerical Simulation

A model of a cantilever beam having dimensions 24*6 mm^2 cross-sections and length of 180 mm is considered for finite element analysis. ANSYS 16.1 software is used to develop the model. Four different models considered in the present analysis. In the finite element beam model with a solid 20 node 186 element type is used. Material properties include linear elastic, isotropic material with a material density of 2700 $\mathrm{kg/m}^3$. Modulus of elasticity is taken as 70 GPa and poison's ratio is 0.33. The constrained are provided on one end of the cantilever beam with all degrees of freedom. Loading is applied to the free end of the beam in the negative Y direction. The load is varied, starting from 2 to 10 N in the step of 2 N. Table 13.1 shows the symbolic representation of the samples used in this study.

Table 13.1 Samples and symbols

Types of samples used in the analysis	Symbols
Sample without notch	A
Sample with a single notch at the center within the gauge length	B
Sample with double notch within the gauge length	C
Sample with three-notch within the gauge length	D

13.2.3 Strain Gauge Measurement

Strain gauges are the most commonly used strain measurement technique. Strain gauges are resistive sensors used for strain measurement of the structure at the point of contact. Working with strain gauge is based on the principle that change in the physical variable will produce a change in the resistance of the wire. By measuring this change in resistance, change in the physical variable can be predicted. The physical variable in the present work corresponds to strain (ε). Strain gauges are used in a Wheatstone bridge, which consists of four resistors in an electrical circuit. One of the resistors is replaced with a strain gauge (quarter bridge), and the resulting circuit can be used to measure the strain of the cantilever beam. Wheatstone bridge used in present work for strain measurement using strain gauge. Strain values can be obtained by using Eq. 13.5.

$$\varepsilon = \frac{-4V_r}{\text{G.F}(1 + V_r)}\left(1 + \frac{R_L}{R_g}\right) \tag{13.5}$$

But due to certain measuring uncertainties under the effect of electric and magnetic fields, it gives random errors that affect its accuracy. Proper functioning of the strain gauge also depends on the contact of strain gauge with the specimen, which requires the precise technical experience. Please do not change the margins of the template as this can result in the footnote falling outside the printing range.

13.3 Experiment

Gwinstek digital storage oscilloscope GDS-1000B Series with an output frequency range of 1–2000 kHz is used for data storage. A Helium–Neon (He–Ne) laser source with a wavelength of 1556 nm is used to provide a necessary signal through fiber optic cable. Pan with negligible weight is used for static loading at the free end of the beam. The aluminum samples having surfaced mounted strain gauge and optical fiber fixed at the center gauge length are used for testing of the cantilever beam structure. An optic to electric converter is used for the necessary conversion of an optical signal coming from the fiber cable to an electrical signal for necessary detection. Notch as per standard is made at different locations on the beam. Four different sets of samples

Table 13.2 Test sample details

Sample parts	Specifications
Beam (L*B*H)	Aluminum (180*24*6 mm^3)
Optical fiber	Cross-section area = 2000 μ refractive index = 1.45
Strain gauge	120 Ω
Quarter wheatstone bridge	Resistance = 120 Ω each, excitation voltage = 3 V

(samples A, B, C, D) with different notch locations and numbers are made having three samples in each set. Details of the test sample are given in Table 13.2.

13.3.1 LASER and Noise

The important characteristic to address in a transmission system is the noise performance of the system. There are different sources of noise in a communication system. The initial source of noise arises from the circuit itself, which is known as internal noise. The significant effect of noise is observed at higher frequencies, and laser diodes operate at comparatively high frequencies. Therefore, it became important to recognize noise in high-frequency transmission. Fluctuations in light intensity of a LASER output can be carried as noise in transmission signals. Reflection noise is a LASER noise which occurs due to the Fresnel reflection phenomenon. The output signal from a LASER carries a specific wavelength, satisfying both gain and phase condition between two reflecting LASER regions. But due to finite spectral width in the output of a LASER, noise is generated which is known as mode partition noise. From the above discussion, it can be concluded that the practical use of strain calculation may lead to some errors due to noise in the signals. There exist various methods to remove noise data using filters and other software. These signals can be successfully used to determine strain values using neural network technology.

13.4 Neural Network

Artificial neural network (ANN) is an intelligence-based processing technique that is made up of interconnected processing units called neurons which transmits a weighted function of its input to the preceding layer of the neuron. An ANN includes three layers: input, hidden, and output layer. Variables are given as input in the input layer, which is connected to a hidden layer for processing. The hidden layer contains an activation function that calculates the weight and bias of the variable for exploring the effect of predictors upon the target variables. The output layer is used for predicting the results with error estimation. Backpropagation algorithm initially, the input is propagated to the hidden layer; it propagates the sensitivity back to reduce error. At the end of the process, it updated the weights and bias. In the present work, two different training functions, namely Levenberg–Marquardt backpropagation and Bayesian regularization backpropagation, are compared for the approximation of strain values. To compare the performance of both algorithms different errors like root mean square error (RSME) and mean square error (MSE) between real and predicted data is analyzed.

13.4.1 Network Evaluation

The proposed ANNs undergo training, validation, and testing phases of 1000 combinations of data. Stratification of data for testing and training is done using the holdout cross-validation method. Out of 1000 data, 667 data was used for training, and the remaining 333 data was used for testing. Comparison of the LM and BR algorithm is done for sample A. In the present analysis, even the number of neurons starting from 2 to 30 is considered for both the algorithms. Method of selection of algorithm and the minimum number of neurons in the hidden layer is based on the minimum error criteria regarding RMSE and MSE. A comparison of both errors related to LM and BR algorithms analyzed. The formulas for error criteria are as follows:

$$\text{RMSE} = \frac{1}{2}\sum_{i=1}^{N}\left(Y_i' - Y_i\right)^2 \tag{13.5}$$

$$\text{MSE} = \sum_{i=1}^{N}\frac{\left(Y_i' - Y_i\right)^2}{N} \tag{13.6}$$

The objective is to devise the criteria that estimate the number of hidden neurons as a function of input neurons to develop a model for strain prediction. Output of error histogram of both errors shows that minimum error for the LM algorithm is obtained by using 2 neurons in the hidden layer but for the BR algorithm its 10 neurons which give the minimum error. Using a smaller number of neurons in a network has an advantage of the reduction in computing time, but in this study, the error is considered as the bases of neuron selection. Therefore, errors of both LM at 2 neurons and BR at 10 neurons are considered.

Figure 13.1 shows that error using 10 neurons in the BR algorithm is much smaller compared to using 2 neurons in the LM algorithm. Hence, fixation of the number of hidden neurons is done based on minimum error estimation. The BR backpropagation algorithm is selected for training the neural network to verify the strain values.

Regression value is used in this study to evaluate the performance of the artificial neural system. Regression analysis of various samples under different loads corresponding to test and train state is shown in Fig. 13.2. Regression values are low for

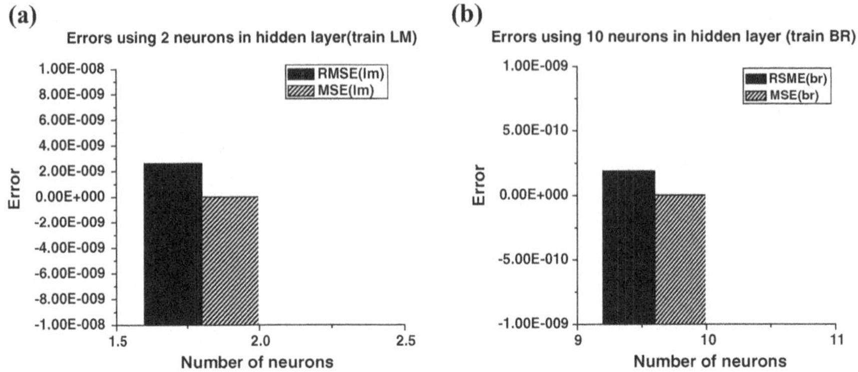

Fig. 13.1 **a** Error histogram for Levenberg–Marquardt backpropagation algorithm with 2 neurons; **b** error histogram for Bayesian regularization backpropagation algorithm with 10 neurons

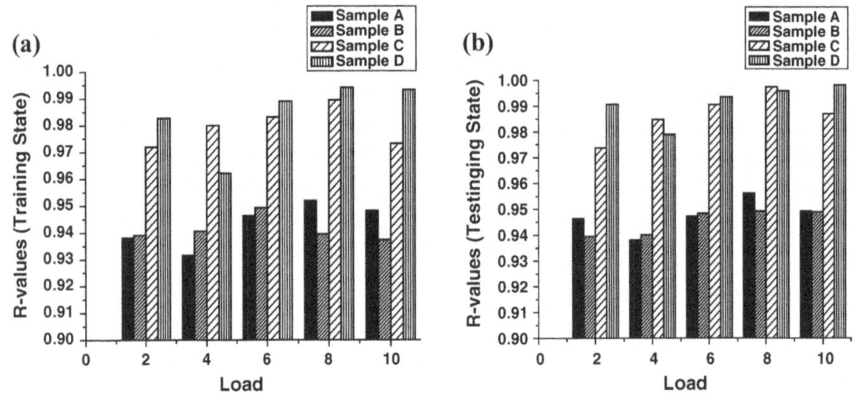

Fig. 13.2 Regression histogram showing variation in sample type at different loads. **a** Training R-value histogram; **b** testing R-value histogram

both train and test samples, but it increases on increasing the load. All regression values are well within the acceptable range.

Plots of the regression values related to train and test sets are shown in Fig. 13.2. At low load as 2 N and 4 N regression, value is low corresponding to both training and testing sets. But the value of regression increases to an acceptable range with an increase in load. The objective of the above least squares estimation and regression analysis is to develop a model for strain determination using optical parameters.

13.5 Results and Discussion

Samples having variation in number and location of notches were tested to obtain strain variations for comparison. Loading was applied using the weight pan method at the extreme free end of the sample. Different loads used for loading include 2, 4, 6, 8, and 10 N. Experimental strain values were obtained as output using a designed neural network and conventional strain gauge. Neural network analysis was performed at 1000 epochs as further training at higher epoch's results in high error variation. Bayesian regularization backpropagation algorithm based on minimum error criteria is used in the neural network. This algorithm works on the probabilistic interpretation of network parameters. It involves a probability distribution of network weights. Based on minimum error criteria and considering the four different errors, the Bayesian regularization backpropagation algorithm has been chosen over the Levenberg–Marquardt backpropagation algorithm.

Four different methodologies used in this work include strain gauge measurement (sg), finite element analysis (fea), analytical solution (Haslach), and neural network analysis (nn). Comparative axial strain plots under different loading are shown in Fig. 13.3. The plot of sample A shows that an analytical solution (by Haslach) underpredicts the strain values at 2 N and overpredicts values at 6 and 8 N load. In

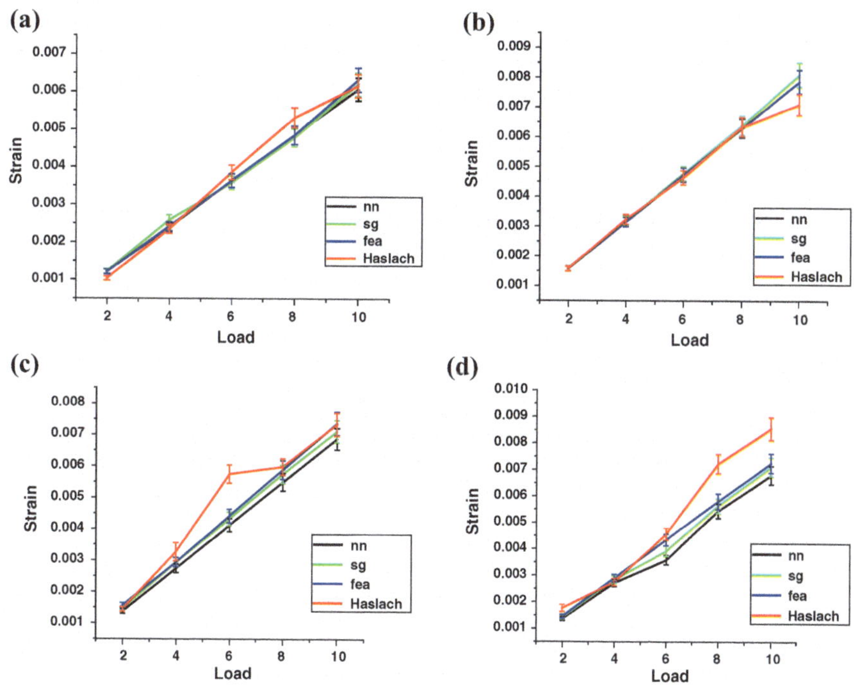

Fig. 13.3 Percentage error in strain relative to finite element strain values **a** sample A; **b** sample B; **c** sample C; **d** sample D

sample B, strain gauge overpredicts the strain value but underpredicts the same at 10 N load. For sample C, analytical solution overpredicts the strain values at 4, 6, and 8 N load. In sample D over prediction of strain values are observed corresponding to 2, 6, 8, and 10 N. Strain values using an analytical solution (by Haslach) corresponding to different samples under different static load provides an overfit and underfit predictions. Similarly, strain gauge axial strain values show a close prediction with finite element analysis. Axial strain prediction by using proposed neural network technology provides the best accuracy with the least error.

13.6 Conclusion

The present work proposes a systematic procedure for the selection of neural network architecture related to structural health monitoring. The selection of transfer function and the minimum number of neurons in the hidden layer based on minimum error criteria are proposed. Two algorithms, namely Levenberg–Marquardt backpropagation and Bayesian regularization backpropagation, were tested with the number of hidden neurons ranging from 2 to 30. Two basic errors (MSE and RMSE) were analyzed. In the present study, a smart structural health monitoring technique using optical parameters with an artificial neural network is proposed. The present work proposes a systematic procedure for the selection of neural network architecture related to structural health monitoring. The selection of transfer function and the minimum number of neurons in the hidden layer based on minimum error criteria are proposed. The relative percentage error of various samples shows that neural network prediction is more accurate compared to the analytical solution. This proposed smart structural health monitoring proves that under static loading conditions on cantilever beam structure, strain values can be predicted with reasonable accuracy using optical data from surface-mounted fiber optic sensors.

References

1. Chan, T.H.T., Yu, L., Tam, H.Y., Ni, Y.Q., Lin, S.Y., Chung, W.H., and Cheng., L. K.: Fiber Bragg grating sensors for structural health monitoring of Tsing Ma bridge: background and experimental observation. Eng. Struct. **28**, 648–659 (2006) (Elsevier Ltd.)
2. Her, S.C., Yang, C.M.: Dynamic strain measured by Mach-Zehnder interferometric optical fiber sensors. Sensors **12**, 3314–3326 (2012)
3. Regez, B., Sayeh, M., Mahajan, A., Figueroa, F.: A novel fiber optics based method to measure very low strains in large scale infrastructures. Measurement **42**, 183–188 (2009) (Elsevier Ltd.)
4. Kadhim, R.A.: Strain measurement by using phase modulated fiber optic sensors technology. Diyala J. Eng. Sci. **8**, 27–41 (2015)
5. Lunia, A., Isaac, K.M., Chandrashekhara, K., Watkins, S.E.: Aerodynamic testing of a smart composite wing using fiber-optic strain sensing and neural networks. Smart Mater. Struct. **9**, 767–773 (2000)
6. Gholamzadeh, B., Nabovati, H.: Fiber optic sensor. Int. J. Electron. Commun. Eng. **8**, 6 (2008)

7. James, S.W., Tatam, R.P.: Optical fiber long-period grating sensors: characteristics and application. Meas. Sci. Technol. **14**, R149–R161 (2003)
8. Heredia, A.S., Aguilar, A.P.M., Ocampo, A.M.: Laser-beam reflection for steel stress–strain characterization. Measurement **113**, 92–98 (2018) (Elsevier Ltd.)
9. Butter, C.D., Hocker, G.D.: Fiber optics strain gauge. Appl. Opt. **17**, 2867–2869 (1978)
10. Haslach Jr., H.W., Sirkis, J.S.: Surface-mounted optical fiber strain sensor design. Opt. Soc. Am. Appl. Opt **30**(28) (1991)

Chapter 14
Downtime Reduction of End of Line Pump Testing Machine

Aishwarya Hadgaonkar, Sagar Sapkal and Aman Arora

Abstract In manufacturing industries, maximum rate of production is the main focus. To get maximum production rate, machine should be available for maximum period of time. Therefore, decrease in downtime of machine which leads to productivity improvement is reported in this paper. In a well-known pump manufacturing company, end of line pump testing machine is identified as bottleneck machine and downtime reduction is attempted on this machine. Fault tree analysis is carried on collected data from the industry, and root causes for all the major problems responsible for downtime are found out. After thorough analysis, solutions to these problems are suggested. By implementation of these solutions, reduction in downtime, in frequency of occurrence of problem, and in number of rejections is achieved. The proper implementation of filter changing, alignment of bed and changing of connectors as per decided frequency leads to decrease in downtime and hence increase in productivity. Thus, by implementing these solutions after thorough evaluation and by suggesting proper maintenance schedule most of the problems associated with end of line pump testing machine are resolved. The percentage reduction in downtime is reported.

Keywords Bottleneck machine · Downtime · Fault tree analysis · Maintenance services

14.1 Introduction

In manufacturing industry, in order to have a greater production rate a machine should be available for maximum period of time. This means that downtime of machine should be decreased so as to get maximum production rate. With a combined effort, correct equipment, and good atmosphere, uptime of a machine in a manufacturing

---

A. Hadgaonkar · S. Sapkal (✉)
Walchand College of Engineering, Sangli, India

A. Arora
Cummins Technologies India Pvt. Ltd., Phaltan, India

S. Yadav et al. (eds.), *Proceedings of International Conference in Mechanical and Energy Technology*, Smart Innovation, Systems and Technologies 174,
https://doi.org/10.1007/978-981-15-2647-3_14

plant can be significantly increased [1]. The given industry which is producing components required for engine assembly is facing problem of maximum downtime. It is necessary to reduce the bottleneck for improving productivity. Fault tree analysis (FTA) is applicable for carrying out the work. This analysis method is mainly used in the field of manufacturing engineering to understand how failure takes place, to identify the best ways to reduce risk, or to determine rate of failure [2]. In the given industry for assembly of a pump, it goes through many operations. After assembling all parts and fittings, the whole assembled product comes to the station called as end of line (EoL) pump testing station. At this station, functional testing of a pump is carried out. There are many problems identified due to which downtime gets increased and hence bottleneck of this machine needs to be removed.

In view of the above study, the literature regarding uptime improvement, overall equipment effectiveness (OEE) improvement is surveyed and followed. Nwanya et al. [1] have given optimization of machine downtime in the plastic manufacturing and they mentioned that pit stop contributes 25.64% of planned downtime duration. Dai and Vyatkin [2] presented a cloud-based decision support system (DSS) for self-healing in distributed automation systems using FTA. By running multiple parallel simulations of control software with real-time inputs ahead of system time, faults could be detected and corrected automatically using autonomous industrial software agents. They have given the solutions to reduce the downtime. Daniewski et al. [3] also reported the results of an analysis of indicators describing the effectiveness of actions taken and repairs made by the maintenance service in a food industry. They concluded that maintenance service action plays very important role in improving production and reducing downtime. Chandrakar and Kumar [4] presented reduction of breakdowns in food processing plants through failure analysis. The reason for breakdown is also analyzed and some of the tools of root cause analysis like why-why analysis and fishbone diagram are implemented to identify actual cause of breakdown. Fakorede et al. [5] gave productivity increase by optimum utilization of machines and manpower energy. From the analysis, it is concluded that optimum utilization of machine leads to increase in productivity. Gupta and Garg [6] presented OEE improvement by total productive maintenance (TPM) implementation. Gupta and Vardhan [7] also presented optimizing OEE, productivity and production cost for improving sales volume in an automobile industry through TPM. Puvanasvaran et al. [8] presented overall equipment efficiency (OEE) improvement using time study in an aerospace industry. Maynards' Operation Sequencing Technique (MOST) is adopted and OEE improvement is mentioned. Vijayakumar and Gajendran [9] mentioned that OEE of the injection molding process is increased from 61 to 81% through the implementation of availability, better utilization of resources and high-quality products.

14.2 Methodology

FTA is applied for downtime reduction of identified bottleneck machine. For EoL pump testing machine, which is identified as bottleneck machine, the categorization of problems is done based on machine, procedure, and material. The total downtime of each problem related to machine, related to procedure and related to material is reported in Tables 14.1, 14.2, and 14.3, respectively. To find a top event, i.e., major problems causing downtime, Pareto analysis is carried out. From Fig. 14.1, it is clear that to reduce 70–80% downtime, first seven major problems are required to be focused. The causes are found out for each problem and fault tree is developed. The fault tree for the downtime is given in Fig. 14.2.

14.3 Investigation Action Plan

The investigation action plan is developed for the causes found out from the above-mentioned fault tree.

For TP3 Failure, the major problem is in improper fitting of all connectors of a pump. The cone-shaped HP connector as shown in Fig. 14.3a should be properly inserted in the slot and tightened with required torque. To avoid the failure, differential pressure filter as shown in Fig. 14.3b should be cleaned properly as per the decided frequency.

In case of TP1 Failure, to get required mass flow all connectors as shown in Fig. 14.4a should be fixed properly. Lube connector as shown in Fig. 14.4b should be properly inserted in the slot to avoid leakage. Also, the air bubble should be removed to get required flow.

To avoid the issue of machine automatic shutdown, proper alignment of pallet connectors should be done with machine connectors. And the focus should be given to the level of oil in the tank. The connectors and oil tank level is shown in Fig. 14.5.

In case of calibol oil and lube oil top up, the filters are used before entry of oil into the pump and fuel dispenser is used instead of hand pump to avoid time for filling. The hand pump and fuel dispenser are shown in Fig. 14.6.

For TP5 Failure, the rubber dampers are used to avoid vibrations and to avoid the disturbance to pressure regulator setting because of the high pressure generated in this test plan. The damaged filter and the damper are shown in Fig. 14.7.

To avoid pallet clamping problem and thus the breakage of guide pin, the torque wrench is used instead of nut runner. The guide pin and torque wrench are shown in Fig. 14.8.

In order to resolve pallet docking problem, the alignment of a bed and conveyor should be done regularly. Shim addition is also made to align permanently to reduce the gap created because of vibrations. The ball conveyor and shim addition are shown in Fig. 14.9.

Table 14.1 Total downtime of problems related to machine

Problems	Downtime (1st month)	Downtime (2nd month)	Downtime (3rd month)
Master pump failed at TP3 (Pallet A)	865	1484	520
Drain connector socket sensor problem	20	–	–
HPP in pressure low	170	–	–
Pallet stuck in machine	33	–	–
HPP drain pressure low	125	145	–
Oil tank temp low	130	–	15
Master pump failed at TP1	85	870	535
HPP drain pressure flow high	244	–	–
Drain pressure high (TP3)	130	–	–
Pump stuck in pallet	30	–	–
Alignment of coupling	270	–	–
Rail pressure sensor zero to shift high	10	40	–
Lube oil pressure low/high	340	205	–
Master failed at LPP inlet pressure low	60	–	–
Master reading not ok	30	115	980
Master motoring 'P' slope high	345	–	–
Master pump motoring flow high	150	15	–
Camshaft position nok	15	60	–
Front door sensor problem	10	–	–
Oil tank temp high	20	10	–
Cyflex software problem	–	50	–

(continued)

Table 14.1 (continued)

Problems	Downtime (1st month)	Downtime (2nd month)	Downtime (3rd month)
Master failed at TP5 (HPP out mass flow low)	90	365	815
Differential fuel filter reading increase	–	15	140
Changing orifice and removing leakage	–	–	225

Table 14.2 Total downtime of problems related to procedure

Problems	Downtime (1st month)	Downtime (2nd month)	Downtime (3rd month)
Calibol oil and lube oil top up	569	515	315
Shim adding in bed (handover to maintenance)	75	–	–
Lube oil not available	30	–	–
Bed alignment (maintenance)	60	–	–
Quality hold for checking of bypass valve	–	60	–
Machine automatic shutdown and IMV not responding	330	346	776
Pallet docking problem	655	85	–

14.4 Results and Discussion

The data before and after implementation of solutions are collected from the industry. The percentage reduction in downtime, percentage reduction in frequency of occurrence of problem, and percentage reduction in number of rejections are calculated and are reported in Table 14.4.

Figures 14.10, 14.11, and 14.12 show the comparison between before and after implementation of the suggested solutions, in terms of downtime, frequency of occurrence of the problem, and number of rejections, respectively.

The component gets rejected because of machine problems and not because of procedural or material problem. Machine automatic shutdown, oil top up, pallet

Table 14.3 Total downtime of problems related to material

Problems	Downtime (1st month)	Downtime (2nd month)	Downtime (3rd month)
Pallet A/B trails(maintenance/ME)	174	80	–
HPC connector leakage	92	211	115
GTU gear and pulley adjusting	30	30	–
IMV broken	90	15	10
LPP outlet connector broken	20	15	40
Pallet gear damage	75	–	–
Guide pin broken	–	472	–
Pallet clamping problem (clamping bolt damage)	–	545	–
Scanner cable damage/scanner not working	–	–	85
HP fitting not available	–	–	380

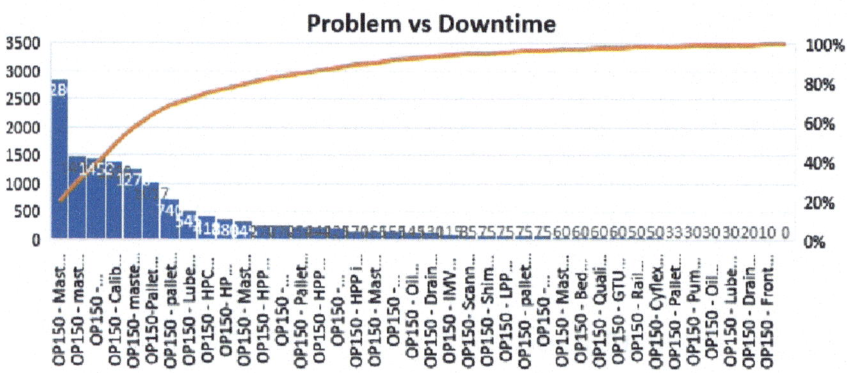

Fig. 14.1 Pareto chart of problem versus downtime

clamping problem/guide pin broken and pallet docking problem causes no rejection since these are not machine problems. Hence, these data show zero rejections for these problems for both, before and after implementation.

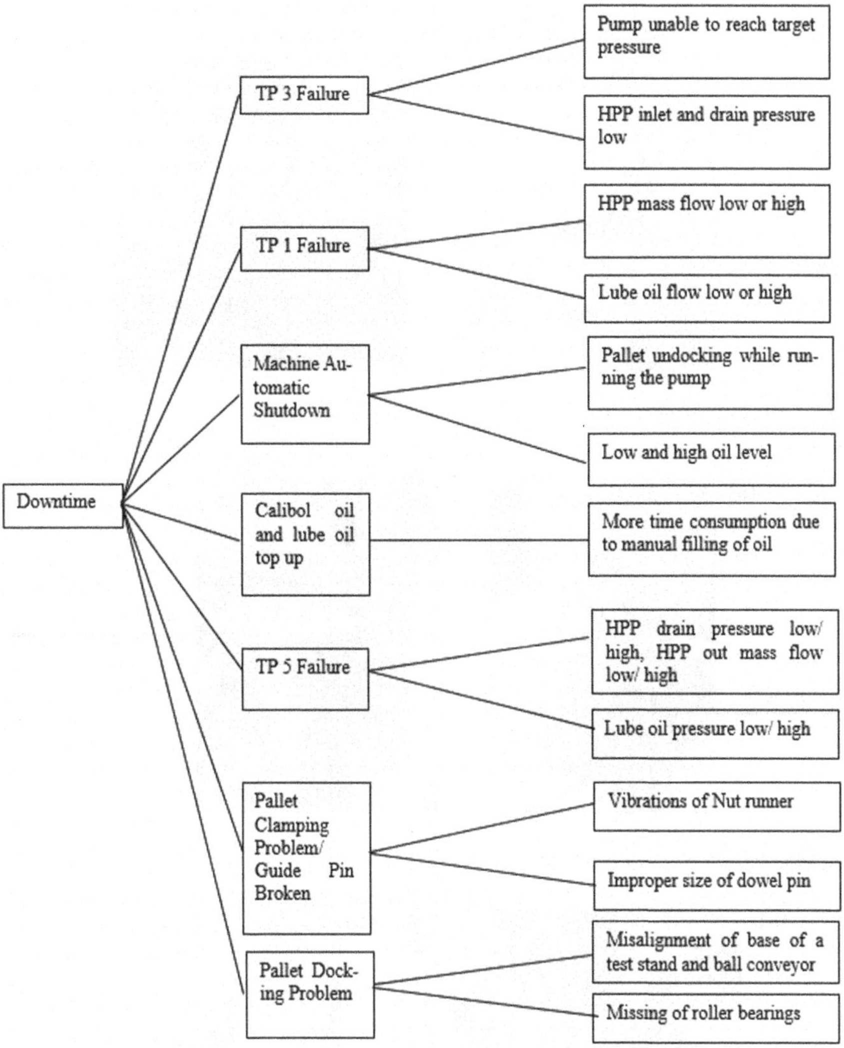

Fig. 14.2 FTA for downtime

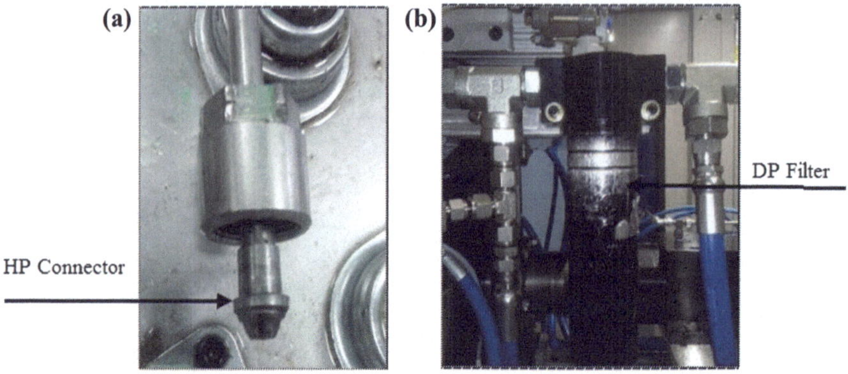

Fig. 14.3 **a** Cone-shaped HP connector **b** differential pressure filter

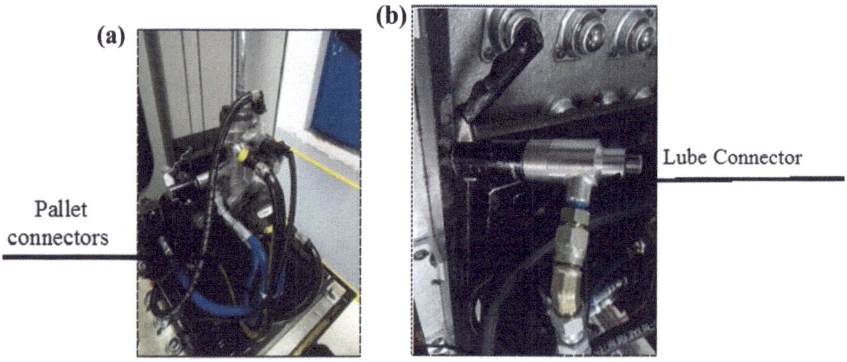

Fig. 14.4 **a** Connectors of pallet **b** lube connector

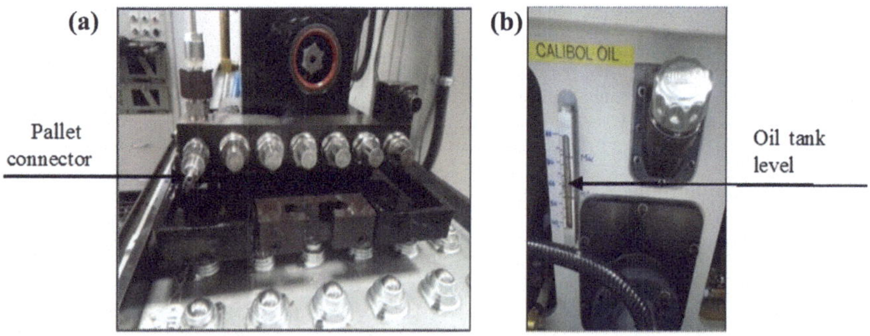

Fig. 14.5 **a** Pallet connectors **b** oil tank level

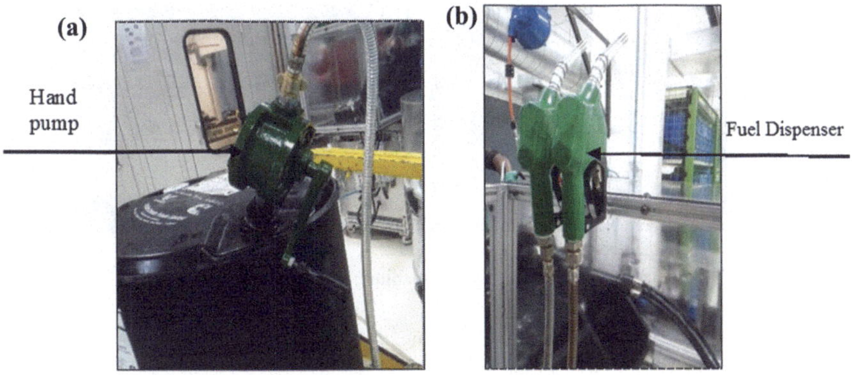

Fig. 14.6 **a** Hand pump for oil filling **b** fuel dispenser

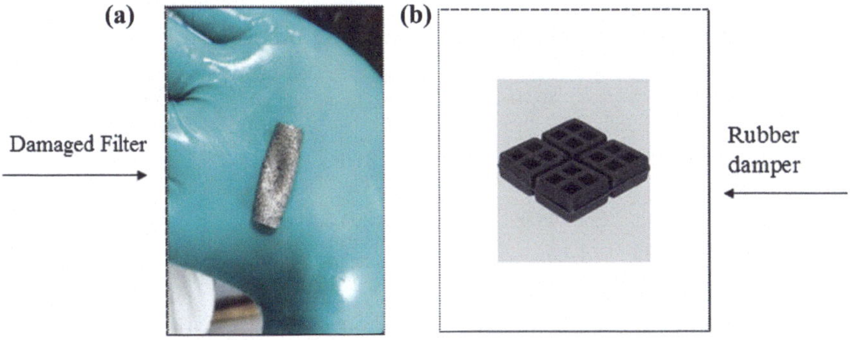

Fig. 14.7 **a** Damaged filter **b** rubber damper

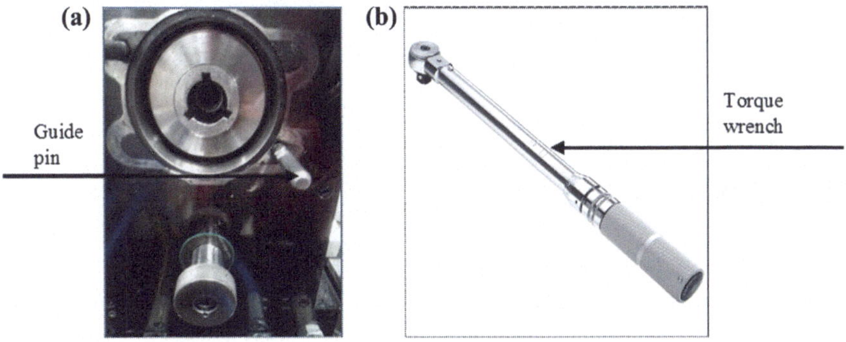

Fig. 14.8 **a** Guide pin **b** torque wrench

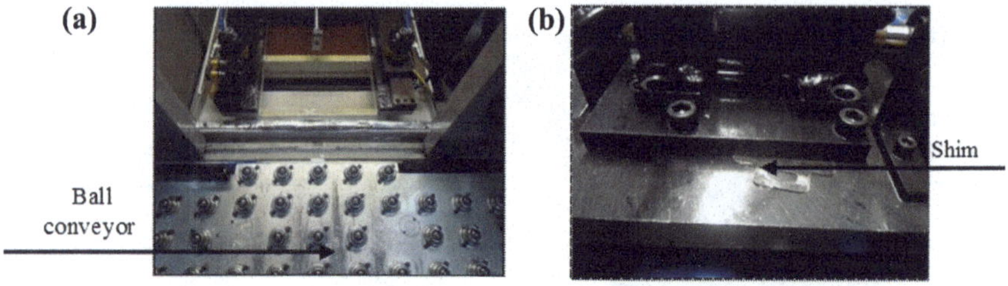

Fig. 14.9 **a** Ball conveyor and base of test stand **b** shim addition

Table 14.4 Percentage reduction in downtime, frequency of problem and number of rejections

S. No.	Major problems	% Reduction in downtime	% Reduction in frequency of problem	% Reduction in number of rejections
1	Master pump failed at TP3	19.97	39.13	11.90
2	Master pump failed at TP1	86.04	82.35	53.96
3	Machine automatic shutdown	100	100	0
4	Calibol oil and lube oil top up	95.99	93.40	0
5	Master failed at TP5	66.14	22.22	70.37
6	Pallet clamping problem/guide pin broken	98.72	66.67	0
7	Pallet docking problem	48.51	25	0
Average percentage reduction		73.62	61.25	45.41

14.5 Conclusion

The downtime of given pump testing machine is reduced by applying FTA. The solutions for major problems causing downtime are suggested and implemented. The average reduction in downtime is 73.62%, the average reduction in frequency of occurrence of the problem is 61.25%, and the average reduction in number of rejections is 45.41%. Thus by applying FTA and Pareto chart and by analyzing the major problems, proper solutions are suggested for reducing the downtime of the bottleneck machine. After implementation of the solutions, significant reduction in all the performance measures is observed which in turn leads to improved productivity.

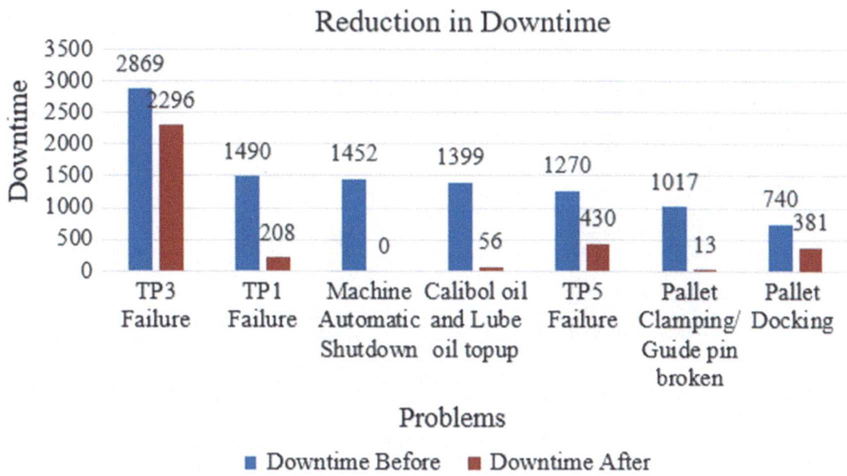

Fig. 14.10 Comparison between reductions in downtime

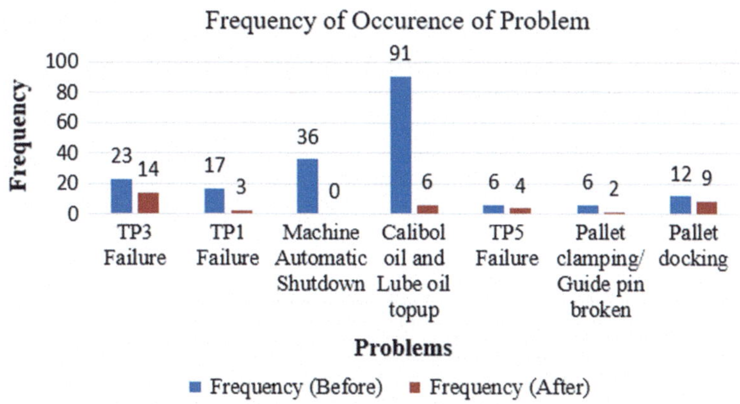

Fig. 14.11 Comparison between frequencies of problems

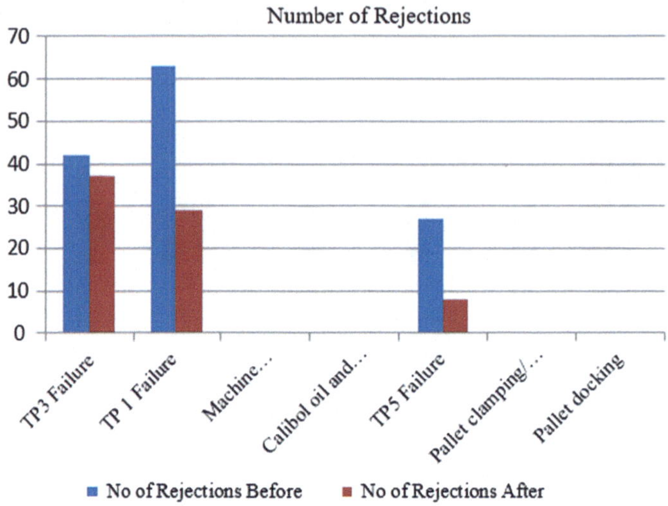

Fig. 14.12 Comparison between numbers of rejections

Acknowledgements The authors are grateful to Cummins Technologies India Pvt. Ltd. for providing the facility and related support for this work.

References

1. Nwanya, S., Udofia, J., Ajayi, O.: Optimization of machine downtime in the plastic manufacturing, production and manufacturing, 1–12 (2017)
2. Dai, W., Vyatkin, V.: A cloud-based decision support system for self-healing in distributed automation systems using fault tree analysis. IEEE Trans. Indus. Inf. 989–1000 (2018)
3. Daniewski, K., Kosicka, E., Mazurkiewicz, D.: Analysis of the correctness of determination of the effectiveness of maintenance service actions. Manag. Product. Eng. Rev. 20–25 (2018)
4. Chandrakar, J., Kumar, R.: Reduction of breakdowns in food processing plants through failure analysis. Int. J. Adv. Eng. Res. Stud. 212–214 (2015)
5. Fakorede, D., Babatunde, A., Ovat, F.: Productivity increase by optimum utilization of machines and manpower energy. Int. J. Eng. Res. Dev (2014)
6. Gupta, A., Garg, R.: OEE improvement by TPM implementation. Int. J. IT Eng. Appl. Sci. Res. (IJIEASR) 115–123 (2012)
7. Gupta, P. Vardhan, S.: Optimizing OEE, productivity and production cost for improving sales volume in an automobile industry through TPM. Int. J. Product. Res. 1–13 (2015)
8. Puvanasvaran, A.P., Mei, C.Z., Alagendran, V.A.: Overall equipment efficiency improvement using time study in an aerospace industry. Proc. Eng. 271–277 (2013)
9. Vijayakumar, S., Gajendran, S.: Improvement of overall equipment effectiveness (OEE) in injection molding process industry. J. Mech. Civ. Eng. 47–60 (2014)

Chapter 15
Modelling of Sustainable Manufacturing Barriers in Pharmaceutical Industries of Himachal Pradesh: An ISM-Fuzzy Approach

Anbesh Jamwal, Rajeev Agrawal, Sumit Gupta, Govind Sharan Dangayach, Monica Sharma and Md. Aktar Zahid Sohag

Abstract In the present time, global competition among the manufacturing firms has forced all the industries to adopt new manufacturing system different from traditional manufacturing system. Adoption of sustainable manufacturing has become a critical issue in the present manufacturing scenario for Indian pharmaceutical industries. At present, pharmaceutical industries among the world are interested to become sustainable in order to cover all three aspects environmental, social and economic. There are number of practices which have been done in pharmaceutical industries among the world to achieve the sustainability. The present study focuses on the development of fuzzy-ISM model for barriers which affect the sustainability in pharmaceutical industries of Himachal Pradesh. In the first stage of this study, survey methodology is adopted for the data collection from pharmaceutical industries of Himachal Pradesh, and after that, fuzzy-ISM model is developed.

Keywords Sustainable manufacturing · Lean manufacturing practices · Fuzzy methodology · Interpretive structural modelling

15.1 Introduction

At present, manufacturing is the backbone of today's industrial society. Government policies among the world have forced all industries to adopt new manufacturing system which do not harm the environment during any stage of manufacturing which is termed as sustainable manufacturing [1]. In India, pharmaceutical industries are

A. Jamwal (✉) · R. Agrawal · G. S. Dangayach · M. Sharma
Department of Mechanical Engineering, Malaviya National Institute of Technology, J.L.N. Marg, Jaipur, Rajasthan 302017, India

S. Gupta
Department of Mechanical Engineering, A.S.E.T., Amity University, Noida, Uttar Pradesh, India

Md. A. Z. Sohag
Department of Mechanical Engineering, Alakh Prakash Goyal Shimla University, Shimla, Himachal Pradesh, India

S. Yadav et al. (eds.), *Proceedings of International Conference in Mechanical and Energy Technology*, Smart Innovation, Systems and Technologies 174,
https://doi.org/10.1007/978-981-15-2647-3_15

devoted to invent the new medicines which allow the patients to live healthier and longer with the minimum environmental depletion during the production [2]. The concept of sustainability in the production of pharmaceutical goods is not new in India. In the past few years, pharmaceutical industries of India have focused more on improving the quality, minimizing the waste in both the manufacturing areas and in research and development [3]. Pharmaceutical industries are not only driven by cost reduction of products but also driven by awareness of increasing the environmental sustainability during the manufacturing processes [4]. Sustainable manufacturing focuses on the use of processes which do not harm the environment, employees and consumers during and after the manufacturing [5]. Sustainable manufacturing emerged in earlier of 1970s which was largely motivated by environmental incidents and depletion of natural resources [6]. Gupta and Dangayach [7] suggested three pillars of sustainability, i.e. social, economic and environmental which are modelled and incorporated latterly for the sustainable development in manufacturing practices. Sustainable manufacturing is also known as green manufacturing, environmental conscious manufacturing and environmental responsible manufacturing, Sustainable manufacturing focuses on waste management, recycling, conservation of resources, water supply, pollution control, environmental protection and other related issues [8]. Sustainable manufacturing emphasizes on the design, manufacturing, delivery, use and disposal of product with the minimal environmental impact during all the stages. In the present time, global competition has forced industries to make environmentally friendly products with economic feasibility [9]. To improve the environmental performance in pharmaceutical industries, sustainable manufacturing is the visible solution at present time [10]. Many pharmaceutical industries are considering environmental management as an integral part of economic development [11]. The present study focuses on the modelling of barriers with the help of fuzzy interpretive structural modelling which affect the sustainability in pharmaceutical industries of Himachal Pradesh.

15.2 Literature Review

The idea of sustainability is accepted as common wish at present time in the world, but still there is guidance required for the implementation of its practices in the industries because of its complexity in consideration of dimension of sustainability [12]. Manufacturing plays an important role in the development of our industries, but it consumes lot of energy and resources and emits a lot of emission to our environment [13]. Today, manufacturing is the source for all the goods for living, for production, for our transportation system, for health and for human safety; manufacturing is the foundation of our life at present time [14]. Thus, the implementation of sustainability concepts in the manufacturing will help to achieve sustainability in industries. Table 15.1 shows the various connotations of sustainable manufacturing in literature during last five years.

Table 15.1 Connotations of sustainable manufacturing in literature review during the last five years

Author	Year	Connotation
[15]	2014	Manufacturing plays an important role in the development of industries by increasing the value of product while minimizing the environmental impact which contribute to sustainability
[16]	2015	Sustainable manufacturing has become a critical issue at present time. Achieving sustainability is the need of industries due to diminishing non-renewable sources of energy and government regulation related to environment, health and safety as well as demand of environmentally friendly products by customer
[17]	2016	Environment concerns have boost up the need of sustainable practices in industries, but at the same time there is need of achieving superior performance from these practices. Integration of big data with sustainable manufacturing can help to achieve this performance
[18]	2017	Claimed that there is a wide deviation from the core understanding of sustainable manufacturing. There are number of issues which are associated with sustainable manufacturing are still not clear
[19]	2018	Assess the impact of sustainable manufacturing in leather industries of Bangladesh and concluded that leather industry is contributing remarkably in the growth of country but does not have a good image due to environmental pollution. It is found that circular economy most affect the implementation of sustainable manufacturing practices in Bangladesh
[20]	2019	Investigated the effectiveness of TP strategies in achieving the sustainable development in manufacturing industries of India and find that innovative capability and corporate strategies highly affect the sustainability in Indian industries

Sustainable manufacturing plays an important role in the development of pharmaceutical industries of Himachal Pradesh. However, adoption of a new manufacturing system over existing one can never be an easy task, as it requires funds, time and commitments.

The adoption of sustainable manufacturing in pharmaceutical industries of Himachal Pradesh is influenced by some key barriers which are discussed below:

1. **Lack of support from top-level management (B1)**: The neglection shown by top-level management to change the existing environment with no proper information system demotivates the middle-level management and workers. There is a need for top-level management to understand the importance of achieving sustainability.
2. **Power shortage (B2)**: Lack in adequate power supply is a still bottleneck in the growth of pharmaceutical industries which arises due to higher power costs and chronic shortages. Production and distribution of power supply should be balanced which will help in the growth of the industries.
3. **Higher costs (B3)**: At present, higher costs are still a major issue in many Indian industries which arises due to the improvement in processes and products.

Investors show less interest in investing the sustainable technology when there are higher investment costs and less return on investment.

4. **Negative attitudes of industries towards sustainability projects (B4)**: Lack of proper knowledge and training about sustainable practice in industries is the main reason for negative attitudes of the industries towards sustainability projects.
5. **Lack of funds for green projects in industries (B5)**: Uneven distribution of funds without considering the government policies and environment norms is the main reason behind lack of funds for green projects in industries. Lower or longer return on investment on sustainable projects makes tough to implement the green projects for many SMEs of Himachal Pradesh.
6. **Lack of performance benchmarks (B6)**: It is difficult for industries to assess their sustainability performance in the absence of practical parameter and guidelines. So, industries and academia should collaborate with each other for the development of sustainable framework.
7. **Lack of awareness in local customers about green products by industries (B7)**: At present, industries are depending on customer demands. Industries are customer driven. Industries are designing and manufacturing their products as desired by customer. So, knowledge in customers regarding green products is a key factor which influences the sustainability.
8. **Lack of awareness towards sustainability concept (B9)**: In most industries, top management does not have any idea about sustainability practices at the shop floor level. Also, there is a big deviation from core sustainability concepts in industries. This resists the implementation of sustainability in Indian industries.
9. **Lack of awareness programmes conducted locally by industries (B8)**: In India, industries are facing difficulties in the adoption of sustainable manufacturing because of market pressure. Industries need to train their workers to cope up with the difficulties which are at the present time.

15.3 Research Methodology

Sustainable manufacturing barriers which affect the sustainability in pharmaceutical industries of Himachal Pradesh are found out with the discussion with industry experts and academia. Further, data collection is done by questionnaire survey in which a set of questions is framed and brainstorming sessions are done with both industry people and academia. The collected data is framed with the help of fuzzy-ISM technique.

15.3.1 Introduction to Fuzzy-ISM

Interpretive structural modelling is an interactive learning process in which relationship between direct and indirect elements is grouped together and structured into a systematic model. Interpretive structural modelling is used to identifying the relationship among the number of factors which influence an output [21]. This technique starts with the identification of factors which influence the output of the problem and further it extends to a group solving technique [22]. Several steps followed for the assessment of sustainable manufacturing barriers which affect the sustainability of pharmaceutical industries of Himachal Pradesh are shown in Fig. 15.1.

Linguistic scale taken for present study is shown in Tables 15.2 and 15.3.

The defuzzification is done by the equation:

$$M = A1 + 2A2 + 2A3 + A4/6 \tag{15.1}$$

By Eq. (15.1), defuzzification value matrix is obtained which is shown in Table 15.4.

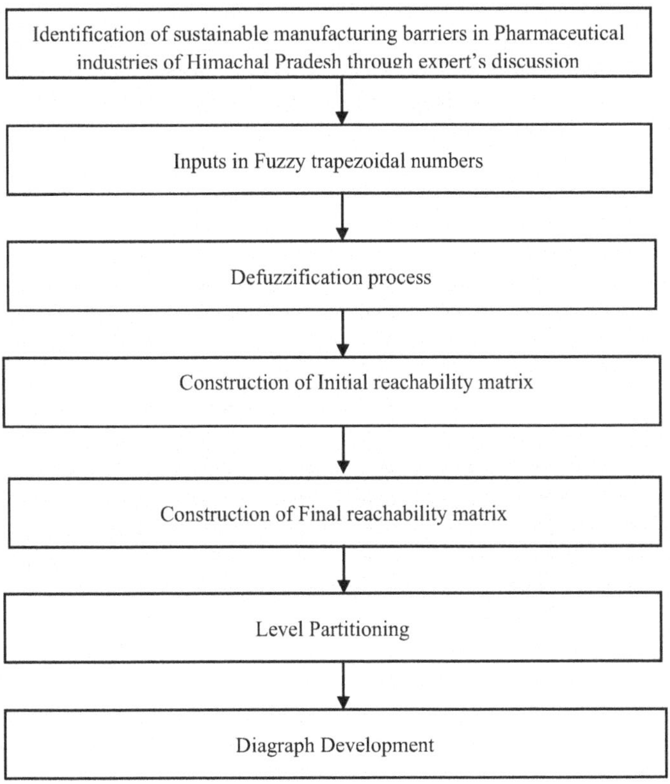

Fig. 15.1 Methodology adopted in the present study

Table 15.2 Linguistic scale

Linguistic value	Linguistic term
(0.0, 0.010, 0.20)	Very less related (VLR)
(0.10, 0.20, 0.30)	Less related (LR)
(0.20, 0.30, 0.40, 0.50)	Medium less related (MLR)
(0.40, 0.50, 0.60)	Medium related (MR)
(0.50, 0.60, 0.70, 0.80)	Medium high related (MHR)
(0.70, 0.80, 0.90)	High related (HR)
(0.80, 0.90, 1)	Very high related (VHR)

Table 15.3 Fuzzy input matrix

Barrier code	BR1	BR2	BR3	BR4	BR5	BR6	BR7	BR8	BR9
BR1	–	VHR	MHR	MLR	MLR	LR	VLR	VLR	MLR
BR2	LR	–	MLR	VLR	VLR	MLR	LR	MLR	LR
BR3	MLR	MHR	–	LR	MLR	VLR	VLR	LR	MLR
BR4	MR	VHR	VHR	–	HR	MLR	MR	MHR	VHR
BR5	LR	VLR	MLR	VLR	–	VLR	LR	VLR	MLR
BR6	VLR	MLR	LR	MLR	LR	–	MHR	MLR	LR
BR7	MLR	MHR	VHR	HR	HR	MHR	–	VHR	MLR
BR8	LR	VLR	VLR	LR	HR	LR	LR	–	LR
BR9	VLR	MLR	LR	MLR	VLR	VHR	MLR	VLR	–

Table 15.4 Defuzzification value matrix

Barrier	BR9	BR8	BR7	BR6	BR5	BR4	BR3	BR2	BR1
BR1	0.35	0.07	0.07	0.20	0.35	0.35	0.65	0.93	x
BR2	0.20	0.35	0.20	0.35	0.07	0.07	0.35	x	0.20
BR3	0.35	0.20	0.07	0.07	0.35	0.20	x	0.65	0.35
BR4	0.93	0.65	0.50	0.35	0.80	x	0.93	0.93	0.50
BR5	0.35	0.07	0.20	0.07	x	0.07	0.35	0.07	0.20
BR6	0.20	0.35	0.65	x	0.20	0.35	0.20	0.35	0.07
BR7	0.35	0.93	x	0.65	0.80	0.80	0.93	0.65	0.35
BR8	0.20	x	0.20	0.20	0.80	0.20	0.07	0.07	0.20
BR9	x	0.07	0.35	0.93	0.07	0.35	0.20	0.35	0.07

Table 15.5 Initial reachability matrix

Barrier	BR9	BR8	BR7	BR6	BR5	BR4	BR3	BR2	BR1
BR1	0	0	0	0	0	0	1	1	1
BR2	0	0	0	0	0	0	0	1	0
BR3	0	0	0	0	0	0	1	1	0
BR4	1	1	1	0	1	1	1	1	1
BR5	0	0	0	0	1	0	0	0	0
BR6	0	0	1	1	0	0	0	0	0
BR7	0	1	1	1	1	1	1	1	0
BR8	0	1	0	0	1	0	0	0	0
BR9	1	0	0	1	0	0	0	0	0

Initial reachability matrix:

Formatting of IR matrix is done by the use of Table 15.4 and equation below (Table 15.5):

$$M = \left\{ \frac{0,\ M \leq 0.45}{1,\ M \geq 0.45} \right\} \tag{15.2}$$

After checking the transitivity of matrix, initial reachability matrix is converted into final reachability matrix which is shown in Tables 15.6 and 15.7.

Diagraph Development

See Fig. 15.2.

Table 15.6 Final reachability matrix

Barrier	BR9	BR8	BR7	BR6	BR5	BR4	BR3	BR2	BR1
BR1	0	0	0	0	0	0	1	1	1
BR2	0	0	0	0	0	0	0	1	0
BR3	0	0	0	0	0	0	1	1	0
BR4	1	1	1	1*	1	1	1	1	1
BR5	0	0	0	0	1	0	0	0	0
BR6	0	0	1	1	0	0	0	0	0
BR7	1*	1	1	1	1	1	1	1	1*
BR8	0	1	0	0	1	0	0	0	0
BR9	1	0	0	1	0	0	0	0	0

Table 15.7 Level portioning

Reachability set	Antecedent set	Interaction set	Level
BR1, BR2, BR3	BR1, BR4, BR7	BR1	3
BR2	BR1, BR2, BR3, BR4, BR7	BR2	1
BR2, BR3	BR1, BR3, BR4, BR7	BR3	2
BR1, BR2, BR3, BR4, BR5, BR6, BR7, BR8, BR9	BR4, BR7	BR4, BR7	4
BR5	BR4, BR5, BR7, BR8	BR5	2
BR6, BR7	BR4, BR6, BR7, BR9	BR6	2
BR1, BR2, BR3, BR4, BR5, BR6, BR7, BR8, BR9	BR4, BR7	BR4, BR7	4
BR5, BR8	BR4, BR7, BR8	BR8	3
BR6, BR9	BR4, BR7, BR9	BR9	3

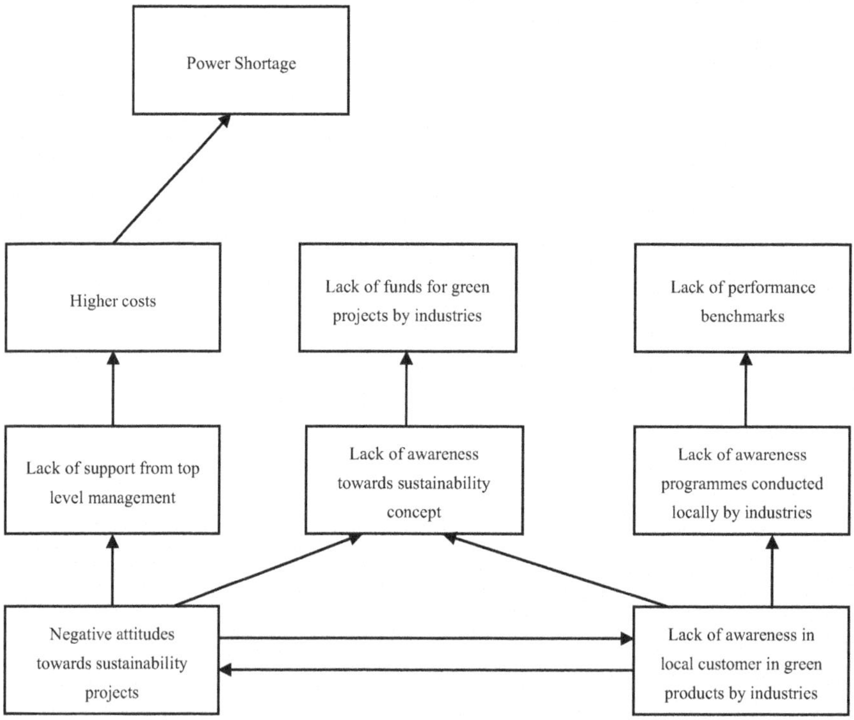

Fig. 15.2 Fuzzy-ISM model for barriers affecting the sustainability in pharmaceutical industries of Himachal Pradesh

15.4 Results and Discussion

Sustainability plays an important role in the development of any manufacturing industry. The barriers which affect the sustainability in pharmaceutical industries of Himachal Pradesh are identified with the help of discussion with industry experts and academia. Further, fuzzy-ISM approach is used to modelling the barriers. There are four levels found after level portioning. At the fourth level, negative attitudes towards sustainability projects and lack of awareness of local customer in green products are found which mostly affect the other barriers have high driving powers. At the third level, lack of support from top-level management, lack of awareness towards sustainability concepts and lack of awareness programmes conducted locally are found. At the second level, higher costs, lack of funds for green projects and lack of performance benchmark are found which have very less effect on sustainability in pharmaceutical industries. At the first level, power shortage is found which has maximum dependence power and lower driving power. Fuzzy-ISM modelling shows that sustainability in pharmaceutical industries is mostly affected by negative attitude of industries towards sustainability projects and lack of awareness of local customer in green products. However, Government of Himachal Pradesh is supporting the sustainability projects in Himachal Pradesh and also providing the financial aid to industries. Still industries need to work on these two barriers to overcome this problem.

Pharmaceutical industries of Himachal Pradesh have a big contribution in the GDP of India. Mostly, pharmaceutical industries are in Himachal Pradesh because its climate advantage. Climate of Himachal Pradesh is suitable for pharmaceutical products production. Pharmaceutical industries consume lot of energy and emit lot of pollution which affect the sustainability of industries. In the present paper, there are nine key barriers are found out in the discussion with industry experts and academia which affect the sustainability of pharmaceutical industries. Fuzzy-ISM modelling is used to find the relationship between the barriers. It is found that sustainability of pharmaceutical industries is mostly affected by negative attitude of industries towards sustainability projects and lack of awareness of local customer in green products. Industries should focus on these two problems which can be eliminated by proper training and knowledge about sustainability trends around the globe. Further, this study can be extended by finding the enablers of sustainability in pharmaceutical industries of Himachal Pradesh.

15.5 Conclusion

As per discussion with industry experts and data collected through different industries, nine barriers were found out which resist the implementation of lean manufacturing systems over the traditional manufacturing system. Hybrid fuzzy-ISM was adopted to assess the level and relationship between barriers. It is found in the study that lack of top management commitment and bad supplier is the main issue among most of the industries. As these two barriers have most driving power and less dependence power. If the industries want to implement sustainable manufacturing practices, then industries should have to overcome from these two barriers. Lack of technology advancement is found as less driving power barrier.

References

1. Singh, P.L., Sindhwani, R., Dua, N.K., Jamwal, A., Aggarwal, A., Iqbal, A., Gautam, N.: Evaluation of common barriers to the combined lean-green-agile manufacturing system by two-way assessment method. Adv. Ind. Prod. Eng., pp. 653–672. Springer, Singapore (2019)
2. Gautam, N., Ojha, M.K., Swain, P., Aggarwal, A., Jamwal, A.: Informal investigation of fourth-party and third-party logistics service providers in terms of indian context: an AHP approach. Adv. Ind. Prod. Eng., pp. 405–413. Springer, Singapore (2019)
3. Sharm, P., Jamwal, A., Aggarwal, A., Bhardwaj, S., Sood, R.: Major challenges in adoption of RFID for Indian SME's. Int. Res. J. Eng. Technol **5**, 2247–2255 (2018)
4. Jamwal, A., Aggarwal, A., Gupta, S., Sharma, P.: A study on the barriers to lean manufacturing implementation for small-scale industries in Himachal region (India). Int. J. Intell. Enterpr. **6**(2–4), 393–407 (2019)
5. Gupta, S., Dangayach, G.S., Singh, A.K., Rao, P.N.: Analytic hierarchy process (AHP) model for evaluating sustainable manufacturing practices in Indian electrical panel industries. Proc. Soc. Behav. Sci. **189**, 208–216 (2015)
6. Gupta, S., Dangayach, G.S., Singh, A.K.: Key determinants of sustainable product design and manufacturing. Proc. CIRP **26**, 99–102 (2015)
7. Gupta, S., Dangayach, G.S.: Sustainable waste management: a case from indian cement industry. Braz. J. Oper. Prod. Manag. **12**(2), 270–279 (2015)
8. Dangayach, G.S., Deshmukh, S.G.: Manufacturing strategy: literature review and some issues. Int. J. Oper. Prod. Manag. **21**(7), 884–932 (2001)
9. Mishra, P., Dangayach, G.S., Mittal, M.L.: An Ethical approach towards sustainable project Success. Proc. Soc. Behav. Sci. **25**, 338–344 (2011)
10. Ciriminna, R., Pagliaro, M.: Green chemistry in the fine chemicals and pharmaceutical industries. Org. Process Res. Dev. **17**(12), 1479–1484 (2013)
11. Hong, P., Kwon, H.B., Jungbae Roh, J.: Implementation of strategic green orientation in supply chain: an empirical study of manufacturing firms. Eur. J. Innov. Manag. **12**(4), 512–532 (2009)
12. Meena, M.L., Dangayach, G.S., Bhardwaj, A.: Measuring quality of work life among workers in handicraft industries of Jaipur. Int. J. Ind. Syst. Eng. **17**(3), 376–390 (2014)
13. Mathur, A., Mittal, M.L., Dangayach, G.S.: Improving productivity in Indian SMEs. Prod. Plann. Contr. **23**(10–11), 754–768 (2012)
14. Jain, J., Dangayach, G.S., Agarwal, G., Banerjee, S.: Supply chain management: Literature review and some issues. J. Stud. Manuf. **1**(1), 2 (2010)
15. Dornfeld, D.A.: Moving towards green and sustainable manufacturing. Int. J. Precis. Eng. Manuf. Green Technol. **1**(1), 63–66 (2014)
16. Amrina, E., Vilsi, A.L.: Key performance indicators for sustainable manufacturing evaluation in cement industry. Proc. CIRP **26**, 19–23 (2015)
17. Dubey, R., Gunasekaran, A., Childe, S.J., Wamba, S.F., Papadopoulos, T.: The impact of big data on world-class sustainable manufacturing. Int. J.Adv. Manuf. Technol. **84**(1–4), 631–645 (2016)
18. Moldavska, A., Welo, T.: The concept of sustainable manufacturing and its definitions: a content-analysis based literature review. J. Clean. Prod. **166**, 744–755 (2017)

19. Moktadir, M.A., Rahman, T., Rahman, M.H., Ali, S.M., Paul, S.K.: Drivers to sustainable manufacturing practices and circular economy: a perspective of leather industries in Bangladesh. J. Clean. Prod. **174**, 1366–1380 (2018)
20. Singla, A., Ahuja, I.S., Sethi, A.S.: An examination of effectiveness of technology push strategies for achieving sustainable development in manufacturing industries. J. Sci. Technol. Pol. Manag. **10**(1), 73–101 (2019)
21. Gorane, S.J., Kant, R.: Modelling the SCM enablers: an integrated ISM-fuzzy MICMAC approach. Asia Pacific J. Market. Logist. **25**(2), 263–286 (2013)
22. Khatwani, G., Singh, S.P., Trivedi, A., Chauhan, A.: Fuzzy-TISM: a fuzzy extension of TISM for group decision making. Global J. Flexib. Syst. Manag. **16**(1), 97–112 (2015)

Chapter 16
Computational Analysis for Thermal Behaviour of Different Types of Steels in Grinding Process

Puneet Kumar Gupta and N. P. Yadav

Abstract This paper includes computational analysis for thermal behaviour of different types of steel as workpiece materials during the grinding process. The three-dimensional model was included to analyse the thermal behaviour on the workpiece surface during the grinding process. The computational fluid dynamics approach is used to simulate the results. A comparative analysis has been done for carbon steel, cast steel and carbon nickel steel for different depth of cut at constant speed of grinding wheel. The effect of water-based and oil-based coolant (kerosene) is also analysed to understand the temperature variation on the workpiece surface. The maximum temperature on the surface increases with increased in depth of cut, and the higher temperature was observed in carbon nickel steel.

Keywords Thermal · Temperature · Grinding process · CFD

16.1 Introduction

The grinding is a process, where abrasive particles are used as a cutting medium to produce the smooth surfaces and fine tolerances. The process of machining is used where a high degree of accuracy and close tolerance are required. The grinding wheel consists of abrasive particles called grains, which take part in cutting action. There is extremely high energy expenditure per unit volume which is required to remove the material. Most of the energy is converted into heat, due to which high temperature is produced in the grinding zone. Therefore, high temperature is generated in grinding zone because of that the possibility of thermal damage to the workpiece such as burning, phase transformation, softening, unfavourable residual tensile stresses and cracks increases. So, the analysis of the thermal behaviour of grinding process is very important. Malkin et al. [1] have developed a thermal model and concluded that in the grinding process total energy depends on chip formation, sliding and ploughing energy components. Most of the ploughing and sliding energies are conducted to the workpiece as heat, while 55% chip formation energy is conducted to the workpiece. Ueda et al. [2] studied the thermal behaviour of abrasive grains in grinding radiation

P. K. Gupta (✉) · N. P. Yadav
Bundelkhand Institute of Engineering & Technology, Jhansi, India

S. Yadav et al. (eds.), *Proceedings of International Conference in Mechanical and Energy Technology*, Smart Innovation, Systems and Technologies 174,
https://doi.org/10.1007/978-981-15-2647-3_16

sensor, where infrared radiation sensors utilizing fibre optics have found workpiece temperature near the ground surface. Torrance [3] described that the workpiece burn is the major thermal damage during grinding. He explained this with the grinding of plain carbon and alloy steel. Tarasov [4] explained the bluish temper colours that are formed during grinding of steel which indicates the workpiece burn and oxide layer formation. Malkin and Cook [5] have discussed that the workpiece burn leads to increase in forces, workpiece surface to deteriorate and rate of wheel wear increases. Torrance et al. [6] have explained workpiece burn of bearing steels which have an adverse effect on fatigue life, which is due to the formation of untempered martensite. Lavine and Jen [7] developed a model and explained that the workpiece temperature depends on different parameters and can help in the optimization of the grinding process in achieving the higher removal rate of material with no thermal damage. Seven non-dimensional parameters are explained on which non-dimensional temperature depends. Guo and Malkin [8] have analysed the energy partition in grinding. They reported the thermal composite that is formed by the abrasive grains and grinding fluid flowing in the pores of wheel and moves at the wheel velocity relative to the grinding zone. The fraction of energy partition to the workpiece is simulated by equating the temperature of the workpiece surface and the composite surface at each point along the grinding zone. They also discussed the variation of energy partition in the grinding zone. Liao et al. [9] have discussed the importance of temperature consideration in grinding and the thermal behaviour at the interface of the abrasive grain and workpiece and also discussed shear plane between the workpiece and the chip. They concluded that under general condition, the rate of flow of the coolant is an important factor to analyse the thermal boundary layer of the fluid in the ground zone. Jin and Cai [10] studied three different thermal models for deep cut and analysed the temperature variation in ground part on using moving plane heat source. Moulik et al. [11] formulated a finite element model, for calculating the temperatures and stresses when moving source of heat is used. For this, hardened steel and plane moving heat source is used to analyse the thermal stresses. Mamalis et al. [12] discussed a model for the simulation of grinding of hardened steels with aluminium oxide wheels. For this task, the implicit FEM code MARC is used. For this, the transient and nonlinear model is used to analyse the maximum temperature and the temperature field on the ground workpiece. Jin and Stephenson [13] have studied the three-dimensional finite element transient heat transfer simulations for high efficiency deep grinding. The analytical models' results have been satisfied with the experimental measurements. Morgan et al. [14] discussed different methods of applications of grinding fluid and requirement of its quantity for any grinding process. A computational fluid dynamics model (CFD) is developed, and results are compared with experimental methods. Stefan Mihic et al. [15] discussed the CFD models to analyse the behaviour of the flow of fluid and how the heat transfer phenomenon takes place in a grinding process. For this, a three-dimensional model is used, and computational fluid dynamics approach is used. Gupta and Yadav [16] studied the behaviour of different grinding fluids in heat transfer during grinding process using computational fluid dynamics approach. They described the effect of water-based and oil-based coolants in heat transfer phenomenon during the grinding process. The present work focuses the computational fluid dynamics analysis for thermal behaviour of different materials during the grinding process through ANSYS software [17]. The mathematical modelling was done, and the results are predicted by the validation of the present model. The validation is done with Stefan et al. [15] for maximum change in temperature. The effect of water-based and oil-based coolant for at variable depth of cut is analysed to understand the temperature variation on the surface of workpiece.

16.2 Mathematical Modelling

16.2.1 Momentum Equation

The flow of coolant injected during the grinding process is estimated by considering the steady, incompressible flow. The Reynolds-averaged Navier–Stokes equations have been used for the flow calculation. Therefore, the momentum equation for the grinding process can be written as follows [15–17]:

$$S_i = -\left(\frac{\mu}{\alpha} \cdot v_i + C_2 \cdot \rho \cdot |v| \cdot v_i\right) \tag{16.1}$$

where S_i represents the source term for the i-th (x, y or z) direction, $|v|$ represents the magnitude of velocity, α represents the permeability, C_2 is constant and termed as inertial resistance factor, and μ is the fluid viscosity.

16.2.2 Energy Equation

The standard energy transport equation was used as discussed by Stefan Mihic et al. [15] in the contact zone. Similar predictions are also discussed in [16, 17]. In the contact zone, the conduction flux uses an effective conductivity of fluid and abrasive particles as mentioned in Eq. (16.2):

$$\nabla \cdot \left(\vec{v}\left(\rho_f \cdot E_f + p\right)\right) = \nabla \cdot \left[k_{\text{eff}} \cdot \nabla \cdot T - \left(\sum_i h_i \cdot J_i\right) + \left(\bar{\bar{\tau}} \cdot \vec{v}\right)\right] + S_f \tag{16.2}$$

16.2.3 Heat Generation Modelling

The heat generated in grinding is caused due to the grains which have rubbing action with workpiece. So, for the elementary heat source, the strength Q_{1Z} can be written as [15] in Eq. (16.3):

$$Q_{1z} = F_{vz} \cdot v_t \cdot t_z \tag{16.3}$$

where $F_{vz} = \frac{F_{vm}}{N_a \cdot A_k}$ represents the main grinding force (F_{vm} represents the total main grinding force, N_a represents the number of active grains per unit of surface of grinding wheel, A_k represents the surface area of contact between wheel and grinded part), v_t represents the grinding wheel circumferential velocity, and $t_z = \frac{\sqrt{a \cdot D_t}}{V_t}$ represents the time period in which grain cuts the workpiece (a represents the grinding depth, and D_t is the diameter of wheel). The final equation for the total heat strength is obtained as [15] in Eq. (16.4):

$$Q' = \frac{v_p}{v_t} \cdot a^{(3/2)} \cdot D_t^{(1/2)} \cdot k_{vm} \tag{16.4}$$

where v_p is the workpiece velocity and k_{vm} is the specific main grinding resistance, calculated as $k_{vm} = \frac{C_k}{\varepsilon_k \sqrt{A}}$. Various parameters for specific main grinding resistance are given in Tables 16.1 and 16.2.

16.2.4 Turbulence Modelling

In this study, the turbulent model is based on Reynold stress model. It is also discussed by Stefen Mihic et al. [15], and the equation is represented by Eq. (16.5).

Table 16.1 Parameters C_k and ε_k for common materials [15]

Materials	C_k	ε_k
Carbon steel	$1.936 * 10^8$	6.1
Chrome nickel steel	$6.263 * 10^8$	10.4
Cast steel	$2.196 * 10^8$	6.7

Table 16.2 Different properties of grinding fluids [17]

	Water-based coolant	Kerosene oil
Density (Kg/m^3)	998.2	780
Specific heat (J/Kg K)	4182	2090
Conductivity (W/m K)	0.6	0.149
Viscosity (Kg/m s)	0.00103	0.0024

$$\frac{\delta}{\delta t}\left(\rho\cdot\overline{u_i^{'}\cdot u_j^{'}}\right)+\frac{\delta}{\delta x_k}\left(\rho\cdot u_k\cdot\overline{u_i^{'}\cdot u_j^{'}}\right)=\frac{\delta}{\delta x_k}\Big[\rho\cdot\overline{u_i^{'}\cdot u_j^{'}\cdot u_k^{'}}$$
$$+\overline{\rho\cdot(\delta_{kj}\cdot u_i^{'}+\delta_{kj}\cdot u_j^{'})}\Big]+\frac{\delta}{\delta x_k}\left[u\cdot\frac{\delta}{\delta x_k}(\overline{u_i^{'}\cdot u_j^{'}})\right]$$
$$-\rho\cdot\left(\overline{u_i^{'}\cdot u_j^{'}}\frac{\delta u_j}{\delta x_k}+\overline{u_i^{'}\cdot u_j^{'}}\cdot\frac{\delta u_i}{\delta x_k}\right)-\rho\cdot\beta\cdot\left(g_i\cdot\overline{u_j^{'}\cdot\theta}+g_j\cdot\overline{u_i^{'}\cdot\theta}\right)$$
$$+\overline{\rho\cdot\left[\left(\frac{\delta u_i^{'}}{\delta x_j}+\frac{\delta u_j^{'}}{\delta x_i}\right)\right]}-2\cdot\mu\cdot\overline{\frac{\delta u_i^{'}}{\delta x_k}\cdot\frac{\delta u_j^{'}}{\delta x_k}}$$
$$-2\cdot\rho\cdot\Omega_k\cdot\left(\overline{u_j^{'}\cdot u_m^{'}}\cdot\varepsilon_{\mathrm{ikm}}+\overline{u_i^{'}\cdot u_m^{'}}\cdot\varepsilon_{\mathrm{jkm}}\right)+s_{\mathrm{user}} \tag{16.5}$$

where $\frac{\delta}{\delta x_k}\left(\rho\cdot u_k\cdot\overline{u_i^{'}\cdot u_j^{'}}\right)$ is the convection term, $-\frac{\delta}{\delta x_k}\left[\rho\cdot\overline{u_i^{'}\cdot u_j^{'}\cdot u_k^{'}}+\overline{\rho\cdot(\delta_{kj}\cdot u_i^{'}+\delta_{kj}\cdot u_j^{'})}\right]$ is the turbulent diffusion term, $\frac{\delta}{\delta x_k}\left[u\cdot\frac{\delta}{\delta x_k}\left(\overline{u_i^{'}\cdot u_j^{'}}\right)\right]$ is the laminar diffusion term, $-\rho\cdot\left(\overline{u_i^{'}\cdot u_j^{'}}\frac{\delta u_j}{\delta x_k}+\overline{u_i^{'}\cdot u_j^{'}}\cdot\frac{\delta u_i}{\delta x_k}\right)$ is the stress production term, $-\rho\cdot\beta\cdot\left(g_i\cdot\overline{u_j^{'}\cdot\theta}+g_j\cdot\overline{u_i^{'}\cdot\theta}\right)$ represents the buoyancy production term, $\overline{\rho\cdot\left(\frac{\delta u_i^{'}}{\delta x_j}+\frac{\delta u_j^{'}}{\delta x_i}\right)}$ is the pressure strain term, $-2\cdot\mu\cdot\overline{\frac{\delta u_i^{'}}{\delta x_k}\cdot\frac{\delta u_j^{'}}{\delta x_k}}$ is the dissipation term, and $-2\cdot\rho\cdot\Omega_k\cdot\left(\overline{u_j^{'}\cdot u_m^{'}}\cdot\varepsilon_{\mathrm{ikm}}+\overline{u_i^{'}\cdot u_m^{'}}\cdot\varepsilon_{\mathrm{jkm}}\right)$ is the production by system rotation term.

16.3 Numerical Simulation

The flow domain consists of a rotating grinding wheel which is rotating with 68 m/s tangential velocity in the clockwise direction as given in Table 16.3, the upper layer of workpiece, the grinding zone, a nozzle from where grinding fluid is coming and the shield. The diameter of rotating grinding wheel is 340 mm, the radius of the

Table 16.3 Parameters for the simulation

Parameters	Values of parameters
Inlet velocity of fluid	20 m/s
Wheel velocity (m/s)	68
Workpiece speed	60 mm/min
Depth of cut (mm)	0.5, 0.7, 0.9, 1.1
Diameter of wheel	340 mm
Grinding fluid	Water-based and kerosene
Workpiece material	Carbon steel, cast steel and carbon nickel steel

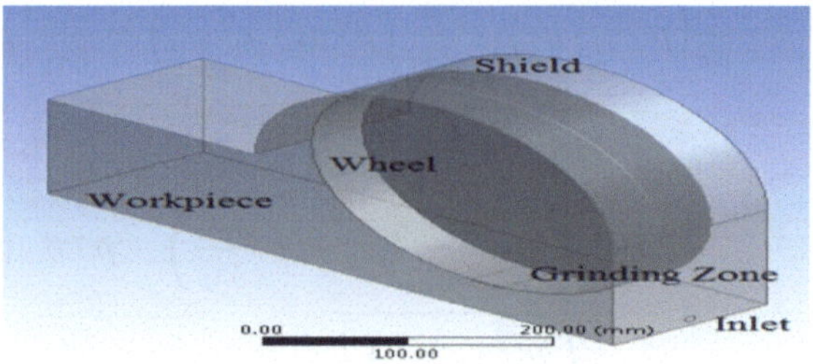

Fig. 16.1 Geometry of grinding process

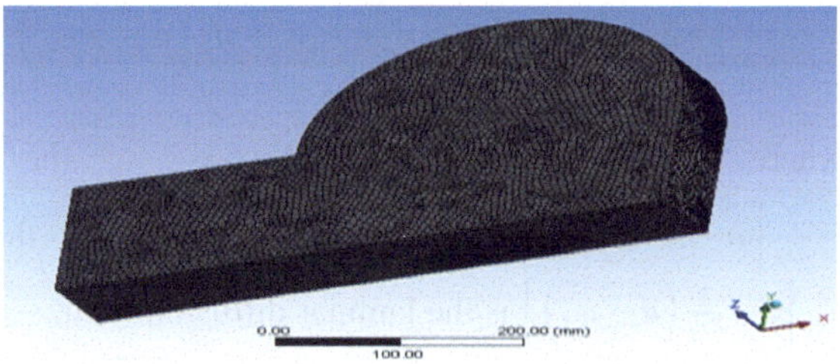

Fig. 16.2 Meshed geometry

shield is 180 mm, the area of workpiece is 580 mm * 150 mm, and the area of the grinding zone is 50 mm * 30 mm. The grinding depths are taken as 0.5, 0.7, 0.9 and 1.1 mm. The size of the nozzle is 5 mm. The width of the wheel is 50 mm. The analysis was done for three different workpiece materials, carbon steel, cast steel and carbon nickel steel. The water-based coolant and kerosene oil of same inlet velocity 20 m/s are used as a coolant to study heat transfer phenomenon in different materials for different depth of cut. The required properties of the grinding fluids are given in Table 16.2. The geometry of the grinding process is shown in Fig. 16.1. Figure 16.2 represents the meshed geometry. The grid size for this model is selected for the number of elements 520,473 by grid-independent test. The average element quality is 0.831. The validation of present study has been done with Stefan et al. [15] by comparing the maximum change in temperature.

16.4 Result and Discussion

16.4.1 Variation of Maximum Grinding Temperature for Different Workpiece Materials at Various Depth of Cut

Figures 16.3, 16.4, 16.5 and 16.6 represent contour of total temperature in grinding zone at different depth of cut for different workpiece materials when kerosene oil is used as grinding fluid. Figure 16.3 is showing the variation of total temperature for

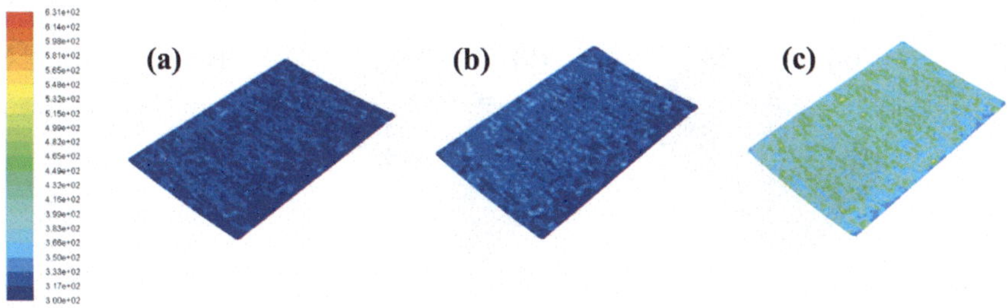

Fig. 16.3 Contour for total temperature in grinding zone at 0.5 mm depth of cut for **a** carbon steel, **b** cast steel, **c** carbon nickel steel

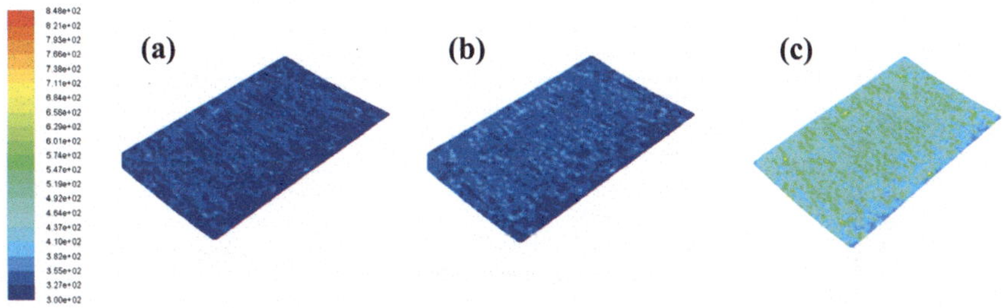

Fig. 16.4 Contour for total temperature in grinding zone at 0.7 mm depth of cut for **a** carbon steel, **b** cast steel, **c** carbon nickel steel

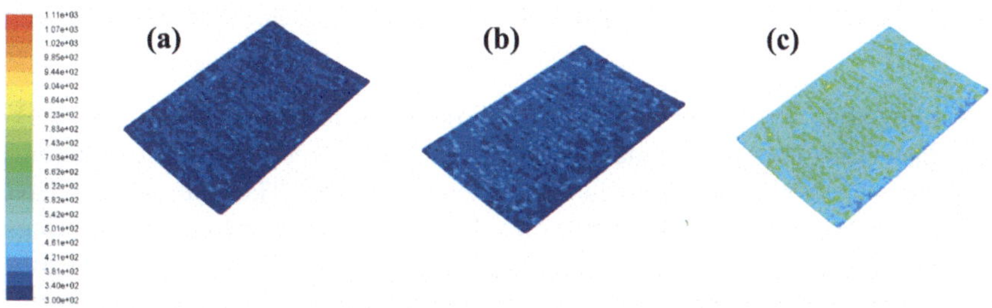

Fig. 16.5 Contour for total temperature in grinding zone at 0.9 mm depth of cut for **a** carbon steel, **b** cast steel, **c** carbon nickel steel

0.5 mm depth of cut, Fig. 16.4 for 0.7 mm, Fig. 16.5 for 0.9 mm and Fig. 16.6 for 1.1 mm depth of cut.

The variations of temperature at the surface of workpiece increase with increase in depth of cut. The variation of maximum temperature with the depth of cut on the surface of the workpiece is given in Fig. 16.7.

It is observed that maximum temperature in carbon steel is least, whereas it is maximum in the case of carbon nickel steel. The maximum grinding temperature

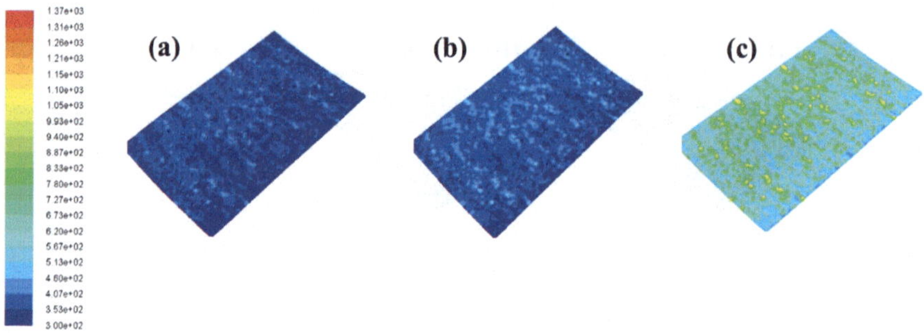

Fig. 16.6 Contour for total temperature in grinding zone at 1.1 mm depth of cut for **a** carbon steel, **b** cast steel, **c** carbon nickel steel

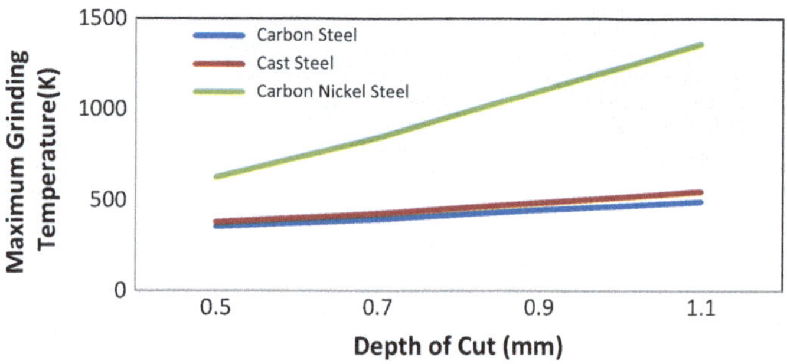

Fig. 16.7 Variation of maximum grinding temperature with depth of cut for different materials

in grinding zone depends on specific main grinding resistance offered by the workpiece materials. The specific main grinding resistance depends on various thermal properties of materials as shown in heat generation model. The materials that have less specific main grinding resistance offer less heat flux. Therefore, the maximum grinding temperature will be less; whereas, if the value of specific grinding temperature is more, it means it resists the heat transfer through the materials and heat flux and maximum grinding temperature will be more. The specific grinding temperature depends on parameters given in Table 16.1. Figure 16.7 represents the variation of maximum grinding temperature with different depth of cut. It shows that maximum grinding temperature increases with increase in depth of cut because of increase in heat zone area. It also represents that maximum grinding temperature in the case when carbon steel and cast steel used as workpiece is less as compared to case when carbon nickel steel is used as workpiece materials. This is because of high specific resistance in carbon nickel steel as compared to cast steel and carbon steel. The carbon steel has least specific grinding temperature.

16.4.2 *Variation of Maximum Grinding Temperature for Different Grinding Fluids at Different Depth of Cut*

The grinding fluids can be classified into two categories, oil-based coolants and water-based coolants. The variations of temperature on the surface of workpiece for water-based coolants and oil-based coolant (kerosene) are further studied for carbon nickel steel as workpiece material because the maximum temperature with cutting depth is higher than other two materials as discussed in Fig. 16.7. Figures 16.8, 16.9, 16.10 and 16.11 represent the contour of total temperature at different depth of cut for two different coolants for carbon nickel steel as workpiece material.

Figure 16.12 shows the variation of maximum grinding temperature at different depth of cut for two different coolants. It is observed that water-based coolants have better cooling capacity as compared to oil-based coolant because of its better thermal properties. The water-based coolant has high thermal conductivity, high density, high specific heat and lower value of viscosity. It can also be observed that

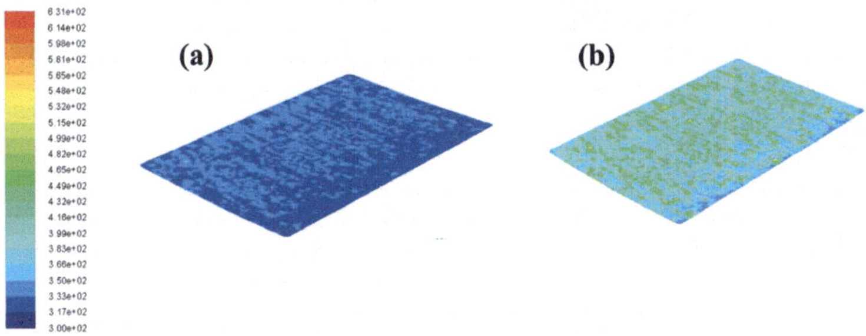

Fig. 16.8 Contour of total temperature at 0.5 mm depth of cut for carbon nickel steel for **a** water-based coolant, **b** kerosene

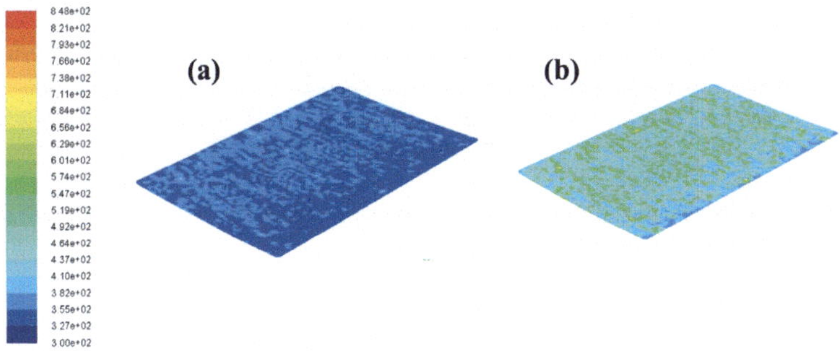

Fig. 16.9 Contour of total temperature at 0.7 mm depth of cut for carbon nickel steel for **a** water-based coolant, **b** kerosene

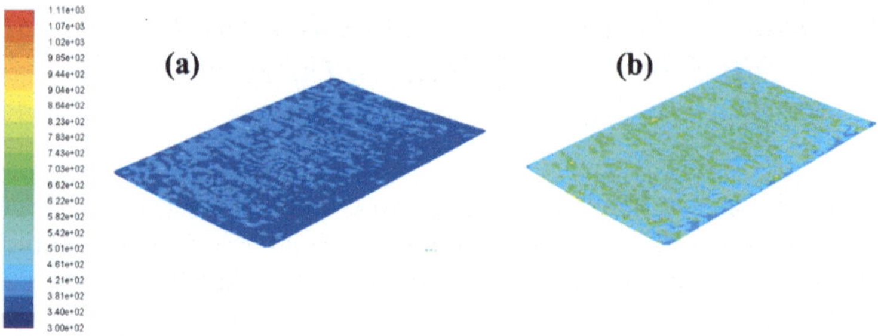

Fig. 16.10 Contour of total temperature at 0.9 mm depth of cut for carbon nickel steel for **a** water-based coolant, **b** kerosene

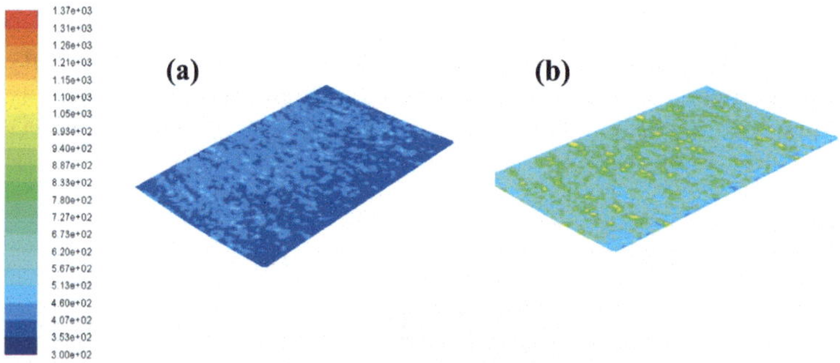

Fig. 16.11 Contour of total temperature at 1.1 mm depth of cut for carbon nickel steel for **a** water-based coolant, **b** kerosene

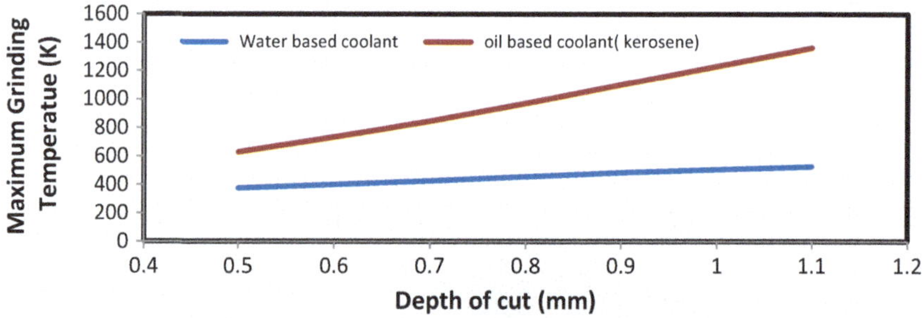

Fig. 16.12 Variation of maximum grinding temperature with depth of cut for different coolant when carbon nickel steel is used as workpiece material

in the case of carbon nickel steel, there may be chance of thermal damage at high grinding temperature. In that situation, water-based coolant can be used. Therefore, it is advisable to use water-based coolant when high degree of cooling is required.

16.5 Conclusions

This paper focused the computational fluid dynamics analysis for thermal behaviour of different types of steels as workpiece materials during the grinding process through ANSYS software. The three-dimensional model was included to analyse the variation of temperature on the surface of workpiece during the grinding process, and results are expected by validation of the present model. The maximum temperature on the surface of the workpiece increases with increase in depth cut for carbon steel, cast steel and carbon nickel steel at constant speed of grinding wheel. The higher temperature was observed in carbon nickel steel. The higher value of specific grinding resistance leads to higher heating capacity and hence higher maximum grinding temperature. The water-based coolant was better if a high degree of cooling is required because of its better thermal properties as compared to oil-based coolants.

References

1. Malkin, S., Anderson, R.B.: Thermal aspects of grinding: part 1: energy partition, part 2: surface temperature and work piece burn. ASME J. Eng. Ind. **96**, 1177–1191 (1974)
2. Ueda, T., Hosokawa, A., Yamamoto, A.: Studies of temperature of abrasive grains in grinding—application of infrared radiation pyrometer, ASME. J. Eng. Ind. **107**, 127–133 (1984)
3. Torrance, A.A.: Metallurgical effects associated with grinding. In: Proceedings of the 19th International Machine Tool Design and Research Conference, pp. 637–644
4. Tarasov, L.P.: Some metallurgical aspects of grinding. Mach. Theor. Pract. ASM **2**, 409–464 (1950)
5. Malkin, S., Cook, N.H.: The wear of grinding wheels part I attritious wear. Trans. ASME J. Eng. Ind. **93**, 1120–1128 (1971)
6. Torrance, A.A., Stokes, R.J., Howes, T.D.: The effect of grinding conditions on the rolling contact fatigue life of bearing steel. Mech. Eng. **10**, 63–73 (1983)
7. Lavine, A.S., Jen, T.C.: Thermal aspects of grinding: heat transfer to workpiece, wheel, and fluid. ASME J. Heat Transf. **113**, 296–303 (1991)
8. Guo, C., Malkin, S.: Analysis of energy partition in grinding. ASME J. Eng. Ind. **117**, 55–61 (1995)
9. Liao, Y.S., Luob, S.Y., Yanga, T.H.: A thermal model of the wet grinding process. J. Mater. Process. Technol. **101**, 137–145 (2000)
10. Jin, T., Cai, G.Q.: Analytical thermal models of oblique moving heat source of deep grinding and cutting. J. Manuf. Sci. Eng. ASME **123**, 185–190 (2001)
11. Moulik, P.N., Yang, H.T.Y., Chandrasekar, S.: Simulation of thermal stresses due to grinding. Int. J. Mech. Sci. **43**, 831–851 (2001)
12. Mamalis, A.G., Kundrak, J., Markopoulos, A.: Numerical simulation for the temperature fields and the heat affected zones in grinding. Prod. Syst. Inform. Eng. **1**, 3–16 (2003)
13. Jin, T., Stephenson, D.J.: Three dimensional finite element simulation of transient heat transfer in high efficiency deep grinding. CIRP Ann. Manuf. Technol. **53**, 259–262 (2004)
14. Morgan, M.N., Jackson, A.R., Wua, H., Baines-Jones, V., Batako, A., Rowe, W.B.: Optimisation of fluid application in grinding. CIRP Ann. Manuf. Technol. **57**, 363–366 (2008)

15. Mihic, S., Cioc, S., Marinescu, I., Weismiller, M.: Detailed study of fluid flow and heat transfer in the abrasive grinding contact using computational fluid dynamics method. J. Manuf. Sci. Eng. ASME **135**, 041002-1 (2013)
16. Gupta, P.K., Yadav, N.P.: Numerical simulation for the temperature behaviour on workpiece surface for different working fluids during grinding process. IJARSE **6**, 638–647 (2017)
17. Ansys, Inc.: FLUENT 15.0 user's guide. Canonsburg, PA (2013)

Chapter 17
Experimental Investigations of Thermal Contact Conductance for Tool–Sample Interface

Mohammad Asif and Mohd Nadeem

Abstract There are many practical applications where two metals are kept in contact, such as metal forming, metal cutting, fin-tube heat exchangers, electronic packaging, metal casting, I C engines, nuclear reactors, and bolted joints. The study of thermal contact conductance (TCC) is essential for the thermal and mechanical analysis required for the suitable design of a system. In this work, the thermal contact conductance has been studied for metal cutting applications. In metal cutting operations, a tool comes in contact with a metal to be cut, and the heat is generated at the contact interface. Hence, axial heat flow experiments have been carried out to evaluate the thermal contact conductance at the interface of tool and metal, placed in line contact. The upper specimen is essentially a metal cutting tool made of high-speed steel while the lower specimen is of mild steel, a common metal to be machined. Steady state approach has been employed to estimate the thermal contact conductance at the interface of tool and metal, under atmospheric conditions. The experiments have been carried out with different surface finishes for each specimen pairs, under varying loads and varying interface temperatures to study the effect of surface properties, contact pressure, temperature and surface deformations on thermal contact conductance. Further, the normalized results of TCC at the interface of tool and metal have been compared with theoretical models. Moreover, the results have been presented in the form of a suitable correlation which would be beneficial for the design of these systems for the present range of working parameters.

Keywords Thermal contact conductance · Metal cutting · Tool–sample interface · Steady state approach

M. Asif (✉) · M. Nadeem
Mechanical Engineering Department, Aligarh Muslim University, Aligarh, India
e-mail: masif@zhcet.ac.in

S. Yadav et al. (eds.), *Proceedings of International Conference in Mechanical and Energy Technology*, Smart Innovation, Systems and Technologies 174,
https://doi.org/10.1007/978-981-15-2647-3_17

17.1 Introduction

Precise knowledge of TCC has been found to be crucial in many engineering applications such as electronic packaging, gas turbine cooling, internal combustion engine, heat exchangers, metal forming processes, and sample–tool interface in metal cutting operations. In metal cutting process, when the tool comes in contact with the sample, high temperatures are generated in the cutting zones due to large plastic deformations of the sample material. Due to high temperature in the tool rake face, tool wears which lead to deformation of tool and hence tool life is affected [1]. Therefore, the determination of maximum temperature in the tool during metal cutting is very essential which is linked with the estimation of the thermal contact conductance at the interface of tool and sample metal. Engineering surfaces are never perfectly flat microscopically; therefore, when two surfaces are kept in contact, most of the heat flow lines pass through a limited number of actual contact spots. Hence, the thermal contact resistance (TCR) is the resistance to heat flow due to the actual area of contact being only a small fraction of the nominal or apparent area. The converse of TCR is said to be thermal contact conductance (TCC) defined as follows:

$$\text{TCC} = \frac{q}{\Delta T} \tag{17.1}$$

where 'q' is the flux, and ΔT is the temperature drop across the interface. In fact, the heat transfer across the interface can take place by the combination of three different modes, i.e. conduction through the individual micro-asperities, convection or conduction through the interstitial medium inside the gap, and by the radiation across the gap. It was reported in the pertinent literature that for the interface temperature less than 400 °C, radiation conductance may be neglected [2].

Experimental work is one of the most significant parts of research on TCC since it considers real conditions and provides actual results as compared to theoretical works. Therefore, many researchers have been doing experimental work on TCC. Jeevanashankara et al. [3] performed an experimental investigation on TCC of flat metallic contacts by specimens made of stainless steel, mild steel, brass, aluminium, and copper in the loading range from 5 to 25 kg with different interstitial fluid such as air, aniline, neem oil, castor oil, glycerine, and glycerine-oil mixture. Mcwaid and Marschall [4] estimated the thermal contact conductance for ten pairs of surfaces in flat contact under vacuum environment for contact pressures between 0.16 and about 6.9 MPa for aluminium 6061-T6 and 304 stainless steel. Milanez et al. [5] carried out an experimental study for measuring the thermal contact conductance between two SS 304 specimens in flat contact under vacuum environment with contact pressure varied from 0.0158 to 3.0 MPa in both ascending and descending levels and the mean interface temperature ranged from 15 to 60 °C. Wahid et al. [6] performed an investigation on stainless steel AISI 304 pairs in flat contact having effective RMS roughness of 2.88, 5.4, 14.3, 21.2 μm and mean interface temperatures 40–100 °C, for contact pressure of 0.43 MPa in a vacuum environment. Sunil Kumar and Ramamurthi [7] examined the thermal contact conductance between aluminium and stainless steel flat contacts for low interface temperatures in the range of 50–300 K. Singhal et al. [8] developed an experimental facility to estimate thermal contact conductance for aluminium, brass, copper, and stainless steel in surface contact with contact pressure ranged from 0.5 to 5.5 MPa. The thermal contact conductance has been studied with interface pressure, surface finish, and types of materials. Yuncu [9] studied the thermal contact conductance as a function of nominal contact pressure experimentally in the atmospheric conditions. A total number of 26 pairs of steel, copper, brass, and aluminium test pieces in flat contact were examined having different surface roughness over a range of contact pressures 0–16 MPa. Xing et al. [10] studied the thermal contact conductance for high interfacial temperatures (320–550 °C) and contact pressure in the range of 2.96–15.68 MPa. Misra and Nagaraju [11] performed experimental studies to study the effect of thermal stresses on thermal contact conductance for (OFHC) Cu in flat contacts at contact pressures of 0–1.0 MPa and interface temperature of 20–100 °C. The experiments have been performed in three different environments (vacuum, argon, and nitrogen). Liu et al. [12] carried out experiments to estimate the thermal contact conductance on two brass columns at room temperature under the interfacial pressure of 0.166–2.636 MPa. Wahid and Madhusudana [13] performed experiments on stainless steel flat contact under a vacuum (3 Pa) to estimate the solid spot conductance and gap conductance with interstitial gases He, Ar, N_2, CO_2 and mixtures of He and Tariq and Asif [14] have done experimental study on TCC for stainless steel, brass, and copper materials under vacuum environment. The pressure was varied in the range of 0.6–15 MPa while temperature range was 30–100 °C. In another important work, Asif and Tariq [15] have established a general correlation and various specific correlations for TCC which can be used for a range of parameters under vacuum conditions. Jam and Fard [16] determined the thermal contact conductance at the tool–chip interface in metal cutting using an inverse methodology based on the comparison of finite element analysis results and experimental results from literatures. Norouzifard and Hamedi [17] determined the thermal contact conductance in the tool–chip contact area in the machining operation by inverse thermal solution and infrared thermography. Further, the effect of cutting parameters such as cutting speed and feed rate have also been studied.

From the literatures, it is evident that thermal contact conductance (TCC) plays a vital role for the metal cutting application, and therefore, the study of TCC at the interface of tool and metallic sample is of utmost importance. Hence, in this work, the experimental investigations of thermal contact conductance for tool–sample interface have been carried out to evaluate its TCC. The upper specimen has been taken essentially a metal cutting tool made of high-speed steel (HSS) while the lower specimen is of mild steel (MS), a common metal to be machined. The type of contact for metal cutting is indeed a line contact which is studied experimentally and theoretically in very few literatures. Axial heat flow experiments have been performed on HSS–MS pairs with different surface finishes for each specimen pairs under varying contact pressures and temperatures to study the effect of various parameters on thermal contact conductance.

17.2 Experimentation

A simple experimental facility has been established based on an axial heat flow method to estimate the thermal contact conductance under atmospheric conditions.

17.2.1 Experimental Set-Up

The main components of the experimental set-up include heating system, cooling system, loading system, supporting structure, insulations (axial and radial), and the temperature measurement system. The two specimens have been kept in contact with the non-contacting ends heated from the upper side and cooled from the lower side to achieve downward axial heat flow. Heating system consists of a heating block which is made of a high conductivity material, i.e. copper that can act as a uniform heat supplier to the specimen. Two cartridge-heating rods each of 150 W has been inserted in the block for the generation of high density heat flux. The applied heat flux is controlled manually with the help of dimmerstat, connected with voltmeter and ammeter.

To extract the heat in a downward direction axially, a proper cooling system has been arranged. A cylindrical cooling block made of copper is used to make efficient cooling in which circular groves are made for the flow of chilled water. The chilled water is supplied in the cooling block by a freezer. Two insulating blocks of nylon fibre are placed at the top of heating block and below the cooling block to reduce the axial heat loss. In addition, a very good insulation made of glass wool is covered around the heater and specimens to minimize the radial heat loss. Ten copper constantan (T-Type) thermocouples of 0.5 mm diameter and at least 1 m long have been used for the axial temperature measurements in both the specimens. To perform the experiment for various loading conditions, a proper loading system has been arranged. It comprises a screw jack, load cell, and a load indicator to measure the load applied to the contacting specimens. A block diagram and a pictorial view of the complete experimental system are shown in Fig. 17.1.

17.2.2 Experimental Procedure

Before the start of experiments, all the surfaces are properly cleaned with acetone, and all the surfaces except the test interface are applied with a highly conducting silicone thermal paste to reduce the thermal contact resistance of other interfaces. Subsequently, the loading mechanism, heater, and freezer have been started. The temperatures at each location of specimens are measured by the thermocouples connected with temperature indicator. The system is kept under running condition until the specimens attain the steady state temperature. Steady state is considered to be

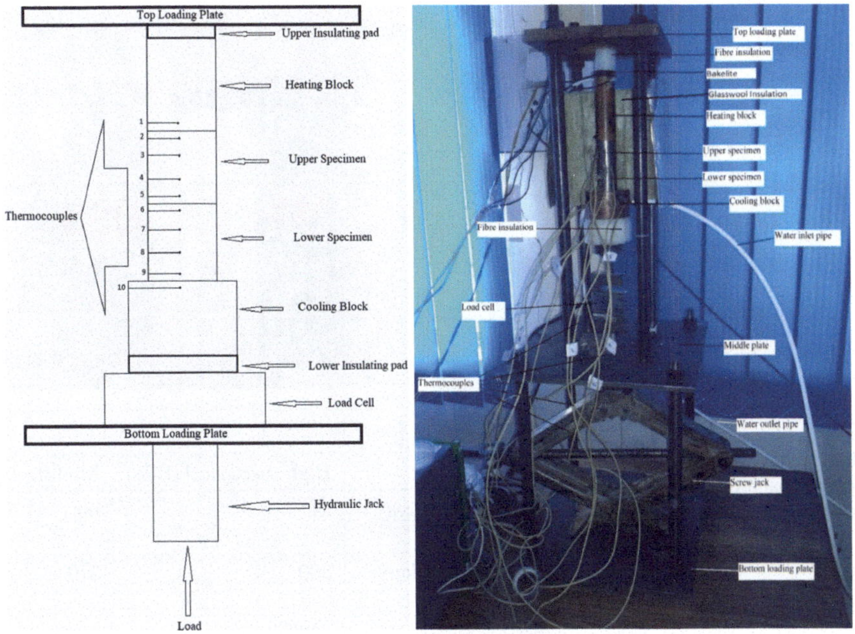

Fig. 17.1 Block diagram and a pictorial view of the experimental set-up

achieved when the temperature difference at each location is 0.1 °C in 30 min. Thereafter, the applied load is increased to the next level, and again the temperature readings are taken in the similar manner up to the steady state. The procedure is repeated in the same way for all set of the experiments. Each specimen's pair has been tested under three loads 15, 25, and 40 kg (contact pressures = 0.3, 0.5, 0.8 MPa), while mean interface temperature ranges between 35 and 40 °C.

17.2.3 Specimen Characteristics

The upper specimen is a rectangular block of dimension 45 × 12 × 8 mm, made of high-speed steel (HSS), which is essentially a cutting tool of lathe machine (Fig. 17.2), while lower specimen is 35 mm height and 10 mm diameter cylinder made of mild steel, a common metal of wide applications (Fig. 17.2). The specimen pairs constitute line contacts joining together. Three different surface finishes have been prepared on mild steel specimens using different grade of emery papers so as to generate the various effective surface finishes of specimen pairs. The thermo-mechanical properties of the materials of the specimens have been presented in Table 17.1.

Fig. 17.2 Samples of high-speed steel and mild steel

Table 17.1 Thermo-mechanical properties of the materials in contact [18]

Material	High-speed steel (HSS)	Mild steel
Thermal conductivity '*k*' (W/mK)	21	60
Modulus of elasticity '*E*' (GPa)	200	210
Viker's hardness '*H*' (MPa)	8500	1400
Poisson's ratio '*ν*'	0.28	0.303

17.3 Data Analysis

Transient temperature data has been recorded in both the specimens in contact at different axial locations up to the steady state. A well-validated steady state methodology has been employed to estimate the thermal contact conductance at the interface of specimens. In this approach, heat flux at the interface and axial temperature data in both the specimens at steady state is utilized, and TCC has been calculated by using the following Eq. (17.1) as: Thermal contact conductance, TCC $= \frac{q}{\Delta T}$; here, q is the average heat flux, $q = \frac{(q_1+q_2)}{2}$
where

q_1 is the heat flux through upper specimen, $q_1 = -k_1\left(\frac{dT}{dx}\right)_1$

and

q_2 is the heat flux through lower specimen, $q_2 = -k_2\left(\frac{dT}{dx}\right)_2$.

Here, k_1 and k_2 are the thermal conductivities, and $\left(\frac{dT}{dx}\right)_1$ and $\left(\frac{dT}{dx}\right)_2$ are the temperature gradients for upper and lower specimens, respectively.

Further, 'ΔT' is the temperature drop at the interface of the specimens, which is calculated by using the linear extrapolation of temperature distribution in both the specimens.

17.3.1 Normalization

To examine the combined effect of different parameters on thermal contact conductance (TCC) and also to compare the experimental results with existing theoretical models, a normalization procedure is commonly used in the literatures [14, 15, 19] which is as follows:

Thermal contact conductance is normalized as $h^* = \frac{h\sigma}{mk}$
where 'σ' is the effective root mean square roughness defined as $\sigma = \sqrt{\sigma_1^2 + \sigma_2^2}$
'm' denotes the effective average asperity slope defined as $m = \sqrt{m_1^2 + m_2^2}$
and k is harmonic mean of thermal conductivities of both specimens.

While the pressure is normalized as

$$P_p = \frac{P}{H};\ \textbf{for plastic models and}\quad P_e = \frac{P\sqrt{2}}{E'm}\ \textbf{for elastic models};$$

where 'P' is the applied pressure evaluated on the basis of nominal contact area, 'H' is the Vicker's hardness of the material of the specimen having lower hardness in the specimen pair, and E' is reduced modulus of elasticity of the specimens' material given by

$$\frac{1}{E'} = \frac{1 - \nu_1^2}{E_1} + \frac{1 - \nu_2^2}{E_2}.$$

RMS roughness 'σ' and average asperity slope 'm' have been calculated by using the correlations as $\sigma \approx 1.25 \cdot R_a$ and $m = 0.1664 \cdot (R_a)^{0.4}$ [20].

17.4 Results and Discussion

The experiments have been carried out on high-speed steel (HSS) and mild steel (MS) kept in contact for three specimens' pairs ($R_a = 0.62, 0.92, 2.02\ \mu m$). Each specimen's pair has been applied under three loads 15, 25, and 40 kg (contact pressures = 0.3, 0.5, 0.8 MPa). The effect of interface temperature has also been studied by varying the applied heat flux. Results of TCC have been presented graphically, and the effect of various parameters on TCC has been discussed in the upcoming sections.

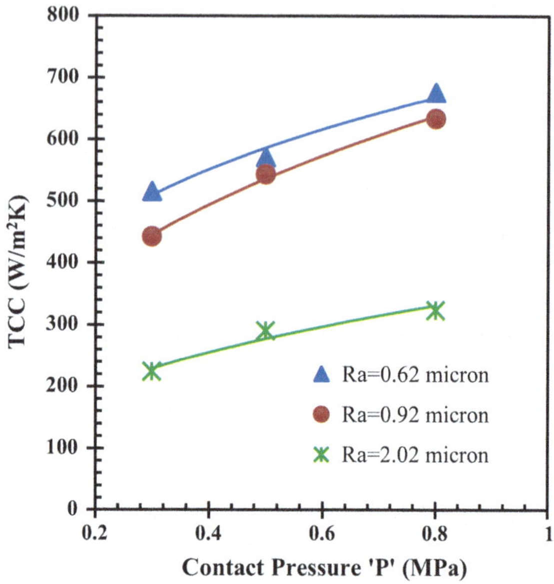

Fig. 17.3 Variation of TCC with pressure

17.4.1 *Effect of Contact Pressure*

Figure 17.3 shows the variation of TCC at the interface of HSS–MS with the contact pressure. From Fig. 17.3, it has been observed that TCC is increased with pressure for all of the test surfaces. Indeed, contact pressure rise results in the deformation of micro-asperities leading to the increase of actual contact area hence increase in TCC. Further, it has been found that TCC values of HSS–MS contact range from 225 to 677 W/m^2 K for the present surfaces under the set pressure range. Moreover, the percentage increase in TCC is found to be 31% at lowest load and 44% at highest load with 2.7 times of contact pressures. The higher increase of TCC is observed at higher load which shows more deformation at higher loads.

17.4.2 *Effect of Surface Finish*

Figure 17.4 presents the variation of TCC with the surface finish. From the figure, it may be observed that the thermal contact conductance increases with the improvement in surface finishes. Essentially, the decrease of surface roughness results in the change of distribution and size of microscopic contact spots. Hence, two surfaces in contact will get more number of micro-asperities in contact leading to the increase in the actual contact area and hence thermal contact conductance. From Fig. 17.4, it is evident that the percentage improvement in TCC is 96–130% with 70% improvement

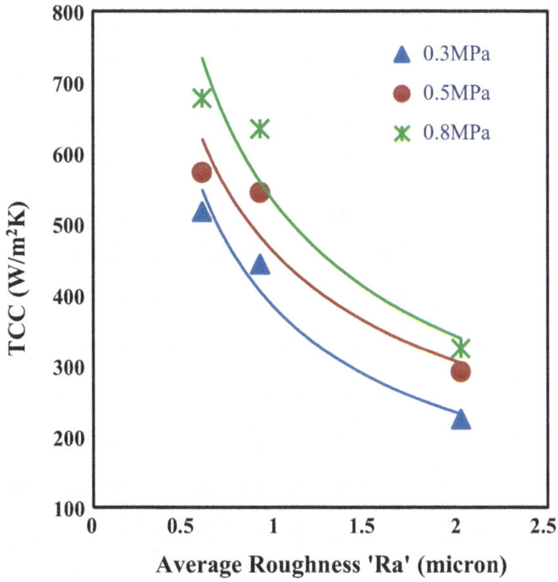

Fig. 17.4 Variation of TCC with roughness

in surface finish. Further, it has been observed that the percentage increase of TCC with the surface finish is higher at lower load which shows less deformation at lower loads.

17.4.3 Effect of Temperature

The TCC has been evaluated varying the mean interface temperatures by changing the heat input to the specimens. However, mean interface temperature range is low (35–40 °C) due to the experimental limitations. Figure 17.5 shows the variation of TCC for HSS–MS contact with the temperature. From the figure, it is evident that the TCC increases with the temperature. It may be attributed to the change in surface deformation which causes the change in actual contact area and hence TCC. However, the effect of temperature on TCC is not very significant for the present temperature range. It is observed that rise in TCC is 15% with approx. 1 °C rise in temperature.

17.4.4 Non-dimensional Correlations

In Fig. 17.6a, results of TCC in normalized form ($h^* = h\sigma/mk$) have been plotted against dimensionless pressures $P_p\left(P_p = P/H\right)$ and compared with Mikic plastic

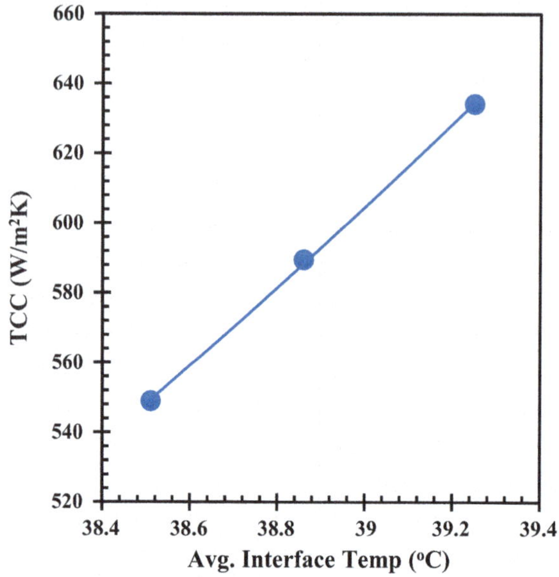

Fig. 17.5 Variation of TCC with temperature ($R_a = 0.92$ μm)

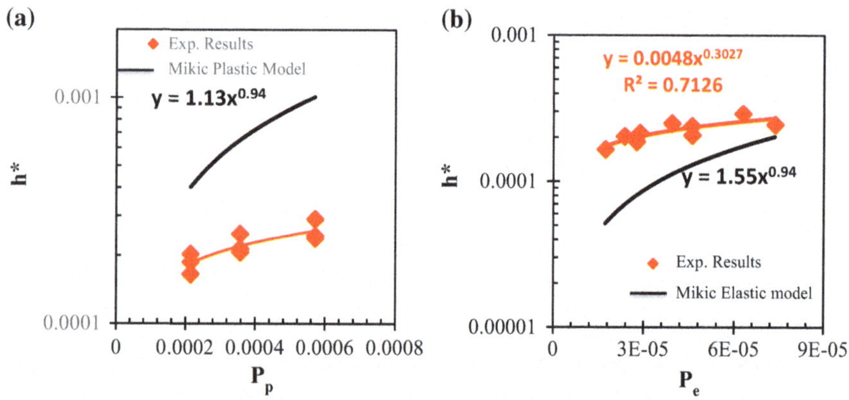

Fig. 17.6 Normalized TCC with dimensionless pressure and comparison with (**a**) Mikic plastic model, (**b**) Mikic elastic model

model [19] $\left[h^* = 1.13\left(P_p\right)^{0.94}\right]$ which is a theoretical model developed on the basis of plastic deformation theories. From Fig. 17.6a, it is evident that the experimental results of TCC do not agree with the plastic-based theoretical model. Indeed, theoretical model is developed for nominally flat metallic contact while the present results are for line contact. In addition, it shows that deformation in the present case is not fully plastic. Further, the normalized results of TCC have been shown in

Fig. 17.6b with dimensionless pressures P_e $\left(P_e = \frac{P\sqrt{2}}{E'm}\right)$ and compared with Mikic elastic model [19] $\left[h^* = 1.55(P_e)^{0.94}\right]$ which is a theoretical model developed on the basis of elastic deformation.

From Fig. 17.6b, it has been observed that experimental results do not show good agreement with the elastic model but closer than plastic models. From the above results, it may be concluded that the deformations of present contact surfaces under loads are neither plastic nor elastic. However, results lie in between the two models, hence deformations might be considered to be elastoplastic for the present case. Eventually, Fig. 17.6 shows the combined effect of all the parameters which are used to normalize TCC and nominal pressure P, i.e. σ, m, k, E and H. Hence, the results of normalized plots may be correlated with the surface characteristics and thermo-mechanical properties of the materials. Therefore, the results of TCC for HSS–MS contacts are represented in the form of correlation as $h^* = 0.004(P_e)^{0.302}$ with 0.712 R^2 value.

17.5 Conclusions and Scopes of Future Work

The study of thermal contact conductance (TCC) has been performed on high-speed steel and mild steel (HSS–MS) contact, which is essentially a contact found in metal cutting operations. The experimental investigations based on axial heat flow method have been carried out on HSS–MS pairs with different surface finishes for each specimen pairs under varying contact pressures and temperatures to study the effect of various parameters on thermal contact conductance. The following conclusions have been drawn from the experimental investigations:

- Overall, the TCC has been observed to increase with the rise of contact pressure.
- The percentage increase in TCC is found to be 31–44% with 2.7 times of pressure.
- TCC has been found to improve with surface finish. The percentage improvement in TCC is 96–130% with 70% improvement in surface finish.
- The effect of temperature on TCC is not significant. The rise in TCC is found to be 15% with approx. 1 °C rise in temperature.
- The experimental results do not show good agreement with the plastic and elastic models. Hence, the deformation of contact surfaces might be considered to be elastoplastic for the present case.
- The results of TCC for HSS–MS contacts are represented in the form of a suitable correlation as $h^* = 0.004(P_e)^{0.302}$ with 0.712 R^2 value. The present correlation might be useful for the thermal analysis of tool–sample contact for the design of metal cutting systems.

17.5.1 Scope of Future Works

The future work might be directed towards the study of TCC for different types of tool–sample pairs such as stainless steel tool and carbide tool and different sample metals to be cut. Further, study on TCC should be done for higher interface temperature range. Moreover, contact pressure and interface temperature should be related to cutting parameters like cutting speed, feed rate and the effect of cutting parameters be studied on TCC.

References

1. Abukhshim, N.A., Mativenga, P.T., Sheikh, M.A.: Investigation of heat partition on high speed turning of high strength alloy steel. Int. J. Mach. Tools Manuf **45**, 1687–1695 (2005)
2. Madhusudana, C.V.: Thermal contact conductance. Springer, New York (1996)
3. Jeevanashankara, Madhusudana C.V., Kulkarni, M.B.: Thermal contact conductance of metallic contacts at low loads. Appl. Energy **35**, 151–164 (1990)
4. Mcwaid, T., Marschall, E.: Thermal contact resistance across pressed metal contacts in a vacuum environment. Int. J. Heat Mass Transf. **35**, 2911–2920 (1992)
5. Milanez, F.H., Culham, J.R., Yovanovich, M.M.: Experimental study on thermal contact conductance of bead blasted SS 304 at light loads. In: 40th AIAA Aerospace Sciences Meeting, pp. 14–17. Reno, NV (2002)
6. Wahid, S.M.S., Madhusudana, C.V., Leonardi, E.: Solid spot conductance at low contact pressure. Exp. Therm. Fluid Sci. **28**, 489–494 (2004)
7. Sunil Kumar, S., Ramamurthi, K.: Thermal contact conductance of pressed contacts at low temperatures. Cryogenics **44**, 727–734 (2004)
8. Singhal, V., Litke, P.J., Black, A.F.: An experimentally validated thermo-mechanical model for the prediction of thermal contact conductance. Int. J. Heat Mass Transf. **48**, 5446–5459 (2005)
9. Yuncu, H.: Thermal contact conductance of nominaly flat surfaces. Heat Mass Transf. **43**, 1–5 (2006)
10. Xing, L., Zhang, L.W., Zhang, X.Z., Yue, C.X.: Experimental investigation of contact heat transfer at high temperature based on steady-state heat flux method. Exp. Heat Transf. **23**, 107–116 (2010)
11. Misra, P., Nagaraju, J.: An experimental study to show the effect of thermal stress on thermal contact conductance at sub-mega Pascal contact pressures. J Heat Transf **132**, 094501-1-4 (2010)
12. Liu, J., Feng, H., Luo, X., Hu, R., Liu, S.: A simple setup to test thermal contact resistance between interfaces of two contacted solid materials. In: 11th International Conference on Electronic Packaging Technology and High Density Packaging, Xi'an, China, pp. 16–20 (2010)
13. Wahid, S.M.S., Madhusudana, C.V.: Gap conductance in contact heat transfer. Int. J. Heat Mass Transf. **43**, 4483–4487 (2010)
14. Tariq, A., Asif, M.: Experimentally investigation of thermal contact conductance for nominally flat metallic contact. Heat Mass Transf. **52**, 291–307 (2015)
15. Asif, M., Tariq, A.: Correlations of thermal contact conductance for nominally flat metallic contact in vacuum. Exp Heat Transf **29**, 1–29 (2016)
16. Jam, J.E., Fard, V.N.: A novel method to determine tool-chip thermal contact conductance in machining. Int. J. Eng. Sci. Tech. **3**, 8491–8501 (2011)
17. Norouzifard, V., Hamedi, M.: Experimental determination of the tool-chip thermal contact conductance in machining process. Int. J. Mach. Tools Manuf. **84**, 45–57 (2014)

18. CRC Hand Book: The CRC Hand Book of Mechanical Engineering, 2nd edn. CRC Press, Taylor & Francis Group, New York (2004)
19. Mikic, B.B.: Thermal contact conductance; theoretical considerations. Int. J. Heat Mass Transf. **17**, 205–214 (1974)
20. Antonetti, V.W., Whittle, T.D., Simons, R.E.: An approximate thermal contact conductance correlation. J. Electr. Packag. **115**, 131–134 (1993)

Chapter 18
An Investigative Study on Motorbike Through CFD Solar

Meeta Sharma and Danish Hussain

Abstract The aerodynamics of the vehicle changes with the structure of the vehicle which must be painstakingly considered while its design. In any case, the streamlined improvement experiences from the timeliest stages, in the design upgrade program, and remains up to full creation level, until the moment that the detail is progressed. The wind tunnel estimations keep on being the most well-known and broadly utilized methodology. The iterative simulation techniques through computational fluid dynamics (CFD) with high-end computers are used for aiding and reducing the trial tests. Proposed work is focusing on the investigative analysis of a motorbike for different models. Through the CFD analysis, various motorcycle models and radiator profiles are examined for the different flow conditions. Further, the effect of specific fuel consumption for the different models is also examined in which the conventional motorbike is having 0.642% change in fuel efficiency than radiator motorbike. The comparison of the aerodynamic drag acting on conventional motorcycle, radiator motorcycle, and superbike using computational fluid dynamics (CFD) estimated that the superbike model having lowest amount of drag coefficient of 11.89% by basic conventional model. The radiator model at 60° has lowest coefficient of drag around 63.65%, i.e., lower than the basic model of the radiator. Therefore, through this CFD analysis the airflow and advancements in the design of an efficient motorbike in terms of better riding ability, control, comfort, and low fuel consumption are observed.

Keywords CFD model · Conventional bike · Superbike · Radiator bike

18.1 Introduction

The computational fluid dynamics (CFD) simulation environments have become an essential method for the automotive industry's aerodynamic designs. The aerodynamic design is governed by various crucial factors—such as optimization of the engine cooling system and reduction in the drag force. All these design parameters are limited by continuously changing technological limitations. The most important

M. Sharma (✉) · D. Hussain
Amity University Uttar Pradesh, Sector-125, Noida, India

S. Yadav et al. (eds.), *Proceedings of International Conference in Mechanical and Energy Technology*, Smart Innovation, Systems and Technologies 174,
https://doi.org/10.1007/978-981-15-2647-3_18

concern with the air-cooled engine is that the cooling depends on vehicle motion and surrounding temperature.

According to the requirements, designers focus upon aerodynamically good design, comfort, and stability at all speed. The aerodynamic design of the vehicle affects the efficiency, maximum speed, and fuel consumption of the motorcycle which is caused by the aerodynamic resistance. The aerodynamic resistance or drag acting on the vehicle is around 90% of the total resistance at racing speeds [1–3].

The CFD analysis assumes a noteworthy job in vehicle structure [4]. For instance, to analyze the 3D flow traits inside motorbikes it is used as an important tool, which is respectably difficult and genuinely exorbitant to get tentatively [5, 6]. In addition to it, CFD has been used to improve specific structured parts, to increase energy efficiency, and to decide the effect of neighborhood geometrical modifications of the motorbike on the basis of streamlined flow pattern [7, 8].

The air-cooled engines of the motorbikes' radiator resolve the overheating-related issues of the engine. At the same time, it also increases the overall drag acting on the vehicle which affects the overall fuel consumption. Therefore, radiator modeling is important parameter to be analyzed and evaluated.

This work investigates the performance of the turbulent airflow conditions and radiator design of the motorbike. The effects of aerodynamic forces such as lift and drag coefficients are examined for the three cases—conventional bike, radiator motorbike, and superbike using computational fluid dynamics (CFD). The effect of drag on the radiator motorbike model is observed by changing the radiator curvature angles. The optimal design of the motorcycle is studied by methods for a product that understands by implicating the Navier–Stokes conditions for the continuum by repeating the fluid dynamics at the sub-atomic dimension.

18.2 Motorbike Modeling

The present research simulation work was carried out on a sportive Harley Davidson XLH 53C BI motorbike with rider model as shown in Fig. 18.1. The motorbike geometry (Fig. 18.1) includes the nomenclature of the assumed in the existing work. The coordinate system defined such as, where the direction of travel is negative z-direction, x-axis denotes the lateral direction and y-axis denotes the vertical direction. The baseline dimensions for the motorbikes length, height, and width are 2.275, 1.165, and 0.825 m, respectively. This motorbike model is tested for the velocities range as mentioned in Table 18.1.

Fig. 18.1 View of the conventional motorbike

Table 18.1 Constraints for measuring aerodynamic parameters

Parameter	Input sections	User input
Model properties	Viscous and turbulent	Modified $k - \varepsilon$ turbulence model
Materials	Fluid	Air
Boundary conditions	Velocity inlet	Magnitude = 30 m/s
	Turbulent intensity = 2%, at a length scale of = 0.08 m	
Reference values	Area	Lifting area for lift Frontal area for drag
	Density	1.225 kg/m^3
	Length	0.08 m
	Temperature	293.15 K
	Velocity	10–30 m/s
Monitors	Forces	Monitors
Calculation activities	Iteration count	250 iterations

18.2.1 Modification in Motorbike

With the help of SOLIDWORKS Software 2018, open-source CAD software, the shape of the motorbike geometry was altered as shown below.

Case 1: In this model, the porous radiator is attached to the conventional motorbike and tested under similar circumstances as of the motorbike (Fig. 18.2).

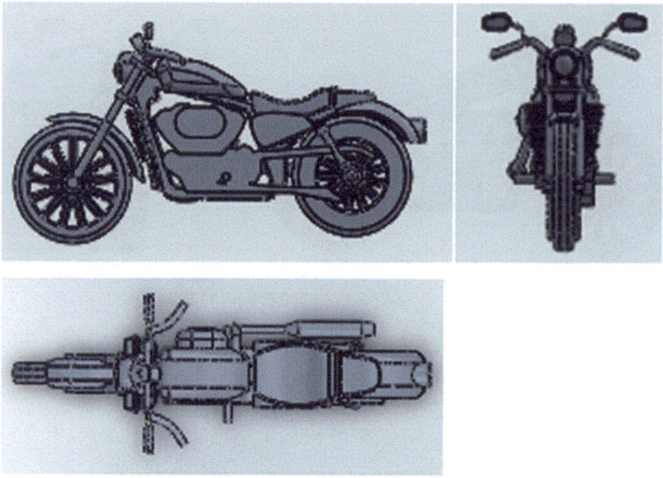

Fig. 18.2 View of radiator motorbike

Case 2: In another model, it is converted to a superbike model from the conventional motorbike model by covering from both sides. The superbike model is tested under similar conditions (as described in Table 18.1) (Fig. 18.3).

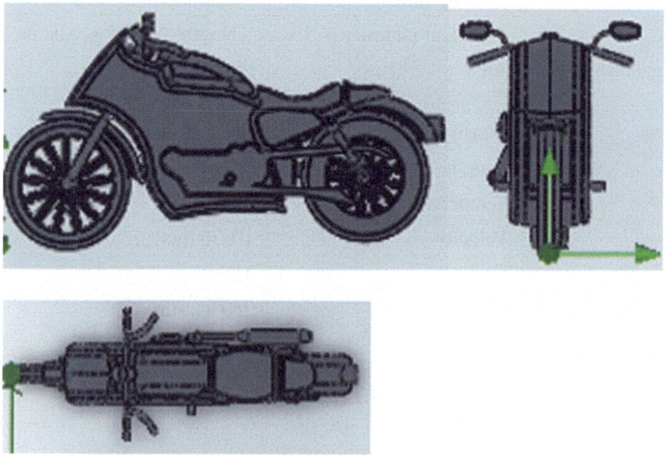

Fig. 18.3 View of the superbike design model

18.2.2 Numerical Technique Used

The open-source finite volume CFD package SOLIDWORKS (Version 2018) is chosen to resolve the incompressible flow equation of related iteration. The flow around a motorbike exposed to front wind is controlled by highly 3D turbulent flow configurations. It uses a revised $k - \varepsilon$ dual equation turbulence model to simulate correctly a comprehensive range of turbulence states in association with immersed boundary Cartesian meshing methods. This allows for precise flow field resolution with low cell mesh densities. For more information on immersed boundary meshes observed by Ooi et al [9] and their work at Stanford University with general motors on CFD simulations techniques are more useful in upfront engineering design studies.

As per the fluid being studied and the flow conditions being observed, any fluid flow situation can be viewed. According to fluid examined under the given condition, the flow situation is summarized by Schlichting (1959) as given below:

- A laminar flow, i.e., a flow without any disturbances.
- A turbulent flow, i.e., a flow regime considered by arbitrarily 3D vortices and intensive mixing.
- A transitional flow, i.e., between laminar and turbulent flow regions.

18.2.3 Boundary Conditions and Used Computational Domain

A comprehensive computational area is used for investigative analysis and is given as follows:

- The computational domain has an inlet and an outlet.
- At the inlet, the velocity is 30 m/s (constant) and the flow is in the opposite direction of travel, Z.
- The whole system is under 1 bar.
- Z-dimension from start to end was set as 9 m. The distance of the exit plane from the motorbike surface is chosen so that it will not affect the flow around the motorbike.
- And there is the no-slip condition which is used on each face of the computational domain (Fig. 18.4).

18.2.4 Radiator Modeling

Radiators are no longer a rare sight on motorcycles, earlier it was only four wheelers we used to witness "Radiators" but with the evolution of motorcycle industry and

Fig. 18.4 Computational domain of the conventional motorbike

the production of big capacity engines, the radiators have become a common sight. And the radiator greatly contributes to overall drag which is 12.7% (radiator shape and cooling drag combined) by Van Dijck [10] (Fig. 18.5).

Using detailed geometry of a radiator, a porous radiator has been designed by means of SOLIDWORKS software. As we can see in Fig. 18.3, there are two portions in it: Above one is water radiator and the below one is the oil radiator. The radiator geometry was put under the simulation iterations for different inlet speeds, and the coefficient of drag is measured.

The radiator geometry was created with exact fin dimensions, fin density, and with the required fin shape. As we know by adding radiator with a motorbike, it will

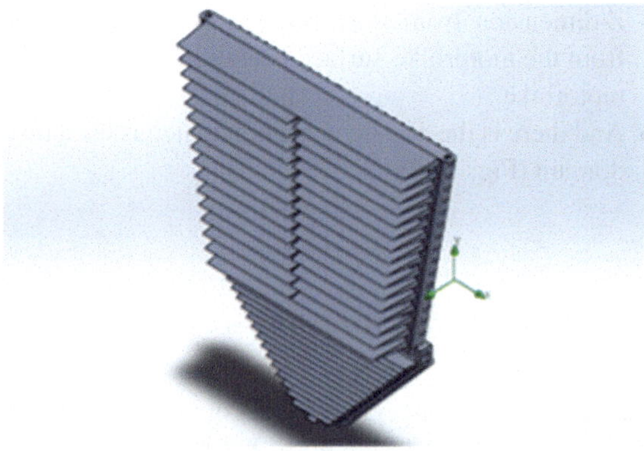

Fig. 18.5 Original oil and water radiator

increase the drag force acting on it. So, the radiators are redesigned by changing the curvature angle to decrease the drag force as shown below (Table 18.2).

Table 18.2 List of the radiator at different curvature angle

1. Radiator at 20° curvature angle	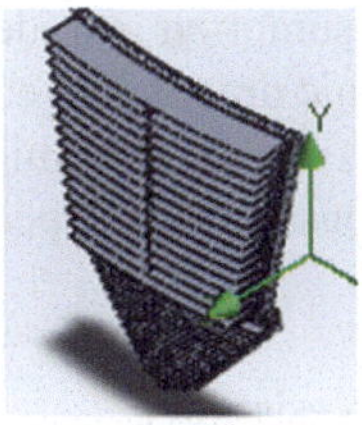	Frontal area: 0.10316 m^2
2. Radiator at 40° curvature angle	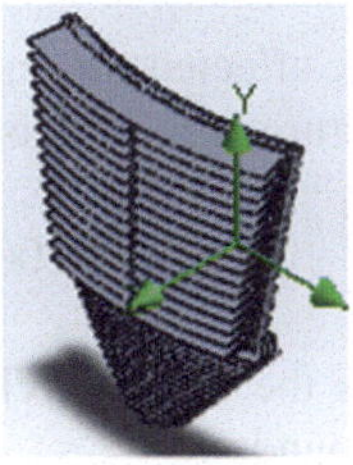	Frontal area: 0.10127 m^2
3. Radiator at 60° curvature angle	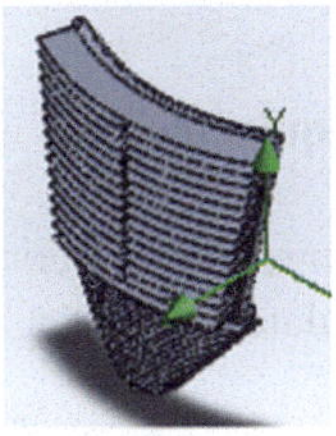	Frontal area: 0.0995135 m^2
4. Radiator at 80° curvature angle	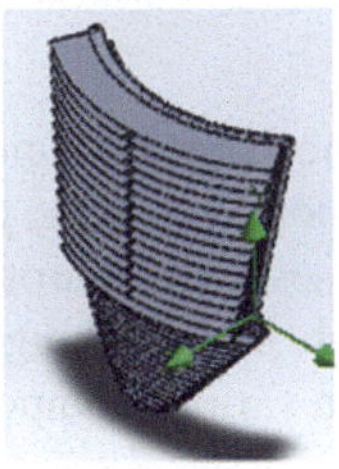	Frontal area: 0.09706 m^2
5. Radiator at 100° curvature angle	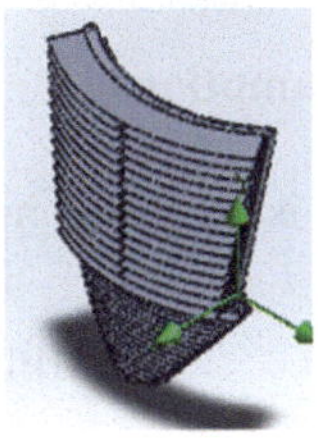	Frontal area: 0.09441 m^2

18.2.5 Fuel Consumption Estimation Model

In an automotive development process, the reduction in fuel consumption is a primary parameter for the manufacturer. An ideal vehicle body design requires, decrease in drag force, fuel consumption and ideal driving performance. The important aspect for lower aerodynamic drag is fuel economy. As long as concerns for fuel economy rise and fuel costs go up, aerodynamic drag will have to develop. The several factors, on which the fuel consumption depends, are mentioned as follows:

18.2.5.1 Rolling Resistance

Rolling resistance is affected through numerous parameters such as air pressure, tread depth, temperature, and side force.

$$F_R = (0.0041 + 0.000041V)\mathrm{W} \tag{18.1}$$

where "V" indicates the relative vehicle speed (in miles/hour), and "W" is gross vehicle weight (in pounds).

18.2.5.2 Power Requirement

The engine total power required is corrected for the power train mechanical efficiency (η_m) and can be calculated as:

$$P = \frac{(F_D + F_R + F_G + F_I) \cdot V}{\eta_m} + P_E \tag{18.2}$$

Even though the mechanical efficiency of the power train is determined by important system of the vehicle such as the gear/transmission system and the wheel bearing condition, it can be assessed to be around 83% for motorcycle under noble conditions. For calculation purpose, the climbing resistance and inertial resistance are neglected. The power loss due to various components in use is assumed as 900 W.

18.2.5.3 Fuel Consumption

The fuel consumption is estimated in kilometer per liter as:

$$\mathrm{KMPL} = \frac{V \cdot \eta_t \cdot E_F}{P \cdot 10^6} \tag{18.3}$$

where η_t is maximum thermal efficiency engine specified, which can go up to 40%, and E_F is the energy content in the fuel (45.8×10^6).

18.3 Results and Discussions

18.3.1 Streamline Flow for Motorbikes

Figure 18.6 describes the airflow variations for the three cases: conventional bike, radiator bike, and superbike. These variations show the streamline flow around the motorbike for the selected models in which air velocity variations can be observed around the motorbike. These variations significantly affect aerodynamic forces and performance of the motorbike.

18.3.2 Effect of Drag Force with Vehicle's Velocity

The effect of drag force on velocity is explained in Fig. 18.7. This graphical result shows that with the increase in speed, there is a linear variation in coefficient of drag. The conventional bike is having lowest drag force because of its shape, and by adding radiator into it, the shape becomes complex because air has to pass through the fins and vents due to which there is a pressure variation in front and back of the radiator. The superbike frontal area is higher than other bikes which increases the drag force.

18.3.3 Effect of Lift Force with the Vehicle's Velocity

Figure 18.8 shows that the lift force is highest in radiator bike in comparison with conventional bike and superbike. It is observed lowest in conventional bike because lift area of the bike is low which creates a low-pressure difference among the top and lower of the motorcycle.

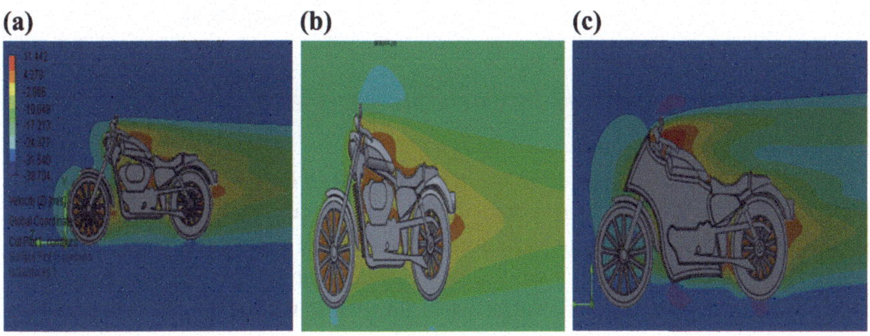

Fig. 18.6 **a** Velocity variation over conventional motorbike, **b** Velocity variation over radiator motorbike, **c** Velocity variation over superbike

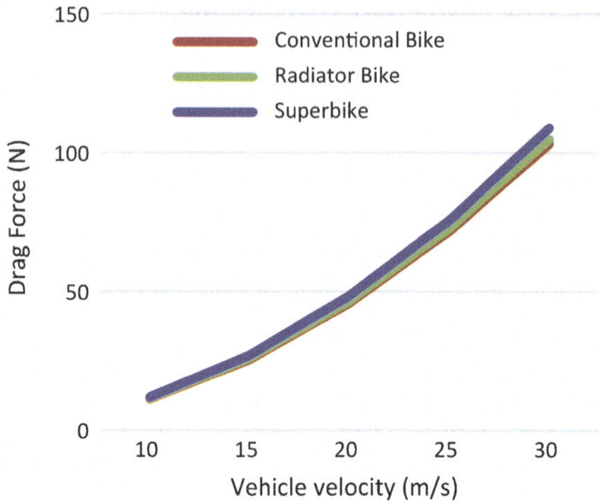

Fig. 18.7 Variation of drag force versus velocity

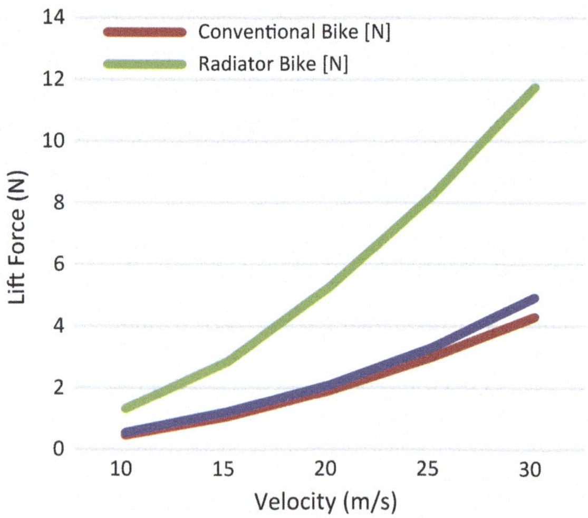

Fig. 18.8 Variation of lift force versus velocity

18.3.4 Effect of Change in Curvature Angle of the Radiator

Figure 18.9 describes the variation of the coefficient of drag with radiator curvature angle. It is observed from the graphical result that by changing the curvature angle of the radiator, there is significant variation in the value of the coefficient of drag. The radiator at 0° shows highest value of drag coefficient and the radiator at 60° lowest value. Because of the pressure variation in the front and backside of the radiator, there is different value for different angled radiator.

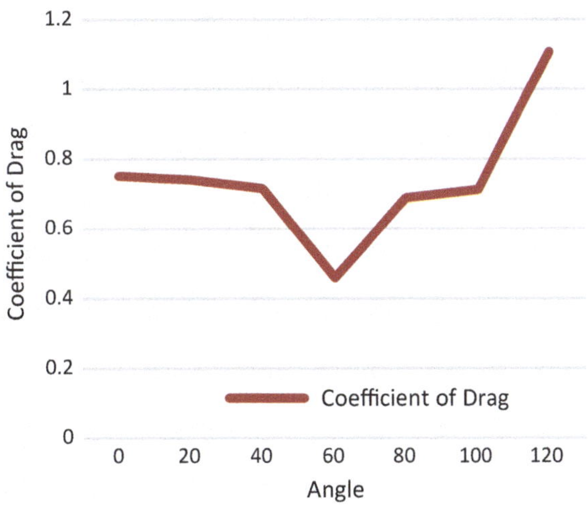

Fig. 18.9 Variation of the coefficient of drag versus angle

18.3.5 *Fuel Consumption Estimation*

There are several factors on which the fuel consumption depends on which graphs have been plotted which are as follows:

18.3.5.1 Rolling Resistance

Figure 18.10 shows the rolling resistance along with velocity variation for the three chosen cases. It is revealed from the results that the rolling resistance is directly

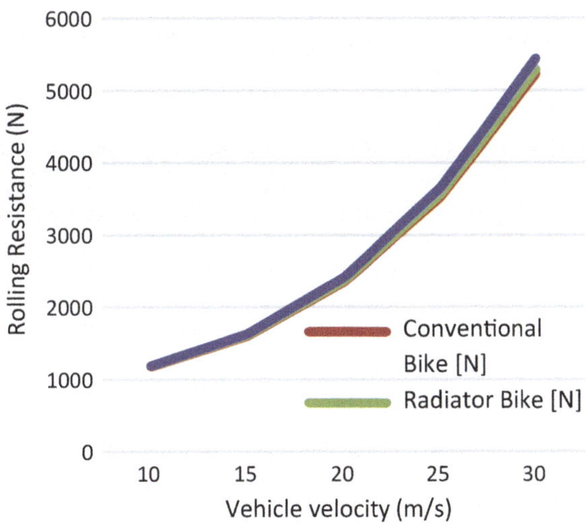

Fig. 18.10 Variation of rolling resistance versus velocity

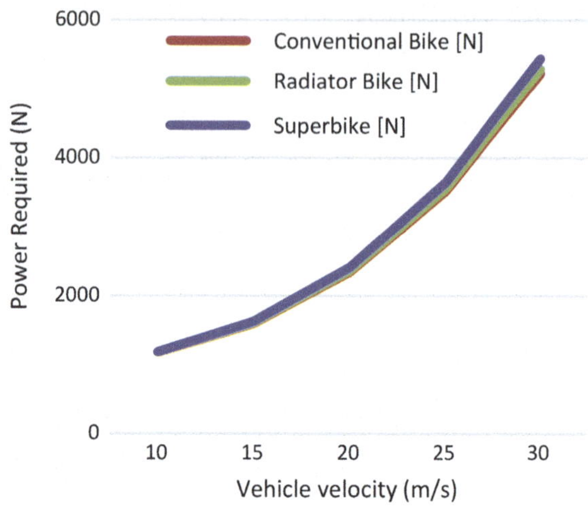

Fig. 18.11 Variation of power required versus velocity

proportional to the vehicle velocity and vehicle gross weight. The minor difference is observed between the three cases due to the small difference in the motorbikes gross weight.

18.3.5.2 Power Requirement

Based on the different assumptions and concerns, the power needed to overcome aerodynamic and rolling resistance is given in Fig. 18.11. It is observed that the power requirement is linearly increasing for higher velocities. The superbike model needs highest power requirement rest than other models of the bike.

18.3.5.3 Fuel Consumption

Figure 18.12 reveals the variation in fuel consumption with the vehicle's velocity. The above graphical representation shows that the fuel economy is highest in the conventional bike, in the range of 12–17 m/s speed. The mileage of the bike decreases with higher speed of the vehicle. The percentage variations in fuel economy on radiator motorbike and superbike with respect to conventional bike are 1.19% and 3.90%, respectively.

18.4 Conclusions

Following conclusions are made by investigating motorbike with the help of SOLIDWORKS CFD solver.

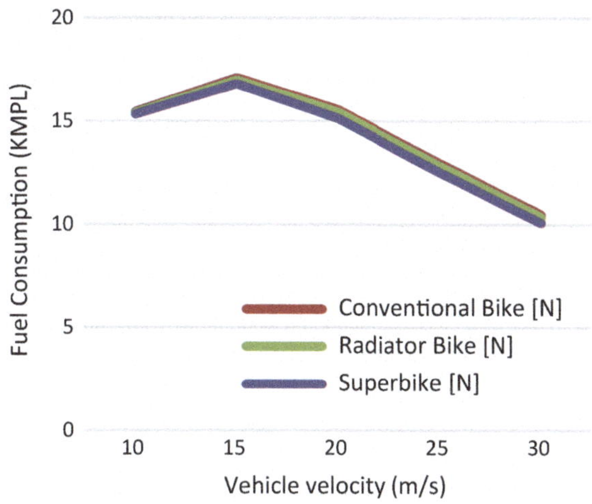

Fig. 18.12 Variation of fuel consumption versus velocity

- The drag coefficient of a superbike model is higher than other models because more streamlined shape is provided to run under high-speed conditions.
- The application of radiator at 60 is having the lowest coefficient of drag than other models of radiator which is done by using the two-equation $k - \varepsilon$ turbulence model, and this represents an overall drag reduction of 63.65% just by changing the curvature angle of the radiator.
- The fuel efficiency is highest in the conventional motorcycle than other models and good to drive in the range of 12–17 m/s speed.
- The use of CFD as a faster and less expensive tool to carry out analysis on aerodynamic objects before their production is very vital.
- Modified $k - \varepsilon$ turbulence model can be relied upon when analyzing aerodynamic shapes under subsonic speeds.
- Shape change optimization studies can be carried out effectively with the use of SOLIDWORKS finite volume CFD solver, thus improving the drag coefficient in the radiator.

References

1. Grappe, G., Candau, R., Belli, A., Rouillon, J.D.: Aerodynamic drag in field cycling with special reference to the Obree's position. Ergonomics **40**(12), 1299–1311 (1997)
2. Kyle, C.R., Burke, E.R.: Improving the racing bicycle. Mech. Eng. **106**(9), 34–45 (1984)
3. Lukes, R.A., Chin, S.B., Haake, S.J.: The understanding and development of cycling aerodynamics. Sports Eng. **8**, 59–74 (2005)
4. Angeletti, M., Sclafani, L., Bella, G., Ubertini, S.: The role of CFD on the aerodynamic investigation of motorcycles. In: SAEtransactions, pp. 1103–1111 (2003)
5. Chu, L.-M., Chang, M.-H., Hsu, H.-C., Chien, W.-T., Liu, C.-H.: Simulation and experimental measurement of flow field within four-stroke motorcycle engines. J. Chin. Soc. Mech. Eng. **29**(2), 149–158 (2008)

6. Gentilli, R., Zanforlin, S., Frigo, S.: Numerical and experimental analysis on a small GDI, stratified charge, motorcycle engine. Paper presented at the Proceedings of the 8th Biennial Conference on Engineering Systems Design and Analysis, Turin, Italy (2006)
7. Takahashi, Y., Kurakawa, Y., Sugita, H., Ishima, T., Obokata, T.: CFD analysis of airflow around the rider of a motorcycle for rider comfort improvement (No. 2009-01-1155). SAE Technical Paper (2009)
8. Watanabe, T., Okubo, T., Iwasa, M., Aoki, H.: Establishment of an aerodynamic simulation system for motorcycle and its application. JSAE Rev. **24**(2), 231–233 (2003). https://doi.org/10.1016/s0389-4304(03)00003-1
9. Ooi, A., Iaccarino, G., Durbin, P.A., Behnia, M.: Reynolds averaged simulation of flow and heat transfer in ribbed ducts. Int J Heat Fluid Flow **23**(6), 750–757 (2002).
10. Van Dijck, T.: Computational evaluation of aerodynamic forces on a racing motorcycle during high-speed cornering (No. 2015-01-0097). SAE Technical Paper (2015)

Chapter 19
5S Housekeeping Lean Technique by Using System Dynamics Method

Richa Sharma, Shubham Sharma and Megha Sharma

Abstract The 5S housekeeping lean technique is considered to improve the plant performance and optimize the overall productivity. The current study examines the active significance of the 1S (sorting) in the industry using system dynamics process. In order to measure the performance output effect of 1S, a system dynamic approach is used for a manufacturing sector presented in the form of the case study. The aim of the system dynamic process was to analyze the outcome of plant performance prior to the implementation of the 5S lean philosophy. The manufacturing sector which is new to introduce the implementation of lean tools can easily replica ethereal-life results by using SD (system dynamics) approach. The current study demonstrates a methodology based on a simulation that ease the standardized effective improvement, implementation and sustaining of 5S lean initiative to improve quality, streamline production, and optimize efficiencies.

Keywords 5S · System dynamics · Housekeeping · Japanese tool · Lean manufacturing · Quality management

19.1 Introduction

The current market scenario demands agility when it comes to production and various techniques, and innovative processes are introduced which have been proven to increase the efficiency of the production systems. This gives room for instability in the implementation of the adequate tool as continuous improvement is difficult to achieve with respect to the architecture of the system and operating system parameters, and hence, the competition along with profit gain operation is a major concern to most of the industries on a global level [1].

The practices in association with the lean manufacturing are required to holistically implement in an organization to embrace the various benefits which were

R. Sharma (✉) · M. Sharma
Department of Mechanical Engineering, Amity University, Noida, Uttar Pradesh 201313, India

S. Sharma
Amity University, Noida, Uttar Pradesh, India

S. Yadav et al. (eds.), *Proceedings of International Conference in Mechanical and Energy Technology*, Smart Innovation, Systems and Technologies 174,
https://doi.org/10.1007/978-981-15-2647-3_19

intended [2]. The audit is done by using 5S lean tool for the organizational performance linked with the lean practices being implemented. The importance of top management and leadership in the introduction and implementation of lean manufacturing plays a significant role [3]. Four organizational factors (human capital, performance, time orientation, and risk appetite) and two environmental factors (risk and stability) are considered to have an effect on leader's choice of transactional versus transformational styles [4].

5S improves the quality of the production by reducing processes with non-value time consumption, and thus increasing the overall productivity of the system [5]. It has been considered as an overall successful tool for enhancing efficiency with or without integrating it with other available lean improvement tools to drastically minimize the transition time [6, 7]. 5S makes the factory a cleaner place, the safety at workplace and the product quality is increased, the problems are easy to detect and prevent, waste and costs are reduced, the product or service fulfills the customer needs in the most efficient manner [8]. Implementing the 5S methodology requires organization and safety of the working process, properly marking and labeling of the working place, and audits to establish the work in progress and to maintain the improved activities [9]. 5S is an effective tool for improvement of organizational performance, regardless of organization type, size, its production, or its service. Consequently, 5S techniques would strongly support the objectives of organization to achieve continuous improvement and higher performance [10].

However, most studies dealing with organizational agility have focused on the characteristics of agile organizations rather than on the underpinning organizational capabilities [11, 12].

19.2 System Dynamics Methodology

System dynamics (SD) is an approach which is used to find out or understands the reasons and effect connection present in the system. SD is a concoction of abstraction tools that help us to facilitate the construction of the complex systems [13]. SD utilizes the simulation to check the dynamic system response [14].

SD is not a new tool to solve the complex problems and helps in decision making. It has used since the 1950s. SD can be used to increase productivity and examine a system of any industry [15]. The research relating to the use of SD within manufacturing systems is noticeably high and an increasing number of researches are related to the lean philosophy. Introduction of system dynamics is preferred to control the qualitative cost and expenses [16]. System dynamics is used as a lean assessment tool to assess and improve lean performance for a print packaging manufacturing system [17]. The author [18] introduced a system dynamics (SD) modeling tool due to its dynamic characteristic and its advantages, such as the possibility of integrating qualitative factors and defining learning loops which can help management to track the effect of changing each incorporated value-added (VA) and non-value-added (NVA) activity on each other's cost as well as on the quality costs in real time.

System dynamics is a virtual modeling methodology which consists of a diagram interlinking stock and flow (SFD). Stocks are accumulations that characterize the system (expressed in terms of the number of items or quantities) [19], inventory, and works-in-process being the appropriate examples of the same (WIP). In the current article, we use the system dynamic approach to examine the effect of sorting activities in a case study of manufacturing industry.

19.3 Design of SD Model

In the present study, the SD model is designed with the help of AnyLogic software. The dynamics of the system related to the variables constitute by the flow and stock variables of the SD model [4]. In the medium- and small-scale industries, the major problems are due to the variations in the demand and system internal problems, for example, failure of any component and unreliable scheduling. In this article, the textile sector-related manufacturing system is presented.

It is very easy to understand the dynamics used in this model. Lead time is calculated in the comparison of targeted lead time. If the delay is more, then there is a need for more efforts to reduce or compensate delay in the lead time with a target level. Due to the delay, the time spent for the sorting activities are reduced because the employees are focused only to compensate the delay, not on the sorting activities. The items are left without sorting and there is an increase in the stock of things unsorted. Due to this, there is the chance of disturbance in finding the items for reuse and so the time for setup or repair also increases. This results in an increased manufacturing lead time. This is the initial problem for manufacturing system. Hence, the present case study uses the SD to connect with dynamic changes of the sorting issue.

19.4 Parametric Equations of System Dynamics Model

To understand the relationship between the cause and effect variables, parametric equations are used. These equations are used to simulate the SD model. The equations used for the SD model are given in Table 19.1. The value for the order inlet rate is like that of the normal production quantity. Hence, the situation becomes critical and crucial as no availability is left. Remaining equations are based on a very simple concept. For example, if the time required in finding tools is more, setup time is also increased. This results in nonlinearity introduced for example “Gap in Lead Time” and “Lead Time Pressure.”

This analysis represents the combination of lead time pressure with supplementary work hours to compensate the lead time pressure. Considering Fig. 19.1, there is minimal pressure on the system till the gap is less than 3 days. Upon increment of the gap to 9 days, a 40% increase in the pressure is noticed, i.e., more work hours are

Table 19.1 List of SD model critical parameters

Elements of SD model	Formula
Order entry rate	Value = 20 orders
Order backlog	d(order backlog)/di = order entry rate − throughput initial orders = 27 orders
Throughput	Production capacity/manufacturing cycle time
Lead time	Order backlog/throughput
Target lead time	Constant, value = 2 days
Gap in lead time	Lead time − target lead time
Lead time pressure	Gap pressure (gap in lead time)
Percentages corners cut	Lead time pressure
Effect of cut on capacity	Percentage corners cut
Normal production capacity	Value = 20 orders
Production capacity	Effect of cut on capacity * normal production capacity
Effect of cut on sorting	Percentages corners cut
Normal sorting capacity	Value = 20 orders
Normal sort time	Value = 1 day
Sort time	Normal sort time * effect of cut on sorting
Sorting backlog	Rate of items for sorting − rate of sorting initial value = 27
Rate of sorting	Sorting capacity/normal sort time
Time used in finding tools	(Sorting backlog/ rate of sorting)/rate of items for sorting
Time to setup	Normal time to setup + time used in finding
Normal time to setup	Value = 0.05
Time to repair	Time used in finding + normal time to repair
Normal time to repair	Value = 0.8
Breakdown frequency	Value = 0.4
Manufacturing cycle time	Normal manufacturing cycle time + time to repair + time to setup

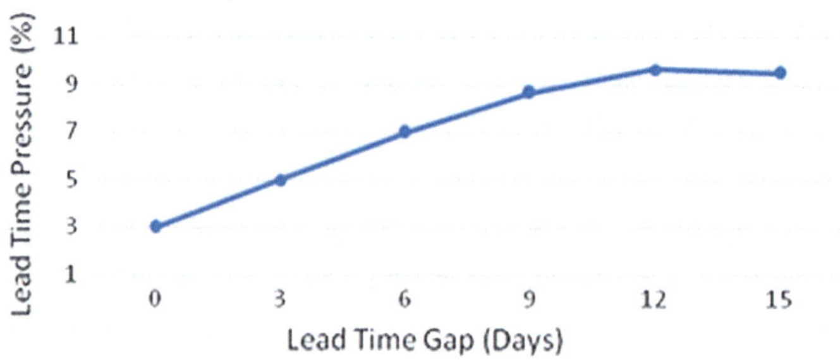

Fig. 19.1 Nonlinear graphical relationship between lead time pressure and lead time gap

required, to compensate the lead time gap, is approximately 40%. If the gap exceeds 15 days, the pressure increases to up to 60% but maintains this level. It is assumed that the system pressure will not exceed 60% as it is the permissible limit up to which the system can stretch itself.

19.5 Exploratory Validation of the SD Model

The simulation of SD model is carried out for 20 days, and they are displayed in Fig. 19.3. The 20 days duration was selected keeping in mind the enough time required for the analysis of the effect of changes.

It is clear from Fig. 19.2, over time, the daily order completion increased due to a considerable reduction in the time spent in locating tools for equipment setup or repair. A steep drop in throughput is noticed from 18 to 15 orders completed per day after 10 days and saturates for the next 10 days. A similar trend can be seen in the real system during the season of high demand where throughput decreases rapidly and maintains between 13 and 16 orders completed per day which is the minimum system requirement.

The simulation model specifies the uplifting trend over the 20 days period to about 0.4 days. This realizes the actual trend as the orders are increased and the items are misplaced on the shop floor. In the actual system, on one day, a similar value was recorded. Every shop floor individual, 30 to be precise, spends 4–6 min for finding the tools, and the summation of this is roughly around 2.5 h of the per day production time.

Manufacturing lead time ranges from 1 day to 10 days upon increment. This trend is mostly observed with an increase in the market demand additional working hours were added to the daily schedule to avoid the increase in manufacturing lead time beyond 15 days.

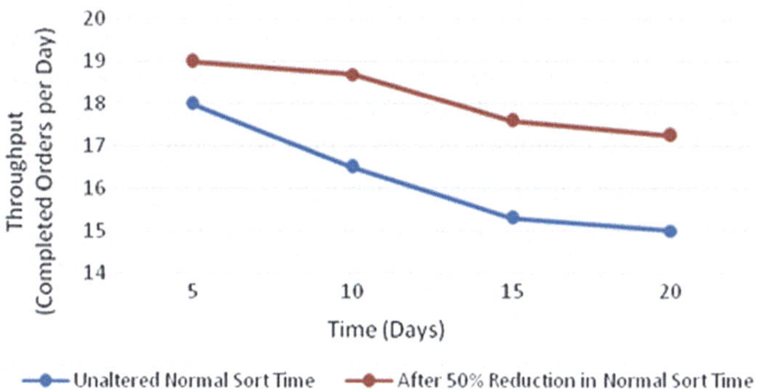

Fig. 19.2 Augmentation of the system in terms of throughput and time before and after normal sort time reduction

The parameter linked with the present SD model is normal time to repair, normal production capacity, and normal sort time. By modifying these parametric variables value, the simulation can describe the detailed behavior of the system in real-life conditions. The normal sort time variable is taken as a critical parameter of the 5S practice of the system.

Normal sort time outlines the summation of various time taken to categorize the items. In the present scenario, the average time taken by a worker to sort an item is 8–10 min. Similarly, the per day sorting requirement is about 50 different sorting. At the cessation of the day (8 h shift), about 470–500 min are spent on sorting activities. The time spent on one item for sorting is not considered but if it calculated for 8 h shift it has a meaningful sense. This suggests the desired approach to minimize the normal sort time and improve the positive effect of sorting on the system.

From the figure mentioned below, it can be noticed that the simulation results in an increment in the throughput in the initial few days before the values drop. As a result of an increase in throughput, there is an increase in the efficiency level of sorting which enhances the system before an increase in job order numbers. On the 20th day, the throughput value is 17.25 completed per day orders when collated to around 15 completed orders for the present situations. A favorable influence on the manufacturing lead time is noted which has further improved at the end of the simulation with the enhancement of normal sort time due to improved throughput and optimized sorting time. The total time spent on finding out the items reduces from 0.4 to about 0.2 work hours per day. An upgrade in time spent on locating items is around 50%. As a result, there is more time to workers for doing value-added activities and the time spent on finding items, machine setup, and repair decreases.

19.6 Prologue to an Improvised 5S Tactic

In order to improve the sorting problems, the SD was applied to reduce the normal sorting time step by step. The justification for the improvement of the amount of normal sorting time required was desired strongly. The reason for that is that the reduction in the time taken for sorting may not considerably enhance the throughput while the reduction of the normal sort time may require training, new shelves, and a little effort. It is clear from Fig. 19.3, when the normal time is reduced by around 80%, the throughput increased by nearly about 50%. This result justifies improving the normal sorting time.

Figure 19.3 shows the anticipation of the throughput improvement by around 15–20% with a decrease in normal sorting time by 50%. Based on the above evaluation along with the views of the supervisors in the organization, the initial tactic was to converge a 20% reduction and expect a 5% rise in throughput which would shift from 15.12 to 15.7 completed orders per day to analyze the decremental effects of the tool. The top management of supervisors expects that 20% reduction can be

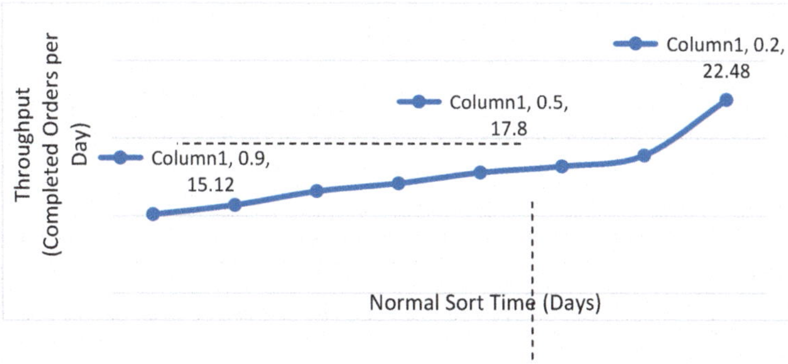

Fig. 19.3 Variation in throughput with incremental refinement in normal sort time generated using SD simulation

done possibly within the short period and response can be used to prove the need for further improvements. The papers having standardized times and procedures for sorting passed on each shelf. There was a higher probability of a quicker and easier item location.

There was training to utilize the standardized time and routine for an individual sorting activity. There are also some sorting and sweeping activities during the changeover [5]. There were shelves and compartments which were clearly labeled with different signs and images as per requirements like part, tool, and items. Maps were also placed at the all storage entry section showing the arrangements of items. The time taken to implement the improvements was approximately 2 weeks. The cost for shelves arrangement is not major because the shelves were already in place. To change the system performance in the sorting behavior, all the other variables remain intact and unaffected. After the two weeks of implementation, an investigation was facilitated to assess the positive fluctuation in the throughput of the system and the time taken to sort items. The simulation generated results were juxtaposed to the real-life results as shown in Table 19.2. The achievement of a reduction in time was rewarded because of an improvement in sorting activities and that reduces the additional man-hours during high demand season.

Table 19.2 Comparison of simulation results to actual life circumstances

Parameter	Real-life results	Theoretical result (simulation generated)	Calculated variance
Normal sort time (days)	1	1	0
Throughput (orders/day)	15.3	16.67	1.37

19.7 Conclusion and Future Scope

The core idea behind the development of the SD model was to simulate the relationship between throughput of the system and 5S tool implementation and indicate the mutual variation as a result of improvement. The implemented methodology showcases the correlation between the improvement inculcated in the 5S technique and its impact on the performance of the manufacturing system. The theoretical assumptions and hypothesis are justified prior to the validation of the SD model which clarifies the improvement and positive optimization of the manufacturing system when the lean philosophy is improved.

The applied methodology not only depicts the throughput improvements but also highlights the optimistic improvements in the setup and repair time taken which are the two fundamental elements of just-in-time (JIT) manufacturing and total productive maintenance (TPM).

This tool can yield beneficial results when implemented in manufacturing plants with different lean philosophies implemented and at different levels of integration and intricacy. 5S lean tool has applications in various fields. 5S could be applied to healthcare facilities regardless of locations. It could be not only a tool for health workers and facility managers but also a strategic option for policymakers [20]. The 5S has the potential to improve client satisfaction at resource-poor health facilities and could, therefore, be recommended as a strategic option for improving the quality of healthcare service in low- and middle-income countries [21].

References

1. Ablanedo-Rosas, J., Alidaee, B., Moreno, J., Urbina, J.: Quality improvement supported by the 5S, an empirical case study of Mexican organizations. Int. J. Prod. Res. **48**(23), 7063–7087 (2010). https://doi.org/10.1080/00207540903382865
2. Agrahari, R., Dangle, P., Chandratre, K.: Implementation of 5S methodology in the small scale industry: a case study. Int. J. Sci. Technol. Res. **4**(4):180–187 (2015). ISSN 2277-8616
3. Alefari, M., Salonitis, K., Xu, Y.: The role of leadership in implementing lean manufacturing. Proc. CIRP **63**, 756–761 (2017). https://doi.org/10.1016/j.procir.2017.03.169
4. Baškarada, S., Watson, J., Cromarty, J.: Balancing transactional and transformational leadership. Int. J. Organ. Anal. **25**(3), 506 (2017). https://doi.org/10.1108/IJOA-02-2016-0978
5. Al-Aomar, R.A.: Applying 5S lean technology: an infrastructure for continuous process improvement. World Acad. Sci. Eng. Technol. **59**, 2014–2019. 10.1.1294.4667 (2011)
6. Ali, R.M., Deif, A.M.: Dynamic lean assessment for takt time implementation. Proc. CIRP **17**, 577–581 (2014). https://doi.org/10.1016/j.procir.2014.01.128
7. Bayo-Moriones, A., Bello-Pintado, A., Merino-Díaz de Cerio, J.: 5S use in manufacturing plants: contextual factors and impact on operating performance. Int. J. Qual. Reliab. Manag. **27**(2), 217–230 (2010). https://doi.org/10.1108/02656711011014320
8. Veres, C., et al.: Case study concerning 5S method impact in an automotive company. Proc. Manuf. **22**, 900–905 (2018). https://doi.org/10.1016/j.promfg.2018.03.127
9. Filip, F., Mărăscu-Klein, V.: The 5S lean method as a tool of industrial management performances. In: IOP Conference Series: Materials Science and Engineering, vol. 95, p. 012127 (2015). https://doi.org/10.1088/1757-899x/95/1/012127

10. Ghodrati, A., Zulkifli, N.: The Impact of 5S implementation on industrial organizations performance. Int. J. Bus. Manag. Invent. **2**(3), 43–49. ISSN (ONLINE)-2319-8028 (2013)
11. Appelbaum, S.H., Calla, R., Desautels, D., Hasan, L.: The challenges of organizational agility part 1. Ind. Commercial Train. **49**(1), 6–14 (2017). https://doi.org/10.1108/ict-05-2016-0027
12. Appelbaum, S.H., Calla, R., Desautels, D., Hasan, L.: The challenges of organizational agility part 2. Ind. Commercial Train. **49**(2), 39–74 (2017). https://doi.org/10.1108/ICT-05-2016-0027
13. Chapman, C.D.: Clean house with lean 5S. Qual. Progr. **38**(6), 27–32 (2005)
14. Chuanjie, X.: Research on implementation plan of 5S management in university library. In: Proceedings of the 2nd International Conference on Systems Engineering and Modeling (ICSEM–13), Advances in Intelligent Systems Research, pp. 574–577. Published by Atlantis Press, Paris, France (2013)
15. Patel, C.V., Thakkar, D.: Review on implementation of 5S in various organization. Int. J. Eng. Res. Appl. **4**(3), 774–779. ISSN 2248-9622 (2014)
16. Baškarada, S., Watson, J., Cromarty, J.: Balancing transactional and transformational leadership. Int. J. Organ. Anal. **26**, 331 (2017). https://doi.org/10.1108/ijoa-05-2017-1163
17. Omogbai, O., Salonitis, K.: A lean assessment tool based on systems dynamics. Proc. CIRP **50**, 106–111 (2016). https://doi.org/10.1016/j.procir.2016.04.169
18. Khataie, A.H., Bulgak, A.: A cost of quality decision support model for lean manufacturing: activity-based costing application. Int. J. Qual. Reliab. Manag. **30**(7), 751–764 (2013). https://doi.org/10.1108/IJQRM-Jan-2011-0016
19. Patel, C.V., Thakkar, H.: A case study: 5 s implementation in ceramics manufacturing company. Bonfring Int. J. Ind. Eng. Manag. Sci. **4**(3), 132–139 (2014). https://doi.org/10.9756/bijiems.10346
20. Kanamori, S., et al.: Applicability of the 5S management method for quality improvement in health-care facilities: a review. Trop. Med. Health **44**, 21 (2016). https://doi.org/10.1186/s41182-016-0022-9
21. Kanamori, S., et al.: Impact of the Japanese 5S management method on patients and caretakers satisfaction: a quasi-experimental study in Senegal. Global Health Action **9**, 32852 (2016). https://doi.org/10.3402/gha.v9.32852

Chapter 20
Comparison-Based Analysis of Travel Time Using Support Vector Regression

Vivek Agrawal, J. N. Singh, Ashish Negi and Sanjay Kumar

Abstract In this application of a machine learning performance, for the forecast of the time, the vehicle will be spent among to the different positions in a near about section. The foundation of guess will be knowledge process which highlighted on the past data about the actions of vehicles in deliberation, a set of semantic variables to acquire approximate time exactly. According to researcher propose a city broad and synchronized model for the judgment of journey period of some street (Represent as a series of linked road segment) in the authentic instance in a city base on the Global Positioning System trajectory of vehicle arriving in present moment of period and more than a period of the past and map data source. In this research, we apply support vector regression (SVR) for travel time predictions and evaluate the result to different techniques of travel time prediction method by means of real road traffic information. In this paper, we relate support vector regression (SVR) for travel time forecast and contrast the result to former baseline travel time prediction method by factual highway traffic information. Our aim of is to use SVR analyst to can decrease appreciably both relation mean bugs and root mean square bugs of predict journey period.

Keywords Support vector regression (SVR) · Global positioning system (GPS) · Machine learning (ML) · Support vector machine (SVM) · Root mean error (RME) · Root-mean-square error (RMSE)

V. Agrawal (✉) · J. N. Singh
Uttrakhand Technical University, Dehradun, India

A. Negi
Galgotia University, Greater Noida, India

S. Kumar
G.B. Pant Engineering College, Pauri Garhwal, Uttarakhand, India

S. Yadav et al. (eds.), *Proceedings of International Conference in Mechanical and Energy Technology*, Smart Innovation, Systems and Technologies 174,
https://doi.org/10.1007/978-981-15-2647-3_20

20.1 Introduction

Travel time is largely a fundamental and raw element of advanced traveller systems. Travel time can help us predict and plan our schedules and departure time accordingly. Travel time is one of the greatest necessities in today's tight schedule scenario. It can be utilized in shipping arrangement, design and operations, and assessment. Mainly, travel time statistics are vital pre-trip and en route data in higher traveller knowledge systems. They are very helpful for drivers and travellers in decision making or schedules making. With accurate travel time forecast, a route guidance method knows how to recommend best possible routes or caution of probable traffic blocking to users, on the basis of which they can then choose the finest leaving period or approximate their likely coming time. Travel time computation depends upon the motor vehicle pace, the traffic stream, the possession and such factors that are extremely dependent on climate situation and traffic incidents. These qualities make the prediction of travel time more and more composite, thereby making it further complex to get to the finest correctness. However, the weekly, seasonal and daily data will be analysed at a large scale and the various patterns will be drawn out of them. The daily data will help to distinguish between rush hour data and slow traffic data time. On a different side, the weekly travel time data will help to distinguish between weekday and weekend traffic data. Similarly, the seasonal data will help to distinguish between the traffic states and the travel times between the summer and the winter seasons, because the formation of support vector machine (SVM) theory been an exhaustive study on SVM for categorization and regression [1–3]. SVM is relatively agreeable from a speculative aim analysis and can have more prospective and performance in realistic applications. In SVM, results assured global minima, and on the other side, ERM can merely position local minima. In support of illustration, in the preparation of neural networks procedure, whichever figure of local minima can be obtained, which do not guarantee for deliberation of global minima. In addition, SVM is extra adaptive to compound system and can agreement through contaminated statistics effortlessly. This characteristic of SVM enhances the capability of simplification that is the bottleneck of its antecedent, the neural network method. In recent times, a purpose of SVM to time series forecasting, called support vector regression (SVR), has furthermore revealed several breakthrough and possible presentation, like (Andersen 1999) [4], forecasting of electricity price, assessment of power expenditure and rebuilding of chaotic systems. Apart from for traffic flow prediction, yet, there is hardly any SVR results on time series examination for ITS. In view of the fact that there are lot of winning results of time-varying application through SVR prediction, it is an appealing feature in the subsequent study of applying SVR for travel time model. In this paper, we bring into play SVR for forecast of travel time for highway users. It demonstrates that SVR is appropriate to journey period forecast along with outperforms numerous preceding methods.

In this paper, we will predict the travel times for particular road segments, and correspondingly, try to minimize the error percentage of our predictions as compared to other prediction and classification models. Basically, we are going to use support vector regression model to predict and classify the various travel time values. In Sect. 20 of this paper, we will be developing an understanding of the support vector regression algorithm. Further, in the next section, we will be discussing the procedure and methodology of predicting these travel time values. Later, we will be comparing our results to the earlier models and predictions, finally, ending the paper with problem statement and the future works and improvements required.

20.2 Support Vector Regression

Support vector machines sustain linear the same as nonlinear regression methods that we can refer to as SVR. As an alternative to try to fit the biggest promising road among two modules while restraining boundary violation, SVR seeks to fit as a lot of amount of instances feasible on the road despite the fact that warning border violations. The breadth of the lane is guarded by means of a hyperparameter epsilon. Support vector regression performs linear regression in an advanced dimensional breathing space. In support vector regression, every statistics in the training set represents the measurement it possesses. As soon as we calculate the kernel among a test point and a training set point, the resultant assessment gives the coordinate of test point in the advanced dimensional space. When we calculate that vector, it can be used to carry out a linear regression.

There is a requisite of training set: $T = (X, Y)$ which covers the area of concern and is accompanied by means of solutions scheduled on that area.

The SVM job is to estimate the function, used within the production procedure of the training set

$$F(X) = Y$$

In a classification problem, the vectors X are used to describe a hyperplane so as to separate the two dissimilar classes in the result.

These vectors are used to carry out linear regression. The vectors nearby to the test point are referred like support vectors. We can evaluate our function anywhere so any vectors could be closest to our test evaluation location (Fig 20.1).

20.2.1 Methodology

Building a SVR

1. Collect a training set $T = \{X, Y\}$
2. Decide a kernel in addition to its parameters and some regularization necessary.
3. Form the correlation matrix, K.

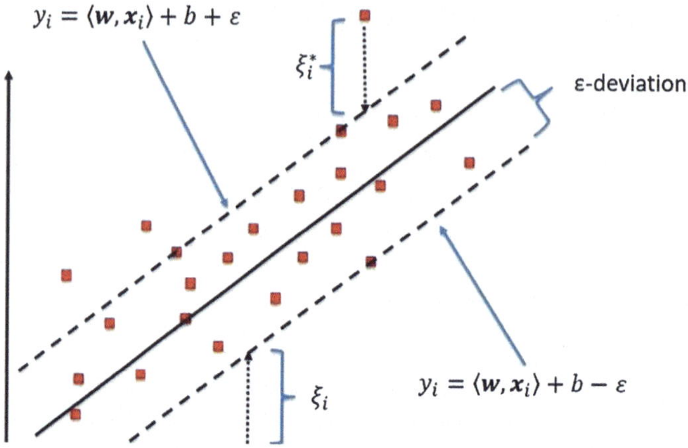

Fig. 20.1 Support vector regression

4. Teach your mechanism, accurately or roughly, to obtain retrenchment coefficients $a = \{ai\}$.
5. Use those coefficients, create your estimator $f(X, a, x) = y$.

Next step is to choose a kernel

1. Gaussian
2. Regularization
3. Noise.

Correlation Matrix

$$K_{i,j} = \exp\left(\sum_k \theta_k \left|x_k^i - x_k^j\right|^2\right) + \in \delta_{i,j}$$

The main part of the algorithm $k\alpha = y$.

y is the vector of values corresponding to the training set, and K is the correlation matrix. A is a set of unknowns we need to solve for

$$\alpha = k^{-1} y^{-1}$$

- Once the parameters are known, form the estimator
- We use the coefficients that we found during the optimization step and the kernel we started off with.
- To estimate the value y for a test point, x, compute the correlation vector k,
- $y = ak$

$$K_{i,j} = \exp\left(\sum_k \theta_k \left|x_k^i - x_k^j\right|^2\right)$$

In simple words, SVR has a dissimilar regression aim in comparison of linear regression. In linear regression, we are trying to minimize the error between the prediction and data. In SVR, our objective is to make certain that errors do not go beyond the threshold.

20.3 Related Work

Wu et al. [5] apply support vector regression (SVR) for travel time forecast and contrast the results among former baseline travel time calculation methods by means of actual highway traffic statistics. Because support vector machines have extra capability in terms of generalization and assurance to achieve global minima for a particular set of training facts, it is supposed to facilitate support vector regression carry out fine in context of time series study. With contrast towards former baseline predictors, the results of SVR predictor illustrate that both the RME and RMSE of predicted travel times can be condensed considerably. The possibility of executing SVR in travel time prediction is verified and proves the applicability and well presentation of SVR for analysis of traffic data. Following a sequence of experiments, a bunch of SVR parameters are proposed that is used in forecast of travel time.

Bang et al. [6] conducted abundant studies intended for the assessment of journey period on the foundation of records obtained from loop detectors. The study is directed on the presentation estimation of a bunch of assessment technique—algorithms which calculate approximately route travel period by means of precise pace statistics through double-loop detectors applications, like display travel times on variable note symbols.

They initially projected the LMM scheme to practice travel time data as of survey vehicles, which captures the description of traveller trips for the period of both peak and non-peak hours, and moreover allow estimating percentiles of travel times, based upon ground truth statistics. During deeply crowded peak time, travel time distribution during a little time is minute. After that, presentation-based assessment is prepared on behalf of three travel time assessment algorithms which utilize pace statistics by loop detectors. Afterwards, it was initiated that while road journey period is comparatively small and the change through free-flow to utmost blocking is sluggish, difference with dissimilar assessment algorithms are non considerable and the immediate tour period may be taken in favour of straightforwardness.

Cho et al. [1] address the trouble of forecasting tour period. Because predictors, four techniques are offered and single idea, 'days-in-a-week' outcome, intended to get better the presentation is prepared. Like exposed within assessment results through the way from Univ. Avenue to San Francisco on highway 101 North, allowing for 'days-in-a-week' reduce forecast mistake considerably. Moreover, TVLR (d) is recommended as a constructive system with its excellent presentation and cleanness.

To additional recover the correctness of forecasting, it can be used with other models resembling linear model with link flow and possession. It furthermore attempts to believe former factors which influence journey period like traffic statistics for the crossing highways. Even though TVPS perform fine in weekdays, its presentation is difficult in weekend. For TVLR, presentation is poorer through relative errors more than 10%. This matter can be addressed by means of techniques taking into consideration days-in-a-week effect.

Wang et al. [2] proposed that a citywide and instantaneous replica for approximating journey period of some lane (represented as a series of linked street segment) in actual instance within a town, relying upon GPS trajectories of vehicles arriving in existing period and over a time of record and map statistics source. Although this is an extremely essential job in numerous traffic monitoring and direction-finding systems, the difficulty still persists due to subsequent three challenges. One is data scarcity difficulty, i.e., numerous street segments are travelled by non-GPS-equipped vehicles. In majority of the cases, individual cannot locate a trail closely traversing an enquiry lane also. Second, for the portion of a lane through trajectories, there is numerous behaviour of applying these trajectories to approximate the equivalent journey period. Third, we require to immediately answer users' query that might happen in some division of a specified town.

To deal with these challenges, they have modelled diverse traveller's journey period on dissimilar street segments in dissimilar period slots through a three-measurement tensor. Pooled among geospatial, temporal and chronological contexts educated from trajectories and record statistics, they pile up the tensor's absent data with a context conscious tensor disintegration technique.

Wu et al. [3] relate support vector regression (SVR) for travel time forecast and apply the results for comparison with former baseline methods using actual road traffic statistics. Because of better simplification capability and assurance of global minima of support vector machines for known training statistics, it is supposed that SVR will offer enhanced presentation intended, for instance, sequence examination. After comparing, results demonstrate, SVR predictor can notably decrease both RME and RMSE of predicted journey period. It confirmed the applicability of SVR in journey period prediction and proves the performance of SVR is fine in favour of traffic data study.

Perreira et al. [7] explore the use of a machine learning system in the direction of prediction of time taken by a vehicle connecting two points within an approximated region. The forecast is based on past statistics regarding the travels perform by vehicles having into consideration a bunch of semantic variables for getting approximate period exactly. The paper in addition shows a testing with real-world statistics. Even though the work is introductory, results were agreeable.

The travel time assessment involves an amount of variables that make the explanation further difficult. Change in traffic behaviour very much manipulates the approximate time of the vehicle. Variables in consideration are the vehicle driver

Table 20.1 Comparison table

y_test	Simple linear	Multiple linear	Support vector
913.64	841.3	821.255	875.336
832.444	826.999	826.999	871.785
800.733	875.552	929.005	873.63
829.517	1025.64	977.533	870.034
893.12	826.999	666.639	872.793

itself, the vehicle's description, the vehicle weight, vehicle identifier, if the day is previous to a vacation. The method used in paper has not so far completely tested with real-time statistics, but initial results proved to be realistically acceptable.

Xhang et al. [8] conducted an experiential assessment of journey period evaluation method lying on single-loop detector statistics. The point to worry is the regression technique lying on an instinctive stochastic replica as planned, and the usual method of using an individuality describing pace and possession with the hypothesis of an ordinary vehicle length. The study is modified in the boundaries forced by every accessible data set. In addition, they have introduced a number of variations of the regression technique and provide examples which propose instructions for upcoming task to more recover the regression system.

20.4 Comparison of All Alogorithms (y_test)

See Table 20.1.

20.5 Proposed Model

Our support vector regression model is performed in the machine learning's integrated development environment namely Annaconda. We use the spider version of it because it contains the maximum number of tools required for the same. Our test and development environment basically make use of these three libraries:

NumPy:

NumPy is a general-purpose array-processing package. It provides a high-performance multidimensional array object and also tools to work upon with these arrays.

It is a basic package for scientific computation through Python. It contains a variety of character together with following main points:

- A powerful N-dimensional array object
- Sophisticated (broadcasting) functions
- Tools for integrating C/C++ and Fortran code
- Useful linear algebra, Fourier transform and random number capabilities.

In addition to its clear technical uses, NumPy can be used as a capable multidimensional box of basic statistics. Arbitrary data types can be defined by means of Numpy that allow NumPy to seamlessly and rapidly put together with a broad range of database.

Matplotlib.pyploy:

Matplotlib is a Python library used to create 2D graphs and plots by means of python scripts. It has a unit named pyplot that make stuff simple intended for plotting by giving attribute to manage line styles, font properties, formatting axes, etc.

It chains a broad diversity of graphs and plots that is—histogram, bar charts, power spectra, error charts, etc. It is used with NumPy to give an atmosphere which is an efficient open-source substitute for MATLAB. It can as well utilize among graphics toolkits like PyQt and wxPython.

Pandas:

Pandas is an open-source, BSD-licensed Python library that provides high-performance, user-friendly statistics structures and information investigation kit for the Python programming language. Python with Pandas is used in a broad series of fields together with educational and commercial domains including finance, economics, statistics, analytics, etc.

Pandas is a conventional Python package for statistics science, and with fine cause: it offers powerful, expressive and flexible data structures which compose statistics manipulation and examination easy, among numerous previous stuff. The DataFrame is one of these structures (Table 20.2).

The prediction results were obtained and the error percentages were found to be decreasing, i.e. 6.67%, of error contrary to the previous results of our conventional predictors, i.e. 7.21% of error.

Table 20.2 Final result comparison

Simple linear	Multiple linear	Support vector
3.8883	6.1783	4.1924
5.1372	5.1372	7.3040
0.2200	6.3384	9.1037
17.8850	12.3557	2.2759
5.2468	2.3620	6.4913

20.6 Conclusion

Support vector machine and support vector regression have verified their achievement in time series examination and statistical education. On the other hand, slight task has made for transfer statistics investigation. In this paper, we interrogate the possibility of applying support vector regression in travel time forecast. Following many experiments, we suggest a bunch of SVR parameters which can guess travel times finely. Results demonstrate that the SVR forecaster considerably outperforms the previous baseline predictors. Compared to previous baseline predictors, our results show that the SVR predictor can decrease notably both RME and RMSE of predicted travel times. This evidences the applicability of support vector regression in traffic data examination. Also, the graph for the SVR shows the results more accurate when compared to simple linear regression and multiple linear regressions.

References

1. Cho, Y., Kwac, J.: Department of Electrical Engineering Stanford University, Stanford, CA, 94305, USA
2. Wang, Y., Zheng, Y., Xue, Y.: Microsoft research, No. 5 Danling Street, Haidian District, Beijing 100080, China 2. College of Computer Science, Zhejiang University 3 Department of Computer Science, Cornell University. Travel Time Estimation of a Path using Sparse Trajectories
3. Wu, C.H., Member, IEEE; Ho, J.M., Member, IEEE: Lee, D.T., Fellow, IEEE: Travel-time prediction with support vector regression
4. Andersen, T.G., Bollerslev, T.: Forecasting financial market volatility: Sample frequency vis-a-vis forecast horizon. Elsevier (1999)
5. Wu, C.H., Ho, J.M., Lee, D.T.: Travel-time prediction with support vector regression. IEEE Trans. Intell. Transp. Syst. **5**(4), 276–281 (2004). https://doi.org/10.1109/tits.2004.837813
6. Jeff, X., Ban, X.J., Li, Y., Skabardonis, A., Margulici, J.D.: Performance evaluation of travel-time estimation methods for real-time traffic applications. J. Intell. Transp. Syst. **14**(2), 54–67 (2010)
7. Pereira, L.: MasieroTeCGraf-PUC–RJ RuaMarquês de São Vicente, 225 Rio de Janeiro—Brazil +55 21 3527-2503 leone@tecgraf.puc-rio.br TeCGraf -PUC–RJ RuaMarquês de São Vicente, 225 Rio de Janeiro—Brazil +55 21 3527-2508 tilio@tecgraf.puc-rio.br Travel Time Prediction using Machine Learning
8. Zhang, X., Rice, J., Bickel, P.: Department of Statistics, University of California at Berkeley, Empirical comparison of travel time estimation methods (1999)

Chapter 21
Predictions of Minimum Fluid Film Thickness of Journal Bearing Using Feed-Forward Neural Network

Sunil Kumar, Vijay Kumar and Anoop Kumar Singh

Abstract Neural network approach plays a significant role in the performance predictions of mechanical systems due to its fast response and ability to solve complex problems. In present work, feed-forward neural network model is developed for the predictions of minimum fluid film thickness of journal bearing. Literature data is used for the validation of the model. The results obtained by neural network model are highly accurate and precise. The accuracy of the model is achieved through gradient descent algorithm. The decrease in fluid film thickness is observed with increase in external applied radial load. The predictions are made within and out of the prescribed range for radial load as input parameter. This model is best suited for the predictions of performance characteristics of journal bearings.

Keywords Neural network · Journal bearing · Fluid film thickness · Feed-forward neural network

21.1 Introduction

Artificial intelligence (AI) is widely used these days for the solution of complex problems. The complexity in the interaction of input and output parameters make the problems difficult to solve. Due to the advancement in computational technologies, it is easy and less time consuming to handle such issues. Neural network (NN) approach is highly recommended to handle such complex issues. Multilayer neural network with single hidden layer is considered for predictions in this work [1–3].

Journal bearings are used as moving mechanical assemblies to carry large radial loads at high rotational speed. Lubricant flows in the clearance space of journal and bearing. Holes, pockets or slots are provided in the bearing shell for the intake of lubricant flow [4]. The flow of lubricant is controlled by restrictors such as orifice, capillary compensated or constant flow valve restrictor [5, 6]. An appropriate thickness of lubricant film is required to avoid the surface-to-surface contact. The

S. Kumar (✉) · V. Kumar · A. K. Singh
Chitkara University Institute of Engineering and Technology, Chitkara University, Rajpura, Punjab, India
e-mail: sunil.sharma@chitkara.edu.in

S. Yadav et al. (eds.), *Proceedings of International Conference in Mechanical and Energy Technology*, Smart Innovation, Systems and Technologies 174,
https://doi.org/10.1007/978-981-15-2647-3_21

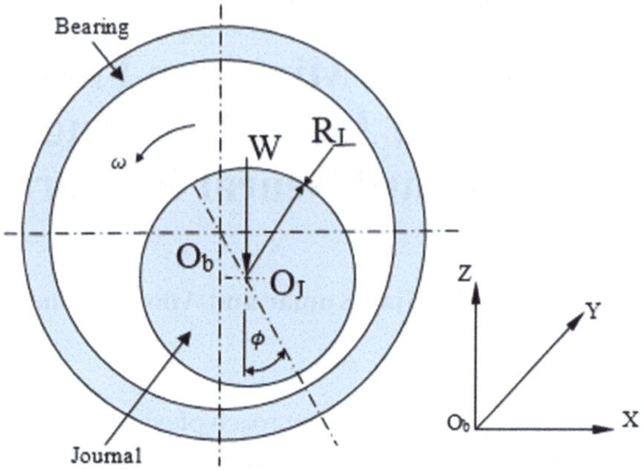

Fig. 21.1 Schematic diagram of journal bearing

variation in fluid film thickness is caused due to hydrodynamic action. The focus of the present work is on the variation of minimum fluid film thickness ($\bar{h}_{min}$) with respect to the applied radial load ($\bar{w}$). To achieve the performance characteristics of journal bearings, theoretical and experimental methods are available. The schematic diagram of journal bearing is shown in Fig. 21.1. In this figure, W, R_J, O_b, O_J, ω_J and ϕ represent radial load, journal radius, bearing centre, journal centre, journal angular velocity and attitude angle.

Feed-forward neural network model with one hidden layer is developed for the predictions of minimum fluid film thickness. The desired output for this supervised model is collected from the literature [7] and used as the training data for the model. The variations in the desired output and model output are minimised using back propagation algorithm. This algorithm uses gradient descent approach to minimise the error. In this approach, the associated weights are updated in many iterations till the convergence criteria achieved [8].

21.2 Journal Bearing Analysis

Journal bearings have been analysed by many researchers to determine their static and dynamic performance characteristics [9, 10]. Finite element method (FEM) is widely used for theoretical analysis. The governing Reynolds equation in non-dimensional form for flow field in journal bearing is given as Eq. (21.1) [11];

$$\frac{\partial}{\partial \alpha}\left(\bar{h}^3\overline{F}_2\frac{\partial \bar{p}}{\partial \alpha}\right)+\frac{\partial}{\partial \beta}\left(\bar{h}^3\overline{F}_2\frac{\partial \bar{p}}{\partial \beta}\right)=\Omega\left[\frac{\partial}{\partial \alpha}\left\{\left(1-\frac{\overline{F_1}}{\overline{F_0}}\right)\bar{h}\right\}\right]+\frac{\partial \bar{h}}{\partial \bar{t}} \quad (21.1)$$

In this equation, α and β shows circumferential and longitudinal coordinates, respectively. $\overline{F}_1$, $\overline{F}_2$ and $\overline{F}_3$ are the viscosity functions. $\bar{p}$, Ω and $\bar{h}$ are the pressure, speed parameter and film thickness, respectively. This governing equation can be modified depending upon the fluid model used. Finally, the performance characteristics can be determined by using appropriate relations.

Micropolar fluid model is most popular for FEM analysis of journal bearing. Micropolar fluid is the non-Newtonian lubricant whose performance in journal bearing depends upon the characteristic length of fluid and coupling number. The present neural network model is trained for micropolar lubricated pocket hybrid journal bearing. The modified form of this equation for micropolar fluid model can be written as Eq. (21.2) [7];

$$\frac{\partial}{\partial\alpha}\left(\frac{\bar{h}^3}{12\bar{\mu}}\varnothing_m\frac{\partial\bar{p}}{\partial\alpha}\right)+\frac{\partial}{\partial\beta}\left(\frac{\bar{h}^3}{12\bar{\mu}}\varnothing_m\frac{\partial\bar{p}}{\partial\beta}\right)=\frac{\Omega}{2}\frac{\partial\bar{h}}{\partial\alpha}+\frac{\partial\bar{h}}{\partial\bar{t}} \tag{21.2}$$

In this equation, $\varnothing_m$ is the micropolar function and $\bar{\mu}$ is the non-dimensional dynamic viscosity.

Fluid film thickness can be obtained by the expression given in Eq. (21.3).

$$\bar{h}=1-\overline{X_J}\cos\alpha-\overline{Z_J}\sin\alpha \tag{21.3}$$

where $\overline{X_J}$ and $\overline{Z_J}$ are the journal centre coordinates.

Minimum fluid film thickness in journal bearing is the static performance characteristic which varies with externally applied radial load.

21.3 Neural Network Algorithm

Neural network is the mapping system based on the human nervous system. It consists of an input layer, output layer and number of hidden layers. The number of neurons in input and output layers depends upon the associated problem but the number of neurons in hidden layer can be decided heuristically. In feed-forward neural network, the neurons in one layer are connected to the neurons of another layer by weights as shown in Fig. 21.2. x_{i_0} are the input parameters which are connected to the neurons of hidden layer by weights $w_{i_1 i_0}$. The response of the hidden layer v_{i_1} is connected to the output layer by weights $w_{i_2 i_1}$. The response of the output layer is y_{i_2}. i_0, i_1 and i_2 are the indices for the input, hidden and output layer neurons.

The response from the hidden layer and output layer is calculated by Sigmoid activation function. The desired output in the proposed neural network model is in the range of 0–1; hence, sigmoid activation function is selected. Also, in most cases of journal bearing analysis, the same activation function was used by researchers. In the present work, there is no need of normalisation or standardization of input and output dataset because the values are already in the prescribed range. Equations (21.4) and (21.5) show the hidden layer response and the output layer response, respectively.

$$v_{i_1}=\frac{1}{1+e^{-h_{i_1}}} \tag{21.4}$$

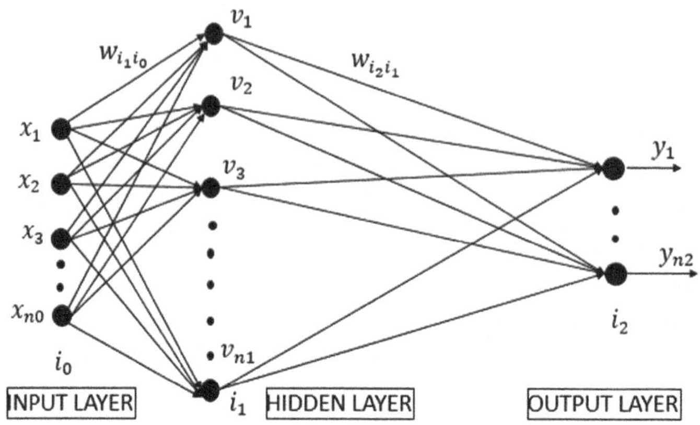

Fig. 21.2 Schematic diagram of feed-forward neural network

$$y_{i_2} = \frac{1}{1 + e^{-h_{i_2}}} \tag{21.5}$$

h_{i_1} and h_{i_2} are the input values to the hidden layer and output layer, respectively. These are given by Eqs. (21.6) and (21.7);

$$h_{i_1} = \sum_{i_0=1}^{n_0} w_{i_1 i_0} x_{i_0} \tag{21.6}$$

$$h_{i_2} = \sum_{i_1=1}^{n_1} w_{i_2 i_1} v_{i_1} \tag{21.7}$$

Mean square error is calculated by using relation as Eq. (21.8);

$$\text{Error} = \frac{1}{2} \sum_{i_2=1}^{n_2} \left(y_{i_2}^d - y_{i_2} \right)^2 \tag{21.8}$$

where $y_{i_2}^d$ is the desired output and y_{i_2} is the model output. Weights are updated by relations shown in Eqs. (21.9) and (21.10):

$$w_{i_2 i_1}(\text{new}) = w_{i_2 i_1}(\text{old}) + \eta \delta_{i_2} v_{i_1} \tag{21.9}$$

$$w_{i_1 i_0}(\text{new}) = w_{i_1 i_0}(\text{old}) + \eta \delta_{i_1} x_{i_0} \tag{21.10}$$

δ_{i_2} is the error back propagated from output layer to hidden layer and δ_{i_1} is the error back propagated from hidden layer to input layer.

The present work is to develop the neural network model to predict $\bar{h}_{\min}$ values for the journal bearing. These values are compared with the literature [7] to validate the program. Neural network diagram for journal bearing is shown in Fig. 21.3. Two neurons for input layer (first for load and second for bias), six neurons for hidden layer and one neuron for output layer are considered in present NN model. Initially, the weights are incorporated as random numbers and the bias as +1 for all the neurons of hidden layer. Number of layers and neurons are selected heuristically in the present work. Model is trained by selecting different number of neurons in hidden layer each time. When the convergence of error is achieved in minimum number of iterations, the proposed model is selected for the predictions.

The parameters for the neural network predictor are presented in Table 21.1. η is the learning rate. n_i, n_h and n_o are the neurons in input, hidden and output layer, respectively. Iterations performed for the convergence of the error are represented by N.

Scilab is used as coding software for neural network model development. The code for weight initialization is presented as follows;

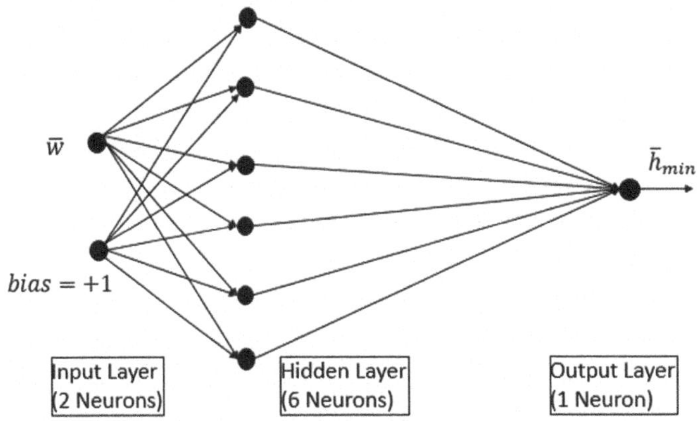

Fig. 21.3 Neural network diagram for journal bearing

Table 21.1 Neural network architecture

Neural network	η	n_i	n_h	n_o	N	Activation function
	1	2	6	1	18,000	Sigmoid

Table 21.2 Updated weights and bias between input and hidden layer neurons

S. No.	Weights	Bias
1	−1.3778589	−0.2236416
2	−1.3221158	−0.2229135
3	−1.2219427	−0.1930271
4	−1.2669601	−0.2133452
5	−1.2829547	−0.2142233
6	−1.3317057	−0.2194876

$$function[NN] = \underline{nn\_initialization}(\boldsymbol{Ni}, \boldsymbol{Nh}, \boldsymbol{No})$$
$$NN.whi = (2 * rand(\boldsymbol{Nh}, \boldsymbol{Ni}) - 1) * 0.01;$$
$$NN.woh = (2 * rand(\boldsymbol{No}, \boldsymbol{Nh}) - 1) * 0.01;$$
$$NN.Ni = \boldsymbol{Ni};$$
$$NN.Nh = \boldsymbol{Nh};$$
$$NN.No = \boldsymbol{No};$$
$$endfunction$$

The Scilab code for sigmoid activation function initialization is presented as follows;

$$function[\boldsymbol{y}, \boldsymbol{h}, \boldsymbol{v}, \boldsymbol{y1}] = \underline{nn\_sigmoid\_fwd}(\boldsymbol{NN}, \boldsymbol{x})$$
$$\boldsymbol{h} = \boldsymbol{NN}.whi * \boldsymbol{x};$$
$$\boldsymbol{v} = 1./(1 + exp(-\boldsymbol{h}));$$
$$\boldsymbol{y}1 = \boldsymbol{NN}.woh * \boldsymbol{v};$$
$$\boldsymbol{y} = 1./(1 + exp(-\boldsymbol{y1}));$$
$$endfunction$$

The updated weights and bias values between input and hidden layer neurons are listed in Table 21.2 and updated weights between hidden layer and output layer neurons are listed in Table 21.3.

The complexity of the neurons connectivity can be easily handled by Scilab program with very fast learning rate.

Table 21.3 Updated weights and bias between hidden and output layer neurons

S. No.	Weights
1	0.8774820
2	0.8416375
3	0.7653444
4	0.8015796
5	0.8112383
6	0.8451896

21.4 Results and Discussion

The neural network model is trained by considering radial load ($\bar{w}$) as input and minimum fluid film thickness ($\bar{h}_{\text{min}}$) as desired output from literature data [7]. The variation of $\bar{h}_{\text{min}}$ with respect to $\bar{w}$ is shown in Fig. 21.4 for the neural network model and the modelled results are compared with desired output.

The design and operating parameters for capillary compensated, micropolar lubricated pocket entry hybrid journal bearing are as follows;

Restrictor design parameter $(\overline{C}_{s_2}) = 0.5$, aspect ratio $\bar{a}_b = 0.14$, speed parameter $(\Omega) = 1$, characteristic length $(l) = 20$ and coupling number $(N_c^2) = 0.5$.

The decrease in $\bar{h}_{\text{min}}$ is observed with the increase in $\bar{w}$. It is clear from this graph that the modelled values are very close to the desired values. The prescribed range for radial load ($\bar{w}$) is 0.1–1.

Predictions for $\bar{h}_{\text{min}}$ are made by neural network model within the prescribed range (Table 21.4) and out of the range (Table 21.5). The predicted $\bar{h}_{\text{min}}$ values for NN model are compared with FEM values from the graph shown in Fig. 21.4. The error is calculated for predicted model output values. The accuracy of the model is very high because the error is less than 0.3%.

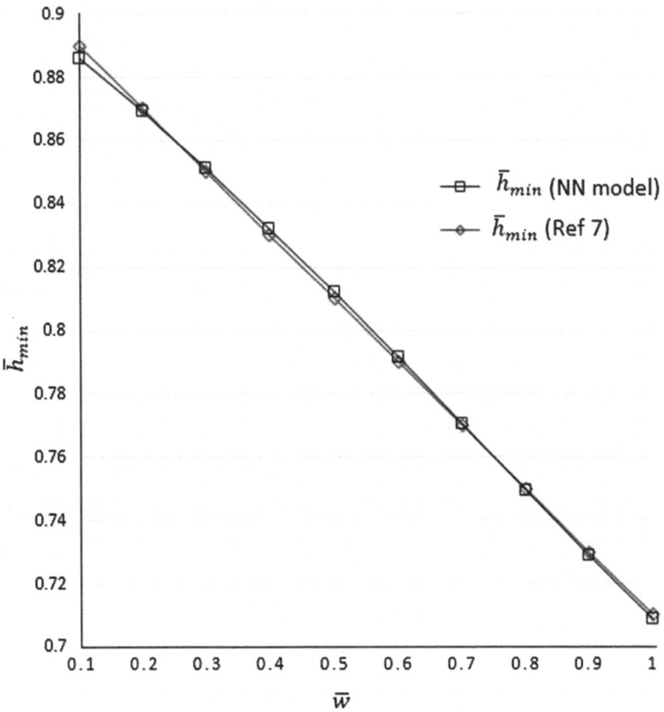

Fig. 21.4 $\bar{h}_{\text{min}}$ versus $\bar{w}$ (desired and neural network model values)

Table 21.4 Neural network predictions within range

S. No.	$\bar{w}$	$\bar{h}_{min}$ (ref 7)	$\bar{h}_{min}$ (NN model)	Error (%)
1	0.25	0.86	0.8605	0.06
2	0.35	0.84	0.8419	0.23
3	0.45	0.82	0.8222	0.27
4	0.55	0.80	0.8017	0.21
5	0.65	0.78	0.7808	0.10
6	0.75	0.76	0.7598	0.03
7	0.85	0.74	0.739	0.14
8	0.95	0.72	0.7187	0.18

Table 21.5 Neural network predictions out of range

S. No.	$\bar{w}$	$\bar{h}_{min}$ (NN model)
1	0.03	0.8965
2	0.05	0.8936
3	1.25	0.6630
4	1.50	0.6244
5	1.75	0.5934

The decrement in the error with respect to number of iterations is shown in Fig. 21.5. The error is converged in 18,000 iterations. The model is trained for different input and output data. The proposed data is sufficient because the predictions which are made using the trained model are very precise and accurate.

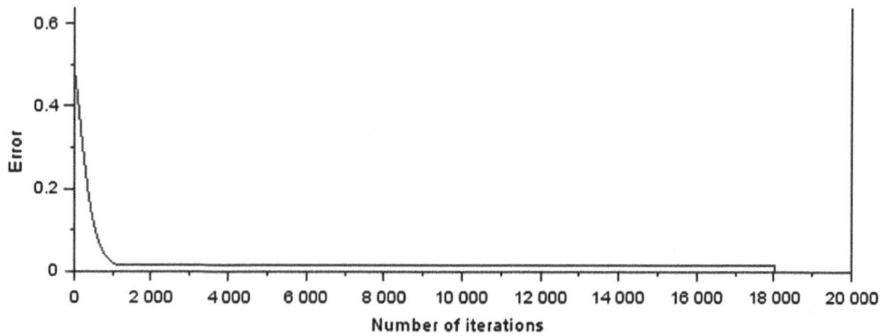

Fig. 21.5 Error convergence graph for neural network model

21.5 Conclusions

In this paper, a feed-forward neural network model is developed for analysing and predicting minimum fluid film thickness of the journal bearing. As shown in the results, the predicted values of $\bar{h}_{\min}$ are very close to the desired values with error <0.3%. The decrease in $\bar{h}_{\min}$ observed as a result of increase in $\bar{w}$. This model can be considered as a best predictor for the behaviour of journal bearings due to its high accuracy. The Delta rule is used as error back propagation. Due to the fast learning rate and the ability to solve complex problems, this neural network model is best suitable for modelling performance characteristics of journal bearings.

References

1. Sinanoğlu, C.: A neural predictor to analyse the effects of metal matrix composite structure (6063 Al/SiCp MMC) on journal bearing. Ind. Lubr. Tribol. **58**(2), 95–109 (2006)
2. Sinanoğlu, C.: Design of neural model for analysing journal bearings considering effects of transverse and longitudinal profile. Ind. Lubr. Tribol. **61**(3), 132–139 (2009)
3. Patel, P.M., Prajapati, J.M.: A review on artificial intelligent system for bearing condition monitoring. Int. J. Eng. Sci. Technol. **3**(2), 1520–1525 (2011)
4. Garg, H.C., Sharda, H.B., Kumar, V.: On the design and development of hybrid journal bearings: a review. Tribotest **12**(1), 1–19 (2006)
5. Ram, N.: Effect of couple stress lubrication on symmetric hole-entry hybrid journal bearing. Tribol. Online **12**(2), 58–66 (2017)
6. Ram, N., Sharma, S.C.: Analysis of orifice compensated non-recessed hole-entry hybrid journal bearing operating with micropolar lubricants. Tribol. Int. **52**, 132–143 (2012)
7. Sharma, S.C., Rajput, A.K.: Effect of geometric imperfections of journal on the performance of micropolar lubricated 4-pocket hybrid journal bearing. Tribol. Int. **60**, 156–168 (2013)
8. Svozil, D., Kvasnicka, V., Pospichal, J.: Introduction to multi-layer feed-forward neural networks. Chemometr. Intell. Lab. Syst. **39**(1), 43–62 (1997)
9. Dowson, D.: A generalized Reynolds equation for fluid-film lubrication. Int. J. Mech. Sci. **4**(2), 159–170 (1962)
10. Fowles, P.E.: A simpler form of the general Reynolds equation. J. Lubr. Technol. **92**(4), 661–662 (1970)
11. Kucinschi, B.R., Fillon, M., Fre, J., Pascovici, M.D.: A transient thermoelastohydrodynamic study of steadily loaded plain journal bearings using finite element method analysis. J. Tribol. **122**(1), 219–226 (2000)

Chapter 22
Application of Taguchi Method to Optimize the Surface Roughness During Face Milling of Rolled Steel (AISI 1040)

Kulwinder Singh, Anoop Kumar Singh and K. D. Chattopadhyay

Abstract Surface roughness has a significant role in the life cycle of mechanical components. Numerous secondary finishing operations are performed to achieve the desired roughness value. The aim of this study is to obtain better surface finish by optimizing the input parameters at primary stage of face milling operation. Effect of machining parameters (cutting velocity, feed rate, and depth of cut) and machining strategies (rolling direction, cutter offset, soaking time) are taken into consideration. Taguchi's L_{27} orthogonal array is used for the design of experiments (DOE). Face milling of rolled steel (AISI 1040) is performed on vertical milling centre (HURCO-VM10) using titanium nitride (TiN)-coated carbide inserts. Surface roughness (SR) is evaluated in terms of signal-to-noise ratio (SNR). Analysis of means (ANOM) is applied on SNR to find the optimal levels of each input parameter. Moreover, significance level and percentage contribution of parameters is calculated using analysis of variance (ANOVA). Feed rate followed by cutting velocity is found the major contributors. Confirmation test is carried out to evaluate the amount of improvement and validation of predicted model. Improvement of 8.41 dB in SR is observed.

Keywords Taguchi · Analysis of means (ANOM) · Analysis of variance (ANOVA) · Surface roughness (SR) · Face milling

22.1 Introduction

Under various manufacturing process, machining process has its own significance to generate a finish product with good surface finish and size in tight tolerance. Machining is a metal removal process, used to shape the final product. Cutting tool is used to remove the material in the form of chips. During cutting, rotary and linear motions are given to cutting tool and workpiece, respectively, depending upon machining operation. In vertical milling machine, rotary cutting tool fixed on vertical spindle fed against the workpiece to remove the material. Parts of different size

K. Singh (✉) · A. K. Singh · K. D. Chattopadhyay
Chitkara University Institute of Engineering and Technology, Chitkara University, Rajpura, Punjab, India
e-mail: kulwinder.singh@chitkara.edu.in

S. Yadav et al. (eds.), *Proceedings of International Conference in Mechanical and Energy Technology*, Smart Innovation, Systems and Technologies 174,
https://doi.org/10.1007/978-981-15-2647-3_22

and shapes can be manufactured on milling machine. Different types of operations such as face milling, end milling, slot milling, drilling, and boring are performed on milling machine [1]. Face milling operation is commonly used operation in various industrial sectors such as automobile, aerospace, press tools, and shipbuilding. Mostly, medium carbon steel (AISI 1040) is used in these sectors to make moderate hard and tough component. It is equal to EN-08 grade steel and comes under a group of medium carbon steel. Top and bottom plates of die sets are also made by this material. Hardening followed by tempering is performed on component after machining. Surface roughness affects the life of part in many ways such as wear and tear, resistance to corrosion, and fatigue strength. Among all quality characteristics, the achievement of desired surface roughness is a big challenge for machining sector. Mostly, additional polishing operation is performed on the part to fulfil the customer requirement. Polishing operations have significant contribution to manufacturing cost [2]. Thus, it is required to control the surface roughness at primary stage of manufacturing. To achieve desired surface roughness, the optimization of process parameters is must. On the basis of experience, it is a difficult task to set the machine at optimum settings without disturbing other quality characteristics.

Currently, many types of optimization techniques (mathematical or statistical) are used to optimize the input parameters. In this study, Taguchi method is employed to optimize the surface roughness (SR) during face milling. As per Taguchi philosophy, it is a loss to society if the outcomes are deviating from the target. Taguchi quality loss is measured in terms of signal-to-noise ratio (SNR) [3, 4]. Taguchi method is used worldwide for the optimization of input parameters. First of all, input parameters along with their levels are selected on the basis of experience and literature review. Suitable orthogonal array is selected on the basis of number of parameters and their levels in the design of experiments (DOE). Taguchi proposed a partial factorial design technique to reduce the number of experiments. Also, the repetition of experiments is recommended (minimum three times) to minimize the variability in outcomes. Analysis of means (ANOM) is used to calculate the mean value of each level for a particular parameter. Further, optimum level of each parameter is selected as an optimum setting. Moreover, analysis of variance (ANOVA) is applied on data values to get the information of significant parameters along with its contribution in variation [5–9]. Further, optimum settings of all process parameters for roughness reduction may decline tool life or edge quality. To overcome this issue, multi-objective optimization techniques are recommended as applied in Najiha et al. [10]. Various metaheuristic optimization techniques such as ant colony optimization, particle swarm optimization, genetic algorithm, and simulated annealing can be used for the optimization of process parameters as reported in Radiša et al. [11]. For process optimization, performance selection index (PSI) method is also proposed by Petković et al. [12]. However, more study can also be done for process parameters optimization.

22.2 Experimental Method

22.2.1 Experimental Set-up

In this research work, rolled steel (AISI 1040) is chosen as work material because of its vast applications in the manufacturing industry. It is a medium carbon steel with the chemical composition shown in Table 22.1. Specimens of size 48 × 48 × 12 mm are prepared, and experiments are performed as per L_{27} orthogonal array. The vertical milling machine (HURCO-VM10) shown in Fig. 22.1 is used for face milling operation. Schematic diagram of face milling process is shown in Fig. 22.2. Cutter of diameter 80 mm with equally spaced seven numbers of titanium nitride (TiN)-coated carbide inserts is used for face milling. SR in terms of arithmetic average roughness (Ra) is measured. Surface roughness test is conducted with portable surface roughness tester of MITUTOYO, Model: SJ-201P. To avoid the error, average of three numbers of measurements at different positions is determined.

Table 22.1 Chemical composition of material

Material	Chemical composition (wt%)				
	C	Mn	Si	S	P
AISI 1040	0.42	0.9	0.31	0.05	0.05

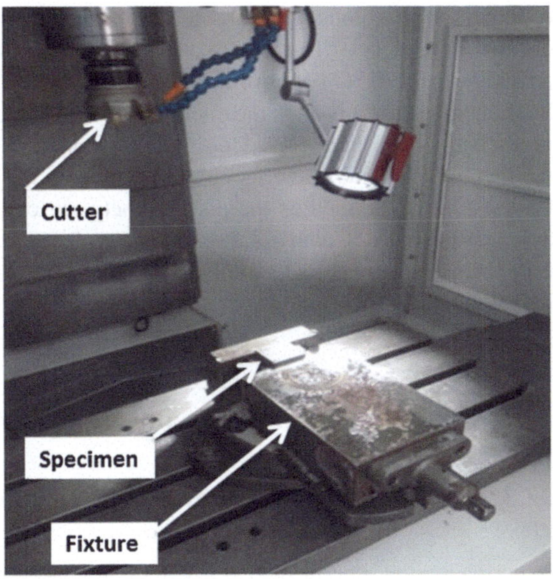

Fig. 22.1 Experimental set-up of face milling

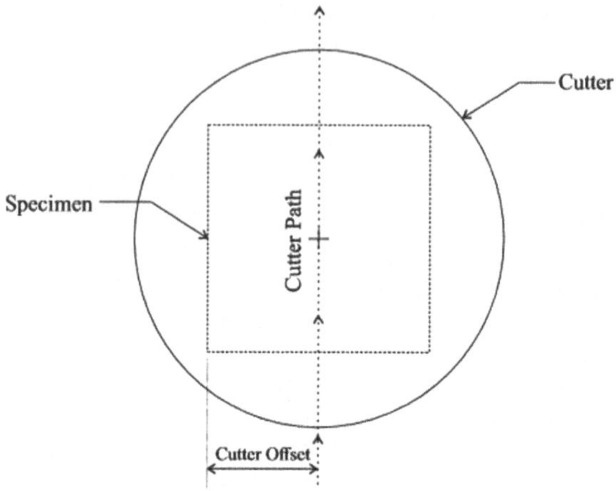

Fig. 22.2 Schematic diagram of face milling selection of input parameters

22.2.2 Selection of Input Parameters

Machining parameters such as cutting velocity, feed per teeth, and depth of cut are the main input parameters in metal removal process. Apart from machining parameters, other three parameters (rolling direction, cutter offset, and soaking time in recovery) are also selected under machining strategy. These parameters have a significant role on various quality characteristics such as surface finish, MRR, and burr growth [13]. Rolled steel behaves like anisotropic material. Strength of rolled sheets varied w.r.t. rolling direction (along and across). Moreover, stresses are generated in the material due to rolling process. Thus, to relieve the stresses, recovery process (low temperature annealing) at 170 °C is performed before machining by varying the soaking time. Soaking time depends upon the thickness of workpiece. Generally, one hour per inch of thickness is taken in consideration [14]. Range of soaking time is fixed based on the literature review. The effect of cutter path on the machinability is also studied. Cutter offset is defined as the distance between cutter rotational axis and central axis of workpiece. Cutter rotational axis is placed from left edge of workpiece at different distances [15]. Detail of input parameters along with level values is shown in Table 22.2.

22.3 Experimental Results and Discussion

22.3.1 Evaluation of Signal-to-Noise Ratio (SNR)

Experimentation is performed as per L_{27} orthogonal array shown in Table 22.3. Each experiment is replicated three times. After machining, the SR (Ra) value of each specimen is measured by using surface roughness tester. To determine the quality

Table 22.2 Detail of face milling input parameters

Code	Input parameter	Level		
		1	2	3
A	Cutting velocity (m/min)	100	150	200
B	Feed per teeth (mm/teeth)	0.05	0.1	0.15
C	Depth of cut (mm)[a]	0.4	0.8	1.2
D	Rolling direction (°)	0	45	90
E	Cutter offset (mm)	12	24	36
F	Soaking time (mm)	30	45	60

[a]Depth of cut is completed in multi-passes with each cut of 0.2 mm

loss, dispersion of data is measured in terms of mean square deviation (MSD). So, all the outcome values are converted into MSD using Eq. (22.1). Then, SNR for each experiment is calculated using Eq. (22.2). Results of SR and SNR are shown in Table 22.3.

$$\mathrm{MSD} = \frac{1}{n}\sum_{k=1}^{n} y_k^2 \tag{22.1}$$

$$\mathrm{SNR} = -10 * \log(\mathrm{MSD}) \tag{22.2}$$

where y represents the SR value of kth replication and n represents the number of replications.

22.3.2 Analysis of Means (ANOM)

In ANOM, mean value of SNR for each input parameter level is calculated to study the effect of input parameters on SR. Results for optimum level are listed in Table 22.4. The effects of individual parameter on SR are also shown in Fig. 22.3. Range of maximum and minimum SNR reveals the contribution of particular parameter in variance. The optimum combination of input parameters with levels is identified as $A_1B_1C_1D_2E_1F_1$. The levels of input parameters for optimum SR value are determined as cutting velocity at level 1, feed per teeth at level 1, depth of cut at level 1, rolling direction at level 2, cutter offset at level 1, and soaking time at level 1.

22.3.3 Analysis of Variance (ANOVA)

ANOVA is applied to identify the significance level of each input parameters along with its contribution. The outcomes of ANOVA are shown in Table 22.5. All the non-

Table 22.3 L_{27} orthogonal array with results

Trial No.	Input parameters						SR (μm)			SNR (dB)
	A	B	C	D	E	F	*R*1	*R*2	*R*3	
1	1	1	1	1	1	1	1.18	1.19	1.18	−1.46
2	1	1	1	2	2	2	1.25	1.26	1.27	−2.01
3	1	1	1	3	3	3	1.41	1.40	1.42	−2.98
4	1	2	2	1	2	3	2.11	2.10	2.10	−6.46
5	1	2	2	2	3	1	1.85	1.86	1.87	−5.39
6	1	2	2	3	1	2	2.07	2.09	2.08	−6.36
7	1	3	3	1	3	2	2.35	2.36	2.36	−7.45
8	1	3	3	2	1	3	2.34	2.33	2.35	−7.38
9	1	3	3	3	2	1	2.62	2.64	2.63	−8.40
10	2	1	3	3	3	2	2.07	2.09	2.08	−6.36
11	2	1	3	1	1	3	1.65	1.66	1.65	−4.37
12	2	1	3	2	2	1	1.75	1.78	1.76	−4.93
13	2	2	1	3	1	1	2.32	2.33	2.34	−7.35
14	2	2	1	1	2	2	2.33	2.35	2.34	−7.38
15	2	2	1	2	3	3	2.17	2.18	2.19	−6.77
16	2	3	2	3	2	3	3.26	3.27	3.26	−10.27
17	2	3	2	1	3	1	2.73	2.74	2.73	−8.73
18	2	3	2	2	1	2	2.71	2.73	2.72	−8.69
19	3	1	2	2	2	3	1.99	2.01	1.99	−6.01
20	3	1	2	3	3	1	2.27	2.29	2.28	−7.16

(continued)

Table 22.3 (continued)

Trial No.	Input parameters						SR (μm)			SNR (dB)
	A	B	C	D	E	F	*R*1	*R*2	*R*3	
21	3	1	2	1	1	2	1.85	1.87	1.85	−5.37
22	3	2	3	2	3	2	2.77	2.76	2.76	−8.83
23	3	2	3	3	1	3	3.10	3.11	3.12	−9.86
24	3	2	3	1	2	1	2.71	2.72	2.73	−8.69
25	3	3	1	2	1	1	2.62	2.63	2.61	−8.37
26	3	3	1	3	2	2	3.16	3.17	3.18	−10.02
27	3	3	1	1	3	3	3.07	3.08	3.06	−9.74

Table 22.4 Results of ANOM

Input parameter	Level			Optimum level
	1	2	3	
A	**−5.32**	−7.21	−8.23	A_1
B	**−4.52**	−7.45	−8.78	B_1
C	**−6.23**	−7.16	−7.36	C_1
D	−6.63	**−6.49**	−7.64	D_2
E	**−6.58**	−7.13	−7.05	E_1
F	**−6.72**	−6.94	−7.09	F_1

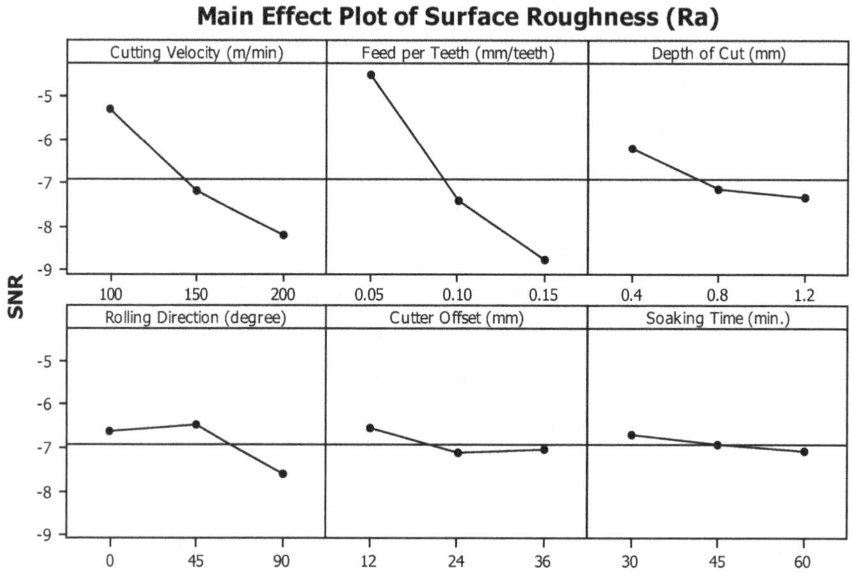

Fig. 22.3 Main effect plot of surface roughness (Ra)

Table 22.5 Results of ANOVA

Input parameter	Degree of freedom	Sum of square	Mean sum of square	*F* ratio (old)	*F′* ratio (new)	New sum of square	Percentage contribution
A	2	39.23	19.62	73.23	58.99	38.57	26.66
B	2	85.88	42.94	160.30	129.12	85.21	58.90
C	2	6.58	3.29	12.27	9.89	5.91	4.09
D	2	6.99	3.49	13.05	10.51	6.32	4.37
E	Pooled	–	–	–	–	–	–
F	Pooled	–	–	–	–	–	–
Error	18	5.99	0.33			8.65	5.98
Total	26	144.66	5.56			144.66	100.00

Table 22.6 Results of confirmation test

	Process parameters of 16th trial	Optimal process parameters		Confidence interval (range)
		Predicted	Experimented	
Level	$A_2B_3C_2D_3E_2F_3$	$A_1B_1C_1D_2E_1F_1$	$A_1B_1C_1D_2E_1F_1$	
Avg. SR (μm)	3.26	1.22	1.24	1.09–1.37
SNR (dB)	−10.27	−1.79	−1.86	−0.81 to −2.78

Improvement in SNR = 8.41 dB

significant parameters are pooled in error to increase the effectiveness of ANOVA. Among significant parameters, feed per teeth is contributing highest with 58.90%, cutting velocity with 26.26%, depth of cut with 4.09%, and rolling direction with 4.37%. Interaction effects of input parameters are neglected because error has minor contribution of 5.98%.

22.3.4 *Confirmation Test*

Finally, in parametric design of Taguchi, confirmation test is performed on the basis of optimal settings to validate the predicted model as shown in Eq. (22.3).

$$\Psi \text{opt} = \Psi m + \sum_{i=1}^{n} (\Psi m_i - \Psi m) \tag{22.3}$$

where Ψopt is the predicted optimum SR value, Ψm is the overall mean, Ψm_i is the mean of optimum level for ith input parameter and n is the number of input parameter. Results of confirmation test are shown in Table 22.6. The confidence interval range for predicted SR is calculated at 95% confidence. SR is improved by 8.41 dB when the optimum input parameters ($A_1B_1C_1D_2E_1F_1$) are used instead of 16th trial (lowest SNR) input parameters ($A_2B_3C_2D_3E_2F_3$). A result of confirmation test validates the success of Taguchi method in optimization of surface roughness.

22.4 Conclusion

In this research work, Taguchi method is applied to optimize the input parameters for minimization of SR. All the parameters are framed in L_{27} orthogonal array for experimentation. ANOM and ANOVA are used to determine the optimum level, significant parameter, and percentage contribution of input parameters. Results reveal that surface roughness of 1.24 μm can be obtained during face milling of rolled steel

(AISI 1040) by using optimum settings of process parameters. Additionally, the following points are concluded from the study:

1. ANOM reveals that cutting velocity of 100 m/min, feed per teeth of 0.05 mm/teeth, depth of cut of 0.4 mm, rolling direction of 45 degree, cutter offset of 12 mm, and soaking time of 30 min are the optimum settings to minimize the surface roughness.
2. Cutting velocity, feed rate, depth of cut, and rolling direction are found significant parameter in ANOVA.
3. Cutter offset and soaking time are found non-significant process parameters.
4. The most dominated parameters are feed rate followed by cutting velocity with contribution of 58.90% and 26.66%, respectively.
5. An improvement of 8.41 dB is achieved during confirmation test at 95% confidence level.
6. Surface roughness of 1.24 μm is achieved with optimum process parameters.

References

1. Ghosh, A., Malik, A.K.: Manufacturing science. East-West Press Private Limited, New Delhi (2009)
2. Kalpakjian, S.: Manufacturing Engineering and Technology ,3rd edn. Reading: Addison-Wesley (1995)
3. Ross, P.J.: Taguchi Techniques for Quality Engineering. McGraw-Hill, New York (1988)
4. Montgomery, D.C.: Design and Analysis of Experiments. John Wiley and Sons, Singapore (1991)
5. Singh, A.K., Kumar, S., Singh, V.P.: Optimization of parameters using conductive powder in dielectric for EDM of super Co 605 with multiple quality characteristics. Mater. Manuf. Processes **29**(3), 267–273 (2014)
6. Parashar, V., Purohit, R.: Investigation of the effects of the machining parameters on material removal rate using Taguchi method in end milling of steel grade EN19. Mater. Today Proc. **4**(2), 336–341 (2017)
7. Rao, R.V., Pawar, P.J.: Parameter optimization of a multi-pass milling process using non-traditional optimization algorithms. Appl. Soft Comput. **10**(2), 445–456 (2010)
8. Gaitonde, V.N., Karnik, S.R., Davim, J.P.: Selection of optimal MQL and cutting conditions for enhancing machinability in turning of brass. J. Mater. Process. Technol. **204**(1–3), 459–464 (2008)
9. Fedai, Y., Kahraman, F., Kirli Akin, H., Basar, G.: Optimization of machining parameters in face milling using multi-objective Taguchi technique. Tehnički glasnik **12**(2), 104–108 (2018)
10. Najiha, M.S., Rahman, M.M., Kadirgama, K.: Parametric optimization of end milling process under minimum quantity lubrication with nanofluid as cutting medium using Pareto optimality approach. Int. J. Automot. Mech. Eng. **13**(2), 3345–3360 (2016)
11. Radiša, R., Dučić, N., Manasijević, S., Marković, N., Ćojbašić, Ž.: Casting improvement based on metaheuristic optimization and numerical simulation. Facta Univ. Ser. Mech. Eng. **15**(3), 397–411 (2017)
12. Petković, D., Madić, M., Radovanović, M., Gečevska, V.: Application of the performance selection index method for solving machining MCDM problems. Facta Univ. Ser. Mech. Eng. **15**(1), 97–106 (2017)

13. Aoda, Y.F., Jabur, L.S.: Effect of the rolling direction and draft on some of the mechanical properties for the medium carbon steel. Int. J. Sci. Res. **3**(12), 2425–2431 (2014)
14. Upadhyay, G.S., Upadhyay, A.: Material Science and Engineering. Viva Books, UK (2007)
15. Olvera, O., Barrow, G.: Influence of exit angle and tool nose geometry on burr formation in face milling operations. Proc. Inst. Mech. Eng. Part B J. Eng. Manuf. **212**(1), 59–72 (1998)

Chapter 23
Evolution of Microstructure Through Various Techniques of Severe Plastic Deformation

Krishna Mohan Agarwal, R. K. Tyagi, Bhuwan Gupta and Abhishek Singhal

Abstract Severe plastic deformation (SPD) is a structuring technique used to manufacture nanomaterials which are used to shape nanoscale structures. Various techniques of SPD are used for grain refinement which has capability of enhancing mechanical properties of material and its alloys. Various SPD methods like ECAP, HPT, ARB, etc., have their own advantages and disadvantages. Comparison of various techniques has been studied in the current research paper. In all important SPD methods, expert coverage of current achievements and future developments in the industry and research directions including both batch and continuous processes has been provided and a concise introduction to the field of mass production of nanomaterials for nanotechnology using the latest examples from the research. It is suitable for various engineering and industrial applications. It has been concluded that ECAP and HPT are better processes for grain refinement.

Keywords Severe plastic deformation · Grain refinement · High-pressure torsion · Cyclic extrusion · Multi-axial forging and equal channel angular pressing

23.1 Introduction

Severe plastic deformation is defined as the manufacturing technique in which the initial parameters are not significantly changed while machining the material into required shape. SPD is a process which is used for refinement of grains by decreasing the grain size and grain boundary. Refinement in grain size leads to the enhancement of mechanical properties, and microstructure becomes more homogenous. There are many types of SPD processes, i.e., high-pressure torsion (HPT), cyclic extrusion, multi-axial forging (MAF) and equal channel angular pressing (ECAP) [1–8].

In SPD, the grain boundaries of the material to be processed are considered and are changed according to the required mechanical properties. The specialty of the SPD is that imposing high strains in the material helps in changing the mechanical properties without any significant changes in its initial dimensions and exceptionally high grain refinement is achieved [8–11].

K. M. Agarwal (✉) · R. K. Tyagi · B. Gupta · A. Singhal
Mechanical Engineering Department, Amity University, Uttar Pradesh Noida 201303, India

S. Yadav et al. (eds.), *Proceedings of International Conference in Mechanical and Energy Technology*, Smart Innovation, Systems and Technologies 174,
https://doi.org/10.1007/978-981-15-2647-3_23

Now, to measure the strain in the samples, the Hall–Petch Eq. (23.1) is used which is described as follows:

$$\sigma_x = \sigma_y + kd^{-1/2} \tag{23.1}$$

where σ_x is friction stress, σ_y is applied stress, d is the grain size and k is a constant of yielding. Hall–Petch equation states that the strength of the material has a hyperbolic relation while decreasing the grain size of the material [11–14].

Thus, SPD has a very large potential in the manufacturing industry since the strength of material is very important factor; thus, SPD is used for grain refinement. In automobile, it is useful for making of chassis and other body parts. In the field of medical, biomaterials play an important role which can be enhanced by SPD. In defense industry, SPD is used in the manufacturing of missile and parts requiring high strength [15–20].

The various techniques of SPD are discussed in this paper and compared based on different tests that were done on different samples being processed by various SPD processes and compared their microstructures and their grain sizes to determine which of these processes are best for manufacturing ultra-fine grains (UFGs).

23.2 Methods of Severe Plastic Deformation

There are various techniques to perform severe plastic deformation. Each method has its own advantages and disadvantages. Further, this paper discusses each method in detail and later compared all the techniques on the basis of grain refinement.

23.2.1 High-Pressure Torsion

High-pressure torsion (HPT) was given by Percy Bridgman [2–4]. In this process, the sheet of the material is kept between two anvils and some compressive stress usually in GPa is applied on the material to deform in its required dimensions.

Figure 23.1 shows schematic diagram of HPT which shows how it is performed. One of the anvils is twisted in either of the direction, and the other anvil is kept constant so that proper compressive stress could be applied on the material to obtain the desired results with accurate dimensions. For more appropriate results, the die used is in the shape of the ring and not in the circular shape. Due to heavy strains realized in the HPT, it is possible to obtain ultra-fine grain (UFG) materials.

The shear strain imposed on the sample is given by the relation in Eq. (23.2):

$$\varepsilon_N = \frac{2\pi Nr}{h\sqrt{3}} \tag{23.2}$$

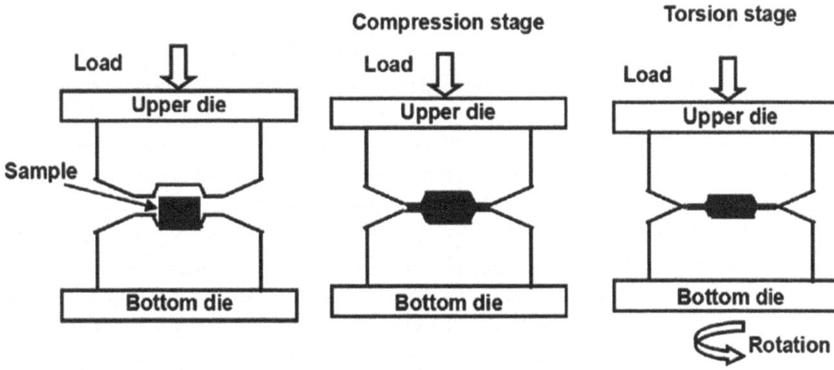

Fig. 23.1 Schematic diagram of HPT

where N is the number of HPT turns, r is the radial distance from the center of the disk and h is the disk thickness. Moreover, using the relationship $\varepsilon = \gamma/\sqrt{3}$ equivalent Von Mises strain imposed on the disk is given by the equation. The main advantage of this process is that it leads to much more refined grains as compared to other techniques. The main disadvantage is that the sample should be of circular type, whereas many industries require sample in sheet or wired form, HPT requires high amount of applied pressure and sample should be in the range of 35 mm in diameter. The strain developed is directly proportional to the distance from the disk center due to which an inhomogeneous distribution of microstructure across the diameter is observed [3].

23.2.2 Equal Channel Angular Pressing

Equal channel angular pressing (ECAP) was given by Segal [6] with its associates in the Minsk Institute in the then Soviet Union. In this process, a plunger is used to apply pressure on the material. It consists of a special die having two special channels meeting at an oblique angle known as the channel angle Φ. The angle made by the cross section of the two channels at the convergence is known as the corner angle Ψ. The strain is developed on the material at the intersection of the two channels. Figure 23.2 shows schematic diagram of ECAP which depicts how the process is executed.

$$\varepsilon_N = \frac{N}{\sqrt{3}}\left[2\cot\left(\frac{\phi}{2}+\frac{\varphi}{2}\right)+\varphi\cos\mathrm{ec}\left(\frac{\phi}{2}+\frac{\varphi}{2}\right)\right] \tag{23.3}$$

Equation 23.3 gives the value of equivalent strain obtained while performing ECAP. Here, the number of passes is N, Φ is channel angle and Ψ is corner angle. The advantage of ECAP is that maximum grain refinement is observed. Other advantage

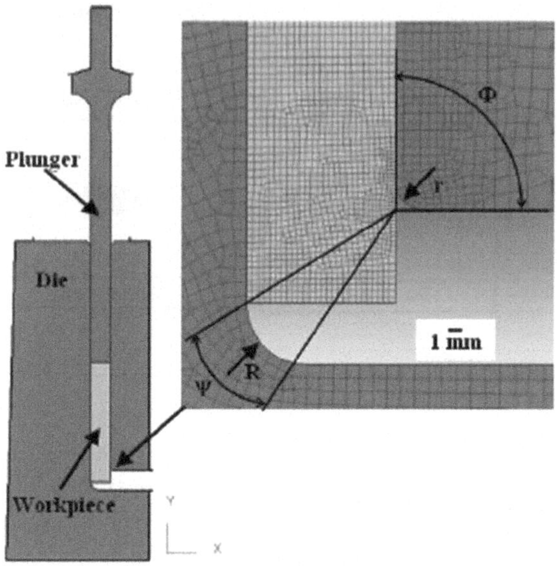

Fig. 23.2 Schematic diagram of ECAP [4]

is that unlike other conventional SPD processes, the dimension of billet does not change with the number of passes [27].

23.2.3 Cyclic Extrusion and Compression

In this process, the material is passed through a cylindrical die of diameter d_0 and then is passed through the series of the cylinders with the same or similar dimensions but has different diameters of the dies in order to refine the grain boundary of the material. The process thus includes compression and extrusion and is repeated several times until the desired properties and thickness are obtained. One of the features is that the material is preserved while having being passed through the dies and there is significant strain observed in the material (Fig. 23.3).

The equivalent strain accumulated is approximately given by Eq. (23.4)

$$\varepsilon_n = 4n \ln\left(\frac{D}{d}\right) \quad (23.4)$$

where the diameter of the chamber is D, diameter of the channel is d and n is the number of cycles of deformation. The strain developed is directly proportional to the number of passes and natural log of D/d. A high hydrostatic pressure is imposed as the billet is compressed from both ends in the CEC process. Special stressed tools are required to impose high load of extrusion–compression; otherwise, the life of the tool will be short. This process is considered best for processing of soft materials

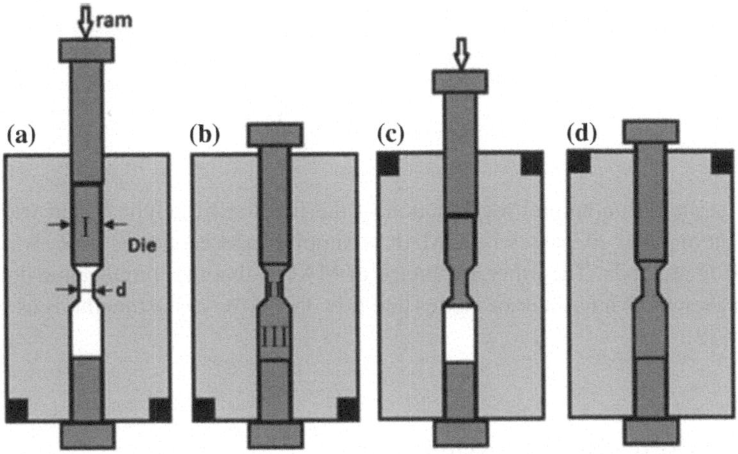

Fig. 23.3 Schematic diagram of CEC [15]

like aluminum alloy. However, the forward extrusion may be canceled due to strain imposed in backward extrusion.

23.2.4 *Multi-axial Forging*

In Fig. 23.4, the MAF principle is shown. It is the process in which the given material is orthogonally forged in *X*, *Y* and *Z* directions. This process is repeated certain number of times as per the desired condition of work. More the number of repetitions, higher the grain refinement. The homogenous strain created in MAFs is comparatively less than that observed in ECAP and HPT. However, the result of process starts at elevated temperatures and also the specific loads are relatively low on tooling square measure. Thus, to get nano-structured fragile materials this methodology is often used.

Fig. 23.4 Schematic diagram of multi-axial forging [9]

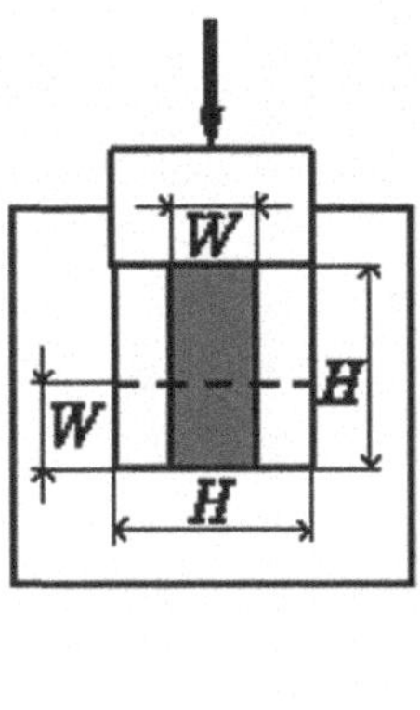

The strain being imposed on the workpiece in multi-axial forging is given by Eq. (23.5)

$$\varepsilon_n = n\frac{2}{\sqrt{3}}\ln\left(\frac{H}{W}\right) \tag{23.5}$$

The strain ε is a natural log function of the fraction of height H and width W, where the number of passes is n. MAF is simplest and economic process among other SPD methods. The other advantage of MAF is that the initial shape does not change even after n number of passes and it is one of the easiest methods as no die is used [22].

23.3 Results and Discussion

An exhaustive study has been done on each process on grain refinement and later concluded the optimal process for highest grain refinement to ultimately see the effect in mechanical properties of material.

23.3.1 *Based on HPT*

In the research, Zhilyaev [24] investigated the microstructures of commercial purity aluminum and found that there is less grain refinement in the initial stages but after HPT the grains got significantly refined after the number of passes. The grain size was found to be reduced from ~0.2–0.5 mm to 3 μm at one revolution. The following were the results of grain refinement. Figure 23.5 shows clearly that initially there is very less homogeneity and are more dislocations, after two passes there is significant amount of homogeneity and after eight passes, even after reducing the scale of observation there is high homogeneity in grain structure [23].

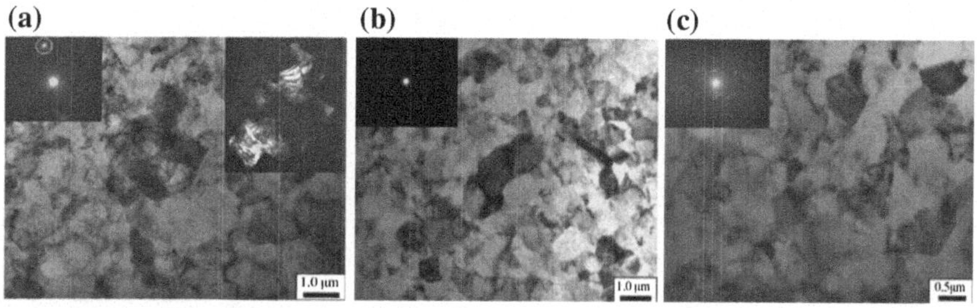

Fig. 23.5 **a** TEM image of initial sample $P = 1$ GPa without HPT, **b** TEM image after 2 revolutions of HPT ($P = 1$ GPa), **c** TEM image after 8 revolutions ($P = 1$ GPa) [24]

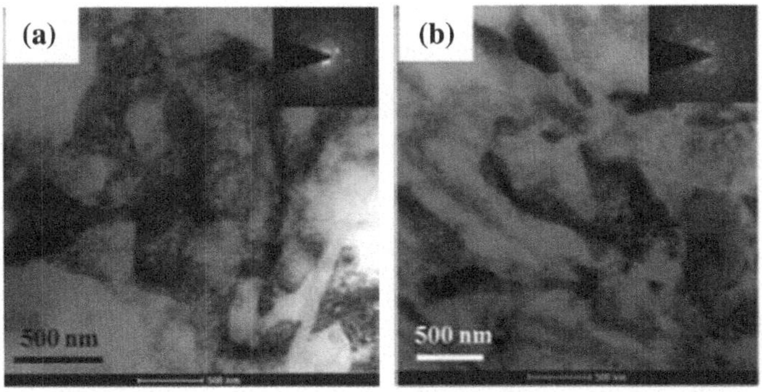

Fig. 23.6 TEM microstructure of Al 6063 alloy; **a** MAF 2 passes and **b** MAF 4 passes [25]

23.3.2 Based on Multi-axial Forging

In the research, Hussaina [25] had observed the grain refinement in multi-axial forging. The sample was Al 6063 alloy. The MAF was performed at 77 k. The author concluded that after $N = 4$ there was development of low grain boundaries and the structure is reduced to ~300 nm. By analyzing Fig. 23.6, in first figure there is high dislocation and in Fig. 23.6b there are less dislocations and more homogeneity in grain structure with grain boundaries closer to each other [21].

23.3.3 Based on Cyclic Extrusion

In the research, Geaman [26] has taken a sample of Al–Si–Cu alloy and has done cyclic extrusion on it to observe the change in grain refinement. It has been concluded that after certain number of passes approximately $N = 4$ the grain boundary reduced to 130 nm. By studying the results of research done by Virgil, in Fig. 23.7a, the structure is non-homogenous. After each pass, the structure is getting refined which is clearly visible and after 15 cycles the structure is highly refined.

23.3.4 Based on ECAP

A scanning electron microscope was used to analyze the microstructure of Al alloy AA6063 whose composition is Al—94.85%, C—2.73%, O—1.32%, Mg—0.67%, Si—0.43% (percentage is by weight). The die used for ECAP technique was made up of H13 tool steel with channel section $d = 20$ mm, channel angle $\Phi = 135°$, corner angle $\Psi = 20°$ and route B_C in which the specimen was rotated 90° in clockwise direction in succeeding number of passes at room temperature of 27 °C. The size of

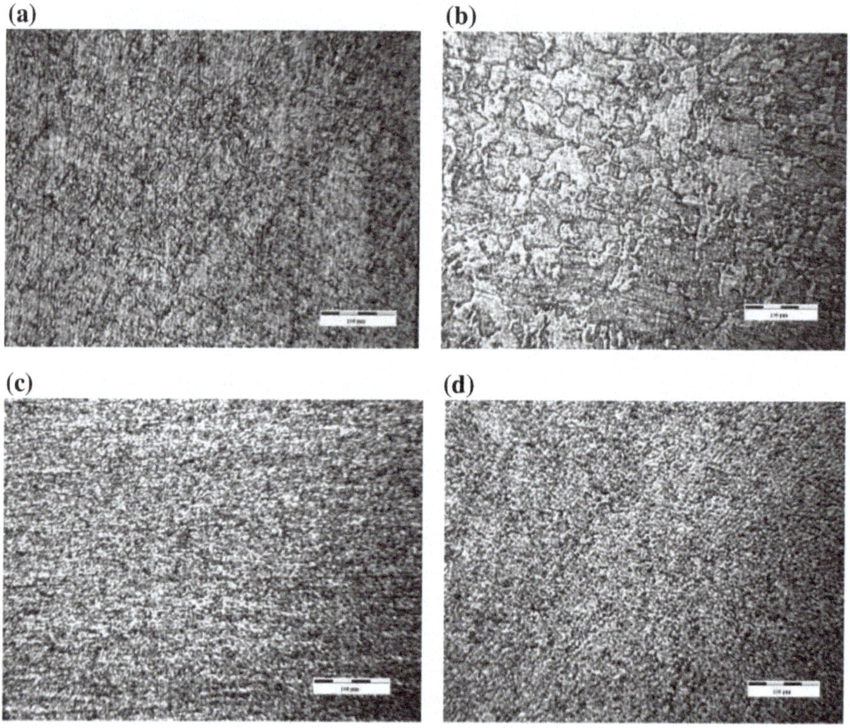

Fig. 23.7 Microstructures after **a** 2, **b** 4, **c** 10 and **d** 15 cycles [26]

sample used was 20-mm diameter and length 100 mm. By observing the results of SEM, the average grain size reduced from 456 to 361 nm (Fig. 23.8).

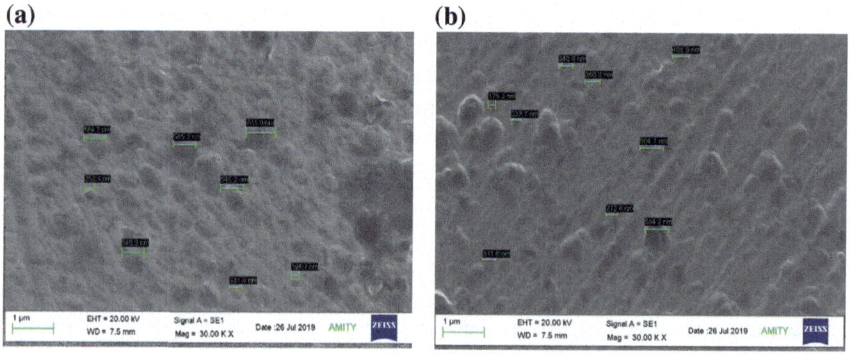

Fig. 23.8 Microstructure **a** after 0 pass and **b** after 3 passes

23.4 Conclusion

The study of various techniques of severe plastic deformation was conducted in this research paper, and it has been observed in depth giving an overview into all the methods. One of the main objectives of SPD is to get grain refinement which improves the mechanical properties of material. SPD has many different techniques, and each process has advantages and disadvantages. From the study, comparing each method with other it is concluded that equal channel angular pressing is an effective and one of the best techniques of severe plastic deformation as in this method the grain refinement is more than other techniques. In results of SEM, the image clearly depicts the change in average grain size as there is drastic reduction of grain size from 456 to 361 nm which can be utilized to enhance the strength for further processing of materials for various engineering and technological applications.

References

1. Olejnik L., Rosochowski, A.: Methods of Fabricating Metals for Nano-Technology. Bulletin of the Polish Academy of Sciences Technical Sciences (2005)
2. Valiev, R.Z., Estrin, Y., Horita, Z., Langdon, T.G., Zehetbauer, M.J., Zhu, Y.T.: Producing bulk ultrafine-grained materials by severe plastic deformation. JOM (2006)
3. Valiev, R.Z., Islamgaliev, R.K., Alexandrov, I.V.: Bulk Nanostructured Materials from Severe Plastic Deformation. Institute of Physics of Advanced Materials, Ufa State Aviation Technical University, 12 K. Marx Street, 450000 Ufa, Russian Federation (1999)
4. Nakashima, K., Horita, Z., Nemoto, M., Langdon, T.G.: Influence of channel angle on the development of ultrafine grains in equal channel angular pressing, Pergamon. Acta Mater. **46**(5) (1998)
5. Langdon, T.G., Furukawa, M., Nemoto, M., Horita Z.: Using equal-channel angular pressing for refining grain size. JOM (2000)
6. Segal V.M.: Engineering and commercialization of equal channel angular extrusion. Mater. Sci. Eng. A 386/1-2269-276 (2004)
7. Agarwal K.M., Tyagi, R.K., Chaubey, V.K., Dixit, A.: Comparison of different methods of severe plastic deformation for grain refinement. IOP Mater. Sci. Eng. (2019)
8. Werenskiold, J.C., Roven H.J.: Strain Measurement in the ECAP Process. Norwegian University of Science and Technology (2001)
9. FaramarzDjavanroodi, Mahmoud Ebrahimi& Jamal F. Nayfeh.Tribological and mechanical investigation of multi-directional forged nickel(2019)
10. Beygelzimer, Y., Kulagin, R., Estrin, Y., Toth, L.S., Kim, H.S., Latypov, M.I.: Twist extrusion as a potent tool for obtaining advanced engineering materials (2017)
11. Sunil, B.R.: Repetitive corrugation and straightening of sheet metals (2015)
12. Bhovi, P.M., Patill, D.C., Kori, S.A., Venkateswarlu, K., Huange, Y., Langdon, T.G.: A comparison of repetitive corrugation and straightening and high-pressure torsion using an Al–Mg–Sc Alloy (2016)
13. Pardis, N., Talebanpour, B., Ebrahimi, R., Zomorodian, S.: Cyclic expansion-extrusion (CEE), a modified counterpart of cyclic extrusion-compression (CEC) (2011)
14. Agarwal, K.M., Tyagi, R.K.: Investigation of mechanical properties of alloys processed after equal channel angular pressing. AGU Int. J. Eng. Technol. (AGUIJET) **5**, e-ISSN: 2455-0442, p-ISSN: 2455-6734 (2017)

15. Esmaeili, F., Mehri, F., Broomand, S.R.: A modified counterpart of cyclic extrusion-compression: experimental study and dislocation density-based finite element modeling (2016)
16. Sheikh, H., Ibrahimi, R.: Investigation on texture evolution during cyclic expansion–extrusion (CEE) technique using crystal plasticity finite element modelling
17. Satapathy, S.: Cyclic extrusion of recycled high-density polyethylene/banana fiber/fly ash cenosphere biocomposites: thermal and mechanical retention properties (2019)
18. Bhov, P.M., Patil, D.C., Kori, S.A., Venkateswarlu, K., Huange, Y., Langdon, T.G.: A comparison of repetitive corrugation and straightening and high-pressure torsion using an Al–Mg–Sc alloy (2016)
19. Edalati, K., Horita, Z.: A review on high-pressure torsion (HPT) from 1935 to 1988 (2016)
20. Agarwal, K.M., Tyagi, R.K., Dixit, A.: Theoretical analysis of equal channel angular pressing method for grain refinement of metals and alloys. Mater. Today Proc. (2019)
21. Liao, C.-C., Hsu, C.-C., Huang, J.-H.: Deformation mechanism of forging tool for multi-stage forming of deep groove ball bearing (2018)
22. Kampen, D., Richter, J., Blohm, T., Knust, J., Langner, J., Stonis, M., Behrens, B.-A.: Design of a genetic algorithm to preform optimization for hot forging process (2019)
23. Ding, L., Kerber, M., Kunisch, C., Kaus, B.J.P.: Plastic yielding of glass in high-pressure torsion apparatus (2018)
24. Zhilyaev, A.P., Oh-ishi K., Langdon, T.G., McNelley, T.R.: Microstructural evolution in commercial purity aluminum during high-pressure torsion (2005)
25. Hussain, M., Nageswararao, P., Singh, D., Jayaganthan, R., Singh, S.: Comparative study of Microstructure and Mechanical properties of Al 6063 alloy Processed by Multi axial forging at 77 K and cryorolling (2014)
26. Geaman, V., Radomira, I., Stoicanescu, M., Popaa, I.: Grain refinement in Al–Si–Cu alloy during cyclic extrusion (2012)
27. Shaeri, M.H., Shaeri, M., Ebrahimi, M., Salehi, M.T., Seyyedin, S.H.: Effect of ECAP temperature on microstructure and mechanical properties of Al–Zn–Mg–Cu alloy (2016)

Chapter 24
Effect of Input Parameters on Burr Formation During Milling Operation: A Review

Kulwinder Singh, Anoop Kumar Singh and K. D. Chattopadhyay

Abstract Burr formation on workpiece during machining is a well-established problem faced by the manufacturing industry. Growth of burr and its formation is observed in all types of machining operations. Burr is required to be removed from the workpiece before its assembly, dispatch or delivery known as deburring operation. Presence of burr on workpiece edge leads to physical injury, assembly problem, malfunctioning, wear and cost to component. This paper reviews the research work in the field of burr formation under milling operation. Role of various input parameters affecting burr formation such as work material, cutting tool, machining parameters, machining strategies, and miscellaneous parameters is described. Numerous researchers proposed different concepts to control the burr formation. The outcomes of the detailed study are presented in a summary sheet. Results show that burr formation in milling operation can be minimized by selecting optimum input parameters.

Keywords Milling operation · Burr formation · Deburring

24.1 Introduction

Burr formation is one of the main issues faced by machining sector. In machining, this phenomenon is also popular as an edge quality of the machined surface. Mostly all the manufacturing industries face this issue. To remove burr, secondary operation, i.e., deburring is done on component. Gillespie [1] reported that deburring operation may cost up to 30% of manufacturing cost for small size, complex, and precision component. Many issues such as malfunctioning, reduced part life, assembly problem, and physical injuries are reported if complete deburring is not done. Burr may be controlled by understanding the burr formation mechanism in a systematic manner. By understanding the burr reduction phenomenon, the cost of component may be reduced and service life of component may be increased. Due to technological development in the last century, lot of new materials is developed. Each material

K. Singh (✉) · A. K. Singh · K. D. Chattopadhyay
Chitkara University Institute of Engineering and Technology, Chitkara University, Rajpura, Punjab, India
e-mail: kulwinder.singh@chitkara.edu.in

S. Yadav et al. (eds.), *Proceedings of International Conference in Mechanical and Energy Technology*, Smart Innovation, Systems and Technologies 174,
https://doi.org/10.1007/978-981-15-2647-3_24

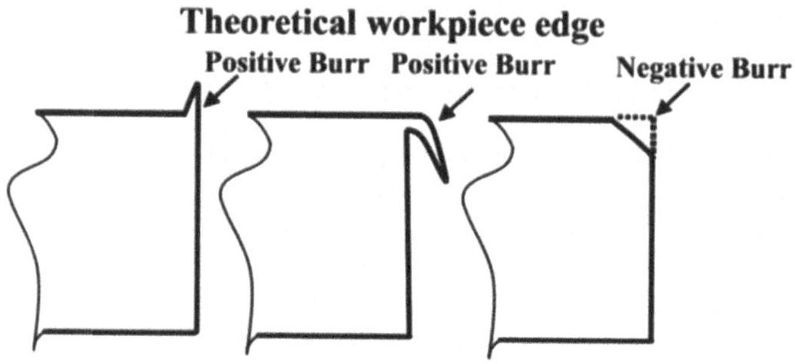

Fig. 24.1 Types of burr [6]

behaves in a different manner and generates different types of burr after machining. So, researchers defines burr in their own perspectives as discussed below:

24.1.1 Burr Description

24.1.1.1 Burr Definition

In engineering drawing, geometrical shape of component is drawn without any edge conditions except chamfering, radius and filleting, etc. But for the specific jobs, functional and safety concerns, such as sharp edge, burr-free edge, etc., are required to be mentioned in drawing. As per ISO 13715 [2], if edge of workpiece is overhang greater than zero, then it will be considered a burred workpiece. Schafer [3] reported that part of a job which is generated on the surface or edge lies outside the required geometry is burr. In German standard DIN 6784 [4], similar type of burr definition can be found, i.e., workpiece edge with overhang portion is known as burred edge. It is only limited up to edge of workpiece but Schafer [3] includes surface as well. Burr is an unwanted projection of material generated on the edge of workpiece due to plastic flow while shearing as reported by Ko and Dornfeld [5]. Moreover, Gillespie [6] defined the burr on the basis of theoretical intersection of two surfaces after machining. Undesired material, beyond the intersection of two surfaces, is known as positive burr and inside the intersection of two surfaces is known as negative burr shown in Fig. 24.1.

24.1.1.2 Deburring

Deburring is the burr removal process to prepare the edge of component. Therefore, it is considered as a secondary operation. Amount of deburring depends upon the burr size along with customer requirements. Different deburring techniques are used by industries such as mechanical, thermal, chemical, and electrical as reported by

Guo and Dornfeld [7]. Robotic arms also preferred for deep and complex zone in deburring.

24.1.1.3 Cost to Company

Deburring cost is an important component during cost estimation of finished product. Many factors contribute to deburring cost such as burr value, deburring technique, machining operation, material properties, complexity, and applications of parts. It may vary from 2 to 14% of manufacturing cost for automotive parts depending upon above mentioned factors. For highly precise and sensitive components such as medical equipment or aerospace industry components, this figure may rise up to 30% [1].

24.2 Parameters Affecting Burr Formation

Scientists proposed many theories on parameters affecting burr formation. Still, it requires a detail investigation on it. Burr minimization is a big challenge to researchers so that deburring cost can be reduced. Because of complex interaction among governing parameters, it is difficult to eliminate the burr. As per Gillespie and Blotter [8], burr can be minimized but it cannot be completely eliminated. Machining parameters, shape of component, tool geometry, and workpiece material are the main four parameters responsible in burr formation [9]. Available literature is reviewed thoroughly to identify the role of a particular parameter in burr formation and contribution of each parameter is discussed as below.

24.2.1 Workpiece

24.2.1.1 Mechanical Properties

To predict the burr formation mechanism, mechanical properties like yield strength, hardness, ultimate tensile strength, and work hardening are found most significant parameters as revealed by Aurich et al. [10]. It is observed that ductility is directly proportional to positive burr formation. Larger burr forms on ductile material until the deformation is restricted by using some supporting arrangement. Mostly, automotive and aerospace parts are manufactured by ductile material which enhances the burr growth [11]. Magnesium alloy is investigated under face milling operation [12]. Brittle material behaves completely reverse to ductile material. Negative burr, i.e., material is removed from the edge of workpiece, is generated on brittle material [13]. Additionally, elasticity promotes the back cutting phenomenon. Material gets recovered after machining by front insert, then again machined by back insert known as back cutting. Rangarajan and Dornfeld [14] revealed that back cutting also affects the burr formation mechanism. Moreover, grain size also affects the burr formation mechanism. In comparison, of normal grain size, less burr is found in fine grain specimen, as reported by Kumar et al. [15].

24.2.1.2 Size and Shape

Size and shape of the workpiece play an important role in burr generation. Demand of microsize component is increasing rapidly because of technological development. Small cutter (less than 1 mm in diameter) is used to perform micromilling. Complex and small-sized components are manufactured by micromilling operation. Deburring of microcomponent is a difficult and time-consuming task, and so, it is essential to manufacture burr-free component in micromilling. Under the investigation of workpiece edge angle, it is reported that edge angle plays an important role in burr formation. Machining on sharp edge (less than 90° angle) generates thin and long burr as compared to edge with 90° or more than 90° angle [16].

24.2.2 Cutting Tool

24.2.2.1 Material and Coating

Cutting tool material has a major role in burr formation mechanism. Balduhn and Dornfeld [17] proposed a database for selection of cutting insert materials. As per Tripathi and Dornfeld [18], burr-free milled surface may be generated by using high-speed diamond tool. Hard coating on cutting inserts is preferred to improve machinability. Razak et al. [19] reported that coated carbide inserts are harder and tougher than uncoated which further helps to improve machinability. Multilayer hard coating increases tool life [20]. Coating acts as a thermal barrier among workpiece, chip and tool insert because of low heat transfer rate. This also helps in reducing friction force as per Singh et al. [21]. Performance of various hard coated and uncoated cutting tools is investigated during micromachining of Ti6Al4V alloy [22]. Outcomes show that uncoated tool generates more burr as compared to coated tools. Among different coatings, PCD coating is observed best in tool life while machining of nickel silver alloy as observed by Swain et al. [23]. Nickel-based superalloy is also investigated by using uncoated and TIALN-coated cutting tool. TIALN-coated tool generates small burr as compared to uncoated tool [24], but Olvera and Barrow [25] reported that there is a negligible effect of coating on burr size.

24.2.2.2 Tool Geometry

Avila and Dornfeld [26] observed that burr size highly depends upon tool geometry. Bansal [27] noticed that least burr with good surface quality can be attained by shaping tool geometry with negative radial and positive axial rack angles. Gillespie and Blotter [8] reported that built-up edge is also responsible for burr formation. Positive rake angle helps to control built-up edge and also to minimize the burr size. Further, Kishimoto [28] observed that burr can also be reduced by reducing lead angle, nose radius and increasing axial rake angle of cutting tool. Niknam and Songmene [29] studied high-speed slot milling and found that tool nose radius affects exit milling burr. Large exit bottom burr is generated by higher nose radius and less burr on exit up side.

24.2.2.3 Tool Wear

The tool wear intensifies burr growth. In milling, back rack and flank face of the cutting tool directly come in contact with chip and machined surface, respectively. Because of continue interface, the formation of crater and flank wear occurs that leads contact area, cutting forces, stresses, and temperature between surface and cutting tool. Worn tool affects shear plane and material deformation flow as compared to sharp tool [30]. Choi et al. [31] also reported that worn tool promotes burr formation at the entry and exit of the machined surface.

24.2.3 Machining Parameters

Machining parameters such as cutting speed, feed rate, and axial depth of cut are the key parameters in material removal process. Optimum settings of machining parameters can be used as a convenient tool to control burr growth. Small burr and good surface finish can be achieved by using high cutting speed on ductile material such as aluminum. But for face milling of magnesium, low speed is suggested to achieve less burr with good surface quality [27]. Rangarajan [32] investigated that surface gets harden while high-speed machining because of higher cooling rate on surface, which helps to reduce burr height. In orthogonal cutting of ductile materials, burr height can be reduced by increasing cutting speed, noticed by Ko and Dornfeld [33]. Moreover, in machining of aluminum alloy, burr can be minimized with high feed rate, revealed by Jones and Furness [34]. Olvera and Borrow [35] presented that depth of cut is the most significant parameter for burr height at tool exit edge. In face milling of aluminum alloy, depth of cut and tool feed rate are found dominated factors for secondary burr formation. Moreover, in micromilling, Chern and Dornfeld [36] found that burr growth is directly proportional to axial depth of cut. As per Wan et al. [37], burr formation can be reduced by minimizing uncut chip thickness up to cutting edge radius. Further if it reduced more, again burr starts to increase. Chip size effect, on burr formation in micromilling, is studied by Zhang et al. [38]. Mohid and Rahim [39] investigated the least burr at high feed rate during milling of ductile material using microball end mill. Chern [40] reported that depth of cut and feed rates are dominated factors to generate secondary burr. Tiabi [41] has proposed a matrix of significant cutting parameters against different types of burr in ranking order. For each type of burr, order of parameters with respect to significance is changed.

24.2.4 Machining Strategy

24.2.4.1 In-Plane Exit Angle

In machining plane, the angle between cutting velocity vector and edge of machined surface is known as in-plane exit angle. Olvera and Barrow [35] observed that in-plane exit angle is an important parameter to minimize exit burr. It also affects the size and shape of the burr. Moreover, Chern [24] studied that shape and size of exit burr (burr at tool exit edge) is changed when the exit angle altered. Exit angle from 30° to 180° with an increment of 30° is altered to observe different types of burr. As per Luo et al. [42], during face milling of aluminum alloy, the least burr is generated at exit angle 76° and 118° while the largest burr is formed at 90°.

24.2.4.2 Tool Path

The path followed by cutting tool on machined surface is known as tool path. Burr is formed due to engagement of cutting tool and machined surface while machining. As per Niknam et al. [43], burr formation can be reduced by implementing proper tool path planning. Path of cutting tool should be planned in such a way so that least burr may be produced.

24.2.5 Miscellaneous Parameters

24.2.5.1 Cooling and Lubrication

Temperature affects the mechanical properties of material. Rangarajan [32] observed that high-speed machining hardens the surface because of high cooling rate which leads to low burr height. Material hardness changes with temperature. At low temperature, material surface gets harden. So, it is tried to control the burr formation by reducing the surface temperature, especially at the machined part edge. Przyklenk [16] applied dry ice on the machined edge to minimize burr. Burr size and tool wear can be minimized by using coolant in machining as reported by Shefelbine and Dornfeld [44]. Tiabi [41] observed that lubrication also helps to reduce friction which further leads to burr formation. But Aurich et al. [45] reported that few materials get harden due to lubrication which makes deburring difficult.

24.2.5.2 Chamfering and Heat Treatment

Chamfering is a pre-machining operation, performed on the edge of work piece. In this operation, negative burr is generated in advance, on the edge of work piece to compensate the positive burr as described in Tiabi [41]. Moreover, Gillespie [46] proposed many methods to control burr formation such as hard machining, laser treatment, localized mechanical, chemical and thermal treatment. Niknam et al. [43] studied the effect of heat treatment on burr size during drilling of aluminum alloys.

24.2.5.3 Clamping

Clamping force on the edge of workpiece also affects the burr formation. In micromilling, support at the edge of workpiece is recommended to eliminate burr [47]. Instant adhesive is also used as a supporting material to the edge of workpiece. It improves the edge rigidity against deformation. Deformation further transferred to the edge of supporting material.

24.2.5.4 Optimization of Process Parameters

As per Medeossi et al. [48], burr formation can be minimized by modeling and optimization of its process parameters. Study of literature shows that each parameter has its own significance in minimizing the burr formation. But still, battle against burr formation is continued because optimum setting of each factor is difficult to control. Summary on input parameters affecting burr formation is listed in Table 24.1.

24.3 Conclusion and Future Scope

Burr on edges of workpiece during milling operation is a critical issue to many manufacturing industries. In this review paper, effect of various input parameters on burr formation is discussed. The detailed study is presented in the tabular form as a summary. This study concludes that burr formation can be controlled by selecting optimum input parameters during milling operation. Many approaches are adopted by research scholars to control this issue but still more work is required in the direction of following areas of grain size and its direction, multiresponse optimization, heat treatment, and cryogenic cooling.

Table 24.1 Summary of literature review

S. No.	Parameter zone	Parameter	Effect on burr formation
1	Workpiece	Mechanical properties	Mechanical properties have a vital role in burr growth. Larger burr generates on ductile material and negative burr generates on brittle materials. Moreover, small burr found on fine grain specimen comparatively normal grain size [10, 11, 15]
		Workpiece edge angle	Machining of sharp edges (<90°) generates thin and large burr as compare to edge of 90° or > 90° [16]
2	Cutting tool	Material and coating	Uncoated tool generates more burr as compare to coated tool. Carbide with hard coating is proposed to reduce burr generation [18, 19]
		Tool geometry	Burr growth is highly depends upon tool geometry. It can be controlled by increasing axial rake angle and decreasing nose radius of cutting tool [26, 28]
		Tool wear	In milling, formation of crater and flank wear occurs that leads contact area, cutting forces, stresses and temperature which further promotes burr growth [30]
3	Machining parameters	Depth of cut	Depth of cut is the most significant factor in burr growth. Less amount of axial depth of cut helps to control burr. Burr formation can be reduced by minimizing uncut chip thickness up to cutting edge radius [36, 37]
		Cutting velocity	To minimize burr, high cutting velocity is preferred for ductile material whereas for brittle material, low cutting velocity is proposed [27, 32]

(continued)

Table 24.1 (continued)

S. No.	Parameter zone	Parameter	Effect on burr formation
		Feed rate	High feed rate is used to reduce burr on ductile material [34]
4	Machining strategies	In-plane exit angle	In machining plane, angle between cutting velocity vector (V) and edge of machined surface is known as In-plane exit angle. The least burr generates at 76° and 118° of in-plane exit angle whereas larger burr generates at 90° [42]
		Tool path	Burr growth can be controlled by proper planning of tool path depending upon the size and shape of component [43]
5	Miscellaneous	Cooling technique	High cooling rate helps to control the burr growth because material hardness changes with temperature. It also helps to increase tool life [32]
		Clamping	Special clamping techniques such as temporary support on edge help in burr minimization [47]
		Surface treatment	Localized mechanical, thermal and chemical treatments are proposed to control burr [46]
		Chamfering	Chamfering is a pre-machining operation, i.e., negative burr is generated on edge to control burr growth [41]

References

1. Gillespie, L.K.: Deburring precision miniature parts. Precis. Eng. **1**(4), 189–198 (1979)
2. International Standard, ISO 13715: Technical drawings—edges of undefined shape—vocabulary and indications (2000)
3. Shafer, F.: Product design influences on deburring. In: Technical Paper MR75-483, Society of Manufacturing Engineers (SME), Dearborn, MI (1975)
4. German Institute for Standardization, DIN 6784: Edges of workpieces—concepts, indications on drawings (1982)

5. Ko, S.L., Dornfeld, D.A.: A study on burr formation mechanism. J. Eng. Mater. Technol. **113**(1), 75–87 (1991)
6. Gillespie, L.K.: The battle of the burr: new strategies and new tricks. Manuf. Eng. (USA) **116**(2), 69–70 (1996)
7. Guo, Y.B., Dornfeld, D.A.: Finite element modeling of burr formation process in drilling 304 stainless steel. J. Manuf. Sci. Eng. **122**(4), 612–619 (2000)
8. Gillespie, L.K., Blotter, P.T.: The formation and properties of machining burrs. J. Eng. Ind. **98**(1), 66–74 (1976)
9. Wang, G.C., Zhang, C.Y.: Mechanism of burr formation in milling. Key Eng. Mater. **259**, 278–281 (2004)
10. Aurich, J.C., Dornfeld, D., Arrazola, P.J., Franke, V., Leitz, L., Min, S.: Burrs—analysis, control and removal. CIRP Ann. Manuf. Technol. **58**(2), 519–542 (2009)
11. Shaw, M.C.: Metal Cutting Principles. Oxford University Press, New York (1984)
12. Matuszak, J., Zaleski, K.: Effect of milling parameters upon burr formation during AZ91 HP magnesium alloy face milling. In: Swic, A., Lipski, J. (eds.) New Materials and IT Technologies in Production Engineering. Lublin Scientific Society, Lublin (2011)
13. Kim, J., Dornfeld, D.A.: Development of an analytical model for drilling burr formation in ductile materials. J. Eng. Mater. Technol. **124**(2), 192–198 (2002)
14. Rangarajan, A., Dornfeld, DA.: Back cutting and tool wear influence on burrs in face milling-analysis and solutions. In: Consortium on Deburring and Edge Finishing. University of California, Berkeley, CA (2004)
15. Kumar, P., Kumar, M., Bajpai, V., Singh, N.K.: Recent advances in characterization, modeling and control of burr formation in micro-milling. Manuf. Lett. **13**, 1–5 (2017)
16. Przyklenk, K.: Abrasive flow machining: a process for surface finishing and deburring of workpieces with a complicated shape. In: Niku-Lari, A. (ed.) Advances in Surface Treatments: Technology—Applications—Effects. Pergamon Press (1986)
17. Balduhn, A., Dornfeld, D.A.: Model of a burr expert system. In: Consortium on Deburring and Edge Finishing. University of California, Berkeley, CA (2003)
18. Tripathi, S., Dornfeld, D.A.: Review of geometric solutions for milling burr prediction and minimization. Proc. Inst. Mech. Eng. Part B J. Eng. Manuf. **220**(4), 459–466 (2006)
19. Razak, N.H., Rahman, M.M., Kadirgama, K.: Experimental study on surface integrity in end milling of HASTELLOY C-2000 super alloy. Int. J. Automot. Mech. Eng. **9**, 1578–1587 (2014)
20. Panda, A., Sahoo, A.K., Panigrahi, I., Rout, A.K.: Investigating machinability in hard turning of AISI 52100 bearing steel through performance measurement: QR, ANN and GRA study. Int. J. Automot. Mech. Eng. **15**(1), 4935–4961 (2018)
21. Singh, T., Singh, P., Dureja, J.S., Dogra, M., Singh, H., Bhatti, M.S.: A review of near dry machining/minimum quantity lubrication machining of difficult to machine alloys. Int. J. Mach. Mach. Mater. **18**(3), 213–251 (2016)
22. Chen, W., Teng, X., Zheng, L., Xie, W., Huo, D.: Burr reduction mechanism in vibration-assisted micro milling. Manuf. Lett. **16**, 6–9 (2018)
23. Swain, N., Venkatesh, V., Kumar, P., Srinivas, G., Ravishankar, S., Barshilia, H.C.: An experimental investigation on the machining characteristics of Nimonic 75 using uncoated and TiAlN coated tungsten carbide micro-end mills. CIRP J. Manufact. Sci. Technol. **16**, 34–42 (2017)
24. Chern, G.L.: Experimental observation and analysis of burr formation mechanisms in face milling of aluminum alloys. Int. J. Mach. Tools Manuf **46**(12–13), 1517–1525 (2006)
25. Olvera, O., Barrow, G.: An experimental study of burr formation in square shoulder face milling. Int. J. Mach. Tools Manuf **36**(9), 1005–1020 (1996)
26. Avila, M.C., Dornfeld, D.A.: The face milling burr formation mechanisms and minimization strategies at high tool engagement. In: Consortium on Deburring and Edge Finishing. University of California, Berkeley, CA (2004)
27. Bansal, A.: Comprehensive approach to burr prediction. In: LMA, Annual Reports, UC Berkeley, pp. 18–24 (2001)
28. Kishimoto, W.: Study of burr formation in face milling-conditions for the secondary burr formation. Bull. Japan Soc. Prec. Eng. **15**(1), 51–53 (1981)

29. Niknam, S.A., Songmene, V.: Factors governing burr formation during high-speed slot milling of wrought aluminum alloys. Proc. Inst. Mech. Eng. Part B J. Eng. Manuf. **227**(8), 1165–1179 (2013)
30. De Souza, A.M., Sales, W.F., Ezugwu, E.O., Bonney, J., Machado, A.R.: Burr formation in face milling of cast iron with different milling cutter systems. Proc. Inst. Mech. Eng. Part B J. Eng. Manuf. **217**(11), 1589–1596 (2003)
31. Choi, G.S., Wang, Z.X., Dornfeld, D.A. and Tsujino, K.: Development of an intelligent on-line tool wear monitoring system for turning operations. In Proceedings of USA–Japan Symposium on Flexible Automation. A Pacific Rim Conference ISCIE Kyoto, pp. 683–690 (1990)
32. Rangarajan, A.: Priority based tool path planning for face milling. In: Masters Report, Department of Mechanical Engineering, University of California, Berkeley, California (2001)
33. Ko, S.L., Dornfeld, D.A.: Analysis and modeling of burr formation and break out in metal. In Proceedings of the Symposium on Mechanics of Deburring and Surface Finishing Processes. ASME Winter Annual Meeting, pp. 79–92 (1989)
34. Jones, S.D., Furness, R.J.: An Experimental Study of Burr Formation for Face Milling 356 Aluminum. Transactions-North American Manufacturing Research Institution of SME, pp. 183–188 (1997)
35. Olvera, O., Barrow, G.: Influence of exit angle and tool nose geometry on burr formation in face milling operations. Proc. Inst. Mech. Eng. Part B J. Eng. Manuf. **212**(1), 59–72 (1998)
36. Chern, G.L., Dornfeld, D.A.: Burr/breakout model development and experimental verification. J. Eng. Mater. Technol. **118**(2), 201–206 (1996)
37. Wan, Z., Li, Y., Tang, H., Deng, W., Tang, Y.: Characteristics and mechanism of top burr formation in slotting micro channels using arrayed thin slotting cutters. Precis. Eng. **38**(1), 28–35 (2014)
38. Zhang, T., Liu, Z., Xu, C.: Influence of size effect on burr formation in micro cutting. Int. J. Adv. Manuf. Technol. **68**(9–12), 1911–1917 (2013)
39. Mohid, Z., Rahim, E.A.: Chip pattern, burr and surface roughness in laser assisted micro milling of Ti6Al4V using micro ball end mill. J. Mech. Eng. Sci. **12**(1), 3410–3430 (2018)
40. Chern, GL.: Analysis of Burr Formation and Breakout in Metal Cutting. In: Ph.D. Thesis. University of California at Berkeley, Berkeley, CA (1993)
41. Tiabi, A.: Burr formation and finishing of pieces. In: Doctoral Dissertation. School of higher technology, University of Quebec, Montreal, QC, Canada (2010)
42. Luo, M., Liu, G., Chen, M.: Mechanism of burr formation in slot milling Al-alloy. Int. J. Mater. Prod. Technol. **31**(1), 63–71 (2008)
43. Niknam, S.A., Wygowski, W., Balazinski, M., Songmene, V.: Milling burr formation and avoidance. In: Paulo Davim J. (ed.) Machinability of Advanced Materials. London, UK: ISTE Wiley (2014)
44. Shefelbine, W., Dornfeld, D.: The effect of dry machining on burr size. In: Consortium on Deburring and Edge Finishing. Laboratory for Manufacturing and Sustainability, University of California, Berkeley, CA (2004)
45. Aurich, J.C., Sudermann, H., Bil, H.: Characterization of burr formation in grinding and prospects for modelling. CIRP Ann. Manuf. Technol. **54**(1), 313–316 (2005)
46. Gillespie, L.K.: Deburring and Edge Finishing Handbook. Society of Manufacturing Engineers, Dearborn, Michigan (1999)
47. Kou, Z., Wan, Y., Cai, Y., Liang, X., Liu, Z.: Burr controlling in micro milling with supporting material method. Procedia Manuf. **1**, 501–511 (2015)
48. Medeossi, F., Sorgato, M., Bruschi, S., Savio, E.: Novel method for burrs quantitative evaluation in micro-milling. Precis. Eng. **54**, 379–387 (2018)

Chapter 25
Investigation on Effects of Deposition Characteristics on Weld Quality During PCGMAW

B. P. Agrawal, P. K. Ghosh, Sudhir Kumar Singh and S. N. Satapathy

Abstract The quality of weld consists of mechanical and metallurgical characteristics of the joint. This quality specifically in thick section welding depends on deposition characteristics. The deposition characteristics include area of fusion (A_F) and area of weld deposit (A_W), deposition form factor (FF), fraction of base metal fusion per unit mass of bead deposition (F_{BMF}), dilution % (D_L) and bead toe angle (B_{TA}). Application of better deposition characteristics in welding results in higher quality of joint. To obtain quality further better, the industries are using advanced techniques of pulsed current gas metal arc welding (PCGMAW) with narrow groove. The application of PCGMAW complicates the selection of pulse parameters used. In this investigation, influence of different combination of pulse parameters on deposition characteristics governing the quality of weld has been studied. Also, the pulse parameters producing better characteristics and quality have been suggested to be used by industries.

Keywords PCGMAW · Pulse parameters · Weld pool · HSLA steel · Thermal cycle

25.1 Introduction

The rising thickness of pieces and usage of new materials in application demands utilization of multipass welding. For a given groove size, the quality of the weld joint is influenced by deposition characteristics. This quality will be in terms of mechanical and metallurgical characteristics. The deposition characteristics comprise FF, F_{BMF}, D_L and B_{TA}. Commonly, different fusion welding processes are mostly utilized for assembling of thicker sections [1–3]. These fusion welding processes administer the quality of joint dependent on their influence on weld thermal behaviour and deposition characteristics [3–5]. The different hitches of fusion welding of SMAW,

B. P. Agrawal (✉) · S. K. Singh · S. N. Satapathy
School of Mechanical Engineering, Galgotias University, Greater Noida, U. P. 201308, India

P. K. Ghosh
Siddharth Group of Institutions, Puttur, Chittoor District, A.P 517583, India

S. Yadav et al. (eds.), *Proceedings of International Conference in Mechanical and Energy Technology*, Smart Innovation, Systems and Technologies 174,
https://doi.org/10.1007/978-981-15-2647-3_25

GTAW and SAW processes can be overcome through GMAW process welded with metal transfer of spray mode [6–8]. This mode is achieved at higher welding current with higher energy input of heat worsening the quality. This is diminished by pulse current gas metal arc (PCGMA) welding process [9–11]. Here, pulsation of current among lower and higher value is applied dropping effective current, energy input and superior manoeuvering of heat distribution. In PCGMAW process, additional parameters evolved are pulse current (I_p), base current (I_b), pulse time (t_p) and pulse frequency (f) [12–14]. The utilization of correct pulse parameters for higher quality is required, the miscellany of which is challenging. This is diminished by the summarized effect of pulse parameters as [15, 16] $\left[\phi = \left(\frac{I_b}{I_p} \times f \times t_b\right)\right]$ where $t_b = \frac{1}{f} - t_p$, base current duration. Energy input in the form of heat (Ω) to weld, arc voltage (V), welding speed (S), weld isotherm and metal transfer behaviour also control the quality of weld [17, 18]. Consequently, the effects of interdependence pulse parameters and mean current I_m managing the quality of deposition characteristics of weld in PCGMAW are scares. Thus, in this investigation, the outcome of applying various grouping of parameters of PCGMAW on quality of deposition characteristics is studied. Such study and observation may be very much useful for industries [19, 20].

25.2 Conduction of Experimentation

The welding was done with GMAW and PCGMAW on 16 mm thickness steel made of high-strength low-alloy (HSLA) steel plate using 1.2 mm diameter filler wire. The welding was done with reverse polarity and shielding of commercial pure argon (99.98%) at a flow rate of 18 l/min. The fixture used is shown in Fig. 25.1. The copper backing plate is used to support the molten metal during welding. The nozzle was

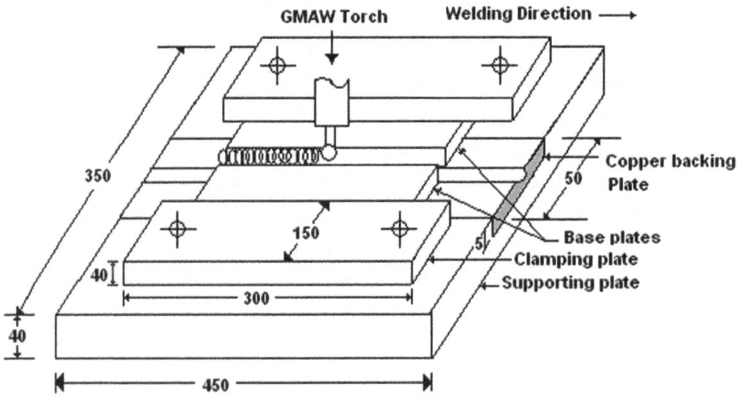

Fig. 25.1 Schematic diagram of welding fixture

kept at a distance of 15 mm from plate. The V and I_m at varied pulse parameters are given in Table 25.1. The samples were polished and etched in 2% nital solution. The deposition characteristics of weld were studied by graphical method (Fig. 25.2).

$$\mathrm{FF} = \frac{W_b}{H_r} \tag{25.1}$$

$$\%\mathrm{Dilution} = \frac{A_F}{A_F + A_W} \times 100 \tag{25.2}$$

The thermal characteristics of the weld of PCGMA welding process are studied with respect to heat input (W), total heat transferred to the weld pool (QT) and weld pool temperature (TWP) using equations discussed [15, 16].

The weld pool temperature (T_{WP}) of PCGMAW processes has been predicted

$$T_{\mathrm{WP}} = (T_{\mathrm{arc}}) + (T_{\mathrm{filler}})$$

where, T_{arc} and T_{filler} are temperature rise at the point in weld pool owing to arc heating and heated metal droplets as reported by author [18, 19].

$$F_{\mathrm{BMF}} = \frac{A_F}{A_w} \tag{25.3}$$

25.3 Outcomes and Discussions

Parameters have substantial impact on the deposition characteristics, the knowledge of which may be useful in getting desired quality.

25.3.1 Deposition Characteristics of GMAW Weld Bead

The deposition characteristics referred as the FF, A_F, D_L and B_{TA} enhance with increase of Ω Fig. 25.3a–d. The increase of FF with the enhancement of Ω has been accredited to escalation of energy transmitted to the weld resulting in the increase of the average temperature and fluidity of weld pool. Higher fluidity accompanied by higher average temperature of molten metal is responsible for more fusion of the filler as well as base metal resulting in higher bead width and FF. The increment of A_F with the increase of Ω has happened due to enhancement of metal transfer per unit length of the weld creating higher temperature of the weld pool resulting in comparatively higher base metal fusion which leads to higher A_F. The variation of D_L may have happened due to variation in area of base metal fusion and area of

Table 25.1 Various combination of pulse parameters and corresponding energy input of molten metal and arc to weld by PCGMAW process

Arc voltage (V)	Total energy input (Ω) (kJ/cm)	Welding speed (S) (cm/min)	Mean current (I_m) (A)	I_{eff} (A)	ϕ	Pulse parameters					Energy input of molten metal and arc	
						I_p	I_b	f (Hz)	t_b (sec)	t_p (sec)	Q_f (J/s)	Q_{AW} (J/s)
28 ± 1	8.20 ± 0.3	31.7	220 ± 3	278	0.04	388	35	102	3.88	4.03	5376	1927
	12.1 ± 0.2	21.4		259	0.08	357	67	104	3.87	4.01	4981	1835
	13.4 ± 0.5	19.3		247	0.15	332	125	107	3.97	3.66	4725	1786
				234	0.24	295	164	106	4.07	3.51	4576	1760
	8.20 ± 0.3	34.6	240 ± 2	298	0.04	410	37	123	3.16	3.51	5765	2137
	12.1 ± 0.2	23.3		288	0.08	377	84	125	2.73	3.55	5372	2042
	13.4 ± 0.5	21.1		257	0.15	350	121	124	3.47	3.17	5122	1989
				242	0.25	316	164	126	3.84	3.01	4989	1966
	8.20 ± 0.3	38.2	265 ± 4	323	0.04	421	47	144	2.38	3.35	6214	2400
	12.1 ± 0.2	25.8		296	0.08	393	75	142	2.66	3.27	5811	2298
	13.4 ± 0.5	23.2		275	0.15	360	146	143	2.97	2.98	5590	2249
				268	0.25	322	190	145	2.55	2.88	5509	2234

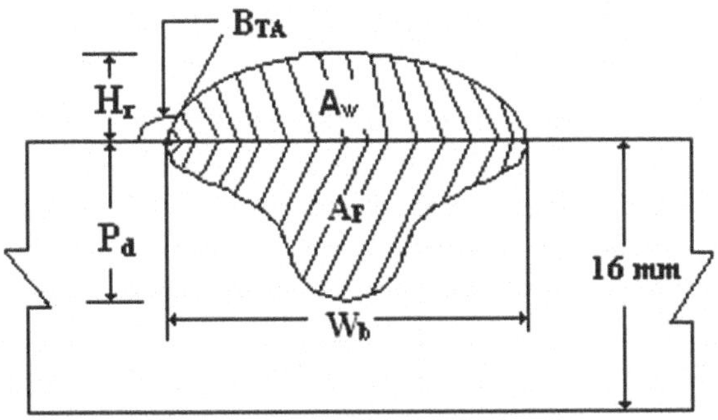

Fig. 25.2 Diagram showing deposition characteristics of weld

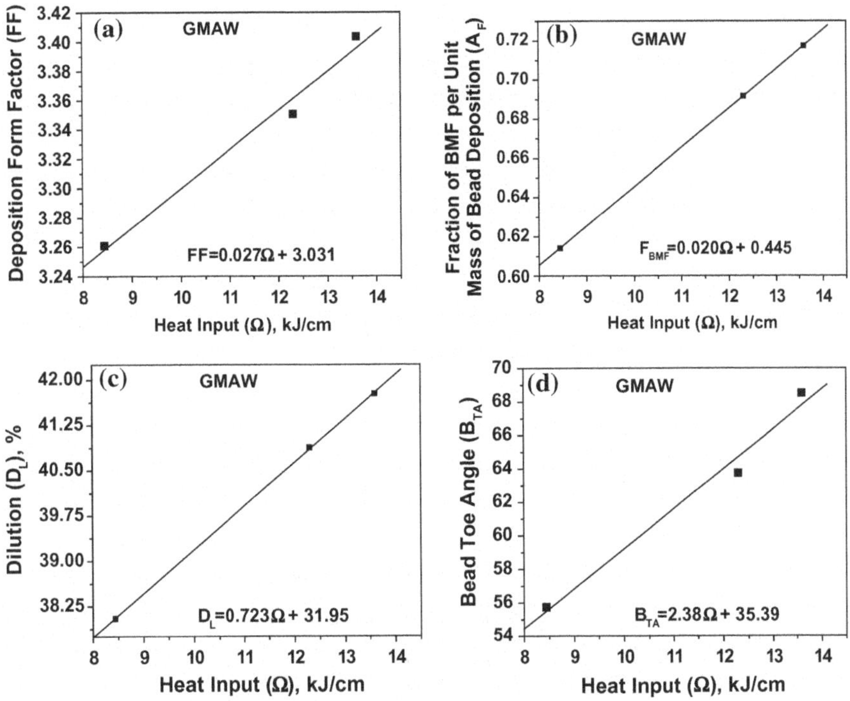

Fig. 25.3 Influence of Ω on deposition characteristics of GMA weld bead

bead deposition. The enlargement of B_{TF} with the enhancement of energy input may have primarily caused by the enhancement of molten metal transferred due to lower welding speed.

25.3.2 Deposition Characteristics of P-GMAW Weld Bead

The FF diminishes with the increase of ϕ (Fig. 25.4). This is due to the reduction of bead width and increment of bead height with the increase of ϕ. The increase of Ω and growth of mean I_m improves FF. This has occurred due to the similar type of variation of bead width and bead height. In welding higher value of deposition form factor is desired. The empirical correlations of FF with the ϕ and Ω are

$$\text{At } I_m = 220 \pm 3\,\text{A} \quad \text{FF} = 0.057\,\Omega - 4.948\phi + 0.096\,\Omega\phi + 4.701 \tag{25.4}$$

$$\text{At } I_m = 240 \pm 2\text{A} \quad \text{FF} = 0.066\,\Omega - 5.247\phi - 0.130\,\Omega\phi + 4.665 \tag{25.5}$$

$$\text{At } I_m = 265 \pm 4\,\text{A} \quad \text{FF} = 0.079\,\Omega - 6.027\phi + 0.231\,\Omega\phi + 4.578 \tag{25.6}$$

At a given Ω and I_m, the increase of ϕ reduces the fraction of base metal fusion per unit mass of bead deposition, F_{BMF} (Fig. 25.5). This has happened due to reduction

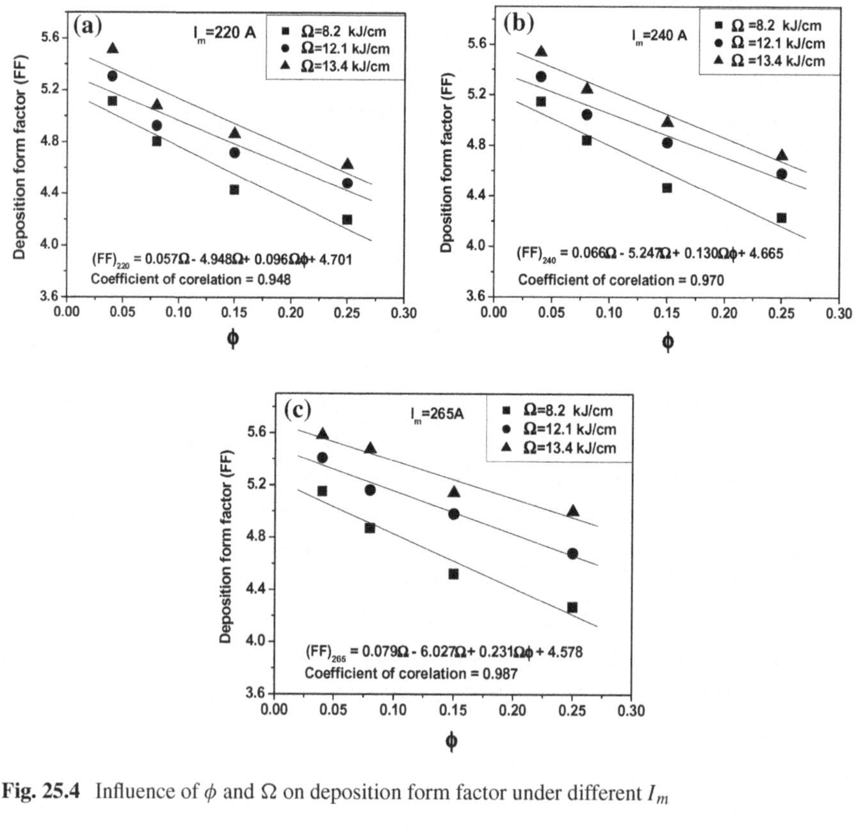

Fig. 25.4 Influence of ϕ and Ω on deposition form factor under different I_m

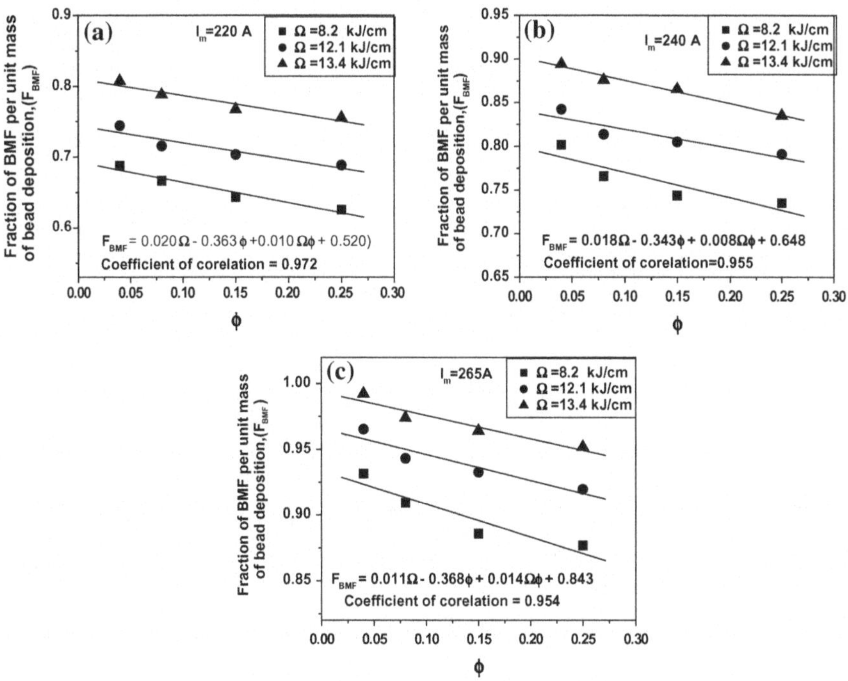

Fig. 25.5 Influence of ϕ and Ω on F_{BMF} under different I_m

of energy transmission to weld with the increase of ϕ, which leads to the reduction in area of the base metal fusion. The figure also depicts that the increase of Ω and the enhancement of I_m increase F_{BMF}, which may have happened due to increase of rate of heat transfer and increased momentum of the transferring droplets giving rise to comparatively higher depth of cavity [10] where the molten metal transfer its heat to the weld pool resulting in increased melting of the base material.

$$\text{At } I_m = 220 \pm 3\,\text{A} \quad F_{\mathrm{BMF}} = 0.020\,\Omega - 0.363\phi + 0.010\,\Omega\phi + 0.520 \tag{25.7}$$

$$\text{At } I_m = 240 \pm 2\,\text{A} \quad F_{\mathrm{BMF}} = 0.018\,\Omega - 0.343\phi + 0.008\,\Omega\phi + 0.648 \tag{25.8}$$

$$\text{At } I_m = 265 \pm 4\,\text{A} \quad F_{\mathrm{BMF}} = 0.011\,\Omega - 0.368\phi + 0.014\,\Omega\phi + 0.843 \tag{25.9}$$

Variation in dilution (D_L) of weld metal largely affects its chemical and metallurgical properties which may consequently affect mechanical characteristics of weld. Enhancement of ϕ decreases the D_L significantly (Fig. 25.6). The figure also shows that the increase of Ω and the increase of I_m enhance the D_L. This variation in D_L as a function of ϕ, Ω and I_m may be the result of the combined influence of both amount of original metal fused and amount of molten metal transferred from filler

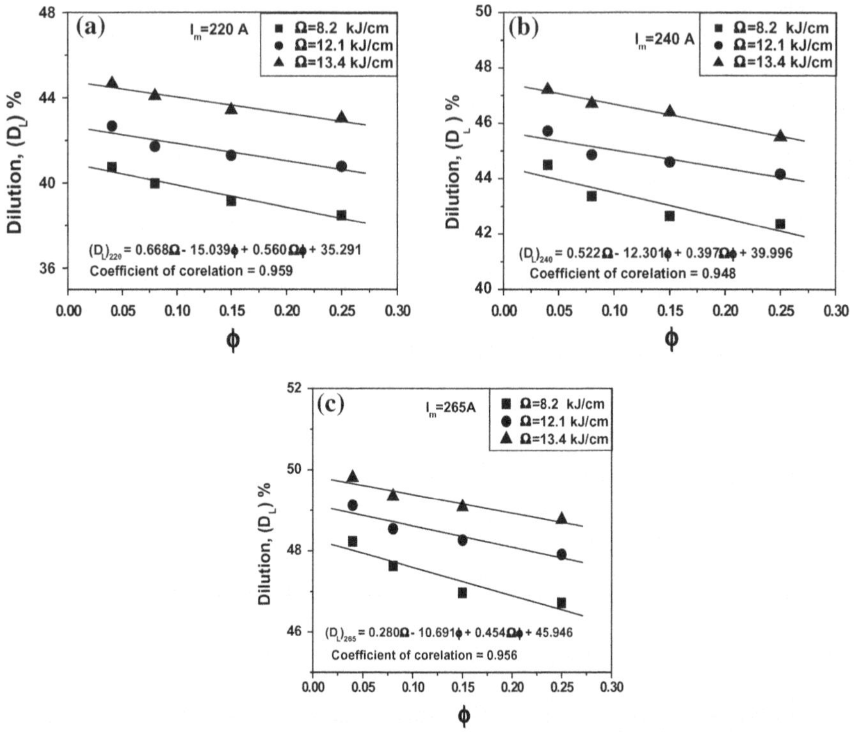

Fig. 25.6 Influence of ϕ and Ω on dilution % (D_L) under different I_m

wire, respectively. Relationship of D_L with the ϕ and Ω developed at different I_m is as follows.

$$\text{At } I_m = 220 \pm 3\,\text{A} \quad D_L = 0.668\,\Omega - 15.039\phi + 0.560\,\Omega\phi + 35.291 \tag{25.10}$$

$$\text{At } I_m = 240 \pm 2\,\text{A} \quad D_L = 0.522\,\Omega - 12.301\phi + 0.397\,\Omega\phi + 39.996 \tag{25.11}$$

$$\text{At } I_m = 265 \pm 4\,\text{A} \quad D_L = 0.280\,\Omega - 10.691\phi + 0.454\,\Omega\phi + 45.946 \tag{25.12}$$

The increase of ϕ reduces the B_{TA}. It is also understood that increase of Ω and increase of I_m diminish the B_{TA} (Fig. 25.7). This type of variation in B_{TA} with ϕ, Ω and I_m is attributed to the variation in average weld pool temperature arises due to the variation in the energy transported to the weld affecting fluidity of deposited weld and the rate of deposited metal to the weld. A higher fluidity of weld metal provides more spreading and thus results in less bead toe angle, while a higher rate of deposited metal to the weld gives rise to lower bead toe angle. A comparatively higher bead toe angle is desirable for the production of a weld joint free from lack of groove wall fusion. The empirical correlations of B_{TA} with the ϕ and Ω developed

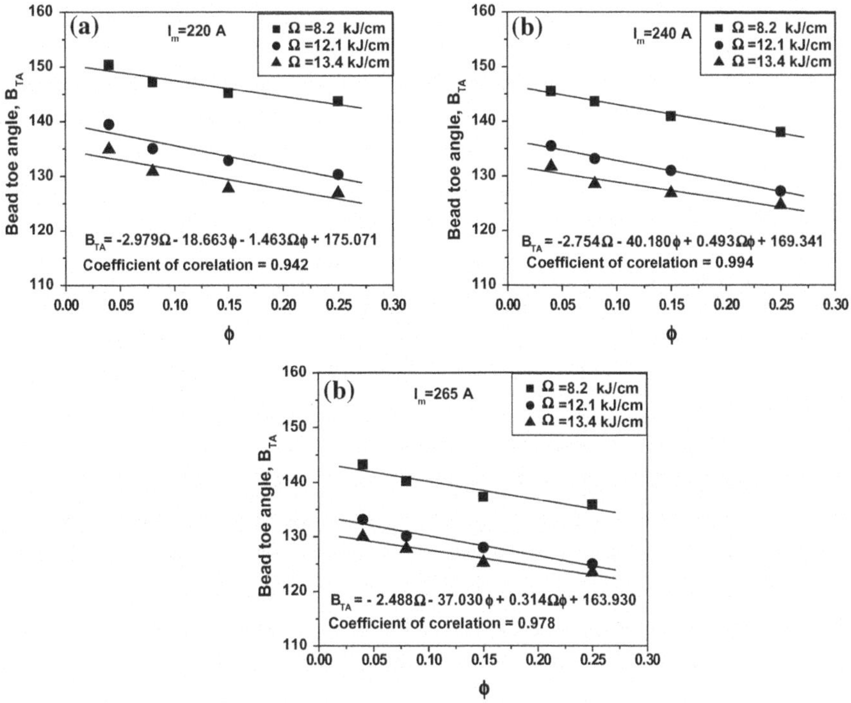

Fig. 25.7 Effect of ϕ and Ω on bead toe angle under different I_m

at different I_m are as follows.

$$\text{At } I_m = 220 \pm 3\,\text{A} \quad B_{\text{TA}} = -2.979\,\Omega - 18.663\phi - 1.463\,\Omega\phi + 175.071 \tag{25.13}$$

$$\text{At } I_m = 240 \pm 2\,\text{A} \quad B_{\text{TA}} = -2.754\,\Omega - 40.180\phi + 0.493\,\Omega\phi + 169.341 \tag{25.14}$$

$$\text{At } I_m = 265 \pm 4\,\text{A} \quad B_{\text{TA}} = -2.488\,\Omega - 37.030\phi + 0.314\,\Omega\phi + 163.930 \tag{25.15}$$

25.4 Conclusions

From the present investigation, the following can be concluded

1. The deposition characteristics are administered by parameters of PCGMAW. Therefore, by proper selection of parameters of PCGMAW, the quality of weld can be controlled.

2. A relatively higher value of ϕ, lower values of I_m and Ω produce better properties and quality of weld joint.
3. These parameters can be used to get a higher quality of weld joint by industries.

References

1. Zhaoa, Y., Lee, P.-S., Chung, H.: Effect of pulsing parameters on drop transfer dynamics and heat transfer behavior in pulsed gas metal arc welding. Int. J. Heat Mass Transf. **129**, 1110–1122 (2019)
2. Kumar, R., Chattopadhyaya, S., Agrawal, B.P., Kumar, S., Devkumaran, K., Arvindan, S.: Effects of types of GTAW including superior GTAW-PC with Superimposed HF current on mechanical and metallurgical properties of super duplex stainless steel weld joints. Mater. Res. Express IOP Sci. **6**(7), 076572 (2019). https://doi.org/10.1088/2053-1591/ab16fe
3. Anant, R., Dahiya, J.P., Agrawal, B.P., Ghosh, P.K., Kumar, R., Kumar, A., Kumar, S., Kumar, K.: SMA, GTA and P-GMA dissimilar weld joints of 304LN stainless steel to HSLA steel; part-1: thermal and microstructure characteristics. Mater. Res. Exp. IOP Sci. **5**(9) (2018)
4. Agrawal, B.P., Ghosh, P.K.: Thermal modeling of multi pass narrow gap pulse current GMA welding by single seam per layer deposition techniques. Mater. Manuf. Processes. **25**(11), 1251–1268 (2010). Tailor and Francis
5. Anant, R., Dahiya, J.P., Agrawal, B.P., Ghosh, P.K., Kumar, R., Kumar, A., Kumar, S., Kumar, K.: SMA, GTA and P-GMA dissimilar weld joints of 304LN stainless steel to HSLA steel; part-2: hot corrosion kinetics. Mater. Res. Express IOP Sci. **5**(9) (2018)
6. Agrawal, B.P., Ghosh, P.K.: Influence of thermal characteristics on microstructure of pulse current GMA weld bead of HSLA steel. Int. J. Adv. Manuf. Technol. **77**(9–12), 1681–1701 (2015)
7. Ghosh, P.K., Agrawal, B.P.: Advanced technique of extra narrow groove welding of thick steel section using pulse current gas metal arc welding process. In: International Conference on Global Trends in Joining, Cutting and Surfacing Technology, 2011, Chennai, India, 17–22 July 2011
8. Kulkarni, S.G., Agrawal, B.P., Ghosh, P.K., Ray, S.: Critical aspects of pulse current GMA welding of stainless steel influencing metallurgical characteristics. Int. J. Microstruct. Mater. Prop. Inderscience **12**(5/6), 363–390 (2017)
9. Wua, Kaiyuan, Caoa, Xuanwei, Yina, Tong, Zenga, Min, Lianga, Zhuoyong: Metal transfer process and properties of double-wire double pulsed gas metal arc welding. J. Manuf. Processes **44**, 367–375 (2019)
10. Agrawal, B.P., Ghosh, P.K.: Assembling of thick section HSLA steel with one seam per layer multi pass PC-GMA welding producing superior quality. J. Braz. Soc. Mech. Eng. Springer Link **39**(12), 5205–5218 (2017)
11. Ghosh, P.K., Agrawal, B.P.: Extra narrow gap gas metal arc welding of thick high strength low alloy steel. In: The Second South East European IIW International Congress, Welding-High-Technology in 21st century, Sofia, Bulgaria, pp. 168–173, 21–24 Oct 2010
12. Baia, P., Wanga, Z., Hua, S., Maa, S., Lianga, Y.: Sensing of the weld penetration at the beginning of pulsed gas metal arc welding. J. Manuf. Processes **28**(1), 343–350 (2017)
13. Agrawal, B.P., Chauhan, A.K., Kumar, R., Anant, R., Kumar, S.: GTA pulsed current welding of thin sheets of SS304 producing superior quality of joint at high welding speed. J. Braz. Soc. Mech. Eng. Springer Link **39**(11), 4667–4675 (2017)
14. Roshith, P., Arivarasu, M., Arivazhagan, N., Srinivasan, A., Phani Prabhakar, K.V.: Investigations on induced residual stresses, mechanical and metallurgical properties of CO_2 laser beam and pulse current gas tungsten arc welded SMO 254. J. Manuf. Processes **44**, 81–90 (2019)

15. Zhao, Y., Shi, X., Yana, K., Wang, G., Jia, Z., He, Y.: Effect of shielding gas on the metal transfer and weld morphology in pulsed current MAG welding of carbon steel. J. Mater. Process. Technol. **262**, 382–391 (2018)
16. Agrawal, B.P., Ghosh, P.K.: Characteristics of extra narrow gap pulse current gas metal arc weld of HSLA steel produced by single seam per layer deposition technique. J. Mater. Eng. Perform. **26**(3), 1365–1381 (2017)
17. Wang, Q., Cong, B., Yang, M.: Output characteristic and arc length control of pulsed gas metal arc welding process. J. Manuf. Processes **29**, 427–437 (2017)
18. Ghosh, P.K., Kulkarni, S.G., Agrawal, B.P.: High deposition pulse current GMAW can change current scenario of thick wall pipe welding. In: International Conference on Pressure Vessel and Piping, ASME-2009, paper no. PVP 2009-775549, vol. 6, pp. 1755–1760, Prague, Czeck Republic, Materials and Fabrication, Parts A and B, 26–30 July 2009
19. Zhao, Y., Chung, H.: Influence of power source dynamics on metal and heat transfer behaviors in pulsed gas metal arc welding. Int. J. Heat Mass Transf. **121**, 887–899 (2018)
20. Kumar, R., Anant, R., Ghosh, P.K., Kumar, A., Agrawal, B.P.: Influence of PC-GTAW parameters on micro structural and mechanical properties of thin AISI 1008 steel joints. J. Mater. Eng. Perform. Springer Link **25**(9), 3756–3765 (2016)

Chapter 26
Prediction of Thermo-Mechanical Properties of MWCNT-Reinforced GFRP and Its Thermo-Elastic Response Analysis in Laminated Composite Plate

Mesfin Kebede Kassa, Ananda Babu Arumugam and Bhim Singh

Abstract This paper presents the prediction of thermal and mechanical properties of multi-walled carbon nanotubes (MWCNT) reinforced with glass fiber-reinforced polymer (GFRP) composite material and its thermo-elastic response analysis in integrated composite plate. A micromechanics modeling approach based on the Eshelby–Mori–Tanaka method and rule of mixture is used to consider the effect of MWCNT agglomeration in the matrix and Chamis model to consider the influence glass fiber reinforcement in the three-phase composite. Thermo-elastic behavior of laminated composite plate enhanced with CNT was simulated by adopting high-order shear deformation theory (HSDT) with displacement field having seven degrees of freedom using finite element formulation with nine-noded Lagrangian plate element by developing FEM codes using MATLAB®. The current results of FEM model are compared with the results available in the literature to verify the effectiveness of developed model. Various parametric studies were conducted to investigate the effect of MWCNT agglomeration, width-to-thickness ratio, side-to-breadth ratio, and weight fraction of MWCNT on the thermo-elastic responses of laminated hybrid composite plate. It has been observed that agglomeration and weight fraction of MWCNT have a significant influence on the thermo-elastic response of laminated composite plates.

Keywords MWCNT · GFRP · Micromechanics · Agglomeration · Weight fraction · HSDT · FEM · Thermo-elastic response

26.1 Introduction

Due to the anisotropy and inhomogeneous behavior of MWCNTs, an important consideration required in the analysis of MWCNT-reinforced hybrid composite structures. Numerous research work were reported for estimating the effective elastic properties of CNT-reinforced polymers analytically, employing an equivalent continuum model of CNT and considering the agglomeration effect and taking CNT

M. Kebede Kassa · A. B. Arumugam (✉) · B. Singh
Department of Mechanical Engineering, SET, Sharda University, Greater Noida, India
e-mail: ananda.babu@sharda.ac.in

S. Yadav et al. (eds.), *Proceedings of International Conference in Mechanical and Energy Technology*, Smart Innovation, Systems and Technologies 174,
https://doi.org/10.1007/978-981-15-2647-3_26

as micro-inclusion based on the Eshelby–Mori–Tanaka approach [1, 2]. Determination of mechanical properties of three-phase CNT-reinforced GFRP and structural response of laminated composite structures reported by [2] propose a numerical model involving two stages for determining mechanical properties and showed the change in natural frequency of the structure by varying agglomeration parameter. The thermo-elastic behavior of the laminated composite plate was indeed investigated by [3]. The present study addresses the prediction of thermal and mechanical properties of MWCNT-reinforced GFRP and its thermo-elastic response analysis in integrated laminated composite plate structure with the help of suitable FEM code developed in MATLAB®. The computed results of the proposed model are validated in the available literature, and finally parametric study of thermo-elastic response of MWCNT-reinforced GFRP-laminated composite plate is reported.

26.2 Material Modeling of the Three-Phase Composite

In this section, micromechanics approach is adopted for modeling of thermal and mechanical properties of three-phase composite. The modeling approach is categorized into two successive steps. Firstly, modeling of MWCNT/Epoxy (EP) based on Eshelby's inclusion theory [4] and Mori–Tanaka's method [5], represented as nanocomposites and secondly, modeling of the enriched nanocomposite with the glass fiber based on Chamis model, represented as hybrid composite material. Each straight fiber of the MWCNT is assumed as an equivalent long fiber with transversely isotropic elastic properties as documented by Odegard [6]. Since the MWCNT is transversely isotropic, the nanocomposite can also be treated as transversely isotropic. The stiffness matrix components of MWCNT can be expressed as hill's constants [7]. The values of the hill's elastic moduli for MWCNT are presented in Table 26.1 [1].

In order to predict the effective elastic and thermal properties of nanocomposite, three different cases of agglomeration models are considered as presented in Fig. 26.1. In the figure, spherical shapes are the areas with high concentration of MWCNT represented as MWCNT micro-inclusion inclusions.

The total volume fraction c_r of the agglomerated MWCNT as presented in Eq. 26.1.

$$c_r = c_r^{\mathrm{INC}} + c_r^m \tag{26.1}$$

Table 26.1 Mechanical properties of MWCNTs in terms of Hill's elastic moduli [1]

Geometry of MWCNT	k_r (GPa)	l_r (GPa)	$m_r = p_r$ (GPa)	n_r (GPa)
MWCNT (5, 5)–(10, 10)	1984.203	900.530	732.703	2210.538
MWCNT (8, 8)–(16, 16)	2443.688	898.447	471.743	1463.653

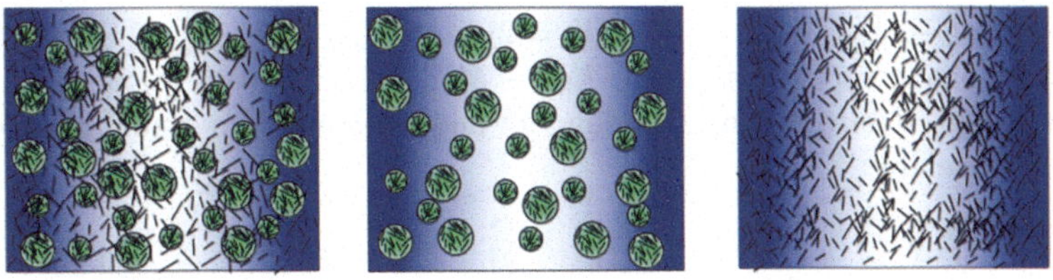

Fig. 26.1 MWCNT agglomeration model, case: **a** partially agglomerated $\lambda < \mu, \lambda < 1$, case: **b** completely agglomerated $\lambda = \mu = 0.3$, and case: **c** null agglomerated $\lambda < \mu, \mu <= 1$

where c_r^{INC} and c_r^m are represented as the volume fractions of MWCNTs dispersed in the concentrated regions and in the matrix, respectively.

The agglomeration of nanotubes governed by the two parameters λ and μ presented in Eq. 26.2.

$$\lambda = \frac{c_{\text{INC}}}{c}, \mu = \frac{c_r^{\text{INC}}}{c_r} \tag{26.2}$$

where c_{INC} is the volume of the sphere micro-inclusion in the nanocomposite. The parameter λ describes the volume fraction of inclusions in relation to the total volume c. On the other hand, the parameter μ describes the volume fraction of the nanotubes dispersed in inclusions relative to the total volume of the nanotubes.

The effective mechanical properties of the nanocomposite with inclusions are evaluated by means of Mori–Tanaka approach [2].

$$\left.\begin{array}{l} K_{\text{INC}}^{\circ} = K_m + \frac{(\delta_r - 3K_m\alpha_r)c_r\mu}{3(\lambda - c_r\mu + c_r\mu\alpha_r)}, \; K_{\text{out}}^{\circ} = K_m + \frac{(1-\mu)(\delta_r - 3K_m\alpha_r)c_r}{3(1-\lambda-c_r(1-\mu)+c_r(1-\mu)\alpha_r)} \\ G_{\text{INC}}^{\circ} = G_m + \frac{(\eta_r - 2G_m\beta_r)c_r\mu}{2(\lambda - c_r\mu + c_r\mu\beta_r)}, \; G_{\text{out}}^{\circ} = G_m + \frac{(1-\mu)(\eta_r - 2G_m\beta_r)c_r}{3(1-\lambda-c_r(1-\mu)+c_r(1-\mu)\alpha_r)} \end{array}\right\} \tag{26.3}$$

where the expression of the variables δ_r, α_r, η_r, β_r, the bulk modulus $K_{m,}$ and shear modulus G_m of the nanocomposite are presented in [8].

The effective bulk modulus K° and the effective shear modulus G° of the composite enriched by MWCNT both in the spherical concentrated region and scattered in the matrix are derived as:

$$\begin{aligned} K^{\circ} &= K_{\text{out}}^{\circ} + \left[1 + \frac{\lambda\left(\frac{K_{\text{INC}}^{\circ}}{K_{\text{out}}^{o}} - 1\right)}{1 + (1-\lambda)\left(\frac{K_{\text{INC}}^{\circ}}{K_{\text{out}}^{o}} - 1\right)\varphi}\right], \\ G^{\circ} &= G_{\text{out}}^{\circ} + \left[1 + \frac{\lambda\left(\frac{G_{\text{INC}}^{\circ}}{G_{\text{out}}^{o}} - 1\right)}{1 + (1-\lambda)\left(\frac{G_{\text{INC}}^{\circ}}{G_{\text{out}}^{\circ}} - 1\right)\psi}\right] \end{aligned} \tag{26.4}$$

where $\varphi = (1+\upsilon^{\circ}_{\text{out}})/3(1-\upsilon^{\circ}_{\text{out}})$ and $\psi = 2(4-5\upsilon^{\circ}_{\text{out}})/15(1-\upsilon^{\circ}_{\text{out}})$. The Poisson's ratio of the nanocomposite, $\upsilon^{\circ}_{\text{out}} = (3K^{\circ}_{\text{out}} - 2G^{\circ}_{\text{out}})/2(3K^{\circ}_{\text{out}} + G^{\circ}_{\text{out}})$

Finally, the effective Young's modulus E° and Poisson's ratio ν° of the nanocomposite are given by:

$$E^{\circ} = \frac{9K^{\circ}G^{\circ}}{3K^{\circ} + G^{\circ}}, \nu^{\circ} = \frac{3K^{\circ} - 2G^{\circ}}{6K^{\circ} + 2G^{\circ}} \tag{26.5}$$

Further, the thermal expansion coefficients of nanocomposites with respect to MWCNT agglomeration is obtained according to the rule of mixture. The thermal expansion coefficients of the concentrated region $\alpha^{\circ}_{\text{in}}$ and in the nanocomposite $\alpha^{\circ}_{\text{out}}$ are given in Eq. 26.6:

$$\left.\begin{aligned} \alpha^{\circ}_{\text{in}} &= \tfrac{\eta c_r}{\lambda}\alpha_r + \tfrac{\lambda-\eta c_r}{\lambda}\alpha_m \\ \alpha^{\circ}_{\text{out}} &= \tfrac{c_r(1-\eta)}{1-\lambda}\alpha_r + \tfrac{1-\lambda-c_r(1-\eta)}{1-\lambda}\alpha_m \end{aligned}\right\} \tag{26.6}$$

where α_r and α_m are the coefficients of thermal expansions (CTE) of MWCNTs and EP, respectively.

The effective CTE of the nanocomposite are given in Eq. 26.7.

$$\left.\begin{aligned} \alpha^{\circ}_1 &= \tfrac{\lambda\alpha^{\circ}_{\text{in}}E^{\circ}_{\text{in}}+(1-\lambda)\alpha^{\circ}_{\text{out}}E^{\circ}_{\text{out}}}{\lambda E^{\circ}_{\text{in}}+(1-\lambda)E^{\circ}_{\text{out}}} \\ \alpha^{\circ}_2 &= \lambda\alpha^{\circ}_{\text{in}}(1+\nu^{\circ}_{\text{in}}) + (1-\lambda)\alpha^{\circ}_{\text{out}}(1+\nu^{\circ}_{\text{out}}) - \alpha_1(\lambda\nu^{\circ}_{\text{in}} + (1-\lambda)\nu^{\circ}_{\text{out}}) \end{aligned}\right\} \tag{26.7}$$

where $\nu^{\circ}_{\text{in}} = (3K^{\circ}_{\text{in}} - 2G^{\circ}_{\text{in}})/2(3K^{\circ}_{\text{in}} + G^{\circ}_{\text{in}})$, $E^{\circ}_{\text{in}} = (9K^{\circ}_{\text{in}}G^{\circ}_{\text{in}})/2(3K^{\circ}_{\text{in}} + G^{\circ}_{\text{in}})$ $\nu^{\circ}_{\text{out}} = (3K^{\circ}_{\text{out}} - 2G^{\circ}_{\text{out}})/2(3K^{\circ}_{\text{out}} + G^{\circ}_{\text{out}})$, $E^{\circ}_{\text{out}} = (9K^{\circ}_{\text{out}}G^{\circ}_{\text{out}})/2(3K^{\circ}_{\text{out}} + G^{\circ}_{\text{out}})$,

α°_1 and α°_2 are longitudinal and transverse CTE of the nanocomposites

The density of the nanocomposite ρ° can be obtained by rule of mixture as:

$$\rho^{\circ} = (\rho_r - \rho_m)V_r + V_m \tag{26.8}$$

The mechanical and thermal properties of the three-phase composite consist of glass fibers evaluated according to the Chamis model [9]. The property associated with the nanocomposite is referred with the superscript "o".

$$\left.\begin{aligned} E_1 = E^f_1 V_f + E^{\circ}V^{\circ}, E_2 = E_3 &= \tfrac{E^{\circ}}{1-V_f(1-E^{\circ}/E^f_2)}, G_{23} = \tfrac{G^{\circ}}{1-V_f(1-G^{\circ}/G^f_{12})} \\ \nu_{12} = \nu_{12} &= \nu^f_{12}V_f + \nu^{\circ}V^{\circ}, \nu_{23} = \tfrac{E_2}{2G_{23}} - 1 \\ \alpha_1 = \tfrac{E^f_1\alpha^f_1 V_f + E^{\circ}\alpha^{\circ}V^{\circ}}{E^f_1 V_f + E^{\circ}V^{\circ}}, \alpha_2 &= \alpha^f_2\sqrt{V_f} + (1-\sqrt{V_f})(1+V_f\nu^{\circ}E^f_1/E_1)\alpha^{\circ}_2 \end{aligned}\right\} \tag{26.9}$$

where E_1 and E_2 are the Young's moduli and α_1 and α_2 are the thermal expansion coefficients in the longitudinal and transverse directions, respectively, and G_{12} is the shear modulus of the composite material. V_f and V° are the volume fraction of the glass fiber and the nanocomposite.

26.3 Mathematical Modeling of Thermo-Elastic Response Analysis of Laminated Composite Plate Hybrid with MWCNT

Laminated composite plate is considered for the development of finite element model to study the thermo-elastic response under thermal and mechanical loading conditions. The length '*L*', breadth '*B*' and each layers with identical thickness 't' of the plate are considered for the analysis. The displacement field is chosen based on HSDT in order to formulate strain energy equations. The constitutive equations are formulated by including the thermo-elastic material model to relate the stress–strain fields. The nine-noded Largragian plate element is chosen and substituted in the strain energy equation to formulate the stiffness and force matrices.

26.3.1 Kinematics

In this paper, the displacement field is assumed based on the higher-order shear deformation theory (HSDT) and presented as:

$$\left.\begin{array}{l} U_2(x,y,z) = u_2(x,y) + z\beta_y(x,y) - \frac{4z^3}{3h^2}\psi_y(x,y) \\ U_2(x,y,z) = u_2(x,y) + z\beta_y(x,y) - \frac{4z^3}{3h^2}\psi_y(x,y) \\ U_3(x,y,z) = w(x,y) \end{array}\right\} \tag{26.10}$$

where U_1 and U_2 are the in-plane deformations and U_3 is the transverse deflection of the plate. u_1 and u_2 are the in-plane deformations of mid-plane. β_x, β_y and Ψ_x, Ψ_y are the slopes and higher-order terms of the plate with respect to x and y direction respectively.

The in-plane strains are given by

$$\begin{Bmatrix} \varepsilon_x \\ \varepsilon_y \\ \gamma_{xy} \end{Bmatrix} = \begin{Bmatrix} \varepsilon_x^0 \\ \varepsilon_y^0 \\ \gamma_{xy}^0 \end{Bmatrix} + z\begin{Bmatrix} \chi_x^b \\ \chi_y^b \\ \chi_{xy}^b \end{Bmatrix} + z^3\begin{Bmatrix} \chi_x^s \\ \chi_y^s \\ \chi_{xy}^s \end{Bmatrix}, \begin{Bmatrix} \gamma_{yz} \\ \gamma_{xz} \end{Bmatrix} = \begin{Bmatrix} \gamma_{yz}^0 \\ \gamma_{xz}^0 \end{Bmatrix} + z^2\begin{Bmatrix} \chi_{yz}^2 \\ \chi_{xz}^2 \end{Bmatrix} \tag{26.11}$$

where ε_x^0, ε_y^0 and γ_{xy}^0 are the in-plane strains and χ_x^b, χ_y^b, χ_{xy}^b χ_x^s, χ_y^s, χ_{xy}^s, χ_{yx}^2 and χ_{xz}^2 are the mid-plane curvatures corresponding to x-a and y axis.

26.3.2 Constitutive Equations

The stress–strain relationships in the coordinates of the laminate of a kth layer are:

$$\begin{Bmatrix} \sigma_x \\ \sigma_y \\ \tau_{xy} \\ \tau_{yz} \\ \tau_{xz} \end{Bmatrix}^{(k)} = \begin{bmatrix} \overline{Q}_{11} & \overline{Q}_{12} & \overline{Q}_{16} & 0 & 0 \\ \overline{Q}_{12} & \overline{Q}_{22} & \overline{Q}_{26} & 0 & 0 \\ \overline{Q}_{16} & \overline{Q}_{26} & \overline{Q}_{66} & 0 & 0 \\ 0 & 0 & 0 & \overline{Q}_{44} & \overline{Q}_{45} \\ 0 & 0 & 0 & \overline{Q}_{45} & \overline{Q}_{55} \end{bmatrix}^{(k)} \left[\begin{Bmatrix} \varepsilon_x \\ \varepsilon_y \\ \gamma_{xy} \\ \gamma_{yz} \\ \gamma_{xz} \end{Bmatrix} - \begin{Bmatrix} \alpha_x \\ \alpha_y \\ \alpha_{xy} \\ 0 \\ 0 \end{Bmatrix} \Delta T \right]^{(k)} \tag{26.12}$$

$$[Q_{ij}]^{(k)} = [T][Q_{ij}]^{(k)}[T]^{\mathrm{T}} \tag{26.13}$$

where ΔT is the change in temperature Q_{ij} are the transformed material constants, T is the Transformation matrix. The thermal expansion co-efficient vector can be transformed into global coordinate as:

$$\begin{Bmatrix} \alpha_x \\ \alpha_y \\ \alpha_{xy} \end{Bmatrix} = \begin{bmatrix} m^2 & n^2 & -2mn \\ n^2 & m^2 & 2mn \\ -2mn & 2mn & m^2 - n^2 \end{bmatrix} \begin{Bmatrix} \alpha_1 \\ \alpha_2 \\ \alpha_3 \end{Bmatrix} \tag{26.14}$$

where $m = \cos\theta$ and $n = \sin\theta$ in which θ is the angle between the global x-axis and the local x-axis of each layer.

The stress resultants are obtained as:

$$\begin{Bmatrix} N \\ M^b \\ T^s \end{Bmatrix} = \begin{bmatrix} A_{ij} & B_{ij} & E_{ij} \\ B_{ij} & D_{ij} & F_{ij} \\ E_{ij} & F_{ij} & H_{ij} \end{bmatrix} \begin{Bmatrix} \varepsilon \\ \chi^b \\ \chi^s \end{Bmatrix} \tag{26.15}$$

where $N = \{N_x, N_y, N_{xy}\}^{\mathrm{T}}$, $M^b = \{M_x^b, M_y^b, M_{xy}^b\}^{\mathrm{T}}$, $T^s = \{T_x^s, T_y^s, T_{xy}^s\}^{\mathrm{T}}$ $\varepsilon = (\varepsilon_x^0, \varepsilon_y^0, \gamma_{xy}^0\}^{\mathrm{T}}$, $\chi^b = \{\chi_x^b, \chi_y^b, \chi_{xy}^b\}^{\mathrm{T}}$, $\chi^b = \{\chi_x^b, \chi_y^b, \chi_{xy}^b\}^{\mathrm{T}}$.

26.4 Finite Element Formulation

The nine-noded Lagrangian isoparametric element with seven degrees of freedom $\{u_i\} = \{u_{1i}, u_{2i}, w_i, \beta_{xi}, \beta_{yi}, \psi_{xi}, \psi_{yi}\}$ at each node is considered for formulating equivalent single-layer laminated composite plate.

$$\{d\} = \sum_{i=1}^{9} [N_i]\{u_i\} (i = 1, 2, 3, \ldots 9) \tag{26.16}$$

where N_i is the shape functions of the element associated with the node 'i'

The generalized linear-strain displacement matrix expressed by

$$\{\varepsilon\} = [B]\{d\} \tag{26.17}$$

The total potential energy for an element 'e' can be obtained by Eq. 26.17

$$\Pi^e = U^e - W^e \tag{26.18}$$

where U^e is the internal strain energy and W^e is the external virtual work

The internal strain energy for an element can be written as

$$U_1^e = \frac{1}{2}\{d\}^{\mathrm{T}}[k_1^e]\{d\} \tag{26.19}$$

Internal strain energy due to transverse deformation

$$U_2^e = \frac{1}{2}\{d\}^{\mathrm{T}}[k_2^e]\{d\} \tag{26.20}$$

where k^e is the stiffness matrix d is the displacement

The element stiffness matrix computed as:

$$k_1^e = \int_{-b}^{b}\int_{-a}^{a}[B_i(x,y)]^{\mathrm{T}} \times \begin{bmatrix} A_{ij} & B_{ij} & E_{ij} \\ B_{ij} & D_{ij} & F_{ij} \\ E_{ij} & F_{ij} & K_{ij} \end{bmatrix} \times [B_i(x,y)]\mathrm{d}x\mathrm{d}y$$

$$k_2^e = \int_{-b}^{b}\int_{-a}^{a}[B_i(x,y)]^{T} \times \begin{bmatrix} A_{ij}^s & D_{ij}^s \\ D_{ij}^s & F_{ij}^s \end{bmatrix} \times [B_i(x,y)]\mathrm{d}x\mathrm{d}y$$

$$\left(A_{ij}, B_{ij}, D_{ij}, E_{ij}, F_{ij}, H_{ij}\right) = \int_{h/2}^{h/2} Q_{ij}\left(1, z, z^2, z^3, z^4, z^6\right)\mathrm{d}z(i, j = 1, 2, 6)$$

$$\left(A_{ij}^s, D_{ij}^s, , F_{ij}^s\right) = \int_{h/2}^{h/2} Q_{ij}\left(1, z^2, z^4\right)\mathrm{d}z(i, j = 5, 4) \tag{26.21}$$

The total stiffness matrix:

$$k_e = k_1^e + k_2^e \tag{26.22}$$

The first variation of the work done on the laminate by external loads for an element given by:

$$\delta W_e = \delta u^{\mathrm{T}} f_e \tag{26.23}$$

The equivalent nodal mechanical load vector given by:

$$f_e = \iint \{N\}^{\mathrm{T}}\{F\}\mathrm{d}x\mathrm{d}y, \{F\} = [0, 0, q, 0, 0, 0, 0]^{\mathrm{T}} \tag{26.24}$$

where q is the transverse pressure loading

The equivalent nodal thermal load is obtained by:

$$f_e^N = \iint [B]^{\mathrm{T}}\{F^N\}\mathrm{d}x\mathrm{d}y \tag{26.25}$$

where $\{F^N\}^{\mathrm{T}} = \left\{N_x^N, N_y^N, N_{xy}^N, M_x^N, M_y^N, M_{xy}^N, S_x^N, S_y^N, S_{xy}^N\right\}$.

The stress, moment, and shear resultants can be obtained as:

$$\begin{Bmatrix} N_x^N, M_x^N, S_x^N \\ N_y^N, M_y^N, S_y^N \\ N_{xy}^N, M_{xy}^N, S_{xy}^N \end{Bmatrix} = \sum_{k=1}^{N} \int_{z_k}^{z_{k+1}} \begin{Bmatrix} \alpha_x \\ \alpha_y \\ \alpha_{xy} \end{Bmatrix} (1, z, z^3)\Delta T \mathrm{d}x\mathrm{d}y\mathrm{d}z \tag{26.26}$$

Equilibrium equations for static analysis using FEM expressed as:

$$[k]\{d\} = \{F\} + \left\{F^N\right\} \tag{26.27}$$

26.5 Result and Discussion

In this section, the effectiveness of the present developed FEM model is assessed using numerical examples available in the literature. Subsequently, the various parametric studies on laminated MWCNT/GFRP composite plates are presented to broaden the knowledge on the thermo-elastic response.

26.5.1 Numerical Validation

The laminated composite plate with (0°/90°/0°) ply orientation subjected to uniformly varying transverse pressure load $q_0 = 100$ is considered with following material properties and end conditions as presented in Reddy [10]:

Table 26.2 Non-dimensional transverse deflection of three-layered simply supported symmetric laminated (0°/90°/0°) plate subjected to transverse pressure loading $q_0 = 100$

References	$a/h = 100$	$a/h = 50$	$a/h = 20$	$a/h = 10$	$a/h = 4$
Reddy [10]	0.6705	0.6838	0.7760	1.0900	2.9091
Present	0.6701	0.6824	0.7679	1.0628	2.8815
Error	0.05	0.02	1.04	2.4	0.9

$$E_1 = 25E_2, G_{12} = G_{13} = 0.5E_2, G_{23} = 0.2E_2, \nu_{12} = \nu_{23} = \nu_{13} = 0.25$$

(a) Loading conditions
Uniform transverse mechanical loading (UDL): $P = P_0$
Thermal loading conditions: $\Delta T = 300\,\text{K}$
(b) The various non-dimensional parameters used within this section presented as

$$\bar{w} = w \times \left(\frac{100E_2h^3}{q_0b^4}\right)\bar{w} = w \times \left(\frac{10T_2h^3}{q_0b^4}\right)$$

(c) Boundary condition: All four edges of the plate are simply supported

$$u_1 = w = \beta_x = \psi_x = 0@y = 0, b; u_2 = w = \beta_y = \psi_y = 0@x = 0, a$$

The normalized deflections of obtained from present FEM are compared with Reddy [10] presented in Table 26.2. It was observed that the error is very less. This is the evident that the developed FEM model is valid for analysis of the static responses of laminated composite plates.

26.5.2 A Parametric Study

26.5.2.1 Effect of Agglomeration and Weight Fraction of MWCNT

The effect of agglomeration and the weight fraction of MWCNT on the normalized deflections of laminated plate subjected to uniformly varying mechanical and thermal load under all edges simply supported are studied. The thermo-mechanical properties of MWCNT/GFRP are determined based on the individual properties listed in Table 26.3 by adopting various models discussed in the previous section. The weight percentage of MWCNT is fixed at 2.5% of the EP is considered for the study of normalized deflection of plate over b/h and a/b.

Three cases are set by varying the agglomeration parameters to consider the agglomeration effect such as **Case A**: Complete agglomeration $\lambda = 0.3,\ \mu = 1$ **Case B**: Partial agglomeration $\lambda = 0.3,\ \mu = 0.7$ **Case C**: Null agglomeration $\lambda = 0.3,\ \mu = 0.3$.

Table 26.3 Mechanical and thermal properties of the reinforcing fiber and the matrix

Mechanical properties		
E-glass fiber (reinforcing fiber) [11]		Polymer matrix (epoxy) [11]
Young modulus	$E_1^f = E_2^f = 73.084$ GPa	$E_m = 3.5$ GPa
Shear modulus	$G_1^f = G_2^f = 30.130$ GPa	–
Poisson's ratio	$\nu_{12}^f = \nu_{23}^f = 0.22$	$\nu_m = 0.34$
Density	$p^f = 2491$ kg/m^3	$p^m = 1200$ kg/m^3
Thermal properties at room temperature		
MWCNT [12]	$\alpha_r = 2.8 \times 10^{-5}$/°C	
Epoxy [13]	$\alpha_m = 2.5 \times 10^{-6}$/°C	
Glass fiber [13]	$\alpha_1^f = \alpha_2^f = 5.2 \times 10^{-6}$/°C	

The bending responses of laminated MWCNT/GFRP plate with complete, partial, and null agglomeration with respect to breadth-to-thickness and material length-to-breadth ratio are shown in Fig. 26.2a, b, respectively. It has been observed that the transverse central deflections are sensitive to both breadth-to-thickness and material length-to-breadth ratio of the laminated composite plate structure.

The variation of transverse central deflection under uniform transverse load with respect to the weight fraction of MWCNT for different agglomeration model is shown in Fig. 26.3a with $a/b = 1$ and $b/h = 50$. The variation of transverse central deflection with respect to the weight fraction of MWCNT model under thermal load with $T = 373$ K, $a/b = 1$ and $b/h = 50$ for different agglomeration models is shown in Fig. 26.3b.

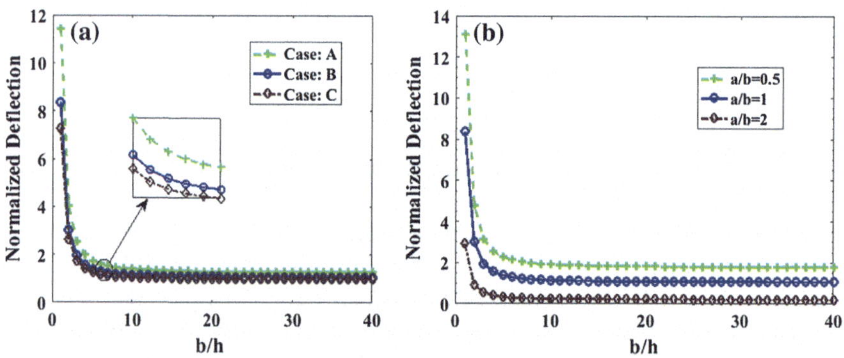

Fig. 26.2 **a** Non-dimensional central deflection ($\bar{w} = w \times E_m h^3 \times 10^2/(p_o b^4)$) versus aspect ratio for simply supported [0°/90°/0°] laminate under uniformly transverse pressure load. **b** The effect of side-to-thickness ratio on non-dimensional central deflection of simply supported three-layer [0°/90°/0°] laminated plate under uniform transverse load

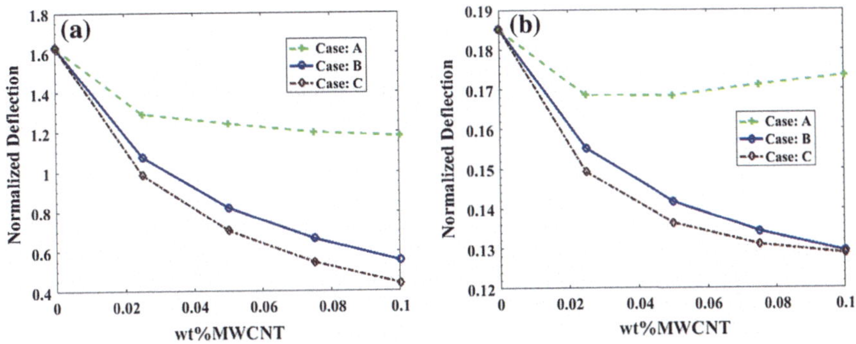

Fig. 26.3 **a** Non-dimensional deflection of simply supported three-layer [0°/90°/0°] laminated plate under uniform transverse load. **b** Non-dimensional deflection ($\bar{w} = w \times h \times 10/(\alpha_m \times T_2 \times b^2)$) of simply supported three-layer [0°/90°/0°/90°] laminated plate under thermal load $\Delta T = 300$ K

26.6 Conclusion

The present work focuses on the prediction of thermal and mechanical properties of MWCNT-reinforced GFRP and its thermo-elastic response analysis in laminated composite plate subjected to uniform transverse pressure load and uniform thermal load. Mechanical and thermal properties are evaluated using a micromechanics approach and nine-noded finite element formulation developed using MATLAB® for thermo-elastic response analysis based on HSDT with displacement field having seven degrees of freedom. The results obtained from developed FEM were validated with the available literature and various parametric studies have been simulated such as influence of MWCNT weight fraction, *a*/*b* and *b*/*h* on the static transverse deflection of hybrid composite plate under thermo-mechanical loading. It was observed that the transverse deflection of a simply supported MWCNT-reinforced GFRP-laminated composite plate increased significantly as the weight fraction of MWCNT increased to a certain limit and become saturated. Moreover, the central transverse deflection significantly varying for various agglomeration cases of MWCNT considered the complete agglomerated MWCNT reaches to the limit value as soon as the other agglomeration cases in both loading conditions.

Acknowledgements This study was carried out under financial support of SERB, DST Project No: ECR/2018/000827. The authors acknowledges for extending the lab facility.

References

1. García-Macías, E. et al.: Multiscale modeling of the elastic moduli of CNT-reinforced polymers and fitting of efficiency parameters for the use of the extended rule-of-mixtures. Compos. Part B Eng. **159**, 114–131 (2019)
2. Tornabene, F. et al.: Multiscale approach for three-phase CNT/polymer/fiber laminated nanocomposite structures. Polym. Compos. **40**(S1), E102–E126 (2019)
3. Adhikari, B., Singh, B.N.: Thermo-elastic analysis of laminated functionally graded CNT plates. AIAA Scitech 2019 Forum (2019)
4. Eshelby J.D.: Proc. Roy. Soc. London A Maths. Phys. Sci. 241, 376 (1957)
5. Mori, T., Tanaka, K.: Acta Metall. **21**, 571 (1973)
6. Odegard, G.M., Clancy, T.C., Gates, T.S.: Modeling of the mechanical properties of nanoparticle/polymer composites. Polymer **46**(2), 553–562 (2005)
7. Hill, R.: J. Mech. Phys. Solids **12**, 213 (1964)
8. Shi, D.-L. et al.: The effect of nanotube waviness and agglomeration on the elastic property of carbon nanotube-reinforced composites. J. Eng. Mater. Technol. **126**(3), 250–257 (2004)
9. Chamis, C.C.: Simplified composite micromechanics equations for hygral, thermal and mechanical properties (1983)
10. Reddy, J.N., Liu, C.F.: A higher-order shear deformation theory of laminated elastic shells. Int. J. Eng. Sci. **23**(3), 319–330 (1985)
11. Tornabene, F., Fantuzzi, N., Bacciocchi, M., Viola, E.: Compos. Part B Eng. **89**, 187 (2016)
12. Dong, C.: Mechanical and thermo-mechanical properties of carbon nanotube reinforced composites. Int. J. Smart Nano Mater. **5**(1), 44–58 (2014)
13. Sideridis, E.: Thermal expansion coefficients of fiber composites defined by the concept of interphase. Compos. Sci. Technol. **51**(3), 301–317 (1994)

Chapter 27
Tolerance Estimation of Different Cavity Geometries Produced by CO_2 Laser

Shrikant Vidya, K. S. Srikanth, Altaf H. Tarique and P. Mathiyalagan

Abstract In this article, the investigation on the tolerances achievable on two cavities, circular and straight channels, are presented. In the case of circular cavities, circularity has been evaluated while in the case of straight channels, straightness has been evaluated, respectively. The investigation begins by fabrication of cavities in series of three having a nominal dimension on polymethyl methacrylate (PMMA), also known as acrylic using CO_2 laser. After that, the Olympus optical microscope was used to examine the geometry of the fabricated microcavities and their tolerances have been evaluated after analyzing these geometries by profile projector and Olympus Analysis Software. It was found that the average circularity tolerance of circular micro-geometries was estimated as 1.34 μm and the average straightness tolerance of microchannels was estimated as 8.66 μm.

Keywords CO_2 laser · Microgeometries · Circularity · Straightness · Profile projector

27.1 Introduction

Metrology plays a vital role in the modern era of micromanufacturing where tolerancing acts like a bridge between the process chain and the product specification. The proper allocation of tolerances can lead to a reduction in costs and the number of rejects or the amount of rework required on components.

Polymers are becoming more important day by day for the production of microfluidic devices as compared to conventional materials due to their low price, simple fabrication methods, and better suitability for single use, especially in medical applications. Among various microfabrication methods for polymers, such as hot embossing [1], injection molding [2], and micromilling [3], infrared laser ablation techniques have gained much importance due to their property of evaporating substrate material

S. Vidya (✉) · K. S. Srikanth · A. H. Tarique · P. Mathiyalagan
School of Mechanical Engineering, Galgotias University, Greater Noida, Uttar Pradesh, India

S. Vidya
Department of Mechanical Engineering, Delhi Technological University, New Delhi, India

S. Yadav et al. (eds.), *Proceedings of International Conference in Mechanical and Energy Technology*, Smart Innovation, Systems and Technologies 174,
https://doi.org/10.1007/978-981-15-2647-3_27

with the application of heat with laser beam only. The high fabrication speed and the peculiar ability of infrared laser systems to change design make it a standout tool for micromanufacturing. The high radiance of laser light [4] together with the characteristics of polymers create the possibilities for the production of microfluidic devices by laser ablation.

The CO_2 laser was selected as the infrared laser for machining purpose. CO_2 lasers typically emit light with a wavelength of 10.6 μm and possess overall efficiencies of approx. 10–13%. In CO_2 lasers, a gas mixture of helium, nitrogen, CO_2, and other additives stimulated by electrical gas discharge is utilized to generate the laser beam. During this process, energy gets transferred from nitrogen molecules to the CO_2 molecules leading to transition of CO_2 molecules from upper energy level to a lower energy accompanied by photon release, ultimately resulting in laser beam emission. The CO_2 molecules return to the ground state after colliding with the helium atoms which is used for another cycle.

In CO_2 laser, most of the literature dealt with process optimization, physical behavior, materials, etc. [5–10]. Many researchers have done experimental and theoretical analysis of CO_2 laser micromachining on PMMA [11–17]. To the best of our knowledge, this is the first report to estimate parts' tolerances or tolerance of geometries produced by CO_2 laser. This investigation studies the tolerances achievable on two cavities: circular and microchannels.

27.2 Experimental Details

The experiments were performed on the 30 W CO_2 laser system, the desktop VLS2.30 (Fig. 27.1), manufactured by Universal Laser System, USA. It is a cost-effective and compact entry-level laser machine for the purpose of laser material processing and

Fig. 27.1 CO_2 laser processing system setup

Table 27.1 Material properties of specimen

Properties	Value	Unit
Yield tensile Strength	75	MPa
Density	1190	Kg/m^3
Modulus of elasticity	3.03	GPa
Processing temperature	156	°C

prototype development. It can accommodate a material processing envelope of 16″ × 12″ × 4″ and can be equipped with a laser cartridge having a maximum power of 30 W.

The polymer material used as specimen was acrylic sheet which is 4 mm thick, having a width of 30 mm and a length of 40 mm. In this direction, Table 27.1 lists the material properties of specimen. At first, a CAD drawing of required microstructures with appropriate dimensions was made using CAD software, and the CAD data are processed in the CO_2 laser. Then, the desired product is directly fabricated in the acrylic sheet by the CO_2 laser of laser beam diameter of 300 μm. The dimensions of the microcavities machined and the machining parameters are shown in Tables 27.2 and 27.3, respectively.

The geometry of machined microcavities was analyzed through Olympus optical microscope. The circularity of circular geometries was evaluated after analyzing these geometries by profile projector, Nikon made while the straightness tolerance of microchannels has been calculated on the basis of data obtained through Olympus analysis software. Figure 27.2 shows the microscopic images of machined microstructures.

Table 27.2 Dimensions of microstructures

Structure	Dimension (μm)		Pitch (μm)	Depth (μm)
	Length	Width/diameter		
Micro channel	8000	300	600	200
Circular		300	600	200

Table 27.3 Machining parameters

Parameters	Value
Laser power	0.75 W
Scanning speed	10 mm/s
PPI	500

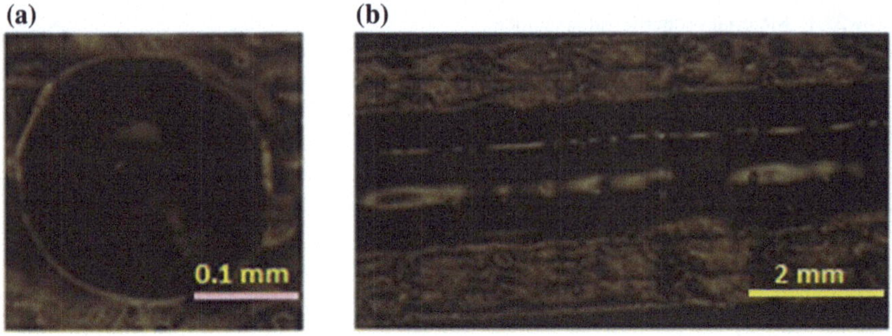

Fig. 27.2 Microscopic images (5×) of micro structures: **a** circular cavity and **b** microchannel

27.3 Results and Discussion

The shape and form of a component can be controlled more precisely using estimation of geometric tolerances. Geometrical tolerances are useful in addition to dimensional tolerances. Geometric tolerances indicate the maximum deviation that is allowed in form or position from true geometry. Reference features—planes, lines, or surfaces must be recognized to specify the geometric tolerances.

27.3.1 Straightness (for Microchannels)

Straightness tells us the tolerance zone within which the considered element or axis lies representing its extent to be called as a straight line [18].

The assessment line for straightness is:

$$y = y_0 + l_0 x \tag{27.1}$$

The deviation is represented mathematically in the form of:

$$e_i = y_i - (y_0 + l_0 x_i) \tag{27.2}$$

The solutions in the form of least square (LS) are given as:

$$l_0 = \frac{\left(N \sum x_i y_i\right) - \sum x_i \sum y_i}{\left(N \sum x_i^2\right) - \left(\sum x_i\right)^2} \tag{27.3}$$

$$y_0 = \frac{\left(\sum y_i - l_0 \sum x_i\right)}{N} \tag{27.4}$$

Table 27.4 Dimensions of microchannel at different points

X (μm)	Y (measurement) (μm)	X (μm)	Y
0	300.06	4800	300.07
800	300.02	5600	300.03
1600	300.05	6400	300.02
2400	300.06	7200	300.09
3200	300.05	8000	300.01
4000	300.01		

Straightness tolerance:

$$h_{\text{straightness}} = \frac{\left|e_{\text{MAX}(+)}\right| + \left|e_{\text{MAX}(-)}\right|}{\sqrt{1 + l_0^2}} \quad (27.5)$$

In this experiment, three microchannels of length 8000 microns were made in patterned form and this channel was divided into 10 regions equally spaced and their microscopic views and dimensions were taken with the help of Olympus Analysis Five software. Table 27.4 shows the measurement data (Y-coordinates) for fixed intervals of X for microchannel. The straightness tolerance of each channel was calculated separately and then their average value was reported. The straightness tolerance is calculated using the above procedures listed by Eqs. (27.1), (27.2), (27.3), (27.4), and (27.5).

For channel 1

$$\begin{aligned} l_0 = {} & [10 \times (800 \times 300.02 + 1600 \times 300.05 + 2400 \times 300.06 + 3200 \times 300.05 \\ & + 4000 \times 300.01 + 4800 \times 300.07 + 5600 \times 300.03 + 6400 \times 300.02 \\ & + 7200 \times 300.09 + 8000 \times 300.01] - [(800 + 1600 + 2400 + 3200 \\ & + 4000 + 4800 + 5600 + 6400 + 7200 + 8000)(300.02 + 300.05 \\ & + 300.06 + 300.05 + 300.01 + 300.07 + 300.03 + 300.02 + 300.09 \\ & + 300.01)]/[10 \times (800^2 + 1600^2 + 2400^2 + 3200^2 + 4000^2 + 4800^2 + 5600^2 \\ & + 6400^2 + 7200^2 + 8000^2) - (800 + 1600 + 2400 + 3200 \\ & + 4000 + 4800 + 5600 + 6400 + 7200 + 8000)^2] \end{aligned}$$

So, $l_0 = -0.0000076$

$$\begin{aligned} y_0 = {} & [(300.02 + 300.05 + 300.06 + 300.05 + 300.01 + 300.07 \\ & + 300.03 + 300.02 + 300.09 + 300.01) + 0.0000076(800 \\ & + 1600 + 2400 + 3200 + 4000 + 4800 + 5600 + 6400 + 7200 + 8000)] \end{aligned}$$

$$y_0 = 300.374$$

Then, the corresponding deviation (e_i) is calculated as:

$$e_1 = 300.02 - (300.374 - 0.0000076 \times 800) = -0.348$$
$$e_2 = 300.05 - (300.374 - 0.0000076 \times 1600) = -0.678 \text{ (Max.}-\text{ve error)}$$
$$e_3 = 300.06 - (300.374 - 0.0000076 \times 2400) = -0.29 \text{ (towards Max.} + \text{ve error)}$$
$$e_4 = 300.05 - (300.374 - 0.0000076 \times 3200) = -0.665$$
$$e_5 = 300.01 - (300.374 - 0.0000076 \times 4000) = -0.660 \text{ (Max.}-\text{ve error)}$$
$$e_6 = 300.07 - (300.374 - 0.0000076 \times 4800) = -0.634$$
$$e_7 = 300.03 - (300.374 - 0.0000076 \times 5600) = -0.677$$
$$e_8 = 300.02 - (300.374 - 0.0000076 \times 6400) = -0.671$$
$$e_9 = 3.09 - (300.374 - 0.0000076 \times 7200) = -0.595$$
$$e_{10} = 300.01 - (300.374 - 0.0000076 \times 8000) = -0.303 \text{ (Max.} + \text{ve error)}$$

Hence, $e_{\max(+)} = 0.296$, and $e_{\max(-)} = 0.678$
Therefore, straightness tolerance:

$$h_{\text{straightness}} = \frac{|e_{\text{MAX}(+)}| + |e_{\text{MAX}(-)}|}{\sqrt{1 + l_0^2}} = \frac{0.296 + 0.678}{\sqrt{1 + (-0.0000076)^2}} = 0.974\,\mu\text{m}$$

Similarly for *channel* 2,

$$h_{\text{straightness}} = 1.764\,\mu\text{m}$$

And for *channel* 3,

$$h_{\text{straightness}} = 1.293\,\mu\text{m}$$

So, the average straightness tolerance is 1.34 μm.

The microchannel walls for channel 1 were largely straight, while for the second channel, the walls have greater slope than the third one. The starting and end points of the laser beam stroke determine the degree of inclination of the sloped surface walls. This may be due to the Gaussian nature of the beam.

The slope of the energy beam profile generally replicates on the PMMA surface with small deviation due to threshold energy requirement. Whereas, in channel 1, the start and end of the stroke are either absorbed or reflected and rest of the travel of stroke is transmitted to PMMA via slot width resulting in flat and vertical wall. The state of stress during laser interaction also plays a key role in the determination of channel shape. There might be many thermal cracks around the channels which may affect the channel cross section and reduce the straightness of the channel.

27.3.2 Circularity (Roundness) (for Circular Geometries)

Circularity is defined by the zone of tolerance enclosed by two concentric circles within which each circular element of the surface should lie down. Roundness error can be defined in the forms of least square circle (LSC), minimum circumscribed circle (MCC), maximum inscribed circle (MIC), and minimum zone circle (MZC), and in the present article, an attempt has been made to evaluate the circularity error using MZC approach (Fig. 27.3). In MZC method, roundness error is measured by using two circles as reference where one circle circumscribes the roundness profile and the other circle inscribes the roundness profile. The difference of diameters of these two circles is employed to estimate the roundness error.

Roundness error = $d_1 - d_2$.
Where d_1—circle drawn outside the profile and d_2—circle drawn inside the profile.
The error of three circular geometries is calculated separately and average value is considered.
For geometry 1, $d_1 = 304$ µm and $d_2 = 297$ µm.
Roundness error = 304 – 297 = 7 µm.
For geometry 2, $d_1 = 309$ µm and $d_2 = 298$ µm.
Roundness error = 309 – 298 = 11 µm.
For geometry 3, $d_1 = 305$ µm and $d_2 = 297$ µm.
Roundness error = 305 – 297 = 8 µm.

Therefore, the average circularity tolerance of circular microgeometries is estimated as 8.66 µm.

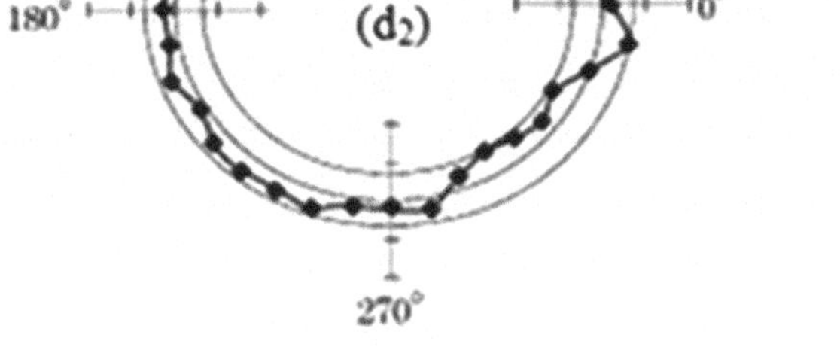

Fig. 27.3 Minimum zone circle method to find roundness error

27.4 Conclusions

An infrared laser micromachining is an outstanding tool for fabricating microfluidic structures in polymers. The shape and form of a component can be controlled more precisely using estimation of geometric tolerances. Geometrical tolerances are useful in addition to dimensional tolerances. Geometric tolerances indicate the maximum deviation that is allowed in form or position from true geometry. Reference features—planes, lines, or surfaces must be recognized to specify the geometric tolerances. The accuracy of microfabricated structures are greatly influenced by geometric tolerances as they indicate the extent of deviation of structure from their true shape.

In the present paper, an attempt has been made to evaluate geometric tolerances on two cavities: circular and straight channels. In case of circular cavities, circularity has been evaluated while in the case of straight channels, straightness has been evaluated, respectively. The Olympus optical microscope was used to examine the geometry of the fabricated microcavities and their tolerances have been evaluated after analyzing these geometries by profile projector and Olympus Analysis Software. It is found that the average circularity tolerance of circular microgeometries is 1.34 μm and the average straightness tolerance of microchannels is 8.66 μm. It shows that the laser micromachining is capable of producing highly accurate microstructures.

Acknowledgements Shrikant Vidya is thankful to the scientists and staffs of Micro Systems Technology Laboratory at CSIR-CMERI, Durgapur for providing the facilities to carry out the research work.

References

1. Gerlach, A., Knebel, G., Guber, A.E., Heckele, M., Herrmann, D., Muslija, A., Schaller, Th: Microfabrication of single-use plastic microfluidic devices for high-throughput screening and DNA analysis. Microsyst. Technol. **7**, 265–268 (2002)
2. Rötting, O., Ropke, W., Becker, H., Gartner, C.: Polymer microfabrication technologies. Microsyst. Technol. **8**, 32–36 (2002)
3. McKeown, P.: From micro- to nano-machining-towards the nanometre era. Sens. Rev. **16**, 4–10 (1996)
4. Ready, J.F.: Industrial Application of Lasers, 2nd edn. Academic, San Diego (1997)
5. Jensen, M.F., Noerholm, M., Christensen, L.H., Geschke, O.: Microstructure fabrication with a CO_2 laser system: characterization, fabrication of cavities produced by raster scanning of the laser beam. Lab. Chip **3**, 302–307 (2003)
6. Prakash, S., Kumar, S.: Fabrication of microchannels: a review. Proc. Inst. Mech. Eng. Part B J. Eng. Manuf. **229**, 1273–1288 (2015)
7. Romoli, L., Tantussi, G., Dini, G.: Experimental approach to the laser machining of PMMA substrates for the fabrication of microfluidic devices. Opt. Lasers Eng. **49**, 419–427 (2011)
8. Prakash, S., Kumar, S.: Experimental and theoretical analysis of defocused CO2 laser microchanneling on PMMA for enhanced surface finish. J. Micromech. Microeng. **27**, 025003 (2017)
9. Nayak, N.C., Lam, Y.C., Yue, C.Y., Sinha, A.T.: CO_2-laser micromachining of PMMA the effect of polymer molecular weight. J. Micromech. Microeng. **18**, 095020 (2008)

10. Imran, M., Rahman, R.A., Ahmad, M., Akhtar, M.N., Usman, A., Sattar, A.: Fabrication of microchannels on PMMA using a low power CO_2 laser. Laser Phys. **26**, 096101 (2016)
11. Mohammed, M.I., Alam, M.N.H.Z., Kouzani, A., Gibson, I.: Fabrication of microfluidic devices: improvement of surface quality of CO_2 laser machined poly(methylmethacrylate) polymer. J. Micromech. Microeng. **27**, 015021 (2017)
12. Snakenborg, D., Klank, H., Kutter, J.P.: Microstructure fabrication with a CO_2 laser system. J. Micromech. Microeng. **14**, 182 (2004)
13. Rauf, M.M., Shahid, M., Durrani, Y.A., Khan, A.N., Hussain, A., Akhter, R.: Cladding of Ni–20Cr coatings by optimizing the CO_2 laser parameters. Arab. J. Sci. Eng. **41**(6), 2353–2362 (2016)
14. Klank, H., Kutter, J.P., Geschke, O.: CO_2 laser micromachining and back-end processing for rapid production of PMMA based microfluidic systems. Lab. Chip **2**, 242–246 (2002)
15. Huang, Y., Liu, S., Yang, W., Yu, C.: Surface roughness analysis and improvement of PMMA-based microfluidic chip chambers by CO_2 laser cutting. Appl. Surf. Sci. **256**, 1675–1678 (2010)
16. Mohammed, M.I., Desmulliez, M.P.Y.: The manufacturing of packaged capillary action microfluidic systems by means of CO_2 laser processing. Microsyst. Technol. **19**, 809–818 (2013)
17. Becker, H., Gaertner, C.: Polymer microfabrication methods for microfluidic analytical applications. Electrophoresis **21**, 12–26 (2000)
18. Hanumaiah, N., Ravi, B.: Rapid tooling form accuracy estimation using region elimination adaptive search based sampling technique. Rapid Prototyping J **13**(3), 182–190 (2007)

Chapter 28
Microstructural and Photocatalytic Performance of $BaZr_xTi_{1-x}O_3$ Ceramics

Umesh Mishra, K. S. Srikanth, Shrikant Vidya, Spandan Shukla, Parveen and Savant Maurya

Abstract In this study, the effect of zirconium addition of ($BaZr_xTi_{1-x}O_3$) on microstructural, dielectric and photocatalytic characteristic level has structure composition $x = 0$–0.15 in detail. All photocatalytic experiments show that the synthesized catalysts had a high photocatalytic activity with Zr addition toward synthetic dye MO under visible light improving the photocatalytic production rate because the band gap shifts toward visible region with increase in x. The ferroelectricity of the samples was done by *P–E* loops, and all samples exhibited well-saturated hysteresis loops. The dielectric measurements indicate the decrease of phase transition temperature from 393 K to near room temperature which is beneficial for many dielectric applications.

Keywords Photocatalysis · Dielectric · Ferroelectric

28.1 Introduction

Interest in lead-free ferroelectric ceramics has increased over the period of time owing to the need to find a suitable replacement of its lead-based counterparts, namely lead zirconate titanate ($Pb(Zr,Ti)O_3$, PZT) which contains ~60 wt% of toxic lead. In this context, as their replacement, these three well-known ferroelectrics based on $BaTiO_3$, $KNbO_3$, and $Na_{1/2}Bi_{1/2}TiO_a$ in pure and modified forms have been explored with renewed interest in the last decade [1–3]. In another study by Yao et al., a giant piezoelectric coefficient (d_{33}) of 697 pC/N has been found in $BaTiO_3$–11$BaSnO_3$ which is highest reported value till date [4–7]. There are various researches going on in the lead-free ceramics, and their main motive is to develop further the various types of cations whose substitution at the Ba and/or Ti sites could enhance the performance of base composition for many dielectric applications, similar to that of Zr, Hf, and Sn. All these findings indicate the worldwide exploration on $BaTiO_3$-based materials for replacement of lead-based materials for various dielectric applications.

For many photovoltaic and photocatalytic applications, these ferroelectric materials are used due to their spontaneous polarization at room temperature. Nasby

U. Mishra · K. S. Srikanth (✉) · S. Vidya · S. Shukla · Parveen · S. Maurya
Department of Mechanical Engineering, Galgotias University, Greater Noida, India

S. Yadav et al. (eds.), *Proceedings of International Conference in Mechanical and Energy Technology*, Smart Innovation, Systems and Technologies 174,
https://doi.org/10.1007/978-981-15-2647-3_28

and Quinn first reported $BaTiO_3$ for water splitting with significant photostability, and then, several potential heterogeneous semiconductor materials like ZnO, CdS, WO_3, Fe_2O_3, Bi_2WO_6, BiOCl, $LaMnO_3$, and $knbO_3$ are explored for enhanced photocatalytic and photovoltaic performance [8–10]. However, these materials create hindrance for the practical applications as a visible active photocatalyst in spite of superior performance. This is because of existence of larger band gap in similar materials. Among the $BaTiO_3$ (BT)-based relaxor ferroelectrics, $BaTiO_3$ doped with Zr garnered researchers' scrutiny due to both application and fundamental interests. The dopant ion Zr can enter into $BaTiO_3$ lattice as Zr^{3+} or Zr^{4+}. The ionic radii of Zr^{3+} are almost similar to the ionic radii of Ba^{2+} ion which makes Zr^{3+} to occupy Ba site. Similarly, Zr^{4+} has an ionic radius almost equal with that of Ti^{4+} which will substitute Ti sites. Further, it is a well-known fact that the Zr^{4+} ions are more stable than Zr^{3+} ions will occupy Ti site. Therefore, it is important to study the site occupancy of Zr in $BaTiO_3$ perovskite lattice [11, 12]. To justify this, few studies had been made which will reveal the substitution of Zr ions can take place at both sites (A and B) in the $BaTiO_3$ perovskite unit cell when concentration of Zr < 8% results in completely different characteristics [13–15].

In this context, parent material $BaZr_xTi_{1-x}O_3$ with ($x = 0$) is selected, Zr ($x =$ 0 M, 0.10 M, 0.12 M, and 0.15 M) concentrations are introduced into perovskite lattice and their photocatalytic activity under visible light has been studied; these compositions are well explored from piezoelectric point of view which is another application of such ferroelectric perovskites.

28.2 Experimental Procedure

28.2.1 Synthesis of $BaZr_xTi_{1-x}O_3$

The ferroelectric ceramics $BaZr_xTi_{1-x}O_3$ ($x = 0$, 0.1, 0.12, and 0.15) was prepared using solid-state reaction. The high purity of reagent powders was used as starting materials like barium carbonate [$BaCO_3$, (99% pure)], titanium dioxide [TiO_2, (99% pure)], and zirconium oxide [ZrO_2, (99.99% pure)]. The materials were weighed according to their stoichiometric ratios, and after that the ball milling of materials was done for 10 h in acetone (wetting agent) to have physical homogeneity. After that, the mixture was dried and calcination was done at 1000 °C up to 4 h.

Further hydraulic press of pressure (3–4 ton cm^{-2}) was applied to make disk-shaped pellets of 12-mm diameter and 1-mm thickness. Sintering was done at 1400 °C for 5 h, and density was measured using Archimedes principle.

28.2.2 Material Characterization

The X-ray diffraction (XRD) gives the phase formation of sintered ceramics (Rigaku Smart Lab, Japan), which is having a range starting from 20° and ending at 80°. The rate at which scanning was carried out is fixed to 2°/min. The surface morphology for our specimens which were sintered at 1400 °C was analyzed employing scanning electron microscopy (FE-SEM). UV–Vis spectroscopy of the prepared specimen was done at room temperature using Shimadzu-2450 UV–Vis spectrophotometer. To ascertain the ferroelectric nature of our samples, we have taken hysteresis loops by keeping polarization in *Y*-axis and electric field along *X*-axis (*P*–*E*) at 50 Hz. These measurements were carried out at different temperature values and various electric fields.

28.2.3 Photocatalytic Study

To check out the photocatalytic degradation activity of $BaZr_xTi_{1-x}O_3$ ferroelectric specimens, a distinctive azo methyl orange (MO) dye solution acted as a water pollutant under the visible light irradiation. To start with the experiment, we first took 0.1 g of our prepared specimen and vessel sonication was done having 20 ml of MO. Further, the dye was kept in dark under constant stirring for 30 min so that it stabilizes. The experiments to check the photocatalytic performance were started as soon as the reaction vessel was put under visible light irradiation with 150 W/420 nm halogen lamp. We also circulated cold water during the reaction process around the vessel. Fixed aliquots of degraded MO sampling were collected to ascertain the reaction kinetics at a set time of intervals for each $BaZr_xTi_{1-x}O_3$ ferroelectric ceramics. The ceramic powders were removed from the collected MO samples by centrifugation at 7000 rpm and then analyzed for the change in MO concentration through Shimadzu-2450 UV–Vis spectrophotometer by measuring the peak intensity at 450 nm.

28.3 Result and Discussion

Figure 28.1 shows the XRD patterns of the $BaZr_xTi_{1-x}O_3$ ceramics for various Zr contents ($x = 0$, 0.1, 0.12, and 0.15) sintered at 1400 °C for 5 h. The lack of any secondary peaks indicates that the solid-state reaction has occurred in the specified sintering conditions. The diffraction peaks of the undoped ceramic correspond well to T-symmetry. However, the diffraction peaks around 46°–48° merge into a single peak, when Zr is introduced into the lattice of $BaTiO_3$, which shows that the host phase undergoes an obvious phase transition. Apart from the structural phase transition, we also noticed with increase in *x* content a gradual shift of diffraction peak

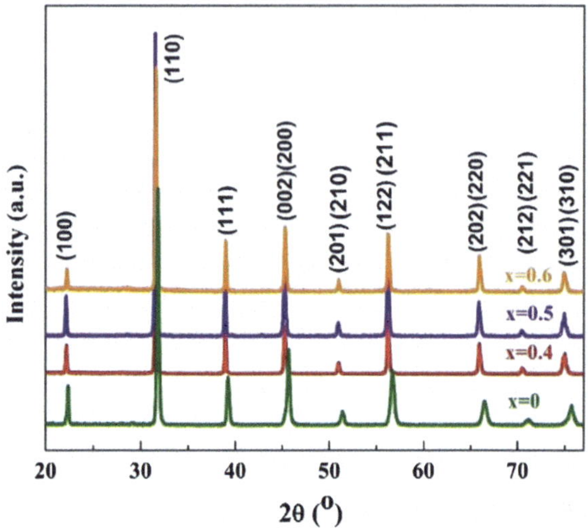

Fig. 28.1 XRD of all samples

toward lower angle. It was due to the replacement of larger ion Zr^{4+} ($r = 0.87$ Å) for Ti^{4+} ($r = 0.6507$ Å) at the B site of perovskite structure. Therefore, XRD is used only to confirm the phase because the main focus of this article is to explore these materials for specific properties as discussed in the above sections. However, a detailed discussion on XRD analysis was required to understand the structural evolution of these materials.

The scanning electron microscope (SEM) images of all the composition states exhibit well-interconnected grains without major voids or anomaly, having a bimodal grain size distribution (larger grains coexist with smaller grains) as shown in Fig. 28.2. Further increase of larger grains on the expense of smaller grains takes place when doped with Zr. Then using Archimedes principle, the densities of $BaZr_xTi_{1-x}O_3$ sintered ceramics were found and the relative densities were all >95% for all compositions.

28.3.1 UV Spectroscopy

The effect of Zr doping on the optical properties of $BaTiO_3$ is shown in Fig. 28.3. There was no significant absorption observed for $BaTiO_3$ specimen in the wavelength range of 400–800 nm as presented in Fig. 28.3.

With the introduction of Zr, the absorbance of visible light range increased with increase in content of Zr in parent $BaTiO_3$. The Tauc plot shows the band gap energy of $BaTiO_3$ and Zr-doped $BaTiO_3$ as shown in Fig. 28.3. As we increase the amount of Zr, the band gap shifted toward the visible region. The band gap value for $BaTiO_3$ and $BaZr_xTi_{1-x}O_3$ ($x = 0.15$) was observed as 3.23 eV and 2.24 eV, respectively.

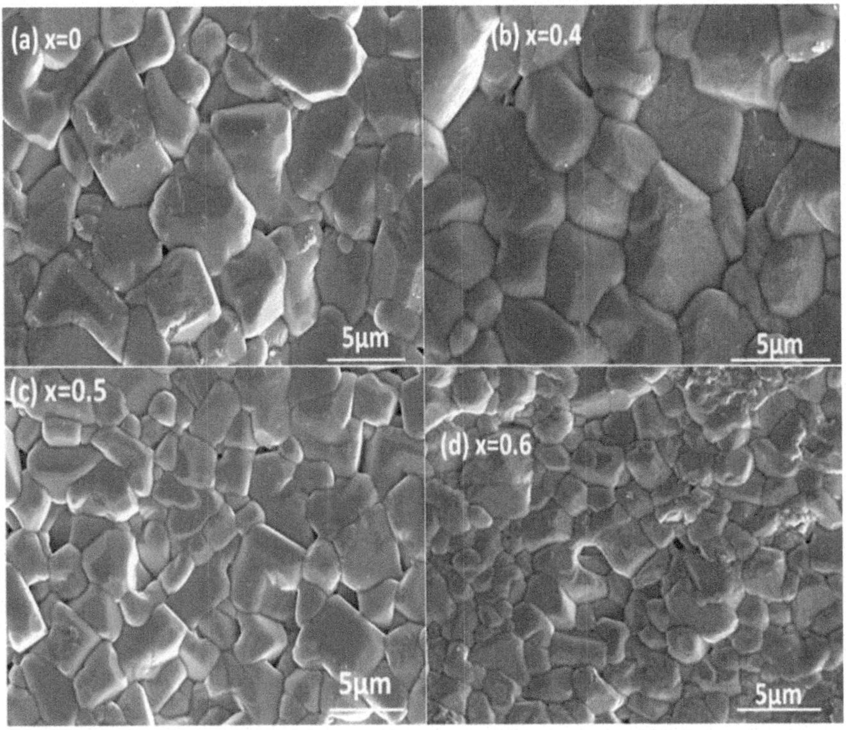

Fig. 28.2 SEM of all samples

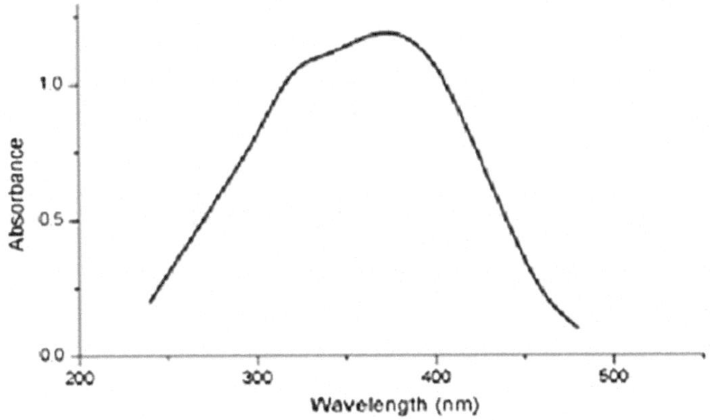

Fig. 28.3 UV spectroscopy of all elements

28.3.2 Hysteresis Loop

In Fig. 28.4, the *P–E* hysteresis loops of Zr-doped $BaTiO_3$ ceramics at constant electric field at 24 kV/cm and at room temperature were shown. All the components show the well-saturated loops which was almost similar to ferroelectric-like loops. The remnant polarization (P_r) and coercive field (E_c) values are ~7 $\mu C/cm^2$ and ~3.5 kV/cm, respectively, for base composition $x = 0$. Hysteresis loops become

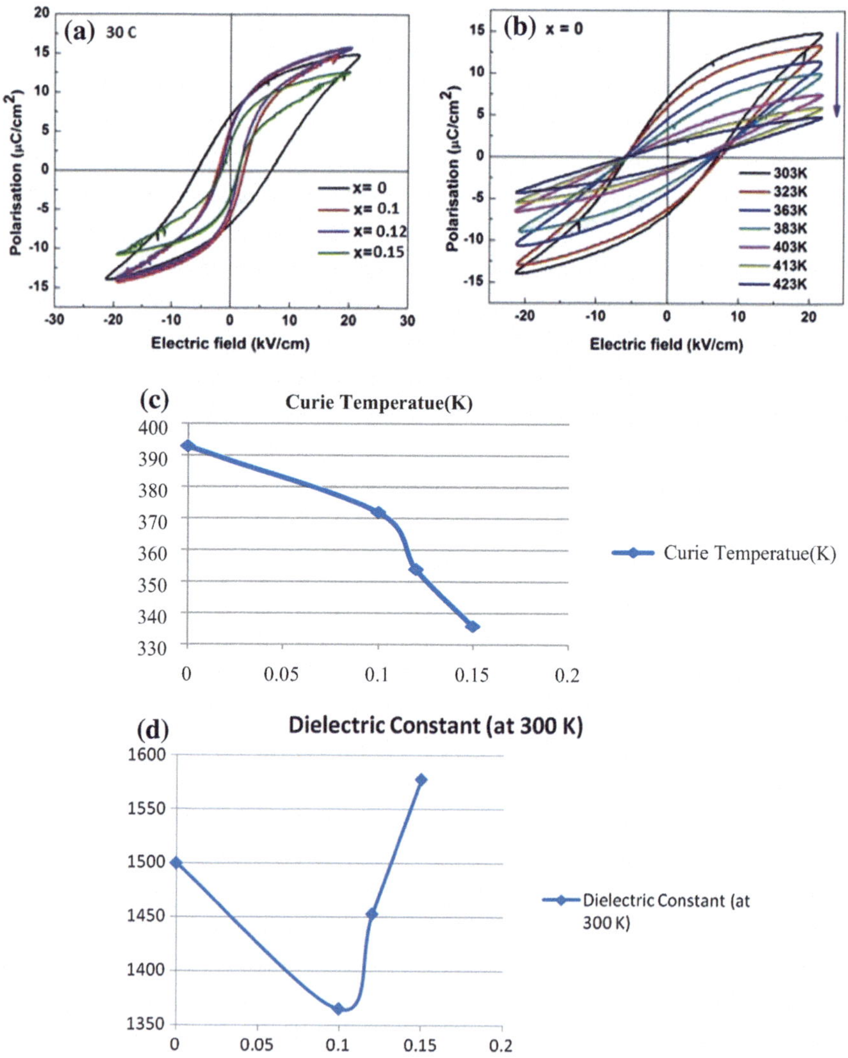

Fig. 28.4 **a** P-E loop at room temperature, **b** $x = 0$ at various temperature, **c** and **d** dielectric measurements

thinner and thinner as the content of Zr increases accompanied by a fast drop of E_c. Such a low value of E_c suggests that the compositions are soft with respect to electric field. The dielectric measurements are shown in Fig. 28.4c, d.

28.3.3 Photocatalytic Study

Under the visible light, BZT catalyst was examined to analyze the photocatalytic degradation of methyl orange dye. After the adsorption equilibrium, the photodegradation experiments were conducted. Figure 28.3 represents the variation in adsorbance spectra at 25 °C of an aqueous solution of methyl orange dye in the presence of BZT catalyst ($x = 0.15$) irradiated by a visible light source. It is very clear from Fig. 28.3 that the peak corresponding to absorption of investigated dye was seen at ~463 nm and the intensity of peak decreases rapidly with the prolonged exposure to time. The orange solution of dye [at time (t) = 0 min] turned into colorless solution after exposure to 120 min as shown in Fig. 28.5.

The photocatalytic deprivation of our used dye using the manufactured Zr-added BZT catalysts was the key measurement in the present article. After the equilibrium condition is reached in dark, the photodegradation trials were initiated with the wavelength of light under visible region with λ > 423 nm. Figure 28.5 shows the distinctive demonstration of the alteration in the absorbance of MO dye of known concentration when the sample $BaZr_xTi_{1-x}O_3$ ceramic ($x = 0.15$) is present. The degradation in MO dye was confirmed by the change in color of MO to colorless from orange color and also photocatalytic performance in $BaZr_xTi_{1-x}O_3$ ceramics under the visible radiation with some duration of time (minimum 2–3 h) as shown in Fig. 28.5.

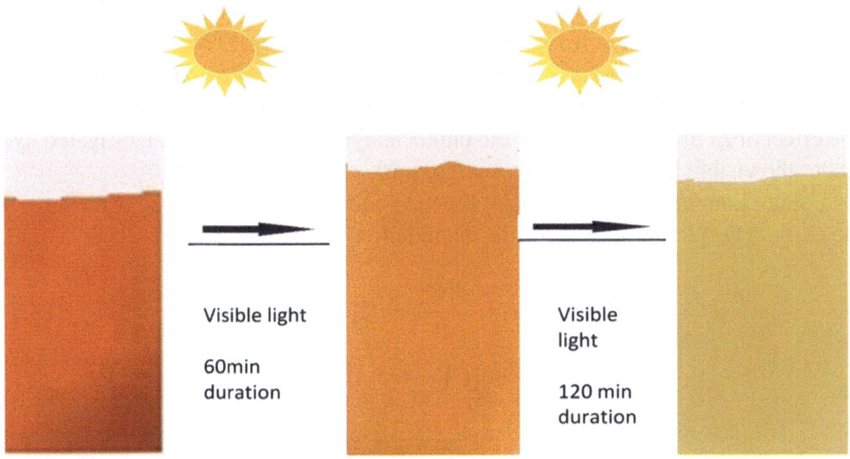

Fig. 28.5 Photocatalysis of sample in visible light

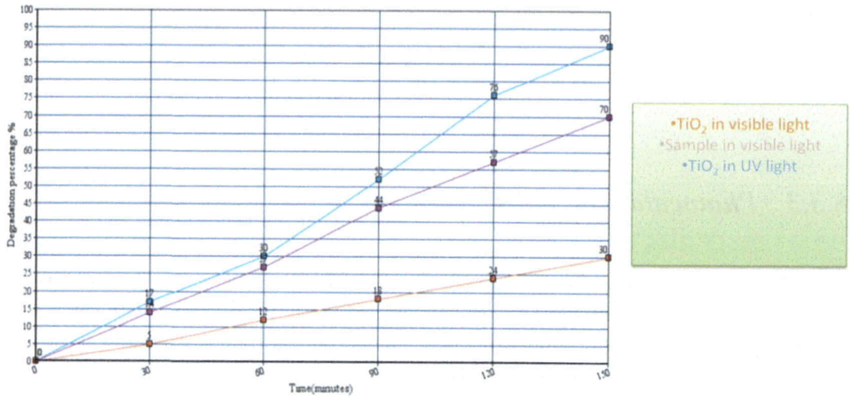

Fig. 28.6 Degradation % as a function of time for samples and compared with TiO_2

The degradation of MO dye for $BaZr_xTi_{1-x}O_3$ ceramics is represented in Fig. 28.6. The efficiency (denoted by $D\%$) after conducting the degradation is calculated for all samples with Eq. (28.1):

$$D\% = \left[\frac{C_0 - C_t}{C_0}\right] \times 100 \tag{28.1}$$

where C_0 means dye's starting concentration and C_t means after few time concentration.

The control experiments for bare MO indicated almost negligible degradation of the pollutant (not shown here), while for $BaZr_xTi_{1-x}O_3$ ceramics showed MO degradation almost in 160 min. Control experiments indicate the photocatalytic activity in $BaZr_xTi_{1-x}O_3$ ceramics and ruling out any other degradation mechanism. For pure BZT catalyst, the dye molecules degraded up to merely 30% whereas it doubled after Zr addition (>15% for $x = 0.15$). The degradation efficiency (D%) of the synthesized catalysts was in the order: $x = 0.15 > x = 0.12 > x = 0.1 > x = 0$. Also, there was no degradation of pollutant observed under dark. Thus, Fig. 28.6 clearly represents the effect of Zr doping in BZT for the photocatalytic degradation of investigated dye under the visible light irradiation ($\lambda > 420$ nm).

Using Langmuir–Hinshelwood kinetic equation, the kinetic behavior of $BaZr_xTi_{1-x}O_3$ photocatalysts for the degradation of MO was studied.

$$-r = \frac{\mathrm{d}[C]}{\mathrm{d}t} = K[C] \tag{28.2}$$

$$\int_{[C]_0}^{[C]_t} \frac{\mathrm{d}[C]}{[C]} = \int_{t=0}^{t=t} K\,\mathrm{d}t \tag{28.3}$$

$$\ln \frac{[C]_t}{[C]_0} = -Kt + \text{constant} \tag{28.4}$$

where the C_0 and C_t stand for amount of pollutant present (MO) after time $t = 0$ and $t =$ minutes. The slope of $\ln(C_t/C_0)$ versus degradation time (t) plots gives the pseudo-first-order kinetic rate constant K. We have plotted this graph for the degradation of MO dye using $BaZr_xTi_{1-x}O_3$ ceramics (not shown here). The pollutant MO shows first-order decay for the visible light catalysis using $BaZr_xTi_{1-x}O_3$ ceramics. The kinetic rate constants were calculated as k_1 (0.249×10^{-2} min^{-1}), k_2 (0.371×10^{-2} min^{-1}), k_3 (0.5×10^{-2} min^{-1}), and k_4 (0.592×10^{-2} min^{-1}) for $(1 - x)$BCT$-x$BST ceramics with $x = 0, 0.1, 0.12$, and 0.15 M concentrations, respectively.

Catalyst	Pollutant	% Degradation	K (min^{-1})	Time	References
$BiFeO_3$ nanoparticles	MO	90	–	16 h	[16]
$Bi_2Fe_4O_9$ nanoparticles	MO	2.5	–	24 h	[16]
$Bi_{20}TiO_{32}$	MO	75	–	180 min	[17]
Bi_2WO_6	MO	–	0.00013	120 min	[18]
C_3N_4	MO	–	0.006	120 min	[19]
BNKLBT	MO	75	0.007	150 min	[19]
BNKLBT	Estriol	85	0.0056	30 min	[16]
TiO_2 (under visible light)	Estriol	–	0.0012	200 min	[20]
BCT under visible light	MO	70	0.006	150 min	[21]
BZT ($x = 0.15$)	MO	70	0.00521	120 min	Present work

28.4 Conclusion

1. The investigation of microstructural, photocatalytic performance along with dielectric of $BaZr_xTi_{1-x}O_3$ ceramics with $x = 0$–0.15 was carried out.
2. The dense samples with interconnected grains were observed by SEM which does not have any major voids.
3. The Zr addition resulted in decrement of Curie temperature which is confirmed from dielectric measurements. From the study, it was concluded that the phase transition of $BaTiO_3$ was diffuse nature.
4. The degradation of dye molecules was found to be nearly 40% which is doubled after adding Zr (>70% for $x = 0.15$).

5. According to the catalytic experiments, the manufactured catalysts are photocatalytically active with zirconium addition to the dye methyl orange (MO) when these samples are subjected to visible light irradiation.

References

1. Fujishima, A., Honda, K.: Electrochemical photolysis of water at a semiconductor electrode. Nature **238**(5358), 37–38 (1972)
2. Bahnemann, D.: Photocatalytic water treatment: solar energy applications. Sol. Energy **77**(5), 445–459 (2004)
3. Malato, S., Fernández-Ibáñez, P., Maldonado, M.I., Blanco, J., Gernjak, W.: Decontamination and disinfection of water by solar photocatalysis: recent overview and trends. Catal. Today **147**(1), 1–59 (2009)
4. Chong, M.N., Jin, B., Chow, C.W., Saint, C.: Recent developments in photocatalytic water treatment technology: a review. Water Res. **44**(10), 2997–3027 (2010)
5. Chakrabarti, S., Dutta, B.K.: Photocatalytic degradation of model textile dyes in wastewater using ZnO as semiconductor catalyst. J. Hazard. Mater. **112**(3), 269–278 (2004)
6. Mills, A., Davies, R.H., Worsley, D.: Water purification by semiconductor photocatalysis. Chem. Soc. Rev. **22**(6), 417–425 (1993)
7. Hoffmann, M.R., Martin, S.T., Choi, W., Bahnemann, D.W.: Environmental applications of semiconductor photocatalysis. Chem. Rev. **95**(1), 69–96 (1995)
8. Rajeshwar, K., Osugi, M.E., Chanmanee, W., Chenthamarakshan, C.R., Zanoni, M.V., Kajitvichyanukul, P., Krishnan-Ayer, R.: Heterogeneous photocatalytic treatment of organic dyes in air and aqueous media. J. Photochem. Photobiol. C: Photochem. Rev. **9**(4), 171–192 (2008)
9. Shang, M., Wang, W., New, H.X.: Bi_2WO_6 nanocages with high visible-light-driven photocatalytic activities prepared in refluxing EG. Cryst. Growth Des. **9**(2), 991–996 (2009)
10. Li, Y., Yao, S., Xue, L., Yan, Y.: Sol–gel combustion synthesis of nanocrystalline $LaMnO_3$ powders and photocatalytic properties. J. Mater. Sci. **44**(16), 4455–4459 (2009)
11. Mu, Q., Zhang, Q., Wang, H., Li, Y.: Facile growth of vertically aligned BiOCl nanosheet arrays on conductive glass substrate with high photocatalytic properties. J. Mater. Chem. **22**(33), 16851–16857 (2012)
12. Gonzalez, C.P., Schileo, G., Murakami, S., Khesro, A., Wang, D., et al.: Continuously controllable optical band gap in orthorhombic ferroelectric $KNbO_3$-$BiFeO_3$ ceramics. Appl. Phys. Lett. **110**, 172902 (2017)
13. Drew, K., Girishkumar, G., Vinodgopal, K., Kamat, P.V.: Boostingfuelcellperformance with a semiconductor photocatalyst: TiO_2/Pt–Ru hybrid catalyst for methanol oxidation. J. Phys. Chem. B **109**(24), 11851–11857 (2005)
14. Kim, H., Kim, J., Kim, W., Choi, W.: Enhanced photocatalytic and photoelectrochemical activity in the ternary hybrid of CdS/TiO_2/WO_3 through the Cascadal electron transfer. J. Phys. Chem. C **115**(19), 9797–9805 (2011)
15. Pelaez, M., Nolan, N.T., Pillai, S.C., Seery, M.K., Falaras, P., Kontos, A.G., Dunlop, P.S., Hamilton, J.W., Byrne, J.A., O'shea, K., Entezari, M.H.: A review on the visible light active titanium dioxide photocatalysts for environmental applications. Appl. Catal. B: Environ. **125**(21), 331–349 (2012)
16. Gao, F., Chen, X.Y., Yin, K.B., Dong, S., Ren, Z.F., Yuan, F., Yu, T., Zou, Z.G., Liu, J.M.: Magnetic and photocatalytic behaviors of Ca Mn co-doped $BiFeO_3$ nanofibres. Adv. Mater. **19**, 2889 (2007)
17. Ruan, Q.J., Zhang, W.-D.: Facile synthesis of nanostructured BiOI microspheres with high visible light-induced photocatalytic activity. J. Phys. Chem. C **113**, 4168 (2009)

18. Cheng, H., Huang, B., Dai, Y., Qin, X., Zhang, X., Wang, Z., Jiang, M.: Band gap engineering design for construction of energy levels well matched semiconductor junction with enhanced visible light driven photocatalytic activity. J. Solid State Chem. **182,** 2274 (2009)
19. Tian, Y., Chang, B., Lu, J., Fu, J., Xi, F., Dong, X.: Green photocatalytic synthesis and application toward the degradation of organic pollutants. ACS Appl. Mater. Interfaces **5**, 7079 (2013)
20. Li, G., Puddu, V., Tsang, H.K., Gora, A., Toepfer, B.: Visible light-induced photocatalytic and antibacterial activity. Appl. Catal. B Environ. **99**, 388 (2010)
21. Srikanth, K.S., Vaish, R., Hooda, M., Kushwaha, H.: Structural and photocatalytic performance of ferroelectric ceramics. Mat. Sci. Semicond. Process. (2018)

Chapter 29
Air Conditioning System Using Solid Desiccant (Silica Gel) as Dehumidifier

Kapil Rajput, Shrikant Vidya, Manoj Kumar Shukla and P. Mathiyalagan

Abstract In this paper, we have arranged different mechanical components to produce air conditioning effect by using solid desiccant material (silica gel) as a dehumidifier. This assembled equipment was tested at various atmospheric temperatures and coolant conditions for monitoring the effect. On room temperatures with relative humidity which consumes very less power to operate with no such harmful effect on climate except releasing hot water, which may be used for any thermodynamic process. The overall performance of the proposed work was checked with proper insulation. The entire system runs by using solar PV as input energy since the system comprises of few DC power operated equipments. In our experiment, we took different temperatures of water, such as natural water from bore well, chilled water and direct ice to produce cooling effect for a room of smaller volume (0.9126 ft^3) made up of acrylic boards. It has composite water tanks containing sands inside it which were used to generate cold water by natural evaporation process along with two water-cooled radiators for exchanging heat between air and cold water for this air conditioning system. From experiments, it has been found that cold water having temperature ranges between 24 and 29 °C provide a good result (COP—0.04) over other water inlet conditions such as applying chilled water and simply putting ice into the water tanks. Also, this type of air conditioning system has many advantages over other conventional types as it is not using any kind of chemical refrigerants and high energy-consuming device with less mechanical components by which mechanical losses, energy consumption with harmful effect to the environment reduces up to a certain limit. This concept can be used for rural and remote areas where a significant energy crisis is observed.

Keywords Desiccants · Evaporator · Humidity · Psychrometry

29.1 Introduction

India is a tropical country and more than 80% of Indian subcontinental area falls under warm humid or composite climatic zone. These climatic zones are characterized by

K. Rajput · S. Vidya (✉) · M. K. Shukla · P. Mathiyalagan
School of Mechanical Engineering, Galgotias University, Greater Noida, Uttar Pradesh, India

S. Yadav et al. (eds.), *Proceedings of International Conference in Mechanical and Energy Technology*, Smart Innovation, Systems and Technologies 174,
https://doi.org/10.1007/978-981-15-2647-3_29

high annual average temperatures and high humidity. Solar thermal system, which matches the cooling demand profile perfectly and can provide a substantial part of the energy needed for air conditioning [1–20], is thus thought to be an efficient way to alleviate these crises. Solar assisted desiccant [21, 22] cooling is one such encouraging system, given the fact that solar energy is abundant renewable resources to meet the cooling load requirements. Among existing solar thermally driven cooling processes, solid rotary desiccant cooling [23] has been recognized as a promising technology. A desiccant is a natural or synthetic substance that is capable of showing a strong attraction for water vapor. Desiccant cooling systems are basically open [24] cycle systems, using water as a refrigerant in direct contact with air. The thermally driven cooling cycle is a combination of evaporative cooling with air dehumidification through the desiccant absorber.

29.2 System Description

According to Fig. 29.1, air from upper room passes to the lower side of the room through a heat exchanger mounted over a fan, where due to cold water interaction, the air gets cooled without an increase in humidity. The air then passes through the desiccant tube where silica gel absorbs the moisture present inside the air creating a slight increase in air temperature. This low temp and low humid air again pass through another heat exchanger filled with cold water coming from the same composite mud pots, where temp of water is maintained low as compare to the surrounding air by the natural evaporation method. Thus, again the temp of output air lowers to its

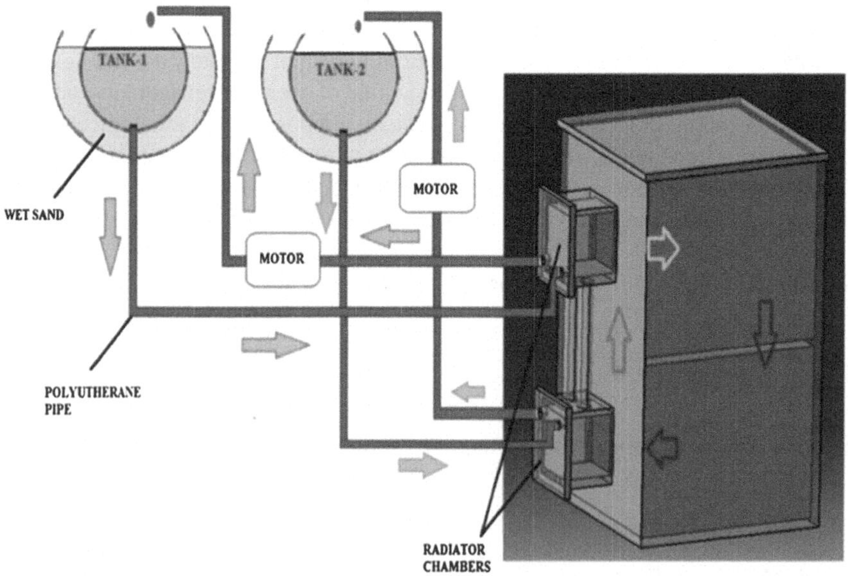

Fig. 29.1 Setup diagram

maximum value keeping low humidity inside the flowing air. This process continues in a cyclic way to make the entire room cooler. In the tank, we have used sand to speed up the rate of evaporation process. Experimentally, it is found that near about 3° to 4° temperature differences are maintained due to this process.

29.3 Mathematical Modeling

Since the entire system carries only two DC operated small radiator fans along with two recirculating pumps which consumes a total of 12 W power which may run by employing small solar PV panels, it is a great advantage in this project. For calculating COP of the system, total power absorbed along with heat absorbed by flowing water is calculated by empirical formulas as below.

Total heat absorbed by the cold water which is circulating outside the radiator pump along with small fans absorbs DC power of 12 W each. Heat released to cold water is circulated inside radiator.

$$\partial Q = mC_{\mathrm{p}}\partial_{\mathrm{T}} \tag{29.1}$$

m Mass flow rate of water
C_{p} Specific heat of water
∂_{T} Change in temperature
∂Q Heat rejected to cold water

$$\mathrm{COP} = \frac{\partial Q}{W_{\mathrm{in}}} \tag{29.2}$$

For calculating work input and heat removed, the best result is taken for consideration to determine the maximum possible cop by this system. So, putting all the calculated values COP is determined as,

(i) By using natural evaporative cooling

$$\mathrm{COP} = \frac{0.0235 \times 4.187 \times 8.8}{12} = 0.072 \tag{29.3}$$

(ii) By using ice

$$\mathrm{COP} = \frac{0.0235 \times 4.187 \times 7}{12} = 0.057 \tag{29.4}$$

(iii) By using cold water only

$$COP = \frac{0.0235 \times 4.187 \times 5}{12} = 0.04 \quad (29.5)$$

From the above calculation, it is clear that using natural evaporative cooling batter COP can be achieved, and from our experiment, it is about 0.072 which is maximum.

29.4 Results and Discussion

According to Fig. 29.2, it is clear that due to application of natural evaporative cold water which is generally available at bore well or tube well. Also, it is measured during hot weather of 39–42 °C; there is a difference in temperature of 12–16 °C is maintained in between these. By applying this water at 21 °C, we are able to achieve the lowermost temp of room as 26.2 °C at 1:30 P.M. where the ambient temp was at 38.9 °C. Also by using desiccant humidity decreases to 53% from its initial value of 56.7%.

By using ice, there is a drastic effect on room temperature and humidity. It was our misconception that using ice would give better achievement of room temp along with humidity which is shown in Fig. 29.3. But it is found that due to lowering the room temp below dew point temp of air, the moisture starts to condense at the contact surface of radiators and get added to the flowing air simultaneously due to rapid increase of moisture in air which again increases its temp while passing through solid desiccants. So, using ice is not suitable for this project.

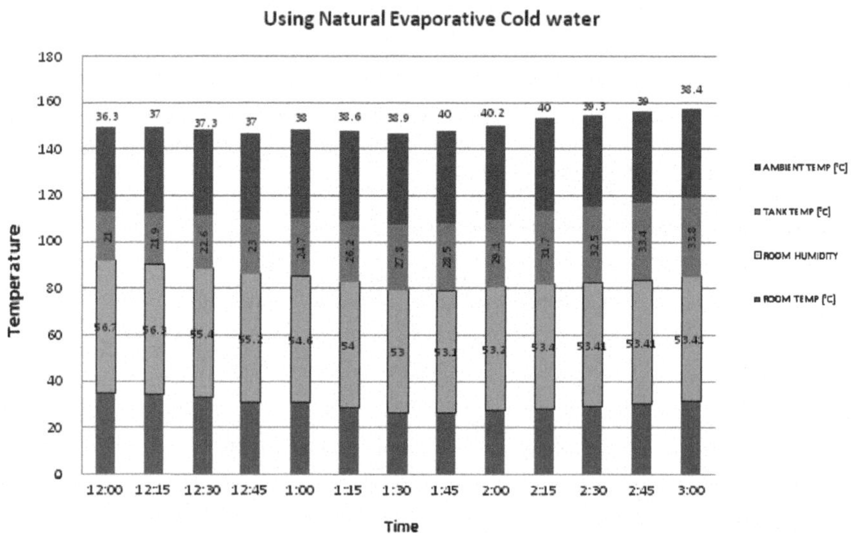

Fig. 29.2 Effects due to natural evaporative water in pot

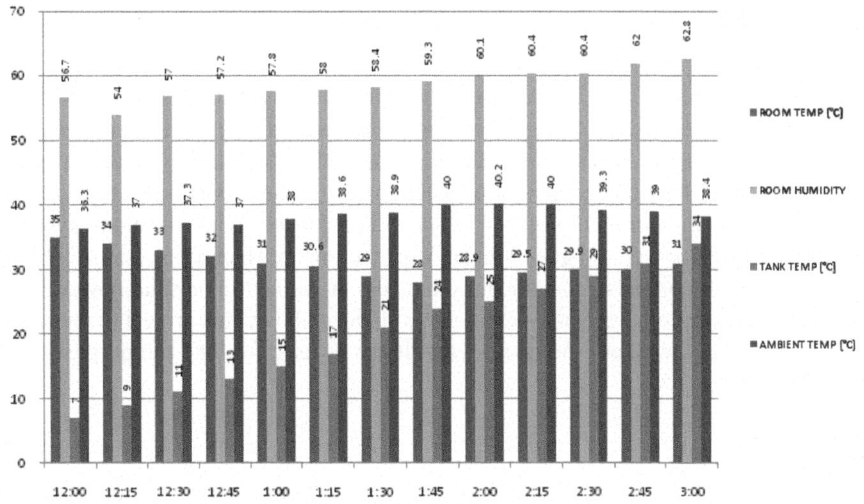

Fig. 29.3 Effects of using ice in pot

In Fig. 29.4, the temp and humidity achieved by using cold water initially at 24 °C give required moisture content that is 54%, which is the lowest one from our experiment but the output temperature what we achieved, i.e., 29 °C is maintained for only 15–30 min, which is not desirable from experimental point of view.

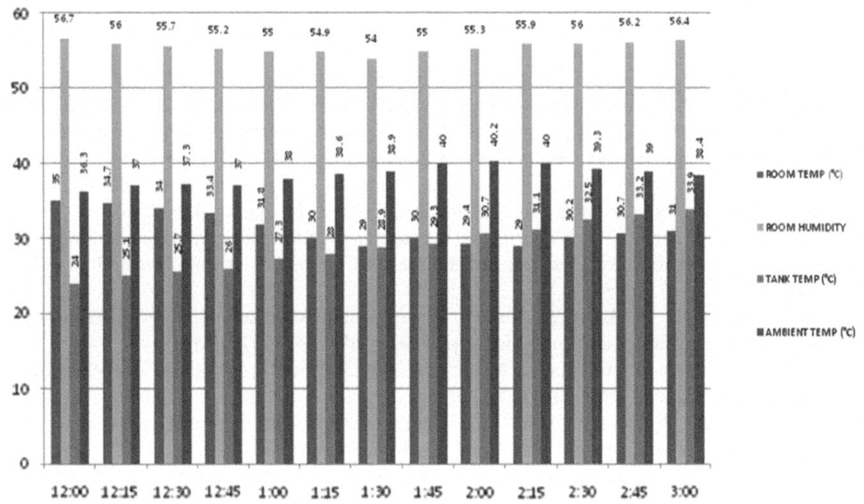

Fig. 29.4 Effects due to cold water only in pot

29.5 Conclusions

From the above three types of analysis, it is clear that using natural evaporative cold water (bore well, tube well, underground sump), etc. provides better result over direct ice and cold water which is about 26–28 °C along with 54% relative humidity which is maintained for 45–60 min. Also, this system provides the maximum COP of 0.072 as compared with other methods such as ice and cold water. Since the entire system requires very less power along without any conventional refrigeration systems which are composed of compressor and heat exchangers, it can be run on solar power with zero effect on atmosphere. Some research is still needed over this project to increase its performance as well as efficiency so that it can be used for long run purposes.

References

1. Pangavhane, D.R., Sawhney, R.L., Sarsavadia, P.N.: Design, development and performance testing of a new natural convection solar dryer. Energy **27**, 579–590 (2002)
2. Gao, W., Lin, W., Liu, T., Xia, C.: Analytical and experimental studies on the thermal performance of cross-corrugated and flat-plate solar air heaters. Appl. Energy **84**, 425–441 (2007)
3. Henderson, D., Junaidi, H., Muneer, T., Grassie, T., Currie, J.: Experimental and CFD investigation of an ICSSWH at various inclinations. Renew. Sustain. Energy Rev. **11**, 1087–1116 (2007)
4. Varola, Y., Oztopb, H.F.: A comparative numerical study on natural convection in inclined wavy and flat-plate solar collectors. Build. Environ. **43**, 1535–1544 (2008)
5. Esen, H.: Experimental energy and exergy analysis of a double-flow solar air heater having different obstacles on absorber plates. Build. Environ. **43**, 1046–1054 (2008)
6. Tang, R., Sun, Z., Li, Z., Yu, Y., Zhong, H., Xia, C.: Experimental investigation on thermal performance of flat plate collectors at night. Energy Convers. Manage. **49**, 2642–2646 (2008)
7. Ozgen, F., Esen, M., Esen, H.: Experimental investigation of thermal performance of a double flow solar air heater having aluminium cans. Renew. Energy **34**, 2391–2398 (2009)
8. Akpinar, E.K., Kocyiğit, F.: Energy and exergy analysis of a new flat plate solar air heater having different obstacles on absorber plates. Appl. Energy **87**, 3438–3450 (2010)
9. Akpinar, E.K., Kocyiğit, F.: Experimental investigation of thermal performance of solar air heater having different obstacles on absorber plate. Int. Commun. Heat Mass Trans. **37**, 416–421 (2010)
10. Alta, D., Bilgili, E., Ertekin, C., Yaldiz, O.: Experimental investigation of three different solar air heaters energy and exergy analysis. Appl. Energy **87**, 2953–2973 (2010)
11. Tanda, G.: Performance of solar air heater ducts with different types of ribs on the absorber plate. Energy **36**, 6651–6660 (2011)
12. Yang, M., Wang, P., Yang, X., Shan, M.: Experimental analysis on thermal performance of a solar air collector with a single pass. Build. Environ. **56**, 361–369 (2012)
13. Hernández, A.L., Quiñoneza, J.E.: Analytical models of thermal performance of solar air heaters of double-parallel flow and double-pass counter flow. Renew. Energy **55**, 380–391 (2013)
14. Benli, H.: Experimentally derived efficiency and exergy analysis of a new solar air heater having different surface shapes. Renew. Energy **50**, 58–67 (2013)
15. Chabane, F., Moummi, N.: Heat transfer and energy analysis of a solar air collector with smooth plate. Eur. Phys. J. Appl. Phys. **66**, 10901 (2014)

16. Bahrehmand, D., Ameri, M.: Energy and exergy analysis of different solar air collector systems with natural convection. Renew. Energy **74**, 357–368 (2015)
17. Saxena, A.: Design and thermal performance evaluation of a novel solar air heater. Renew. Energy **77**, 501–511 (2015)
18. Kabeela, A.E., Khalil, A., Shalaby, S.M., Zayed, M.E.: Experimental investigation of thermal performance of flat and v-corrugated plate solar air heaters with and without PCM as thermal energy storage. Energy Convers. Manage. **113**, 264–272 (2016)
19. Sultan, M., El-Sharkawy, Ibrahim I., Miyazaki, T., Saha, B., Koyama, S.: An overview of solid desiccant dehumidification and air conditioning systems. Renew. Sustain. Energy Rev. **46**, 16–29 (2015)
20. Daou, K., Wang, R.Z.: Desiccant cooling air conditioning: a review. Renew. Sustain. Energy Rev. **10**(2), 55–77 (2006)
21. Ubertini, S., Desideri, U.: Design of a solar collector for year-round climatization. Renew. Energy **28**, 623–645 (2003)
22. Koyuncu, T.: Performance of various design of solar air heaters for crop drying applications. Renew. Energy **31**, 1073–1088 (2006)
23. Ammari, H.D.: Mathematical model of thermal performance of a solar air heater with slats. Renew. Energy **28**, 1597–1615 (2003)
24. Karim, M.A., Hawlader, M.N.A.: Performance evaluation of a v-groove solar air collector for drying applications. Appl. Therm. Eng. **26**, 121–130 (2006)

Chapter 30
Methane–Diesel Dual Fuel Engine: A Comprehensive Review

Ashwin Misra, Mukesh Yadav, Ayush Sharma and Ghanvir Singh

Abstract A methane–diesel dual fuel engine is a relatively new internal combustion technology which uses both methane and diesel fuels for its operation. Methane–diesel dual fuel engines are basically modified conventional compression ignition engines that use methane (90–95%) as the primary fuel and diesel (5–10%) as the secondary fuel. These engines have good thermal efficiencies at higher load values although the performance is relatively lesser during partial load conditions due to the poor charge utility. The most considerable advantage of using this technology is the substantial reduction in emissions in response to the alarming rate of increase in global warming. This article reviews the research done by past researchers in order to study the combustion process in a methane–diesel dual fuel engine and its emissions (CO, hydrocarbons (HC), NO(x)). The emission measurements and results from an engine dynamometer test are analysed and discussed. A perspective on knocking by varying the primary fuel (methane, LPG, natural gas) is also presented. The characteristics and trends are compared to the conventional diesel engine and its future prospects as a viable alternative to the conventional compression ignition engines.

Keywords Methane · Dual fuel engine · Emission · Alternative fuel

30.1 Introduction

Hazardous concerns relating to the environment and the alarming increase in depletion of petroleum sources have led worldwide research to search for substitutes for conventional internal combustion technology and petro-fuels. Large usage of fossil fuels is depleting our resources exponentially and polluting the environment. Thus, it is required to improve awareness regarding the frugal use of the present resources and gradually switch to alternatives, which are eco-friendly [1–3]. Lately, biodiesel/fatty acid methyl ester has achieved interest as a viable substitute, because of the very

A. Misra (✉) · M. Yadav · A. Sharma · G. Singh
Department of Mechanical Engineering, Delhi Technological University, New Delhi, Delhi 110042, India
e-mail: ashwin_bt2k16@dtu.ac.in

S. Yadav et al. (eds.), *Proceedings of International Conference in Mechanical and Energy Technology*, Smart Innovation, Systems and Technologies 174,
https://doi.org/10.1007/978-981-15-2647-3_30

similar fuel characteristics to diesel. It is currently made from uneatable, nutritious or livestock fats [4]. It is not nutritious. As an outcome, in many parts of the world, uneatable oil resources are gaining momentum worldwide, especially wasteland that is not suited to agriculture [5]. Nonetheless, biodiesel from edible oils is being discussed as a result of a long-term food supply [6] questions regarding feedstock. The use of alternative gaseous fuels like methane, LPG is a good way to stop the use of conventional fuels. The advantage of gaseous fuels is the absence of mineral impurities, consistent quality and convenience [7]. It is the main component of natural gas. It also has the least carbon-to-energy ratio amongst all fossil fuels along with a low flame temperature [8, 9]. The relative wider availability of methane on the earth makes it a cheap and viable fuel, though its storage poses difficulties due to its gaseous state under regular conditions for pressure and temperature. Thus, it is a highly economical fuel with considerable advantages over fossil fuels [10–12].

Compression ignition engines are prevalent because of their advantageous performance characteristics such as high compression ratio, high natural torque capability, high thermal efficiency. Regulations have been imposed on emissions and are bound to tighten even more in the future. Conventional diesel engines are therefore not suitable for the future. This problem will be solved by dual fuel engines. A duel fuel combustion engine utilizes reactivity controlled compression ignition (RCCI). The process of combustion in such a dual fuel engine is complex because it combines spark ignition and compressor ignition motor features [13–16]. Dual fuel engines have more advantages than the conventional SI or CI engines. In the latter engines, when the load is high, the emissions are high, and the lowest possible efficiency is accomplished. The advantage of dual fuel engines, compared to the conventional SI or CI engines, is more efficiency obtained during high-load conditions, lower toxicity of the emissions from the engine, minor smoke and fewer particulate emissions, hence more environment-friendliness. One disadvantage of the dual fuel engine is that the efficiency of this engine at part loads is secondary to that of the standard SI or CI engines. This publication is an empirical study and a comparison of the various parameters of engine analysis. This article examines combustion and emission characteristics with distinct load conditions and then affiliates the results obtained with emission and efficiency parameters of the engine used.

30.2 Combustion Process in a Methane–Diesel Dual Fuel Engine

The combustion process is only slightly different from the typical SI and CI engines for the methane–diesel dual fuel engine. A air–methane mixture, recognized as primary fuel, is taken in during the intake stroke. During the compression stroke, the temperature and pressure of this primary fuel is increased by compression. Once the compression stroke ends, i.e. when the piston reaches the top dead centre, a minute amount of diesel fuel is pumped into a tank of primary fuel combustion under high pressure. This diesel fuel, intended for ignition, is termed pilot fuel. Since the mixing of the diesel fuel is improper, a plethora of flame fronts are generated, as the pilot fuel ignites at several places within the compressed volume [17–20]. This causes a smooth and rapid initiation of the combustion process.

Altering the volume of the primary fuel, that is, the air–methane mixture, will alter the engine's power output [21]. Depending on the engine's operating conditions and its design parameters, the quantity of pilot fuel, that is, the diesel fuel, is adjusted.

The amount of diesel fuel utilized in this engine is roughly 4–10% of that used in a purely CI engine [22, 23] (Fig. 30.1).

This dual fuel engine's combustion cycle consists of the following five stages, in sequence [26]:

1. Pilot fuel ignition delay.
2. Pilot fuel premixed combustion.
3. Primary fuel ignition delay.
4. Primary fuel rapid combustion.
5. After-burning.

30.3 Analysis

30.3.1 The Combustion Process

The amount of oxygen drawn during the intake stroke is less for the methane–diesel dual fuel engine, which results from methane drawn with the intake charge, when compared with pure diesel engine. The primary fuel solution has a temperature of

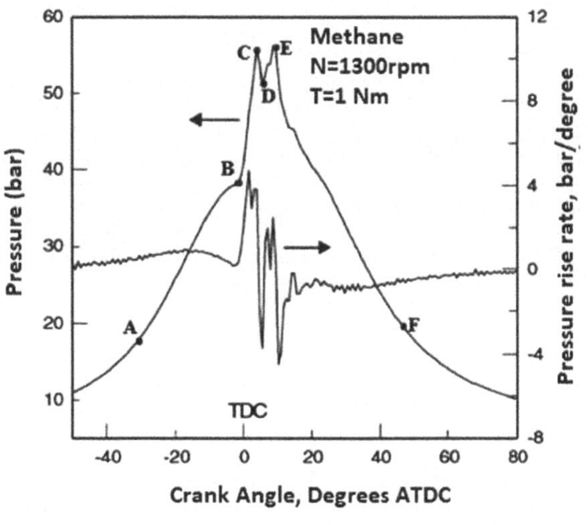

Fig. 30.1 Pressure and pressure rise rate versus crank angle for the methane–diesel dual fuel engine [24, 26]

about 600 °C post compression. When a minute of pilot (diesel) fuel is pumped, the pilot premixed combustion (BC) starts and triggers absolute low explosion. For all heat released at this point, the pilot fuel is essentially responsible.

The primary ignition delay (CD) is a result of the compressed primary fuel undergoing a chemical reaction. There is a slow decline in pressure in this stage.

The unprompted ignition of the pilot fuel marks the commencement of the stage of rapid combustion (DE), which is very volatile [27, 28]. A small quantity of both the primary fuel and pilot fuel explodes, which cause the growth of pressure inside the cylinder. Maximum pressure is achieved in the cylinder once majority of the primary fuel (present in a large quantity) and pilot fuel (present in a small quantity) has combusted [29].

As the rapid pressure rise comes to a conclusion, the after-burning stage commences and continues until the piston moves towards TDC, till the opening of the exhaust valve(s). In this stage, there is a gradual burning of remnants of the primary fuel and the diesel fuel [30–32].

The chief combustion stages can be also observed on the rate of heat released (ROHR) graph for a methane–diesel dual fuel [33].

30.3.2 Performance

30.3.2.1 Rate of Heat Release (ROHR)

With the increasing methane ratio in the medium torque and low torque (respectively Figs. 30.2 and 30.3), the combustion maximum pressure decreases. It can be noted that the dual fuel ROHR system will decrease at low load and the ROHR value will decrease in the presence of increase in methane. The auto-ignition temperature of CH_4 (methane) and the presence of CO_2 (carbon dioxide) affect ROHR quality; as in the case of medium loads and high load due to high-temperature conditions, the two-fuel approach initiates ignition earlier.

30.3.2.2 Brake Power

A lower methane level, as shown in Fig. 30.2, decreases the braking strength of a small but significant number. From this, the energy performance of the brakes can be derived with a drop in methane content [34–36].

30.3.3 Emissions

According to [37, 38], emission measurements and characteristics of different types of engine loads were based on a national standards 13-mode test procedure. The

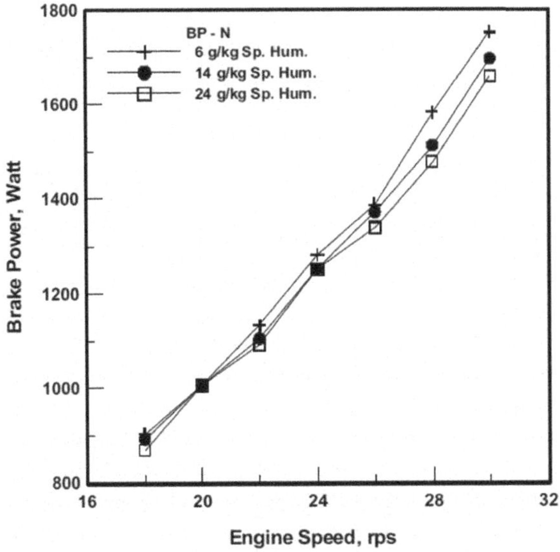

Fig. 30.2 Brake power versus engine speed for different methane ratios

Mode	Load	Speed (rpm)
+	Idle	600
2	10%	2000
3	25%	2000
4	50%	2000
5	75%	2000
6	100%	2000
7	Idle	600
8	100%	3000
9	75%	3000
10	50%	3000
11	25%	3000
12	10%	3000
13	Idle	600

Fig. 30.3 Procedure testing modes (13)

engine is tested in sequential stationary modes on an engine dynamometer [39, 40]. Over the 13 modes, samples are measured for particulate matter emissions. Using methane as a primary fuel, it was found that there was no reduction in nitrogen oxide emissions in mode 8, whilst slightly it reduced on all modes as shown in Figs. 30.3 and 30.4 [41–42].

This result is likely due to the increase in the lag of ignition. Hydrocarbon emissions are shown in Fig. 30.5 and are extremely high. Most clear is the fact that diesel–methane hybrid fuel at high loads has lower reaction temperatures [44, 45]. Increased methane content is mostly responsible for an increased ignition delay at lower loads (Fig. 30.6).

Fuel usage is small, resulting in a sudden increase in the concentration of carbon monoxide (CO) at a weak mixture. This amount is multiple times that usually found for the same load conditions in diesel engines. Compared to diesel engines, the density of exhausted gas smoke is much lower; the picture has always been uphill, with growing concentrations of carbon monoxide as shown in Fig. 30.7. One can assume that not all fuel has been consumed at high speed; unburned fuel can be

Fig. 30.4 Impact of the NO_x pollution on load

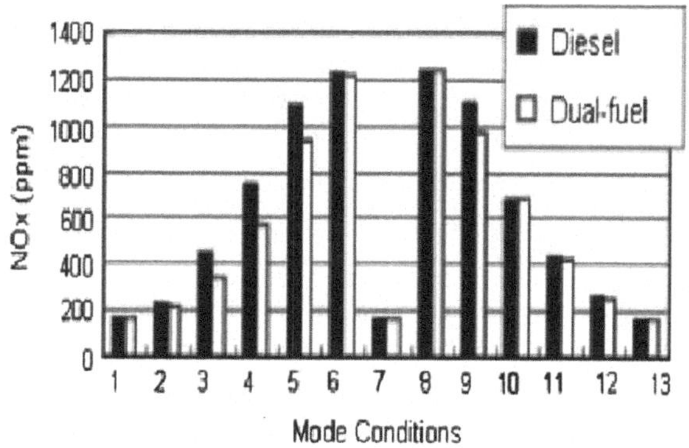

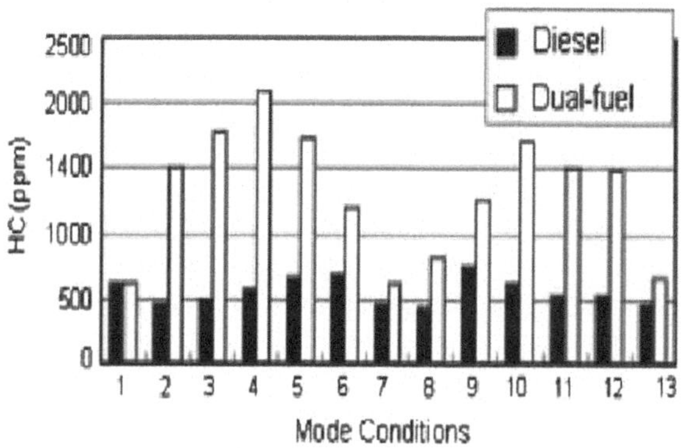

Fig. 30.5 Variation in exhaust gas composition of a dual fuel engine with the overall load mixing strength at 1000 rpm, with methane as a pilot fuel [46]

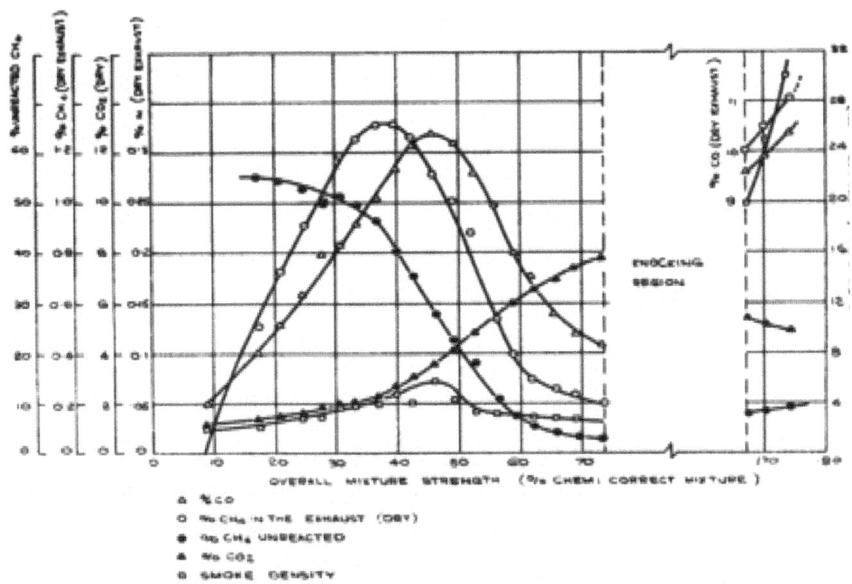

Fig. 30.6 Effect of load on HC emission

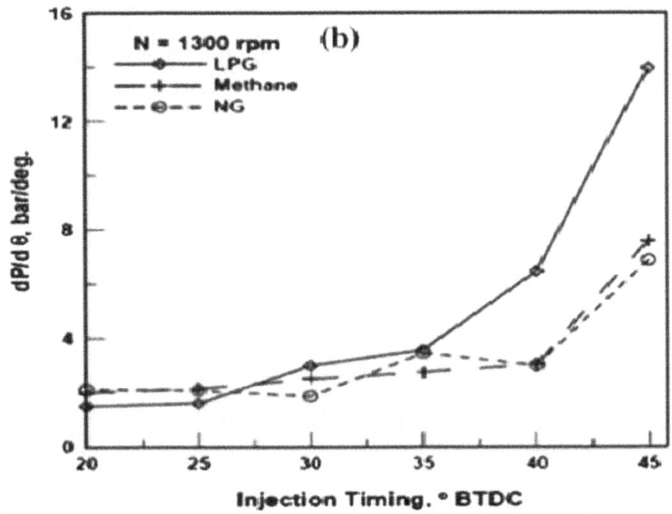

Fig. 30.7 Effects of pilot fuel mass on knocking

extracted in order to deal with exhaust gas. The spike in thermal efficiency, peak pressure and nitrogen oxide emissions for all the engine load values studied is due to an improvement in the traditional range of diesel injection timing. Nevertheless, when the timing for diesel injection transitions from medium to high-load low-speed circumstances, carbon dioxide emissions do not change significantly, whereas unburned methane emissions are increasing [42, 47, 48].

30.4 Knocking

At high-load situation, dual fuel technology engines are subjected to increase the knocking phenomenon. Flame fronts that spread from different ignition hubs do not stretch to all the parts of the cylinder and therefore leave some of the disseminated gaseous fuel unaffected. Several partly oxidized compounds are formed at the boundaries of the central combustion area. As the fuel containing gas concentrations surpass a threshold, a faster rate of heat release along with a pressure rise is cleared by the flame fronts. Further increase in the strength of the mixture can ultimately lead to knocking. This phenomenon was studied using different primary fuels, and graphs were plotted for different injection timeframes of pilot fuel versus pressure differential with respect to Δ. Pressure increase rate, dp/dθ, increases as pilot diesel injection, as shown in the Fig 30.7 [49, 50].

Oxygen enrichment is a profitable and easy way to reduce the blow-on trend of high-load dual fuel methane–diesel engines. By increasing the oxygen concentration inside the cylinder, as can be seen in Fig. 30.8, around 19% decrease in the maximum pressure rise rate is achieved.

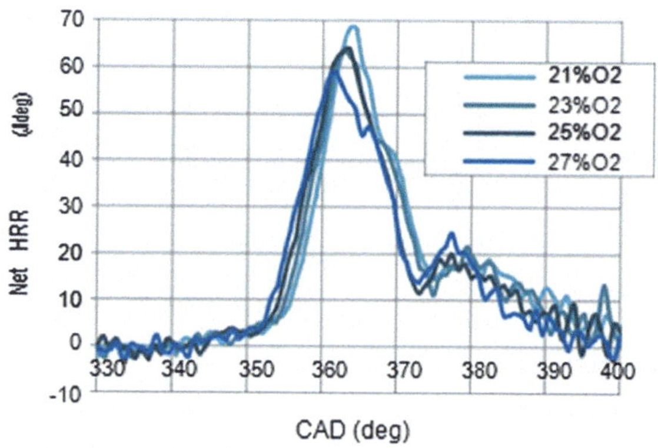

Fig. 30.8 Effects of oxygen enrichment on the heat release rate (HRR) pattern of such an engine [51]

30.5 Conclusion

When the load is low, relative performance of conventional engines as juxtaposed to dual fuel engine is marginally higher. However, the performance can be increased, and at higher loads, it becomes more efficient than a conventional CI engine. By the modification of the conditions of intake charge, improvement in the low load of the component performance and emission can be seen. Comparing the conventional diesel fuel engine with the dual fuel technology engine, the observation made was that the emissions of nitrogen oxide of the dual fuel engine are approximately 200 ppm less than the CI engine. The hydrocarbon emissions of the dual fuel engine were observed to be more than the CI engine. CO emissions were lesser for dual fuel engines in the modes used more often. Power obtained and fuel consumption can be varied with different techniques such as oxygen enrichment and also by varying the injection timings of the combustion process. With the shift in the injection times of pilot fuel and oxygen add-ons, the knocking produced can be varied. Exhausted gas analysis of a dual fuel engine revealed a large number, when fed to the engine above or below those limited levels, of the fuel gas outlast through combustion. The operating conditions and the fuel used are cohesive with these limits [52].

References

1. Lata, D.B., Misra, S., Medhekar, A.: Effect of hydrogen and LPG addition on the efficiency and emissions of a dual fuel diesel engine. Int. J. Hydrogen Energy **37**, 6084–6096 (2012). https://doi.org/10.1016/j.ijhydene.2012.01.014
2. Murthy, K., Madhwesh, K., Shrinivasa, N.: Influence of injection timing on the performance of dual fuel compression ignition engine with exhaust gas recirculation. Int. J. Eng. Res. Dev. **1**, 36–42 (2012)
3. Lata, D.B., Misra, A.: Theoretical and experimental investigations on the performance of dual fuel diesel engine with hydrogen and LPG as secondary fuels. Int. J. Hydrogen Energy **35**(21) (2010)
4. Khan, O., Khan, M.E., Yadav, A.K., Sharma, D.: The ultrasonic-assisted optimization of biodiesel production from eucalyptus oil. Energy Sour. Part A: Recovery Utilization Environ. Effects **39**, 1–9 (2017). https://doi.org/10.1080/15567036.2017.1328001
5. Hirkude, J.B., Padalkar, A.S.: Performance optimization of CI engine fuelled with waste fried oil methyl ester-diesel blend using response surface methodology. Fuel **119** (2014)
6. Atabani, A.E., Silitonga, A.S., Ong, H.C., Mahlia, T.M.I., Masjuki, H.H., Badruddin, I.A., Fayaz, H.: Non-edible vegetable oils: a critical evaluation of oil extraction, fatty acid compositions, biodiesel production, characteristics, engine performance and emissions production. Renew. Sustain. Energy Rev. **18** (2013)
7. Rakopoulos, C.D., Kyritsis, D.C.: Comparative second-law analysis of internal combustion engine operation for methane, methanol, and dodecane fuels. Energy **26**(7) (2001)
8. Zheng, J., Wang, J., Zhao, Z., Wang, D., Huang, Z.: Effect of equivalence ratio on combustion and emissions of a dual-fuel natural gas engine ignited with diesel. Appl. Thermal Eng. **146** (2019)
9. Yousefi, A., Guo, H., Birouk, M.: Effect of swirl ratio on NG/diesel dual-fuel combustion at low to high engine load conditions. Appl. Energy **229** (2018)

10. Kaleemuddin, S., Rao, A.: Development of dual fuel single cylinder natural gas engine an analysis and experimental investigation for performance and emission. Am. J. Appl. Sci. **6** (2009). https://doi.org/10.3844/ajas.2009.929.936
11. Prakash, G., Ramesh, A., Anwar, B.S.: An approach for estimation of ignition delay in a dual fuel engine. SAE 1999-01-0232 (1999)
12. Karim, G., Jones, W., Raine, R.: An examination of the ignition delay period in dual fuel engines. SAE Technical Paper 892140 (1989)
13. Tira, H.S., Herreros, J.M., Tsolakis, A., Wyszynski, M.L.: Characteristics of LPG-diesel dual fuelled engine operated with rapeseed methyl ester and gas-to-liquid diesel fuels. Energy **47**(1) (2012)
14. Oguma, M., Goto, S., Sugiyama, K., Kajiwara, K., Mori, M., Konno, M., Yano, T.: Spray characteristics of LPG direct injection diesel engine (2003). https://doi.org/10.4271/2003-01-0764
15. Vasu, R., Ramakrishnan, E., Ramesh, A., Nagalingam, B., Gopalakrisnan, K.V.: Measurement and control of particulate emission in four stroke single diesel engine using LPG. In: Proceedings of the XV National Conference I.C. Engine and Combustion (1997)
16. Badr, O., Karim, G.A., Liu, B.: An examination of the flame spread limits in a dual fuel engine. Appl. Therm. Eng. **19**(10), 1071–1080 (1999)
17. Leermakers, C.A.J., Berge, B.V.D., Luijten, C.C.M., de Goey, L.P.H., Jaasma, S.: Direct injection of diesel-butane blends in a heavy duty engine. SAE 2011-01-2400 (2011)
18. Karim, G.A.: Combustion in gas fueled compression: ignition engines of the dual fuel type. J. Eng. Gas Turbines Power (2003)
19. Papagiannakis, R.G., Rakopoulos, C.D., Hountalas, D.T., Rakopoulos, D.C.: Emission characteristics of high speed, dual fuel, compression ignition engine operating in a wide range of natural gas/diesel fuel proportions (2010)
20. Ganesan, S., Ramesh, A.: Investigation on the use of water–diesel emulsion in a LPG-diesel dual fuel engine (2001)
21. Abu-Jrai, A., Tsolakis, A., Theinnoi, K., Cracknell, R., Megaritis, A., Wyszynski, M.L.: Effect of gas-to-liquid diesel fuels on combustion characteristics, engine emissions, and exhaust gas fuel reforming. Comparative study. Energy Fuels (2006)
22. Nathan, S.S., Nagarajan, G.: An innovative application of isomerisation of LPG by Al2O3/Pt catalyst for diesel engine in dual fuel operation. SAE (2003)
23. Nagarajan, G., Nathan, S.S.: A new approach for isomerised LPG-diesel dual fuel engine by two different isomerisation catalysts acidic AL_2O_3 vs AL_2O_3/Pt, an experimental study. SAE (2002)
24. Tiwari, D.R., Sinha, P.G.: Performance and emission study of LPG diesel dual fuel engine. Int. J. Eng. Adv. Technol. (IJEAT) (2014)
25. Vijayabalan, P., Nagarajan, G.: Performance, emission and combustion of LPG diesel dual fuel engine using glow plug (2009)
26. Shah, A., Thipse, S.S., Tyagi, A., Rairikar, S.D., Kavthekar, K.P., Marathe, N.V., Mandloi, P.: Literature review and simulation of dual fuel diesel-CNG engines. SAE (2011)
27. Murthy, K., Madhwesh, N., Srinivasa Rao, B.R.: Investigations on the combustion parameters of a dual fuel diesel engine with hydrogen and LPG as secondary fuels (2012)
28. Crookes, R.J., Nazha, M.A.A., Kiannajad, F.: A comparison of ignition and emissions characteristics for alternative diesel fuels and emulsions. In: International Mechanical Engineering Seminar (1990)
29. Kalghatgi, G.: Auto-ignition quality of practical fuels and implications for fuel requirements of future SI and HCCI engines. SAE Technical Paper (2005)
30. Arapatsakos, C., Karkanis, A., Katirtzoglou, G., Pantokratoras, I.: Liquid petroleum gas (LPG) and natural gas (NG) as fuels on diesel engine—dual fuel engine. Recent Adv. Fluid Mech. Heat Mass Transf. (2018)
31. Boretti, A.: Conversion of a heavy duty truck diesel engine with an innovative power turbine connected to the crankshaft through a continuously variable transmission to operate compression ignition dual fuel diesel-LPG. Fuel Process. Technol. (2013)

32. Papagiannakis, R.G., Hountalas, D.T., Rakopoulos, C.D.: Theoretical study of the effects of pilot fuel quantity and its injection timing on the performance and emissions of a dual fuel diesel engine. Energy Convers. Manage. **48**, 2951–2961 (2007)
33. Situmorang, O., Sembiring, R., Kawai, R., Ambarita, H.: Performance, rate of heat release, and combustion stability of dual-fuel mode in a small diesel engine. Energy Sci. Eng.; Kalghatgi, G.: Auto-ignition quality of practical fuels and implications for fuel requirements of future SI and HCCI engines. SAE Technical Paper (2019)
34. Wagemakers, A.M.L.M., Leermakers, C.A.J.: Review on the effects of dual-fuel operation, using diesel and gaseous fuels, on emissions and performance. SAE (2012)
35. Ahmad, N., Babu, M.K.G., Ramesh, A.: Experimental investigations of different parameters affecting the performance of a CNG—diesel dual fuel engine. SAE 2005-01-3767 (2005)
36. Chandra, R., Vijay, V.K., Subbarao, P.M.V., Khura, T.K.: Performance evaluation of a constant speed IC engine on CNG, methane enriched biogas and biogas. Appl. Energy **88**(11) (2011)
37. Qi, D.H., Bian, Y.Z.H., Ma, Z.H.Y., Zhang, C.H.H., Liu, S.H.Q.: Combustion and exhaust emission characteristics of a compression ignition engine using liquefied petroleum gas–diesel blended fuel. Energy Convers. Manage. (2006)
38. Rao, G.A., Raju, A.V.S., Rao, C.V.M., Rajulu, K.G.: Effect of LPG content on the performance and emission characteristics of a diesel-LPG dual fuel engine. Appl. Sci. **8**(5), 825 (2018)
39. Wattanavichien, K.: Spray and combustion visualization of LPG-PME dual fuelling an IDI compression ignition engine. In: 3rd Regional Conference on Mechanical and Aerospace Technology (2011)
40. Kim, Y.J., Kim, K.B., Lee, K.H.: The spray characteristics of a liquid phase LPG port injection type injector for a remodeled diesel engine. SAE 2009-01-1879 (2009)
41. Luft, S., Michalczewski, A.: Analysis of chosen parameters of dual fuel CI engine fuelled with propane-butane gas as a main fuel. SAE 2002-01-2234 (2002)
42. Luft, S.: The influence of regulating parameters of dual fuel compression ignition engine fuelled with LPG on its maximum torque, overall efficiency and emission. SAE 2001-01-3264 (2001)
43. O'Neal, G.B., Storment, J.O., Waytulonis, R.W.: Control of diesel exhaust emissions in underground coal mines—single cylinder engine experiments with modified and non-conventional fuels, SAE SP-495. Diesel Combustion and Emission 1981 Part 3, Int. Off-Highway Meeting & Exp., Milwaukee, WI (1981)
44. Goldsworthy, L.: Combustion behavior of a heavy duty common rail marine diesel engine fumigated with propane (2012)
45. Saanum, I., Bysveen, M., Hustad, J.: Study of particulate matter, NO_x and hydrocarbon emissions from a diesel engine fuelled with diesel oil and biodiesel with fumigation of hydrogen, methane and propane. SAE Technical Paper Series, 2008-01-1809 (2008)
46. Karim, G.A.: A review of combustion processes in the dual fuel engine—the gas diesel engine. Prog. Energy Combust. Sci. **6**(3) (1980)
47. Sahoo, B.B., Sahoo, N., Saha, U.K.: Effect of parameters and type of gaseous fuel on the performance of dual-fuel gas diesel engines—a critical review. Renew. Sustain. Energy Rev. (2009)
48. Yousefi. A., Guo, H., Birouk, M.: Effect of diesel injection timing on the combustion of natural gas/diesel dual-fuel engine at low-high load and low-high speed conditions. Fuel **235** (2019)
49. Lounici, M.S., Benbellil, M.A., Loubar, K., Niculescu, D.C., Tazerouta, M.: Knock characterization and development of a new knock indicator for dual-fuel engines. Energy (2017)
50. Suga, T., Kitajima, S., Fujii, I.: Pre-ignition phenomena of methanol fuel (M85) by the post-ignition technique. SAE Paper 892061 (1989)
51. Abdelaal, M.M., Rabee, B.A., Hegab, A.H.: Effect of adding oxygen to the intake air on a dual-fuel engine performance, emissions, and knock tendency. Energy **61** (2013)
52. Moore, N.P.W., Mitchell, R.W.S.: Combustion in dual fuel engines. In: Joint Conference on Combustion, ASME/Institution of Mechanical Engineers (1955)

Chapter 31
Comparative Study of the Straight Capillary Tube for CO_2 and R22 Refrigerants

Pravin Jadhav, Neeraj Agrawal and Ajit Mane

Abstract This paper indicates a comparative study of a straight capillary tube, by considering an adiabatic, homogenous, and unchoked flow model. The model is developed for a CO_2 and R22 refrigerants by employing the basic principles of fluid and thermodynamics. Thermodynamic properties of R22 and CO_2 are obtained from property code REFPROP and CO2PROP, respectively. The result of the model is validated with earlier published results. The variation in mass flow rate at different geometric factors of the tube has been evaluated. The change in tube length is calculated at the given mass flow rate, tube diameter, and surface roughness. The major change in the performance of the capillary tube is observed for the tube diameter, while less variation is observed with surface roughness. The mass flow rate of CO_2 is nearly 8–9 times of R22 refrigerant at different geometric configurations. The capillary tube length of CO_2 is about 1.4 times of R22 refrigerant. This study is useful to the design of the straight capillary tube with CO_2 and R22 refrigerants.

Keywords Capillary tube · CO_2 · Mass flow rate · R22

Nomenclature

A	Cross-sectional area of capillary tube (m^2)
C_p	Specific heat (kJ kg^{-1} K^{-1})
d	Capillary tube diameter (mm)
f	Friction factor (–)
G	Mass flux (kg m^{-2} s^{-1})
h	Specific enthalpy (kJ kg^{-1})
L	Capillary tube length (m)
m	Mass flow rate (kg s^{-1})

P. Jadhav (✉) · N. Agrawal
Dr. B. A. Technological University, Lonere, Raigad, Maharashtra, India

A. Mane
Annasaheb Dange College of Engineering & Technology, Ashta, Sangli, Maharashtra, India

S. Yadav et al. (eds.), *Proceedings of International Conference in Mechanical and Energy Technology*, Smart Innovation, Systems and Technologies 174,
https://doi.org/10.1007/978-981-15-2647-3_31

P	Pressure (bar)
Re	Reynolds number, $\mathrm{Re} = \rho V d/\mu$ (–)
T	Temperature (K)
V	Velocity (m s^{-1})
x	Dryness fraction (–)

Greek Symbol

ε	Internal surface roughness (mm)
ΔT_{sub}	Degree of subcooling (K)
μ	Dynamic viscosity (Pa s)
v	Specific volume (m^3 kg^{-1})
ρ	Density (kg m^{-3})

Subscripts

1–4	Capillary tube state points
C	Capillary
evpt	Evaporator
g	Saturated vapor
gc	Gas cooler
i	Element
k	Condenser
l	Saturated liquid
tp	Two phase
cr	Critical
sat	Saturation
max	Maximum
min	Minimum

31.1 Introduction

The requirement of the air conditioning and refrigeration system in the global market is continuously increasing; however, due to the ecological issues, there is a need to use new technologies for environmentally friendly refrigerants. Owing to ozone depletion potential (ODP) and global warming potential (GWP), the use of CFCs and HFCs have been avoided. CO_2 is the natural refrigerant having good environmentally friendly characteristics like a zero ODP and less GWP [1]. In many more refrigeration applications, 80% of applications employ a vapor compression refrigeration cycle (VCR). In the VCR cycle, an expansion device is very crucial, where the pressure difference across the high side and low side is maintained. There are many expansion devices that may be employed in the VCR cycle like orifice, capillary tube, thermostatic expansion valve, etc. Among these devices, capillary tube is simple, economical, and self-actuating devices; owing to that, it is used in many small capacity refrigeration systems. The design of the capillary tube may be done by selecting proper tube dimensions for the desired mass of the refrigerant. The required refrigerant flow rate across the capillary tube may be achieved by selecting various combinations of tube length and internal diameter and surface roughness. As shown in Fig. 31.1, the transcritical CO_2 cycle is nothing but half cycle above the critical point and half cycle below the critical point. Similarly, as shown in figure, the VCR cycle with R22 refrigerant is subcritical cycle. In transcritical cycle, flow through a tube is more intricate. There is a need to do compare the transcritical and subcritical cycle.

Much more research has been done on the capillary tubes. Jabaraj et al. [2] developed an experimental setup with a straight capillary tube. The mass flow rates of the adiabatic capillary tube were measured at different inlet conditions and tube geometries for R22 and M20 (blend of R600a and R290). The mass flow rates of R22 are lesser than those of M20 by 3.825%. Based on the experimental result of R22 and M20, a nondimensional correlation was developed to predict the mass flow rate using the Buckingham pi theorem. Agrawal and Bhattacharyya [3] carried out experimental studies for a CO_2 heat pump employing a capillary tube as an expansion device for simultaneous water heating and cooling. Tests were conducted with two stainless steel capillary tubes having specifications of $d = 1.71$ mm, $L = 2.95$ m, $\varepsilon = 3.92$ mm, $d = 1.42$ mm, $L = 1.0$ m, and $\varepsilon = 5.76$ mm. System performance is significantly influenced by gas cooler water inlet temperature, whereas the effect of water flow rate on system performance is modest. An optimum charge is also recorded at which the system yields the best COP with a capillary tube. Performance

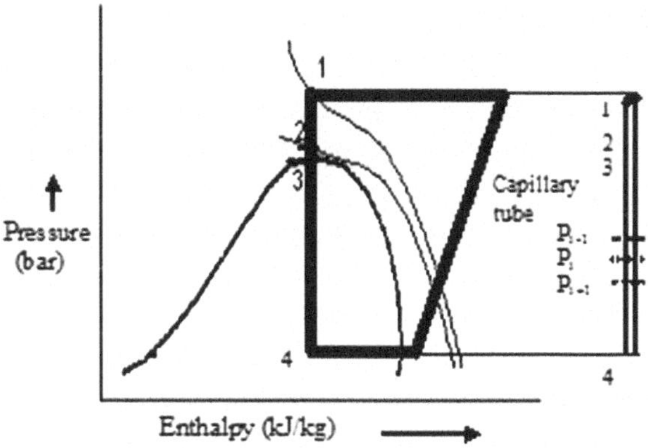

Fig. 31.1 *P–H* chart of the transcritical CO_2 cycle indicating various flow regions

deterioration was more severe at an undercharged condition than an overcharged condition. Zhou and Zhang [4] presented a theoretical and experimental study for a helical capillary tube using R22 refrigerant. Similar to the helical tube, the study had been done with a straight capillary tube. Afterward, the results of the helical tube (coil diameter 40 mm) were compared with the straight tube. For a similar working situation, approximately 10% less mass flow rate was observed with the helical tube. Jadhav and Agrawal [5] developed the numerical model for the straight and spiral capillary tubes using CO_2 and R22. For nearly same operating situation, the length and mass flow rate of straight tube were larger than spiral tube. Similarly, refrigerant flow rate and tube length with CO_2 refrigerant were significantly more than the R22. Agrawal and Bhattacharya [6] conducted a numerical study for a straight capillary tube using a transcritical CO_2 system. A thermodynamic study had been carried for the system. They reported that cooling capacity is more at low evaporator temperature. Jadhav et al. [7–9] studied a numerical model for the helical capillary tube. These works mainly focus on the flow behavior of choked and unchoked flow situations. At various operating and geometric conditions of capillary tube, the flow behavior was reported. Similar to that, system behavior was observed at different operating conditions and capillary tube specifications. The coil diameter played a key role in flow characterization of helical capillary tube. Bansal and Wang [10] had done a numerical study for the straight capillary tube. The study was carried out for R22, R134a, and R600a refrigerant for choked and unchoked flow conditions. The choked and unchoked behavior was efficiently presented by graphical representation.

Comparatively less work was found in the open literature on transcritical CO_2 and subcritical R22 with a straight capillary tube. The objective of this study is to compare the transcritical CO_2 and subcritical R22 system. The influence of various capillary tube dimensions on the mass flow is presented, which is directly concerned with system performance. This study is useful for selecting proper dimensions of the straight capillary tube for a given mass flow rate.

31.2 Mathematical Model

The model is developed for a CO_2 and R22 refrigerants by employing the basic principles of fluid and thermodynamics by employing for assumptions

(i) Throughout the tube, the tube diameter is constant.
(ii) Internal surface roughness is the same throughout the tube.
(iii) Adiabatic and one-dimensional flow in the tube.
(iv) Subcritical flow is homogeneous flow.
(v) The thermodynamic cycle is achieved equilibrium.
(vi) Very less entrance loss.

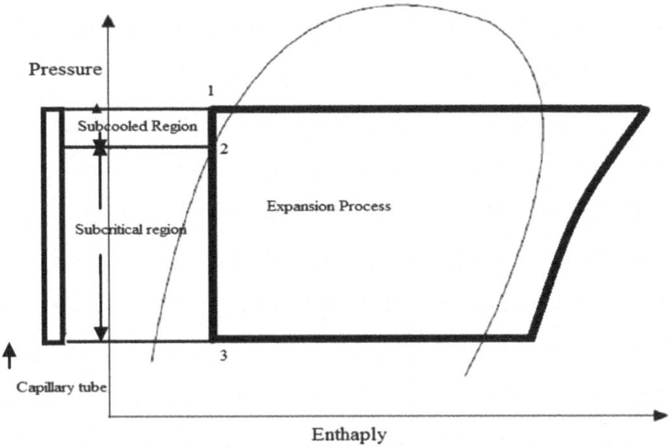

Fig. 31.2 Subcritical R22 cycle with various flow regions in the capillary tube

Fundamental principles of mass, momentum, and energy are used to design the model. The whole capillary tube is divided into small discretized elements to sense the variation in fluid characterizes. In the CO_2 transcritical system, three flow regions are found in capillary tube, viz. supercritical, transcritical, and subcritical flow regions. As shown in Fig. 31.1, CO_2 transcritical cycle, the supercritical, and transcritical flow regions are single phase, while the subcritical flow region is the two-phase region. Similarly, as shown in Fig. 31.2 R22 cycle, the subcooled region is a single phase, while the subcritical phase is the two-phase region. For simplicity, the mathematical model of a single phase and two phases is developed for the CO_2 and R22 refrigerants by employing their properties. The capillary length is computed by summation of all discretized elemental lengths of various flow regions.

31.2.1 Supercritical, Transcritical, and Subcooled Single-Phase Flow Region

Fundamental principles of mass and energy are applied for an element.

$$\frac{AV}{v} = \text{Constant } G = \frac{m}{A} = \frac{V}{v} \tag{31.1}$$

$$dh + \frac{dv^2G^2}{2} = 0 \tag{31.2}$$

The fundamental principle of momentum is used as and solved using Eqs. 31.1 and 31.2

$$-dP - \frac{f_{sp} V G dL}{2d} = GdV \quad dL = \frac{2d}{f_{sp}}\left(\frac{v}{dv} - \frac{dp}{vG^2}\right) \tag{31.3}$$

The Churchill [11] equation is employed to calculate the f_{sp} as

$$f_{sp} = 8\left[\left(\frac{8}{\text{Re}}\right)^{12} + \left(A^{16} + B^{16}\right)^{-\frac{3}{2}}\right]^{\frac{1}{12}} \tag{31.4}$$

where

$$A = 2.457 \ln\left(\frac{1}{\left(\frac{7}{\text{Re}}\right)^{0.9} + \frac{0.27\varepsilon}{d}}\right), \quad B = \frac{37,530}{\text{Re}}, \ \text{Re} = \frac{Gd}{\mu}$$

31.2.2 Subcritical Two-Phase Flow Region

Likewise, the single-phase model, fundamental principles of mass, momentum, and energy are used to design the two-phase flow region. The properties of the two-phase fluid are calculated by considering the quality of refrigerant. The continuity–momentum equation is applied for elemental area is calculated as

$$dL = \frac{2d}{f_{tpm}}\left(\frac{\nu_m}{d\nu} - \frac{dP}{\nu_m G^2}\right) \tag{31.5}$$

where f_{tpm} and ν_m are average friction factor and average specific volume of the refrigerant. Lin [12] friction factor and Churchill are employed to compute f_{tpm}

$$f_{tpm} = Ø_{tp} f_{sp}\left(\frac{\nu_{sp}}{\nu_{tp}}\right) \tag{31.6}$$

$$Ø_{tp} = \left(\frac{\left(\frac{8}{\text{Re}_{tp}}\right)^{12} + \left(A_{tp}^{16} + B_{tp}^{16}\right)^{-\frac{3}{2}}}{\left(\frac{8}{\text{Re}_{sp}}\right)^{12} + \left(A_{sp}^{16} + B_{sp}^{16}\right)^{-\frac{3}{2}}}\right)^{\frac{1}{12}} \left(1 + x\left(\frac{\nu_g}{\nu_l} - 1\right)\right) \tag{31.7}$$

here $A_{tp} = 2.457 \ln\left(\frac{1}{\left(\frac{7}{\text{Re}}\right)^{0.9} + \left(\frac{0.27\varepsilon}{D}\right)}\right)$, $B_{tp} = \frac{37,530}{\text{Re}}$ and $\text{Re} = \frac{GD}{\mu_{tp}}$

31.3 Step-by-Step Solution

The equations of the single-phase flow region and two-phase flow regions are employed and solved by an iterative technique to calculate the capillary tube length. Thermodynamic properties of R22 and CO_2 are obtained from property code REFPROP and CO2PROP, respectively. The property code CO2PROP is developed by employing Helmholtz free energy function [13]. In the current scenario, alternate refrigerants are found out for R22; still, in this study, main focus is given to the comparison of transcritical and subcritical systems. The comparison has been done by comparing the length of the tube for subcritical R22 and transcritical CO_2 systems for a given mass flow rate. Similarly, the mass flow rate at given length is compared for both refrigerants. For calculation of the length of straight capillary tube for both refrigerants, firstly, calculate the length at single and two phases by considering mass momentum and energy equation. Afterward, addition of the lengths is the total length of the tube (as shown in Fig. 31.3). The mass flow rate is calculated for given length of the tube. Firstly, the mass flow rate is guessed for given capillary tube length, considering mass flow rate and calculating length. Secondly, compare the computed length with the given length. If it has the deviations, then assume another mass and repeat the same step until the given length is equal to the calculated length. At that time, we achieve the desired mass flow for a given length (Fig. 31.4).

31.4 Result and Discussion

Flow through straight is numerically studied for CO_2 and R22 cycles.

For CO_2, the study is conducted at $P_{gc} = 100\,\text{bar}$, $T_{gc} = 313\,\text{K}$, $T_{ev} = 273\,\text{K}$ $\varepsilon = 0.00576\,\text{mm}$. Whereas, for the R22 cycle, $\Delta T_{sub} = 3\,°\text{C}$ $P_k = 16.33\,\text{bar}$ $\varepsilon = 0.00576\,\text{mm}$. For calculation of the length, mass flow rate is considered as 0.015 kg/s. However, for determining the mass flow rates, the length of the tube considered as 2 m. The result of the model is validated with Jabaraj et al. [2] and Agrawal and Bhattacharya [3]. The agreement is good as shown in Figs. 31.5 and 31.6. The capillary tube performance varies with the geometric specification of the capillary tube: tube length, internal diameter, and surface roughness.

Tables 31.1 and 31.2 indicate the influence of these specifications on unchoked flow conditions for CO_2 and R22 refrigerants. The flow characteristic of tube is determined by mass flow rate at various factors. The variation in mass flow rate is determined with respect to $L = 2\,\text{m}, d = 1.4\,\text{mm}, \varepsilon = 5.76 \times 10^{-3}\,\text{mm}$, and $T_{ev} = 273\,\text{K}$ as shown in Tables 31.1 and 31.2. For 20% increase in length, the refrigerant mass flow rate reduces about 7.9% and 8.1% in CO_2 and R22, respectively. As the tube length increases, the pressure drops in tube resulting in decrease in the mass flow rate. A very large change in mass is seen due to a variation in the tube diameter; for 20% increase in diameter, refrigerant flow rate rises about 61.6% and 57% in CO_2 and R22 refrigerants, respectively.

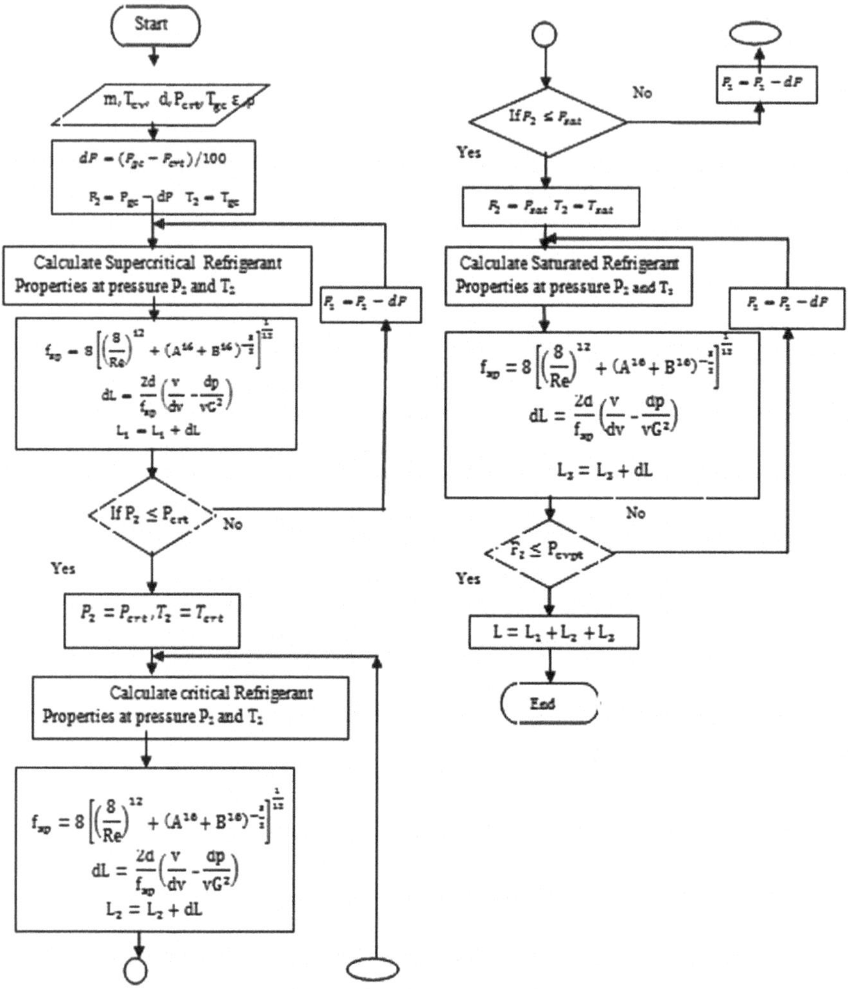

Fig. 31.3 Flow chart for calculation of length for given mass flow rate

For specific tube length, when diameter increases, constraint to flow reduces resulting in an increase in the refrigerant flow rate. A very small change in the mass flow rate is seen; as the internal surface roughness increases by 20%, refrigerant flow rate decreases about 1.4% and 1.9% in CO_2 and R22, respectively. As the roughness increases, constraint due to roughness increases resulting in a decrease in mass flow rate. In designing the capillary tube, the total length of the capillary tube is calculated for a given mass flow rate. Similarly, the length of the tube is calculated by considering the mass flow rate, tube diameter, capillary tube length, and surface roughness. Tables 31.3 and 31.4 indicate the influence of geometric factors on unchoked flow conditions of CO_2 and R22 refrigerants. The variation in

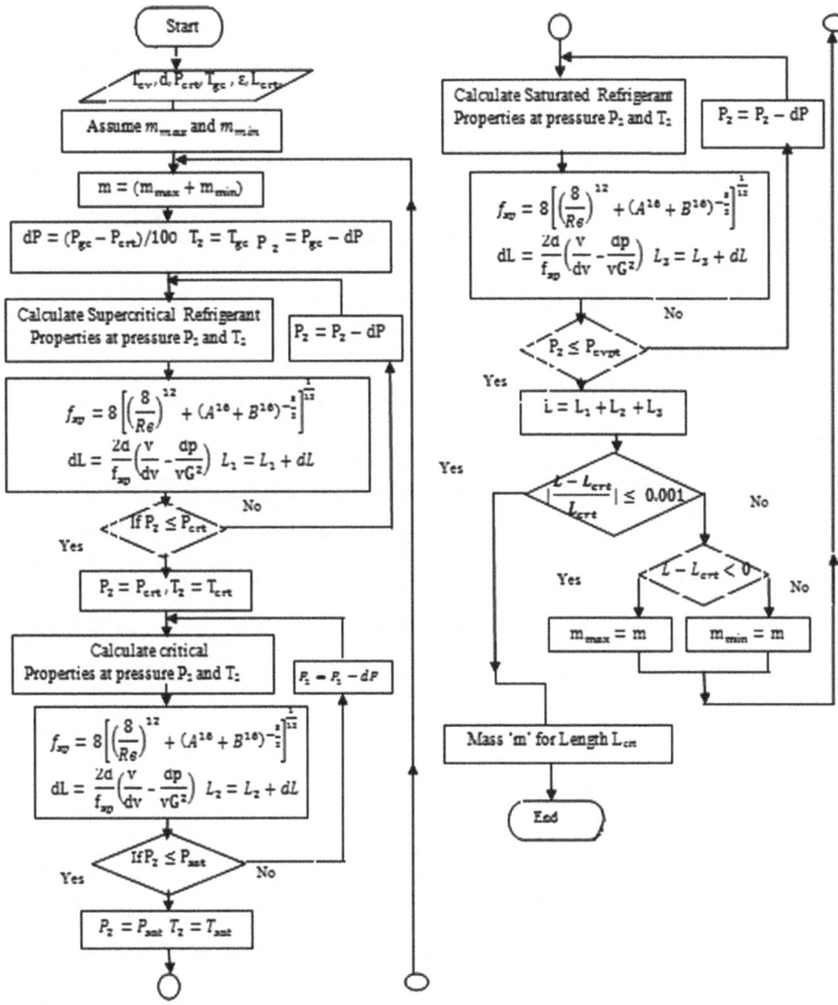

Fig. 31.4 Flow chart of mass calculation for given length

capillary length is determined with respect to $L = 2\,\text{m}$, $d = 1.4\,\text{mm}$, $\varepsilon = 5.76 \times 10^{-3}\,\text{mm}$, and $T_{ev} = 273\,\text{K}$ as shown in Tables 31.3 and 31.4.

For a 20% rise in mass flow rate, the capillary tube length decreases by nearly 32% and 50% in CO_2 and R22, respectively. For specific tube diameter, as the mass flow rate increases, the pressure drop increases results in a decrease of the capillary tube length. A major change in capillary tube length is seen owing to a change in the tube diameter. For a 20% increase tube diameter, the capillary tube length increases about 3 and 3.47 times in CO_2 and R22 refrigerants, respectively. For same length tube, as the tube diameter increases, throttling of refrigerant decreases that increases the tube length. A very small change in the tube length is noted; for the

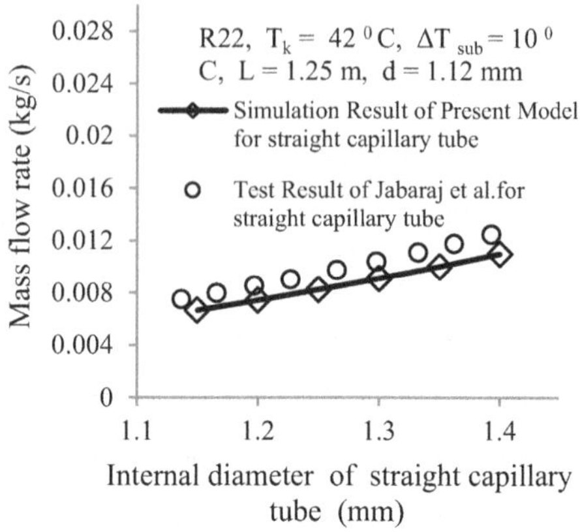

Fig. 31.5 Validation of the present model results with Jabaraj et al. [2] for R22

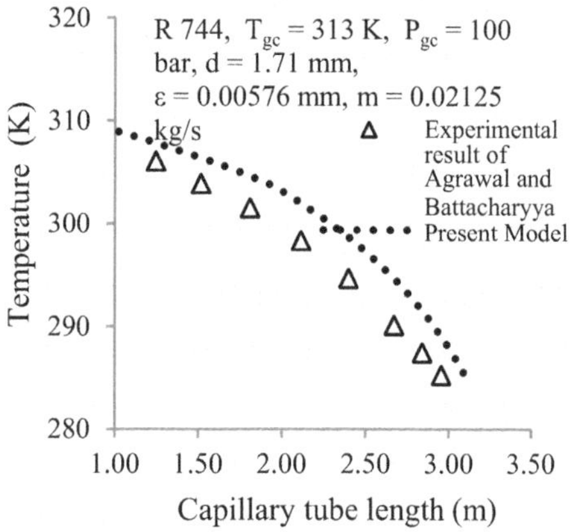

Fig. 31.6 Validation of the present data with Agrawal and Bhattacharya [3] for CO_2

Table 31.1 Percentage variation in mass for various geometric configurations for CO_2 refrigerant

S. No.	Geometric configuration of capillary tube			Mass flow rate, m (kg/s $\times 10^{-4}$)	% Change in	% Change in mass flow rate
	L (m)	d (mm)	$\varepsilon \times 10^{-3}$ (mm)			
1	2	1.4	5.76	177	$\Delta L = 20\%$	−7.9
	2.4	1.4	5.76	163		
2	2	1.4	5.76	177	$\Delta d = 20\%$	61.6
	2	1.68	5.76	286		
3	2	1.4	5.76	177	$\Delta \varepsilon = 20\%$	−1.7
	2	1.4	6.91	174		

Table 31.2 Percentage change in mass for various geometric configurations for R22 refrigerant

S. No.	Geometric configuration of capillary tube			Mass flow rate, m (kg/s $\times 10^{-5}$)	% Change in	% Change in mass flow rate
	L (m)	d (mm)	$\varepsilon \times 10^{-3}$ (mm)			
1	2	1.4	5.76	742	$\Delta L = 20$	−8.1
	2.4	1.4	5.76	682		
2	2	1.4	5.76	742	$\Delta d = 20$	57.0
	2	1.68	5.76	165		
3	2	1.4	5.76	742	$\Delta \varepsilon = 20$	−1.9
	2	1.4	6.91	728		

Table 31.3 Percentage change in length for various geometric configurations for CO_2 refrigerant

S. No.	Geometric configuration of capillary tube			Capillary tube length L (m)	% Change in	% Change in length
	m (kg/s $\times 10^{-3}$)	d (mm)	$\varepsilon \times 10^{-3}$ (mm)			
1	15	1.4	5.76	2.85	$\Delta m = 20$	−32
	18	1.4	5.76	1.95		
2	15	1.4	5.76	2.85	$\Delta d = 20$	3 times
	15	1.68	5.76	7.6		
3	15	1.4	5.76	2.85	$\Delta \varepsilon = 20$	−5
	15	1.4	6.91	2.714		

Table 31.4 Percentage change in length for various geometric configurations for R22

S. No.	Geometric configuration of capillary tube			Capillary tube length L (m)	% Change in	% Change in Length
	m (kg/s $\times 10^{-3}$)	d (mm)	$\varepsilon \times 10^{-3}$ (mm)			
1	15	1.4	5.76	0.33	$\Delta m = 20$	50
	18	1.4	5.76	0.17		
2	15	1.4	5.76	0.33	$\Delta d = 20$	3.47 times
	15	1.68	5.76	1.17		
3	15	1.4	5.76	0.33	$\Delta \varepsilon = 20$	−4.7
	15	1.4	6.91	0.32		

20% increase in surface roughness, tube length decreases by 5% and 4.7% in CO_2 and R22, respectively; the resistance to flow increases as the roughness increases resulting in decrease in tube length Fig. 31.7 indicates a comparison of the mass flow rate of CO_2 and R22 refrigerants at the various combinations of geometric

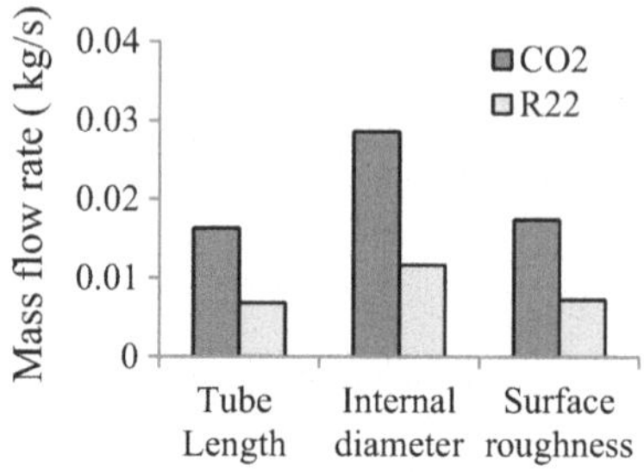

Fig. 31.7 Comparison of the mass flow rate of CO_2 and R22 refrigerants at different geometric configurations

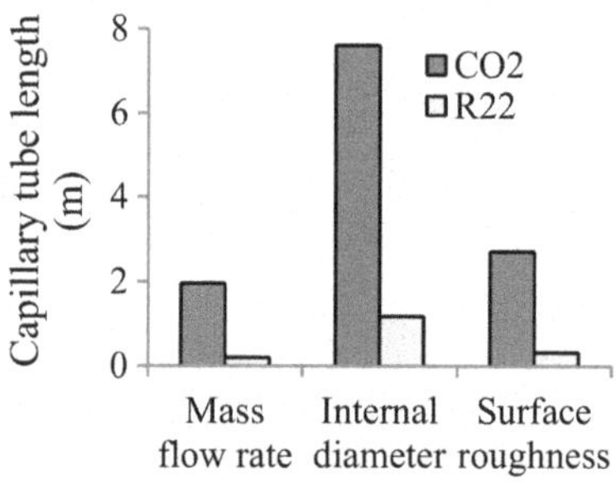

Fig. 31.8 Comparison of capillary tube lengths of CO_2 and R22 at the some physical parameters

configuration. At $L = 2.4$ m, $d = 1.4$ mm, and $\varepsilon = 6.91 \times 10^{-3}$ mm, mass flow rate of CO_2 is almost 1.4 times of R22. Since the operating pressure of the R744 is larger than the R22, the mass flow rate of CO_2 refrigerant is almost 1.4 times the R22 at different combinations of length, diameter, and surface roughness. Figure 31.8 indicates the comparison of capillary tube of CO_2 and R22 refrigerants at different combinations of geometric configuration. For $m = 0.018$ kg/s, $d = 1.4$ mm, and $\varepsilon = 5.76 \times 10^{-3}$ mm, tube length employing CO_2 is about 9 times of R22. Similarly, at $m = 0.015$ kg/s, $d = 1.68$ mm, and $\varepsilon = 5.76 \times 10^{-3}$ mm, capillary tube length employing CO_2 is about 8 times of R22. For $m = 0.015$ kg/s, $d = 1.4$ mm, and $\varepsilon = 6.91 \times 10^{-3}$ mm, capillary tube length employing CO_2 is about 8 times of R22. Since the tube length of the capillary tube is different at various combination of tube geometry (tube length, tube diameter, and surface roughness), the tube length of CO_2 is 8–9 times of R22 refrigerant.

31.5 Conclusion

This paper presents a comparative study on the straight capillary tube for CO_2 and R22. The capillary tube flow behavior is studied. The influence of tube length, internal diameter, and surface roughness on mass flow rate is reported for CO_2 and R22. As the tube length increases by 20%, the mass flow rate decreases by nearly 7.9% and 8.1% in CO_2 and R22, respectively. The key element, which largely influenced the mass flow rate, is the tube diameter; for 20% increase in diameter, refrigerant flow rate increase about 61.6% and 57% in CO_2 and R22, respectively. A very small change in the mass flow rate is seen; as the internal surface roughness increases by 20%, refrigerant flow rate decreases about 1.4% and 1.9% in CO_2 and R22, respectively. A change in capillary tube length is observed for the variation in tube diameter, surface roughness, and refrigerant mass flow rate. For a 20% increase in mass flow rate, the tube length decreases by nearly 32% and 50% in CO_2 and R22, respectively. A large change in capillary tube length is seen owing to a variation in the tube diameter of the tube. For 20% increase in tube diameter, the capillary tube length increases nearly 3 and 3.47 times in CO_2 and R22 refrigerants, respectively. A negligible change in the tube length is seen for the variation in surface roughness; for the 20% increase in roughness, tube length decreases by 5% and 4.7% in CO_2 and R22, respectively. In all the observation, it is clearly observed that the mass flow rate of CO_2 is nearly 8–9 times of R22 refrigerant at different geometric configurations. The capillary tube length of CO_2 is about 1.4 times of R22 refrigerant. This study is useful for selecting the proper geometric parameter of the straight capillary tube for CO_2 and R22 refrigerants.

References

1. Lorentzen, G.: Revival of carbon dioxide as a refrigerant. Int. J. Refrig. **17**(5), 292–300 (1994)
2. Jabaraj, D.B., Kathirvel, A.V., Lal, D.M.: Flow characteristics of HFC407C/HC600a/HC290 refrigerant mixture in adiabatic capillary tubes. Appl. Therm. Eng. **26**(14–15), 1621–1628 (2006)
3. Agrawal, N., Bhattacharyya, S.: Experimental investigations on adiabatic capillary tube in a transcritical CO_2 heat pump system for simultaneous water cooling and heating. Int. J. Refrig. **34**, 476–483 (2008)
4. Zhou, G., Zhang, Y.: Numerical and experimental investigations on the performance of coiled adiabatic capillary tubes. Appl. Thermal Eng. **26**, 1106–1114 (2006)
5. Jadhav, P., Agrawal, N.: A comparative study in the straight and a spiral adiabatic capillary tube. Int. J. Ambient Energy (2018). https://doi.org/10.1080/01430750.2017.1422146
6. Agrawal, N., Bhattacharyya, S.: Adiabatic capillary tube flow of carbon dioxide in a transcritical heat pump cycle. Int. J. Energy Res. **31**, 1016–1030 (2007)
7. Jadhav, P., Agrawal, N., Patil, O.: Flow characteristics of helical capillary tube for transcritical CO_2 refrigerant flow. Energy Procedia **109C**, 431–438 (2017)
8. Jadhav, P., Agrawal, N.: Numerical study on choked flow of CO_2 refrigerant in helical capillary tube. Int. J. Air-Conditioning Refrig. **26**(03), 1850027 (2018). https://doi.org/10.1142/S201013251850027X
9. Jadhav, P., Agrawal, N.: Study of homogenous two phase flow through helically coiled capillary tube. Adv. Sci. Eng. Med. **10**, 1–5 (2018)
10. Bansal, P., Wang, G.: Numerical analysis of choked refrigerant flow in adiabatic capillary tubes. Appl. Therm. Eng. **24**, 851–863 (2004)
11. Churchill, S.W.: Friction factor equations spans all fluid-flow regimes. Chem. Eng. **84**(24), 91 (1977)

12. Lin, S.: Local frictional pressure drop during vaporization of R-12 through capillary tubes. Int. J. Multiph. Flow **17**(1), 95–102 (1991)
13. Sarkar, J., Bhattacharyya, S., Ramgopal, M.: Optimization of transcritical CO_2 heat pump cycle for simultaneous cooling and heating applications. Int. J. Refrig. **27**(8), 830–838 (2004)

Chapter 32
Renewable Energy in Bangladesh: Current Status and Future Potentials

Md. Aktar Zahid Sohag, Priyanka Kumari, Rajeev Agrawal, Sumit Gupta and Anbesh Jamwal

Abstract In the last few years, the demand of energy over the globe is increasing at the rapid rate due to the rapid growth in industrialization in developing countries. The global energy scenario is changing day by day, and developing nations like Bangladesh and India are contributing a great share in the development of energy scenario. In Bangladesh, renewable energy share is only 3% out of its total energy consumption and production and now Government of Bangladesh is promoting the renewable energy projects in order to promote the renewable energy trend and sustainability in Bangladesh. Being a developing country, the electric power required in Bangladesh has been expanding at an essentially high rate. This requirement is fulfilled by the continued development in GDP as of late. The electricity generation of Bangladesh is now fulfilling the demand of electricity of its neighbor countries as well as its small towns. One-third of power generation in Bangladesh depends on the expensive fossil fuels which are imported from other countries and developed nations. The electricity generation in Bangladesh from fossil fuels affects the sustainability of nation and also results in high carbon emissions which is still a cause of worry. The rest 65% of power generation in Bangladesh depends on the natural gas resources which is depleting at the very fast rate, and there will be no future reservation of these resources. There is a need of renewable sources of energy which can fulfill the power demands of towns as well as industries of Bangladesh. Many studies have been done on the factors, drivers and barriers in renewable energy in Bangladesh. This study focuses on the current scenario and future scopes of renewable energy in Bangladesh. An extensive literature survey is taken for this study from the various databases like Elsevier, Springer and Web of Science. Further, future research directions are also

Md. A. Z. Sohag
Department of Mechanical Engineering, Alakh Prakash Goyal Shimla University, Shimla, Himachal Pradesh 171009, India

P. Kumari
Department of Physics, Panjab University, Chandigarh, India

R. Agrawal · A. Jamwal (✉)
Department of Mechanical Engineering, Malaviya National Institute of Technology, J.L.N. Marg, 302017, Jaipur, Rajasthan, India

S. Gupta
Department of Mechanical Engineering, A.S.E.T., Amity University, Noida, India

S. Yadav et al. (eds.), *Proceedings of International Conference in Mechanical and Energy Technology*, Smart Innovation, Systems and Technologies 174,
https://doi.org/10.1007/978-981-15-2647-3_32

given which will motivate the researchers to promote the renewable energy trend in Bangladesh. A brief outline of current electricity generation status and renewable energy source situations in Bangladesh has been presented in this paper.

Keywords Renewable energy · Solar energy · Wind energy

32.1 Introduction

In the present time, increasing global warming and greenhouse effect due to rapid change in industrialization in the developing nations are the primary reasons to explore and adopt the clean energy resources such as renewable energy resources over the non-renewable sources of energy which results in high carbon emission during the power generations [1]. These ecological issues and furthermore the fast consumption of non-renewable energy sources supported the country and different associations to accomplish profoundly efficient and green power plant [2]. Mechanical progression helps in accomplishing a few methods for reaping energy from the renewable sources of energy and utilizes them as the new source of sustainable clean energy which fulfill the present generation need; energy sources are regenerative and do not exhaust after some time [3]. Sustainable energy generation in developing countries helps them to compete with developed countries in terms of emitting less carbon emission from energy production as well as in use of this energy. From 1979 to 2019, CO_2 emission throughout the world from the industries as well as other sources has been expanded from 19.1 billion tons to 33.7 billion tons and is relied upon to additionally increment by 85% inside 2040. The CO_2 outflow has expanded with the expanded utilization of non-renewable energy sources [4]. Over the most recent couple of decades, non-renewable energy source utilization has seen wonderful development. The utilization of primary energy has expanded from 8128 Mtoe in 1990 to 12,932 Mtoe in 2015 [5]. In addition, it is evaluated that 85% of the essential utilization of energy will be provided by petroleum-based sources till 2020. Petroleum-based energy sources provide 85% of the required energy in which 55% of energy is used in transport purposes [6]. The transportation area is the principle shopper of oil and other oil-based commodities. The research objectives of present study are: (a) review of the current energy scenario and research trends in renewable energy in Bangladesh and (b) future research direction to the renewable source's energy and their adoption challenges.

32.2 Economy and Energy Context of Bangladesh

Bangladesh is one of the south Asian countries in the world with the population of 16.47 crores and area of 56,980 m^2. Bangladesh shares its border with the developing countries like India and Myanmar. With its largest seaport Chittagong, Bangladesh shares its goods with other countries [7]. The economic growth rate of Bangladesh is 6.2% in last decade which is expected to increase by 2030, and this is above the economic growth rate than many developing countries over the world [8, 9]. At present time due to the global pressure and polices around the world to save the environment has forced many developing nations to adopt renewable sources of energy over the traditional sources of energy. Bangladesh is now using the maximum of its electricity generation from the renewable sources of energy. It is found that still only 16.9% of power generation around the world is done by the renewable sources of energy [10]. The energy production is still less by renewable sources of energy as compared to the non-renewable sources of energy. The energy production through the natural gas is utilized by industries (16%), power plants (41%), households (8%), hostage control segment (12%) and rest (6%) by the transportation through CNG. Dependency on the natural gas in Bangladesh is increasing at very fast rate which is expanded by 300% from 1990 to 2019. Oil is also the source of the energy in Bangladesh which relies upon the unrefined and refined oil fuels.

32.3 Renewable Energy Sources in Bangladesh

Renewable sources of energy are most economical and clean and can recreate without consumption over the span of time. These sources can support the effect of rapid depletion of non-renewable energy sources and handle the effect that petroleum derivatives have on nature. Although the fact is that the greenhouse gas substance discharge is certifiably not an essential worry for a lower economy country and developing country like Bangladesh, consolidation of renewable energy sources is evident because of rapid depletion of natural resources as well as fossil fuels. Bangladesh has some of renewable energy source possibilities in which sun-oriented and biomass energy conversions are the most encouraging sources of energy at present time. Hydropower energy has restricted potential because of the flat surface and lower accessible heads [11]. Due to lack and insufficient information accessibility, the appraisal of wind control generation is quite difficult in Bangladesh.

Figure 32.1 demonstrates the quantity of introduced renewable innovation measures. Focus to accomplish 6% of the power generation from renewable energy sources inside 2017 by Bangladesh Policy of Renewable Energy (BPRE) could not be accomplished. The offer of the renewable energy source is just 3.9% starting in 2017.

In any case, the legislature is working for the following focus of 11% of the power generation from sustainable sources by the end of 2020.

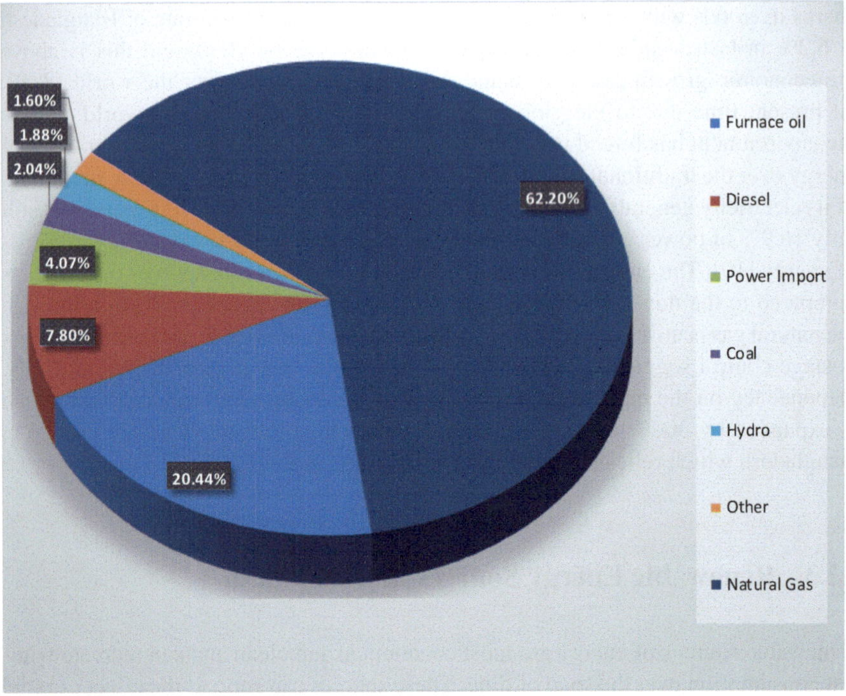

Fig. 32.1 Breakdown of electricity capacity in Bangladesh

32.3.1 Solar Energy in Bangladesh

Solar energy is one of the most plentiful and promising renewable sources of energy asset with higher potential to pick up energy than some other energy sources [12]. It very well may be utilized in two different ways known as the thermal course and the photovoltaic course. In the thermal course, the thermal energy extracted from the sun is utilized for the different purposes like heating of water, room heater, solar lights which is very beneficial for rural areas [13]. The sunlight extracted from solar panels is converted into the useful power which can be utilized in the areas where the supply of electricity is negligible. Bangladesh has great potential of using the solar radiations because of its geographic position as it absorbs 4.6.5 Wh/m^2 of solar radiation everyday which has a great potential to produce the solar energy [14]. Although monthly power production of solar energy is different in each area of Bangladesh, it is found that the maximum power generation is done in the month of March to April and it is minimum from December to January. There are many areas which get maximum power production, but Rajshahi is one of the areas with maximum power production which is normally 1900 kWh/m^2 every year [15]. In spite of the fact that concentrating solar energy power is in the very beginning stage, so the different technologies are growing very quickly in Bangladesh. Solar home system is one of the best solutions at present time [16]. There are around 3.7 million homes introduced around the country which will help the Government of Bangladesh to promote the renewable sources of energy [17].

Infrastructure Development Company Limited (IDCOL) is the main government organization in the segment, and it has begun to work with solar house system since 2004 with the end goal of giving economic energy to the power-denied rural people. It is dealing with an objective to introduce 6 million solar home frameworks with an expected limit of 220 MW continuously 2019 [18]. Under this task, right around 65,000 SHSs are presently being introduced each month coming about 58% yearly increment consistently. Around 180,456 tons of kerosene with an expected estimation of USD 226 million will be supplanted by this project [19]. Non-agricultural lands claimed by the government are being utilized for the solar park undertaking to create clean power. In the urban areas, mechanical and business divisions utilize high-temperature water that is created by natural gas or electric heaters. Consequently, the incorporation of solar water heaters is a need for the government to supplant gas and electric radiators. Hybrid renewable sources of energy are getting popularity in Bangladesh. Hybrid renewable sources of energy can give power to the remote off-matrix areas. There is a continuous project to deliver 7.6 MW control in Hatiya Island of Noakhali utilizing solar–wind–diesel hybrid system [20]. An examination directed by Nandi and Ghosh recommends a breeze PV–battery half and half framework can be utilized as a potential innovation in the remote regions of the country. Solar charging can be an effective option in contrast to regular non-renewable energy sources in the vehicle segment also [21]. So, to accomplish the focus on 10% power from sustainable sources the government and different associations ought to use the solar energy to its maximum capacity.

32.3.2 Wind Energy in Bangladesh

Wind energy is being prevalent everywhere throughout the world because of its alluring commitment on just about zero fuel cost and lower ecological impact than regular sources with guaranteeing everlasting energy supply; additionally, the energy-catching limit of wind generator is higher when contrasted with photovoltaic generator. Regular energy sources adding to environmental change and an unnatural weather change are confined other than depleted [22]. The theoretical measure of electrical power can be evaluated by the accompanying condition. In any case, just 40–47% of this power is essentially accessible for usage.

$$P = \frac{1}{2} \times \rho$$

The yearly normal wind speed at 29 m stature is more than 8 m/s. Normal wind rates of these destinations are much than 7.2 m/s which is, as per NREL (USA), practical for tackling wind control. Various associations of Bangladesh (PDB, BUET,

LGED and so forth.) have directed wind mapping at a few places in particular years. Wind speed is seen to be commonly higher from March to July, and from August to February it is less [23].

Worldwide, wind information and research demonstrate that wind speed not in excess of 8 m/s is not suitable for enormous scale grid-associated electricity creation inside wind parks. From the report, it is discovered that complete wind turbine setup limit in Bangladesh changes from 24 to 56 kW. In this situation, it needs more research and development (R&D) in the field to be directed with thorough examination, in view of geographical setting of Bangladesh.

32.3.3 Hydro-energy in Bangladesh

Globally, small scale of hydropower wound up famous in light of minimal effort, quality and natural advantages. Globally, renewable power source contributed 213 GW of power starting at 2015, where hydropower represents 28% of the absolute energy production. Because of topographical levelness, Bangladesh has limited hydropower potential.

Right when water rate of flow down from a bigger add up to a direction of lower level, water is changed the potential energy into dynamic energy which is used in water turbines to generate the electrical energy. Regardless of the way that hydro-energy is certainly not a thoroughly ideal well-spring of energy, it has two essential primary points diverged from other ordinary sources such as coal, fuel, oil, uranium, oil gas and so on [24].

To start with, the expense of provide and furthermore outflow of greenhouse gas substances both are lower contrasted with those different sources [25]. Figure 32.2 shows the possibility of micro-hydropower sites reached by SRE.

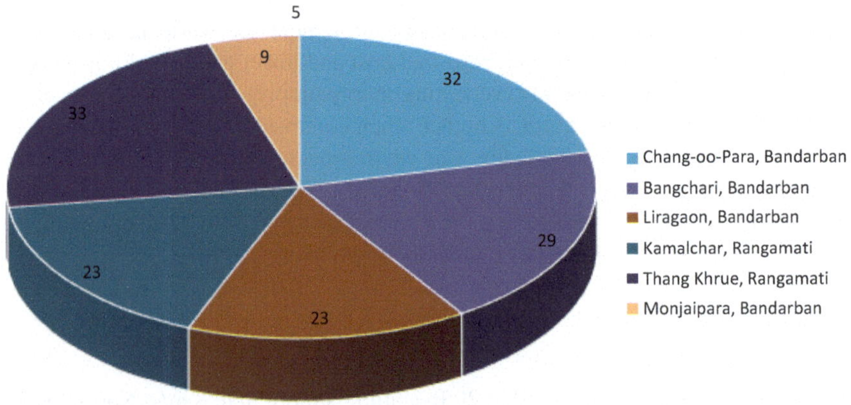

Fig. 32.2 Possibility of micro-hydropower sites researched by SRE

Despite having different conduits, Bangladesh has the obliged ability of hydro-energy because of the non-appearance of sensible head and stream of water. For all intents and purposes, 90% of the streams start from connected country India and Myanmar. Up to 2005, Bangladesh had only one solitary hydroelectric power plant known as Kaptai Hydroelectric Power Plant (KHPP) arranged at the bowl of Karnafuli conduit of full-scale limit 242 MW containing two 39 MW and three 53 MW Kaplan turbines.

Bangladesh Power Development Board (BPDB) is considering the extension of some other 115 MW of point of confinement in this power plant to be kept running generally in the rainy season when wealth water spills out. In 2011, another little scale hydro-control plant of farthest point 48 kW has been presented in Rangamati area [26]. Brahmaputra stream bowl can make 1540 MW and another potential goals the country over, which were found as a proper site for scaled down scale hydro control plant by BPDB and Bangladesh Water Development Board (BWDB), are showed up in Fig. 32.2 Sustainable rural energy (SRE) venture found a couple of zones fitting for scaled downscale hydro-plants in 2010 in Chittagong zones [27]. The hard and fast cutoff of hydro-control was assessed to be 142.4 MW. It addresses the capacity of scaled downscale hydro-control goals researched by SRE. In light of this, BPDB is at present introducing a 60–80 MW water system cum hydro-control plant around there [28].

32.3.4 Biomass Energy in Bangladesh

Biomass energy in Bangladesh is one of the sustainable and supportable energies which extracted from the biowastes covering from animal wastes, crop wastes, city wastes, etc. Biomass energy can be capable to produce heat energy which can be further changed into many different forms of energies by the use of both thermo-chemical and biochemical methods depending upon suitability. This changed over bio energy can be taken care of shipped. The idea of bio-energy in the overall fundamental energy use is 14% beginning at 2016 [29]. Solid biomass is the essential supporter in both overall warmth and power age from biomass energy sources. Horticultural buildups are generally accumulated from the yields, for instance wheat, rice, maize, sugarcane, etc. These buildups can be clearly accumulated from the soil and assembled after the planning of bioplants where the proper utilization of biomass can be taken care [30]. There are many uses of biogas plant such as cooking purposes. It is found that around 93 million tons of wood have been saved after the concept of biomass energy which expanses around the USD 3.1 million. On the other hand, biofuels are the best alternative to solve the carbon emission problems at present time which can be used in transportation and other purposes [31]. Fuels extracted from the rice husk can be used for the cooking purposes which helps to maintain sustainability in the country. Sugarcane plants in Bangladesh are also providing a large amount of energy which helps to solve the fuel problems in Bangladesh. Sixteen cogeneration plants have been started now presented in sugar forms that can create the 51,348 kWh control. Speaks to the power generation limit in 14 cogeneration plants utilizing sugarcane bagasse as fuel in Bangladesh [32].

32.3.5 Geothermal Energy in Bangladesh

With the sustainability issues at the global level, renewable sources of energy such as geothermal energy are becoming popular and adoption of this is growing at the fast rate. As showed by the 2016 yearly U.S. what's more, power production report of global geothermal energy, the presented furthest reaches of geothermal power energy sources is 15,412 MW. USA is one of the largest producers of geothermal power production with a farthest point of 5.15 GW trailed by the Philippines with 2.6 GW and country like Indonesia with 1.96 GW [33]. As there are relatively few thermal gradient destinations in Bangladesh, geothermal energy production can be used to create control [36, 37]. Regardless, it also expects through and through data for the evaluation of geothermal production [34]. Both private and government territory should way to deal with survey the capacity of geothermal energy production with the objective that this benefit can be utilized appropriately [35].

32.4 Future Targets for Renewable Energy in Bangladesh

Bangladesh is now focusing on the renewable sources of energy. Soliciting the legislature has set an earth-shattering system to achieve 41,464 MW by 2035. This course of action outlined out decreased dependence on oil gas and extended the use of coal in country and sustainable power source. The present idea of reasonable power source is simply 5.6%. The guide of the age of intensity from sustainable sources until 2023 has showed up. As a significant part of this game plan, the organization of Bangladesh is tackling different maintainable power source systems. For the national lattice, the legislature has begun a 650 MW sunlight-based power progression program. This program will join the improvement of sun-powered controlled water system siphon overriding diesel water framework siphon, advancement of sunlight-based parks, sun-oriented littler than regular cross-sectional control systems in remote towns, roof sun situated power answer for business, mechanical and private structures and electrification of prosperity center, enlightening establishments, e-centers at the affiliation level, strict associations and railroad stations. Out of this 450 MW, 370 MW will be used for business reason and 170 MW will be used for social organizations. Division smart scattering of 600 MW sun powered power. Development in materials [36] such as metal matrix composites for energy use [37], advanced ceramics [38], high refractory materials [39], sustainable composites [40], natural fiber-based materials [41] and advanced engineering materials [42] has made new research scopes in machining area [43] such as in both traditional machining processes [44] and non-traditional machining processes [45] with the use of non-traditional energy sources.

Further, this investigation can be reached out by finding the hindrances and drivers for the renewable power source appropriation in Bangladesh utilizing the MCDM techniques such as total interpretative structural modeling [46], analytical hierarchy process [47, 48] and hybrid fuzzy-based techniques [49].

32.5 Conclusion

In the present study, it is found that natural gas, coal and refined oil-based sources are the main sources of energy in Bangladesh. IDCOL and some other associations, e.g., LCED, private and public associations and Grameen Shakti are working in the area to promote the renewable sources of energy in the country which also helps to promote the sustainability. There are some research gaps in the adoption of the renewable sources of energy. The renewable sources of energy help Bangladesh to improve their energy as well as economic growth. Apart from this, adoption of renewable sources in country helps to improve the sustainability and reduction of carbon emission.

References

1. Islam, M.R., Islam, M.R., Beg, M.R.A.: Renewable energy resources and technologies practice in Bangladesh. Renew. Sustain. Energy Rev. **12**(2), 299–343 (2008)
2. Islam, A.S., Islam, M., Rahman, T.: Effective renewable energy activities in Bangladesh. Renew. Energy **31**(5), 677–688 (2006)
3. Hossain, A.K., Badr, O.: Prospects of renewable energy utilisation for electricity generation in Bangladesh. Renew. Sustain. Energy Rev. **11**(8), 1617–1649 (2007)
4. Mondal, M.A.H., Denich, M.: Assessment of renewable energy resources potential for electricity generation in Bangladesh. Renew. Sustain. Energy Rev. **14**(8), 2401–2413 (2010)
5. Biswas, W.K., Bryce, P., Diesendorf, M.: Model for empowering rural poor through renewable energy technologies in Bangladesh. Environ. Sci. Policy **4**(6), 333–344 (2001)
6. Mondal, M.A.H., Kamp, L.M., Pachova, N.I.: Drivers, barriers, and strategies for implementation of renewable energy technologies in rural areas in Bangladesh—an innovation system analysis. Energy Policy **38**(8), 4626–4634 (2010)
7. Sarkar, M.A.R., Ehsan, M., Islam, M.A.: Issues relating to energy conservation and renewable energy in Bangladesh. Energy. Sustain. Dev. **7**(2), 77–87 (2003)
8. Islam, M.T., Shahir, S.A., Uddin, T.I., Saifullah, A.Z.A.: Current energy scenario and future prospect of renewable energy in Bangladesh. Renew. Sustain. Energy Rev. **39**, 1074–1088 (2014)
9. Halder, P.K., Paul, N., Joardder, M.U., Sarker, M.: Energy scarcity and potential of renewable energy in Bangladesh. Renew. Sustain. Energy Rev. **51**, 1636–1649 (2015)
10. Anam, K.: Power crisis & its solution through renewable energy in Bangladesh (2011)
11. Hossain, H.Z., Hossain, Q.H., Monir, M.M.U., Ahmed, M.T.: Municipal solid waste (MSW) as a source of renewable energy in Bangladesh: revisited. Renew. Sustain. Energy Rev. **39**, 35–41 (2014)
12. Ahiduzzaman, M., Islam, A.S.: Greenhouse gas emission and renewable energy sources for sustainable development in Bangladesh. Renew. Sustain. Energy Rev. **15**(9), 4659–4666 (2011)

13. Ullah, M.H., Hoque, T., Hasib, M.M.: Current status of renewable energy sector in Bangladesh and a proposed grid connected hybrid renewable energy system. Int. J. Adv. Renew. Energy Res. **1**(11), 618–627 (2012)
14. Urmee, T., Harries, D., Schlapfer, A.: Issues related to rural electrification using renewable energy in developing countries of Asia and Pacific. Renew. Energy **34**(2), 354–357 (2009)
15. Reza, S.E., Nitol, T.A., Abd-Al-Fattah, I.B.N.E.: Present scenario of renewable energy in Bangladesh and a proposed hybrid system to minimize power crisis in remote areas. Int. J. Renew. Energy Res. (IJRER) **2**(2), 280–288 (2012)
16. Ahmed, S., Islam, M.T., Karim, M.A., Karim, N.M.: Exploitation of renewable energy for sustainable development and overcoming power crisis in Bangladesh. Renew. Energy **72**, 223–235 (2014)
17. Mondal, M.A.H.: Implications of renewable energy technologies in the Bangladesh power sector: long-term planning strategies. ZEF (2010)
18. Rahman, M.S., Saha, S.K., Khan, M.R.H., Habiba, U., Hosse, S.M.: Present situation of renewable energy in Bangladesh: renewable energy resources existing in Bangladesh. Glob. J. Res. Eng. (2013)
19. Biswas, M.M., Das, K.K., Baqee, I.A., Sadi, M.A., Forhad, H.M.S.: Prospects of renewable energy and energy storage systems in Bangladesh and developing economics. Glob. J. Res. Eng. (GJRE) **11**(5), 23–31 (2011)
20. Islam, M.: Assessment of renewable energy resources of Bangladesh. In: Electronic Book (2002)
21. Baky, M.A.H., Rahman, M.M., Islam, A.S.: Development of renewable energy sector in Bangladesh: current status and future potentials. Renew. Sustain. Energy Rev. **73**, 1184–1197 (2017)
22. Mondal, M.A.H., Islam, A.S.: Potential and viability of grid-connected solar PV system in Bangladesh. Renew. Energy **36**(6), 1869–1874 (2011)
23. Hossain, M.F., Hossain, S., Uddin, M.J.: Renewable energy: prospects and trends in Bangladesh. Renew. Sustain. Energy Rev. **70**, 44–49 (2017)
24. Uddin, S.N., Taplin, R.: Trends in renewable energy strategy development and the role of CDM in Bangladesh. Energy Policy **37**(1), 281–289 (2009)
25. Habibullah, M., Masjuki, H.H., Kalam, M.A., Rahman, S.A., Mofijur, M., Mobarak, H.M., Ashraful, A.M.: Potential of biodiesel as a renewable energy source in Bangladesh. Renew. Sustain. Energy Rev. **50**, 819–834 (2015)
26. Barua, D.C.: Strategy for promotions and development of renewable technologies in Bangladesh: experience from Grameen Shakti. Renew. Energy **22**(1–3), 205–210 (2001)
27. Bhuiyan, M.M.H., Asgar, M.A., Mazumder, R.K., Hussain, M.: Economic evaluation of a stand-alone residential photovoltaic power system in Bangladesh. Renew. Energy **21**(3–4), 403–410 (2000)
28. Sen, R., Bhattacharyya, S.C.: Off-grid electricity generation with renewable energy technologies in India: an application of HOMER. Renew. Energy **62**, 388–398 (2014)
29. Mondal, A.H., Denich, M.: Hybrid systems for decentralized power generation in Bangladesh. Energy. Sustain. Dev. **14**(1), 48–55 (2010)
30. Martinot, E., Chaurey, A., Lew, D., Moreira, J.R., Wamukonya, N.: Renewable energy markets in developing countries. Annu. Rev. Energy Environ. **27**(1), 309–348 (2002)
31. Mumtaz, A.: U.S. Patent No. 10,215,783. U.S. Patent and Trademark Office, Washington, DC (2019)
32. Mondal, M.A.H.: Economic viability of solar home systems: case study of Bangladesh. Renew. Energy **35**(6), 1125–1129 (2010)
33. Williams, K.R.: U.S. Patent No. 10,288,041. U.S. Patent and Trademark Office, Washington, DC (2019)
34. Nandi, S.K., Ghosh, H.R.: Prospect of wind–PV-battery hybrid power system as an alternative to grid extension in Bangladesh. Energy **35**(7), 3040–3047 (2010)
35. Ahmed, F., Al Amin, A.Q., Hasanuzzaman, M., Saidur, R.: Alternative energy resources in Bangladesh and future prospect. Renew. Sustain. Energy Rev. **25**, 698–707 (2013)

36. Garg, P., Jamwal, A., Kumar, D., Sadasivuni, K.K., Hussain, C.M., Gupta, P.: Advance research progresses in aluminium matrix composites: manufacturing & applications. J. Mater. Res. Technol. (2019)
37. Jamwal, A., Prakash, P., Kumar, D., Singh, N., Sadasivuni, K.K., Harshit, K., Gupta, S., Gupta, P.: Microstructure, wear and corrosion characteristics of Cu matrix reinforced SiC–graphite hybrid composites. J. Compos. Mater. 0021998319832961 (2019)
38. Jamwal, A., Vates, U.K., Gupta, P., Aggarwal, A., Sharma, B.P.: Fabrication and characterization of Al_2O_3–TiC-reinforced aluminum matrix composites. In: Advances in Industrial and Production Engineering, pp. 349–356. Springer, Singapore (2019)
39. Bandil, K., Vashisth, H., Kumar, S., Verma, L., Jamwal, A., Kumar, D., Singh, N., Sadasivuni, K.K., Gupta, P.: Microstructural, mechanical and corrosion behaviour of Al–Si alloy reinforced with SiC metal matrix composite. J. Compos. Mater. 0021998319856679 (2019)
40. Nayim, S.T.I., Hasan, M.Z., Seth, P.P., Gupta, P., Thakur, S., Kumar, D., Jamwal, A.: Effect of CNT and TiC hybrid reinforcement on the micro-mechano-tribo behaviour of aluminium matrix composites. In: Materials Today: Proceedings (2019)
41. Hossain, S., Rahman, M.M., Jamwal, A., Gupta, P., Thakur, S., Gupta, S.: Processing and characterization of pine epoxy based composites. In: AIP Conference Proceedings, vol. 2148, no. 1, p. 030017. AIP Publishing (2019)
42. Kumar, A., Arafath, M.Y., Gupta, P., Kumar, D., Hussain, C.M., Jamwal, A.: Microstructural and mechano-tribological behavior of Al reinforced SiC-TiC hybrid metal matrix composite. In: Materials Today: Proceedings (2019)
43. Sohag, M.A.Z., Gupta, P., Kondal, N., Kumar, D., Singh, N., Jamwal, A.: Effect of ceramic reinforcement on the microstructural, mechanical and tribological behavior of Al-Cu alloy metal matrix composite. In: Materials Today: Proceedings (2019)
44. Nayim, S.T.I., Hasan, M.Z., Jamwal, A., Thakur, S., Gupta, S.: Recent trends & developments in optimization and modelling of electro-discharge machining using modern techniques: a review. In: AIP Conference Proceedings, Sept 2019, vol. 2148, no. 1, p. 030051. AIP Publishing (2019)
45. Jamwal, A., Aggarwal, A., Gautam, N., Devarapalli, A.: Electro-discharge machining: recent developments and trends. Int. Res. J. Eng. Technol. **5**, 433–448 (2018)
46. Jamwal, A., Aggarwal, A., Gupta, S., Sharma, P.: A study on the barriers to lean manufacturing implementation for small-scale industries in Himachal region (India). Int. J. Intell. Enterp. **6**(2–4), 393–407 (2019)
47. Gautam, N., Ojha, M.K., Swain, P., Aggarwal, A., Jamwal, A.: Informal investigation of fourth-party and third-party logistics service providers in terms of Indian context: an AHP approach. In: Advances in Industrial and Production Engineering, pp. 405–413. Springer, Singapore (2019)
48. Sharm, P., Jamwal, A., Aggarwal, A., Bhardwaj, S., Sood, R.: Major challenges in adoption of RFID for Indian SME's. Int. Res. J. Eng. Technol. **5**, 2247–2255 (2018)
49. Singh, P.L., Sindhwani, R., Dua, N.K., Jamwal, A., Aggarwal, A., Iqbal, A., Gautam, N.: Evaluation of common barriers to the combined lean-green-agile manufacturing system by two-way assessment method. In: Advances in Industrial and Production Engineering, pp. 653–672. Springer, Singapore (2019)

Chapter 33
Design and Development of Steam-Based Solar Dryer for Agricultural Products

Masnaji R. Nukulwar and Vinod B. Tungikar

Abstract In the present scenario, agriculture products are processed (dried) in an open atmosphere. The rate of drying is affected by environmental conditions such as relative humidity, temperature, and solar radiations. Under unfavorable weather conditions, the processing time may go beyond a week, which necessitates a requirement to design equipment to assist the drying process to reduce processing time. In this study, a steam-based heating system called as steam-based solar dryer (SBSD) is designed and fabricated. Atmospheric temperature, relative humidity, and solar radiations are the considered parameters. In order to evaluate the feasibility of equipment, pilot experiments were conducted and the efficiency of SBSD is found to be 29.56%. CFD analysis was performed to investigate the effect of non-uniformity of temperature inside the heating space. It was found that assumption of uniform temperature is not appropriate and non-uniform temperature needs to be incorporated for realistic investigation. It is found that there is a temperature variation in the SBSD experimentally and CFD analysis.

Keywords Solar dryer · Drying · Scheffler system · Turmeric drying · Solar thermal application

33.1 Introduction

Curcumin or curcuminoid is the main constitute found in turmeric along with volatile oil, moisture, total ash, acid insoluble ash, crude fiber, bulk index, and granulation. Curcumin is found in the range of 1–6%, and it may vary according to the turmeric. Curing, drying, and polishing size are required after harvesting from the field. Blanching is the first process carried out on the rhizomes. Blanching adds moisture to the rhizomes. Solar dryer plays an important role to remove high moisture from the turmeric. Turmeric is the major spices used in cosmetics, pharmaceutical, textile, and food industries. Curcumin is the main important ingredient in the turmeric. Turmeric is the cash crop in India. Less attendance is given to dry turmeric because of higher moisture content. It can be sold in either dried form or fresh rhizomes [1].

M. R. Nukulwar (✉) · V. B. Tungikar
SGGS Institute of Engineering and Technology, Nanded, Maharashtra 431606, India

S. Yadav et al. (eds.), *Proceedings of International Conference in Mechanical and Energy Technology*, Smart Innovation, Systems and Technologies 174,
https://doi.org/10.1007/978-981-15-2647-3_33

Fossil fuel depletion, energy consumption rate, price of electricity, and CO_2 emission are the main threads of the world. To overcome it, solar energy needs to be used. Solar dish collector is used for high-temperature application compared to the Fresnel reflectors, solar tower, and parabolic trough collector. Solar dish collector is a promising alternative for the solar thermal application [2].

Use of different techniques like thermal storage system improves the effectiveness of the solar dryer. There is significant effects of geometric conditions, heat pump, heat exchanger, air circulating mode, experimental condition, phase change material, and adding of reflectors on solar dryer. Chimney arrangement maintains the airflow velocity by increasing the buoyancy force. Energy consumption can be reduced by increasing air temperature and mass flow rate of air [3].

Phase change material (PCM) is used for the solar cabinet dryer. The CFD analysis was carried out to find the temperature variation inside the solar cabinet dryer. PCM does not affect the quality of produce dried in the solar dryer [4]. Computer vision approach was used to monitor the quality parameter of turmeric-like texture, color, powder behavior, and morphological traits [5].

Thermal performance analysis of evacuated tube-based solar dryer was investigated. The evacuated tube-based solar dryer was developed to dry leaves of plants in the Indian climate condition. Quality of the dried produce inside the solar dryer is better than open sun drying. Solar drying does not have contaminations like birds, flies, microorganism when dried, and insects [6].

Drying kinetics and quality analysis of the mackerels were investigated in the solar-electric hybrid dryer. Two-term and diffuse approach drying models were best fitted for drying of the mackerels. Quality improvement was observed in the solar-electric hybrid dryer as compared with the open sun drying process [7].

Solar drying gives better return to the farmer along with reduced wastage. Direct sun drying method may have poor color value of agricultural production and reduce the price of market. Indirect solar dryer maintains nutrition values. Farmer will get appropriate price after the devoted hard work in crop cultivation. Open sun-dried product cannot directly be sold into the international market due to low quality. Temperature losses in the indirect solar dryer can be minimized by using good-quality insulating material [8].

Simulation model for grapes is developed based on geometric features, specific heat, shrinkage during sorption isotherms, and radiative properties. The same model can be used for the other different products by knowing the above parameter. Simulation model can be used for predicting optimum initial load and drying time [9].

Losses of agricultural product during the drying were estimated 30–40% in developing country. Post-harvesting losses can be reduced by using solar dryer. Solar energy can be used for two main purposes: conversion into electricity and solar thermal application. Solar thermal application includes solar cooking, solar cooling, solar pounds, solar dryer, solar distillation, solar furnace, and solar water heater.

Solar dryer accelerates the drying processes by incorporating the sensible and latent heat [10].

33.2 Materials and Method

33.2.1 SBSD

An SBSD is designed and fabricated at the institute. SBSD is designed for the 10 kg capacity of turmeric. The SBSD consists of drying chamber and solar heater storage system. Figure 33.1 gives the detailed construction of the solar dryer.

Rectangular box is fitted at the bottom of SBSD. Rectangular box is attached at the bottom which acts as a small reservoir. Five trays are placed at equal distance in the drying chamber. Wire mesh trays are placed at equal distance in the drying chamber. Wire mesh trays ensure free passage of air. Steam or hot water is collected

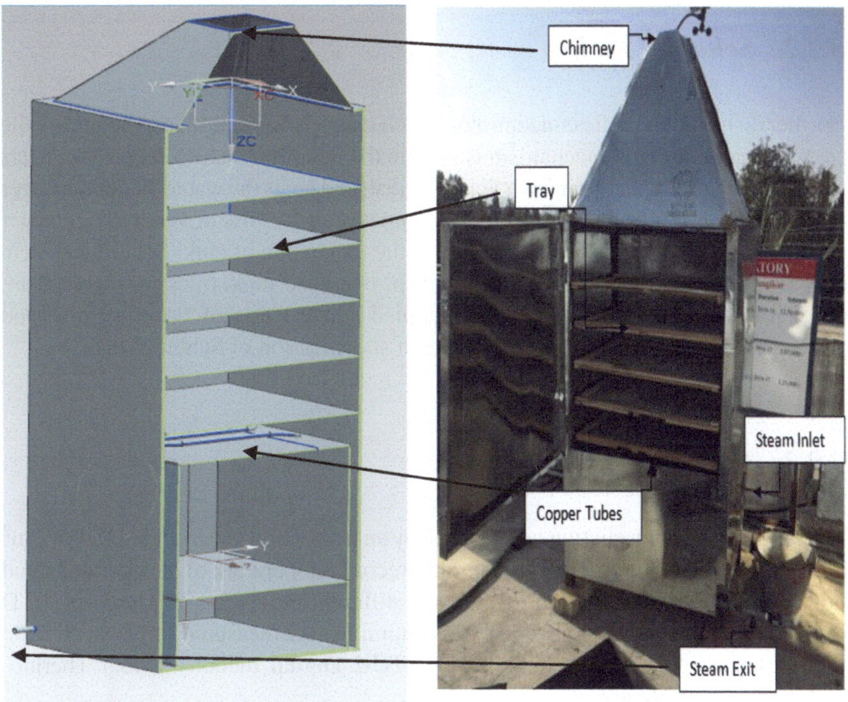

(a) Sectional View of SBSD (b) Illustrated SBSD

Fig. 33.1 SBSD

Table 33.1 Specification of SBSD

Length of dryer	750 mm
Height of dryer	1800 mm
Width of dryer	550 mm
Size of tray	700 mm * 500 mm
Length of copper tube	24,000 mm
Thickness of copper tube	2 mm
Dimension of mild steel box	700 mm * 500 mm * 180 mm
Thickness of insulation	30 mm
Diameter of copper tube	10 mm
Chimney size	10 mm * 10 mm
Insulation type	Glass wool

into the rectangular box. SBSD is insulated by glass wool covered by an aluminum sheet. Baffles are provided to direct the drying air movement (Table 33.1).

33.2.2 *Scheffler Dish*

Scheffler dish is a parabolic concentrator which reflects the sunlight and heat from the sun. Automatic tracking mechanism is used in the Scheffler dish. A receiver collects the reflected sunlight and heat from Scheffler dish and heats the water. Water tank over the receiver is used to refurbish water. It compensates for loss of water. Since water heats, steam is generated and is collected at the upper side of tank. Scheffler system consists of receiver, reflecting parabolic dish, overhead tank, tracking mechanism, and dish stand. Figure 33.2 gives the detailed construction of Scheffler dish and receiver. Table 33.2 gives detailed idea about specification of Scheffler dish.

33.2.3 *Instrumentation*

The intensity of solar radiation is measured by an irradiation meter with ± 0.05 W/m^2 accuracy. *K*-type thermocouples are used to record temperature of steam at inlet and outlet of SBSD. Drying air temperatures at different points are measured by RTD with 0.1 °C resolution. Velocity and relative humidity are measured by Mextech 4208 Digital Thermo-Anemometer and Teesta HTC-2 Digital Indoor/Outdoor Thermo-Hygrometer, respectively.

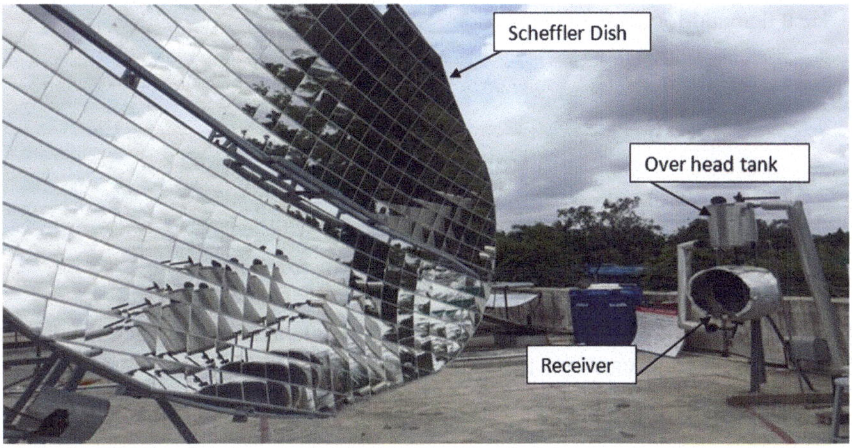

Fig. 33.2 Scheffler dish

Table 33.2 Specification of Scheffler dish

Area	16 m^2
Receiver capacity	18 l
Overhead tank capacity	28.5 l
Average steam flow rate	5 l/h
Material of receiver	Cast iron
Average steam output from system	104 °C
Thickness of receiver	5 mm
Thermic fluid	Water

33.2.4 Design of SBSD

The following assumptions are made while designing new SBSD.

Loading capacity of SBSD $m_f = 10\,\text{kg}$ of turmeric
Drying time for one batch = 20 h
Initial moisture content of turmeric $M_i = 77\%$
Final moisture content of turmeric $M_f = 12\%$.

Mass of water to be removed from turmeric was calculated [11],

$$m_w = m_f * \frac{(M_i - M_f)}{(100 - M_f)} \quad (33.1)$$

$$m_w = 7.16\,\text{kg}$$

Latent heat of vaporization is calculated as 2.43 MJ/kg [11].

Heat demand for drying [11] is

$$Q_{\text{drying}} = m_w * L_{\text{eff}} \tag{33.2}$$

$$Q_{\text{drying}} = 17.39\,\text{MJ}$$

Assumptions for calculating the area of the copper coil are

Efficiency of the solar dryer = 25%
Intensity of solar radiation = 750 W/m^2
According to MNRE Indian Meteorological Department, solar radiation handbook data by solar energy center, energy received per day = 25.12 MJ/m^2.

Area required for heating the air is calculated by the following equation [7]

$$A_c = \frac{Q}{H_T * N_d * \eta} \tag{33.3}$$

where

A_c area of copper coil m^2
Q heat required for drying MJ
H_T total energy absorbed by surface per day
N_d number of days
η efficiency of solar dryer
L_{eff} latent heat of vaporization (MJ/kg).

$$A_c = 0.9274\,\text{m}^2$$

$$\begin{aligned}\text{Surface area used for the rectangular vessel} &= 0.7 * 0.5\\ &= 0.35\,\text{m}^2\\ \text{Surface area required for copper tube} &= 0.9274{-}0.35\\ &= 0.5774\,\text{m}^2\end{aligned}$$

$$\begin{aligned}\text{Perimeter of the copper tube} &= \pi * D\\ &= \pi * 0.0125\\ &= 0.039\,\text{m}\end{aligned} \tag{33.4}$$

$$\begin{aligned}\text{Length of the copper tube} &= \frac{0.5774}{0.039}\\ &= 14.80\,\text{m}\end{aligned}$$

$$\text{Number of turns of copper coil} = \frac{14.80}{0.7} = 21 \text{ turn.}$$

33.2.5 Efficiency of SBSD

The dryer efficiency is evaluated as ratio of the heat demand for drying to energy input to the dryer.

$$\eta_{\text{dryer}} = \frac{m_{\text{w}} L_{\text{eff}}}{m\, c_{\text{p}} \Delta t} \tag{33.5}$$

33.2.6 Experimental Procedure

Scheffler dish and SBSD are installed in the solar laboratory of the institute. Steam generated by Scheffler dish at average temperature 104 °C is circulated through the copper tube of dryer. The steam/hot water is stored in the rectangular box placed at the bottom of SBSD. Atmospheric air coming from downside gets preheated when it comes in contact with steam in the rectangular box. Baffles are provided to guide the air toward copper tubes. Air gains heat when it comes in contact with copper tubes. Heated air moves in upward direction and comes out from the chimney provided at the top of SBSD. RTD is mounted inside the dryer to measure the temperature at different locations.

33.3 Results and Discussion

33.3.1 Climate Variation

Experiments are conducted during February 16, 2019, to February 20, 2019. Figures 33.3 and 33.4 give the variation of atmospheric temperature, solar radiation, and the relative humidity during the experiments. Maximum temperature and solar radiations are observed around 1 p.m.

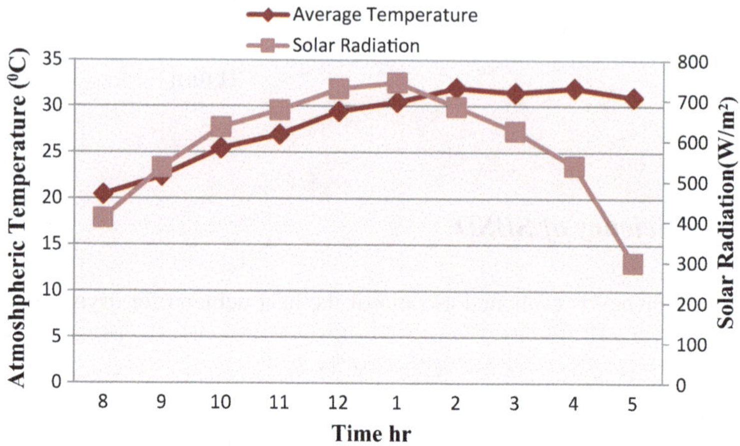

Fig. 33.3 Variation of average atmospheric air temperature and solar radiation

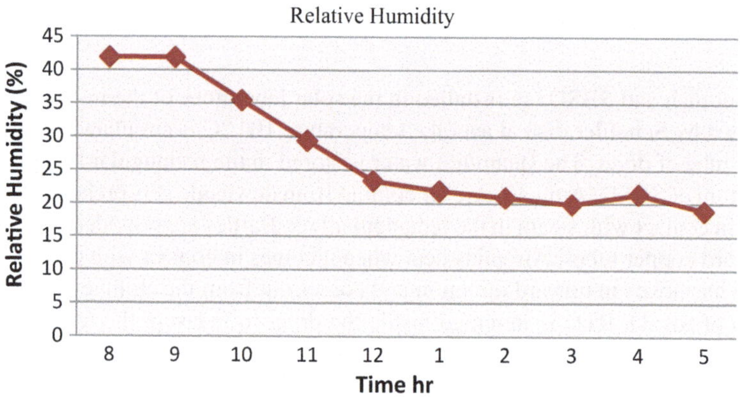

Fig. 33.4 Variation of average relative humidity

33.3.2 Steam Produced and Efficiency of SBSD

Scheffler system is started at 9:00 a.m., and steam is generated at 11:30 a.m. Average steam generated from the Scheffler system is found as 6.85 l/h. Steam generation is measured by direct condensation method. Figure 33.5 shows the steam generated in a day. Maximum steam generation is observed at 2.00 p.m. which could be because of maximum solar radiation during 1.00 p.m. to 1.30 p.m.

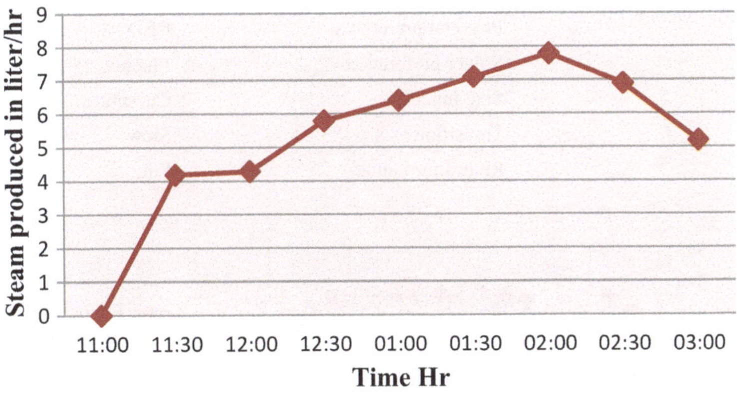

Fig. 33.5 Variation of steam produced

33.3.3 *CFD Analysis*

CFD analysis is performed in the ANSYSTM software 19.2 version. This analysis helps to find the temperature at different locations of the SBSD. Figure 33.6 shows the temperature distribution in the SBSD. Table 33.3 shows the mesh details used to solve the problem. Figure 33.7 shows the temperature distribution in CFD analysis.

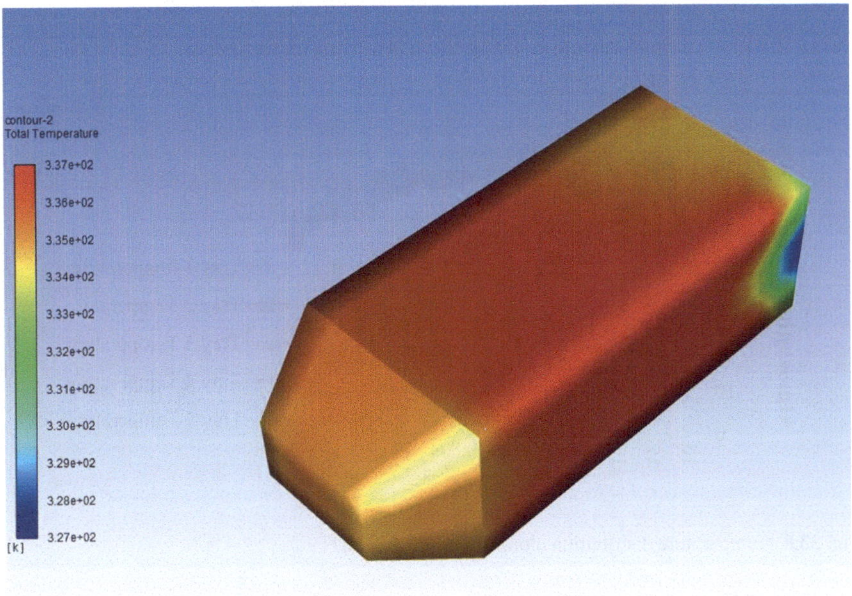

Fig. 33.6 Temperature distribution of SBSD

Table 33.3 Details of meshing

Physical preference	CFD
Solver preference	Fluent
Size function	Curvature
Transition	Slow
Relevance center	Fine

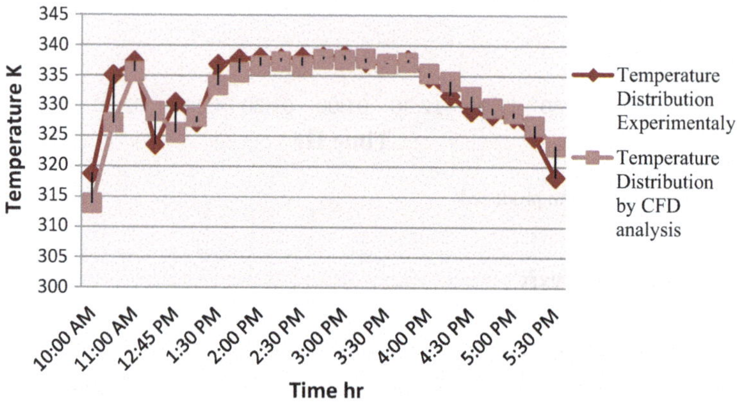

Fig. 33.7 Comparative temperature distribution of SBSD experimentally and CFD analysis

It is found that temperature reduces in upward direction of the SBSD. Figure 33.8 shows the temperature variation along the trays from experiments.

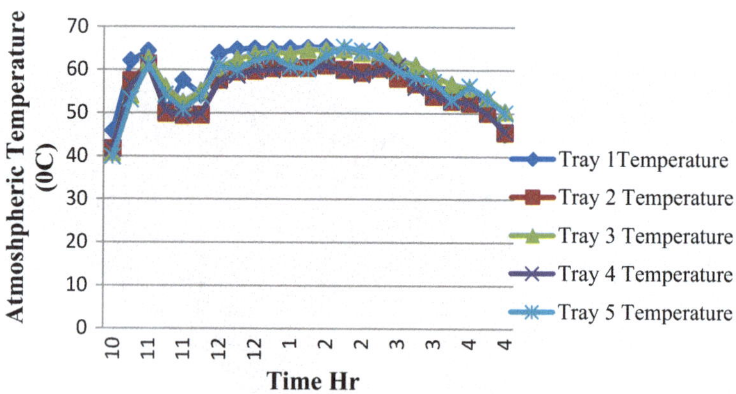

Fig. 33.8 Temperature distribution along the trays of SBSD

33.4 Conclusion

Eco-friendly SBSD working on renewable energy is developed and fabricated. Experimental results show that maximum temperature achieved inside the dryer is 64 °C. CFD analysis is performed and compared to the experimental results. Temperature results are found good agreement for CFD analysis and experiments. Temperature reduces along the vertical direction. Pilot experiments are carried out, and efficiency of the present heating system is found to be 29.56%. The results are encouraging and advocate its application for turmeric drying.

References

1. Karthikeyan, A.K., Murugavelh, S.: Thin layer drying kinetics and exergy analysis of turmeric (*Curcuma longa*) in a mixed mode forced convection solar tunnel dryer. Renew. Energy **128**, 305–312 (2018). https://doi.org/10.1016/j.renene.2018.05.061
2. Stefanovic, V.P., Pavlovic, S.R., Bellos, E., Tzivanidis, C.: A detailed parametric analysis of a solar dish collector. Sustain. Energy Technol. Assess. **25**, 99–110 (2018). https://doi.org/10.1016/j.seta.2017.12.005
3. Sandali, M., Boubekri, A., Mennouche, D.: Improvement of the thermal performance of solar drying systems using different techniques: a review. J. Sol. Energy Eng. **141**, 050802-1-050802-11 (2019). https://doi.org/10.1115/1.4043613
4. Iranmanesh, M., Akhijahani, H.S., Barghi Jahromi, M.S.: CFD modeling and evaluation the performance of a solar cabinet dryer equipped with evacuated tube solar collector and thermal storage system. Renew. Energy **145**, 1192–1213 (2019). https://doi.org/10.1016/j.renene.2019.06.038
5. Sharma, S., Dhalsamant, K., Tripathy, P.P.: Application of computer vision technique for physical quality monitoring of turmeric slices during direct solar drying. J. Food Meas. Charact. **13**, 545–558 (2019). https://doi.org/10.1007/s11694-018-9968-0
6. Singh, P., Vyas, S., Yadav, A.: Experimental comparison of open sun drying and solar drying based on evacuated tube collector. Int. J. Sustain. Energy **38**, 348–367 (2019). https://doi.org/10.1080/14786451.2018.1505726
7. Chavan, B.R., Yakupitiyage, A., Kumar, S.: Mathematical modeling of drying characteristics of Indian mackerel (*Rastrilliger kangurta*) in solar-biomass hybrid cabinet dryer. Dry. Technol. **26**, 1552–1562 (2008). https://doi.org/10.1080/07373930802466872
8. Shrivastava, V., Kumar, A., Baredar, P.: Developments in indirect solar dryer: a review. Dev. Indirect Sol. Dry. Rev. **3**, 67–74 (2014)
9. Ramos, I.N., Brandão, T.R.S., Silva, C.L.M.: Simulation of solar drying of grapes using an integrated heat and mass transfer model. Renew. Energy **81**, 896–902 (2015). https://doi.org/10.1016/j.renene.2015.04.011
10. Liberty, J.T.: Solar crop drying-a viable tool for agricultural sustainability and food security. Int. J. Mod. Eng. Res. **4**, 8–19 (2014). http://www.ijmer.com/papers/Vol4_Issue6/Version-1/IJMER-46010819.pdf
11. Husainy, A.S.N., Pujari, S., Wale, G.: Experimental investigation of mixed mode forced convection solar dryer for turmeric (*Curcuma longa*). Asian Rev. Mech. Eng. **7**(1), 1–6 (2018). ISSN: 2249-6289

Chapter 34
Heat Pipe Fabrication and Performance Evaluation for Different Coolants

Devendra Yadav, Simbal Pal, Saddam Quraishi and Mohammad Farhan

Abstract The heat pipe has higher thermal conductance, which prominent the thermal transport from one point to another point and makes it the most useful in the cooling applications. In this work, the heat pipe is designed and made to cool the electronic devices based on the case study on different parameters (pipe material, working fluid, length, angle, mesh size of wick). The pipe has a length of 30 cm, the internal diameter of 1 cm and a thickness 1 mm. Evaporator, adiabatic and condenser section lengths are 8.5 cm, 6.5 cm and 15 cm, respectively. The coolant filling ratio for water-based heat pipe is 85% of evaporator volume and for acetone-based is 100% by volume of evaporators. Annular fins (5×5 cm^2) of aluminium plates are used in the condenser section to enhance the heat transfer, and fin gap is 1.2 cm. The efficient cooling for heat pipes is observed at 60° of inclination and at a temperature of 60 °C. The maximum heat transfer capability of the wick is maximum at horizontal position.

Keywords Filling ratio · Adiabatic section · Fins · Mesh size · Inclination

Nomenclature

L	Length of the heat pipe
K_w	Permeability of wick
A_w	Cross-sectional area of wick
$\dot{m}$	Mass flow rate of vapour
r_c	Mean effective radius of wick
μ_l	Viscosity of liquid
ρ	Density of the fluid

D. Yadav (✉)
Department of Mechanical Engineering, Galgotias College of Engineering and Technology, Greater Noida, U.P., India

S. Pal · S. Quraishi · M. Farhan
Department of Mechanical Engineering, Axis Institute of Technology and Management, Kanpur, U.P., India

S. Yadav et al. (eds.), *Proceedings of International Conference in Mechanical and Energy Technology*, Smart Innovation, Systems and Technologies 174,
https://doi.org/10.1007/978-981-15-2647-3_34

q Heat flux

34.1 Introduction

The heat pipe is a high thermal conductance device which transfers heat by means of two-phase fluid flow. It transfers heat from a hot end (heat injection) to the cold end (heat exhaust) with minimum temperature differences inside a sealed vessel with the help of a backfilled working medium (or coolant) [1]. The vessel is to be evacuated in order to maintain vacuum inside the vessel. The working fluid flows inside the tube through capillary action with the help of wick. This device uses a combination of evaporation and condensation of the working fluid to transfer heat. It is based on the combination of conductive and convective heat transfer [2]. The basic components of heat pipe are evaporation, adiabatic and condensation regions and wick. The heat pipe picks up the heat from the evaporator [3]. The working fluid in the evaporator region is in the liquid phase which extracts the thermal energy as latent heat of vaporisation and its phase changes from liquid to saturated or superheated vapour. It is easier to change the liquid into a vapour phase under vacuum condition. As pressure increases, the vapour will flow towards the condenser section of heat pipe naturally using the pressure created by the minimum temperature difference travelling via adiabatic section [4]. The working fluid reaches to the condenser section and rejects the contained thermal energy as latent heat of condensation in the condenser section. The heat is dissipated using a heat sink. The working fluid changes its phase from vapour to a liquid state. It returns back towards the hotter section (evaporator section) of the heat pipe with the help of wick structure by utilising the capillary force [5].

The heat pipe used in many fields for the cooling purposes: in compact electronic enclosures, i.e., laptop and desktop computers, microprocessors, audio amplifier components, power supplies, etc., producing a highly efficient cooling effect with compact design due to miniaturisation [6]. The heat pipes are also used in aerospace technology for spacecraft cooling and temperature stabilisation and maintaining the isothermal structures. Its functionality is significant under the adverse solar heating in the field of aerospace due to its lightweight, reliability and no maintenance cost. They are also used for thermal control and heat distribution of satellites and spacecraft [7]. In passive thermal applications, in the field of medical devices in terms of less space required due to its miniaturisation with less environmental effects producing better cooling than pumped liquids and rejecting the heat from electrical and electronics computing hospital devices [8]. Some analysis has done on the effect of filling ratio of working fluid and angle of inclination for both thermosiphon and wick-assisted heat pipe [9, 10]. Some researchers Lips and Lefevre [11] and Carbajal et al. [12] have used thermal and hydrodynamic approaches to understand the variations of temperature, pressure and velocity distribution which is influenced by localised thermal fluxes under transient conditions. Derevyanko et al. [13] have used different wick structural configurations to ensure different heat transfer performance.

In this work, copper is used for making the heat pipe. The main challenging task in manufacturing of heat pipe is to evacuate the vessel to create a vacuum inside the heat pipe. For this, a new manual operated vacuum pump is developed at a very low cost. Using the disposal, working fluid is filled inside the vessel. Cu wire mesh wick structure is selected, and brazing is applied to seal the vessel. The experiments have performed for different coolants such as water and acetone. The performance analysis has done for various evaporator temperatures, heat transfer capability through the wick and a different angle of inclination of the heat pipe.

34.2 Governing Equations

In order to heat pipe function, maximum capillary pumping head must be able to overcome the total pressure drop of fluid in the heat pipe [14].

$$\Delta Pc_{\max} \geq \Delta P_l + \Delta P_v + \Delta P_g \tag{34.1}$$

Liquid pressure drops through the homogeneous wick.
From Darcy's law:

$$\Delta P_l = \frac{\mu_l L_{\text{eff}}}{K_{\text{w(wick)}}} \cdot \frac{\dot{m}}{\rho_l \cdot A_w} \tag{34.2}$$

The pressure drop of vapour

$$\Delta P_v = \frac{128 \mu_v \dot{m} L_{\text{eff}}}{\rho_v \pi D_v^4} \tag{34.3}$$

Pressure difference due to a hydrostatic head

$$\Delta P_g = \rho_l g L \sin(\alpha) \tag{34.4}$$

Capillary pumping head

$$\Delta Pc = \frac{2\sigma_l \cos(\theta)}{r_c} \tag{34.5}$$

Normally, the vapour pressure drop is negligible compared to ΔP_l

$$\Delta Pc = \frac{\mu_l L_{\text{eff}}}{K_{\text{w(wick)}}} \cdot \frac{\dot{m}}{\rho_l \cdot A_w} + \rho_l g L \sin(\alpha) \tag{34.6}$$

From the above equation

$$\dot{m} \cong \left(\frac{2\sigma_l \cos(\theta)}{r_c} - \rho_l g L_{eff} \sin(\alpha) \right) \times \frac{\rho_l K_{w(wick)} A_w}{\mu_l L_{eff}} \tag{34.7}$$

Heat transport capability

$$Q = \dot{m} h_{fg} \tag{34.8}$$

The maximum heat transport capability through wick is given by

$$Q_{max} = \left(\frac{\sigma_l \rho_l h_{fg}}{\mu_l} \right) \left(\frac{A_w K_w}{L_{eff}} \right) \left(\frac{2}{r_c} - \frac{\rho_l g L_{eff} \sin(\alpha)}{\sigma_l} \right) \tag{34.9}$$

34.3 Heat Pipe Fabrication

In present work, the copper pipe has a length of 30 cm which is calculated from the theoretical study. Its internal diameter and thickness are 1 and 0.1 cm, respectively. The length of the evaporator, adiabatic and condenser sections are 8.5, 6.5 and 15 cm, respectively. For the better condensation of vapour, 10 rectangular fins of area $6 \times 6\ cm^2$ are attached in the condensation section. To seal one end of the copper tube, a copper piece with a rectangular shape is brazed. Now a copper plate of suitable size with a hole of 0.5 cm of diameter is used to seal the evaporation section. A copper tube of diameter 0.5 cm is brazed with this hole to introduce passage for filling coolant. Mazumder et al. [10] conducted experiment with varying filling ratios for different coolants, and estimated amount of coolant is to be filled inside the tube for achieving high heat transfer coefficient, reduced thermal resistance and minimum temperature difference between evaporator and condenser sections (the coolant filling ratio for water-based heat pipe is 85% of evaporator volume and for acetone-based is 100% by volume of evaporators). For creating the vacuum inside the tube, set-up is made which has a plastic T-joint connected to the externally attached pipe which is brazed previously to the side of the evaporator. One of the remaining two arms of T-joint which is perpendicular to the central axis of the heat pipe is connected to the vacuum gauge. And another end which is in the same direction of the central axis of the heat pipe is connected to a plastic pipe having a one-way non-returning valve inside the tube and is connected with disposal for creating the vacuum. From this set-up, perfect evacuation inside the heat pipe can be achieved. Cleanness is of prime importance to ensure that no incompatibilities exist in the heat pipe and also make the condition that the wick and wall will be wetted by the working fluid. Also, it checks the leaks. The welds on the heat pipe will be leakproof [15]. Ten aluminium fins are closely fitted and welded along with maintaining 1.2 cm gap between pair of fins in the 15 cm length of condensation section. Teflon is wrapped over the adiabatic section of 6.5 cm length present between the condensation and evaporation sections in order to insulate the adiabatic section. The vacuum-creating arrangement and fabricated heat pipe are shown in Figs. 34.1 and 34.2, and its specifications are listed in Table 34.1.

The evaporator section is made by dipping the evaporator section of the heat pipe into a container of water maintained at constant temperature by a digital temperature controller (W1209). Emulsion rod (2000 W) is used for heating the water. In order to maintain, regulate and indicate the temperature of the water contained in the vessel, there is a direct connection of emulsion rod negative terminal with the AC supply and positive terminal is connected between AC supply and thermoset. J-type thermocouples are connected at the different sections of the heat pipe. Three thermocouples are connected in condensation region for finding a variation of temperature. One thermocouple is connected at the middle of the adiabatic region and one at the middle of the evaporation region. One thermocouple is set free in order to find the temperatures of the tip of fins. All six thermocouples are connected to their respective positive terminals of the digital temperature indicator (DTI). And the negative terminal of the

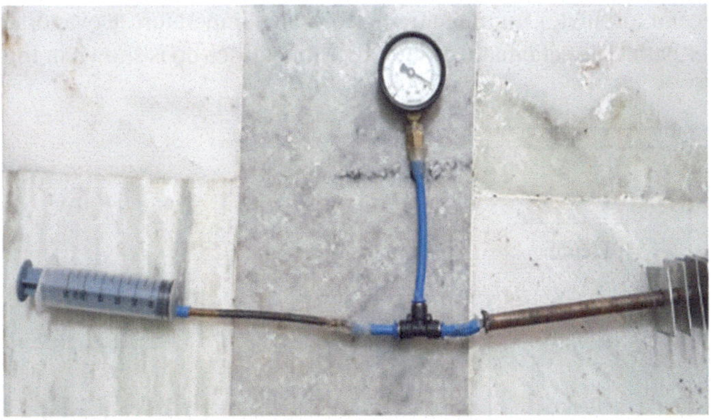

Fig. 34.1 Vacuum creation inside the pipe (by disposal system)

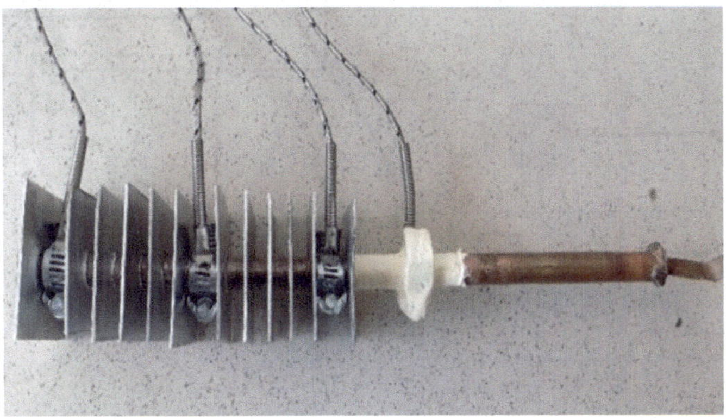

Fig. 34.2 Fabricated heat pipe

Table 34.1 Heat pipe material and design specification

Material	Specifications		
Heat pipe material	Cu	$k = 385$ W/m K	
Fin material	Al	$k = 204.2$ W/m K	
Length (cm)	Evaporator = 8.5 cm	Adiabatic section = 6.5 cm	Condenser = 15 cm
Coolant	Water (at 55 °C)	Volume = 5.6 ml	$\mu = 1.09 \times 10^2$ m Ns/m^2 (milli Newton second/metre square)
	Acetone (at 55 °C)	Volume = 6.5 ml	$\mu = 0.24$ m Ns/m^2 (milli Newton second/metre square)

DTI is connected to the AC supply. In this set-up, thermoset with A/D converter is used in place of turbulator for maintaining constant temperature at evaporator section inside water bath. The schematic of the experimental set-up is shown in Fig. 34.3. To

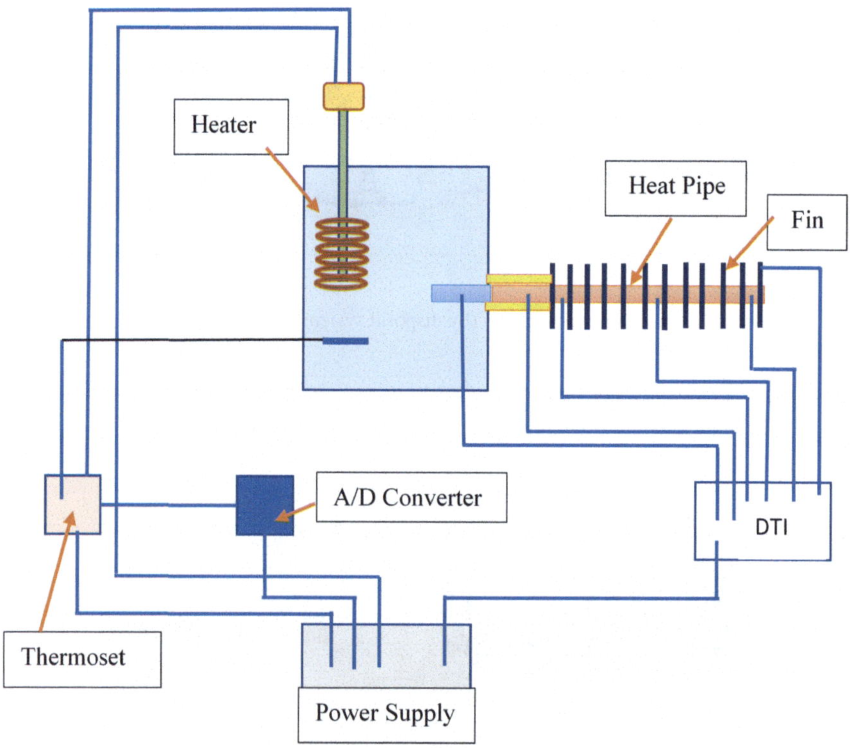

Fig. 34.3 Schematic diagram of the entire collaborated experimental set-up

analyse the effect of angular positions of the heat pipe along with the vessel, an angle measurement set-up was also collaborated in order to measure the angular positions of the heat pipe.

34.4 Results and Discussion

In order to perform the experiment, the evaporator section is kept in contact with a heat supplying zone (the evaporator section in the region which is required to be cooled). There are two approaches as per the principles of heat transfer to supply heat: (1) maintaining constant temperature and (2) supplying constant heat flux. In this work, the temperature of the evaporator section was kept constant. To accomplish this, a container of five litres of capacity filled with water has been taken. An electrical emulsion rod is used for heating water inside the container. And the constant temperature is maintained by using thermoset. Teflon is selected for insulating the adiabatic section due to its high insulating property. Thermosiphon performs better heat rejection in a vertical position due to the effect of gravity which assists in capillary effect. At varying inclinations, effect of gravity is reduced which produces lesser capillary effect. Thus, heat rejection is directly affected. But in case of heat pipes, the wick produces capillary effect. At varying inclinations, effect of gravity changes and the combined effect of wick and gravity at particular inclination decide the returning of the fluid to the evaporator section. Thus, heat rejection and performance of heat pipe varies at different tilt angles [9]. This set-up consists of a stationary arm which is fixed on a wooden base. Another arm is hinged at the other end of the stationary arm at which angle measuring tool is provided for measuring the different angular positions of the heat pipe. For performing the experiment, following components coupled together; (i) the heat pipe arrangement, a container or vessel with water (water bath) for the formulation of evaporator section. (ii) The arrangement which maintains a constant temperature inside the container or vessel. (iii) the temperature measuring and indicating devices. (iv) Angle measuring set-up for varying angular positions of the heat pipe. The container is thoroughly filled with water ensuring that the entire evaporator section is completely surrounded with water outside.

34.4.1 Experimental Analysis on Heat Pipe with Water

As per the theoretical concepts of thermosiphon, the temperature variation along the length of thermosiphon is maximum when it operates at a vertical position. But from the experiment on heat pipe at a different inclination angle of thermosiphon, 60° angle shows the best temperature variation across its length compared to other inclination angles 0°, 30°, 45° and 90°. For this observation, temperature variation across the length at 55 °C of evaporator temperature is shown in Fig. 34.4. So by

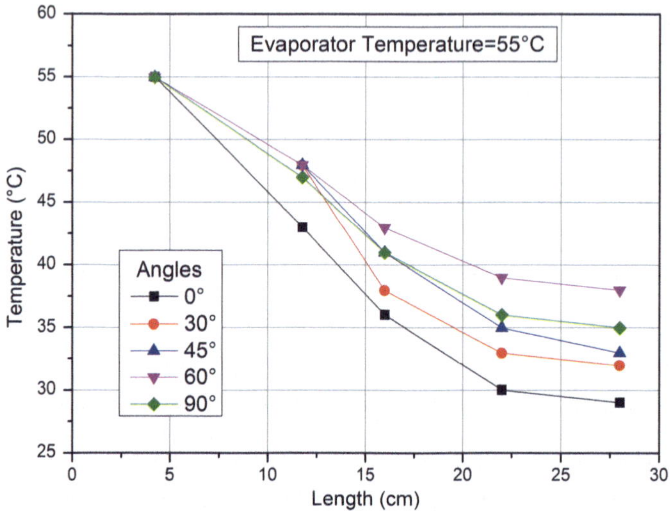

Fig. 34.4 Temperature variation at different heat pipe position (evaporator temperature = 55 °C)

this experimental result, we can say that the efficiency of heat rejection is more in thermosiphon at a 60° angle compared to other angles.

Another comparative analysis is conducted on different evaporator temperature of the thermosiphon (45, 50, 55 and 60 °C) at the vertical position (90° angle), and the temperature variation along the length of the thermosiphon is analysed (Fig. 34.5). The magnitude of temperature along the length is more for 60 °C of evaporator

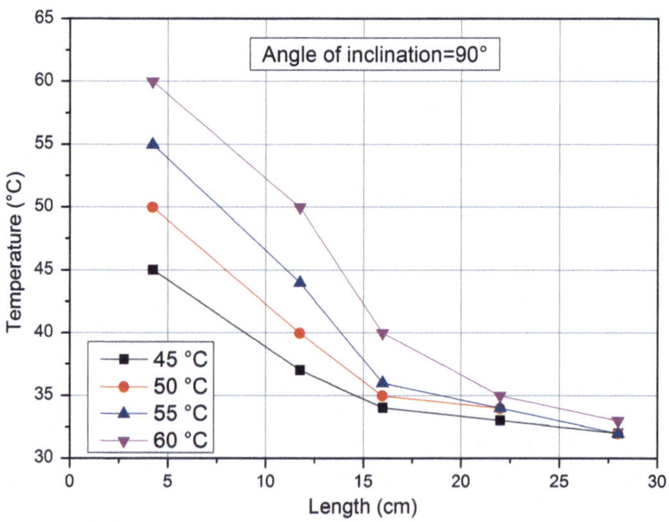

Fig. 34.5 Temperature variation along the length of the heat pipe at vertical position 90°

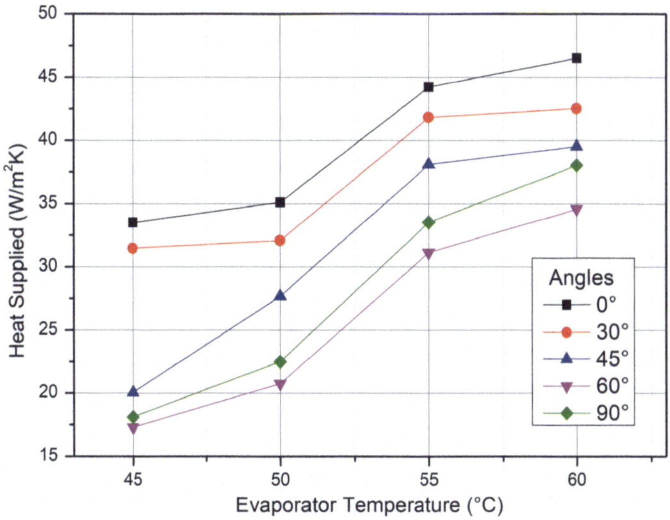

Fig. 34.6 Effect of angle of inclination on heat supplied to the evaporator

temperature. Therefore, heat rejection through the condenser section will be more at this temperature compared to another temperature. And the heat supplied to the evaporator is maximum at the horizontal position (Fig. 34.6).

34.4.2 Experimental Analysis on Heat Pipe with Acetone

As per the theoretical concept, heat pipe shows temperature variation across its length and maximum at horizontal position. From the experimental calculation at different angles of the heat pipe, the temperature variation across the length of the heat pipe at 55 °C of evaporator temperature, 45° angle of the heat pipe shows the maximum temperature difference across the length compared to another inclination angle of the heat pipe (Fig. 34.7). Another experiment is conducted at different evaporator temperature at the vertical position of the heat pipe across its length (Fig. 34.8). By performing different evaporator temperature (45, 50, 55 and 60 °C), the temperature variation is found maximum in 60 °C temperature of the evaporator. So, the heat transfer capability of the heat pipe is maximum at 60 °C temperature.

Using acetone as a working fluid, different experiments have been conducted on it. From the experiment data, the calculation is done on heat supplied to the evaporator section and we have compared this value to the different evaporator temperatures and different inclination angles of the heat pipe (Fig. 34.9). By comparison, heat supplied value in the evaporator section is maximum at 60° angle. The effect of angle and temperature of the evaporator section on maximum heat transfer capability through wick has been analysed by using Eq. 34.9. It is observed that the temperature of the

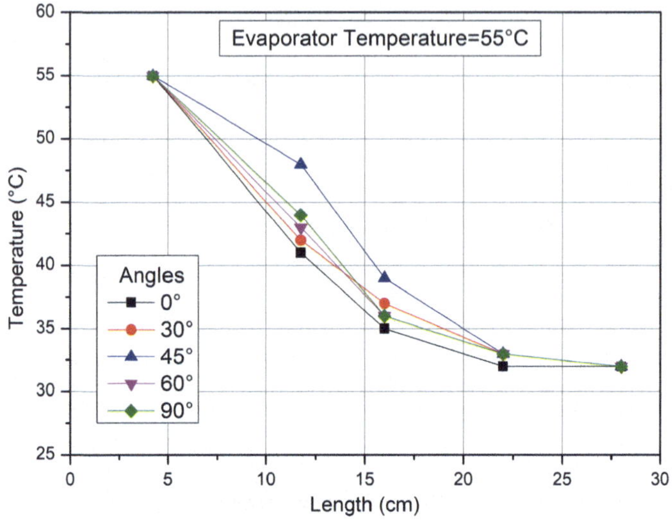

Fig. 34.7 Temperature variation at different heat pipe position (evaporator temperature = 55 °C)

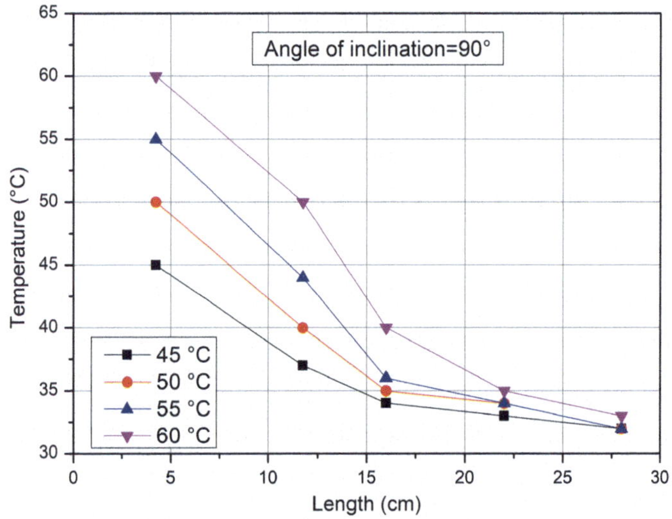

Fig. 34.8 Temperature variation along the length of the heat pipe at vertical position 90°

evaporator section does not affect the heat transfer capability, but the variation is observed for the different positions (angle) of heat pipe as shown in Fig. 34.10. At horizontal position, heat transfer capability is maximum and at vertical position it is minimum.

Normally, the thermal heat transfer capability of the heat pipe is greater than thermosiphon. But in case of our experimental analysis due to large wick diameter,

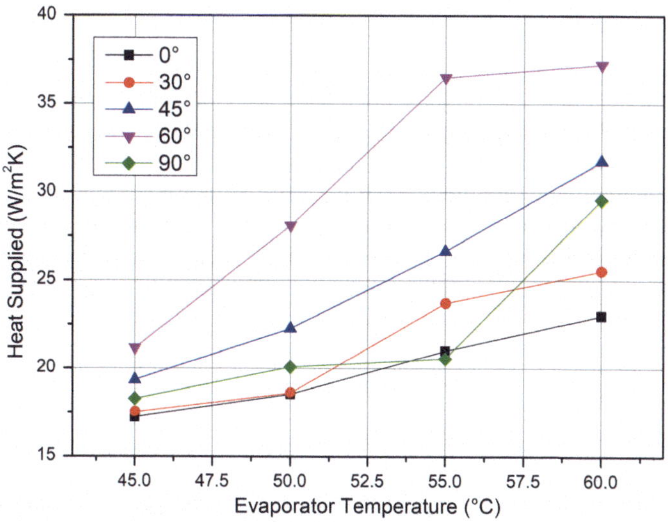

Fig. 34.9 Effect of angle of inclination on heat supplied to the evaporator

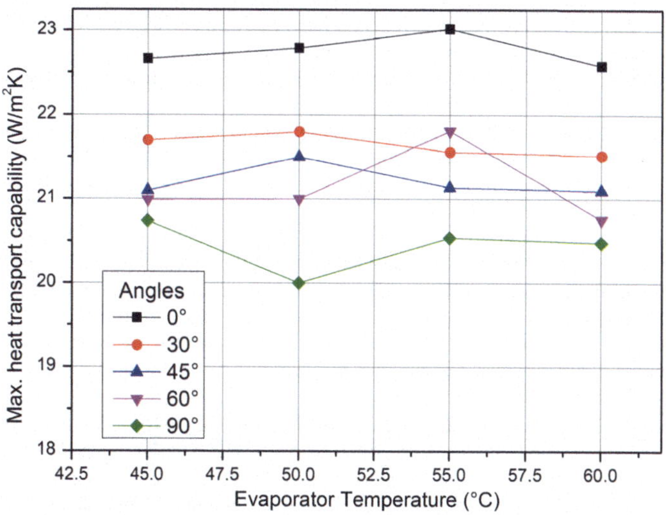

Fig. 34.10 Effect of angle of inclination of the heat pipe in heat transfer capability

there is the availability of smaller porous volume for travelling of working fluid (acetone). Thus, vapour travelling is blocked by wick structure which results as the reduced performance of heat pipe than thermosiphon with water as working fluid. The velocity of the fluid in the porous core must be greater than the velocity in wick configuration in order to resist blockage.

34.5 Conclusion

In this work, heat pipe and thermosiphon have been made at very low cost. A new manual operated vacuum pump is developed for the perfect evacuation of the heat pipe. The fabricated heat pipe performance is analysed for the different coolants, evaporator temperature and the angular positions. Water heat pipe (thermosiphon) has a better performance at 60°. For acetone heat pipe, its heat transfer capability is maximum at horizontal position. Due to the smaller cavity space for vapour transport in acetone heat pipe, it has lower heat transportation capability compared to the water thermosiphon.

References

1. Cotter, T.P.: Theory of Heat Pipes, 37th edn. DTIC Document, Los Alamos Scientific Laboratory, University of California (TID-4500) (1965)
2. Korn, F.: Heat Pipes and Its Applications. Project Report, MVK160 Heat and Mass Transport (May 07, 2008)
3. Panda, K.K., Basak, A., Dulera, I.V.: Design and development of high temperature heat pipes and thermosiphons for passive heat removal from compact high temperature reactor. In: Thorium Energy Conference (ThEC15), vol. 47, no. 25 (2015)
4. Deshpande, A., Patil, V., Patil, R.: Theoretical design of radiator using heat pipes. Int. J. Eng. Res. Technol. **5**(11), 17–23 (2016). ISSN: 2278-0181
5. Vanyasree, G., Ramana, P.V.: Experimental analysis on thermosyphon heat pipe to find heat transfer coefficient. Int. Res. J. Eng. Technol. (IRJET) **4**(8), 129–136 (2017). ISSN: 2395-0056
6. Grooten (Mart), M.H.M.: Towards an Optimal Design of Heat Pipe Equipped Heat Exchanger. Report No-WPC2007.01: 6-48 (2007)
7. Makhankov, A., Anisimov, A., Arakelov, A., Gekov, A., Jablokov, N., Yuditskiy, V., Kirillov, I., Komarov, V., Mazul, I., Ogorodnikov, A., Popov, A.: Liquid metal heat pipes for fusion application. Fusion Eng. Des. **42**(1–4), 373–379 (1998). https://doi.org/10.1016/S0920-3796(98)00216-6
8. Rashidian, B.: Modeling of the heat pipe heat exchangers for heat recovery. In: 2nd WSEAS International Conference on Engineering Mechanics, Structures and Engineering Geology, pp. 114–119 (2015). ISSN: 1790-2769
9. Khan, M.N., Gupta, U., Sinha, S., Singh, S.P., Pathak, S.: Parametric study of the performance of heat pipe—a review. Int. J. Mech. Eng. Technol. (IJMET) **4**(1), 173–184 (2013). ISSN 0976-6340
10. Mazumder, A.K., Akon, A.F., Chowdhury, M.S.H., Banik, S.C.: Performance of heat pipe for different working fluids and filling ratios. J. Mech. Eng. **41**(2), 96–102 (2010)
11. Lips, S., Lefevre, F.: A general analytical model for the design of conventional heat pipes. Int. J. Heat Mass Transf. **72**, 288–298 (2014). https://doi.org/10.1016/j.ijheatmasstransfer.2013.12.068
12. Carbajal, G., Sobhan, C.B., Peterson, C.P., Queheillalt, D.T., Wadley, H.N.G.: Thermal response of flat heat pipe sandwich structure to localized heat flux. Int. J. Heat Mass Transf. **49**, 4070–4081 (2006). https://doi.org/10.1016/j.ijheatmasstransfer.2006.03.035
13. Derevyanko, V., Nesterov, D., Suntsov, S.: Experimental investigation of flat heat pipes to remove high heat fluxes. In: 16th International Heat Pipe Conference, (16th IHPC), Lyon, France (2012)

14. Reay, D.A., Kew, P.A., Mc Glen, R.J.: Heat Pipes Theory, Design and Applications, 6th edn. Elsevier, Amsterdam (2014). https://doi.org/10.1016/C2011-0-08979-2
15. Harris, J.: Modeling, Designing, Fabricating and Testing of Channel Panel Flat Plate Heat Pipe. Report. Utah State University (2008)

Chapter 35
Assessing the Effect of Blending Eucalyptus Oil with Water and Their Impact on Conventional Engine

N. S. Senthur, H. Imamulhasan, M. Maheshkumar and K. Arun

Abstract Intense growth in industrialization and demands of the increasing population constantly resulted in high usage of limited petroleum resources of the earth. In addition, inflexible emission laws compelled manufacturers to look for alternative sources of fuel. This experiment is one such consequence of a similar escapade. The test biodiesel fuels and conventional fuel were individually combusted in a four-stroke conventional Kirloskar diesel engine. The characteristics of engine performance and emission parameters were analyzed. Test fuels taken into consideration were conventional diesel, E20 biodiesel (20% of eucalyptus oil by volume with diesel), and three unique mixtures of eucalyptus-water-diesel emulsions which vary from each other in the ratio of water content. Eucalyptusdiesel-Water Mixture 1 (EWM1), Eucalyptusdiesel-Water Mixture 2 (EWM2), and Eucalyptusdiesel-Water Mixture 3 (EWM3) have 5%, 10% and 15% of water content, respectively. The results portray the advantages of the use of emulsified fuels over conventional fuel and when compared to diesel, EWM3 Eucalyptusdiesel-Water Mixture 3 (EWM3) yielded the highest performance and efficiency and yet being least emissive than other test fuels.

Keywords Biodiesel · Conventional engine · Emulsion · Brake thermal efficiency · Specific energy consumption · Micro-explosion

35.1 Introduction

The diesel engine is a very complex machine to design and operate. Hence, making structural and functional changes to enhance its parameters is a hard task. The efficiency of the conventional diesel engine is low, yet it plays a vital role in various dominating sectors such as transportation, agriculture, construction, and manufacturing sectors. The sole purpose of a diesel engine is to convert chemical energy into purposeful mechanical energy. The two necessary elements required to carry out this task are the fuel and the engine. The fuel has a flexible margin in the field of research and innovation. This is taken as an advantage and is used to influence

N. S. Senthur (✉) · H. Imamulhasan · M. Maheshkumar · K. Arun
Einstein College of Engineering, Tirunelveli, Tamilnadu, India

S. Yadav et al. (eds.), *Proceedings of International Conference in Mechanical and Energy Technology*, Smart Innovation, Systems and Technologies 174,
https://doi.org/10.1007/978-981-15-2647-3_35

the efficiency of the system. In this experiment, diesel is altered by adding various fluids. The mono-alkyl ester derived from vegetable oil constitutes biodiesel [1]. This ensures a wider and more economical alternative source of fuel for diesel engines. Eucalyptus oil is used indirectly as an alternative fuel in the place of diesel for diesel engines, and this is possible due to its similar calorific value when compared to conventional diesel [2]. Transesterification is carried out on vegetable oil to enhance the specific energy consumption. As a consequence, emissions such as carbon monoxide (CO) emissions and hydrocarbon (HC) emissions were minimized. Transesterification converts the esters in the long chain of fatty acids into mono-alkyl esters hence producing biodiesel [3]. Water was included in the fuel mixture to investigate consequences involvement of water during the process of combustions. As oil is immiscible in water, there exist some instability issues while using this fuel, which is overcome by using certain substances like surfactants [4]. The composition of water in water emulsion increases from 5 to 15% as a step of 5% when progressing through EWM1, EWM2, and EWM3. These fuels introduced some interesting and phenomenal processes during combustion. One such process was micro-explosion. It occurs when the water reaches temperatures beyond its boiling point. The water molecules in the oil emulsions begin to heat up quickly than the molecules it coexists with it. Water begins to expand strenuously and splits the molecules within it. This leads to atomization of the oil and better combustion of the same. The vaporized water expands unto thousand seven hundred times of its original volume inside the combustion chamber. The splitting of a single bulk molecule into smaller molecules occurs thereby enhancing the efficiency of combustion [5]. The emissions are lower and cleaner while using oil emulsions instead of diesel oil. But to gain similar performance from the engine the right proportion has to be used. This was found to be 15% of net water content in the fuel and corresponds to EWM3. Emission of particulate matter was lower while compared to combusting oil emulsions. Surfactants are used to preserve the proportions of oil emulsions.

35.2 Materials and Methods

The analysis was carried out by the four-stroke single cylinder Kirloskar diesel engine. Compression ratio of the engine is 17.5:1. Engine has a direct injection compression setup set to an optimal 200 bar. This engine is 5.9 kW of power at 1800 revolutions per minute. The loading conditions were split into four phases ranging from 0 to 100% in steps of 25%. The engine was set to its default configurations and no modifications were made before carrying out the tests. The speed of the engine was set constant for all the test fuels. A piezoelectric sensor was placed inside the engine block to note the pressure (Fig. 35.1; Tables 35.1 and 35.2).

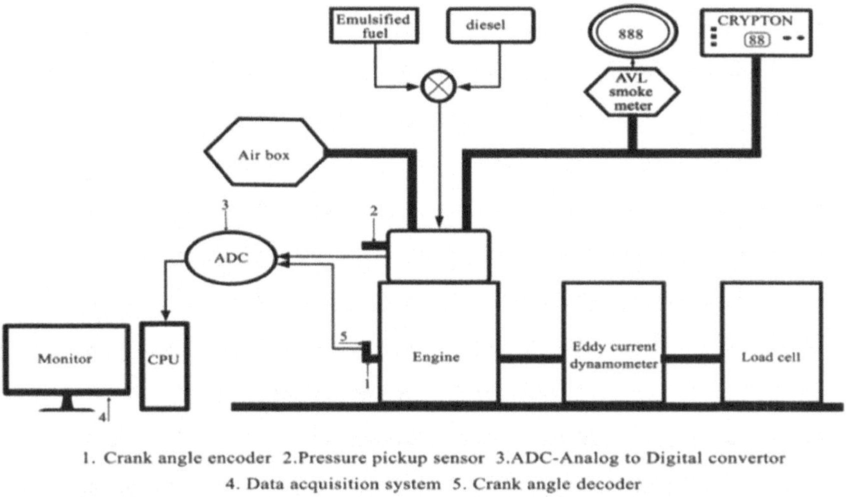

Fig. 35.1 Test engine setup

Table 35.1 Proportion of eucalyptusdiesel-water mixtures

Fuel	Diesel level (%)	Eucalyptus level (%)	Water level (%)	Span 80 level (%)	Tween 80 level (%)
EWM1	74	20	5	0.5	0.5
EWM2	69	20	10	0.5	0.5
EWM3	64	20	15	0.5	0.5

Table 35.2 Diesel and eucalyptusdiesel-water mixtures fuel properties

Fuel	Density (kg/m^3)	Calorific value (MJ/kg)	Flash point (°C)	Fire point (°C)
Diesel	830	45.2	56	64
Eucalyptus oil	890	43.3	34	43
Eucalyptusdiesel-water mixture 1	838	40.51	72	84
Eucalyptusdiesel-water mixture 2	843	38.37	74	85
Eucalyptusdiesel-water mixture 3	844	36.20	75	86

In all the eucalyptus-water emulsions, the proportion of surfactants and eucalyptus oil is 1% and 20%, respectively. The water content in EWM1, EWM2, and EWM3 is 5%, 10%, and 15%, respectively. Due to the immiscibility of water in oil, usage of the mixture without any alterations is unstable.

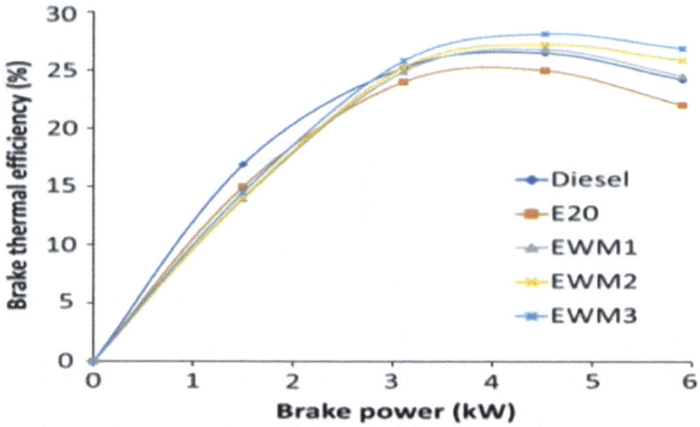

Fig. 35.2 Variation of brake thermal efficiency with brake power

35.3 Results and Discussion

Various physical and thermal properties of dual green fuels of Kanuga oil and Eucalyptus oil and its blends were assessed. The engine effectiveness has been assessed by certain parameters such as thermal efficiency, specific fuel consumption, and exhaust gas temperature.

35.3.1 Brake Thermal Efficiency

The efficiency of the various fuels, in the conventional diesel engine, is shown in Fig. 35.2. The reference fuel—diesel—had a brake thermal efficiency of 26.48%. The test fuel—Eucalyptus-20 biodiesel—was almost as efficient as conventional diesel and missed the mark by 1.16%. The EWM series fuels are more efficient than conventional diesel as near complete combustion takes place due to the phenomenon of micro-explosions [6]. Emulsified fuels—EWM1, EWM2, and EWM3—were 1.15%, 2.87%, and 4.8% more efficient than conventional diesel. Thus, from the analyses, it was found that Eucalyptusdiesel-Water Mixture 3 (EWM3) gave better brake thermal efficiency when compared with other test fuels.

35.3.2 Specific Energy Consumption

The specific energy consumption of all the test fuels used in this experiment is shown in Fig. 35.3. Conventional diesel had a specific energy consumption of 14,868.8 kJ/kWh at full load conditions. Specific energy consumption is increased

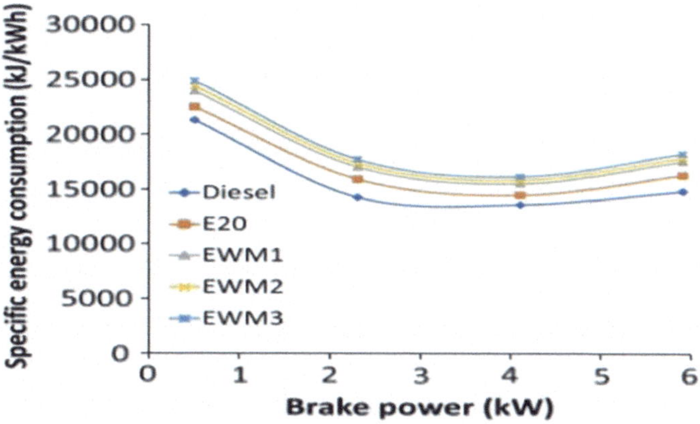

Fig. 35.3 Variation of specific energy consumption with brake power

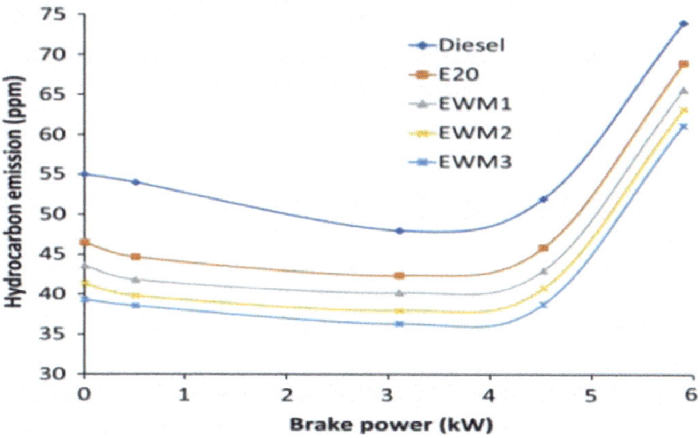

Fig. 35.4 Variation of hydrocarbon emission with brake power

due to the lower heating value of biodiesels [7]. The specific energy consumption of Eucalyptus-20 biodiesel EWM1, EWM2, and EWM3 was found to be 1464.2 kJ/kWh, 2700.84 kJ/kWh, 3052.243 kJ/kWh, and 3410.664 kJ/kWh higher than diesel (Figs. 35.4 and 35.5).

35.3.3 Hydrocarbon Emission

The diesel and Eucalyptus-20 had hydrocarbon emissions of 74 ppm and 59 ppm, respectively. Generally, biodiesels are rich in oxygen content which allows them to have low hydrocarbon content in their emissions [8]. The hydrocarbon emissions of Eucalyptus-20 biodiesel, EWM1, EWM2, and EWM3 are 5 ppm, 8.34 ppm,

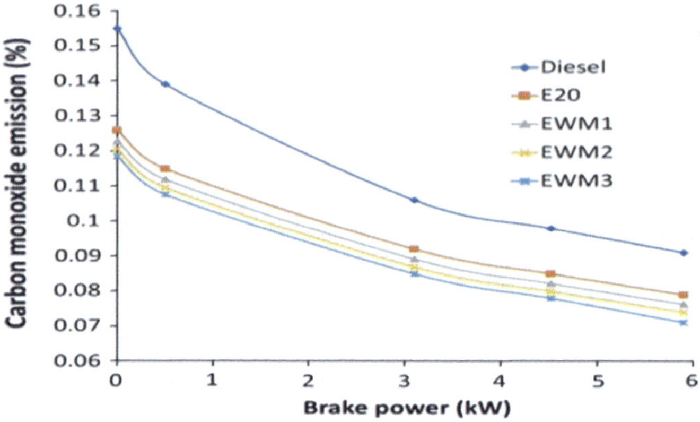

Fig. 35.5 Variation of carbon monoxide emission with brake power

10.69 ppm, and 12.827 ppm, respectively, lower than that of conventional diesel. Hence, the emission of hydrocarbon was the lowest for Eucalyptusdiesel-Water Mixture 3 (EWM3) when compared with other test fuels.

35.3.4 Carbon Monoxide Emission

The carbon monoxide emission of diesel and Eucalyptus-20 biodiesel is 0.0091% and 0.079%, respectively. The abundance of oxygen in biodiesel fuels contributes to the minimal CO emissions [9]. E20 biodiesel, EWM1, EWM2, and EWM3 had a 0.13%, 0.16%, 0.18%, and 0.079% decrement, respectively, in their CO emissions when contrast to the diesel. The emission of carbon monoxide was lowest for Eucalyptusdiesel-Water Mixture 3 (EWM3).

35.3.5 Oxides of Nitrogen Emission

Emission of oxides of nitrogen due to the combustion of the various fuels used in the conventional diesel engine is shown in Fig. 35.6. One of the most prominent causes of the formation of NO_X is due to the Zeldovich mechanism [10]. The NO_X emissions of EWM1, EWM2, and EWM3 are lower than conventional diesel by 9.7%, 12.5%, and 23%, respectively. The emission of NO_X was low for Eucalyptusdiesel-Water Mixture 3 (EWM3) when fed into the conventional diesel engine.

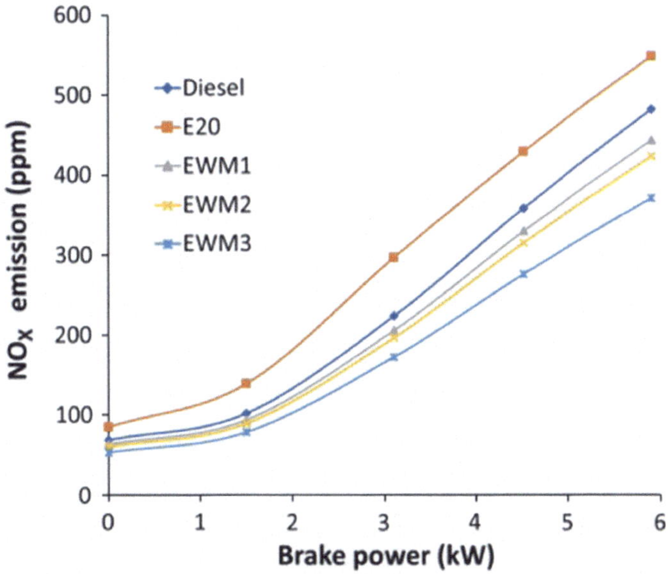

Fig. 35.6 Variation of NO_X emission with brake power

35.3.6 Smoke Opacity Emission

Smoke opacity of various test fuels, when fed into a conventional diesel engine, is shown in Fig. 35.7. Maximum smoke opacity of diesel and Eucalyptus-20 biodiesel are 53.3 HSU and 57.2 HSU, respectively. The smoke opacity emission of E20 saw an increment of 3.9 HSU when compared to conventional diesel. The fuel droplets are further atomized during the course of combustion because of presence the water in emulsion fuel. This results in low smoke opacity emission [11]. The smoke opacity emissions of EWM1, EWM2, and EWM3 are lower than conventional diesel by 1.487 HSU, 2.497 HSU, and 4.527 HSU, respectively. Therefore, emission is lowest for Eucalyptusdiesel-Water Mixture 3 (EWM3) when compared with other test fuels.

35.3.7 Cylinder Pressure Versus the Crank Angle

Cylinder pressure generated while combusting the various fuels during the diesel engine is shown in Fig. 35.8. Eucalyptus-20 had a 2.81% decrement in its peak pressure when compared to conventional diesel. Vaporization of water present in the emulsion fuel provides increment in cylinder pressure [12]. Cylinder pressure of EWM1, EWM2, and EWM3 is higher than diesel by 0.94%, 1.81%, and 2.74%,

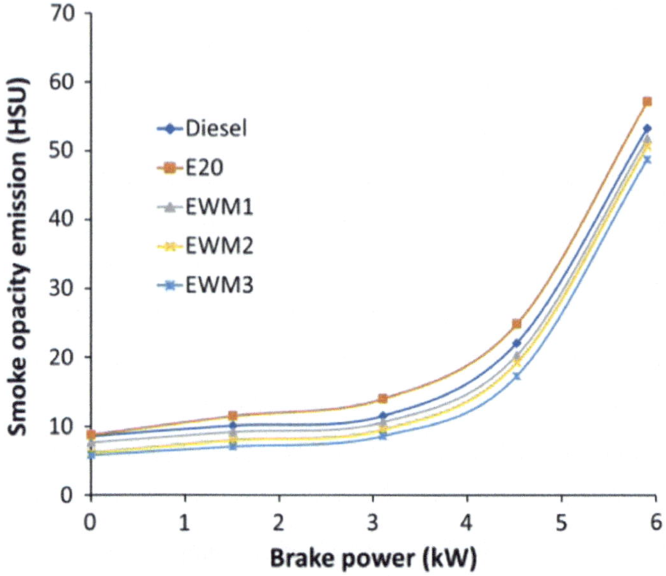

Fig. 35.7 Variation of smoke opacity emission with brake power

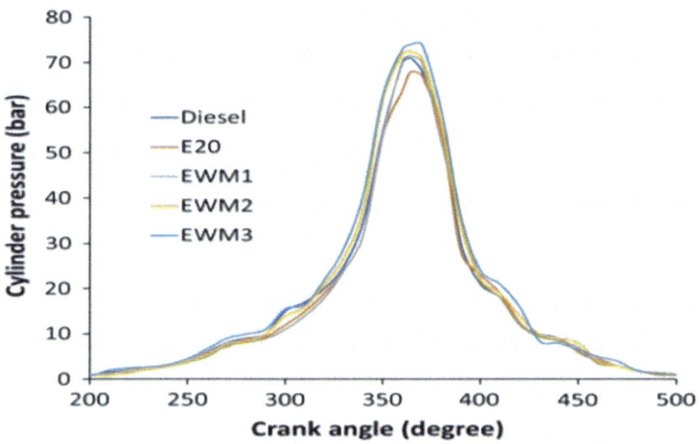

Fig. 35.8 Variation in cylinder pressure with crank angle

respectively. Hence, from the results, it was inferred that the cylinder pressure was high when Eucalyptusdiesel-Water Mixture 3 (EWM3) was fed into the conventional diesel engine.

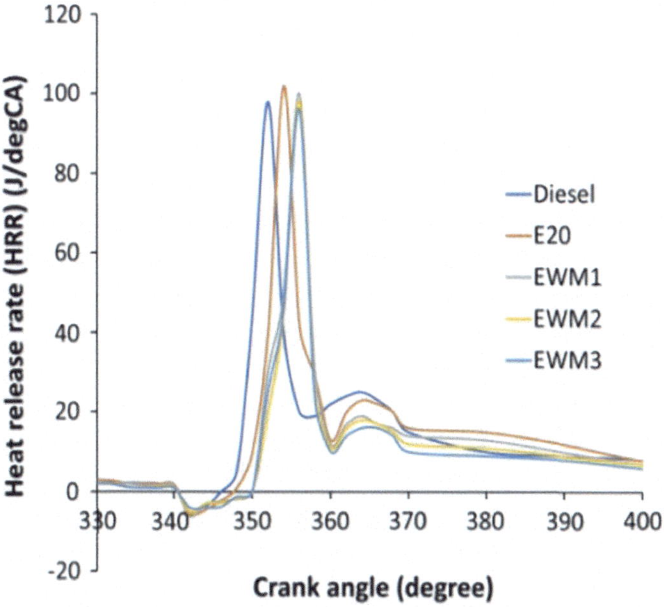

Fig. 35.9 Variation of heat release rate with crank angle

35.3.8 Heat Release Rate

The Diesel, Eucalyptus-20 biodiesel, and a eucalyptusdiesel-water emulsion in the conventional diesel engine heat release rate curve are shown in Fig. 35.9. Maximum heat release rates of the diesel and Eucalyptus-20 biodiesel are 81.3 J/deg CA and 85.3 J/deg CA respectively. Due to a low cetane number of eucalyptus-water emulsion, Eucalyptus-20 biodiesel had a lesser ignition delay when compared to eucalyptus-water emulsions [13]. From the experiments carried out, it was found that EWM2 had the least heat release rate of 78.45 J/deg CA when compared with other test fuel.

35.3.9 Exhaust Gas Temperature

Exhaust gas temperature curves are shown in Fig. 35.10. EWM3 has the least exhaust gas temperature, whereas Eucalyptus-20 biodiesel has the highest exhaust gas temperature. Because of the high viscosity of biofuels, its temperature is higher than conventional diesel [14]. Eucalyptus-20 biodiesel, EWM1, and EWM2 had a 2.77%, 1.66%, and 1.11% increment in their exhaust gas temperatures, respectively, when compared to the reference fuel. Exhaust temperature decreased by 0.83% for

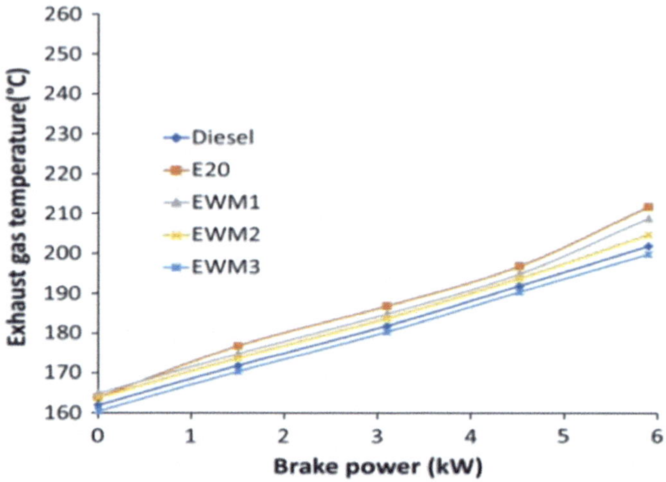

Fig. 35.10 Variation of exhaust gas temperature with brake power

Eucalyptusdiesel-Water Mixture 3 (EWM3), and hence lowest exhaust gas temperature was found for EWM3 when compared with other test fuels in the conventional engine.

35.4 Conclusion

The brake thermal efficiency of entire fuels was compared and EWM3's brake thermal efficiency was higher than diesel and all the other test fuels. The specific energy consumption was higher for eucalyptus-based fuels, Eucalyptus-20 biodiesel. Hydrocarbon emissions were high for the reference fuel and EWM3 emitted the least amount of HC and significantly lesser than conventional diesel. Carbon monoxide emissions were high for diesel and EWM3 produced the slightest decrease in CO emission. EWM3 had the least smoke opacity emissions. EWM3 emitted the least amount of NO_X when compared to other test fuels. EWM3 generated the highest peak pressure within the combustion chamber, while it yielded the least heat release rate among all the test fuels used in this experiment. Subsequently, EWM3 had the least exhaust gas temperature. Hence, EWM3 was found to be superior to all the other test fuels considering its performance.

Acknowledgements The research work was supported by Einstein College of engineering under Anna University, Tamilnadu.

References

1. Forson, F.K., Oduro, E.K., Hammond-Donkoh, E.: Performance of jatropha oil blends in a diesel engine. Renew. Energy **29**, 1135–1145 (2004)
2. Van Gerpen, J.: Green fuel processing and production. Fuel Process. Technol. **86**, 1097–1107 (2005)
3. Kumar, M.S., Ramesh, A., Nagalingam, B.: An experimental comparison of methods to use methanol and jatropha oil in a compression ignition engine. Biomass Bioenergy **25**, 309–318 (2003)
4. Attia A.M., Hassaneen, A.E.: Influence of diesel fuel blended with green fuel produced from waste cooking oil on diesel engine performance. Fuel **167**, 316–328 (2016)
5. Ramadhas, A.S., Jayaraj, S., Muraleedharan, C.: Dual fuel mode operation in diesel engines using renewable fuels: rubber seed oil and coir-pith producer gas. Renew. Energy **33**, 2077–2083 (2008)
6. Nanthagopal, K., Ashok, B., Thundil Karuppa Raj, R.: Influence of fuel injection pressures on *Calophyllum inophyllum* methyl ester fuelled direct injection diesel engine. Energy Convers. Manag. **116**, 165–173 (2016)
7. Yadav, A.K., Khan, M.E., Dubey, A.M., Pal, A.: Performance and emission characteristics of a transportation diesel engine operated with non-edible vegetable oils green fuel. Case Stud. Therm. Eng. **8**, 236–244 (2016)
8. Fazal, M.A., Haseeb, A.S.M.A., Masjuki, H.H.: Green fuel feasibility study: an evaluation of material compatibility; performance; emission and engine durability. Renew. Sustain. Renew. Energy **15**(2), 1314–1324 (2011)
9. Anbarasu, A., Karthikeyan, A.: Performance and emission characteristics of a diesel engine using cerium oxide nanoparticle blended biodiesel emulsion fuel. J. Energy Eng. **142**(1), 04015009 (2016)
10. Annamalai, M., Dhinesh, B., Nanthagopal, K., SivaramaKrishnan, P., Lalvani, J.I., Parthasarathy, M.: An assessment on performance, combustion and emission behavior of a diesel engine powered by ceria nanoparticle blended emulsified biofuel. Energy Convers. Manag. **123**, 372–380 (2016)
11. Arul Gnana Dhas, A., Devarajan, Y., Nagappan, B.: Analysis of emission reduction in ethyne–biodiesel-aspirated diesel engine. Int. J. Green Energy **15**(7), 436–440 (2018)
12. Yang, S.I., Hsu, T.C., Wu, C.Y., Chen, K.H., Hsu, Y.L.: Application of biomass fast pyrolysis part II: the effects that bio-pyrolysis oil has on the performance of diesel engines. Energy **66**, 172–180 (2014)
13. Beld, B.V., Holle, E., Florijn, J.: The use of pyrolysis oil and pyrolysis oil derived fuels in diesel engines for CHP applications. Appl. Energy **102**, 190–197 (2013)
14. Mani, M., Nagarajan, G., Sampath, S.: An experimental investigation on a DI diesel engine using waste plastic oil with exhaust gas recirculation. Fuel **89**, 1826–1832 (2010)

Chapter 36
Investigating Heat Treatment Process for Shear Failed Traction Pinion Supported Using FEA

Saurabh Sarkar and Devendra Shahare

Abstract Stress evaluation in pinions has been a complex area of research, due to its non-linear and non-uniform nature of stress distribution. This work attempts to study a shear failure analysis of forged 17CrMoNi6 grade steel traction pinion associated with an 1156 HP traction motor in the electric locomotive domain. The material of the helical pinion has been inspected using spectroscopy means to determine the grade and composition. Application of FEA software has been incorporated to support the failure results which are verified using analytical model. Stress distribution at the fracture region is analysed based on Lewis and Buckingham equations for static force measurements. A subsequent heat treatment model is also proposed to the inspected material based on the composition and the nature of failure in the pinion. Different stages associated with heat treatment model are elaborately discussed and supported, suggesting different temperature ranges and the desired phase changes to the existing material, using exponential carbon dependence equation, Koistinen and Marburger's equation with detailed composition study.

Keywords Heat treatment · Shear failure · FEA · Exponential carbon dependence equation

36.1 Introduction

The failure associated with any component of a transmission system, especially in the locomotive industry, leads to large casualties in the domain of both the economy and life. To transmit power in the mechanical system through the means of traction pinions have proved to be very effective and efficient means of transmission. These traction pinions incorporate heavy-duty application predominantly in the sub-sector of electric locomotives for passengers and goods transportation, which brings in possibilities of failure at continuous duty. The rail sector, particularly, has to look into the maintenance and cause of such failures associated with these elements. Gear manufacturers have so far been occupied with failures due to high root stresses, high

S. Sarkar (✉) · D. Shahare
Department of Mechanical Engineering, Yeshwantrao Chavan College of Engineering, Hingna Road, Wanadongri, Nagpur, Maharashtra 441110, India

S. Yadav et al. (eds.), *Proceedings of International Conference in Mechanical and Energy Technology*, Smart Innovation, Systems and Technologies 174,
https://doi.org/10.1007/978-981-15-2647-3_36

surface pressures and hardening cracks which strongly impacts on tooth geometry and admissible loads [1]. Shear stress is another parameter that is considered to all gears as it involves mutual rolling surfaces and is most often the originator of initial line pitting along the pitch line and subcase fatigue called spalling on gear teeth [2]. Shear stress runs parallel to the surface at a distance of 0.007 to 0.012 below the case surface, which is an average depth for normal loading conditions, but the actual depth of maximum stress could be deeper. The area most vulnerable, as in the present case, to torsional shear failure, is the transition zone between case and core. The study of shear at the core of pinion shank or traction pinion or a shaft is a bit complex, which occurs along the shear plane, at a direction of 45° from the central axis or longitudinal direction. Most instances of this shear failure have been seen along the shear plane of non-carburized or improper heat-treated shafts and entities. This reveals that microstructure and composition are the key factors controlling such failures which can be ameliorated by adopting a well-studied heat treatment process, wherein a proper understanding of the effect of alloying on martensitic temperature has played immense importance for the development of advanced high-strength steels. Accurate knowledge of martensitic start temperature can aid to ensure that martensite transformation occurs in the most appropriate temperature range during a corresponding quench process [3, 4]. The empirical equations for the martensitic temperature proposed by Steven [5] and Andrews [6] are probably the most widely used. Well known as 'M_s' temperature' it strongly depends on the stability of the austenite parent phase, which is primarily determined by the chemical composition of the steel [5, 6]. Although the influence of the chemical composition on the start of martensite formation in steels has been investigated extensively in the past using empirical models [7–12], the exponential carbon dependence equation stands to be relatively simple and sufficiently accurate for a wide range of carbon contents especially low carbon steels, as in this work. The prediction of martensite starts temperature can be further supported and improvised by using the Koistinen and Marburger's equation at a certain arrest temperature, minimizing the extent of microstructural fatigues.

Literature revealed there was a need for evaluation of concentrated stresses in the material at the threshold to counter these microstructural fatigues. Finite element analysis having a capability to involve complicated tooth geometry is an efficient tool to analyse and determine these stress distributions. Though finite element analysis for deformable bodies is complex and non-linear, literature provides a wide knowledge on the usage of finite element analysis (FEA) in gearing and gear stress evaluation. Chen and Tsay [13] evaluated the contact and bending stresses for a localized bearing contact in a helical gear using finite element analysis. The results provided the influence of gear's design parameter and contact position on gear stress distribution. Hwang et al. [14] presented a contact stress analysis for spur and helical gears between two gear teeth at different contact positions during rotation.

Through literature, it was inferred that the majority of research work concentrated on the prediction of wear in quasi-static conditions mostly based on Archard's model for sliding wear and on dissipated power densities at the mesh [15, 16], and there had been significantly no evident model proposed to cope with failures based on interdisciplinary domain of FEA and heat treatment. In the present work, the authors have proposed a heat treatment process based on the analysis reported by finite element analysis. The heat treatment, supported by exponential carbon equation, provides a probable solution to the shear failure in a helical traction pinion through detailed material investigation to avoid jeopardize through material failure in future.

36.2 Findings

Recognizing the visual signs associated with typical failure modes is integral to analysing and diagnosing gear failures. In this case, a 21 teeth helical traction pinion was investigated, which was manufactured through forged roots. The pinion was operated for transmission of about 850 kW at a proposed gear ratio of 107:21. The dimensions of the pinion were measured as shown in Table 36.1. Pinion's pre-failure and its location with respect to traction motor are shown in Fig. 36.1a, b. Investigation revealed severe damage by total loss of 18 teeth and the rest were badly deformed with cracked finned edges at the tip as shown in Fig. 36.1c. Complete shear failure

Table 36.1 Specification of the failed traction pinion

Number of teeth	21
Module	7.20 mm
Pressure angle	24°
Angle of helix	5.972°
Material	17CrNiMo6
Young's modulus	206 GPa
Yield strength	785 MPa
Ultimate tensile strength	930 MPa
Poisson's ratio	0.3

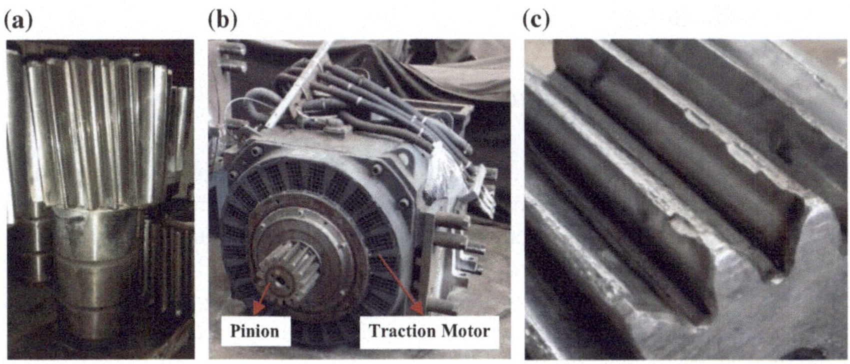

Fig. 36.1 **a** Traction pinion before failure, **b** location of pinion in traction motor, **c** cracked finned edges

Fig. 36.2 Failure of traction pinion through shear at conjoining area

Table 36.2 Composition of failed traction pinion revealed through spectroscopy

Elements (%)	C	Si	Mn	P	S	Cr	Ni	Mo	Nb*
Composition (%)	0.167	0.253	0.527	0.015	0.006	1.656	1.539	0.257	0.008

Remark The Elements marked with an * are OUT OF NOBLE SCOPE

is obtained at the conjoining area or the 'neck' of the traction pinion as shown in Fig. 36.2.

In order to further evaluate the source of the failure, the material of the traction pinion was inspected by spectroscopy. The material was tested in accordance with ASTM-E-415: 2015 standard using Q4-TASMAN; LJRO2/Jr.CCD type spectroscopy from MIA test laboratory and research centre from Nagpur, India. Results revealed the material to be 17CrNiMo6 with the composition shown in Table 36.2.

36.3 Methodology

36.3.1 Finite Element Analysis

Parametric modelling is adopted using a cartesian system to create an involute tooth profile, under the SolidWorks© parametric solid modelling. Finite element method is used to simulate deformation behaviour and stress distribution on the pinion during transmission operation, considering the worst loading conditions using ANSYS Workbench® 15.0.7. Material is considered isotropic with negligible heat generation and thermal stress as necessary assumptions. Material properties of 17CrMoNi6 grade steel as shown in Table 36.1.

Boundary conditions were applied on the helical pinion as shown in Fig. 36.3 based on Lewis and Buckingham model where the pinion shaft is considered as fixed

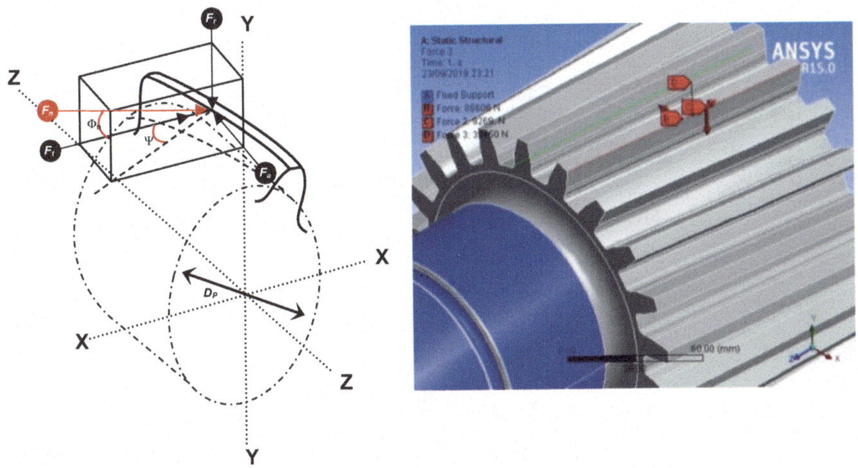

Fig. 36.3 Boundary and loading condition for analysis for the traction pinion

constraint and static forces were applied along the line of action as shown in Fig. 36.6. Linear static forces were applied on the teeth section in axial (F_a), radial (F_r) and tangential (F_t) direction. The magnitude of forces applied has been theoretically evaluated using Eqs. (36.1), (36.2) and (36.3) [17].

$$F_a = F_n \cos \Phi_n \sin \Psi \tag{36.1}$$

$$F_r = F_n \sin \Phi_n \tag{36.2}$$

$$F_t = F_n \cos \Phi_n \cos \Psi \tag{36.3}$$

where, Ψ, Φ_n and F_n are the helix angle, normal pressure angle and resultant force, respectively.

36.3.1.1 Mesh Independency Study

To establish the accuracy of the simulation solution and to keep the computational costs low, the helical pinion was analysed for mesh independence. The grid convergence study was performed by investigating six meshing conditions from M1 to M6 wherein M1 and M2 represent coarse, M3 and M4 medium and M5 and M6 represent fine mesh. Triangle meshing was used for all the six meshing conditions under the same boundary conditions and model discussed in the previous discussion.

The mesh size was reduced in sixfolds from 40 to 1 mm and corresponding interpretation of results is done at each stage. The highest nominal stress obtained at convergence is 1879 MPa at M4 for the above model. M5 and M6 accounted for

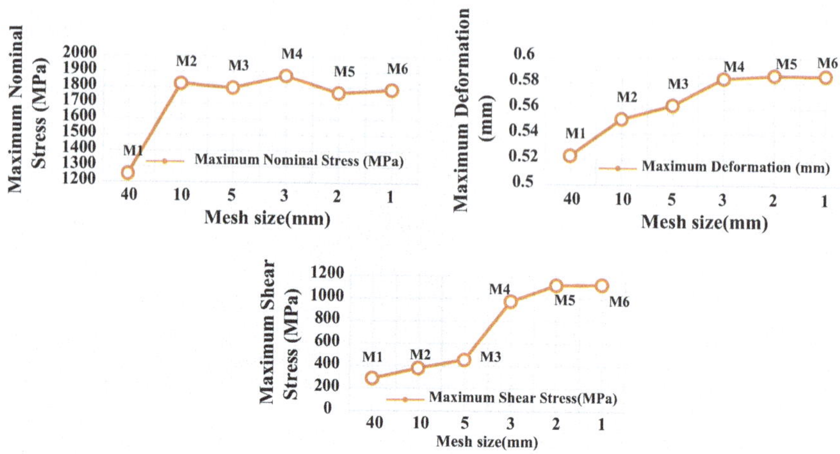

Fig. 36.4 Grid independency test for of all the investigated meshes

0.46% change in the estimated nominal stress values showing grid independency as shown in Fig. 36.4. A similar trend is inferred for maximum deformation and shear stress with 0.13 and 0.3% change in results at meshing condition M5 and M6 showing grid independency. Taking into consideration the computational cost and time incurred during the simulation and the convergence results obtained at meshing condition M5, element size at 2 mm with nodes 452,714 nodes and 277,560 elements is employed for further numerical analysis.

36.3.2 Heat Treatment

This study is primarily focused on proposing carburization, quenching and tempering processes for helical traction pinion to curb the obtained shear failure in future operation. The probable role of an individual alloying element in 17CrNiMo6 on deciding working temperature in each process is discussed, and corresponding conclusions are made on the later text. Data proposed by Gegner [18] showed depreciation of the diffusion of carbon owing to the carbide forming elements Cr, Mn and Mo which reduces activity in austenite, diffusion acceleration and a rise in the carbon activity by Co in the used material. According to Krishtal [19], above 1000 °C, Si decreases the diffusivity of carbon whereas below this temperature it significantly increases the diffusion coefficient [19], an alloy addition of 0.5, 1, 1.5 and 2 wt% Cr at 900 °C approximately corresponds to the carbon diffusion coefficient in binary Fe–C austenite at 868, 837, 810 and 782 °C, respectively. Thus, the effect of 0.5–2 wt% Cr is thus comparable with a temperature reduction of about 30–120 °C at the initial heating stage [20, 21]. The austenite stabilizing alloys, i.e. nickel (Ni) and manganese (Mn), have a potential of lowering the eutectoid temperature, thereby widening the temperature range over which austenite is stable, i.e. γ-phase [22] felicitating carburizing temperature above 800 °C.

36.4 Discussion

Stress distribution at mesh convergence along the teeth has been found to be asymmetrical wherein the maximum stress values are found to occur at the load point. Typical plots are shown in Fig. 36.5 for the failed pinion. This maximum stress value decreases towards the middle up to the root of the subjected tooth under study.

The numerical results extracted from the simulation for the failed pinion reveal maximum von Mises stress is to be 1772.5 MPa at a safety factor is 1.93 at the tip of the tooth section, whereas the shear stress value is approximately 1130 MPa at the adjoining neck region. It is observed that these maximum stress values are inordinate compared to the yield stress of the material (785 MPa) under maximum load which very strongly supports the cause of failure in this case.

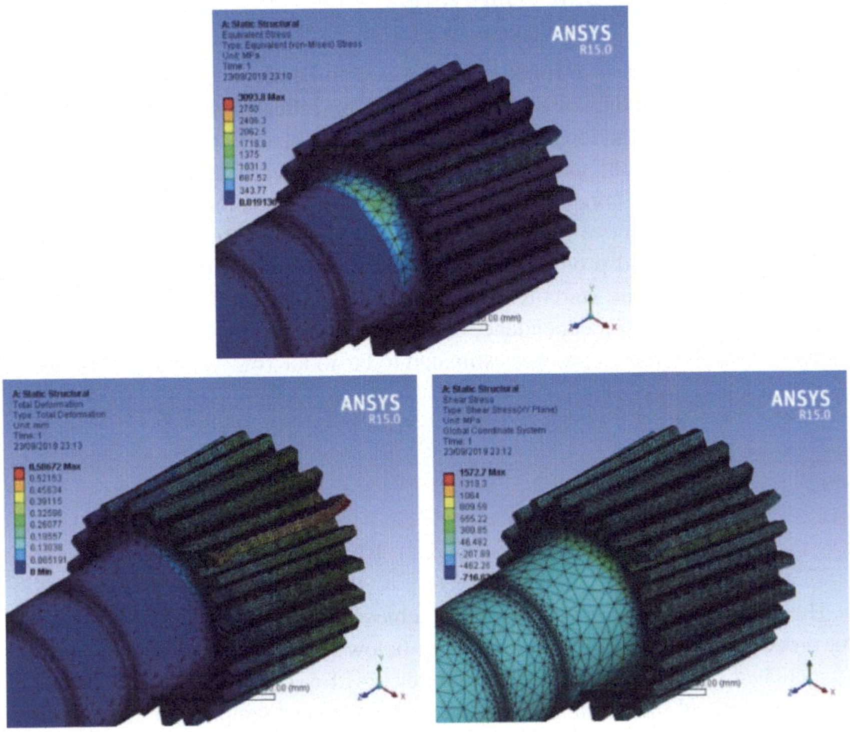

Fig. 36.5 FE contour plot for different stresses at loaded condition

36.4.1 Pre-heating and Gas Carburizing

A conventional pre-heating is proposed to at 480 °C in order to reduce thermal shocks, excess moisture and stress leaks during the carburizing process [23]. Steel 17CrNiMo6 revealed 0.15–0.20% carbon content which makes gas carburization as the preferred heat treatment process [24]. An effective gas carburization to the traction pinion will ensure a 30–50% increment in the load handling capacity of the hardened gear. Based on the heat treatment methodology discussed and have satisfactory carbon diffusion the pinion, steel 17CrNiMo6, is proposed to be carburized below 1000 °C. In addition to this, a substantial presence of chromium content (1.8%) in the material demands a comparative reduction in this temperature (1.8%) by 80–100 °C [19], owing to a gas carburizing temperature close to 900–920 °C. This temperature ensures high diffusivity of carbon assisted by the high silicon content in the steel (0.4%). Based on the model [25], carburization should be performed for about 7–8 h, imparting the desired hardness of about 40 Rc at the core and produce surface carbon concentration approximately amounting to 0.6–0.85 wt% C.

36.4.2 Quenching and Tempering

Uniform martensitic transformation is expected to be achieved through rapid and even oil medium cooling of traction pinion in a vertical position by suppressing ferrite and pearlite reactions, as well as other intermediate reactions such as the formation of bainite [26]. Oil quenching is preferred over water quenching to avoid excess hardness at the core (not more than 40 Rc), which is felicitated by the lower heat transfer coefficient of oil than that of water. This will ensure brittleness does not reach the core and produce risks of cracking and distortion after the heating process. The martensite starts temperature (M_s) is calculated to be about 383 °C using exponential carbon dependence as expressed in Eq. (36.4) and corresponding martensite finish temperature was estimated to be about 168 °C [24, 25, 27–29].

$$M_s(^\circ\mathrm{C}) = 565 - \Sigma K_i x_i - 600\left[1 - \exp(-0.96x_c)\right] \tag{36.4}$$

where

$$\Sigma K_i x_i(^\circ\mathrm{C}) = 31x_{\mathrm{Mn}} + 13x_{\mathrm{Si}} + 10x_{\mathrm{Cr}} + 18x_{\mathrm{Ni}} + 12x_{\mathrm{Mo}} \tag{36.5}$$

The volume fraction of martensite (f) achieved at the end of the above quenching process as a function of undercooling below the martensite start temperature is calculated using Koistinen and Marburger [24] Eq. (36.6), where the difference in actual (T_s) and theoretical reading of martensite start temperature (M_s) is taken approximately 20 °C. 34.783% volume fraction of martensite is expected at the end of the proposed process. The rate parameter (α_m) depends on composition of material and Eq. (36.6) is used to comprehend this value with least discrepancies with experimental data [29].

$$f = 1 - \exp[-\alpha_m(M_s - T_s)] \quad (36.6)$$

$$\alpha_m(*10^{-3}\,K^{-1}) = 27.2 - \Sigma S_i x_i - 19.8[1 - \exp(-1.56x_c)] \quad (36.7)$$

with,

$$\Sigma S_i x_i(°C) = 0.14x_{Mn} + 0.21x_{Si} + 0.11x_{Cr} + 0.08x_{Ni} + 0.05x_{Mo} \quad (36.8)$$

In Eqs. (36.4), (36.5), (36.7) and (36.8) $x_c, x_{Mn}, x_{Si}, x_{Cr}, x_{Ni}$ and x_{Mo} are percentage composition of carbon, manganese, silicon, chromium, nickel and molybdenum, respectively, in the helical pinion (Fig. 36.6).

At the end of the quenching process, post-heat treatment is proposed for the microstructure of pinion to attain equilibrium under the influence of thermal activation [25, 27]. Tempering at a temperature of about 100 °C, below the lower transformation temperature or under lower critical temperature, for a threshold time of 1 h is presented. This ensures a balance of hardness, strength and toughness of the pinion during operation [28]. Tempering would not only impart higher fatigue resistance but would also allow it to relieve such shear stresses at the core and inhibit drastic failure as in this case.

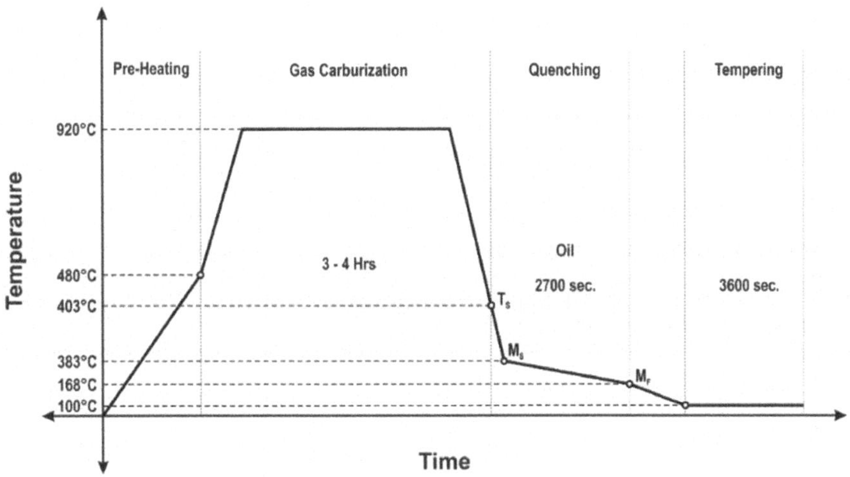

Fig. 36.6 Proposed heat treatment model as function of time and temperature

36.5 Conclusions

- A simulation-driven stress analysis case study is described, which contributes to further research into the utilization of simulation technology for the automotive industry. Maximum shear stress value of 1130 MPa and a maximum deflection value of 0.5867 mm have been obtained at the neck junction. Maximum von Mises stress in the hub is noticed to be 1772.5 MPa which is much above the expected ultimate strength value of the traction material showing traces of potential failure at the teeth.
- A heat treatment model is proposed to curb the shear failure by considerable increment in load handling capacity (up to 30%), increment in carbon % (up to 0.8%) and infusion of hardness (up to 40 Rc) at the core. Exponential carbon dependence equation was evident to provide considerable information regarding % volume of martensite phase formation achieved. This was supported by Koistinen and Marburger's equation stating the 'start' and 'finish' martensitic temperature to be adopted to do so. The proposed model accounts for the effect of most typical substitutional elements to create a benchmark for future application.
- This FEM-based analysis along with heat treatment model can help in reducing turnaround time and cost associated with frequent material change and handling. Additionally, this approach can provide useful information on stresses, deflection and tooth loads that are needed to design strong and quiet transmission parts in the automotive sector. Through this work casualties due to catastrophic failures can be reduced and prove advantageous in maintenance planning.

References

1. Flodin, A., Andersson, S.: Simulation of mild wear in spur gears. Wear **207**, 16–23 (1997)
2. James, A.: Failure Analysis Report. Advanced Engineering and Testing Services. CSIR (1994)
3. Bhadeshia, H.K.D.H.: Bainite in Steels. The Institute of Materials, London (2001)
4. Speer, J.G., Edmonds, D.V., Rizzo, F.C., Matlock, D.K.: Partitioning of carbon from supersaturated plates of ferrite, with application to steel processing and fundamentals of the bainite transformation. Curr. Opin. Solid State Mater. Sci. **8**, 219–237 (2004)
5. Steven, W., Haynes, A.G.: The temperature of formation of martensite and bainite in low-alloy steels. J. Iron Steel Inst. **183**(8), 349–359 (1956)
6. Andrews, K.W.: Empirical formulae for the calculation of some transformation temperatures. J. Iron Steel Inst. **203**, 721–729 (1965)
7. Grange, R.A., Stewart, H.M.: The temperature range of martensite formation. Trans. AIME **167**, 467–490 (1946)
8. Nehrenberg, A.E.: The temperature range of martensite formation. Trans. AIME **167**, 494–498 (1946)
9. Zhao, J.: Continuous cooling transformations in steels. Mater. Sci. Technol. **8**, 997–1004 (1992)
10. Lee, J.K.: Prediction of Tensile Deformation Behavior of Formable Hot Rolled Steels. Posco Technical Research Laboratories Report, Pohang (1999)
11. Gorni, A.A.: Forming and Heat Treating Handbook, pp. 100–102. Sao Vacente, Brazil (2012)
12. Olson, G.B., Cohen, M.: A general mechanism of martensitic nucleation: part III. Kinetics of martensitic nucleation. Metall. Trans. A **7**, 1915–1923 (1976)

13. Chen, Y.C., Tsay, C.B.: Stress analysis of a helical gear set with localized bearing contact. Finite Elem. Anal. Des. **38**, 707–723 (2002)
14. Hwang, S., Lee, J., Lee, D., Han, S., Lee, K.: Contact stress analysis for a pair of mating gears. Math. Comput. Model. **57**(1–2), 40–49 (2013)
15. Brauer, J., Andersson, S.: Simulation of wear in gears with flank interference—a mixed FE and analytical approach. Wear **254**, 1216–1232 (2003)
16. Onishchenko, V.: Tooth wear modeling and prognostication parameters of engagement of spur gear power transmissions. Mech. Mach. Theory **43**, 1639–1664 (2008)
17. Gopinath, K., Mayuram, M.M.: Machine Design II, IITM, Module 2, Lecture 11, 15–17. http://nptel.ac.in/courses/IITMADRAS/Machine_Design_II/pdf/2_11.pdf
18. Gegner, J.: Analytical modeling of carbon transport processes in heat treatment technology of steels. In: Proceedings of 3rd International Conference on Mathematical Modeling and Computer Simulation of Materials Technologies (MMT), Chap. 1, pp. 95–106, Ariel, Israel, 6–10 Sept 2004, College of Judea and Samaria (2004)
19. Krishtal, M.A.: Diffusion Processes in Iron Alloys. Publishing House of the Literature on Ferrous and Nonferrous Metallurgy, Moscow (1963)
20. Blazek, K.E., Cost, J.R.: Carbon diffusivity in iron-chromium alloy. Trans. JIM **17**(10), 630–636 (1976)
21. Hajduga, M., Kučera, J.: Decarburization of Fe–Cr–C steels during high-temperature oxidation. Oxid. Met. **29**(5/6), 419–433 (1988)
22. ASM Handbook, vol. 4, Heat Treating. ASM International, Cleveland (1991)
23. Payares-asprino, M.C., Katsumoto, H., Liu, S.: Effect of martensite start and finish temperature on residual stress. Weld. J. **87**, 279–289 (2008)
24. Koistinen, D.P., Marburger, R.E.: A general equation prescribing the extent of the austenite-martensite transformation in pure iron-carbon alloys and plain carbon steels. Acta Metall. **7**, 59–60 (1959)
25. Bhadeshia, H.K.D.H., Honeycombe, R.W.K.: Steels: Microstructure and Properties. Elsevier Ltd., Amsterdam (2006)
26. van Bohemen, S.M.C.: Bainite and martensite start temperature calculated with exponential carbon dependence. Mater. Sci. Technol. **28**(4) (2012)
27. Bhadeshia, H.K.D.H.: Design of ferritic creep-resistant steel. ISIJ Int. **41**(6), 626–640 (2001)
28. Kumar, R.: Effect of tempering temperature and time strength and hardness of ductile cast iron. IOP Conf. Ser. Mater. Sci. Eng. **75**(01) (2015)
29. Van Bohemen, S.M.C., Sietsma, J.: Martensite formation in partially and fully austenitic plain carbon steels. Mater. Sci. Technol. **25**, 1009–1012 (2009)

Chapter 37
A Review on Processes of Fabrication and Properties of Nano-hybrid Metal Matrix Composites

Mukesh Kumar, R. K. Gupta, Anand Pandey, Rahul Goyal and Ashish Goyal

Abstract This review paper tends to analysis nano-hybrid metal matrix composites with a description of their various fabrication processes as well as their comparative studies and some detailed description of powder metallurgy and stir casting. Owing to the higher specific properties, nano-hybrid metal matrix composites are quite useful in number of industries like aviation, military and automobile, etc., and due to this reason, it is most important topic between researchers. There are a lot of scopes in this field and have a very wide range of possibilities.

Keywords Nano-hybrid · Metal matrix · Composites · NHMMC · Reinforcement

37.1 Introduction

Composites are used for the better mechanical properties and their easy fabrication process. In this regard, metal matrix composites are considered as quite reliable material possessing the properties of low density, high-hardness high-specific strength, better resistance to wear, and also tractable expansion coefficient. It has high prospective to be applied in aerospace technology, automobile, and defence industries because of the existence of Aluminum-centered metal matrix composites (MMCs) [1–4].

Silicon carbide-reinforced aluminum matrix composites have been more focused on among the discontinuous reinforced metal matrix compounds due to intensified mechanical, manageable thermal properties, and less fabrication cost comparatively [5–9].

Engineering combination of reinforcement can be two or more in hybrid composites like as fibers, whiskers, and nanotubes. This can have variety of materials for reinforcement such as (Graphite and SiC), (SiC, Al_2O_3) and (Al_2O_3 and Graphite), and many more. For example, graphite and alumina are used in car engine block as particulates [10–12]. Because of accelerated interfacial area, hybrid metal matrix composites depict better mechanical attributes, the meniscus penetration fault depleted,

M. Kumar (✉) · R. K. Gupta · A. Pandey · R. Goyal · A. Goyal
Department of Mechanical Engineering, Manipal University Jaipur, Jaipur, Rajasthan 303007, India

S. Yadav et al. (eds.), *Proceedings of International Conference in Mechanical and Energy Technology*, Smart Innovation, Systems and Technologies 174,
https://doi.org/10.1007/978-981-15-2647-3_37

and the making of inter-metallic component at interfaces gets reduced [13]. There is a very little resource of research being conducted on hybrid composites based on aluminum. Also investigations on usage of Al alloys with hybrid composition of Al_2O_3 fibers + Al_2O_3 particulates, SiC particles + graphite fibers, carbon short fibers + Al_2O_3 short fibers, Al_2O_3 fibers + SiC particles, BPO_w + WO_3 particles, $ABO_w(Al_{18}B_4A_{33})$ + $BPO_p(BaPbO_3)$, glass fibers + SiC fibers, and ABO_w + SiCp or reinforced aluminum were found [14–19].

Number of nano-metric phases in literature has been paired with matrix metal of different kinds. There is a reference of ceramic compounds (Al_2O_3, SiC, etc.), carbon allotropes, and inter-metallic materials being used to reinforce Al, Cu, Mg, and other many metals and alloys. Carbon nanotubes (CNT) have been given specific predominance due to attributes like electrical conductivity and hardness, as these characteristics deliberate comparatively better mechanical power in the procedure of making the thermal and electrical properties of base material fitter [20–22]. Moreover, MMnCs proved to be very useful in ameliorating yet other related characteristics, such as moisten capacity [23, 24], resistance to wear [25], and rogue conduct [26].

This paper is beneficial for analyzing the theory as well as practical-based background concerned to abundance of MMnCs and the greatest results gained in this field. This paper will describe the properties related to structural and mechanical performance caused by nanoparticle and the synthesis methods.

In order to detect the consequences of both stiff and delicate solid lubricant particle reinforced aluminum matrix, few researchers studied hybrid MMCs [27]. Basavarajappa et al. examined the inclusion of graphite reinforcement in Al2219/15SiCp composites that escalate surface harshness of composites comparative to graphite-free composites. Presence of graphite increases the surface roughness which is acknowledged as the reason of presence of graphite which abrogates the acuminating effect and thus contributes in reducing the coefficient of friction among tool and work piece. Suresh et al. [28] studied the functioning of graphite and in his report said that, the graphite's MRR-reinforced particulate metal matrix composite is superior in comparison to ceramic particle-reinforced composites. It is an easy mechanism to machine Al–SiC–Graphite hybrid composites with relatively less tool wear and maximum MRR by making use of a more mass fraction of graphite. The Al_10% (SiC_Graphite) MMCs has better machinability, flank wear, maximum MRR, minimum surface roughness in each and every cutting condition in comparison with 5 and 7.5% of SiC_Graphite reinforcement [29].

Fundamentally, MMCs are metallic alloys strengthened predominantly with ceramic materials. It is quite common that alloys of Al, Mg, and Ti (light metals) are used as metallic alloys although many metallic alloys have been used [30, 31]. The reason of aluminum being utilized the most metallic alloy as matrix materials in the development of MMCs have been reported [32–34]. In the similar fashion, in literature, there is mention of using interrupted ceramic particulates or whiskers rather than uninterrupted ceramic fibers to generate aluminum matrix composites (AMCs) [35–37]. The biggest problem in front of developing nations is the elevated cost and scarce supply of traditional ceramic reinforcing materials related with the progress of interrupted reinforced aluminum matrix composites (DRAMCs) [38].

Some another compositions related to DRAMC which are subject of attention to researchers are lowly adaptability, lesser breakage firmness and not been able to forecast the corrosion conduct of AMCs [36–40].

37.2 Nano-hybrid Metal Matrix Composite Fabrication Processes

In this section, we will see the comparative study of different types of fabrication processes used for development of nano-hybrid metal matrix composites. It will easier to under these process parameters and their difference by Table 37.1.

There are lots of procedures used for nano-hybrid metal matrix composites fabrication, but in this study, we will discuss two methods, i.e., powder metallurgy and stir casting.

37.2.1 Powder Metallurgy

In order to complete the study by Ravindran et al. [41] prepared four kinds of nanocomposites (Refer Table 37.2). Blend of Al2024 and Gr and SiC powders used were pulverized in a planetary ball mill for manufacturing hybrid Al nanocomposite. The above-mentioned four kinds of powder blend have undergone a planetary ball milling machine of high energy using tungsten balls and vials. The mechanical alloying was performed at 300 rpm using a toluene medium to grind the powder up to 60 h having a ball to powder weight ratio of 10:1 Tungsten balls of a diameter of

Table 37.1 A study on various methods of fabrication

Fabricating methods	Volume fraction (range)	Wear of reinforcement	Expenditure	Shape and size (range)
Powder metallurgy	–	Reinforcement fracture	Expensive	Wide range and restricted by size
Mechanical stirring	0.45–0.70	Small damage	Medium	No size limit
Electromagnetic stirring	0.55–0.85	No damage	Comparatively expensive	No size limit
Squeeze casting	Up to 0.47	Severe damage	Comparatively expensive	Limited up to 2 cm of height (by perform shape)
Spray casting	00.25–0.75	–	Expensive	Limited shape and large size

Table 37.2 Hardness, density, and BHN value (average) of the composites

S. No.	Density of composites (g/cm^3)	Hardness	BHN
I	A12024	2.84	83
II	A12024/5%SiC nanocomposite	2.89	86
III	A12024/5%Gr/5%SiC hybrid nanocomposite	2.85	82
IV	A12024/10%Gr/5%SiC hybrid nanocomposite	2.83	80

10 mm have been used to avoid oxidation or sticking of powders on the wall of the vial [42].

Laser particle size analyzer applied by them to measure the size dispensation of grinded powders. Figure 37.1 exhibits the particle size dispensation of A12024, A12024/5%SiC hybrid nanocomposite, A12024/5%Gr/5%SiC hybrid nanocomposite, A12024/10%Gr/5%SiC/hybrid nanocomposite powders. HITACHI S 3000H is used for scanning electron microscopy (SEM), the powder particles were inspected in morphology point of view.

The SEM images of mixture of hybrid nanocomposite powder can be seen in Fig. 37.2, after different milling times. It is clearly noticeable (Fig. 37.2a) that the as-obtained aluminum particles are comparatively round, but asymmetrical framed aluminum particles are established after 2 h of milling (Fig. 37.2a). On further grinding of the powder of composite for 4 h, flattened aluminum particles like plate are formed (Fig. 37.2b). After 8 h of milling, the Al particles formed were more flattened

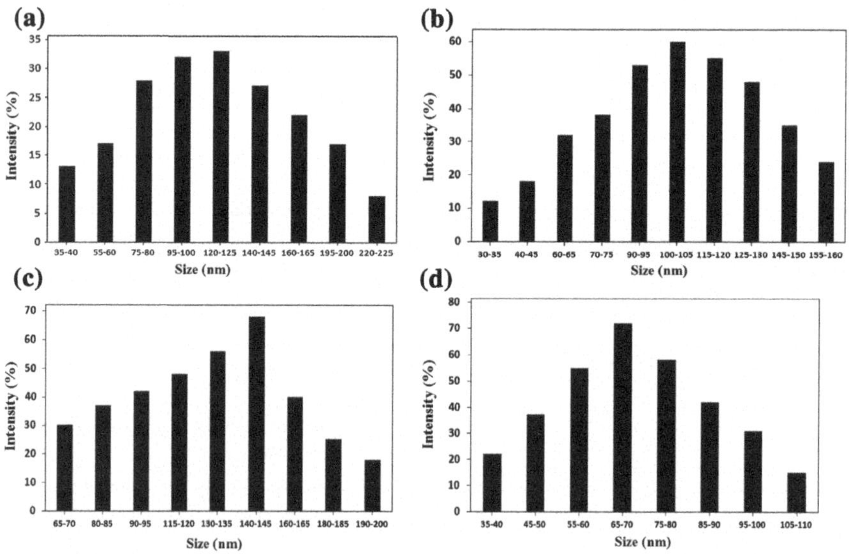

Fig. 37.1 Nano-powders distributions based on different particle size. **a** A12024, **b** A12024/5%SiC, **c** A12024/5%Gr/5%SiC, **d** A12024/10%Gr/5%SiC [41]

Fig. 37.2 Different particle sizes of hybrid composite powder on milling of **a** 2 h, **b** 4 h, **c** 8 h, **d** 16 h, **e** 24 h, **f** 36 h and **g** 40 h [41]

(as shown in Fig. 37.2c) and in milling duration of 16 h, the particles disfigure in a flat shape (Fig. 37.2d). The particles having huge size and flake shape in the 24 h milling due to adaptable quality of the aluminum powder (Fig. 37.2e) and as consequence of 36 h milling there occurs flake-like morphology but, average particle size was decreased (Fig. 37.2f). After 40 h, a minor feature ratio could be notified, and the size distribution has a limited range. As shown in Fig. 37.2a, the symmetric distribution may be evident, describing the equilibrium between welding, and fracture distinctive form of endmost phase of mechanical alloying. The adjacent allotment emerges owing to the appearing of welding and fracture on the sky-high particles [41, 43]. The materials used in the work of Masroor et al. [44] on commercial scale were pure Cu (<75 μm, 99.50%), Cr (<75 μm, 99.50%) and multi walled CNTs (about 10 μm in length, 10–30 nm in diameter) having purity of about 90%. In order to terminate the CNTs lumps, CNTs were treated in ethanol for up to 90 min. There used to be a mechanical grinding of powder mixtures in a considerably ball mill of high energy (PM2400) with desensitized steel vial (volume ~ 300 ml) and 7 balls (dia ~ 20 mm) under argon (Ar) atmosphere.

At first step, Cu–1%Cr solid solution used to be made as the matrix stage. The milling speed was 300 rpm and ball to powder mass ratio (BPR) were 30:1. Toluene (1%) was used as an agent to control process. Milling was performed up to 30 h. This has been established using the X-ray diffraction (XRD) that 20h milling was enough to attain slow state in making of Cu–Cr solid solution. Further step shows, Cu–Cr solid solution alloy and 5% of CNTs was dry milled for 5 h. Three various milling circumstances were examined for disclose consequence of milling energy on distribution of carbon nanotubes and the similar sample decides as dry milling was wet milled in 10 ml of ethanol (99.7% pure) to assess the effectiveness. Further, in order to evaporate ethanol, the milled samples were dried at 80 °C about 1 h. It is evident from these studies that milling time as well as speed presents main role in fabrication of metal matrix composites. All results shown in the SEM were easy to understand and they explain their particle sizes after hours of milling. Another important fabrication process is stir casting and we will discuss it in following section.

37.2.2 Stir Casting

An indigenously developed friction-stir welding machine (40 kN, 11 kW) was brought in use for fabrication of surface composite layer. By using gentle steel backing plate, samples were secured on the hydraulic fixture. The rotational speed at 1000 rpm and persistent diagonal pace at 25 mm/min were secured after various trials. There was a change in the progressive and receded sides of the core metals after friction-stir process qualify. Then, the specimen was brought down to room temperature for every pass and each experiment was performed at same temperature [45] (Fig. 37.3).

In general, the stir casting process deals with melting the matrix material at liquefying temperature in cauldron being chemically dormant to the materials which are further be mixed into it. There may be different types of cauldron but coke fired

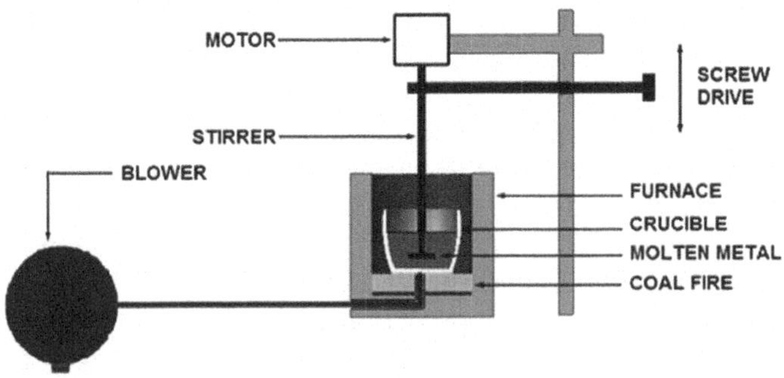

Fig. 37.3 Basic stir casting process [46]

is the most basic type being used. And in terms of furnace, Electric resistance furnace is the most commonly used furnace. To better its amalgamation with the matrix material and to circumvent the mismatching of thermal, it is required to preheat the particulates. Before mixing the particulates, the melt may or may not be stirred. That melted alloy material will depend on cauldron temperature. There is variation based upon different weight percentages (reinforcement) in context to the mixing of the particulates, i.e., either it is two-step or three-step [47].

There was successful fabrication of friction-stir process (FSP), Al/B_4C metal matrix composites. An investigation took place to examine the effectiveness of the FSP passes and reinforcing particle type on the micro stiffness, microstructure, wear properties, and tensile power of the fabricated surface.

1. There was a clear evidence of higher hardness, ultimate strength, fine grain size, and wear rate microstructure in the FSP-used specimens with nano B_4C particles three passes in comparison to the stiffness of the base material, ultimate power, and rate of wear.
2. The micro hardness of the Al/B_4C is comparatively greater than B_4C micro particles. Ultrafine grain size is generated due to existence of nano-sized B_4C particles.
3. In the Aluminum matrix, the portion of nanoparticles becomes uniform comparatively to one pass FSP nanocomposites, by elevating FSP passes and as a result, there was increase in hardness.
4. The addition of B_4C nano-sized particles improved wear properties of the Al5083 alloy in comparison of B_4C micro-sized particle and as a consequence, the wear resistance property of nano SCL was more than unreinforced Al5083 alloy.
5. The usage of nano-sized tensile specimen on stir zone by the FSP is accepted as better than as received Al5083 alloy in terms of mechanical properties [45]. As presented by P. Madhukar et al., the hybrid composites performance is better than nano-sized reinforced based composites and another existing traditional composites specifically when those are fabricated by stir casting route due to fine firmness between inter-particle spacing effects and reinforced particles to increase the tribological and mechanical properties due to invariable and good distribution of nano-ceramic particles were provided [48].

37.3 Properties of NHMMCS

The determination of the hybrid composites properties is by the reinforcement of selection each property of nano-sized particles and the metal matrix. The most common property is the mechanical property due to its adaptability, power, and resistance of impact, stiffness, and toughness of fracture. In order to accelerate the mechanical properties [49–56], ceramic particles are used, such as Alumina, B_4N, SiC, and TiC, and are deteriorated by elevating rice husk ash (RHA), bamboo leaf ash (BLA), fly ash (FA), and mica [57–64].

37.4 Conclusion

This study helps to understand the importance of nano-hybrid metal matrix composites. As after the study, we are able to conclude that there is a large scope to explore more in this field to study and research as quite few researchers have gone through nano-hybrid metal matrix composites. Moreover, their specific properties are very lucrative for our future implementation in leading industries like aerospace, military, and automobile. Further, their comparative study on fabrication process gives the idea to understand about the appropriate method to select for fabrication. There is also mention of powder metallurgy and stir casting as they are very popular in researchers due to their cost and process. In the future, we will study on various machining process used for nano-hybrid metal matrix composites.

References

1. Torralba, J.M., da Costa, C.E., Velasco, F.: P/M aluminum matrix composites: an overview. J. Mater. Process. Technol. **133**, 203–206 (2003)
2. Surappa, M.K.: Aluminium matrix composites: challenges and opportunities. Sadhana-Acad. Proc. Eng. Sci. **28**, 319–334 (2003)
3. Zhang, X., Hu, T., Rufner, J.F., LaGrange, T.B., Campbell, G.H., Lavernia, E.J., et al.: Metal/ceramic interface structures and segregation behavior in aluminum-based composites. Acta Mater. **95**, 254–263 (2015)
4. Guo, B., Zhang, X., Cen, X., Wang, X., Songa, M., Ni, S., Yi, J., Shen, T., Du, Y.: Ameliorated mechanical and thermal properties of SiC reinforced Al matrix composites through hybridizing carbon nanotubes. Mater. Charact. **136**, 272–280
5. Tuber, H., Degischer, H.P., Lefranc, G., Schmitt, T.: Thermal expansion studies on aluminium-matrix composites with different reinforcement architecture of SiC particles. Compos. Sci. Technol. **66**, 2206–2217 (2006)
6. Nam, T.H., Requena, G., Degischer, P.: Thermal expansion behaviour of aluminum matrix composites with densely packed SiC particles. Compos. Part A Appl. Sci. Manuf. **39**, 856–865 (2008)
7. Chawla, N., Deng, X., Schnell, D.R.M.: Thermal expansion anisotropy in extruded SiC particle reinforced 2080 aluminum alloy matrix composites. Mater. Sci. Eng. A **426**, 314–322 (2006)
8. Molina, J.M., Rhême, M., Carron, J., Weber, L.: Thermal conductivity of aluminum matrix composites reinforced with mixtures of diamond and SiC particles. Scr. Mater. **58**, 393–396 (2008)
9. Evans, A., Sanmarchi, C., Mortensen, A.: Study on properties and selection of metal matrix and reinforcement material for composites. In: Proceedings of the 6th European Conference on Genetic Programming, pp. 34–46 (2003)
10. Wong, W.L.E., Gupta, M., Lim, C.Y.H.: Enhancing the mechanical properties of pure aluminum using hybrid reinforcement methodology. Mater. Sci. Eng. A **423**, 148–152 (2006)
11. Ge, D., Gu, M.: Mechanical properties of hybrid reinforced aluminum based composites. Mater. Lett. **49**, 334–339

12. Zhang, X., Geng, L., Wang, G.S.: Fabrication of Al-based hybrid composites reinforced with SiC whiskers and SiC nanoparticles by squeeze casting. J. Mater. Process. Technol. **176**, 141–151 (2006)
13. Sureshbabu, J.S., Nair, P.K., Kang, C.G.: Nano and hybrid aluminum based metal matrix composites: an overview. In: 16th International Conference on Composite Materials, Kyoto, Japan, pp. 1–5 (2007)
14. Feng, Y.C., Geng, L., Zeng, P.Q., Zeng, Z.Z., Wang, G.S.: Fabrication and characteristic of Al-based hybrid composite reinforced with tungsten oxide particle and aluminum borate whisker by squeeze casting. Mater. Des. **29**, 2023–2026 (2008)
15. Fan, G.H., Geng, L., Zeng, Z.Z., Wang, G.S., Zeng, P.Q.: Preparation and characterization of $Al_{18}B_4O_{33}$ + $BaPbO_3$/Al hybrid composite. Mater. Lett. **62**, 2670–2672 (2008)
16. Feng, Y.C., Geng, L., Fan, G.H., Li, A.B., Zeng, Z.Z.: The properties and microstructure of hybrid composites reinforced with WO_3 particles and $Al_{18}B_4O_{33}$ whiskers by squeeze casting. Mater. Des. **30**, 3632–3635 (2009)
17. Kumar, D.R., Narayanasamy, R., Loganathan, C.: Effect of glass and SiC in aluminum matrix on workability and strain hardening behavior of powder metallurgy hybrid composites. Mater. Des. **34**, 120–136 (2012)
18. Guan, L., Geng, L., Zhang, H., Huang, L.: Effects of stirring parameters on microstructure and tensile properties of (ABOw + SiCp)/6061Al composites fabricated by semi-solid stirring technique. Trans. Nonferrous Met. Soc. China **21**, s274–s279 (2011)
19. Muley, A.V., Aravindan, S., Singh, I.P.: Mechanical and tribological studies on nano particles reinforced hybrid aluminum based composite. Manuf. Rev. **2**, 1–15 (2015)
20. Casati, R., Vedani, M.: Metal matrix composites reinforced by nano-particles—a review. Metals **4**, 65–83 (2014)
21. Uddin, S.M., Mahmud, T., Wolf, C., Glanz, C., Kolaric, I., Volkmer, C., Höller, H., Wienecke, U., Roth, S., Fecht, H.: Effect of size and shape of metal particles to improve hardness and electrical properties of carbon nanotube reinforced copper and copper alloy composites. Comp. Sci. Technol. **70**, 2253–2257 (2010)
22. Bakshi, S.R., Lahiri, D., Agarwal, A.: Carbon nanotube reinforced metal matrix composites—a review. Int. Mater. Rev. **55**, 42–64 (2010)
23. Trojanova, Z., Lukac, P., Ferkel, H., Riehemann, W.: Elastic and plastic behavior of an ultrafine-grained Mg reinforced with BN nanoparticles. Mater. Sci. Eng. A **370**, 154–157 (2004)
24. Deng, C.F., Wang, D.Z., Zhang, X.X., Ma, Y.X.: Damping characteristics of carbon nanotube reinforced aluminum composite. Mater. Lett. **61**, 3229–3231 (2007)
25. Shehata, F., Fathy, A., Abdelhameed, M., Mustafa, S.F.: Preparation and properties of Al_2O_3 nanoparticle reinforced copper matrix composites by in situ processing. Mater. Des. **30**, 2756–2762 (2009)
26. Ferkel, H., Mordike, B.L.: Magnesium strengthened by SiC nanoparticles. Mat. Sci. Eng. A **298**, 193–199 (2001)
27. Basavarajappa, S., Paulo Davim, J.: Influence of graphite particles on surface roughness and chip formation studies in turning metal matrix composites. Mater. Res. **16**(5), 990–996 (2013)
28. Suresh, P., Marimuthu, K., Ranganathan, S., Rajmohan, T.: Optimization of machining parameters in turning of Al-SiC-Gr hybrid metal matrix composites using grey-fuzzy algorithm. Trans. Nonferrous Met. Soc. China **24**, 2805–2814 (2014)
29. Siddesh Kumar, N.G., Shiva Shankar, G.S., Basavarajappa, S., Suresh, R.: Some studies on mechanical and machining characteristics of Al2219/n-B_4C/MoS_2 nano-hybrid metal matrix composites. Measurement **107**, 1–11 (2017)
30. Casati, R., Vedani, M.: Mechanical and functional properties of ultrafine grained Al wires reinforced by nano-Al_2O_3 particles. Metals **4**(1), 65–83 (2014)
31. Alaneme, K.K., Olubambi, P.A.: Microstructural analysis and corrosion behavior of Fe, B, and Fe-B-modified Cu-Zn-Al shape memory alloys. J. Mater. Res. Technol. **2**(2), 188–194 (2013)
32. Das, D.K., Mishra, P.C., Singh, S., Pattanaik, S.: Fabrication and heat treatment of ceramic-reinforced aluminium matrix composites—a review. Int. J. Mech. Mater. Eng. **9**(1), 1–15 (2014)

33. Alaneme, K.K., Bodunrin, M.O.: Acta Tech. Corvininesis Bull. Eng. **6**(3) [cited 2014 Aug 25] [Internet]. Available from: http://acta.fih.upt.ro/pdf/2013-3/ACTA-2013-3-18.pdf (2013)
34. Alaneme, K.K., Aluko, A.O.: Fracture toughness (K1C) and tensile properties of as-cast and age-hardened aluminium (6063)–silicon carbide particulate composites. Sci. Iran. **19**(4), 992–996 (2012)
35. Surappa, M.K.: Tribological behavior of Al–Si–SiCp composites/automobile brake pad system under dry sliding conditions. Sadhana **28**(1–2), 319–334 (2003)
36. Kok, M.: Production and mechanical properties of Al_2O_3 particle-reinforced 2024 aluminium alloy composites. J. Mater. Process. Technol. **161**(3), 381–387 (2005)
37. Yigezu, B.S., Mahapatra, M.M., Jha, P.K.: Study on properties and selection of metal matrix and reinforcement material for composites. J. Miner. Mater. Charact. Eng. **1**(4), 124–130 (2013)
38. Oghenevweta, J.E., Aigbodion, V.S., Nyior, G.B., Asuke, F.: J. King Saud Univ. Eng. Sci. [Internet] [cited 2014 Aug 25]. Available from: http://www.sciencedirect.com/science/article/pii/S1018363914000245 (2014)
39. Bhandakkar, A., Prasad, R.C., Sastry, S.M.: Enhancement of surface coating characteristics of epoxy resin by dextran: an electrochemical approach. Int. J. Compos. Mater. **4**(2), 108–124 (2014)
40. Bodunrin, M.O., Alaneme, K.K., Chown, L.H.: Aluminium matrix hybrid composites: a review of reinforcement philosophies; mechanical, corrosion and tribological characteristics. J. Mater. Res. Technol. **4**(4), 434–445 (2015)
41. Ravindran, P., Manisekar, K., Vinoth Kumar, S., Rathika, P.: Investigation of microstructure and mechanical properties of aluminum hybrid nano-composites with the additions of solid lubricant. Mater. Des. **51**, 448–456 (2013)
42. Sijo, M.T., Jayadevan, K.R.: Analysis of stir cast aluminium silicon carbide metal matrix composite: a comprehensive review. Procedia Technol. **24**, 379–385 (2016)
43. Ma, W., Lu, J.: Effect of surface texture on transfer layer formation and tribological behaviour of copper–graphite composite. J. Wear **270**, 218–229 (2011)
44. Masroor, M., Sheibani, S., Ataie, A.: Effect of milling energy on preparation of Cu–Cr/CNT hybrid nano-composite by mechanical alloying. Trans. Nonferrous Met. Soc. China **26**, 1359–1366 (2016)
45. Yuvaraj, N., Aravindan, S.: Fabrication of Al5083/B4C surface composite by friction stir processing and its tribological characterization. J. Mater. Res. Technol. **4**(4), 398–410 (2015)
46. Suresh, S., Moorthi, N.S.V.: Process development in stir casting and investigation on microstructures and wear behavior. Proc. Eng. **64**, 1183–1190 (2013)
47. Annigeri Veeresh Kumar, U.K.G.B.: Method of stir casting of aluminum metal matrix composites: a review. Mater. Today Proc. **4**, 1140–1146 (2017)
48. Madhukar, P., Selvaraj, N., Rao, C.S.P.: Manufacturing of aluminium nano hybrid composites: a state of review. IOP Conf. Ser. Mater. Sci. Eng. **149**, 012114 (2016). https://doi.org/10.1088/1757-899x/149/1/012114
49. Singh, J., Chauhan, A.: Characterization of hybrid aluminum matrix composite for advance applications—a review. J. Mater. Res. Technol. **5**(2), 159–169 (2015)
50. Rajmohan, T., Palanikumar, K., Ranganathan, S.: Manufacturing of aluminium nano hybrid composites: a state of review. Trans. Nonferrous Met. Soc. China **23**, 2509–2517 (2013)
51. Arun Kumar M.B., Swamy R.P.: Flyash and e-glass fiber reinforced hybrid metal matrix composites. ARPN J. Eng. Appl. Sci. **6**(5) (2011)
52. Poovazhagan, L., Kalaichelvan, K., Rajadurai, A., Senthilvelan, V.: Characterization of hybrid silicon carbide and boron carbide nano particles reinforced aluminum alloy composites. Procedia Eng. **64**, 681–689 (2013)
53. Alaneme, K.K., Ademilua, B.O., Bodunrin, M.O.: Mechanical properties and corrosion behaviour of aluminium hybrid composites reinforced with silicon carbide and bamboo leafash. Tribol. Ind. **35**(1), 25–35 (2013)
54. Fatile, O.B., Akinruli, J.I., Amori, A.A.: Microstructure and mechanical behaviour of stir-cast Al-Mg-Sl alloy matrix hybrid composite reinforced with corn cob ash and silicon. Int. J. Eng. Technol. Innov. **4**(4), 251–259 (2014)

55. Wong, W.L.E., Karthik, S., Gupta, M.: Development of hybrid $MgAl_2O_3$ composites with improved properties using microwave assisted rapid sintering route. J. Mater. Sci. **40**, 3395–3402 (2005)
56. Kumar, M., Gupta, R.K., Pandey, A.: A review on fabrication and characteristics of metal matrix composites fabricated by stir casting. IOP Conf. Ser. Mater. Sci. Eng. **377**, 012125 (2018). https://doi.org/10.1088/1757-899x/377/1/012125
57. Kumar, M., Gupta, R.K., Pandey, A.: Study on properties and selection of metal matrix and reinforcement material for composites. In: AIP Conference Proceedings, vol. 2148, p. 030019 (2019). https://doi.org/10.1063/1.5123941
58. Khare, M., Gupta, R.K., Bhardwaj, B.: Development of empirical relationship for surface roughness during the machining of metal matrix composite. IOP Conf. Ser. Mater. Sci. Eng. **402**, 012010 (2018). https://doi.org/10.1088/1757-899x/402/1/012010
59. Khare, M., Gupta, R.K., Ghosh, S.S., Bhardwaj, B.: Effect of reinforcements (B4C & Al_2O_3) and rotational speed on tribological properties of aluminum alloy 7075 hybrid composites through friction stir processing. In: AIP Conference Proceedings, vol. 2148, p. 030008 (2019). https://doi.org/10.1063/1.5123930
60. Srivastava, K.R., Gupta, R.K., Khare, M.: Int. J. Eng. Adv. Technol. (IJEAT) **8**(6). ISSN: 2249-8958 (2019)
61. Khare, M., Gupta, R.K., Goyal, R.: Evaluation of mechanical properties of AA7075/Al_2O_3/Mg hybrid composites. Int. J. Eng. Adv. Technol. (IJEAT). **8**(6). ISSN: 2249-8958 (2019)
62. Singh, H., Haq, M.I.U., Raina, A.: Dry sliding friction and wear behaviour of AA6082-TiB2 in situ composites. Silicon (2019). https://doi.org/10.1007/s12633-019-00237-y
63. Ul Haq, M.I., Anand, A.: Evaluation of mechanical and tribological properties of directionally solidified Al-Si based FG composite. Silicon **10**, 1819 (2018). https://doi.org/10.1007/s12633-017-9675-1
64. Haq, M.I.U., Anand, A.: Dry sliding friction and wear behaviour of hybrid AA7075/Si3N4/Gr self lubricating composites. Mater. Res. Express **5**, 066544 (2018). https://doi.org/10.1088/2053-1591/aacc50

Chapter 38
Variation of Shear Properties in Carbon Fiber Composites as Extensional Forces Applied with Different Fiber Orientation

Joginder Singh and M. R. Tyagi

Abstract A composite can have two types of loading, i.e., axial and off-axial loading. When loading on the carbon fiber reinforced composite (CFRC) is along the longitudinal or transverse direction, then it is known as specially orthotropic condition. In this condition, there is no interaction between the extensional and shear properties. Extensional loading generates only extensional stress and extensional strain. Similarly, shear loading generates only shear stress and shear strain. Therefore, cross-coefficients are zero in such condition. Due to which, there are no distortions in CFRC and this is the ideal conditions for engineering application. In off-axial loading, the loading is not along the longitudinal or transverse direction, then it is known as generally orthotropic condition. In this condition, there is so much interaction between the extensional and shear properties that extensional loading generates both extensional and shearing properties. Similarly, shearing loading generates both shearing and extensional properties. Therefore, both the cross-coefficients, i.e., m_x and m_y are nonzero for this situation. So, there is distortion in the CFRC which is not recommended for engineering application. Here, only shear properties are discussed like shear modulus, cross-coefficients, shear strain, and Poisson's ratio. Behavior of each of the property is discussed with theoretical calculation and graph with fiber orientation.

Keywords CFRC · Off-axis loading · Shear stress · Shear strain · Poisson's ratio · Shear modulus

38.1 Introduction

The properties of CFRC are different in different direction. There is huge difference in the CFRC properties from longitudinal to transverse directions but there is some relation between them. It can also define as axis and off-axis properties. The off-axial properties are predicted by carrying out macroscopic analysis of a lamina using the principles of mechanics and assuming a lamina to be macroscopically homogenous.

J. Singh (✉) · M. R. Tyagi
Department of Mechanical Engineering, Manav Rachna University, Faridabad, Haryana, India
e-mail: joginder@mru.edu.in

S. Yadav et al. (eds.), *Proceedings of International Conference in Mechanical and Energy Technology*, Smart Innovation, Systems and Technologies 174,
https://doi.org/10.1007/978-981-15-2647-3_38

The behavior of CFRC lies between the isotropic and anisotropic materials. A composite can be of unidirectional or multidirectional. Here, only unidirectional CFRC are considered for the simplification of calculation. Almeida et al. [1] performed different test like Iosipescu shear test, double-notched shear, V-notched rail, and short beam shear to know the shear behavior of glass fiber epoxy laminates. Naresh et al. [2] investigated and compared the effect of fiber orientation on carbon and glass epoxy composites subjected to shear and bending. Leopold et al. [3] reviewed the comparison of analytical approaches predicting the compressive strength of carbon and glass fiber reinforced polymers. Ali and Anjaneyulu [4] investigated the effect of unidirectional fiber volume fraction and orientation on the double bone suspension system of composite material. Dhakate et al. [5] developed the unidirectional composite of carbon fiber and phenolic matrix. Fiber volume fraction was selected from the range of 30%, 40%, 50%, and 60%. Vinay et al. [6] studied the effect of reinforcement on the hybrid composite. They used hybrid composite because it can be recycled, low cost, low density, etc. Torres et al. [7] studied the mechanical properties of natural fiber composites because of their sustainability benefits. They found that variability of long natural fibers is same as that of the carbon fibers. Reddy et al. [8] studied the carbon and glass fiber composites for the high strength application. They used unidirectional carbon fiber like T-300 12k and T-700 12k for the fabrication with the help of filament winding technique. Shamsuyeva et al. [9] reviewed the materials, manufacturing, and applications of composites to describe the different mechanical properties like impact, bending, damping, etc. Baran et al. [10] investigated the fabrication methods of composite fibers. As the usage of fiber composite is increasing, it is important to understand the fabrication method so that residual stresses in the composite can be reduced. Wu et al. [11] developed the method of recovering the carbon fibers T700 with the help of $ZnCl_2$. Epoxy degraded at 380 °C in 80 min in the presence of $ZnCl_2$. Ahmed et al. [12] studied the impact properties of the unidirectional composites. They found that unidirectional composites have low ability to absorb impact load.

In this study, elastic properties of a unidirectional carbon epoxy composite like major and minor Poisson's ratio, cross-coefficients, shear strain, and shear modulus are considered with respect to fiber orientation.

38.2 Design of the Specimen

The specimen of CFRC is designed as per the standard of ASTM E8M. Isometric view of the specimen is shown in Fig. 38.1.

Fiber orientation varies from 0° to 90° and in the steps of 5°. At a time, fibers remain at a single angle because CFRC considered is unidirectional (Fig. 38.2).

The required properties for the calculation purpose are shown here in the form of Table 38.1. Each value is selected on the basis of books and literature review [13].

The required parameters for the calculation purpose are shown here in the form of Table 38.2.

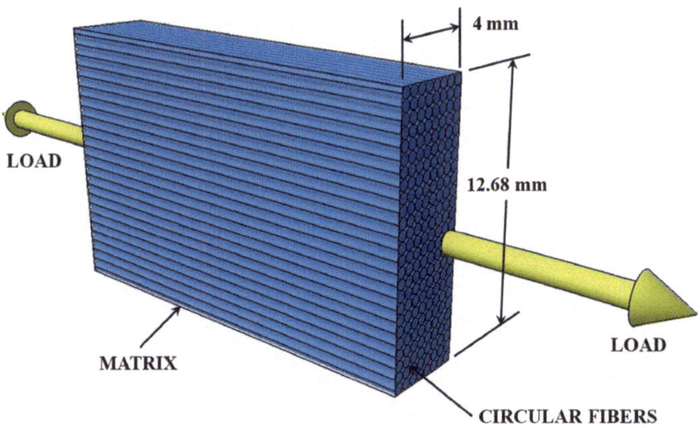

Fig. 38.1 Isometric view of extensional force applied on the specimen of CFRC

Fig. 38.2 Top view of extensional force applied on the specimen of CFRC

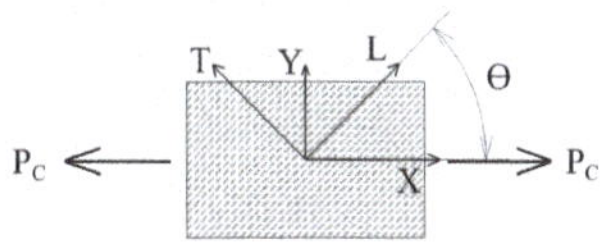

Table 38.1 Properties of carbon fiber and cast epoxy resin

S. No.	Type of properties	Symbols	Values
1	Elastic modulus of fiber	E_f	350,000 N/mm^2
2	Tensile strength of fiber	σ_{uf}	2100 N/mm^2
3	Shear modulus of fiber	G_f	13,000 N/mm^2
4	Shear strength of fiber	τ_f	23 N/mm^2
5	Poisson's ratio of fiber	ν_f	0.20
6	Elastic modulus of epoxy	E_m	3500 N/mm^2
7	Tensile strength of cast epoxy resin	σ_{um}	130 N/mm^2
8	Shear modulus of epoxy	G_m	2000 N/mm^2
9	Shear strength of epoxy	τ_m	6.33 N/mm^2
10	Poisson's ratio of epoxy	ν_m	0.35

Table 38.2 Parameters considered of CFRC based on experimental testing and theoretical calculations

S. No.	Parameters	Symbols	Values
1	Volume fraction of fibers	V_f	0.5
2	Width of specimen	–	12.68 mm
3	Thickness of specimen	–	4 mm
4	Load carried by the CFRC in *X*-direction	P_c	45,400 N
5	Zeta for elastic modulus and shear modulus in transverse direction (according to Halpin–Tsai equation)	ξ	2 and 1

Table 38.3 Values calculated for extensional properties of CFRC

S. No.	Type of properties	Symbols	Values
1	Volume fraction of matrix	V_m	0.5
2	*X*-sectional area of CFRC	A_c	50.72 mm^2
3	*X*-sectional area of fiber and matrix	A_f and A_m	25.36 mm^2
4	Load carried by the fibers in *X*-direction	P_f	44,950.5 N
5	Load carried by the matrix in *X*-direction	P_m	449.505 N
6	Stresses in fibers in *X*-direction	σ_f	1772.5 N/mm^2
7	Stresses in matrix in *X*-direction	σ_m	17.725 N/mm^2
8	Stress in the *L* direction of the CFRC	σ_L	895.11 N/mm^2
9	Elastic modulus of CFRC in *L* direction	E_L	176,750 N/mm^2
10	Iota for elastic modulus in *T* direction	η	0.97
11	Elastic modulus of CFRC in *T* direction	E_T	13,400 N/mm^2
12	Stress in the *T* direction of the CFRC	σ_T	67.86 N/mm^2
13	Iota for shear modulus in *T* direction	η	0.73
14	In-plane shear modulus of a CFRC	G_{LT}	4315.79 N/mm^2
15	Major and minor Poisson's ratio	ν_{LT} and ν_{TL}	0.275 and 0.021
16	Strain in *L* and *T* direction of the CFRC	ϵ_L and ϵ_T	0.0051 and 0.0037
17	In-plane shear stress of the CFRC	τ_{LT}	10.648 N/mm^2
18	In-plane shear strain of the CFRC	γ_{LT}	0.00247

Basic parameters are calculated for Extensional Properties of CFRC are shown in Table 38.3 [14].

38.3 Results and Discussion

38.3.1 Poisson's Ratio (ν_{xy})

The Poisson's ratio 'ν_{xy}' relates the stress in *X*-direction to the strain in *Y*-direction [15].

$$\frac{\nu_{xy}}{E_x} = \frac{\nu_{LT}}{E_L} - \frac{1}{4}\left(\frac{1}{E_L} + \frac{2\nu_{LT}}{E_L} + \frac{1}{E_T} - \frac{1}{G_{LT}}\right)\sin^2 2\theta \tag{38.1}$$

The value of 'ν_{xy}' can be obtained from Eq. (38.1) and shown in Table 38.4. The graph is shown in Fig. 38.3.

Similarly, 'ν_{yx}' is the Poisson's ratio which relates the stress in *Y*-direction to the strain in *X*-direction.

Table 38.4 Final values of ν_{xy}, ν_{yx}, m_x, m_y, γ_{xy}, and G_{xy} of CFRC

$\theta°$	ν_{xy}	ν_{yx}	m_x	m_y	γ_{xy}	G_{xy} (N/mm^2)
0	0.28	**0.02 (LV)**	**0.00 (LV)**	**0.00 (LV)**	0.00247	**4316 (LV)**
5	0.37	0.04	3.30	−1.18	0.00265	4401
10	0.49	0.08	6.30	−2.13	0.00276	4665
15	0.55	0.14	8.72	−2.63	**0.00278 (HV)**	5138
20	0.58	0.21	10.37	−2.54	0.00272	5868
25	**0.59 (HV)**	0.28	**11.12 (HV)**	−1.78	0.00257	6912
30	0.58	0.34	10.95	−0.40	0.00235	8301
35	0.56	0.41	9.94	1.52	0.00205	9926
40	0.54	0.46	8.24	3.76	0.00170	11,381
45	0.50	0.50	6.10	6.10	0.00129	**11,991 (HV)**
50	0.46	0.54	3.76	8.24	0.00084	11,381
55	0.41	0.56	1.52	9.94	0.00037	9926
60	0.34	0.58	−0.40	10.95	−0.00012	8301
65	0.28	**0.59 (HV)**	−1.78	**11.12 (HV)**	−0.00060	6912
70	0.21	0.58	−2.54	10.37	−0.00106	5868
75	0.14	0.55	−2.63	8.72	−0.00149	5138
80	0.08	0.49	−2.13	6.30	−0.00188	4665
85	0.04	0.37	−1.18	3.30	−0.00221	4401
90	**0.02 (LV)**	0.28	**0.00 (LV)**	**0.00 (LV)**	**−0.00247 (LV)**	**4316 (LV)**

LV Lowest Value, *HV* Highest Value

$$\frac{\nu_{yx}}{E_y} = \frac{\nu_{TL}}{E_T} - \frac{1}{4}\left(\frac{1}{E_L} + \frac{2\nu_{LT}}{E_L} + \frac{1}{E_T} - \frac{1}{G_{LT}}\right)\sin^2 2\theta \tag{38.2}$$

The different values of 'ν_{yx}' calculated are shown in Table 38.4. The graph between the Poisson's ratio 'ν_{yx}' and fiber orientation of the CFRC is shown in Fig. 38.4.

38.3.2 Cross-Coefficient ($\mathbf{m_x}$)

Shear strain will be observed when off-axis stress is applied in the CFRC. Cross-coefficient is the bridge between the extensional stress and shear strain. It can be of two types, i.e., cross-coefficient in *X*- and *Y*-directions [16]. After simplification of the equations, the expression will be as follows:

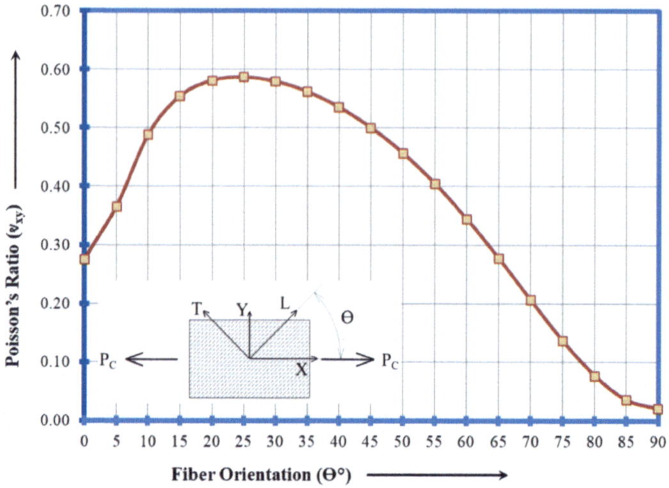

Fig. 38.3 Graph between Poisson's ratio (ν_{xy}) and fiber orientation ($\theta°$)

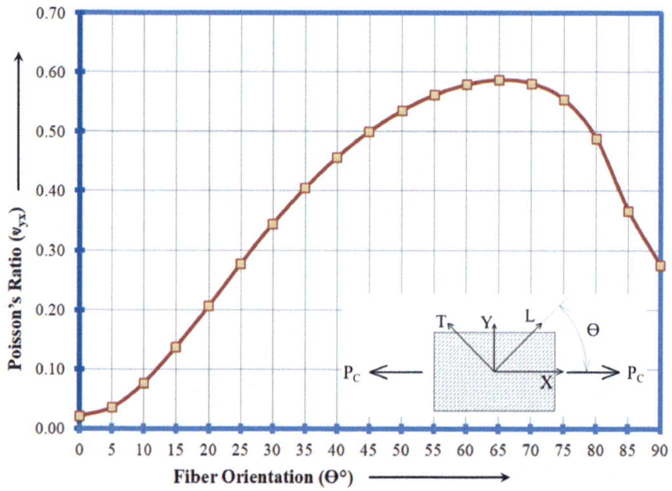

Fig. 38.4 Graph between Poisson's ratio (ν_{yx}) and fiber orientation ($\theta°$)

$$m_x = \sin 2\theta \left[\nu_{\mathrm{LT}} + \frac{E_{\mathrm{L}}}{E_{\mathrm{T}}} - \frac{E_{\mathrm{L}}}{2G_{\mathrm{LT}}} - \cos^2\theta \left(1 + 2\nu_{\mathrm{LT}} + \frac{E_{\mathrm{L}}}{E_{\mathrm{T}}} - \frac{E_{\mathrm{L}}}{G_{\mathrm{LT}}} \right) \right] \quad (38.3)$$

The value of 'm_x' can be obtained from Eq. (38.3) as all other values are shown in Table 38.3. The different values of 'm_x' calculated are shown in Table 38.4. The graph between the cross-coefficient 'm_x' and fiber orientation of the CFRC is shown in Fig. 38.5.

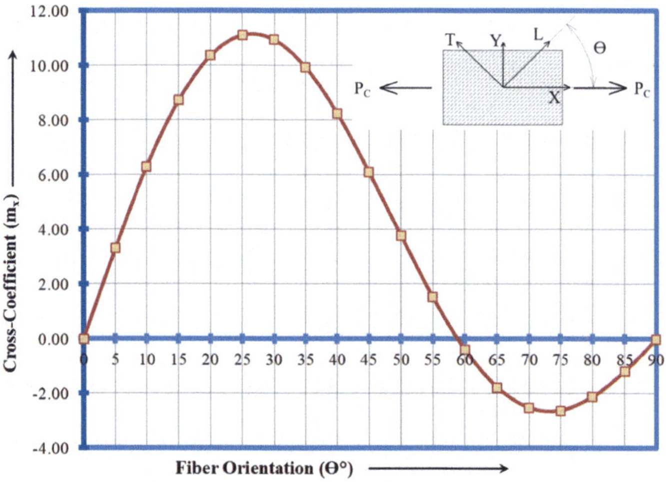

Fig. 38.5 Graph between cross-coefficient (m_x) and fiber orientation ($\theta°$)

Similarly, cross-coefficient 'm_y' can be expressed in the form of shear strain, extensional stress and elastic modulus as shown. After simplification of the equations, the expression will be as follows:

$$m_y = \sin 2\theta \left[\nu_{\mathrm{LT}} + \frac{E_{\mathrm{L}}}{E_{\mathrm{T}}} - \frac{E_{\mathrm{L}}}{2G_{\mathrm{LT}}} - \sin^2\theta \left(1 + 2\nu_{\mathrm{LT}} + \frac{E_{\mathrm{L}}}{E_{\mathrm{T}}} - \frac{E_{\mathrm{L}}}{G_{\mathrm{LT}}} \right) \right] \tag{38.4}$$

The value of 'm_y' can be obtained from Eq. (38.4) as all other values are shown in Table 38.3. The different values of 'm_y' calculated are shown in Table 38.4. The graph between the cross-coefficient 'm_y' and fiber orientation of the CFRC is shown in Fig. 38.6.

38.3.3 Shear Strain (γ_{xy})

Shear strain will be developed whenever fibers are not aligned with load [17]. Shear strain 'γ_{xy}' at an angle can be calculated with the help of E_{X}, E_{L}, E_{T}, ν_{LT}, τ_{LT}, G_{LT}, and θ as shown here.

$$\gamma_{xy} = \tau_{xy} \left[\frac{1}{E_{\mathrm{L}}} + \frac{2\nu_{\mathrm{LT}}}{E_{\mathrm{L}}} + \frac{1}{E_{\mathrm{T}}} - \left(\frac{1}{E_{\mathrm{L}}} + \frac{2\nu_{\mathrm{LT}}}{E_{\mathrm{L}}} + \frac{1}{E_{\mathrm{T}}} - \frac{1}{G_{\mathrm{LT}}} \right) \cos^2 2\theta \right] \tag{38.5}$$

The value of 'γ_{xy}' can be obtained from Eq. (38.5) as all other values are shown in Table 38.3. The different values of 'γ_{xy}' calculated are shown in Table 38.4. The graph between the cross-coefficient 'γ_{xy}' and fiber orientation of the CFRC is shown in Fig. 38.7.

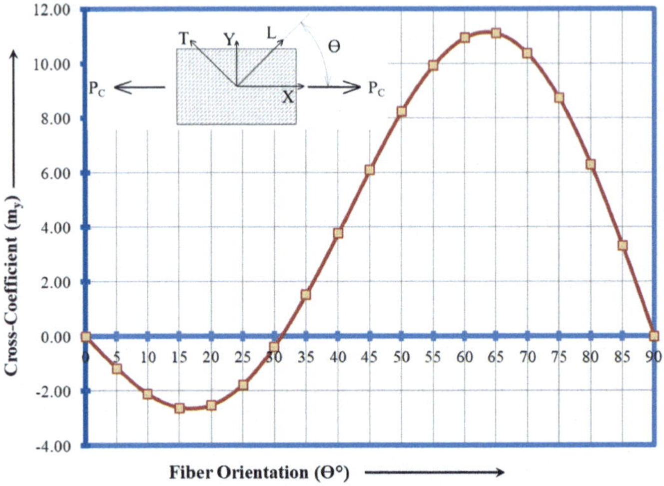

Fig. 38.6 Graph between cross-coefficient (m_y) and fiber orientation ($\theta°$)

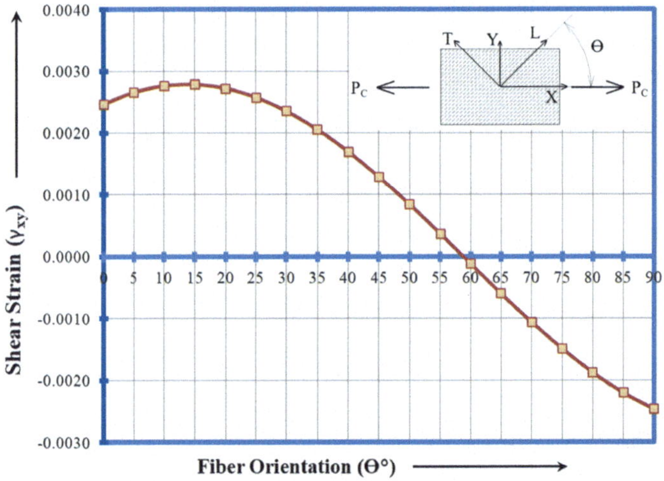

Fig. 38.7 Graph between shear strain (γ_{xy}) and fiber orientation ($\theta°$)

38.3.4 Shear Modulus ($\mathbf{G_{xy}}$)

The ratio of shear stress and shear strain is known as shear modulus. It is also known as the modulus of rigidity [18]. Shear modulus changes as the fiber orientation varies and it can be calculated with the help of E_L, E_T, ν_{LT}, G_{LT}, and θ as shown.

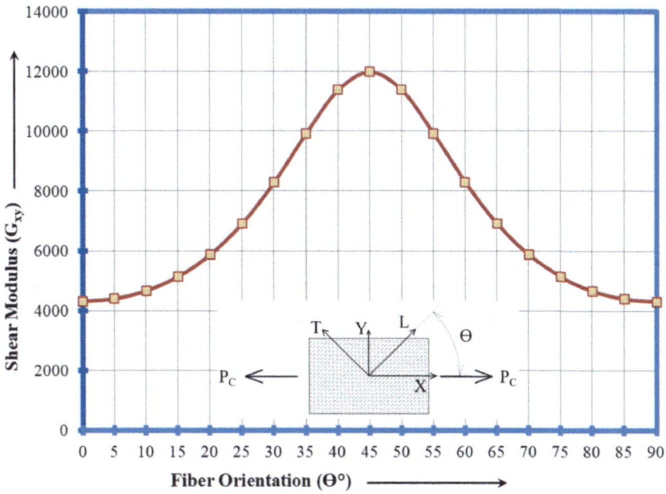

Fig. 38.8 Graph between shear modulus (G_{xy}) and fiber orientation ($\theta°$)

$$\frac{1}{G_{xy}} = \frac{1}{E_{\mathrm{L}}} + \frac{2\nu_{\mathrm{LT}}}{E_{\mathrm{L}}} + \frac{1}{E_{\mathrm{T}}} - \left(\frac{1}{E_{\mathrm{L}}} + \frac{2\nu_{\mathrm{LT}}}{E_{\mathrm{L}}} + \frac{1}{E_{\mathrm{T}}} - \frac{1}{G_{\mathrm{LT}}}\right)\cos^2 2\theta \tag{38.6}$$

The value of 'G_{xy}' can be obtained from Eq. (38.6) as all other values are shown in Table 38.3. The different values of 'G_{xy}' calculated are shown in Table 38.4. The graph is shown in Fig. 38.8.

The different values of ν_{xy}, ν_{yx}, m_x, m_y, γ_{xy}, and G_{xy} of CFRC which are calculated in the above section shown in Table 38.4.

38.3.5 Discussion

- The results indicate that major Poisson's ratio 'ν_{xy}' is higher at 25°, whereas minor Poisson's ratio 'ν_{yx}' is higher at 65°. The highest value of Poisson ratio 'ν_{xy}' is 0.59 at 25° and lowest value is 0.02 at 90°. The 'ν_{xy}' increases from 0° to 25° and then reduces slowly up to 90°. The highest value of Poisson ratio 'ν_{yx}' is 0.59 at 65° and lowest value is 0.02 at 0°. The 'ν_{yx}' increases slowly from 0° to 65° and then reduces slowly up to 90°.
- The cross-coefficients 'm_x' and 'm_y' are zero at $\theta = 0°$ and $\theta = 90°$. The highest value of cross-coefficients 'm_x' and 'm_y' is 11.12 at 25° and 65°, respectively. The cross-coefficient 'm_x' becomes negative beyond 60°. Similarly, the cross-coefficient 'm_y' is negative up to 30°.
- The highest value of shear strain 'γ_{xy}' is 0.00278 at 15° and lowest value is −0.00247 at 90°. The 'γ_{xy}' becomes negative beyond 60°.
- The value of 'G_{xy}' is largest at $\theta = 45°$ and it is symmetric about $\theta = 45°$.

38.4 Conclusions

- Extensional and shear interaction increase under off-axis loading condition. Due to this, distortion in CFRC increases.
- The results were in accordance with the theory and the literature. Finally, all the calculations were considered successful in evaluating shear characteristics of the CFRC.
- Balanced orthotropic lamina is better to use in engineering application. First, because elastic modulus in longitudinal and transverse are equal, i.e., $E_L = E_T$. Second, major and minor Poisson's ratios are equal, i.e., $\nu_{LT} = \nu_{TL}$. So, there will be almost no distortion in CFRC.

References

1. Almeida, J., et al.: Effect of fiber orientation on the shear behavior of glass fiber/epoxy. Mater. Des. 789–795. www.elsevier.com (2014)
2. Naresh, K., et al.: Effect of fiber orientation on carbon/epoxy and glass/epoxy composites subjected to shear and bending, vol. 267, pp. 103–108. ISSN: 1662-9779 (2017)
3. Leopold, C., et al.: Comparison of analytical approaches predicting the compressive strength of fibre reinforced polymers. www.mdpi.com/journal/materials (2018)
4. Ali, M., et al.: Effect of fiber-matrix volume fraction and fiber orientation on the design of composite suspension system. IOP Conf. Ser. **455**(1) (2018)
5. Dhakate, S., et al.: Mechanical properties of unidirectional carbon-carbon composites as a function of fiber volume content. Carbon Sci. **3**(3), 127–132 (2002)
6. Vinay, H., et al.: A review on investigation on the influence of reinforcement on mechanical properties of hybrid composites. IJPAST 39–48. ISSN 2229-6107 (2014)
7. Torres, J., et al.: The mechanical properties of natural fibre composite laminates: a statistical study. Compos. Part A 99–104. www.elsevier.com (2017)
8. Reddy, C., et al.: Mechanical characterization of unidirectional carbon and glass/epoxy reinforced composites for high strength applications. Mater. Today Proc. 3166–3172 (2017)
9. Shamsuyeva, M., et al.: Review on hybrid carbon/flax composites and their properties. Hindawi Int. J. Polym. Sci. 17 p. Article ID 9624670 (2019)
10. Baran, I., et al.: A review on the mechanical modeling of composite manufacturing processes. Arch. Comput. Methods Eng. **24**(2), 365–395 (2017)
11. Wu, T., et al.: Efficient reclamation of carbon fibers from epoxy composite waste through catalytic pyrolysis in molten ZnCl2. Roy. Soc. Chem. Adv. **9**(1), 377–388 (2019)
12. Ahmed, P., et al.: Effect of unidirectional and woven fibers on impact properties of epoxy. Maxwell Scientific Publication Corp., pp. 197–205. ISSN: 2040-7459 (2016)
13. Garoushi, S., et al.: Short fiber reinforced composite: the effect of fiber length and volume fraction. JCDP **7**(5) (2006)
14. Tyagi, S., et al.: Experimental and numerical analysis of tensile strength of unidirectional glass/epoxy composite laminates with different fiber percentage. IJAER **13**(15), 12157–12160. ISSN 0973-4562 (2018)
15. Swolfs, Y., et al.: Fibre hybridisation in polymer composites: a review. Compos. Part A Appl. Sci. Manuf. 181–200 (2014)
16. Abdalla, F.H., et al.: Determination of volume fraction values of filament wound glass and carbon fiber reinforced composites. **3**. ISSN 1819-6608 (2008)

17. Hamza, M.S.: Study the effect of carbon fiber volume fraction and their orientations on the thermal conductivity of the polymer composite materials. Al-Khwarizmi Eng. J. **4**, 80–89 (2008)
18. Flore, D., et al.: Influence of fibre volume fraction and temperature on fatigue life of glass fibre reinforced plastics. AIMS Mater. Sci. 770–795 (2016)

Chapter 39
Static Response Analysis of a Hybrid Beam with Variable Piezo Layer Thicknesses Under Thermal Loads

N. Rahman and M. N. Alam

Abstract Static response analysis of a hybrid beam is presented under thermal loads. The zigzag theory-based 1D finite element model is being used for the analysis. Laminated composite and sandwich beams with surface-bonded piezoelectric layers are considered as the problem beams. Influence of piezoelectric layer thickness on axial and across the thickness distribution of temperature and static response parameters is presented for thermal load cases with sinusoidal axial variation. The problem beams are also modelled as 2D deformable shell in ABAQUS. Sequential heat transfer and static analysis steps are used to obtain the distribution of displacements and stresses under thermal loads. 1D-FE results for clamped-clamped and clamped-free end conditions are matched with 2D-FE results.

Keywords Hybrid beam · ABAQUS · Zigzag theory · Thermal load · Piezo layer

39.1 Introduction

Structural elements consisting of composite and sandwich laminates are often subjected to varying temperature conditions. Since these laminates have dissimilar thermal properties in different layers, the temperature change results in considerable distortion in the shape of structure and the thermal stresses induced in the structure. Thus, response of a hybrid structure in thermal environment is influenced by the adjoining temperature or thermal load. Jonnalagadda et al. [1] presented the results for static response analysis of polyvinylidene-difluoride-layer-integrated composite plates under electro-thermo-mechanical loading. Vel and Batra [2, 3] presented a 3D analytical solution for obtaining the piezoelectric and thermal distortions of laminated plates. Tzou and Ye [4] studied the thermal influence on sensor and actuator of piezoelectric material layers in beam-type precision devices using solid finite element. Kapuria and Alam [5] evolved a new one-dimensional beam element found on coupled efficient layer-wise theory (zigzag theory) for hybrid beams subjected to thermal loads. Xu and Noor [6] used analytical solutions to examine the response of a

N. Rahman (✉) · M. N. Alam
Department of Mechanical Engineering, Zakir Husain College of Engineering and Technology, A. M. U. Aligarh, Aligarh, Uttar Pradesh, India

S. Yadav et al. (eds.), *Proceedings of International Conference in Mechanical and Energy Technology*, Smart Innovation, Systems and Technologies 174,
https://doi.org/10.1007/978-981-15-2647-3_39

cylindrical shell structure consisting of composite laminate subjected to temperature change, electric potential and mechanical load. Raja et al. [7] reviewed fundamentals of piezothermoelasticity and included the effect of temperature in developing a piezothermoelastic triangle shell element. Gu et al. [8] implemented a higher order theory for temperature field and used coupled analysis of hybrid laminates subjected to thermo-electro-mechanical loads. Rahman and Alam [9, 10] used zigzag theory-based finite element model using 1D beam element for laminated hybrid beam under electro-thermo-mechanical load. Giunta et al. [11] used advanced one-dimensional finite elements to model three-dimensional beam for investigating the thermoelastic behaviour of 3D isotropic and laminated beams. Mayandi and Jeyaraj [12] carried out buckling, bending and free vibration analysis of polymer composite beams subjected to thermal load using finite element method.

This work presents the influence of thicknesses of piezoelectric layer on the static response parameters of a hybrid beam under thermal loads. The zigzag theory-based FE model using 1D beam element is being used for the analysis. Laminated composite and sandwich hybrid beams with piezoelectric layers mounted at the exposed planar surfaces of the substrate are considered as the problem beams. All these beam structures are also modelled as 2D deformable shell using commercial finite element package—ABAQUS. Sequential heat transfer and static analysis steps are used to acquire the static response results for beams under thermal loads. 1D-FE results are compared with 2D-FE results for clamped-clamped and clamped-free boundary conditions.

39.2 Finite Element Model of Laminated Hybrid Beam

The zigzag theory-based 1D finite element model [5] for hybrid beam is being used for which the displacements u_x and u_z may be indicated in the form of the generalized displacements $\tilde{u}_1$ and $\tilde{u}_2$ as:

$$\begin{Bmatrix} u_x \\ u_z \end{Bmatrix} = \begin{bmatrix} g_1(z) & O_{1\times 2} \\ O_{1\times 4} & g_2(z) \end{bmatrix} \begin{Bmatrix} \tilde{u}_1 \\ \tilde{u}_2 \end{Bmatrix} + \begin{bmatrix} \overline{S}^{kj}(z) & 0 \\ 0 & \overline{\xi}^j_\theta(z) \end{bmatrix} \begin{Bmatrix} \frac{\partial \theta^j}{\partial x} \\ \theta^j \end{Bmatrix} \tag{39.1}$$

where $O_{1\times 2}$ and $O_{1\times 4}$ are, respectively, 1×2 and 1×4 order zero matrices. The thermal and potential fields are assumed as piece-wise linear between n_θ and n_ϕ number of points, respectively, in the direction of the beam thickness.

The variational equation for hybrid beam is achieved as:

$$\int_0^a \left[\delta\bar{\varepsilon}^{\mathrm{T}}\overline{F} - \delta\tilde{u}_2^{\mathrm{T}} g_{u\phi} - \delta\tilde{u}_2^{\mathrm{T}}\left\{\hat{q}_2 - \hat{q}_4^{l}\right\}^{\mathrm{T}} + \left(\frac{\partial\delta u_{z0}}{\partial x} N_x \frac{\partial u_{z0}}{\partial x}\right)\right]\mathrm{d}x$$
$$- \left[N_x^*\delta u_{x_0}^* + V_x^*\delta u_{z_0}^* - M_x^*\frac{\partial\delta u_{z_0}^*}{\partial x} + P_x^*\delta\psi_0^* + \left(H^{l*} - V_\phi^{l*}\right)\delta\phi^{l*} + R_x^{l*}\frac{\partial\delta\phi^{l*}}{\partial x}\right]\Bigg|_0^a \tag{39.2}$$

where $\bar{\varepsilon}$ is the vector of generalized strains and $\overline{F}$ is the vector of generalized stresses; N_x, M_x, P_x, R_x^l, V_x, V_ϕ^l, H^l and G^l are the beam stress resultants and electric displacement resultants, respectively. Superscript * is used to mark values at the ends.

Finite elements with two and three nodes are considered for interpolating the electromechanical and thermal variables, respectively [5]. Each node of the two-noded element has mechanical and electric potential degrees of freedom which are four and $2n_\phi$ in number, respectively. However, three-noded element has a single temperature degree of freedom at each node.

An element's contribution in the integral of Eq. (39.2) is attained as:

$$T^{\mathrm{e}} = \delta d^{\mathrm{eT}}\left[K^{\mathrm{e}} d^{\mathrm{e}} + P_\theta^{\mathrm{e}} - P^{\mathrm{e}}\right] \tag{39.3}$$

where K^{e} is the element stiffness matrix, d^{e}, P^{e} and P_θ^{e} are element displacement, load and thermal load vectors, respectively.

Thus, after integration, the general equation is achieved as:

$$K^{\mathrm{e}} d^{\mathrm{e}} = \overline{P}^{\mathrm{e}} - \overline{P}_\theta^{\mathrm{e}} \tag{39.4}$$

The elemental contributions are added up, and the equation of the system is finally achieved as:

$$Kd = P - P_\theta \tag{39.5}$$

in which K, d, P and P_θ are the assembled form of the elemental stiffness matrix K^{e}, elemental displacement vector d^{e}, elemental load vector P^{e} and elemental thermal load matrix P_θ^{e}.

39.3 Results and Discussion

The hybrid beam of Ref. [13] is considered for validation of the present formulation. The beam is modelled with simply supported end conditions and analysed for static

Table 39.1 Static response of hybrid beam of Ref. [13] under thermal loading

S. no.	Entity	Thermal load case 1		Thermal load case 2	
		Ref. [13]*	1D-FE	Ref. [13]*	1D-FE
1	$\bar{u}_z(a/2, -h/2)$	−0.84	−0.87	2.07	2.15
2	$\bar{u}_z(a/2, h/2)$	2.40	2.51	1.83	1.95
3	$\bar{\sigma}_x(a/2, -h/2)$	1.09	1.05	−0.67	−0.65
4	$\bar{\sigma}_x(a/2, h/2)$	−1.05	−1.07	−1.35	−1.33
5	$\bar{\sigma}_x^p(a/2, 2/5h)$	−1.08	−1.11	−1.31	−1.32

*Results read from graphs

response under the thermal load cases (1 and 2) considered in Ref. [13]. The results are presented (Table 39.1) with beam span-to-thickness ratio, $S = 5$.

On comparison, it is found that the present 1D-FE results with reference to the analytical results given in Ref. [13] are in a range of 1.9–4.6% for load case 1 and 0.8–6.6% for load case 2.

39.3.1 Response with Variation in Piezoelectric Layer Thicknesses

The static response results under thermal load using 1D-FE model (zigzag theory) and 2D-FE model (ABAQUS) are obtained for hybrid beams (d) and (e) of Ref. [14]. The hybrid beams with integrated piezoelectric layers are modelled for different piezo layer thicknesses varying from $0.1h$ to $0.3h$, keeping the total beam thickness h as same in all cases. The following two thermal load cases are examined for a beam of length l.

Load case 1: $\theta(x, \pm 0.5h) = T_0 \sin\left(\frac{\pi x}{l}\right)$
Load case 2: $\theta(x, 0.5h) = -\theta(x, -0.5h) = T_0 \sin\left(\frac{\pi x}{l}\right)$

The results are presented in non-dimensionalized form given as:

$$\bar{u}_z = 100u_z/hS^2T_0\alpha_T, \quad \bar{\sigma}_x = \sigma_x/\alpha_T Y_0 T_0, \quad \bar{\phi} = \phi d_0/\alpha_T h T_0 \text{ and } \overline{T} = T/T_0$$

where $S = l/h$ is the ratio of span length l to beam thickness h.

2D-FE (ABAQUS) results are obtained by performing sequential heat transfer and static analysis procedures. A quadratic heat transfer quadrilateral element with 8 nodes (DC2D8) is used in the heat transfer step, whereas plane stress elements CPS8R and CPS8RE are used in the static analysis step which follows the previous heat transfer step. The temperature profile acquired in the heat transfer step is applied as predefined field in static analysis.

The temperature distribution through the beam thickness at mid-span of beams is shown in Fig. 39.1 under the two thermal load cases. For all the thicknesses of

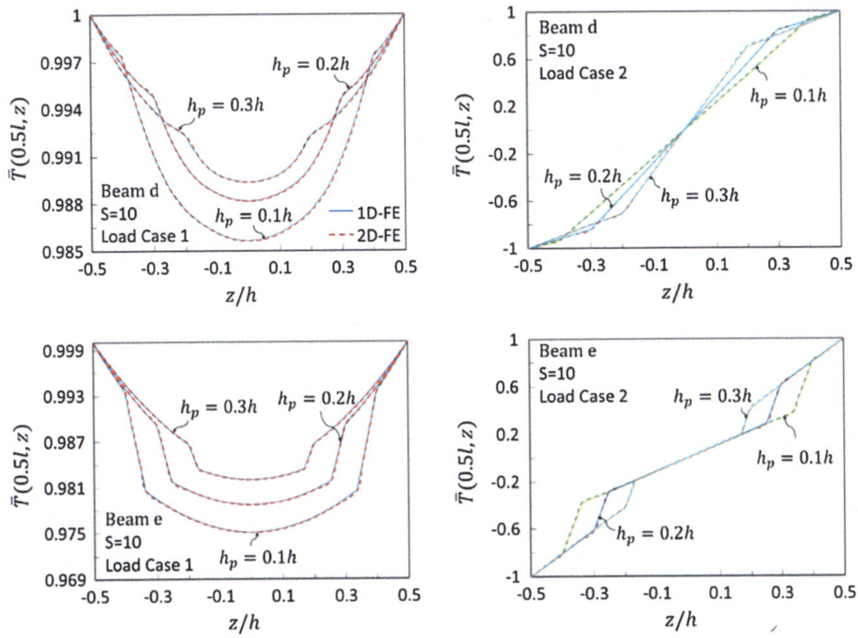

Fig. 39.1 Temperature distribution across the thickness of hybrid beams

piezoelectric layer, the temperature at mid-surface is maximum under load case 1 and zero under load case 2. The 1D-FE results show an excellent match with the 2D-FE results.

Figure 39.2 shows the axial temperature distribution under the two load cases for both the beam configurations at the surfaces of the substrate where piezoelectric layers are bonded. At the upper surface of elastic substrate, the temperature is maximum at the mid-span and the value of temperature is higher for load case 1 in comparison with load case 2. An inverse trend is obtained at the lower surface of the elastic substrate for load case 2, where the temperature is minimum at mid-span. The more is the thickness of piezoelectric layer, the higher is the temperature.

Figure 39.3 shows the deflection profiles of bottom surface of clamped-free hybrid beams (d) and (e) for thermal load case 1, and Fig. 39.4 shows the corresponding distribution of transverse displacement and normal stress across the thickness of beam at mid-span length. Maximum deflection occurs at the mid-span and not at the free end of the beam. Further, the deflection is more when the piezoelectric layer thickness is less, and for the same beam configuration, the deflection is more for span-to-thickness ratio, $S = 10$ in comparison with $S = 20$. The normal stress is maximum in face of the sandwich substrate in beam (e), and the stress value is higher when larger thicknesses of piezoelectric layers across the beam thickness are considered.

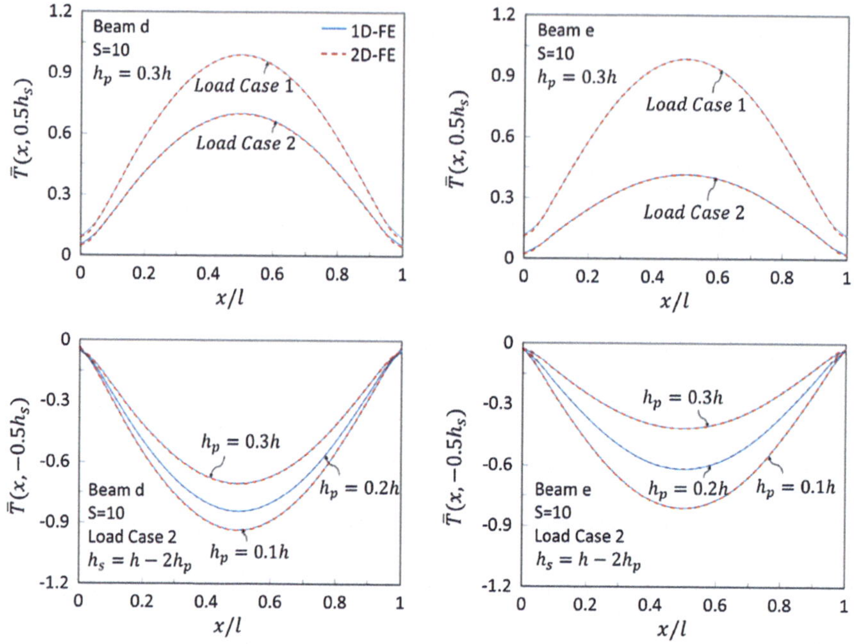

Fig. 39.2 Axial temperature distributions for hybrid beams

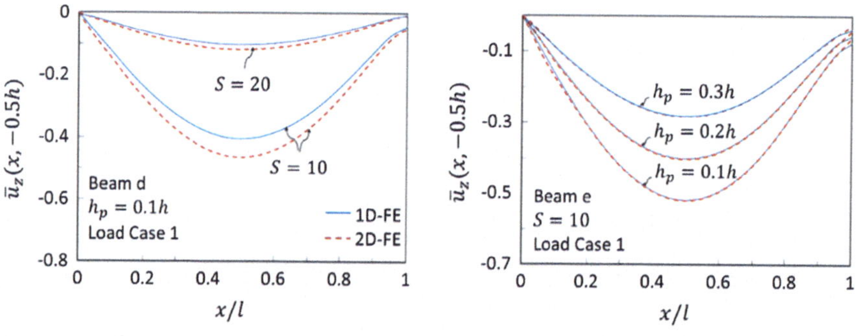

Fig. 39.3 Deflection profiles of clamped-free hybrid beams under load case 1

The deflection is maximum at the free end of clamped-free hybrid beams in thermal load case 2 as shown in Fig. 39.5, and its value increases considerably with increase in piezoelectric layer thickness. Figure 39.6 shows the corresponding distribution of transverse displacement across the thickness of beam at mid-span.

Similar trends are observed for the hybrid beams for thermal load cases considered under clamped-clamped boundary conditions (Figs. 39.7, 39.8, 39.9 and 39.10). The deflection is maximum at mid-span, and the value of deflection increases with decrease in piezoelectric layer thickness (Fig. 39.7) for thermal load case 1. However,

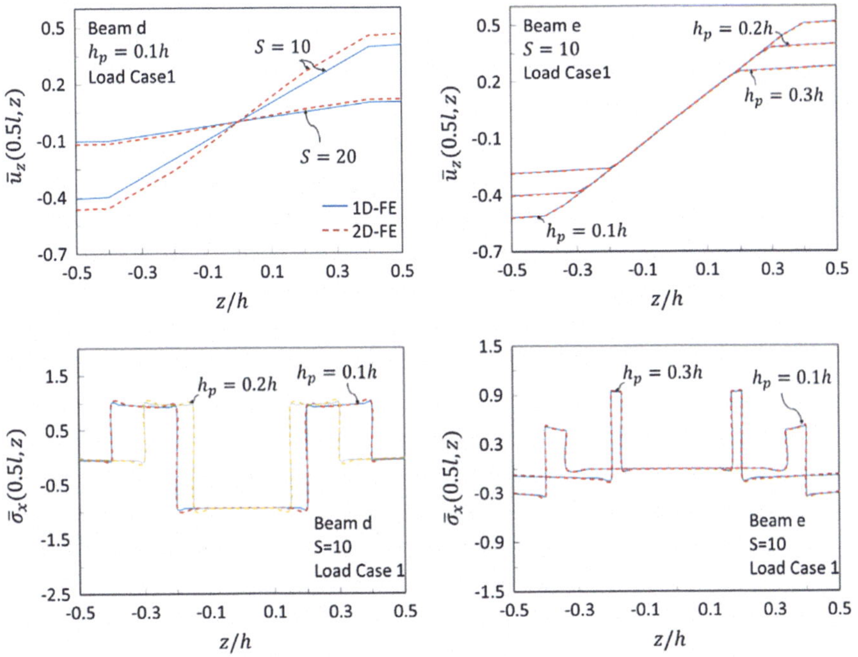

Fig. 39.4 Transverse displacement and normal stress distribution across the thickness of clamped-free hybrid beams

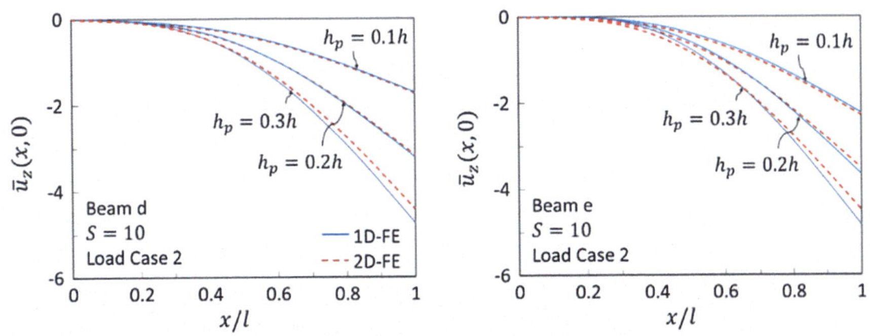

Fig. 39.5 Deflection profiles of clamped-free hybrid beams under load case 2

an inverse trend is observed for load case 2 (Fig. 39.9). Since the laminates under consideration are symmetric laminates, no deflection of mid-plane occurs for thermal load case 1, and it corresponds to thermal stretching problem. Conversely, the mid-plane deflects in load case 2 for either boundary conditions of the hybrid beam. It corresponds to thermal bending problem. The distribution of transverse displacement across the thickness at mid-span for either boundary conditions is symmetric about the mid-plane for load case 2. Similarly, the distribution of normal stresses across the

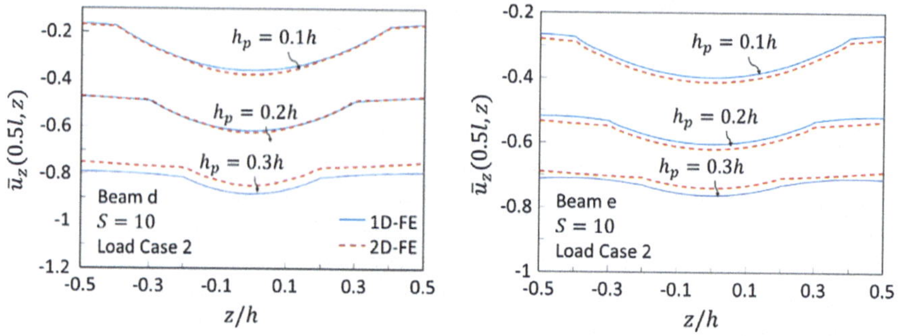

Fig. 39.6 Transverse displacement distribution across the thickness of clamped-free hybrid beams

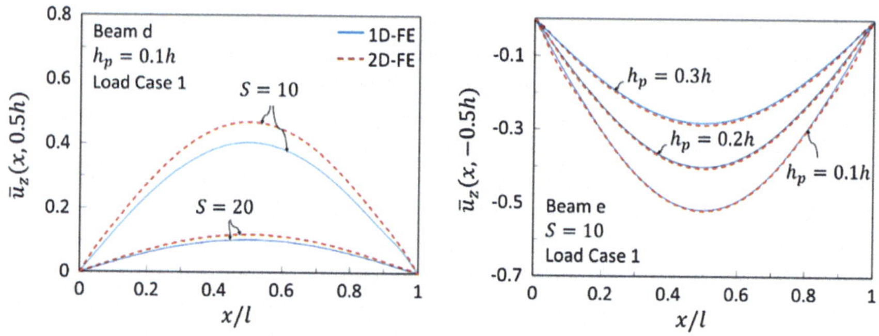

Fig. 39.7 Deflection profiles of clamped-clamped hybrid beams under load case 1

thickness is symmetric with respect to mid-plane for load case 1. The 1D-FE results are in good agreement with the 2D-FE (ABAQUS) results.

39.4 Conclusions

The indistinguishable match of temperature profile captured using one-dimensional and two-dimensional finite element models for clamped-free and clamped-clamped beams concedes that these are precise for moderately thick hybrid beams and give fairly accurate temperature distribution. The identical temperature distribution as obtained for the two beam boundary conditions demonstrates that temperature distribution is unconstrained of the end conditions. Furthermore, the thickness of piezoelectric layers is significant as it results in different structural response of a hybrid beam under thermal loads. These findings may be employed as a reference for conducting experimental analysis and may be used as a basis for modelling of hybrid structures for varied industrial and research applications.

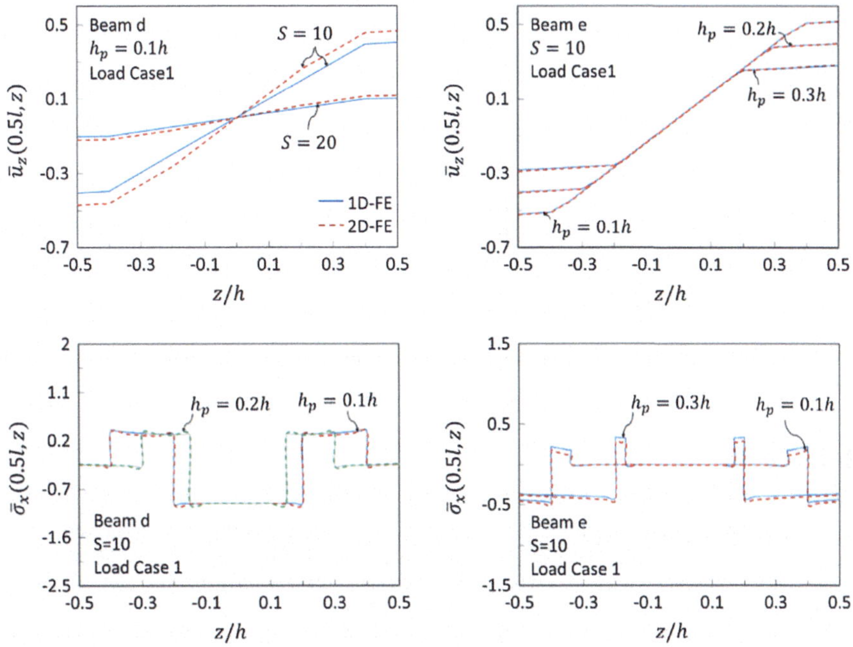

Fig. 39.8 Transverse displacement and normal stress distribution across the thickness of clamped-clamped hybrid beams

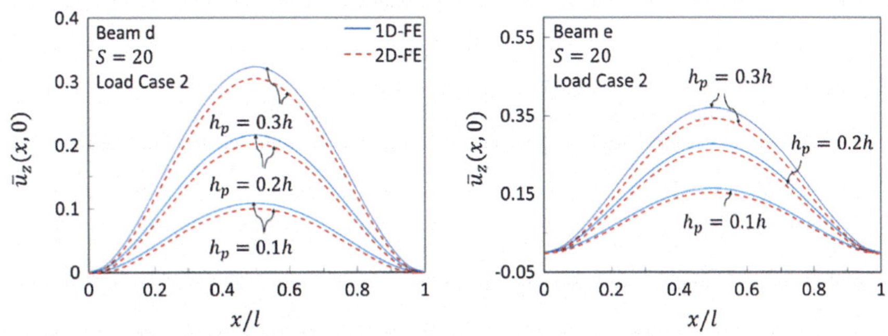

Fig. 39.9 Deflection profiles of clamped-clamped hybrid beams under load case 2

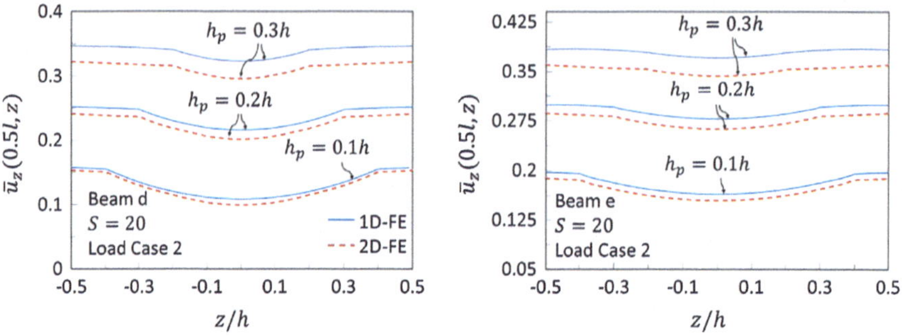

Fig. 39.10 Transverse displacement distribution across the thickness of clamped-clamped hybrid beams

References

1. Jonnalagadda, K.D., Blanford, G.E., Tauchert, T.R.: Piezothermoelastic composite plate analysis using first-order shear deformation theory. Comput. Struct. **51**, 79–89 (1994)
2. Vel, S.S., Batra, R.C.: Cylindrical bending of laminated plates with distributed and segmented piezoelectric actuators/sensors. AIAA J. **38**, 857–867 (2000)
3. Vel, S.S., Batra, R.C.: Generalized plane strain thermopiezoelectric analysis of multilayered plates. J. Therm. Stresses **26**, 353–377 (2003)
4. Tzou, H.S., Ye, R.: Piezothermoelasticity and precision control of piezoelectric systems: theory and finite element analysis. J. Vib. Acoust. **116**, 489–495 (1994)
5. Kapuria, S., Alam, M.N.: Coupled efficient zigzag finite element analysis of piezoelectric hybrid beams under thermal loads. J. Therm. Stresses **29**, 553–583 (2006)
6. Xu, K., Noor, A.K.: Three dimensional analytical solutions for coupled thermoelectroelastic response of multilayered cylindrical shells. AIAA J. **34**, 802–812 (1996)
7. Raja, S., Rohwer, K., Rose, M.: Piezothermoelastic modeling and active vibration control of laminated composite beams. J. Intell. Mater. Syst. Struct. **10**, 890–899 (1999)
8. Gu, H., Chattopadhyay, A., Li, J., Zhou, X.: A higher order temperature theory for coupled thermo-piezoelectric-mechanical modeling of smart composites. Int. J. Solids Struct. **37**, 6479–6497 (2000)
9. Rahman, N., Alam, M.N.: Structural control of piezoelectric laminated beams under thermal load. J. Therm. Stresses **38**, 69–95 (2015)
10. Rahman, N., Alam, M.N.: Thermal buckling analysis of laminated hybrid beam with variable piezoelectric layer thicknesses. IOP Conf. Ser. Mater. Sci. Eng. **404**, 012018 (2018)
11. Giunta, G., De Pietro, G., Nasser, H., Petrolo, M.: A thermal stress finite element analysis of beam structures by hierarchical modelling. Compos. Part B **95**(15), 179–195 (2016)
12. Mayandi, K., Jeyaraj, P.: Bending, buckling and free vibration characteristics of FG-CNT-reinforced polymer composite beam under non-uniform thermal load. J. Mater. Des. Appl. **229**(1) (2015)
13. Kapuria, S., Ahmed, A., Dumir, P.C.: Static and dynamic thermo-electro-mechanical analysis of angle-ply hybrid piezoelectric beams using an efficient coupled zigzag theory. Compos. Sci. Technol. **64**, 2463–2475 (2004)
14. Rahman, N., Alam, M.N.: Effect of piezoelectric layer thickness on static response of a hybrid piezoelectric beam. Adv. Sci. Eng. Med. **10**, 455–459 (2018)

Chapter 40
Kinematics Evaluation of Gait Spatiotemporal Parameters of Transfemoral Amputees While Walking with Jaipur Knee and 3R15 Knee Joints: A Case Study with Indian Population

Pawan Mishra, Sachin Kumar Singh, Vinayak Ranjan, Sonu Singh and Sahil Sujit Deshpande

Abstract After amputation surgery, patients use prosthesis to compensate missing functionality. The performance of the above-knee amputee is significantly reduced than normal subjects. The knee joint plays an important role in overall performance of the prosthesis. In India, a cost-effective solution like a Jaipur knee joint has been widely used. Many researchers have studied knee joints like 3R15 and 3R20; however, the performance of Jaipur knee joint during gait has not been reported with the 3R15 joint yet. Therefore, the objective of this research was to evaluate the comparative variations in spatiotemporal gait parameters of the transfemoral patients with 3R15 and Jaipur knee joints. Gait analysis was performed on 11 unilateral transfemoral amputees. Subjects were longtime users of the Jaipur knee joint. Subjects walked with their comfortable speed to evaluate the gait parameters with six VICON cameras and KINEMATRIX SYSTEM—a 3D motion analysis system. The difference between the performances of the subjects with Jaipur knee and 3R15 joints was associated with the walking speed, which enhanced gait performance with Jaipur joint. The spatiotemporal parameters were compared using the paired t-test method. After the study, it was found that the spatiotemporal parameters of patients were enhanced with the Jaipur knee joint. Subjects reported walking speed of 61.5 m/min with Jaipur knee joint, while the reported speed with 3R15 was 29.8 m/min (p-value = 0.05). Based on the increase in few parameters (p-value < 0.055), the gait symmetry was reported enhanced than that of 3R15 knee joint.

P. Mishra (✉) · S. K. Singh
Department of Mechanical Engineering, Indian Institute of Technology—Indian School of Mines, Dhanbad, India
e-mail: pawan.mishra@bennett.edu.in

V. Ranjan · S. S. Deshpande
Department of Mechanical and Aerospace Engineering, Bennett University, Greater Noida, India

S. Singh
Department of Physical Medicines and Rehabilitation, Army Hospital, R&R, New Delhi, India

S. Yadav et al. (eds.), *Proceedings of International Conference in Mechanical and Energy Technology*, Smart Innovation, Systems and Technologies 174,
https://doi.org/10.1007/978-981-15-2647-3_40

Keywords Amputation · Gait · Prosthetic knee · Spatiotemporal · Transfemoral

40.1 Introduction

Mobility may be defined as the quality of life. A person may lose his/her leg due to a trauma or medical reasons such as cancer or peripheral vascular disease, leading to reduction in their functionality. The leg prosthesis serves the purpose to restore the reduced functionality suffered by the individual and utility of natural biological systems. Unilateral transfemoral amputees require a prosthesis leg to support their weight, delivering a mobility performance similar to a natural leg. Gait techniques are recognized worldwide to quantify the performance of a leg prosthesis. Other prosthetic segments like socket, shank, knee joint and foot are also analyzed through gait analysis. Collado explained Balzac's recommendations for human gait analysis [1]. Ramkrishnan et al. classified and distinguished the bipedal gait asymmetries [2]. Jaroslav et al. examined that how much a knee joint affects the gait symmetry and reported a comparison of different knee joints and their respective pelvic movements [3]. According to a sample survey conducted by the Government of India, there were about one million people that had lost their limbs out of ten million with disabilities, and another four million were polio affected, while others had associated problems. The loss of limb was rampant as the results of accidents, landmine explosions, gunshot wounds in warzone, peripheral vascular diseases, cancer and Buerger's disease. Disabled people constituted about 2% of India's population, of which those with locomotive disabilities were the most prevalent (approximately 56%) [4]. Overall, the proportion of these causes varies in different developing and developed countries. The Jaipur knee joint has been widely used in developing countries [5].

The primary function of a knee unit for above-knee prosthesis is to control and support the prosthetic leg during standing position and the shank in swing position during walking. The prosthetic knee joint plays an important role as a critical component in case of transfemoral amputation. Various studies are reported with computerized or active knee joints that indicate reduced possibility of stumbling and falling, increase in walking velocity of amputee, improvement in adapting any walking speed [6] and better results on staircase [7] and hill slope [8]. Boonstra et al. [8] examined gait performance for transfemoral amputees with Tehlin and 3R20 prosthetic knee joints. It is already studied that the knee joint not only compensates the control of the leg during stance phase but also controls the shank during swing phase of the gait cycle [9]. Apart from the knee joint, the socket, shank, foot and ankle are also the important components which constitute the prosthesis assembly influencing the gait performance. Many researchers have examined a range of knee joint and ankle joint motion during gait [10, 11]. Few studies are reported the effect of prosthetic components like knee unit and ankle on gait performance [12–18]. Sapin et al. [17], who used Proteor's Hydro Cadence System based on single-axis prosthetic knee, compared the walking pattern of transfemoral amputees with other knee joints. Lee and Hong [18] reported how a prosthetic ankle affects mobility on the walking pattern of above-knee amputees with knee joints with a stance phase control. Evaluation of gait performance of 3R20 and 3R15 knee joints has been reported by A. Taheri and Karimi [19] comparing the differences in performance of the subjects using a paired *t*-test.

It is worth mentioning that there is a lack of reported data for gait performance of the transfemoral amputees with a Jaipur knee joint. The Jaipur knee joint, provided by Bhagwan Mahaveer Viklang Sahayta Samiti (BMVSS) Jaipur, is fitted free of cost to the millions of Indian amputees. Due to its widespread use among the Indian population, it was more significant to compare the performance of a Jaipur knee joint with other available knee joints to support clinician as well as to amputees. However, to the author's knowledge, there is no published literature on the comparison of gait performance of the amputees while walking with 3R15 and Jaipur knee joints. Therefore, this study focused to evaluate the relative effectiveness of leg prosthesis with 3R15 and Jaipur knee joints with Indian population data.

40.2 Methods

40.2.1 Subjects

Eleven individuals who had suffered from unilateral transfemoral amputations after a trauma, without any comorbidity, volunteered in this research [20], and details of all subjects are represented in Table 40.1. All participants used the Jaipur knee joint on

Table 40.1 Details of subjects (transfemoral amputees) participated in the study

Subjects	Age (years)	Sex	Reasons	Amputation (years)	Acclimation periods (weeks)
A	27	M	PVD	11	32.4
B	43	M	Trauma	16	11.8
C	57	M	Trauma	18	10.5
D	41	M	PVD	14	12.5
E	37	F	Trauma	9	11.7
F	59	M	Trauma	28	12.4
G	38	M	CA	6	13.7
H	46	F	Trauma	12	13.2
I	34	M	CA	13	14.9
J	65	M	PVD	30	10.6
K	33	M	PVD	8	14.9

Abbreviation CA—cancer; PVD—peripheral vascular disease

a regular basis. For this experiment, subjects wore the 3R15 knee joint daily. Experienced and trained gait laboratory staff supervised this study. Moreover, informed and written consents were obtained from each subject prior to the experimental setup.

These subjects were recruited for research on voluntarily basis. The study structure and steps involved were conferred with the themes. Gait laboratory supervisor confirmed comfortable socket fit, inclusive fitment and ensured sound and proper alignment for stability and comfort to amputees. Clinician examined all subjects for their any current or preexisted injury or laxity, circulatory or neurologic conditions in their back, hip, knee or ankle. This might have influenced their normal gait pattern.

40.2.2 Study Design

In this study, repetitive experiments were conducted, and measurements had been noted. Data were collected in two different sessions. The prosthetic knee joint only was altered in both sessions. In this study, only prosthetic knee joint acted as an independent variable. To remove nonuniformity due to the socket, suspension and prosthetic foot, socket system, suspension system and foot remained same during both trials for each subject. Firstly, all subjects were suggested to walk for five trials with their self-selected speed. Average speed reported with Jaipur knee joint was noted. When the first session ended, the Jaipur knee joint was replaced by 3R15 knee joint. The instructions were given to subjects to use the 3R15 knee joint till the clinician did not find their gait stabilized with the newly replaced knee joint. Average acclimation period was 13.6 weeks, mentioned in Table 40.2. This acclimation period is based on the clinical observation in gait laboratory testing and feedback of patients. Participants were called again to collect data with 3R15 knee joint at gait laboratory. Again, data were collected at comfortable speed. Average speed was considered at the self-selected pace. Also, in the second session, average speed was noted. Thus, all subjects have completed the full necessary requirements of this study with each type of knee joint, respectively.

Table 40.2 Details of eleven participants volunteered in this investigation

Parameters	Side of amputation	Average age (years)	Duration of amputation (years)	Mass (kg)	Height (m)	Acclimation periods (weeks)
Mean value	Right (8), left (3)	46	12–29	51 ± 14.3	1.51 ± 0.1	12.4–14.9

40.2.3 Prosthesis Fitment and Alignment

Prosthesis alignment means to ensure relative positioning and to maintain the correct orientation of the prosthetic components. Alignment of the prosthesis affects patient's comfort level and its function. Improper socket fitting may result in nonuniform distribution at the socket and limb interface. This further may lead to pain, severe tissue damage and discomfort. Besides this, misalignment in prosthesis components may cause difficult flexion or reduction in knee stability. Correct alignment allows for optimal gait, optimal pressure distribution across the stump, optimal stability, optimal control and reduced energy consumption. To achieve this, bench alignment and static alignment were carried out under the direction of experienced and trained technicians at the gait laboratory.

40.2.4 Gait Analysis

Kinematics data were recorded by using six VICON cameras having a frame rate of 60 frames per seconds and used infrared (IR) light-emitting diode strobes. The location of cameras was set such that they recorded the required motion at the hip, knee and ankle joints from both sides. These reflective markers are placed at prominent bony positions to ensure anatomic coordinate systems. The marker locations were recorded and processed with the help of motion capture system for pelvis points, thigh point positions and foot data. The first data set was captured for each subject in the standing and static position to determine the location of the join centers. Ground reaction forces were measured using a strain-gauzed AMTI force plate of dimensions (508 × 460 mm). The force plate was fixed at the center of an 8-m-long walkway which is tuned to a video system. The 3D marker coordinates and AMTI force plate data were utilized as the input to a commercial software program, VICON 612 Data Station, to estimate the 3D joint kinematics. Al-Fatafta et al. studied the repeatability of knee joint's kinematics performance in different planes during stair climbing [21]. Richard et al. estimated knee kinematics using multibody optimization [22]. Nicholas et al. evaluated the kinematics and kinetics parameters of knee for different landing applications [23]. Maria et al. also explained the kinematics measurements of human gait [24]. During gait analysis, Kramer [25] observed that ground reaction force providing a basis for comparison of biomechanical functions in different knee joints. Kramer also highlighted the comparison of loading, safety and patient's effort as shown in Fig. 40.1.

Gait cycle periods were selected and recorded between two successive heel strike events. All intermediate state timings during a gait cycle were articulated as a percentage of the gait cycle, without considering the actual time for a stride, to represent a normalized gait cycle. Subjects were asked to walk comfortably in the forward and reverse direction on the pathway as a trial basis to capture right or left side data. Out of many trials, the best five steps were captured for each right and left legs. The

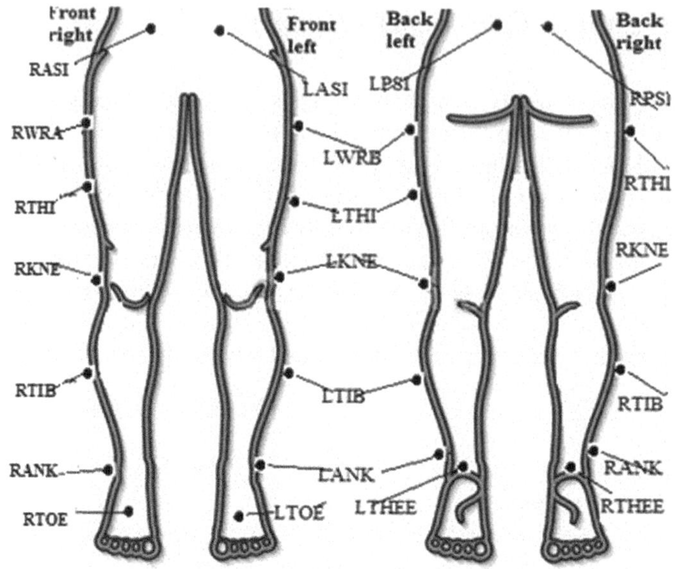

Fig. 40.1 Schematic representation for positioning of marker used in this study

spatiotemporal gait parameters like stump length (the distance between hip joint and the distal end of stump) and stump muscle strength were examined in this study. Kendal suggested this measurement and assigned grades between 0 and 5 [26] and same is represented in Table 40.3.

Besides this, flexor and extensor, muscle strength measurement was done using dynamometer with accuracy of ±1% [27]. Pearson correlation coefficient indicated the spatiotemporal parameters and stump musculature strength. Paired *t*-test was used to find out the difference between walking patterns of the subjects with 3R15 and Jaipur knee joints. Moreover, Shapiro–Wilk test was also used to determine the normal distribution with significant level 0.05.

Table 40.3 Kendal's suggested grades [26]

Grade 0	Grade 1	Grade 2	Grade 3	Grade 4	Grade 5
Complete paralysis	Flicker of contraction present	Active movement with gravity	Active movement against gravity	Active movement against gravity and some resistance	Normal power

40.3 Results

The strength of hip flexor muscle in the prosthetic side and sound side was used, respectively, as 99.05 ± 23.06 N and 109.27 ± 26.98 N. There was not any significant difference between the hip extensor muscle force in prosthetic and intact sides (p-value > 0.05).

Mean values of various spatiotemporal parameters obtained during gait with Jaipur knee and 3R15 knee joints are mentioned in Table 40.5. The average walking speed noted with 3R15 joint was considerably lower than that measured with Jaipur knee, 29.40 ± 0.35 m/min and 61.58 ± 1.7 m/min, respectively (p-value = 0.05). During gait with 3R15 joint, stride length was 26.7 ± 15.8 cm compared to 42.81 ± 16.8 cm for Jaipur joint (p-value = 0.05). Also, while walking with 3R15 joint, a difference has been observed in percentage of stance phase than that with Jaipur knee joint; this was noted as 63.9 ± 5 for 3R15 and 49.6 ± 3 for Jaipur joint as observed from Table 40.5.

During gait, the maximum angle of the knee joint and ankle joint between the prosthetic and normal leg sides is presented in Fig. 40.2 and Fig. 40.3. During the gait analysis, various kinematics parameters for both knee joints are shown in Table 40.6. It is evident that with Jaipur knee, the maximum knee flexion angle was 8.47° ± 5.6°, with respective to 8.95° ± 8.1° for 3R15 (p-value = 0.25). A remarkable difference is observed in maximum angles of the prosthetic knee and ankle joints between the normal leg and prosthetic leg sides as in Figs. 40.2, 40.3 and Table 40.6.

Pearson correlation test resulted a high significant correlation between gait parameters and stump muscle strength ($R = 0.81$).

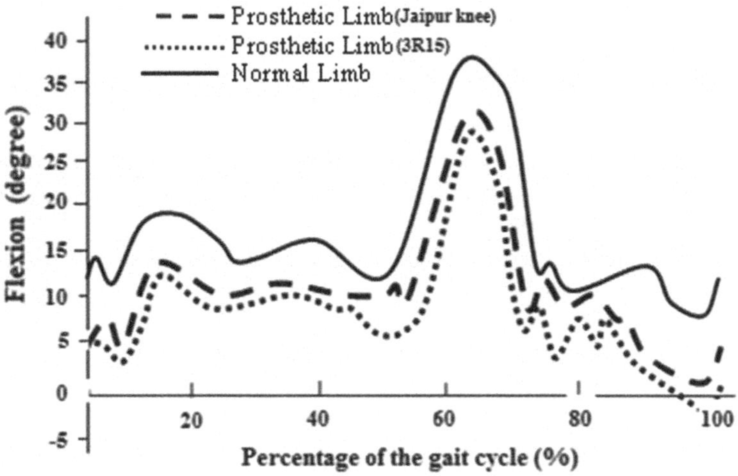

Fig. 40.2 Knee joint angle pattern in the normal side and in prosthetic side with Jaipur knee and 3R15 knee joint

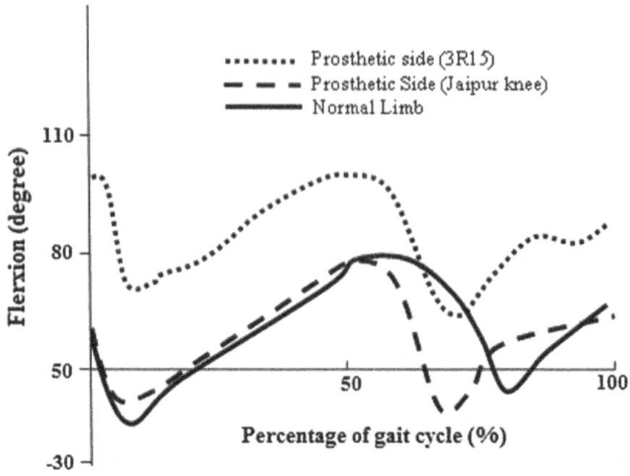

Fig. 40.3 Ankle joint angle pattern in the normal leg and in prosthetic leg sides with Jaipur knee and 3R15 knee joints

40.4 Discussion

The role of knee joint is most vital and imperative in case of above-knee amputation. A variety of mechanisms and components of prosthesis have been developed to enhance the walking and standing performance of the subjects. The knee joint has drawn a great deal of attention from researchers and scientists as compared to another add-on component. A diversified range of knee joints is developed to improve stability in stance phase and mobility in swing phase [16, 17, 28]. A significant improvement took place in conventional knee designs [17]. This is also evident from the result of this research as shown in Tables 40.4, 40.5 and 40.6, as Murray et al. [29] concluded a significant gap between the performance of normal subjects

Table 40.4 Mean value of the stump lengths and strength in both prosthetic and intact side

Parameter	Stump length (cm)	Muscle strength (N)			
		Hip flexor	Hip flexor	Hip extensor	Hip extensor
		Prosthetic side	Intact side	Prosthetic side	Intact side
Mean value ± SD	25.34 ± 5.7	99.05 ± 23.06	109.27 ± 26.98	114.6 ± 50.1	116.08 ± 25.1

Table 40.5 Two types of knee joints and their gait spatiotemporal data

Parameters	Step length (cm)	Cadence (step/min)	Stance (%)	Walking speed (m/min)
Normal side	55.46 ± 26.1	59.39 ± 17.41	57.2 ± 2	65.87 ± 1.01
Jaipur knee	42.81 ± 16.8	71.9 ± 11.17	49.6 ± 3	61.58 ± 1.7
3R15	26.7 ± 15.8	52.2 ± 12.7	63.9 ± 5	29.40 ± 0.35

Table 40.6 Maximum angle: knee and ankle joints; maximum accelerations: hip, knee and ankle joints (in the sound and amputee sides)

Parameters	Knee		Ankle		Max acceleration		
	Stance	Swing	Stance	Swing	Knee	Ankle	Hip
Normal side	48.99 ± 22.1	30.1 ± 14.8	50.95 ± 33.5	80.2 ± 15	7.10 ± 6.3	17.15 ± 12.9	4.4 ± 1.30
Jaipur knee	8.47 ± 5.6	22.7 ± 14.8	57.6 ± 26.9	83.5 ± 20.6	8.6 ± 5.3	25.4 ± 13.1	5.2 ± 1.4
3R15	8.95 ± 8.1	21.8 ± 9.5	50.9 ± 34.2	68.52 ± 12.9	10.20 ± 10.5	25.5 ± 14.0	4.3 ± 1.27

and above-knee amputees. Remarkably, the performance of the normal subjects is considerably higher than that of the amputees.

The cadence mean value (rate at which a person can walk step per minute), stride length (distance between two consecutive points of the initial contact for same foot) and speed while regular walking were observed 110 ± 6 steps/min, 1.4 m and 87 m/min, respectively, [29] with respect to the resulted speed 71.9 steps/min, 1.19 m, 61.58 m/min in walking with Jaipur knee joint and 52.2 steps/min, 0.90 m and 30.01 m/min with 3R15 knee joint, respectively, as indicated in Table 40.5. The results illustrated a noteworthy variation between values of spatiotemporal parameters obtained from both sessions. Farahmand et al. [30] examined the average value of stride length and walking speed 1.003 ± 0.12 m and 66.8 ± 1.96 m/min, respectively, for artificial limb. In other studies, the spatiotemporal gait parameters were not reported with much variation in the artificial and natural limbs [9, 25, 30], while the average values of the gait parameters are found appreciably lower with Jaipur knee joint in comparison with the natural leg side (p-value $= 0.05$).

The stance phase range in the natural leg was 57.2% compared to 49.6% with Jaipur joint as mentioned in Table 40.5. It may be explained that the amputees consumed much time duration in the swing when it compared to the stance phase with Jaipur joint. It indicates that Jaipur knee joint compensates the missing functionality and load bearing restraint resulted by an amputation [31]. The mean speed of the amputee with Jaipur knee joint was observed 61.58 m/min in contrast to 30.01 m/min for the amputee with the 3R15 knee joint. This disparity may be linked with possibility of cadence and larger step length with Jaipur knee joint. The peak of flexion angle of the knee joint in Jaipur knee was near to that in the case of regular walking as shown in Fig. 40.2. Consequently, it may be observed that gait balance with Jaipur knee joint is closer to that of regular walking. Spatiotemporal parameters for the Jaipur knee joint were also considerably improved than that of 3R15 knee joint in swing phase. Similarly, from Fig. 40.3, it is evident that pattern of ankle joint with Jaipur knee is more aligned to normal limb than in the case with 3R15 joint.

During gait, stump muscle strength, for flexor position and extensor positions, responds a momentous effect on the gait performance. Likewise, from Table 40.4, the strength of the stump was approximately the same in flexor and extensor, as in the case of natural leg. In this research, considerable correlation was observed between the stride length, cadence, average walking speed and the muscle strength. Besides, the degree of successful walking with prosthetic knee joints depends on the stump muscle power. It is evident that in surrounding the hip joint, the muscle power was closer to that of natural leg, which resulted in small difference in the values of angular speeds for the both leg joints like hip and knee. Few limitations to be acknowledged about this research are that only kinematics parameters were examined, and this study did not cover the patterns of upper limb motions during investigation.

40.5 Conclusions

It can be concluded that walking speed with single-axis Jaipur knee is found more than that of 3R15 knee joint in this study. More symmetry was observed in gait pattern of above-knee amputees with Jaipur knee joint than that of 3R15 knee joint. The number of subjects was a limitation in this study. For that reason, it is suggested to include a greater number of subjects in further study. It is also suggested that during gait, applied force on the limb and the motion on pelvic joints like constraints should be examined for further study.

Acknowledgement The study was supervised by Indian Institute of Technology—Indian School of Mines (IIT-ISM), Dhanbad, under the Ph.D. program. Gait laboratory support was provided by Research and Referral Army hospital, New Delhi. We acknowledge the motivation from Bhagwan Mahaveer Viklang Samiti (BMVSS), Jaipur. We acknowledge the cooperation of all subjects who participated in the study; without their support, the study was not possible at all. The authors are solely responsible for the contents and declare to present necessarily their personal view only toward the study.

Conflict of Interest The authors declare no conflicts of interest.

Source of Support Nil.

References

1. ColladoVazqueza, S., Carrillo, J.: Balzac and human gait analysis. Neurologia **30**(4), 240–246 (2015)
2. Ramakrishnan, T., Lahiff, C., Reed K.B.: Comparing gait with multiple physical asymmetries using consolidated metrices. Front. Nerorobotics **12**(2) (2018)
3. Jaroslav, U., Jandacka, D., Farana, R., Zahradnik, D., Rosicky, J., Janura, M.: Kinematics of gait using bionic and hydraulic knee joints in transfemoral amputees. ActaGymnica **47**(3), 130–137 (2017)
4. Disabled persons in India: Report No. 485 (58/26/1). National Survey Organisation, Ministry of Statistics and Programme Implementation, Govt. of India (2012)
5. The 50 Best Inventions of 2009: The $20 knee. TIME (2009)
6. Orendurff, M., Segal, A.D., Klute, G.K., Mcdowell, M.L., Pecoraro, J., Czerniecki, J.: Gait efficiency using the c-leg. J. Rehabil. Res. Dev. **43**, 239–246 (2006)
7. Winter, D.A.: The Biomechanics and Motor Control of Human Gait: Normal, Elderly and Pathological, 2nd edn. Waterloo Biomechanics, Waterloo (1991)
8. Boonstra, A.M., Schrama, J.M., Eisma, W.H., Hof, A.L., Fidler, V.: Gait analysis of transfemoral amputee patients using prostheses with two different knee joints. Arch. Phys. Med. Rehabil. **77**, 515–520 (1996)
9. Bae, T.S., Choi, K., Hong, D., Mun, M.: Dynamic analysis of above-knee amputee gait. ClinBiomech **22**, 557–566 (2007)
10. Kadaba, M.P., Ramakrishnan, H.K., Wootten, M.E.: Measurement of lower extremity kinematics during level walking. J. Orthop. Res. **8**, 383–392 (1990)
11. Moisio, K.C., Sumner, D.R., Shott, S., Hurwitz, D.E.: Normalization of joint moments during gait: a comparison of two techniques. J. Biomech. **36**, 599–603 (2003)
12. Lim, P.: Advances in prosthetics: a clinical perspective. J. Phy. Med. Rehab **11**, 13–33 (1997)

13. Kaufman, K.R., Levine, J.A., Brey, R.H., Iverson, B.K., Mccrady, S.K., Padgett, D.J., et al.: Gait and balance of transfemoral amputees using passive mechanical and microprocessor-controlled prosthetic knees. Gait Posture **26**, 489–493 (2007)
14. Jepson, F., Datta, D., Harris, I., Heller, B., Howitt, J., Mclean, J.: A comparative evaluation of the Adaptive knee and Catech knee joints: a preliminary study. J. Prosthet. Orthot. Int. X **32**, 201–218 (2010)
15. Lythgo, N., Marmaras, B., Connor, H.: Physical function, gait, and dynamic balance of transfemoral amputees using two mechanical passive prosthetic knee devices. Arch. Phys. Med. Rehabil. **91**, 1565–1570 (2010)
16. Yang, L., Solomonidis, S.E., Spence, W.D., Paul, J.P.: The influence of limb alignment on the gait of above-knee amputees. J. Biomech. **24**, 981–997 (1991)
17. Sapin, E., Goujon, H., De Almeida, F., Fode, P., Lavaste, F.: Functional gait analysis of transfemoral amputees using two different single-axis prosthetic knees with hydraulic swing-phase control: kinematic and kinetic comparison of two prosthetic knees. Prosthet. Orthot. Int. **32**, 201–218 (2008)
18. Lee, S., Hong, J.: The effect of prosthetic ankle mobility in the sagittal plane on the gait of transfemoral amputees wearing a stance phase-controlled knee prosthesis. Proc. Inst. Mech. Eng. H **223**, 263–271 (2009)
19. Taheri, A., TaghiKarimi, M.A.: Evaluation of the gait performance of above-knee amputees while walking with 3R20 and 3R15 knee joints. J. Res. Med. Sci. **17**(3), 258–263 (2012)
20. Mishra, P., Singh, S., Ranjan, V., et al.: J. Med. Syst. **43**, 55 (2019)
21. Al Fatafta, H., Liu, A., Hutchins, S., Jones, R.: Knee joint kinematics and kinetics during the force plate attachable interlaced stairway climbing (FPAIS) in healthy individuals. Prosth. Orth. Open J. (2017)
22. Richard, V., Lamberto, G., Lu, T.W., Cappozzo, A., Dumas, R.: Knee kinematics estimation using multi-body optimization embedding a knee joint stiffness matrix: a feasible study. Plos One **11**(6), 1–18 (2016)
23. Nicholas, R.H., Rafferty, D., Meleesa, F.W., Simonson, A., Lovalekar, M., Reinert, A., Sell, T.: Lending kinematics and kinetics at the knee during different landing tasks. J. Athl. Train. **52**(12), 1101–1108 (2017)
24. Maria, S., Pablo, C.: Human gait kinematics measurement. Open J. Orthop. **7**, 79–89 (2017)
25. Kramer, K., Sorgel, Varady, P., Brand, A., Augat, P.: Evaluation of different prosthetic knee joint systems by means of gait analysis. Gait Posture **39**(1), S85 (2014)
26. Kendall, F.P., Mccreary, E.K., Provance, P.G.: Muscles: testing and function, with posture and pain. Williams & Wikins, North America (2005)
27. Harkonen, R., Harju, R., Alaranta, H.: Accuracy of Jamar dynamometer. J. Hand Ther. **6**, 259–62 (1993)
28. Su, P.F., Gard, S.A., Lipschutz, R.D., Kuiken, T.A.: Gait characteristics of persons with bilateral transtibial amputations. J. Rehabil. Res. Dev. **44**, 491–501 (2007)
29. Murray, M.P., Drought, A.B., Kory, R.C.: Walking patterns of normal men. J. Bone Jt Surg. **49**, 335–360 (1964)
30. Farahmand, F., Rezaeian, T., Narimani, R., Hejazi Dinan, P.: Kinematic and dynamic analysis of the gait cycle of above-knee amputees. ScientiaIranica **13**, 261–271 (2006)
31. Fitzlaff, G., Sepp, H.: Lower limb prosthetic components: design, function and biomechanical properties. VerlagOrthopadie-Technik (2002)

Chapter 41
Dimensional Variation and Wear Analysis of 3D Printed Pattern for Sand Casting

S. U. Sapkal, I. C. Patil and S. K. Darekar

Abstract Additive manufacturing (AM) technology has proven its importance in the field of foundry technology by means of its use for master pattern preparation and investment casting pattern (lost PLA). Huge research has been reported for the strength of the 3D printed part and its use as functional part for different applications. 3D printed sand casting pattern shows its potential where components of complex shapes are required to be sand cast in batch quantity if it is in development stage or on breakage of pattern during ongoing production of castings. 3D printing by using Fused Deposition Modeling (FDM) is a fast process, so till the complete repair of the broken pattern 3D printed pattern can be used. This study emphasis on dimensional variation and wear analysis of FDM printed pattern for sand casting. For this investigation, materials are tested against pneumatic ramming for the abrasive wear and strength required for the sand casting pattern. Also, to improve the life of 3D printed sand casting pattern, attempts of post-processing are made by two different methods, namely electroplating and abrasive resistant coating.

Keywords Additive manufacturing · Casting pattern · Electroplating · Abrasive resistant coating

41.1 Introduction

Additive manufacturing (AM) technology is used for the manufacturing part for engineering industry like aircraft and automotive products as well as medical industry such as dental restorations and medical implants. Nowadays, this technology, especially the fused deposition modeling (FDM), is found to be a cost-effective method and useful for making patterns due to its advantages such as design freedom, incorporation of intricate features, reduced labor, and speed of manufacturing. With the advancement in technology, filament materials are also evolving and providing more strength that makes the use of 3D printed part for industrial application. Though FDM process is economical, favorable, and easy to use than any other AM process, it has found many limitations in terms of surface quality and part accuracy. Alafaghani et al.

S. U. Sapkal (✉) · I. C. Patil · S. K. Darekar
Walchand College of Engineering, Sangli, India

S. Yadav et al. (eds.), *Proceedings of International Conference in Mechanical and Energy Technology*, Smart Innovation, Systems and Technologies 174,
https://doi.org/10.1007/978-981-15-2647-3_41

[1] studied six parameters and their effect on the dimensional accuracy. They found that building direction, extrusion temperature, and layer height play a significant role to improve the mechanical property and dimensional accuracy. Mohamed et al. [2] investigated the effect of layer thickness, air gap, raster angle, build orientation, road width, and number of contours on the printing speed and material consumption. They found that layer thickness is the most significant parameter which affects the printing speed and feedstock material consumption. Shrinkage is one of the defects detected in the 3D printed part due to which its surface finish gets affected. The results of the research of Sood et al. [3] are very helpful in investigating the life of patterns. They have performed thirty-two experiments considering five variables that are layer thickness, orientation, raster angle, raster width, and air gap. The observations are taken by using the scanning electron microscope (SEM) for adhesion wear, crushing, crack development, pit formation, and abrasion. To improve the properties of the filament material used in 3D printing particularly for FDM process, Kumar and Panneerselvam [4] carried out the research to find out the effect of reinforcement of nylon by glass fiber on the two-body abrasive wear. They concluded that reinforcement significantly increases the strength of the nylon and reduces the wear. Mohamed et al. [5] studied the creep resistant property of the PC-ABS filament material. Their focus was on the raster fill angle, raster air gap, layer thickness, etc. They have developed model for the calculation of creep displacement. Utilizing the result of developed model, they generated response surface and found that lower the raster angle, lower the creep displacement. A higher air gap can also be helpful in reducing the creep displacement. For utilizing the advantages of 3D printing technology, it is essential to know the surface wear of the FDM printed part due to sliding.

Saleh et al. [6] contributed in investigating the effect of electroplating on the mechanical properties such as Young's modulus and impact strength of 3D printed parts. They claimed that electroplating improves the surface finish of the 3D printed parts. Jin et al. [7] developed a theoretical model for the geometrical representation of filament during chemical finishing. They built the specimen by using polylactic acid (PLA) material followed by chemical post-processing with dichloromethane vapors. The chemical profile is observed by a laser microscope, and the results are compared with theoretical results. They noted that chemical polishing not only reduces roughness of the specimen but also improves toughness.

Jain et al. [8] mentioned the cost comparison and quality check between simple sand casting and lost PLA casting. They found that the cost of lost PLA method is more than traditional sand casting, but the accuracy and finishing of the casted part by the lost PLA method are very high compared with the traditional method. For this research, they considered a case study of casting of middle disk of Oldham coupling. They have suggested that to get zero defect casting at one go, it would be better to use the lost PLA method though the cost is high. Singh et al. [9] found very important use of additive manufacturing in the field of biomedical implants. They suggested that the standard part used as implants may not work efficiently and create problems. With the use of the additive manufacturing, we can produce the individual implants with less cost as compared with investment casting method used traditionally. The only problem with the FDM patterns is its surface finish. They tried to improve the surface finish of the prints with the use of acetone vapor bath, electroplating, and abrasive flow polishing. Among all these techniques, they found that vapor bath has enough potential of post-processing that fulfills the required surface finish in the field of biomedical implants. Mustafa et al. [10] considered the parameters' strength and cost, while they studied the effect of cost by means of time and volume.

It has become a practice to use additive manufacturing in foundry technology to build master patterns which lead to less manufacturing time, but it has not commonly observed that 3D printed patterns are used directly as sand casting pattern. Therefore, attempts are made to investigate the wear and dimension of the 3D printed pattern for polymers. In this paper, the research work related to use of different materials and post-processing methods for 3D printed patterns and the effects on dimensional variation are reported. The remainder of the paper is organized as follows: Sect. 41.2 provides the description of materials and parameter selection along with preliminary experiments. Section 41.3 deals with experimentation and measurements. The results are discussed in Sect. 41.4, and the conclusions are mentioned in Sect. 41.5.

41.2 Parameters' Selection and Preliminary Experiments

In FDM, filament of various polymers is used as raw materials which get heated and extruded to create layer-by-layer formation of three-dimensional geometry. Commonly used materials for filaments are acrylonitrile butadiene styrene (ABS) and PLA. Other than these materials, polycarbonate (PC), PC-ABS, nylon, polyethylene terephthalate glycol (PET-G), high impact polystyrene (HIPS), and thermally conductive plastics (TCP) are also available as filament materials. Based on filament properties, price, and availability of printing facility, ABS, PC-ABS, and nylon are selected as filament materials in this study. FDM printing with nylon and PC-ABS is a difficult task as nylon absorbs moisture, and PC-ABS requires controlled surrounding temperature to avoid distortion. On the basis of a literature survey, standards are given by filament manufacturer and 3D printer capabilities; the process parameters are selected and their values are reported in Table 41.1. As per the requirement of filament materials under consideration different values of bed temperature, nozzle temperature and printing speed are selected. All other parameters such as layer thickness, infill percentage, raster angle, raster width, air gap, number of contour, and infill pattern are kept same for all the three filament materials under consideration.

The FDM printer used for pattern printing is as shown in Fig. 41.1. In order to ensure the accuracy of the 3D printer, samples are prepared by setting the parameter values 0.2 mm for layer thickness, 0.4 mm for raster width, and 45° for raster angle. The 3D printed samples are tested under digital microscope as shown in Fig. 41.2.

The measured values of layer thickness, raster width, and raster angle are 0.204 mm, 0.401 mm and 45.5°, respectively. As the maximum variation between set values and obtained values is not more than 2%, the printer is considered to be suitable for pattern making.

Table 41.1 FDM process parameters

No.	Parameter	PC-ABS	Nylon	ABS
1	Bed temperature	110°	65 °C	100 °C
2	Nozzle temperature	275°	260 °C	240 °C
3	Printing speed (mm/s)	50	30	35
4	Layer thickness (mm)	0.2	0.2	0.2
5	Infill percentage (%)	60	60	60
6	Raster angle (°)	45	45	45
7	Raster width (mm)	0.4	0.4	0.4
8	Air gap	0	0	0
9	Number of contours	3	3	3
10	Infill pattern	Honeycomb	Honeycomb	Honeycomb

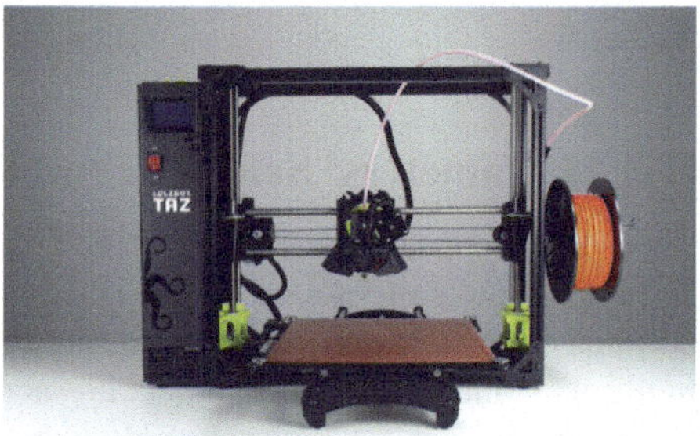

Fig. 41.1 FDM facility

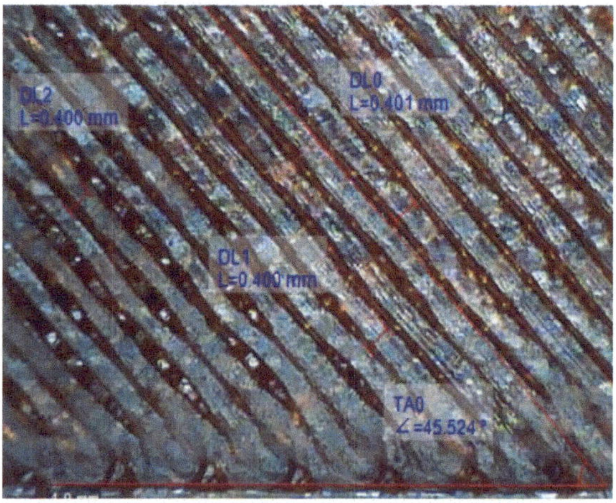

Fig. 41.2 Image captured by digital microscope

41.3 Experimentation and Measurement

In order to test most of the dimensions on various surfaces, the following model as shown in Fig. 41.3 is selected. The tolerance provided on the pattern is ±700 μm.

The 3D model of a sample under consideration is converted into .stl file and imported into FDM printing software. By setting the parameters as mentioned in Table 41.1, the sample patterns are 3D printed for three filament materials under consideration. Figure 41.4 shows three printed sample patterns of nylon, ABS, and PC-ABS materials.

The molds by using these patterns are prepared, and then, ramming operation is carried out in cycle's steps. The pneumatic ramming is performed, and for predicting the life of the pattern, dimensional wear of test pattern is considered. The dimensions of various surfaces of the pattern are measured by using coordinate measuring machine as shown in Fig. 41.5.

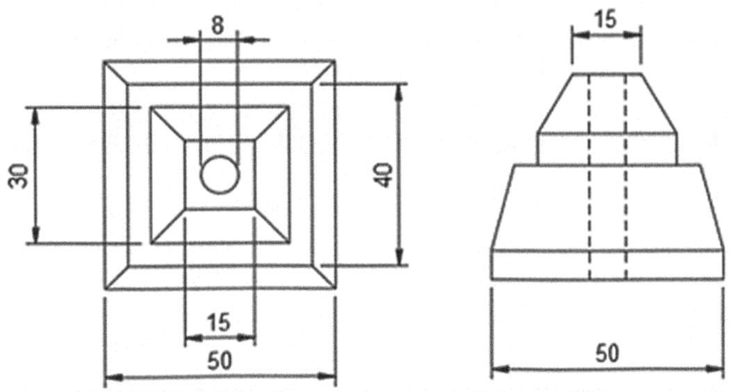

Fig. 41.3 Model of sample pattern

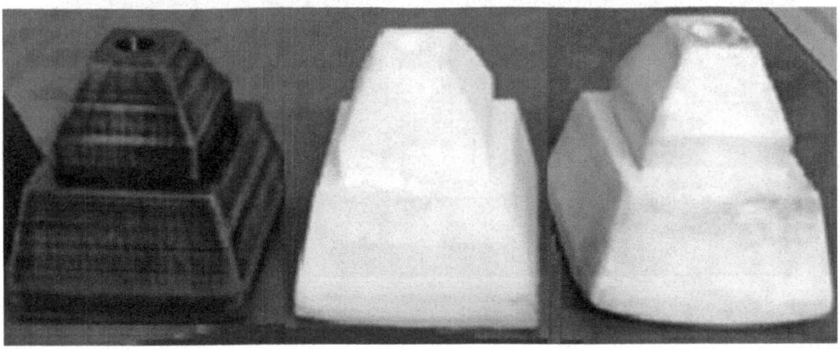

Fig. 41.4 FDM printed pattern

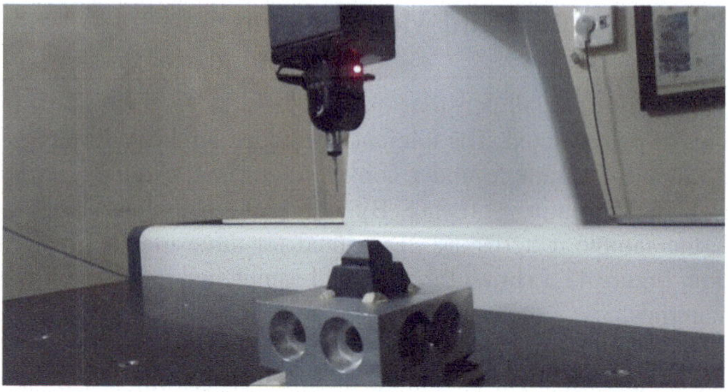

Fig. 41.5 CMM used for measurement

A total 600 ramming cycles are performed, and after an interval of 150 cycles, length of each side a, b, c, and d as shown in Fig. 41.6 are measured on CMM. For measurement, horizontal plane and inclined planes are plotted for all surfaces, and line intersection is marked to get required results.

Further, two samples of ABS material with the same dimensions are printed on FDM. On one sample wear resistant coating and on other sample electroplating is applied as shown in Fig. 41.7.

The thickness of wear resistant coating is about 250 μm, and the thickness of electroplating along with conductive coat is about 350 μm. The pneumatic ramming is carried out on molds of coated and electroplated sample patterns to check the improvement in wear.

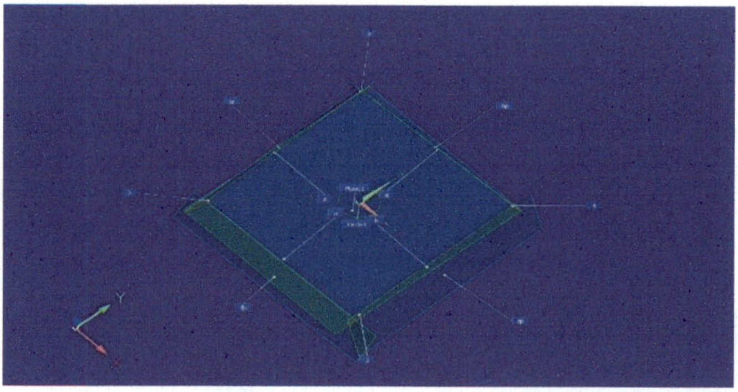

Fig. 41.6 Graphical view on CMM

Fig. 41.7 Coated and electroplated FDM patterns

41.4 Results and Discussions

Along with the change in the dimensions, the change in weight of the pattern is also measured by using a weighing machine for understanding the effect of wear. The average change in dimensions of samples and change in mass of the samples after an interval of 150 ramming cycles up to 600 ramming cycles for ABC, PC+ABS, and nylon materials are reported in Table 41.2.

The average change in dimensions of sample patterns against a number of cycles is plotted as shown in Fig. 41.8. From the graph, it has been observed that difference in wear of pattern in PC-ABS is marginal due to its inner resistant. The difference of wear in nylon samples is higher but shows the least wear conditions among the three materials used. The difference in wear of pattern in ABS is more than PC-ABS, and wear rate is more than nylon.

Table 41.2 Dimensional wear of patterns

Type of patterns	Ramming cycle (nos.)	Average change in dimension (mm)	Change in mass (gm.)	Cumulative wear (%)
ABS	150	0.07335	0.0097	7.33
	300	0.069475	0.0086	14.28
	450	0.056775	0.0036	19.96
	600	0.056025	0.0032	25.56
PC+ABS	150	0.07945	0.0101	7.81
	300	0.06217	0.0078	14.03
	450	0.069125	0.0082	20.94
	600	0.069375	0.0069	27.88
Nylon	150	0.062875	0.0053	6.30
	300	0.057625	0.0037	12.05
	450	0.04257	0.0022	16.31
	600	0.03225	0.0013	18.98

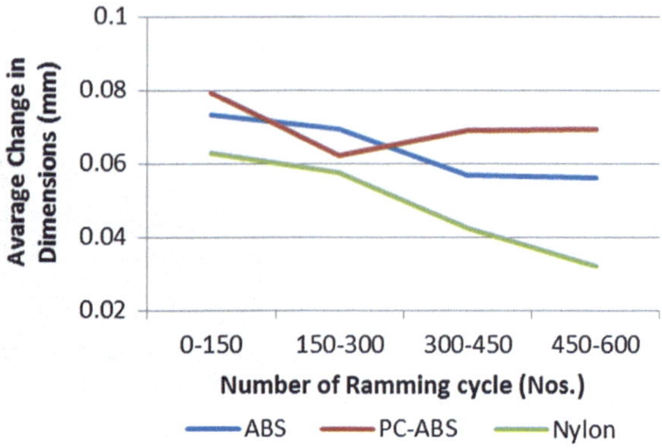

Fig. 41.8 Graph of average dimensional wear

From Fig. 41.8, it is observed that the wear rate of every sample is decreasing with the increase in the number of ramming cycles due to the fact that initially, the patterns are more rough and getting smoother with number of ramming. Variation in wear of ABS pattern is due to its less strength and stair-stepping effect of the process. There is a scope of improving wear of ABS by using post-processing methods. Also, in terms of ease of printing, ABS is a better option for casting pattern.

The attempt is made to reduce the surface wear of the 3D printed patterns of ABS by coating with the abrasive resistant coat and electroplating. Table 41.3 shows the average change in dimensions of samples and change in mass of the samples after an interval of 150 ramming cycles up to 600 ramming cycles for coated ABS and for electroplated ABS. Figure 41.9 shows the plot of average dimensional wear versus number of ramming cycles for all the samples, i.e., patterns of ABS, PC+ABC, nylon, coated ABS, and electroplated ABS.

From the graph, it is shown that the ramming of 600 cycles found very little effect on change in dimension of test samples for coated and electroplated ABS. Both coating and electroplating improve the wear resistance of ABS material. Abrasives

Table 41.3 Dimensional wear of coated and electroplated pattern

Type of pattern	Ramming cycle (nos.)	Average change in dimension (mm)	Change in mass (gm.)	Cumulative wear (%)
Coated	150	0.00795	0.009	0.795
	300	0.0072	0.0003	1.515
	450	0.0081	0.0003	2.325
	600	0.007025	0.0004	3.0275
Electroplated	150	0.002525	0.0019	0.2525
	300	0.00265	0.0017	0.5175
	450	0.001325	0.0014	0.61
	600	0.002375	0.0017	0.8775

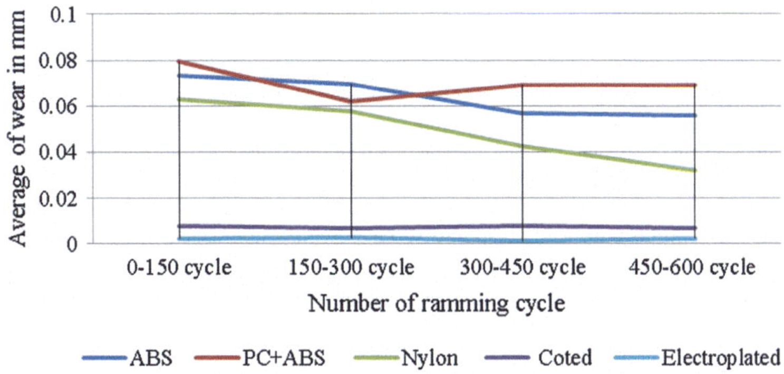

Fig. 41.9 Graphs of dimensional wear of all samples

resistant coating has shown improvement in surface quality of the coated pattern, and wear rate is reduced by approximately 85%. Though it is difficult to electroplate the 3D printed pattern, the results of the average wear are very significant as it reduces wear almost 95%. Hence, it is advisable to use post-processing technique for FDM printed pattern.

41.5 Conclusion

In this paper, the usefulness of FDM patterns for sand casting has been studied. Three different materials are taken for experimental study, and pattern of sand casting are printed by FDM printer with similar input parameters. Dimensional variations of patterns over the period of number of ramming cycles have been recorded, and dimensional wear is analyzed. In addition, post-processing techniques such as abrasive coating and electroplating are applied on the FDM printed pattern of ABS material. Among the three materials used in this investigation, ABS material can be considered as better filament material considering the ease of printing, cost of the printing, and time taken for printing. Though the dimensional wear of PC+ABS and nylon are comparable with ABS material, the coating, as well as electroplating, improves the wear property of ABS. This study can be further explored for checking compressive and impact strength and the improvement in strength of patterns due to post-processing.

References

1. Alafaghani, A., Qattawi, A., Alrawi, B., Guzman, A.: Experimental optimization of fused deposition modeling processing parameters: a design-for-manufacturing approach. In: Procedia Manufacturing, 45th SME North American Manufacturing Research Conference, NAMRC 45, LA, USA, vol. 10, pp. 791–803. Elsevier, USA (2017)
2. Mohamed, O.A., Masood, S.H., Bhowmik, J.L.: Mathematical modeling and FDM process parameters optimization using response surface methodology based on Q- optimal design. Appl. Math. Model. **40**, 10052–10073 (2016)
3. Sood, K., Ohdar, R.K., Mahapatra, S.S.: Experimental investigation on wear of FDM processed part. Adv. Mater. Res. **445**, 883–888 (2012)
4. Kumar, S., Panneerselvam, K.: Two-body abrasive wear behavior of nylon 6 and glass fiber reinforced (GFR) nylon 6 composite. In: Procedia Technology, Global Colloquium in Recent Advancement and Effectual Researches in Engineering, Science and Technology (RAEREST 2016), vol. 25, pp. 1129–1136, Elsevier Ltd. (2016)
5. Mohamed, O.A., Masood, S.H., Bhowmik, J.L.: Experimental investigation of time dependent mechanical properties of PC-ABS prototypes processed by FDM additive manufacturing process. Mater. Lett. **193**, 58–62 (2017)
6. Saleh, N., Hopkinson, N., Hague, R.J.: Effects of electroplating on the mechanical properties of stereo lithography and laser sintered parts. Rapid Prototyp. J. **10**, 305–315 (2004)
7. Jin, Y., Wan, Y., Zhang, B., Liu, Z.Q.: Modeling of the chemical finishing process for polylactic acid parts in fused deposition modeling and investigation of its tensile properties. J. Mater. Process. Technol. **240**, 233–239 (2017)
8. Jain, P., Kuthe, A.M.: Feasibility Study of manufacturing using rapid prototyping: FDM approach. Procedia Eng. **63**, 4–11 (2013)
9. Singh, D., Singh, R., Singh, K.B.: Development and surface improvement of FDM pattern based investment casting of biomedical implants: a state of art review. J. Manuf. Process. **31**, 80–95 (2018)
10. Mostafa, K.G., Montemagno, C., Qureshi, A.J.: Strength to cost ratio analysis of FDM nylon 12 3D printed parts. In: 46th SME North American Manufacturing Research Conference, vol. 26, pp. 753–762 (2018)

Chapter 42
Mechatronic Structure for Forearm Prosthesis with Artificial Skin

Monica Dascalu, David Dragomir, Daniel Besnea, Lidia Dobrescu, Ana Maria Pascalau, Dragos Dobrescu, Eduard Franti, Edgar Moraru and Anca Plavitu

Abstract The paper presents a mechatronic structure for an artificial hand that can be used in several types of prostheses, depending on performance/costs requirements. The data presented in the introductory section justify the importance of the research, since limb amputation is quantitatively significant among medical issues at the global level. The mechanical structure was realized through 3D printing, after it was designed with SolidWorks software package. For better operation flexibility and accuracy, haptic feedback was included using both pressure sensors and "artificial skin" made of Velostat. The motors that command the mobile elements are included in the empty space inside the hand, while the electronics (build around an Arduino board) is embedded in the forearm. The mechatronic structure is light, versatile and can be used in both myoelectric and neural prostheses. The main original contribution of the paper is the haptic feedback using both pressure sensors and Velostat. The work is a part of a multidisciplinary project that will use this structure in a neural prosthesis with neural bio-feedback.

Keywords Mechatronics · Artificial skin · Intelligent prosthesis · Tactile feedback

42.1 Introduction

Limb amputation is one of the medical problems with a global, alarming growth in the last years. World Health Organization and the International Society for Prosthetics and Orthotics declared, a decade ago, that approximately "*50 million people (0.5%–0.8% of world population) have limb loss, 30 million people being located in Africa, Asia and Latin America*" [1]. Advanced Amputee Solutions [2] forecasted

---

M. Dascalu (✉) · D. Dragomir · D. Besnea · L. Dobrescu · A. M. Pascalau · D. Dobrescu · E. Moraru
Politehnica University of Bucharest, Splaiul Independentei Street 313, Bucharest, Romania
e-mail: monica.dascalu@upb.ro

D. Dragomir · E. Franti
IMT Bucharest, Erou Iancu Nicolae Street 32B, Bucharest, Romania

M. Dascalu · E. Franti · A. Plavitu
ICIA, Calea 13 Septembrie Street, Bucharest, Romania

S. Yadav et al. (eds.), *Proceedings of International Conference in Mechanical and Energy Technology*, Smart Innovation, Systems and Technologies 174,
https://doi.org/10.1007/978-981-15-2647-3_42

that the number of people with amputated limbs in the USA will reach 3,600,000 by 2050. "*There are more than 1 million annual limb amputations globally*" [3]. This is due to the large number of traumatic accidents and to various diseases that lead to amputations (diabetes, peripheral blood circulation conditions, bone cancer, congenital illnesses, etc.). In Norway, "*every year, about 25,000 injuries are registered in the Norwegian Labour Inspection Authority's Registry of Work-related Injuries*" [4]. The increasing number of people with amputated hands generates acute global concern and is indicative of a high market demand. "*Every 30 min a limb is lost due to a landmine, every 20 s a limb is lost due diabets*", said Dr. David Armstrong, a professor of surgery at the University of Arizona [5]. Analyses of retail prostheses markets have pointed the deficiencies in the existing devices, which are unable to fully meet critical requirements of users [6] as well as the high expectations from the public towards this kind of products, manifested in the growth of worldwide sales (US$2.8 billion in 2010 [7], US$19.4 billion in 2015 [8]).

42.2 State of the Art

All over the world, in recent years, intelligent prostheses have created a strong interest, which is visible in the ever-increasing number of international research projects, scientific articles in specialized journals worldwide and, last but not least, in commercial products such as the artificial hands with neuronal or myoelectric command. Systematic efforts made by many interdisciplinary specialist research teams lead to significant annual progress that enriches the market with better performing artificial hands. Some centres specialized in devising intelligent prostheses are located in Taiwan, Japan, the USA, Germany and the UK. Nowadays, a great variety of forearm and hand prostheses are available in commercial products, with different levels of speed, force and degrees of freedom, reproducing partially the functional features of the human hand. Some of the best-selling models available currently on the market are the following:

- Myo-prostheses for the forearm ("i-limb ultra") from Touch Bionics [9];
- Exo-prosthetic devices from Otto Bock HealthCare GmbH [10];
- Prostheses for adults and children from Fillauer LLC [11];
- Various other intelligent prosthetic devices for upper limbs produced by smaller manufacturers (RSL Steeper, DEKA, HDT, etc.).

The best-performing hand prostheses on the market (ultra-i-limb (Touch Bionics) [12] and Michelangelo prosthetic hand (Ottobock) [13]) have fingers with only 2 mobile phalanges. The mechatronic structure described in this article has as an element of novelty 3 phalanges per fingers (except the thumb).

Nowadays, fitting the artificial hands on the amputation stump is done by means of a detachable connecting system. New modalities of implanting the connecting system in the stump are being explored but are still in the experimental and research phases [14]. From practical perspective, an artificial hand is expected to act similarly to the limb it replaces, not only in movement possibilities and control, but also regarding the sensory experience [2]. However, the best-performing artificial hands manufactured nowadays are still far from reaching the abilities of a healthy hand.

42.3 The Mechanical Structure of the Prosthesis

The mechanical structure of the prosthesis was designed according to the state-of-the-art performances of similar products. The main characteristics that define the performances of a prosthesis are the number of mobile segments, the number of articulations (joints), the number of degrees of freedom of each articulation, the force, speed and precision of operation. Other important features are the reproducibility of movements and the performances of the control system [15]. Taking into consideration that the actual utility of a prosthesis is given by its ability to execute in real time the movements intended by the patient, the mechanical structure of the prosthesis was conceived as to implement sequences of movements similar to those realized by a healthy hand.

Our model is based on the following assumptions and design options:

1. The prosthesis follows the basic structure of the human hand regarding the mobile elements, mechanical elements and articulations, with 3 phalanges per finger and 2 for the thumb; the palm and the wrist follow a functional similarity of a hinge articulation, not the human skeleton which has a large number of small bones;
2. The thumb was considered to be independent regarding the other fingers; each finger has an individual control which results in independent flexion, adduction and abduction of the fingers and thumb;
3. The mechanical structure of the palm includes the space for the acting elements of the fingers and thumb but also has to allow the attachment of the artificial skin needed for the bio-feedback;
4. The whole structure is modular and articulated, thus offering the possibility to adapt it to different levels of complexity and functionality (and costs), starting from a common basic structure; the modular architecture has another significant advantage: being composed by subsystems and devices that are combined in kinematic couples, it has a clear mechanical model, offering the information for the kinematic and dynamic analysis of the mechanical and operational systems of the prosthesis.

From the kinematic perspective, the mechanical structure of the prosthesis consists of a configuration of rigid elements, connected by rotation and translation articulations [16, 17] as the example in Fig. 42.1. In this model [18], the rotation joints (Ji) are represented by cylinders (the axis of rotation is the axis of the cylinder; for example, J5 realizes the supination/pronation movement); the segments between cylinders are solid rigid connections, and some of them equipped with tactile sensors (Ti). The model was used in the kinematic computations for the mechatronic structure described in this paper. The relative positions of the components give the

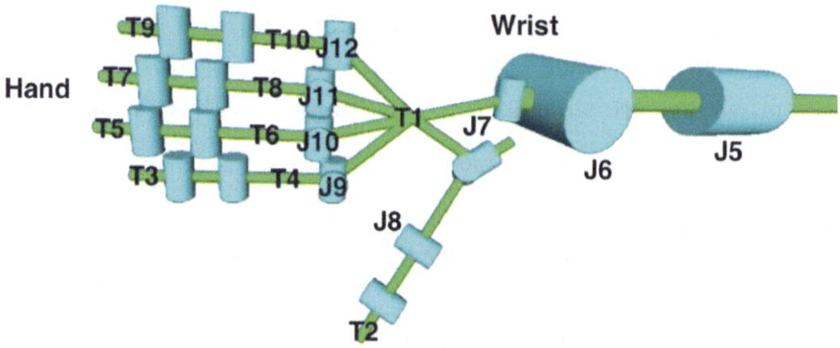

Fig. 42.1 Kinematic chain of an artificial hand [18]

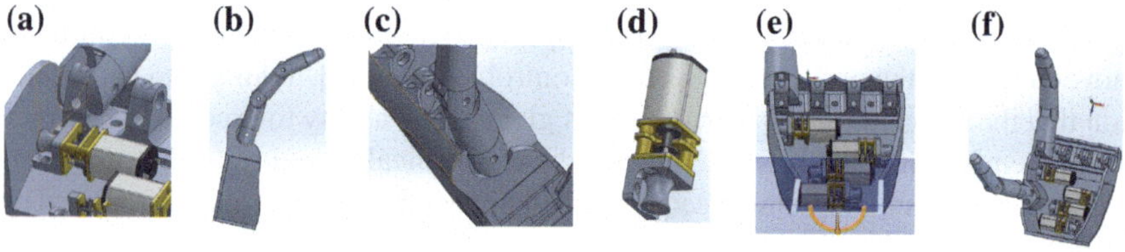

Fig. 42.2 Design of the mechanical structure

global position of the prosthesis and also represent one of its functional conditions [19]. The speed and acceleration of an element are obtained by recurrence from the parameters of the previous element. The sequence of all the transformations associated with each element allows the computation of the general movement parameters of the whole mechanical structure, which is made of the cascade connection of all the elements. Ultimately, the parameters of the terminal elements (last phalanges) can be computed. The 3D design of the prosthesis was done with SolidWorks software package.

The images in Fig. 42.2 represent different views and details of the hand of the prosthesis. A finger with 3 phalanges, side view, can be seen in image a, while its articulation with the palm is represented in image b (the detail is for the index finger). The motors (Fig. 42.2c) are mounted in the space of the palm (Fig. 42.2d, with a detail in Fig. 42.2e for the index). There is one motor for each finger, including the thumb; this motor moves all the articulations of the finger using a plastic cable. The last Fig. (42.2f) shows the interior of the palm with the motor elements, having the thumb and the index attached. The dimension of the structure is given by the dimension of the hand that it replaces but also has to accommodate all the mechanical components inside the palm compartment.

The computer design of the prosthesis allowed also the kinematic modelling and simulation of the behaviour of the mechanical structure. Some results of the kinematic simulation realized with SolidWorks are reproduced in Fig. 42.3:

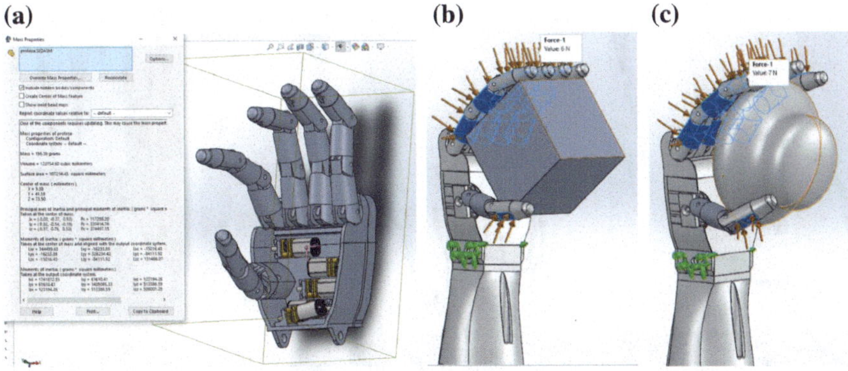

Fig. 42.3 Kinematic simulation for the behaviour of the mechatronic structure

(a) the mechanical characteristics of the structure; the results of the kinematic simulation recommend the use of this structure in a prosthesis for precise manipulation of objects up to 1 kg (the weight of the palm is 194 g);
(b–c) the regular repartition of forces in the contact points for the manipulation of an object with a regular form.

All the mechanical components of the structure were printed with a 3D printer. Technical specifications:

- the printer: 3D MakerBot Replicator [20].
- EasyPrint filament (printing temperature 185–220 °C, melting point 175–179 °C).
- printing material: polylactic acid (PLA), a biodegradable material that can be used for high-resolution 3D printing.

42.4 Implementation of the Command and Control Block of the Prosthesis

The most used method for the control of the movements of a prosthesis is the kinematic control. Although the solving of the equation set in the mathematical model may be complicated, the solution can also result from the physical implementation. No matter the model of a prosthesis, either analogic or digital (numeric), it is a reduction of the actual implementation. The most common models have 3–5 fingers (typically with 2 phalanges), having different numbers of degrees of freedom. The main challenge in the design and operation of such a prosthesis is the decomposition of the natural movements of the human hand in sequences of movements of the kinematic chain and their realization through the mechanical structure. Different fabrication stages are presented in Fig. 42.4. While the mechanical control elements are built in the interior of the hand, under the palm, the electronic circuitry of the

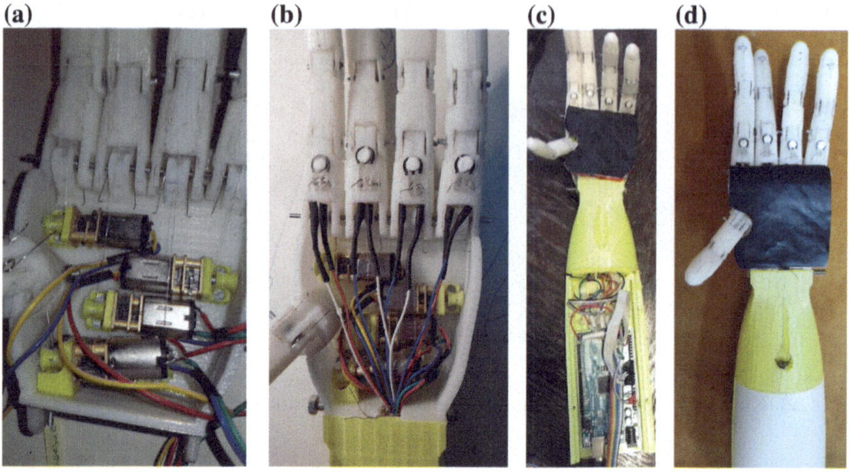

Fig. 42.4 Mechanical structure of the prosthesis. **a** The motors inside the palm for flexion of fingers; **b** pressure sensors mounted on the fingers; **c** the palm is covered with Velostat; **d** the Arduino board is embedded inside the forearm of the prosthesis

command and control block is embedded into the empty space of the prosthesis, in the forearm. The implementation uses an Arduino board for the command of five DC micromotors that move the mobile elements of the artificial hand. The motors which can be seen in Fig. 42.4a, with all the connecting wires in Fig. 42.4b, are powered at 6 V; because of their small dimensions and weight, they are appropriate for this application. The Arduino board commands keep the movement of the mobile elements in their actual range of motion, using a control loop that includes DC motor drivers and Hall sensors as shown in Fig. 42.5. Two neodim micro-magnets are mounted at the extremities of the range of motion for each finger. The rotation of the motor towards the extremities range will modify the intensity of the magnetic field, which

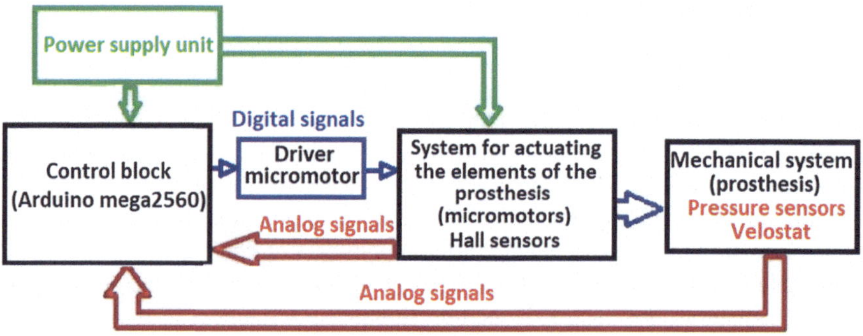

Fig. 42.5 Functional diagram of the prosthesis's command and control including the feedback loops

will trigger a negative feedback loop to stop the movement towards flexion. The outputs of the Hall sensors are connected to the analogic inputs of the Arduino board. The variation of the magnetic field is converted in a 12-digits numerical value. The extension is practically the progressive release of the flexion for each finger. Technical specifications of our implementation: the board: Arduino Mega 2560 v3; the motors: Pololu micromotors with carbon brushes, weight 9.5 g each, dimensions: 10 mm × 12 mm × 26 mm, gear ratio 250:1; the drivers: 1.8A Low Voltage Brushed DC Motor Driver from Texas Instruments; the Hall sensors: Honeywell SS441A.

42.5 Implementation of the Artificial Skin and Pressure Sensors

The palm of the mechanical structure was covered with Velostat acting as artificial skin. Velostat, polymeric foil impregnated with carbon, is a pressure-sensitive material. Its resistance reduces with either flexing or pressure. The electrical resistance of the Velostat varies depending on the mechanical pressure exerted on it (Fig. 42.6). This material has been used in the previous work of the authors in testing an artificial skin structure [21]. Compared to the pressure sensors, the artificial skin has the advantage that it fits in the shape of the palm, thus providing accurate tactile information emerging from the contact with objects in real time. The difference between the pressure sensors and the Velostat is the fact that a sensor is strictly local: it measures the pressure applied on it, while the Velostat is sensible to pressure applied in different regions, thus not requiring precision in touch. The artificial skin signals are monitored by the control block of the sensorial interface which was integrated in the control micro-system of the prosthesis. Pressure sensors were mounted on the fingers of the prosthesis. One sensor of type FSR402 for the first proximal phalange of each finger has been used. The electrical scheme and characteristic are reproduced in Fig. 42.7 from [22].

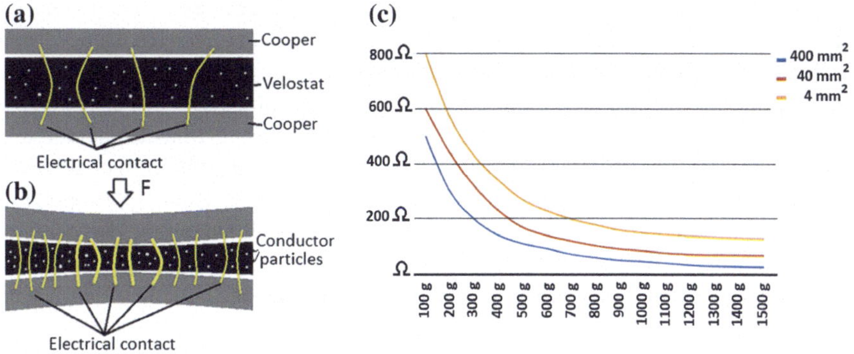

Fig. 42.6 **a, b** Functioning of Velostat [23]; **c** resistance/force characteristics [24]

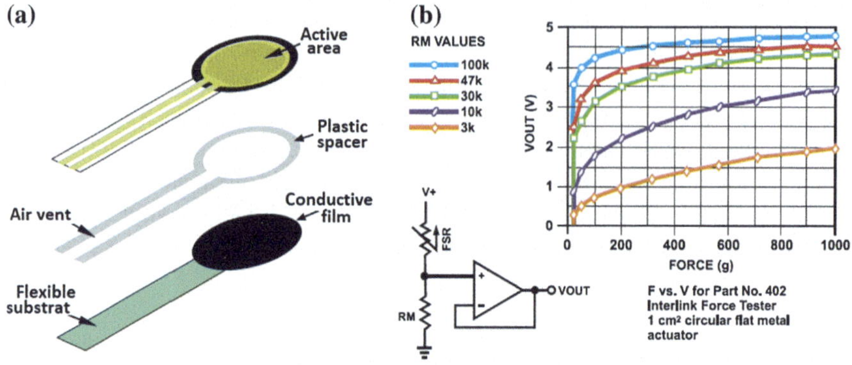

Fig. 42.7 **a** FSR402 pressure sensors; **b** electric schematics and characteristics of the sensors [22]

All the pressure sensors and also the electrical contacts of the Velostat are connected to analog inputs of the Arduino board. The variations in mechanical pressure are therefore processed by the control system in order to give the proper commands during the operation of the prosthesis, in tasks that imply manipulation of objects. Based on the information acquired from the sensors, two distinct feedback levels were implemented: an automated one directly implemented in the prosthesis and another one for bio-feedback, to allow the patient to sense the interactions of the prosthesis with the objects. At the first level of feedback, according to the pattern of the signal map on the artificial skin, the control micro-system commands action or corresponding adjustment of the motor elements in the prosthesis, for adequate manipulation of various objects.

In this way, self-adjustment decisions for position and exerted force will be taken directly at the level of the prosthesis, without requiring conscious intervention by the patient on each motor element. Under circumstances such as performing certain gestures or gripping a fragile object (without destroying it), automatic tuning will act much more rapidly and precisely as compared to the case where only patient feedback is involved (acting as command and supervision). The sensorial interfaces will grant the prosthesis a large degree of adaptability and self-management, so as to enable real-time response to complex stimuli and commands, coming from the user or from the environment where it operates. The functioning of the sensorial feedback with the pressure information acquired on fingers and palm of the prosthesis was tested using the application DataGlove set-up [25, 26] with minor modifications to include the signals from the palm. Figure 42.8 presents the monitor indicators for the variation in mechanical pressure measured from the sensors and Velostat (right). The operator pushes the index and the palm in the first image, the middle finger and the palm in the second image. According to the experiments carried out, we consider that the mechatronic structure is suitable to be used as a prosthesis for a patient with forearm amputation. Depending on the specific situation of the patient, the commands can be obtained either from the muscles of the stump [27] or the peripheral nervous system [28, 29]. The implementation of tactile feedback to the prosthesis can be used in

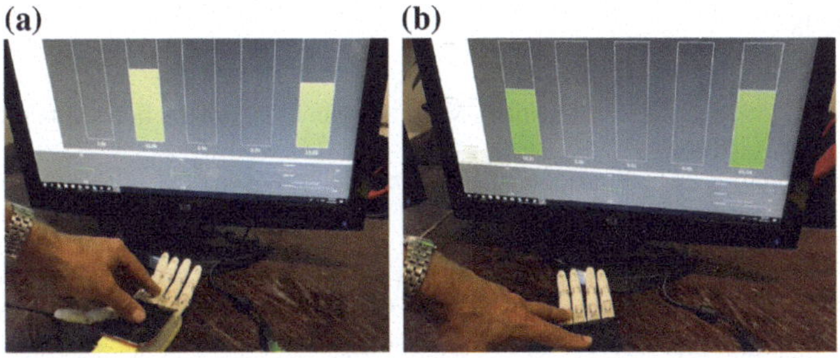

Fig. 42.8 Pressure level applied on fingers (sensors) and palm (Velostat) is monitored with the software application

future research to provide patients with different tactile sensations when handling different objects with the aid of the prosthesis, depending on the roughness of each object [30, 31].

42.6 Conclusions

The paper presents the mechatronic structure of a prosthesis that can be used for patients with forearm amputation (the electronics is embedded inside the artificial forearm). The artificial hand has 5 fingers and 3 phalanges per finger except the thumb, with only 2. Each finger is controlled by one DC motor that moves the finger in flexion. The extension of fingers is realized when the movement command is released. The mechanical skeleton of the prosthesis was printed with a 3D printer, thus allowing personalized design in relation with the patient's physical features (dimensions) and other technical specifications of our project. For better operation of the prosthesis, the structure includes pressure sensors and a Velostat layer as "artificial skin". Thus, the force applied by the motors is adjusted based on the pressure sensed on each finger and on the surface of the palm. This kind of feedback is the main original contribution of the paper. The pressure signals can also be used in a bio-feedback loop to offer direct information to the patient. The control block that processes the myoelectric or neural signals (and the acquisition of the signals) is beyond the aim of this paper. The structure has to be further developed and improved with computing algorithms for different movements and the actual implementation of the operating block, based on physiological signals.

Acknowledgements The work of this paper was done with financial support from ARMIN EEA Grant, EEA-RO-NO-2018-0390, nr. 8/2019.

References

1. Kurichi, J.E., Bates, B.E., Stineman, M.G.: Amputation. International Encyclopedia of Reabilitation, (2010)
2. Ziegler-Graham, K., MacKenzie, E.J., Ephraim, P.L., Travison, T.G., Brookmeyer, R.: Estimating the prevalence of limb loss in the United States: 2005–2050. Arch. Phys. Med. Rehabil. **89**(3), 422–429 (2008)
3. Delzell, E.: Global lessons improve amputation prevention, Low. Extrem. Rev. (June) (2011)
4. Samant, Y., Parker, D., Wergeland, E., Westin, S.: Work-related upper-extremity amputations in Norway. Am. J. Ind. Med. **55**(3), 241–249 (2012)
5. Armstrong, D.G.: Every 30 min a limb is lost due to a landmine, every 20 s a limb is lost due to diabetes (2018)
6. McGimpsey, G., Bradford, T.C.: Limb Prosthetics Services and Devices Critical Unmet Need: Market Analysis. Bioengineering Institute Center for Neuroprosthetics Worcester Polytechnic Institution (2017)
7. Hochman, P.: Bionic legs, i-Limbs, and other super human prostheses you'll envy. Fast Co. Mag. (2010)
8. Global Industry Analysts, Inc.: Orthopedic Prosthetics—A Global Strategic Business Report (MCP-3224) (2012)
9. The i-Limb Ultra-Prosthetic Hand is Designed for Those Who want more from Their Prosthesis, Touch Bionics Inc. and Touch Bionics Limited (2019)
10. Näder, H.G.: Product Development Milestones. Research & development (2019)
11. Rightmire, M.: Introducing the world's first water-resistant, multi-articulating hand, Upp. Extrem. Prosthet. (2019)
12. i-Limb® Ultra: https://www.ossur.com/prosthetic-solutions/products/touch-solutions/i-limb-ultra
13. Michelangelo Prosthetic Hand: https://www.ottobockus.com/prosthetics/
14. Burck, J.M., Bigelow, J.D., Harshbarger, S.D.: Revolutionizing prosthetics: systems engineering challenges and opportunities. Johns Hopkins APL Tech. Dig. **30**(3), 186–197 (2011)
15. Goto, S.: Industrial robotics: theory, modeling and control (force-free control for flexible motion of industrial articulated robot arm). In: Advanced Robotic Systems International, pp. 813–840, Proliteratur Verlag (2007)
16. Roy, N., Newman, P., Srinivasa, S.: Robotics: Science and Systems VIII. The MIT Press (2013)
17. Saha, S.K.: A unified approach to space robot kinematics. IEEE Trans. Rob. **12**(3), 401–405 (1996)
18. Massera, G., Tuci, E., Nolfi, S.: Evolution of Object Manipulation Skills in Humanoid Robots. Laboratory of Autonomous Robotics and Artificial Life (2019). http://laral.istc.cnr.it/res/manipulation/
19. Faizura, W., Tarmizi, B.W., Elamvazuthi, I., Begam, M.: Kinematic and dynamic modeling of a multi—fingered robot hand. Int. J. Basic Appl. Sci. IJBAS—IJENS **09**(10), 61–68 (2009)
20. Perry, M.: Method: A Manufacturing Workstation, Makerbot Education Starter Kit (2019). http://www.makerbot.com/
21. Donțu, O.G., Barbilian, A., Florea, C., Lascar, I., Dobrescu, L., Sebe, I., Scarlet, R., Mihaila, C., Moldovan, C., Patanzica, M., Besnea, D., Grămescu, B., Dobrescu, D., Lazo, V., Firtat, B., Edu, A.: Mechatronic finger structure with pressure-sensitive conductive layer. ROMJIST **20**(2), 139–150 (2018)
22. Electronic Projects Focus, Force Sensing Resistor—How it Works and its Applications, Know all about Force Sensing Resistor Technology. https://www.elprocus.com/force-sensing-resistor-technology/. Last accessed (2019)
23. Martinaitis, A., Daunoraviciene, K.: Low cost self-made pressure distribution sensors for ergonomic chair: are they suitable for posture monitoring? Technol. Health Care **26**(S2), 655–663 (2018)

24. Reps, M.: Pressure Sensor Matrix Mat Project (2019). https://reps.cc/?p=50
25. DG5 VHand 3.0, WiFi Version KIT (2019). http://www.dg-tech.it/vhand3/
26. Chidi, C., Howard, A.: Prospects of implementing a vhand glove as a robotic controller. The Tower **3**, 43–51 (2011)
27. Milea, P.L., Dascalu, M., Franti, E.D., Barbilian, A., Stoica, I.C.: Tactile feedback experiments for forearm prosthesis with myoelectric control. Roman. J. Inf. Sci. Technol. **20**(2), 101–114 (2017)
28. Moldovan, C., Dobrescu, L., Ristoiu, V., Gheorghe, R., Firtat, B., Dinulescu, S., Brasoveanu, C., Ion, M., Neagu, A., Dobrescu, D., Pascalau, A.M., Pogarasteanu, M., Oproiu, A.M., Stoica, I.C., Edu, A.: Cheng Feng Ifrim, Experimental measurements in the acquisition of biosignals from a neuronal cell culture for an exoprosthesis command. Revista de Chimie **69**(10), 2948–2952 (2018)
29. Oproiu, A.M., Lascar, I., Dontu, O., Florea, C., Scarlet, R., Sebe, I., Dobrescu, L., Moldovan, C., Niculae, C., Cergan, R., Besnea, D., Cismas, S., David, D., Muraru, D., Neagu, T., Pogarasteanu, M.E., Stoica, C., Edu, A., Ifrim, C.F.: Topography of the human ulnar nerve for mounting a neuro-prosthesis with sensory feedback. Revista de chimie **69**(9), 2494–2497 (2018)
30. Anand, A., Irfan M., Haq, U., Raina, A., Vohra, K., Kumar, R., Sharma, S.M.: Natural systems and tribology—analogies and lessons. Mater. Today Proc. **4**(4), 5228–5232 (2017)
31. Barnes, C.J., Childs, T.H.C., Henson, B., Southee, C.H.: Surface finish and touch—a case study in a new human factors tribology. Wear **257**(7–8), 740–750 (2004)

Chapter 43
Solution Model of Issues of Property of E-Governance by Using Cloud Computing

Prashanta Kumar Bhuyan and Arundhati Walia

Abstract The citizen of the country has great weakness for their properties. Property may be in various forms as land, plot, flat, etc. All are always very much bother to keep their property safe and secure. At all time, many critical issues are reported to authority at different levels for the solution, but most of the time, these issues are unattended or pending over the years. Due to lack of proper documentation and proof, law and order are unable to solve the problem at many levels. The citizen of the country acquired many unidentified properties over the years, and when government are planning for any fruitful development of that area, they are unable to do it and facing many challenges for the same. This paper aims to suggest a conceptual model with the help of cloud computing to develop a centralized system for the country to solve all the critical issues related to property, and the citizen can able to get their proper information related to their property.

Keywords E-Governance · Cloud computing · Transformation · SaaS · PaaS · IaaS · G2G · G2B · G2C · RTI · ICT

43.1 Introduction

India is a developing country, and the Government of India is taking a lot of pain in all fronts to make the country to become a developed country. E-governance is a major initiative to develop the country and to give sufficient facilities to the citizen to live their live in very simple and comfortable way. Many e-governance projects are successfully lived, and the citizens are enjoying the facility out of it. But again many core area where the Government of India must focus to give the better lifestyle to the citizen. Property issues are very common, and critical issues and citizen with law and order are wasting their quality time to solve the issues. Through e-governance with cloud computing technology, the issues can be minimized to a great extent.

P. K. Bhuyan (✉) · A. Walia
Department of Computer Science and Engineering, HRIT, Ghaziabad, India

S. Yadav et al. (eds.), *Proceedings of International Conference in Mechanical and Energy Technology*, Smart Innovation, Systems and Technologies 174,
https://doi.org/10.1007/978-981-15-2647-3_43

43.1.1 E-Governance

E-governance is the internal government operations with the help of electronic technology [1] to simplify and improve democratic, government and business aspects of governance. Modern information and communication technologies (ICT)such as Internet, LAN, mobiles, etc. by governments to improve the effectiveness, efficiency, service delivery and promote democracy [2]. E-business, e-democracy or e-government are the parts of e-governance.

43.1.2 An E-Governance Model

In an e-governance model, various concepts like government, citizens and business are given major emphasis. Various interaction areas in e-governance model, as shown in Fig. 43.1, are G2C, G2B and G2C which indicates the interaction of common citizeNs, business NGO's with the government and interaction of central government with local state government.

43.1.3 Four Phases of E-Governance

The government initiates its e-governance concepts mostly by delivering online information through various e-governance projects. According to Gartner, e-governance [4] will mature according to the following four phases:

1. Information → Presence
2. Interaction → Intake processes
3. Transaction → Complete transactions

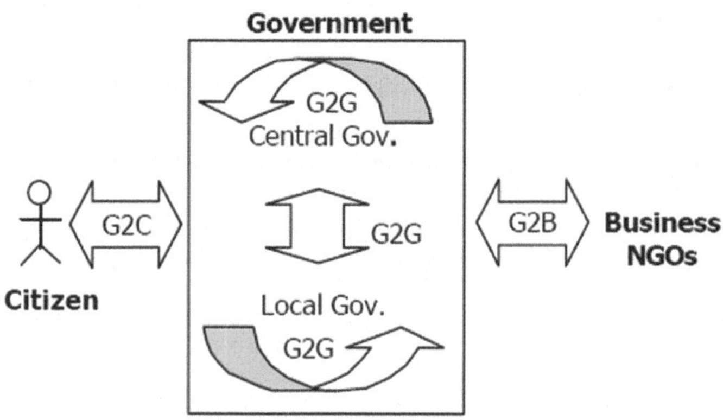

Fig. 43.1 E-governance model [3]

4. Transformation → Integration and change.

In the first phase, e-governance is providing the relevant information to public (G2C & G2B) [5] by the means of websites. (G2G) [6], where government can also disseminate static information with electronic means, such as the Internet. In the *second* phase, the interaction between government and the public (G2C & G2B) [7] is stimulated with various applications where the citizens can interact with government by asking questions via e-mail, use search engines and can download forms and documents. In third phase the complexity of the technology increases (G2C & G2B) as the complete transactions are carried out online without actually going to an office. Various online services are filing income and filing property tax, extending/renewal of licenses, visa and passports and online voting, etc. Moreover, the third phase is more complex because of security and personalization issues like digital (electronic) signatures [8] will be necessary to enable legal transfer of services. In fourth phase all information systems are integrated and the public can get G2C [2] & G2B services at one (virtual) counter [9]. One single point of contact for all services is the ultimate goal. In this phase, cost savings [10], efficiency and customer satisfaction should reach at highest possible level.

43.2 Cloud Computing

Cloud computing is a model for enabling simple, convenient, on-demand network access to a shared pool of configurable computing resources [11], e.g., networks, servers, storage, applications, and services that can be rapidly provisioned and released with minimal management effort or service provider interaction. There are three layers of cloud computing.

43.2.1 Infrastructure as a Service (IaaS)

Infrastructure as a service includes hardware/network with storage facility of the datacenter. Storage area network will be in proper place to fulfill the benefits of cloud computing in e-governance initiatives. Cloud computing architectures are able to provide a common infrastructure for all applications to work which is simple and easy to use and deploy to get maximum productivity.

43.2.2 Platform as a Service (PaaS)

Cloud offers a standard platform in providing different kinds of systems, middleware and integrated systems.

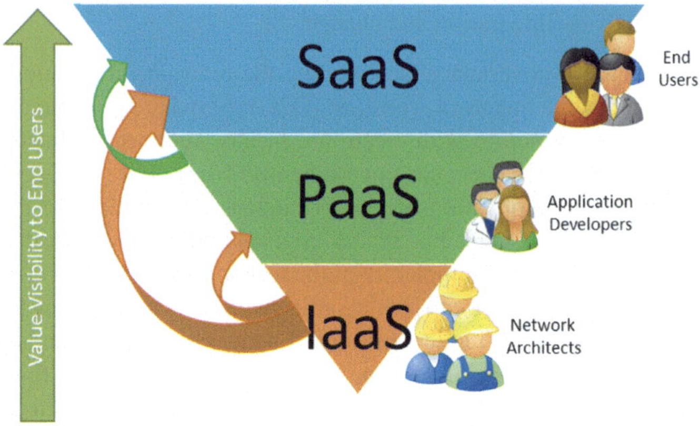

Fig. 43.2 Cloud computing service model [12]

Various platforms are OS provisioning, queuing service, database services, middleware services and workflow Services. Figure 43.2 shows cloud computing service model.

43.2.3 Software as a Service (SaaS)

Cloud computing provides different applications as a service. If a case of new district deciding to move to e-governance solution for some applications for their citizens, the district need not have to purchase applications, hardware and software. Instead, the district can make a request for a particular service from the cloud provider. Applications' instances can then be created for their use. Numerous applications [13] can be provided as standard services, where departments can request and manage. As far as security-related issues are concerned, various key risk factors are identified across the non-listed real estate funds and appreciate the fund managers and investor to take proper decision [14].

By cloud computing technology to the real estate sector, we move from traditional to smart real estate giving high quality of information related to government to bridge the gap between the real estate consumer and stakeholder [15]. Use of cloud computing technology helps real estate to save substantial amount of money since cloud providers handle the cost of infrastructure, upgrades, and license, etc. Real estate owners can work from any location online on any device, thus making life and government simple [16]. Use of cloud supports various tools in real estate sector like e-rooms where lawyers, architects, stakeholders and consultant can contribute in ongoing communication, use of virtual desktop and mobile devices for remote login [17]. In real estate industry, the SaaS technology plays a major role in interacting various software by providing better benefits to user like automatic update, easier to onboard new users of software, etc. [18]. Cloud services when applied to e-governance are very useful to reduce the software, infrastructure and platform cost [19]. Cloud-based e-governance system provides many benefits to the government like reduced cost, distributed data storage, availability of resources [20].

Government is now trying to adopt cloud computing because of non-requirement of infrastructure to offer e-services effectively [21]. Information regarding government should be kept secure. Various reviews were performed on security-related issue by applying certain technology. Unique identification number (UID) has already been implemented in India to identify genuine user.

43.3 Current Issues of Property in India

43.3.1 Issues at Village Level

Property issues are very critical at village level because the citizen is not able to get proper information. They are not able to find the proper way to make their property safe and secure. The biggest problem is transformation of generation, but the property may or may not be transferred to the concerned person at present. Government offices keep the hard copy of the document over the years, and no citizen is taking pain to take any action on it. Many times, it is seen that the property is disputed, and selling or buying of the property is very difficult. Law and order, everyday, is facing many cases related to property, and also, the citizen is not getting the proper way to resolve the issue over the years. Demarcation of property is also a biggest problem.

43.3.2 Issues at City Level

City-level proper planning is made for the citizen to acquire their property, and all properties are properly registered. But, some unauthorized disputed property is again allotted or acquired by the citizen which destroys the beauty of city. Authority faced a lot of problem to resolve the issue, and many times, the life of citizen is not comfortable with law and order. Private agencies many times made confusion to the citizen for buying and selling of property. They also do unauthorized construction to make money. Citizen are unable to get proper information from the authority for their query and always stay in dilemma while doing any activity related to property.

43.3.3 Issues at Metropolitan City Level

Metropolitan city of India is crowded due to many opportunities. Infrastructure is increasing day by day to accommodate more number of citizens. Much real state development is going on to give the better facility to citizen. Private builders are making many multi-stored building. Many agriculture lands are converted to residential. Authority is taking a lot of pain to make the city life more comfortable and to accommodate more citizens in the city. But, there is a big problem to authority as well as citizen to manage many activities manually. Many states are trying to do the record keeping in electronic form. Issues in terms of malfunctioning of property and private builders are making fool of citizen. Authenticity of property can not be known to the citizen in a proper way. Many citizens have multiple property in different states but unable to trace the property. Multiple buyers and sellers of the same property is a critical issue because of manual activity of property system.

43.3.4 Issues at Government Level

States are taking pain to make electronic form of documentation but a huge work to be done to provide the facility to the citizen related to their property issues.

43.3.5 Issues in Law and Order

Everyday, huge cases are coming to authority related to property issue. They are unable to find proper documents to settle down and resolve the issue. Many times, the process is running over the years, and citizen always getting negative result from the authority. As such, there is no centralized system to get the status of property to resolve the critical issues. Lack of man power and office-space with overload work, authority is not able to manage the work in proper time. Both citizen and authority are blaming each other without solving the problem. To manage huge volume of work in manual way in India scenario is itself critical issue.

43.3.6 Issues with Citizen

The citizen is always developing a negative mentality against the authority related to property issue. They do not rely on the system approach. Many citizens are misguided by others to blame the existing system. Many illiterate citizens are easily relying on agents to solve their problem. Corruption is again a critical issue in terms of property selling and buying deal. The citizen has no faith on authority due to delay of work and not getting any proper documentary evidence for their property. The citizens are not interested to know property rule and regulation. They never try to acquire any knowledge related to their property. Generation transformation related to property is again a critical issue for them. Parental property when transferred to the next generation citizen is least bother about the process.

43.4 Conceptual Solution Model

Various issues related to the property can be resolved through e-governance by using cloud computing. When data will available in electronic form, huge data can be managed in simple, reliable and accurate way with minimum time. Report and error analysis will be easier. The model of property system is a centralized system for the country. All property-related data must manage in a centralized way. One storage and access every place at every level can be made. All the existing data must be transferred to electronic form, and the data must be linked with unique ID for the particular citizen. Role-based portal must be created to access the data at citizen level and authority level. Figure 43.3 shows the centralized property system of India.

At high level, all states must take initiative to convert all the existing properties data to electronic form, and for new data, authority should work in portal. Unauthorized property must be identified, and proper system must be followed to resolve the issue. The citizen must get informed about the system, and they must check their property data and also give their opinion in the portal for issues. Many challenges will come at the initial level because many disputes will come. All states and union territory data when ready, they must transferred to a centralized property system of India. State property system can be categorized at three levels as village level, block level, district level.

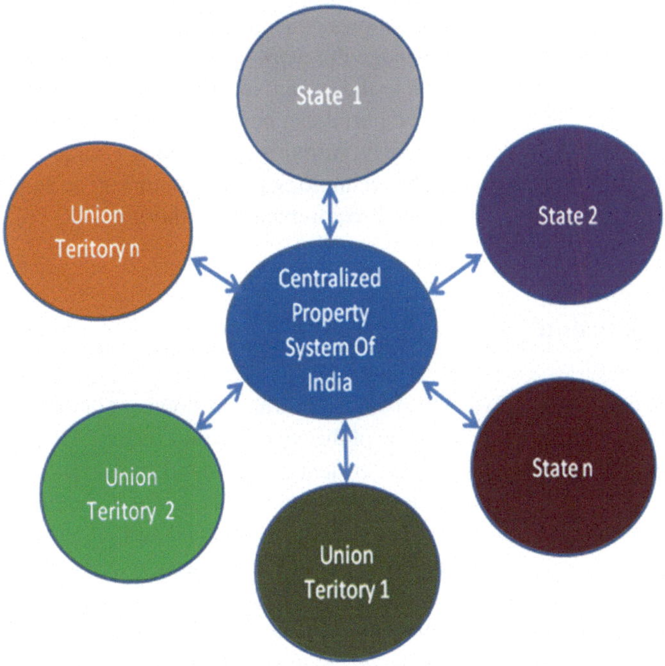

Fig. 43.3 Centralized property system of India

In village level, the cases are very complicated in regard to property. First, deep study of the existing document is required, and meeting must be fixed with the citizen for clarification and interaction of authority, and citizen must be very polite so that citizen will cooperate to do the work. Pending issue must be solved at that level. When all existing data get converted into electronic form, the property portal will able to manage any activity in a simplest form. A block consists of many villages. Data compiled at village level that must be collected to block level. A particular block authority must verify all the village data, and if any mismatch found with the data, then the authority must interact with citizen and get a complete solution. When block-level work is up to mark, we should move for the next level. With increase in level, the work will be completed. A district consists of many blocks. All blocks activity must migrate to district level as administration is better at district level in comparison with village and block level. Many challenges can be faced, and data integration must be done in a proper way as the system is more complicated at this stage. When all three levels are completed, then we have to move forward for the state-level property system. All level migration of data should be stored in a state property system. When all states are ready with their system, they can move for a centralized property system of India. All state data will be integrated to a centralized system.

The centralized property system portal must develop in cloud computing technology as cloud computing technology has many advantages in terms of implementation of an e-governance project. This centralized system consists of state-level system and union territory system. Figure 43.4 shows conceptual model of property system of India. Both systems are able to manage the work of the existing, new and disputed property. Role-based citizen access to the system must be there. The system must be user friendly, and information in the portal can be made crystal clear and simple so that every citizen can understand easily. Administrator rule is a unique role and has control with the overall system. The technology phenomenon can be studied after a great study of the existing system and requirement of authority.

43.5 Conclusion

Developed countries are always focusing on overall development of the citizen. They try to provide the comfortable lifestyle of the citizen in the country. All e-governance projects are playing major rule for betterment of the country's overall development. Developed country like India has to work on major areas of e-governance projects. Many projects like income tax, railway, passport, judicial system, etc., are started, and the citizens are enjoying the benefits out of it. Property issues are very critical, and for the development of the country, centralized e-governance project is a must. Many critical issues can be minimized with the model suggested, and the citizen overhead related to property issues can be also minimized. Right to Information (RTI) policy for citizen can be achieved through e-governance projects. Information communication technology (ICT) has a major role to enhance the lifestyle of citizen

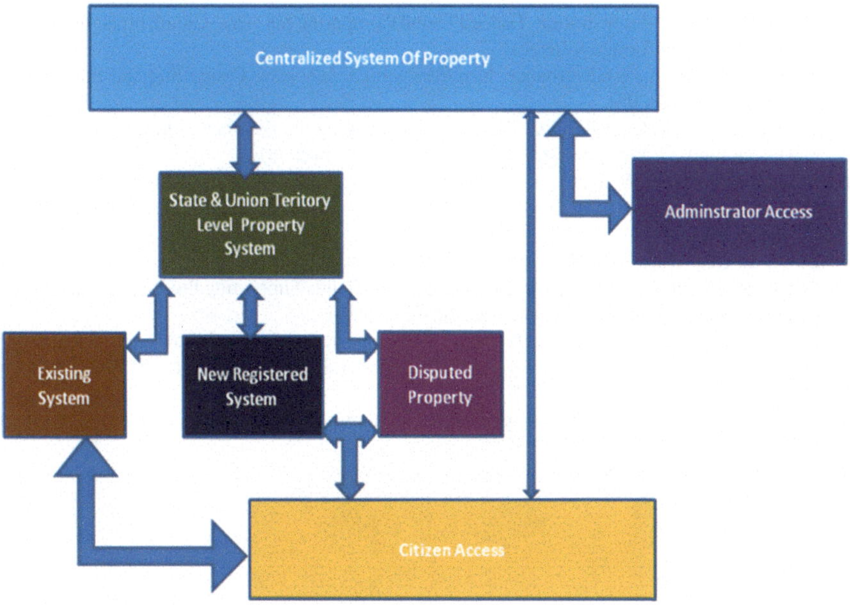

Fig. 43.4 Conceptual model of property system

in the country. The conceptual model can be more enhanced semantically to retrieve the relevant information of a particular property by common citizen using semantic Web access.

References

1. E-Governance-A way to good Governance: The Indian J. Polit. Sci. **69**(1), 43–48 (2008)
2. http://www.egovindia.org/egovernancepaper.doc
3. Misra, D.C., Dhingra, A.: e-Governance Maturity Model. NIC, New Delhi (2002)
4. Saugata, B., Masud, R.R.: Implementing E-Governance Using OECD Model (Modified) and Gartner Model (Modified) Upon Agriculture of Bangladesh. IEEE 1-4244-1551-9/07 (2007)
5. Backus, M.: e-Governance in Developing Countries. IICD Research Brief No 1 (2001)
6. Larsen, B., Milakovich, M.: Citizen relationship management and E-government. Lect. Notes Comput. Sci. **3591**, 57–68 (2005)
7. White Paper on E-Governance Strategy in India
8. http://www.indiaegov.org/knowledgeexchg/egov_strategy.pdf
9. Bhuiyan, S.H.: Modernizing Bangladesh public administration through e-governance: benefits and challenges, Gov. Inf. Q. **28**(1), 54–65 (2011)
10. Bhuiyan, S.H.: Bhoomi, Gyan Ganga, e-governance and the right to information: ICTs and development in India. Gov. Inf. Q. **28**(1) (2011)
11. Dawes S.S.: Governance in the digital age: a research and action framework for an uncertain future. Gov. Inf. Q. **26**(2), 257–264 (2009)
12. Katz, R.N.: The Tower and the Cloud: Higher Education in the Age of Cloud Computing, Educause, (2009)

13. General Services Administration: Federal Cloud Computing Services, Cloud. https://apps.gov/cloud/advantage/information/page.do
14. Cloud computing for e-Governance: http://www.iiit.ac.in/~vasu. Computing and Information Technology 16:4 (2008)
15. Gupta, A., Newell, G., Bajaj, D., Mandal, S.: Identifying the risk factors in Indian non-listed real estate funds. J. Prop. Invest. Financ. **36**(5), 429–453 (2018)
16. Ullah, F., Samad, S., Wang, C.: A systematic review of smart real estate technology: drivers of, and barriers to, the use of digital disruptive technologies and online platforms. Sustainability pp. 1–18 (2018)
17. Cloud for Real Estate: White paper. www.rapidscale.net
18. Mladenow, A., et al.: Mobility for 'Immovable's'—Clouds Supporting the Business with Real Estates Procedia Computer Science 63, 120–127 (2015)
19. Power of Saas for Real Estate Technology: White paper, Rentlytics (2006)
20. Das, R.K., et al.: Adoption of cloud computing in egovernance, CCSIT part III, CCIS 133, pp 161–172 (2011)
21. Smitha, K.K., et al.: Cloud based E-governance system: a survey. Procedia Eng. **38**, 3816–3823 (2012)

Chapter 44
Optimization of Vacuum Impregnation Plant Using Taguchi Methodology

Varun Shekhar and Manthan Jadhav

Abstract The main reason for high life of the electric motor is the process of encapsulation. The process which is used for the encapsulation is known as vacuum pressure impregnation (VPI). This process has helped the end-user to reduce electric motor failures substantially. The following research gives emphasis on the possibilities of the optimization of the VPI plant for better outputs. Six control parameters have been selected for the optimization of the VPI plant, and Taguchi's experimental design has been selected for finding the influencing function of the control parameters.

Keywords Vacuum impregnation · Preheating temperature · Filling time · Thickness of resin

44.1 Introduction

Vacuum pressurized is a process in which two parameters vacuum and pressure are used for the penetration of liquid in various devices. It is mostly used to impregnate an electrical apparatus with insulating resins.

V. Shekhar (✉)
Sangam University, Bhilwara, Rajasthan 311001, India
e-mail: Varun.bhandari@sangamuniversity.ac.in

M. Jadhav
Quality Engineer Aqua Machineries Pvt. Ltd., Ahmadabad 382445, India

S. Yadav et al. (eds.), *Proceedings of International Conference in Mechanical and Energy Technology*, Smart Innovation, Systems and Technologies 174,
https://doi.org/10.1007/978-981-15-2647-3_44

44.2 Problem Definition

44.2.1 Problem

Develop an optimized set of input parameters of the vacuum pressurized impregnation plant, for the Aqua Machineries Pvt. Ltd., for the improvement of the plant and to ensure the optimum use of the resources and to minimize the excessive use of the resin for better cost efficiency of the plant.

44.2.2 Objective of Work

The proposed process specifically aims to:

1. Develop optimum input parameters
2. Reduce the excess amount of the resin
3. Reduce the cycle time of the entire process.

44.3 Literature Review

Sgobba et al. [1] have research on the concept of assessment of vacuum pressure impregnation (VPI) of ITER correction coils. In the ITER magnet system, it consists of set of 18 superconducting correction coils (CC) which are used for solving the errors generally arising from the geometric deviation which is caused due to tolerance and manufacturing tolerance.

Kumar [2] has research on the concept of assessment of impact of impurities in epoxy anhydride vacuum impregnation (VI) resin system. This paper results in gaining new knowledge about the important and significant role played by the quality control parameters in analysing the combined effect of different kinds of impurities that are present in the raw materials of VI resin.

Yu et al. [3] have developed the insulation technology with vacuum pressure impregnation (VPI) for ITER correction coil. ITER magnet system includes 18 correction coils (CC), made with Nb Ti cable-in-conduit 10 Ka conductor, wound into multiple pancakes.

Hubrig [4] has carried out the study on managing coil winding epoxy vacuum impregnation systems at the task level for increased yield, lower cost and extended performance. He has noted out that how compound chemistry enables task-level management of material acceptance, handling and storage, impregnation processing and cure management to increase manufacturing yield, lower unit cost and ensure optimum life cycle performance.

Nakayama [5] has carried out study on global vacuum pressure impregnation insulation system for turbine generators.

44.4 Methodology and Approach

44.4.1 Proposed Methodology

From the survey of the VPI plant, six input parameters have been selected for the optimization of the plant. The input parameters are shown in Table 44.1.

The responses which have to be chosen are impregnated resin thickness, time to fill the body and total process time.

For the above experiment, Taguchi's L18 orthogonal array has been used. The design of the experiment is as given in Table 44.2.

44.5 Results and Analysis

44.5.1 Results

The experiments have been performed over the M270 body. The result is as shown in Table 44.3.

44.5.2 Analysis

In this mathematical calculation, we get the model *F*-value of 6.29 which implies that the model is significant, and due to the noise, there is only a 2.70% chance that a large *F*-value can occur (Table 44.4).

Table 44.1 Levels of input parameter

Operating parameter	Level 1	Level 2	Level 3
Preheating temp (°C)	100	120	140
Preheating time (min)	40	60	80
Vacuum pump pressure (bar)	0.5	0.6	0.7
Submerge time (min)	10	20	30
Air pressure (kg)	1.5	2	2.5
Air pressure time (min)	5	8	10

Table 44.2 Design of experiment

Experiment No.	Preheating temp (°C)	Pre heating time (min)	Vacuum pump pressure (bar)	Submerge time (min)	Air pressure (kg)	Air pressure time (min)
1	100	40	0.5	10	1	8
2	100	60	0.6	20	2	10
3	100	80	0.7	30	3	12
4	120	40	0.6	20	2	12
5	120	60	0.6	30	3	8
6	120	80	0.7	10	1	10
7	**140**	**40**	**0.6**	**10**	**3**	**10**
8	140	60	0.7	20	1	12
9	140	80	0.5	30	2	8
10	100	40	0.7	30	2	10
11	100	60	0.5	10	3	12
12	100	80	0.6	20	1	8
13	120	40	0.6	30	1	12
14	120	60	0.7	10	2	8
15	120	80	0.5	20	3	10
16	140	40	0.7	20	3	8
17	140	60	0.5	30	1	10
18	140	80	0.6	10	2	12

"Values of ""Prob > *F*"" less than 0.0500 give us the result that model term are significant as here E, H are important model terms (Table 44.5).

The "Pred R-Squared" of 0.1947 is not as nearer to the "Adj R-Squared" of 0.7887 as we are normally expecting so in this case the difference is greater than 0.2. This gives us the result that there are high chances of a large block effect or a possible problem with your model and/or data.

So during the mathematical calculation, the things to be considered are model reduction, response transformation, outliers, etc. Generally, every empirical model should be tested by doing confirmation runs.

"Adeq Precision" measures the signal-to-noise ratio. Generally, a ratio greater than 4 is required, and in this case, the ratio of 8.186 indicates an adequate signal. So in this case, the model can be used to navigate the design space.

Table 44.3 Result for (M270)

Experiment No.	Thickness of resin	Filling time
1	0.62	3
2	0.60	3
3	0.53	2
4	0.60	3
5	0.58	3
6	0.65	2
7	0.56	4
8	0.72	3
9	0.61	2
10	0.63	3
11	0.50	3
12	0.71	2
13	0.70	3
14	0.61	3
15	0.59	2
16	0.50	3
17	0.71	3
18	0.65	2

Table 44.4 Anova of thickness (M270)

Source	Sum of squares	*df*	Mean square	*F*-value	*p*-value Prob > *F*
Model	0.073	12	6.100 E−003	6.29	0.0270
C-preheating temp	1.633 E−003	2	8.167 E−004	0.84	*0.4840*
D-preheating time	5.700 E−003	2	2.850 E−003	2.94	*0.1433*
E-vacuum pressure	0.024	2	0.012	12.39	*0.0116*
F-submerge time	9.700 E−003	2	4.850 E−003	5.00	*0.0642*
G-air pressure	5.333 E−004	2	2.667 E−004	0.27	*0.7704*
H-air pressure time	0.032	2	0.016	16.29	*0.0065*
Residual	4.850 E−003	5	9.700 E−004		
Cor. total	0.078	17			

Table 44.5 R-Square table (M270)

Std. Dev.	0.031	R-Squared	0.9379
Mean	0.61	Adj R-Squared	0.7887
C.V. %	5.06	Pred. R-Squared	0.1947
PRESS	0.063	Adeq precision	8.186

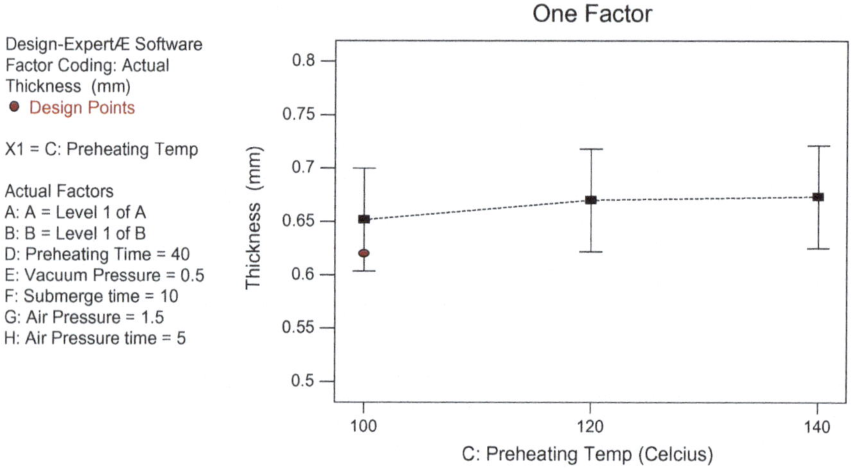

Fig. 44.1 Thickness versus preheating temperature

44.5.3 Effect of Single Parameter on Thickness

Figure shows the effect of preheating temperature over thickness. We can see very little increase in thickness up to 120° of temperature and then remains constant (Figs. 44.1 and 44.2).

The figure above shows the relation between preheating time and thickness. As an increase in preheating time thickness decreases and remains constant (Fig. 44.3).

Relation between vacuum pressure and thickness is shown above. We can see drastically increase in thickness as an increase in vacuum pressure. So we can say that vacuum pressure has a significantly high impact over thickness (Fig. 44.4).

Figure above shows the relation between submerge time and thickness. We can observe drastically decrease in thickness up to level 2 which is 20 min and again increase for the next level which is 30 min (Fig. 44.5).

From the above graph, we can deduce that air pressure has a very low effect over the thickness as there isn't any change over the different levels of air pressure (Fig. 44.6).

One can observe from the above figure that as the air pressure increase thickness will decrease but it decrease significantly over the last level which is 10 min.

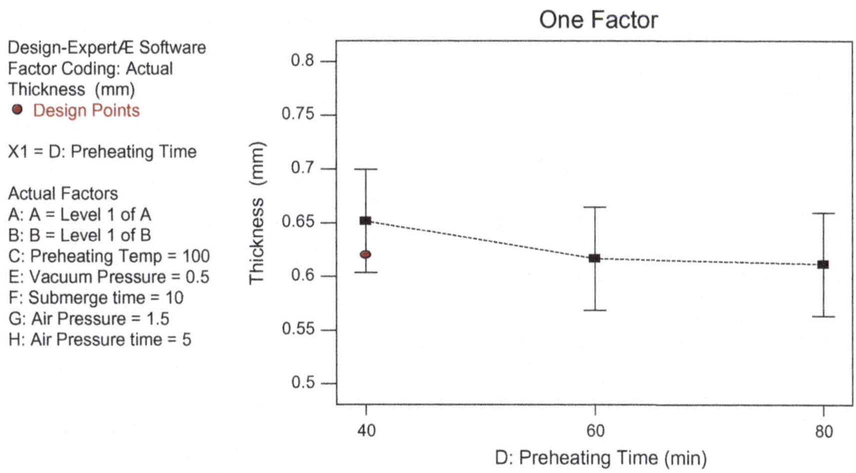

Fig. 44.2 Thickness versus preheating time

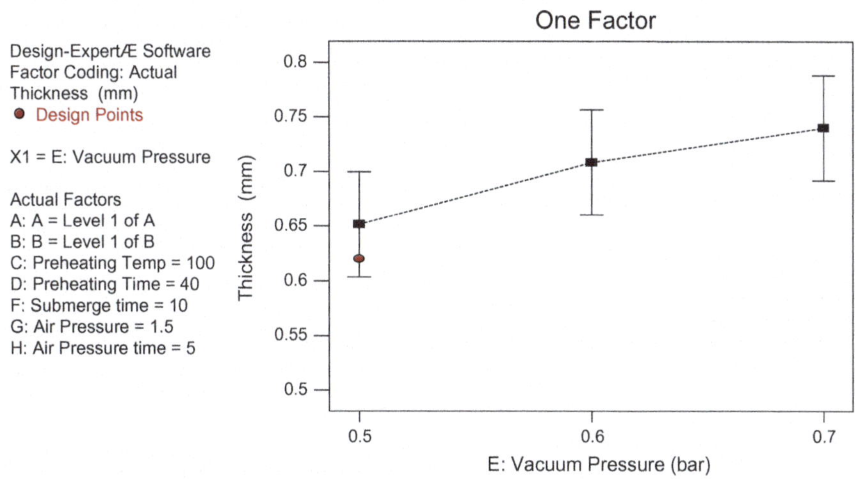

Fig. 44.3 Thickness versus vacuum pressure

44.6 Conclusion

From the analysis and experimentation, we came to know about the possible set of the parameters for the vacuum pressurized impregnation plant to work efficiently. We can see the set of optimum input parameters in Fig. 44.7.

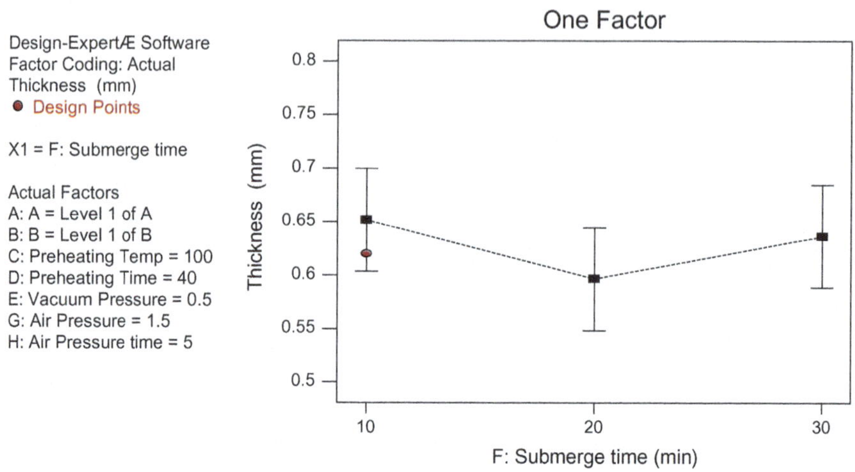

Fig. 44.4 Thickness versus submerge time

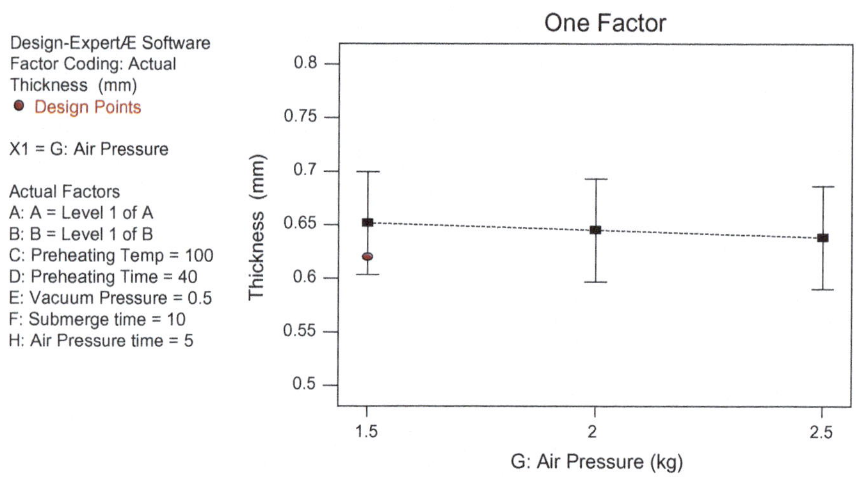

Fig. 44.5 Thickness versus air pressure

With this set of parameters, we can assure the thickness of the resin which has been impregnated to the winding is minimum which is around 0.6 mm and the filling time also should be minimum for each body.

By using these set of parameters, we can assure the maximum use of the resources which is available as well as the minimum waste of the resin material.

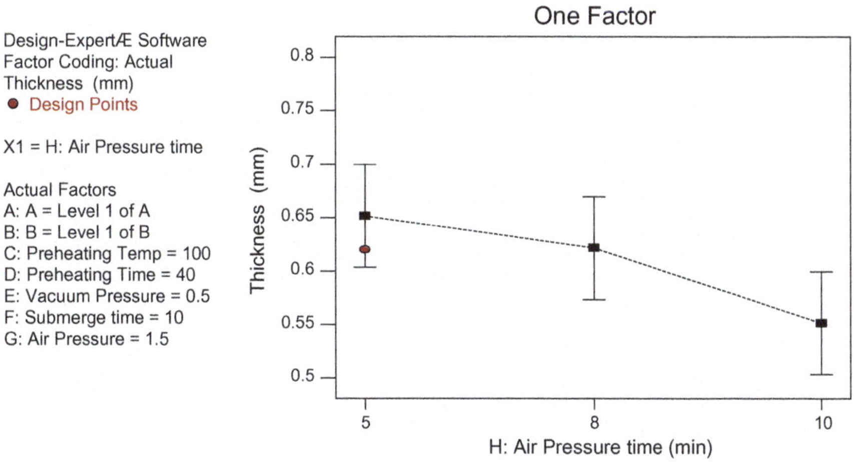

Fig. 44.6 Thickness versus air pressure

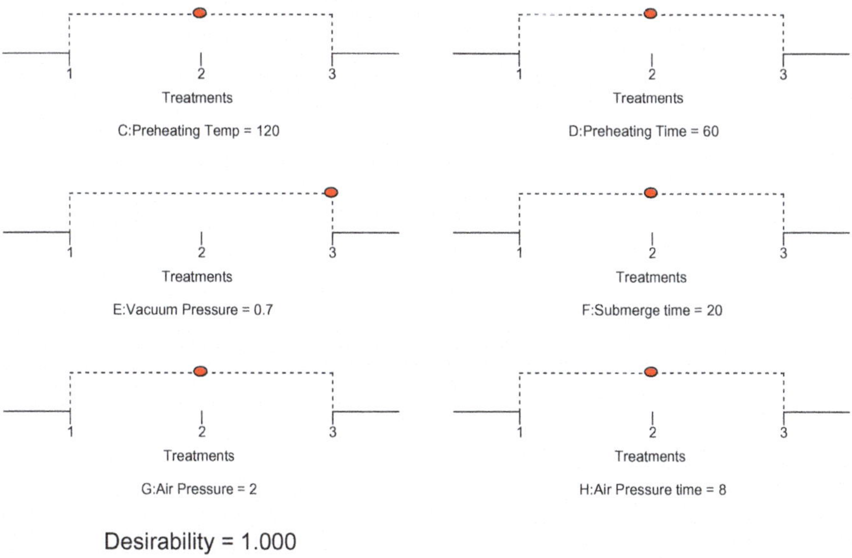

Fig. 44.7 Optimized parameter for M270

References

1. Sgobba, S., Marcinek, D.J., Libeyre, P., Carillon, A.: Advanced examination techniques applied to the assessment of vacuum pressure impregnation (VPI) of ITER correction coils. In: IEEE Transaction on Applied Superconductivity. IEEE, New York (2014)
2. Kumar, V.: Assessment of impact of impurities in epoxy—anhydride vacuum impregnation (VI) resin system. In: IEEE Transaction on Applied Superconductivity. pp. 1947–1954. IEEE, New York (2011)
3. Yu, X., Wu, W., Pan, W., Han, S., Wang, L., Wei, J., Liu, L., Du, S., Zhou, Z., Foussat, A., Libeyre, P.: Development of insulation technology with vacuum-pressure-impregnation (VPI) for ITER correction coil. In: IEEE Transaction on Applied Superconductivity. pp. 503–506. IEEE, New York (2012)
4. Hubrig, J.G.: Managing coil winding epoxy vacuum impregnation systems at the task level for increased yield, lower cost and extended performance. In: Processing of 2005 Particle Accelerator Conference, pp. 4260–4262. IEEE, Tennessee (2005)
5. Nakayama, A.: Development of global vacuum pressure impregnation insulation system for turbine generators. In: Asian International Conference on Dielectrics and Electrical Insulation. IEEE, Tokyo (2002)

Chapter 45
A Recent Study of Thermal Performances of a Circular Tube Using Internal Tape Inserts

Mohammed Zaki Hayat, Prateek Tanwar, Hardik Chauhan, Gopal Nandan and Nafees Alam Wani

Abstract Energy conservation and cost minimization are important aspects of heat exchanging devices. In the heat exchanger, concerted efforts are going to improve the heat transfer augmentation. Researchers tried active as well as passive techniques for it. In the current work research study of different types of twisted tape, twisted wires have been critically examined in recent years. The experimental and numerical aspect has been presented. This will help the system design for the maximization of heat transfer and minimization of the pumping power of the thermal systems.

Keywords Perforated tape · Twisted tape · Augmentation of heat transfer · Passive heat transfer · Swirl flow

45.1 Introduction

The heat exchanger is the most common application of thermal engineering. The use of the heat augmentation technique increases the performance heat exchanger which leads to downsizing of the system as well as cost reduction. The efficiency of it is dropped because the exchange of heat between the solid and the fluid is very less due to high thermal resistance in the solid–fluid interface. To mitigate this problem, several methods are implemented like making the internal walls of the tube rough and many more. The use of twisted tape inserts (TTIs) enhances heat transfer. It has been reported in several applications [1–5]. It is due to the very simple, low cost and simplicity of its installation inside the tube. It was found that certain types of inserts can be very useful inside the tubes which can facilitate convective heat transfer among solid and fluid. The tube with TTI has higher efficiency as compared to without inserts in the tube. The application of inserts generates the swirl flow, which intermixes with the fluids. Intermixing is the primary reason for heat transfer enhancement. This provides continuous swirl flow, which causes the turbulence at the near-wall liquid interface and hence reducing the thermal layer barrier by breaking the sub-layer. Using TTI increases the heat transfer due to augmentation in the turbulence [6–13]. Later, the inserts are modified by twisting the strips, making them perforated,

M. Z. Hayat · P. Tanwar · H. Chauhan · G. Nandan (✉) · N. A. Wani
ME Department, Amity University Uttar Pradesh, Noida, India

S. Yadav et al. (eds.), *Proceedings of International Conference in Mechanical and Energy Technology*, Smart Innovation, Systems and Technologies 174,
https://doi.org/10.1007/978-981-15-2647-3_45

etc., and it was experimentally observed that heat convection has been drastically elevated. There are several factors on which heat transfer depends upon, such as friction penalty, pressure loss. To maximize the heat transfer while minimizing the friction penalty, coil inserts are the better choice. Generally, the material of inserts should be such that it is unreactive to the fluid medium and should be light in weight. The most commonly used materials for the inserts are aluminum and copper. In this paper, various types of inserts are extensively discussed and explained. This plays an important role in heat transfer enhancements [14–33].

45.2 The Various Twisted Tapes Used as Inserts

45.2.1 Twisted Tape

It is a type of tape having a similar length of the tube where it is inserted. It is made by twisting a metallic strip (mostly steel, aluminum, copper) which is highly polished. Insertion of tape heat transfer is enhanced. The parametric study has been done experimentally by varying the pitch and twist ratio of tape. The distance measured along the longitudinal axis between two consecutive coplanar points is referred to as pitch. The ratio of pitch to inner diameter is referred to as a twist ratio. The use of TTI increases the heat transfer which is a positive point, but simultaneously the pressure drop also increases slightly in comparison with the tube without inserts. It was observed that the heat transfer rate is maximum when the clearance between insert and tube wall is taken 1 mm whereas if the clearance is further increased then the heat transfer decreases.

45.2.1.1 Cross-Hollow TTI

This is a type of twisted tape insert in which the twist ratio is kept 3. It consists of central and lateral support structures to fix its parts and to affix it to the tube. The thickness of the supporting structure is kept 1 mm, and the axial length of the support is 2 mm. Also, a ring-shaped gasket is used to prevent the axial movement of the insert as this gasket is clipped at the beginning of the test section [34]. In this, we have considered three cases of hollow width (6, 8 and 10 mm) as shown in Fig. 45.1. Such type of insert produce swirl in the flowing fluid which helps in heat transfer enhancement. The heat transfer reduces when the hollow width is increased due to which the swirl motion decreases leading to the decrease in heat transfer as the high swirl motion leads to proper mixing of the fluid leading to heat transfer enhancement.

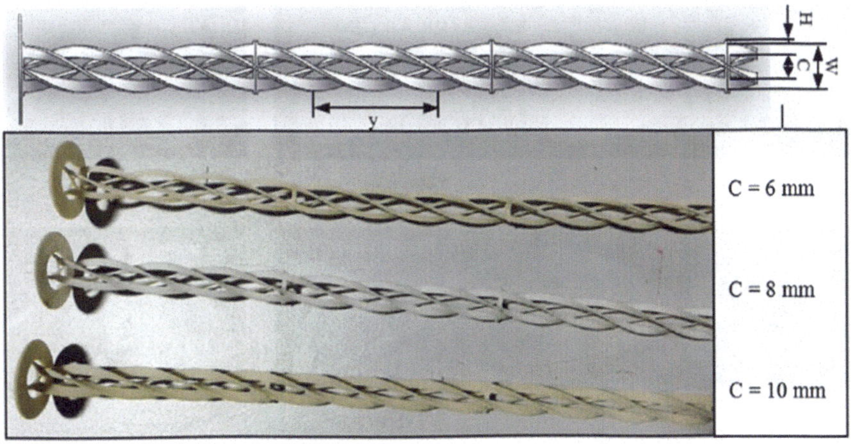

Fig. 45.1 Cross-hollow TTI with different hollow widths [34]

45.2.1.2 TT with a Different Wing Geometry

The TTI was modified by forming wings on the edges of the tapes with different shapes such as triangle, rectangle and trapezoid (Fig. 45.2). To produce an alternate axis, the plane of TTI tape was adjusted to 60° difference relative to the adjacent plane [35]. The wings were made at the edges of tapes at every twist ratio along its longitudinal axis.

45.2.1.3 Perforated Helical Twisted Tapes

The holes formed at the tapes are referred to as perforated TTI. The holes are formed along the tube axis of the tape (Fig. 45.3). The perforated tape damps the pressure loss. The perforated tape works well with a lower friction penalty. Moreover, the results highly fluctuate with the hole diameter and pitch ratio [36]. Experiments were conducted for different sizes of hole diameters on the tape.

45.2.1.4 Peripherally Cut Twisted Tapes

Such type of tapes has been reported by Smith et al. [37] for all types of flow through the tube. The depth of cut, width of cut and their ratio showed heat transfer enhancement (Fig. 45.4).

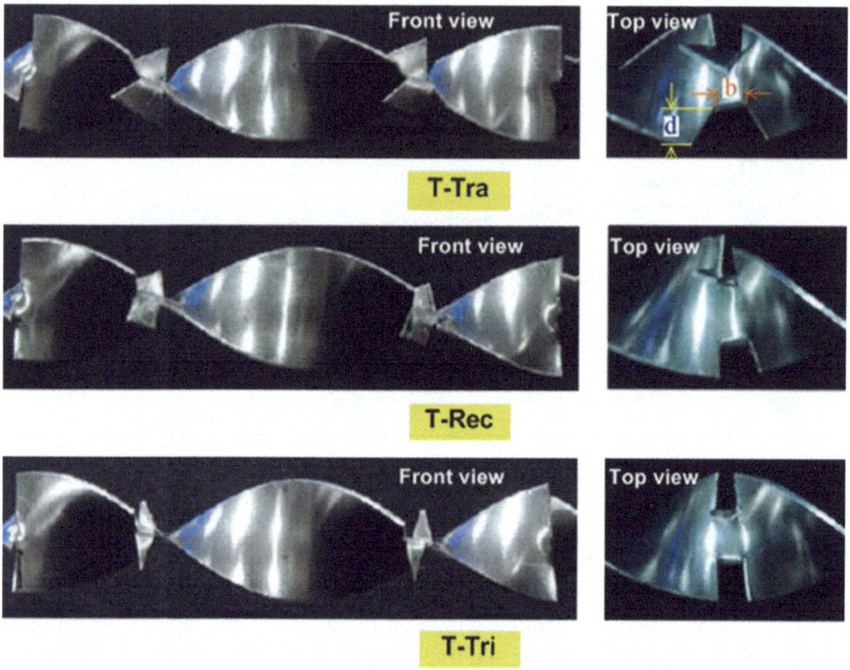

Fig. 45.2 TTI with triangular, rectangular and trapezoidal wings [35]

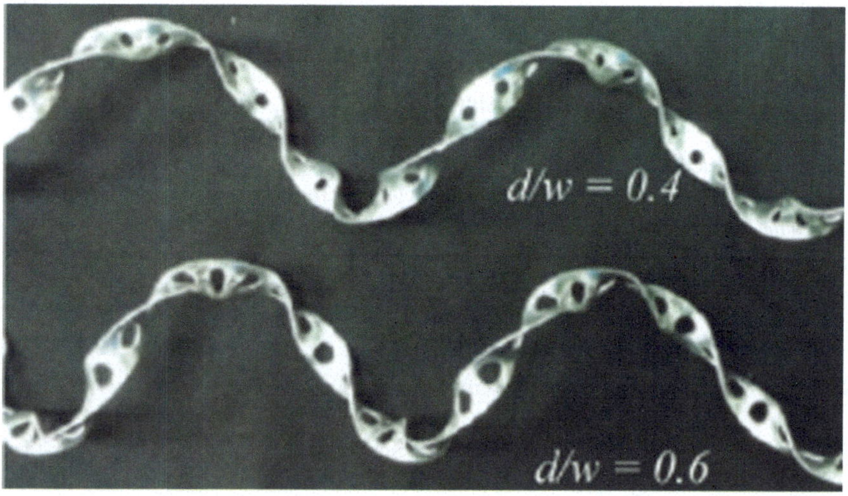

Fig. 45.3 Perforated twisted tapes [36]

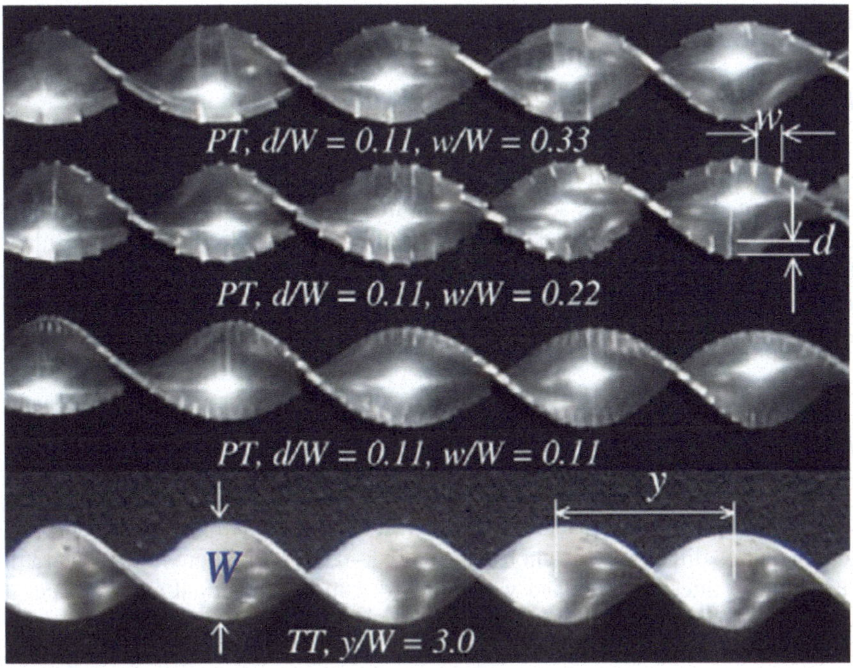

Fig. 45.4 Peripherally cut twisted tapes [37]

45.2.1.5 Helical Screw TTI

These tapes are formed by making helical grooves on the twisted tape. It was found that the left or right helical twisted tapes are better in terms of augmenting the convective heat transfer as compared to the straight-grooved helical twisted tape because it causes the fluid to move in different directions repeatedly and hence ensures efficient mixing. Afterward, it was tested that helical screw tape without core rod is about 2 times more efficient than one with a core rod. The helically twisted tape with internal roughness is also experimentally proved to be highly productive in increasing the heat transfer rate [36]. Several other types of TTI like short-length twisted tape, loose fit twisted tapes, serrated twisted tapes, jagged twisted tapes have been fabricated and studied for performance study.

45.3 Effects of Different Inserts

45.3.1 Effect of Perforated Helical TT Inserts

Nanan et al. [36] used aluminum tapes for the fabrication of perforated twisted tapes and conducted experiments at uniform heat flux. The ratio of a hole of perforation and width of the tape was varied from 0.2 to 0.6 at different hole pitches. The result of the perforation of tapes is not encouraging. The reduction in heat transfer has been reported as helically twisted tapes due to blockages of fluid flow. The turbulence intensity diminished.

45.3.2 Effects of TT-Varied Wing Geometry

Using three different wing shapes, experiments were conducted with a constant twist ratio and three different wing chord ratios of 0.1–0.3, for turbulent flow range [35]. The TT with alternate axes and wings provided the best thermal performance. It was observed that TT with trapezoidal wings offers highest thermal performance factor, followed by tapes with rectangular wings and then by triangular wings. The maximum heat transfer rates were observed with trapezoidal wings. This was due to lager tips, which provide more effective stream disturbances. The disturbances generate stronger turbulence near the tube wall. The maximum thermal performance factor 1.42 is found with trapezoidal wings as compared to the plain tube.

45.3.3 Effects of Cross-Hollow TTI

The results obtained in this study with the hollow width of 6, 8 and 10 mm and the value of Re ranging from 5600 to 18,000 are that it creates swirl motion in the fluid flow which leads to proper mixing of the fluid near walls and central region [34]. For high Re, the heat transfer coefficient reaches maximum value with the TTI with a 6-mm hollow width and reaches the minimum value with 8-mm hollow width. This concludes that heat transfer is maximum at less hollow width which helps in proper mixing of the fluid and enhances the heat transfer.

45.3.4 Effect of Peripherally Cut Twisted Tapes

The peripherally cut twisted tapes were used by Smith et al. [37]. They varied the depth and width of cut on the periphery of tapes. They used three different types of cut. They conducted experiments for depth of cut and width of tape ratio for 0.11,

0.22 and 0.33 for Re ranging from 1000 to 20,000. In higher flow ratio, the ratio of Nu ratio was high up to 12.8 times in the laminar flow region and 2.6 times in the turbulent flow region. The effect of peripherally cut TTI is more effective at low Re. In laminar flow, being lower velocity, turbulence is negligible and correspondingly there will be a low value of Nu. By using tapes, turbulence intensity was magnified, and hence, more Nu is observed as compared to flow in the tube without inserts, whereas in turbulence flow the inherent turbulent intensity was high already. Hence, the use of inserts increases Nu by 2.6 times. The maximum thermal performance ratio is 4.88 at lower value of Re.

45.3.5 Effect of the Helix Angle of the TT

The effect of helix angles 9°, 13°, 17° and 21° in turbulent flow range has been examined [38]. At the lower helix angle, the maximum heat transfer rate was observed. This is about 1.65 times as compared to inserts.

45.4 Conclusions

The enhancement of heat transfer by the use of inserts increases due to an increase in turbulence intensity. The inserts behave as turbulence promoters. This increases the tangential velocity, which breaks the fluids sub-layers at solid–liquid interface. The thickness of the sub-layer offers resistance in the heat flow. In the current study, the geometrical dimension of insets like width, thickness, twist ratio, perforation on the inserts, winglets, winglets with triangular, square, tetrahedron cut has been discussed. The combination of single and pair longitudinal inserts also has been discussed experimentally in the laminar flow and turbulent flow in the circular tubes. Researchers have also used wire types of inserts with varying geometry. Study suggests that triple twisted tapes have the best thermal performance among single, twin and triple cases for same pumping power consumption.

References

1. Chang, S.W., Gao, J.Y., Shih, H.L.: Thermal performances of turbulent tubular flows enhanced by ribbed and grooved wire coils. Int. J. Heat Mass Transf. **90**, 1109–1124 (2015)
2. Bas, H., Ozceyhan, V.: Heat transfer enhancement in a tube with twisted tape inserts placed separately from the tube wall. Exp. Thermal Fluid Sci. **41**, 51–58 (2012)
3. Li, P., Liu, Z., Liu, W., Chen, G.: Numerical study on heat transfer enhancement characteristics of tube inserted with centrally hollow narrow twisted tapes. Int. J. Heat Mass Transf. **88**, 481–491 (2015)

4. Promvonge, P., Eiamsa-ard, S.: Heat transfer behaviors in a tube with combined conical-ring and twisted-tape insert. Int. Commun. Heat Mass Transfer **34**(7), 849–859 (2007)
5. Murugesan, P., Mayilsamy, K., Suresh, S., Srinivasan, P.: Heat transfer and pressure drop characteristics in a circular tube fitted with and without v-cut twisted tape insert. Int. Commun. Heat Mass Transf. **38**(3), 329–334 (2011)
6. Sekhar, T.V.R., Nandan, G., Prakash, R., Muthuraman, M.: Modeling a renewable energy collector and prediction in different flow regimes using CFD. Mater. Today: Proc. **5**(2), 4563–4574 (2018)
7. Sekhar, T.V.R., Nandan, G., Prakash, R., Muthuraman, M.: Investigations on viscosity and thermal conductivity of cobalt oxide-water nano fluid. Mater. Today: Proc. **5**(2), 6176–6182 (2018)
8. Sekhar, T.V.R., Prakash, R., Nandan, G., Muthuraman, M.: Pressure drop characteristics & efficiency enhancement by using TiO_2–H_2O nanofluid in a sustainable solar thermal energy collector. Int. J. Environ. Sustain. Dev. (2018)
9. Vadhera, J., Sura, A., Nandan, G., Dwivedi, G.: Study of phase change materials and its domestic application. Mater. Today: Proc. **5**(2), 3411–3417 (2018)
10. Sekhar, T.V.R., Prakash, R., Nandan, G., Muthuraman, M.: Preparation of Co_3O_4–H_2O nanofluid and application to CR-60 concentrating solar collector. Prog. Ind. Ecol., Int. J. **11**(3), 227–246 (2017)
11. Sekhar, T.V.R., Prakash, R., Nandan, G., Muthuraman, M.: Performance enhancement of a renewable thermal energy collector using metallic oxide nanofluids. Micro Nano Lett. **13**(2), 248–251 (2018)
12. Suresh, A.K., Khurana, S., Nandan, G., Dwivedi, G., Kumar, S.: Role on nanofluids in cooling solar photovoltaic cell to enhance overall efficiency. Materials Today: Proceedings **5**(9), 20614–20620 (2018)
13. Sharma, R., Gupta, A., Nandan, G., Dwivedi, G., Kumar, S.: Life span and overall performance enhancement of solar photovoltaic cell using water as coolant: a recent review. Mater. Today: Proc. **5**(9), 18202–18210 (2018)
14. Krishn, S., Goyal, M., Nandan, G., Kumar, S., Kumar, P., Shukla, A.K.: Pool boiling using nanofluids: a review. In: Lecture Notes in Mechanical Engineering, pp. 325–336. Springer Singapore (2019)
15. Rathour, R.S., Chauhan, V., Agarwal, K., Sharma, S., Nandan, G.: Cooling of solar photovoltaic cell: using novel technique. In: Lecture Notes in Mechanical Engineering, pp. 521–529. Springer, Singapore (2019)
16. Agrawal, T., Ajitkumar, R., Prakash, R., Nandan, G.: Sodium silicide as a hydrogen source for portable energy devices: a review. Mater. Today: Proc. **5**(2), 3563–3570 (2018)
17. Sekhar, T.V.R., Nandan, G., Prakash, R.: Parabolic trough solar collectors. Eur. J. Sci. (EJS) 43–53 (2018)
18. Srivastav, A., Patil, A.K., Nandan, G.: Investigation of heat transfer and friction in rib roughned fin under forced convection. In: 1st International Conference on Research in Science, Engineering & Management (IOCRSEM 2014), pp. 117–125 (2014)
19. Rawat, K.S., Thakur, H.C., Nandan, G.: CFD analysis of a pentagonal rib over absorber plate of a solar air heater. In: 3rd International Conference on Manufacturing Excellence, pp. 191–196, Amity University, Uttar Pradesh, Noida, India (2016)
20. Kumar, S., Nandan, G., Singh, G.K.: Numerical study of natural convection and radiative heat transfer of heated inner cylinder placed inside a isothermally cooled circular enclosure. In: Third International Conference on Manufacturing Excellence, pp. 179–183, Amity University, Uttar Pradesh, Noida, India (2016)
21. Kumar, V., Shrivastava, R., Nandan, G.: Energy saving using phase change material in refrigerating system. In: Third International Conference on Manufacturing Excellence, pp. 184–190, Amity University, Uttar Pradesh, Noida, India (2016)
22. Anamika, Thakur, H.C., Nandan, G.: A review on solar air heater performance using different artificial roughened rib. In: International Conference Emerging Development In Engineering and Technology (ICEDET-2016), pp. 347–349, Shivalik College of Engineering, Dehradun, India (2016)

23. Sekhar, T.V.R., Nandan, G., Prakash, R.: Parabolic trough solar collectors. Eur. J. Sci. (EJS) **1**(1), 43–53 (2018)
24. Nandan, G., Sahoo, P.K., Kumar, R., Chatterjee, B., Mukhopadhyay, D., Lele, H.G.: Experimental investigation of sagging of a completely voided pressure tube of Indian PHWR under heatup condition. Nucl. Eng. Des. **240**(10), 3504–3512 (2010)
25. Nandan, G., Majumdar, P., Sahoo, P.K., Kumar, R., Chatterjee, B., Mukhopadhyay, D., Lele, H.G.: Study of ballooning of a completely voided pressure tube of Indian PHWR under heat up condition. Nucl. Eng. Des. **243**, 301–310 (2012)
26. Nandan, G., Sahoo, P.K., Kumar, R., Chatterjee, B., Mukhopadhyay, D., Lele, H.G.: Experimental investigation of sagging and ballooning in LOCA for Indian PHWR. Eng. Lett. **18**(3) (2010)
27. Kumar, R., Nandan, G., Sahoo, P.K., Chatterjee, B., Mukhopadhyay, D., Lele, H.G.: Ballooning of pressure tube under LOCA in an Indian pressurised heavy water reactor. In: 14th International Heat Transfer Conference, IHTC14, 7, pp. 317–323 (2010)
28. Nandan, G., Majumdar, P., Sahoo, P.K., Kumar, R., Chatterjee, B., Mukhopadhyay, D., Lele, H.G.: Experimental investigation of sagging and ballooning for a completely voided pressure tube of Indian PHWR under heatup condition. World Acad. Sci., Eng. Technol. **61**, 197–204 (2010)
29. Nandan, G., Lele, H.G., Sahoo, P.K., Chatterjee, B., Kumar, R., Mukhopadhyay, D.: Experimental investigation of heat transfer during LOCA with failure of emergency cooling system. In: 5th International Conference on Heat Transfer, Fluid Mechanics and Thermodynamics (2007)
30. Nandan, G., Sahoo, P.K., Kumar, R., Chatterjee, B., Mukhopadhyay, D., Lele, H.G.: Thermomechanical behavior of pressure tube of Indian PHWR at 20 bar pressure. World Acad. Sci., Eng. Technol. **61**, 205–213 (2010)
31. Nandan, G., K.Sahoo, P., Kumar, R., Chatterjee, B., Mukhopadhyay, D., Lele, H.G.: Thermomechanical behavior of Pressure Tube of Indian PHWR at 20 bar Pressure. In: International Conference on Mechanical, Industrial, and Manufacturing Engineering, Cape Town, South Africa (2010)
32. Kumar, R., Nandan, G., Sahoo, P.K., Chatterjee, B., Mukhopadhyay, D., Lele, H.G.: Ballooning of pressure tube under LOCA in an Indian pressurised heavy water reactor. In: 14th International Heat Transfer Conference, pp. 319–323, Washington, DC, USA. American Society of Mechanical Engineers Digital Collection (2010)
33. Shukla, A.K., Sharma, A., Sharma, M., Nandan, G.: Thermodynamic investigation of solar energy-based triple combined power cycle. Energy Sources, Part A: Recovery, Utilization, and Environmental Effects 41(10), 1161–1179 (nov 2018)
34. He, Y., Liu, L., Li, P., Ma, L.: Experimental study on heat transfer enhancement characteristics of tube with cross hollow twisted tape inserts. Appl. Therm. Eng. **131**, 743–749 (2018)
35. Wongcharee, K., Eiamsa-ard, S.: Heat transfer enhancement by twisted tapes with alternate-axes and triangular, rectangular and trapezoidal wings. Chem. Eng. Process. **50**(2), 211–219 (2011)
36. Nanan, K., Thianpong, C., Promvonge, P., Eiamsa-ard, S.: Investigation of heat transfer enhancement by perforated helical twisted-tapes. Int. Commun. Heat Mass Transfer **52**, 106–112 (2014)
37. Eiamsa-ard, S., Seemawute, P., Wongcharee, K.: Influences of peripherally-cut twisted tape insert on heat transfer and thermal performance characteristics in laminar and turbulent tube flows. Exp. Thermal Fluid Sci. **34**(6), 711–719 (2010)
38. Bhuiya, M., Ahamed, J., Chowdhury, M., Sarkar, M., Salam, B., Saidur, R., Masjuki, H., Kalam, M.: Heat transfer enhancement and development of correlation for turbulent flow through a tube with triple helical tape inserts. Int. Commun. Heat Mass Transfer **39**(1), 94–101 (2012)

Chapter 46
Impact of IoT-Enabled Supply Chain—A Systematic Literature Review

Ranjan Arora, Abid Haleem and P. K. Arora

Abstract In this era of high competition, success lies in implementing an efficient supply chain. The problem faced by modern-day supply chain is that it is not proactive. Proactive supply chain helps in attacking the problem well in time, along with reducing all potential threats and risks. Internet of things (IoT) is the solution to this problem as it helps in making supply chain proactive. IoT-enabled supply chain is useful in tracking and tracing products in the entire supply chain along with delivering the right product at the right place at the right time. All this has huge impact on the business because IoT-enabled supply chain ensures a proactive supply chain. Thus, in nut shell it helps in revenue increase, enhancing customer service and providing real-time visibility to the entire supply chain. This paper reviews the existing literature to study and consolidate the impact of IoT-enabled supply chain on organizations. This study will provide a framework to academicians and practitioners working in the industry to focus on IoT-enabled supply chain along with its impact on organization.

Keywords Internet of things (IoT) · Supply chain management · IoT-enabled supply chain

46.1 Introduction

Success of an organization depends on proper functioning of all business functions, but one of the most important components for success is supply chain. This is because organizations have realized that an effective and efficient supply chain is needed in order to be competitive globally. The need of the hour is to effectively implement a

R. Arora (✉)
Consultant, Ghaziabad, Uttar Pradesh, India

A. Haleem
Jamia Millia Islamia, New Delhi, India

P. K. Arora
Galgotias College of Engineering and Technology, Greater Noida, Uttar Pradesh, India

S. Yadav et al. (eds.), *Proceedings of International Conference in Mechanical and Energy Technology*, Smart Innovation, Systems and Technologies 174,
https://doi.org/10.1007/978-981-15-2647-3_46

supply chain strategy so as to reduce operational costs, improve customer satisfaction and increase flexibility. Supply chain management is "the active management of supply chain activities to maximize customer value and achieve a sustainable competitive advantage. Supply chain activities cover everything from product development, sourcing and production to logistics, as well as the information systems needed to coordinate these activities" [1]. The real problem in supply chain is not getting the right information at the right time leading to loss in revenue; excessive inventory; high logistics cost and low customer service [2]. It is important in a supply chain to get the right data well in time so that all potential threats are identified and resolved. The thrust is on reducing all risks to supply chain by making it proactive instead of being reactive. IoT-enabled supply chain helps in connecting all available resources through Internet, and providing real-time data helps mitigate these shortfalls [3]. It allows decision making without human intervention by enabling and integrating communication technologies [4]. Along with it, the advantage is to interconnect and communicate with all devices that are uniquely identified [5]. This is key to success, and a business is bound to succeed which has a strong thrust on proactive supply chain. The literature on IoT-enabled supply chain is less though it is bringing a fundamental change in industries [6]. This has encouraged the authors to study and consolidate the impact of IoT-enabled supply chain.

46.2 Literature Review

Literature review has been done by undertaking following steps:

- Research conducted in IoT-enabled supply chain: This includes going through the existing research.
- Database has been selected, and search criteria are defined.
- Inclusions and exclusions are defined.
- Data are collected, collated and analysed.

46.2.1 Internet of things (IoT)

IoT is defined as

> A "sensor network", which uses RFID and various information sensing devices connected to internet to provide real time information [7].
>
> A grid of objects linked together so that can interact and communicate with each other [8].

Research has been done by various researchers, highlighting the various components of IoT along with the benefits and its impact on business. Below highlighted are some of the key data points on IoT.

- Transmission of data through Internet for smart business decisions [8]
- According to an estimation made by Cisco, more than 18,000 companies rely on our IoT solutions [9].
- "Things" connected to the Internet and creation "Smart" systems [10].
- IoT has gained popularity and acceptance as a standard because of rapid growth of high-speed Internet and smart devices [11].
- Due to resource diversity in IoT, the devices are neither secure nor capable of securing themselves [11].
- Network of devices and machines which interact with each other is called Internet of things (IoT) [12].
- Success of IoT is dependent on deployment of five technologies [12].
  - Wireless sensor networks (WSNs)
  - Middleware
  - Radio Frequency Identification (RFID)
  - Software for IoT applications
  - Cloud computing.
- IoT is also termed as Internet of everything (IoE) as it has the potential to reform the business [13].

46.2.2 IoT-Enabled Supply Chain

Supply chain has been transformed by the implementing IoT, and organizations have benefited by the adoption of IoT-enabled supply chain. As highlighted earlier, it is important to have a supply chain which is proactive and not reactive. IoT enables supply chain to become proactive. Based on the previous research done in IoT-enabled supply chain, the key characteristics are highlighted below:

- IoT is changing the operating mode of supply chain management [7].
- IoT-enabled supply chain helps overcome traditional challenges faced by supply chains such as vulnerability, uncertainty, complexity and cost [14].
- IoT provides a systematic approach to supply chains in order to reduce costs, in efficiencies, and emissions [15].
- IoT increases efficiency in an organizations supply chain [16].
- IoT helps in harmonization between supplier and distributor [8].

46.2.3 Impact of IoT-Enabled Supply Chain

Below listed are some of the previous research done in identifying the impact of IoT-enabled supply chain on organizations

- IoT-enabled supply chain helps in satisfying customer requirements, increasing sales, enhancing customer service and real-time visibility of product delivery [8].
- IoT will play a key role in condition-based maintenance for high capital assets.
- This will ensure improved availability of machine, increase in productivity and guaranteed security [17].
- IoT-enabled supply chain helps suppliers and distributors to update product or material information, so that real-time record updation can take place [8].
- IoT is playing an instrumental role in addressing global challenges like pollution, climate change and disease outbreaks [18].
- IoT is useful in preventive maintenance, process quality, supply chain tracking and tracing by providing real-time data and delivery of the product in the right place at the right time [19].

46.3 Results and Discussion

Work has been done in IoT, and previous researchers have also highlighted the benefits brought to an organization by implementing IoT. The previous research has also highlighted the business functions which will get impacted by making them IoT enabled, but not much work has been carried out in the field of IoT-enabled supply chain. Most of the previous work is in IoT without studying its impact on supply chain. This work has highlighted the importance of IoT. IoT helps in getting things connected to the Internet by creating a "Smart" system. It is primarily a network which helps machines/products/things interact with each other. IoT-enabled supply chain helps in making supply chain real as well as proactive. This further helps in harmonization in the whole chain along with providing a systematic approach. This in turn helps in increasing revenue for the organizations by increasing customer service and preventing stock-outs. It is also visible that this is a new area where a lot of researches have not been done. There is also a scope of studying the impact of IoT-enabled supply chain by using artificial intelligence (AI) and robotics process automation (RPA).

46.4 Conclusion

The main objective of this paper is to study the increase in interest in IoT-enabled supply chain along with the benefits it brings for the organizations. In order to be competitive in this globalized environment, organizations need to implement an effective supply chain strategy. IoT has become a core component of every supply chain strategy implementation, and IoT-enabled supply chain helps in improving performance and operational efficiency. The current research has some limitations which need to be acknowledged. First, the research has taken few research done earlier in this space.

It would be great to undertake this study by taking into account more researches done earlier. The second limitation is that this work is not specific to a particular industry. The future work can be undertaken by taking an industry and doing survey on the benefits of IoT-enabled supply chain. This will involve taking into consideration the statistical tools for further analysis.

Acknowledgements The productive comments and feedback provided by practitioners have helped the authors in quality improvement.

References

1. Council of Supply Chain Management Professionals (CSCMP): CSCMP's Definition of Supply Chain Management https://cscmp.org/CSCMP/Certify/Fundamentals/What_is_Supply_Chain_Management.aspx?WebsiteKey=0b3f453d-bd90-4121-83cf-172a90b226a9 (2017). Accessed on 30 Jan 2019
2. Williams, B.D., Waller, M.A.: Top-down versus bottom-up demand forecasts: the value of shared point-of-sale data in the retail supply chain. J. Bus. Logistics **32**(1), 17–26 (2011)
3. Parry, G.C., Brax, S.A., Maull, R.S., Ng, I.C.: Operationalising IoT for reverse supply: the development of use-visibility measures. Supply Chain Manag.: Int. J. **21**(2), 228–244 (2016)
4. Zhou, L., Chong, A.Y., Ngai, E.W.: Supply chain management in the era of the internet of things. Int. J. Prod. Econ. **159**, 1–3 (2015)
5. Aggarwal, C.C., Ashish, N., Sheth, A.: The internet of things: a survey from the data-centric perspective. In: Managing and Mining Sensor Data, pp. 383–428. Springer, Boston, MA (2013)
6. da Rocha, K.E. Mendes, J.V., de Santa-Eulalia, L.A., Moris, V.A.S.: Adoption of IoT in logistics & supply chain management: a systematic literature review. XXXVII Encontro Nacional De Engenharia De Producao (2017)
7. Min, Z., Yuan, S., Zhenwei, X., Hua, Z.: Applied research on supply chain management based on internet of things. Value Eng. **26**, 028 (2010)
8. Dhumale, R.B., Thombare, N.D., Bangare, P.M.: Supply chain management using internet of things. Int. Res. J. Eng. Technol. (IRJET) **4**(06) (2017)
9. Internet of things (IOT)—Cisco. URL http://www.cisco.com/web/solutions/trends/iot/overview.html (2019)
10. Jilani, M.T., Rehman, M.Z.U., Khan, A.M., Chughtai, O., Abbas, M.A., Khan, M.T.: An implementation of IoT-based microwave sensing system for the evaluation of tissues moisture. Microelectron. J. **88**, 117–127 (2019)
11. Khan, M.A., Salah, K.: IoT security: review, blockchain solutions, and open challenges. Future Gener. Comput. Syst. **82**, 395–411 (2018)
12. Lee, I., Lee, K.: The Internet of Things (IoT): applications, investments, and challenges for enterprises. Bus. Horiz. **58**(4), 431–440 (2015)
13. Sodhro, A.H., Pirbhulal, S., Sangaiah, A.K.: Convergence of IoT and product lifecycle management in medical health care. Future Gener. Comput. Syst. **86**, 380–391 (2018)
14. Abdel-Basset, M., Manogaran, G., Mohamed, M.: Internet of Things (IoT) and its impact on supply chain: a framework for building smart, secure and efficient systems. Future Gener. Comput. Syst. **86**, 614–628 (2018)
15. Accorsi, R., Bortolini, M., Baruffaldi, G., Pilati, F., Ferrari, E.: Internet-of-things paradigm in food supply chains control and management. Procedia Manuf. **11**, 889–895 (2017)
16. Ashton, K.: That 'internet of things' thing. RFiD J. **22**(7), 97–114 (2013)
17. Li, H., Parlikad, A.K.: Social internet of industrial things for industrial and manufacturing assets. IFAC-PapersOnLine **49**(28), 208–213 (2016)

18. Kafle, V.P., Fukushima, Y., Harai, H.: Internet of things standardization in ITU and prospective networking technologies. IEEE Commun. Mag. **54**(9), 43–49 (2016)
19. Sivamani, S., Kwak, K., Cho, Y.: A study on intelligent user-centric logistics service model using ontology. J. Appl. Math. (2014)
20. Dehury, C.K., Sahoo, P.K.: Design and implementation of a novel service management framework for IoT devices in cloud. J. Syst. Softw. **119**, 149–161 (2016)

Chapter 47
Study of a Water-Based Hybrid Solar Photovoltaic Thermal Collector

Shubham Sharma, Ravi Prakash, Sandeep Tiwari and Arun Kumar Tiwari

Abstract The electrical conversion efficiency of a solar photovoltaic panel depends on ambient conditions. The efficiency of the photovoltaic panel (PV) system is affected by the module temperature rise. Operating solar panels at high temperate affects efficiency as well overall life of the panel. The researchers have proposed a design that will provide simultaneity electricity and water at a high temperature which is commonly known as photovoltaic thermal (PVT). This is a very cost-effective solar energy application. However, further concerted studies (experimental and analytical) must are required in the design of new heat collectors having more thermal efficiency and lower cost. Researchers reported the use of air and water as cooling fluids in the PVT system. But water-based systems are practically more effective as compared to air-based systems. The temperature fluctuation in the water-based system is lower than the air-based collectors. In this paper, efforts made by various researchers to study of water-based PVT have been presented, which give direction to further research.

Keywords Solar photovoltaic · Thermal conductivity · Overall efficiency · Electrical conversion efficiency · PVT

47.1 Introduction

The solar energy is converted into other forms (thermal energy and electrical energy) which can be used in the applications. The current is produced by photovoltaic (PV) systems. The photovoltaic technology converts incident solar irradiations into electricity. It is ideal, practical and economical. This is used in household and field applications such as a park, agricultural field, railways stations and airports. Literature suggests that the maximum electrical conversion efficiency of a polycrystalline system is 4–17%. The radiation falling on panel absorbed or reflected by the panel;

S. Sharma (✉) · R. Prakash · S. Tiwari
Amity University, Noida, Uttar Pradesh, India

A. K. Tiwari
IET Lucknow, Lucknow, Uttar Pradesh, India

S. Yadav et al. (eds.), *Proceedings of International Conference in Mechanical and Energy Technology*, Smart Innovation, Systems and Technologies 174,
https://doi.org/10.1007/978-981-15-2647-3_47

thus, the temperature of panel increases [1–5]. The operating temperature of the PV module affects its performance. The temperature of the module depends on several factors like local climatic condition, solar irradiation, losses from the PV module, soiling of panel top and various ambient conditions. As the operating temperature of the panel increases, the conversion efficiency decreases. The life of the panel may decrease if operated at high temperatures. The greenhouse effect, environmental concern and fast-drying out of fossil fuels, researchers are doing concerted efforts to maximize the use of solar PV panels. The PV panel converts solar energy into electricity and backside of the panel an absorber is attached which provides heat energy. By cooling the PVT devices, improvement in the performance is observed. The PVT system gives higher energy as compared to a standard PV system. They are cost-effective [6]. The recent research compilation of this area has been presented by several investigators [7, 8]. Moradi et al. [2] have presented all possible types of PVT collectors explored by researchers as shown in Fig. 47.1. In this paper, the recent development in the PVT technologies has been considered for review by focusing on thermal as well as electrical conversion efficiency.

This will help in engineering system designs and future research endeavors. During the last few years, concerted research has been done in the PVT system worldwide.

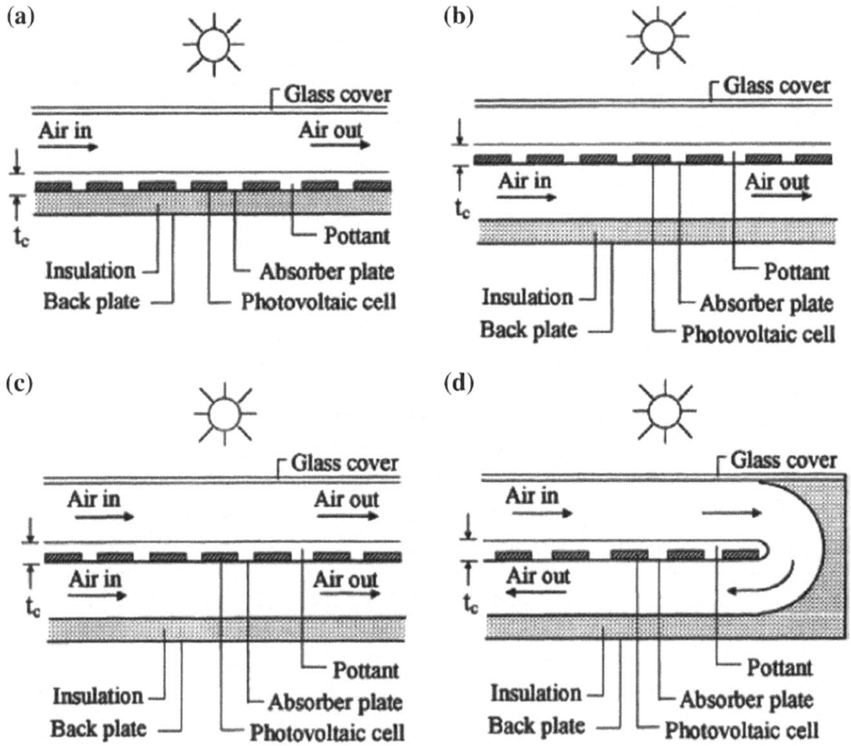

Fig. 47.1 Different configurations for hybrid PVT collector [2]

In the systems air, water and refrigerant have been used for heat collection. Many design principles prosed, developed and studied. The electrical efficiency [9] of the photovoltaic thermal collector is given by Eq. 47.1.

$$\eta_{el} = \eta_o\left[1 - \beta\left(T_{pv} - 298\right)\right] \tag{47.1}$$

where η_o is coefficient for photovoltaic conversion efficiency and β is coefficient for photovoltaic conversion efficiency at reference temperature 298 K. Researchers reported the use of air, water or refrigerant as cooling fluids for heat removal and to cool the solar cells for better electrical conversion efficiency.

High thermal capacity makes the water a suitable media for heat extraction from PV [11]. In the PVT system, the installation of PV modules with the heat absorber plate considerably manages the electrical and thermal performance of the system. The assembly is produced by stacking layers as shown in Fig. 47.2. The well-sealing and tight contact in all layers are maintained. The front casing consists of a 4 mm low glass pane by maintaining a 25 mm air gap over the plate. The lower side of the system is covered by a polyurethane foam insulation. The complete assembly is fixed in an aluminum-alloy frame with proper rubber sealants. The mass flow rate of working fluids in the PVT system plays a very important role. This is beneficial for the cooling of the PV module. The critical mass flow rate of fluid has been obtained. At the optimum fluid flow rate, thermal performance and electrical conversion efficiency have been obtained [12, 13]. The heat extraction from the panel is further improved if water-based nanofluids are circulated through the tube fitted at the back of the panel. But it has certain drawback, as the volumetric fraction of nanofluids is increased, the pressure drop is also increased [7, 14–18]. Hence, the power requirement to maintain the flow will also be increased.

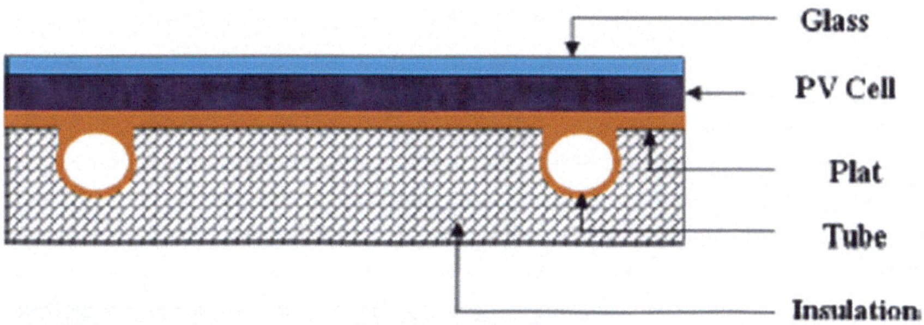

Fig. 47.2 Configuration of a hybrid PVT [10]

47.2 Experimental Observation of the PVT System Using Water

For controlled and accurate measurement, Fudholi et al. [19] developed an indoor test by using 23 halogen lights (0.5 kW) at the laboratory consists and fixed in nine columns to simulate the solar radiation. By doing so, the temperature at the input, outlet and ambient can be measured. The wind velocity, current and voltage, and water flow rate in the PVT systems can be measured properly. For water-based PVT collectors, alternative designs of collectors fabricated using hollow tubes of stainless steel. This has three types of configurations. The design configurations of the absorber plate are (a) web flow, (b) direct flow and (c) spiral flow as shown in Fig. 47.2. Among all three collectors, spiral flow absorber has 65% efficiency and PV efficiency is 13%. The experiments were conducted at a mass flow rate of 0.041 kg/s. The primary energy-saving efficiency is 79–91%. The absorber plate and tube (14 mm outer diameter and 2 mm thickness) were welded before attaching the back of the PV panel. The plate and tube are made of galvanized steel. The spacing between tubes was kept 30 mm. This hybrid collector is more economical. The temperature variation in all sub-layers was recorded to study thermal and electrical efficiency. This system has relatively good heat absorption and lower production cost [10].

Rejeb et al. [20] have numerically investigated the photovoltaic/thermal sheet and tube collector. The increase in the temperature of the PV system decreases electrical efficiency whereas thermal efficiency increases. The inlet temperature of fluids also affects the performance of the panel. The maximum thermal and electrical power recorded in the summer season as compared to season due to the availability of higher solar irradiation. Similarly, sheet and tube collector type of absorber (Fig. 47.3) was used by Khelifa et al. [21] for mathematical modeling.

The energy equations were written, and the coupled differential equations were solved to predict temperatures of all sub-layers of the system as well as coolant. For the heat transfer analysis, modeling was done using *FLUENT* software without considering direct solar radiation. The effects of radiation are applied in terms of heat flux. Numerical results were validated by the experimental results. The panel temperature is dismissed by 15–20% due to the water circulation through the tube.

The complete system performance was analyzed in the laminar and turbulent flow regime [22]. The total energy efficiency is higher for the glazed PVT system

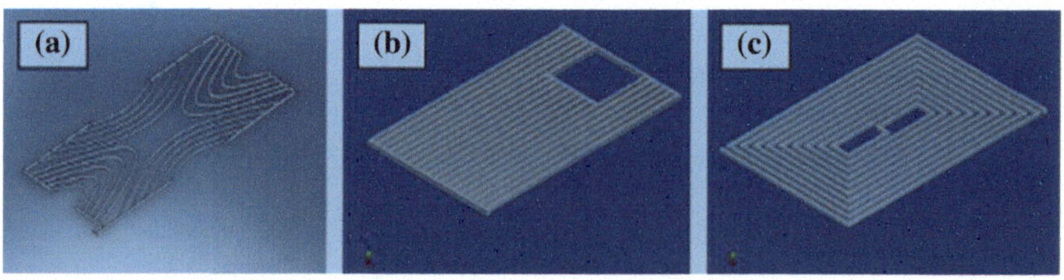

Fig. 47.3 Design configurations of absorber plate [19]

Fig. 47.4 Heat exchanger attachment to the rear side of the PV module [21]

as compared to unglazed systems. In the turbulent flow region, total efficiency is enhanced with a higher diameter of the pipe. Overall, total energy efficiency was maximum in the turbulent region. On the other hand, the exergy efficiency is greater in the laminar flow region. Aste et al. [23] used polycrystalline cells (230 Wp) and attached a roll-bond (Al) heat exchanger which was fixed tightly. The water was used as a coolant with a flow rate of 0.055 kg/s. The hybrid system shows higher overall efficiency as compared to the simple module in terms of primary energy (Fig. 47.4).

Rawat et al. [24] used copper sheets with copper tubes to fix it behind the panel. The lower attachment will act as a heat extractor to the panel, as the water flows through the tubes using a small capacity pump with a discharge rate of 0.002 kg/s (Fig. 47.5). The electrical efficiency was improved, and the thermal efficiency was observed approximately 50% and during experimentation.

The effect of contact conductance between different sub-layers and ohmic loss is very significant. In experimental setup, sheet and tube assembly was used. The variation of the electrical, thermal overall energy and exergy efficiencies are in the range of 12.2%–13%, 18.2%–41.8%, 19%–28.9% and 8.9%–10%, respectively. The heat transfer study has been reported by several researchers [25–45].

Fig. 47.5 Experimental setup of PV and PVT system [24]

47.3 Conclusions

The electrical conversion efficiency of silicon-based solar photovoltaic panel is 4–17%. Some of the incident solar raditions are absorved by the panel. The remanining radiations are reflected back to the envirnonment and thus lost. The conversion efficiency decreases with a rise in the panel increases. This depends on direct normal incidence, its reflection, loss from the module, soiling of the panel, etc. The life of the panel also decreases due to operation at elevated temperatures. The use of photovoltaic thermal system simultaneously provides current and heat energy. Based on the numerical and experimental study, researchers proposed a new design of PVT. This is a very cost-effective solar energy application. Researchers reported the use of air and water as cooling fluids in the PVT system. The panel temperature fluctuation is less in a water-based system, and very high fluctuation is observed with a gas-based (air) system. Among the three approaches, the water-based PVT is more effective as compared to air-based systems.

References

1. Aste, N., del Pero, C., Leonforte, F.: Water flat plate PV thermal collectors: a review. Sol. Energy **102**, 98–115 (2014)
2. Moradi, K., Ebadian, M.A., Lin, C.X.: A review of PV/t technologies: effects of control parameters. Int. J. Heat Mass Transf. **64**, 483–500 (2013)
3. Suresh, A.K., Khurana, S., Nandan, G., Dwivedi, G., Kumar, S.: Role on nanofluids in cooling solar photovoltaic cell to enhance overall efficiency. Mater. Today: Proc. **5**(9), 20614–20620 (2018)
4. Sharma, R., Gupta, A., Nandan, G., Dwivedi, G., Kumar, S.: Life span and overall performance enhancement of solar photovoltaic cell using water as coolant: a recent review. Mater. Today: Proc. **5**(9), 18202–18210 (2018)

5. Rathour, R.S., Chauhan, V., Agarwal, K., Sharma, S., Nandan, G.: Cooling of solar photovoltaic cell: using novel technique. In: Lecture Notes in Mechanical Engineering, pp. 521–529. Springer, Singapore (2019)
6. Lari, M.O., Sahin, A.Z.: Design, performance and economic analysis of a nanofluid-based photovoltaic/thermal system for residential applications. Energy Convers. Manag. **149**, 467–484 (2017)
7. Sekhar, T.V.R., Prakash, R., Nandan, G., Muthuraman, M.: Preparation of Co_3O_4-H_2O nanofluid and application to CR-60 concentrating solar collector. Prog. Ind. Ecol. Int. J. **11**(3), 227–246 (2017)
8. Sekhar, T.V.R., Prakash, R., Nandan, G., Muthuraman, M.: Performance enhancement of a renewable thermal energy collector using metallic oxide nanofluids. Micro Nano Lett. **13**(2), 248–251 (2018)
9. Sobhnamayan, F., Sarhaddi, F., Alavi, M., Farahat, S., Yazdanpanahi, J.: Optimization of a solar photovoltaic thermal (PV/t) water collector based on exergy concept. Renew. Energy **68**, 356–365 (2014)
10. Touafek, K., Khelifa, A., Adouane, M.: Theoretical and experimental study of sheet and tubes hybrid PVT collector. Energy Convers. Manag. **80**, 71–77 (2014)
11. Daghigh, R., Ruslan, M., Sopian, K.: Advances in liquid based photovoltaic/thermal (PV/T) collectors. Renew. Sustain. Energy Rev. **15**(8), 4156–4170 (2011)
12. Ibrahim, A., Othman, M.Y., Ruslan, M.H., Mat, S., Sopian, K.: Recent advances in flat plate photovoltaic/thermal (PV/t) solar collectors. Renew. Sustain. Energy Rev. **15**(1), 352–365 (2011)
13. Ji, J., Han, J., Tai Chow, T., Yi, H., Lu, J., He, W., Sun, W.: Effect of fluid flow and packing factor on energy performance of a wall-mounted hybrid photovoltaic/water-heating collector system. Energy Build. **38**(12), 1380–1387 (2006)
14. Sekhar, T.V.R., Nandan, G., Prakash, R., Muthuraman, M.: Modeling a renewable energy collector and prediction in different flow regimes using CFD. Mater. Today: Proc. **5**(2), 4563–4574 (2018)
15. Sekhar, T.V.R., Nandan, G., Prakash, R., Muthuraman, M.: Investigations on viscosity and thermal conductivity of cobalt oxide- water nano fluid. Mater. Today: Proc. **5**(2), 6176–6182 (2018)
16. Sekhar, T.V.R., Prakash, R., Nandan, G., Muthuraman, M.: Pressure drop characteristics & efficiency enhancement by using TiO_2-H_2O nanofluid in a sustainable solar thermal energy collector. Int. J. Environ. Sustain. Dev. (2018)
17. Sekhar, T.V.R., Nandan, G., Prakash, R.: Parabolic trough solar collectors. Eur. J. Sci. (EJS), 43–53 (2018)
18. Sharma, S., Tiwari, A.K., Tiwari, S., Prakash, R.: Viscosity of hybrid nanofluids: measurement and comparison. J. Mech. Eng. Sci. **12**(2), 3614–3623 (2018)
19. Fudholi, A., Sopian, K., Yazdi, M.H., Ruslan, M.H., Ibrahim, A., Kazem, H.A.: Performance analysis of photovoltaic thermal (PVT) water collectors. Energy Convers. Manag. **78**, 641–651 (2014)
20. Rejeb, O., Dhaou, H., Jemni, A.: A numerical investigation of a photovoltaic thermal (PV/T) collector. Renew. Energy **77**, 43–50 (2015)
21. Khelifa, A., Touafek, K., Moussa, H.B., Tabet, I.: Modeling and detailed study of hybrid photovoltaic thermal (PV/t) solar collector. Sol. Energy **135**, 169–176 (2016)
22. Yazdanifard, F., Ebrahimnia-Bajestan, E., Ameri, M.: Investigating the performance of a water-based photovoltaic/thermal (PV/t) collector in laminar and turbulent flow regime. Renew. Energy **99**, 295–306 (2016)
23. Aste, N., Leonforte, F., Pero, C.D.: Design, modeling and performance monitoring of a photovoltaic thermal (PVT) water collector. Sol. Energy **112**, 85–99 (2015)
24. Rawat, P., Debbarma, M., Mehrotra, S., Sudhakar, K.: Design, development and experimental investigation of solar photovoltaic/thermal (PV/T) water collector system. Int. J. Sci. Environ. Technol. **3**(3), 1173–1183 (2014)

25. Agrawal, T., Ajitkumar, R., Prakash, R., Nandan, G.: Sodium silicide as a hydrogen source for portable energy devices: a review. Mater. Today: Proc. **5**(2), 3563–3570 (2018)
26. Nandan, G., Sahoo, P.K., Kumar, R., Chatterjee, B., Mukhopadhyay, D., Lele, H.G.: Experimental investigation of sagging and ballooning in LOCA for Indian PHWR. Eng. Lett. **18**(3) (2010)
27. Kumar, R., Nandan, G., Sahoo, P.K., Chatterjee, B., Mukhopadhyay, D., Lele, H.G.: Ballooning of pressure tube under LOCA in an Indian pressurised heavy water reactor. In: 14th International Heat Transfer Conference, IHTC, vol. 14(7), pp. 317–323 (2010)
28. Nandan, G., Sahoo, P.K., Kumar, R., Chatterjee, B., Mukhopadhyay, D., Lele, H.G.: Experimental investigation of sagging and ballooning for a completely voided pressure tube of Indian PHWR under heatup condition. World Acad. Sci. Eng. Technol. **61**, 197–204 (2010)
29. Rawat, K.S., Thakur, H.C., Nandan, G.: CFD analysis of a pentagonal rib over absorber plate of a solar air heater. In: 3rd International Conference on Manufacturing Excellence, pp. 191–196. Amity University Uttar Pradesh, Noida, India (2016)
30. Kumar, S., Nandan, G., Singh, G.K.: Numerical study of natural convection and radiative heat transfer of heated inner cylinder placed inside a isothermally cooled circular enclosure. In: Third International Conference on Manufacturing Excellence, pp. 179–183. Amity University Uttar Pradesh, Noida, India (2016)
31. Shukla, A.K., Gupta, S., Singh, S.P., Sharma, M., Nandan, G.: Thermodynamic performance evaluation of SOFC based simple gas turbine cycle. Int. J. Appl. Eng. Res. **13**(10), 7772–7778 (2018)
32. Kumar, V., Shrivastava, R., Nandan, G.: Energy saving using phase change material in refrigerating system. In: Third International Conference on Manufacturing Excellence, pp. 184–190. Amity University Uttar Pradesh, Noida, India (2016)
33. Anamika, Thakur, H.C., Nandan, G.: A review on solar air heater performance using different artificial roughened rib. In: International Conference Emerging Development in Engineering and Technology (ICEDET-2016), pp. 347–349. Shivalik College of Engineering, Dehradun, India (2016)
34. Sekhar, T.V.R., Nandan, G., Prakash, R.: Parabolic trough solar collectors. Eur. J. Sci. (EJS) **1**(1), 43–53 (2018)
35. Nandan, G., Lele, H.G., Sahoo, P.K., Chatterjee, B., Kumar, R., Mukhopadhyay, D.: Experimental investigation of heat transfer during LOCA with failure of emergency cooling system. In: 5th International Conference on Heat Transfer, Fluid Mechanics and Thermodynamics (2007)
36. Nandan, G., Sahoo, P.K., Kumar, R., Chatterjee, B., Mukhopadhyay, D., Lele, H.G.: Thermomechanical behavior of pressure tube of Indian PHWR at 20 bar pressure. World Acad. Sci. Eng. Technol. **61**, 205–213 (2010)
37. Majumdar, P., Chatterjee, B., Nandan, G., Mukhopadhyay, D., Lele, H.G.: Assessment of the code "PTCREEP" for IPHWR pressure tube ballooning study. J. Press. Vessel Technol. Trans. ASME **133**(1) (2011)
38. Nandan, G., Kumar, R., Sahoo, P., Chatterjee, B., Mukhopadhyay, D., Lele, H.: Thermomechanical response of pressure tube during degraded cooling condition event in a pressurised heavy water reactor. In: 19th National & 8th ISHMT-ASME Heat and Mass Transfer Conference. JNTU Hyderabad, India (2008)
39. Nandan, G., Sahoo, P.K., Kumar, R., Chatterjee, B., Mukhopadhyay, D., Lele, H.G.: Experimental investigation of sagging of a completely voided pressure tube of Indian PHWR under heatup condition. Nucl. Eng. Des. **240**(10), 3504–3512 (2010)
40. Nandan, G., Majumdar, P., Sahoo, P.K., Kumar, R., Chatterjee, B., Mukhopadhyay, D., Lele, H.G.: Study of ballooning of a completely voided pressure tube of Indian PHWR under heat up condition. Nucl. Eng. Des. **243**, 301–310 (2012)
41. Kumar, R., Nandan, G., Sahoo, P.K., Chatterjee, B., Mukhopadhyay, D., Lele, H.G.: Ballooning of pressure tube under LOCA in an Indian pressurised heavy water reactor. In: 14th International Heat Transfer Conference, Washington, DC, USA, pp. 319–323. American Society of Mechanical Engineers Digital Collection (2010)

42. Vadhera, J., Sura, A., Nandan, G., Dwivedi, G.: Study of phase change materials and its domestic application. Mater. Today: Proc. **5**(2), 3411–3417 (2018)
43. Krishn, S., Goyal, M., Nandan, G., Kumar, S., Kumar, P., Shukla, A.K.: Pool boiling using nanofluids: a review. In: Lecture Notes in Mechanical Engineering, pp. 325–336. Springer, Singapore (2019)
44. Shukla, A.K., Sharma, A., Sharma, M., Nandan, G.: Thermodynamic investigation of solar energy-based triple combined power cycle. Energy Sources Part A: **41**(10), 1161–1179 (2018)
45. Lamnatou, C., Chemisana, D.: Photovoltaic/thermal (PVT) systems: a review with emphasis on environmental issues. Renew. Energy **105**, 270–287 (2017)

Chapter 48
Kinematic Analysis of Steering Mechanism: A Review

Mubina Shekh, O. P. Umrao and Dharmendra Singh

Abstract Mostly, two-wheel steering (2 WS) systems are used to control the vehicle. But many researchers are working in this area, for a narrow space how a car can take turn or back without any failure occurs. There are different types of drives in a vehicle such as front-wheel, rear-wheel or all-wheel drive (2 and 4 WS). But for the reason of safety, four-wheel steering (4 WS) vehicles termed as Quadra Steering System are being used. In this paper, the features of different models of car steering system used have some drawbacks like failing at high speed, slipping of the tracks, higher turning radius. To overcome these drawbacks, a suitable and appropriate steering system has been proposed and it has been presented here.

Keywords Kinematic analysis · Two-wheel steering (2 WS) · Four-wheel steering (4 WS) · Turning radius

48.1 Introduction

Many time endeavours have been devoted to develop a reliable and safe steering design for the driver, which is easy to operate at busy and narrow road to take a back and U turn. Many researchers are working on wheel drive mechanism of vehicle from many years, so numerous designs are proposed for reducing turning radius by researchers. Arvind et al. [1] said less body lean during fast lane changes and turns; theoretical methods were also found in different literatures. Nityananth said in his paper that steering arrangement is to turn the front wheels using a hand-operated steering wheel, which is positioned in front of the driver [2]. Chaudhary, in his paper, has done the kinematic analysis of the four-wheel steering system. These data are

M. Shekh · O. P. Umrao
Department of Mechanical Engineering, IMS Engineering College, Ghaziabad, India

D. Singh (✉)
Department of Mechanical Engineering, ABES Engineering College, Ghaziabad, India
e-mail: dharmendra0677@gmail.com

S. Yadav et al. (eds.), *Proceedings of International Conference in Mechanical and Energy Technology*, Smart Innovation, Systems and Technologies 174,
https://doi.org/10.1007/978-981-15-2647-3_48

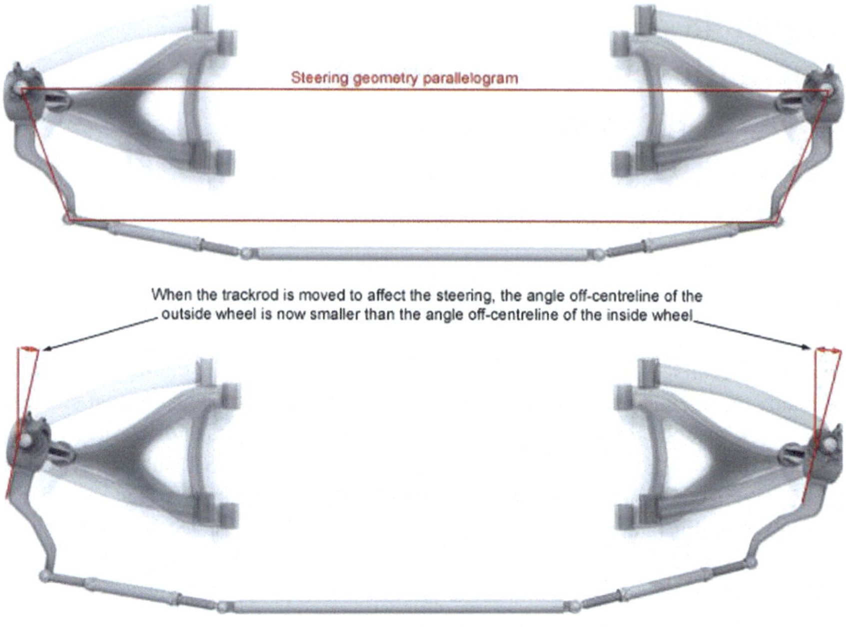

Fig. 48.1 Link arrangement

compared to know the effect of the system on the vehicle in terms of the turning radius [3]. Bhishikar et al. said in his paper that all the four-wheel mechanism of the vehicle used for steering purpose, it is most effective [4]. Abe said that rear wheels turn by an angle to aid the front wheels in steering. Though, in these systems, the rear wheels steered by 20 or 30, their main aim was to support the front wheels rather than steer by themselves [5]. Mistry et al. in his work stated that all vehicles existed for the most part two-wheel controlling framework to control the development of the vehicle whether it is front-wheel drive, back-wheel drive or all-wheel drive [6]. Civeraet et al. [7] said in his project report that to obtain instant rotation centre or angles (according to Ackerman principle), used an articulated trapezium called Jeantaud trapezium as shown in Fig. 48.1.

48.2 Steering Mechanism System

Steering is used for changing the direction of two- or four-wheel axle with reference to the chassis, so as to move the vehicle in any desired path. Steering system is based on the following mechanism—Ackerman's steering mechanism, Davis steering mechanism [8].

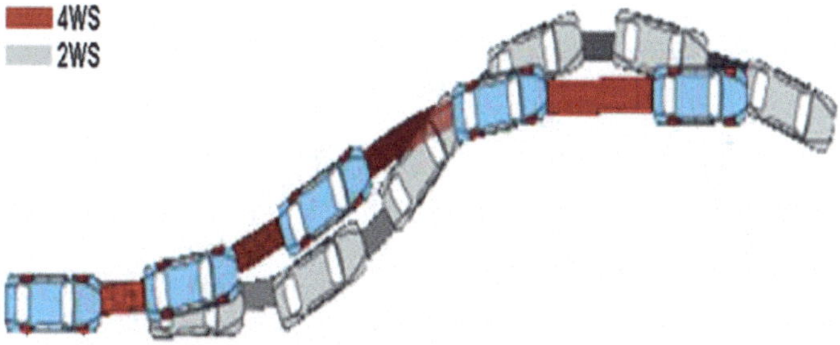

Fig. 48.2 Movement difference between two-wheel steering system and four-wheel steering system [Ref. 3]

48.3 Comparison of Four-Wheel Steering System with Two-Wheel Steering System

In four-wheel directing framework, it is progressively effective and stable on cornering improved guiding responsiveness and accuracy in fast straight-line steadiness as shown in Fig. 48.2. Prominent improvement is fast, simpler, more secure path evolving moves. Littler turning range and tight space required to moves the car at low speed, relative wheel angles and their control and danger of hitting a hindrance is enormously decreased [3, 9].

48.4 Design of Two- and Four-Wheel Steering System

It is to be recalled that both the controlled wheels do not turn a similar way, since the inward wheels travel by a more drawn out separation than the external wheels, as portrayed in Fig. 48.3.

Ackermann two- and four-wheel directing framework geometries are a geometrical game plan of various linkages in the two- and four-wheel guiding arrangement of vehicles intended to take care of the issue of wheels within and outside of a go expecting to follow out circles of various radii. The two- and four-wheel directing framework support focuses are joined by an inflexible bar called the tie pole which is a piece of the two- and four-wheel guiding framework components. In immaculate Ackermann, the middle purpose of the majority of the circles followed by all wheels will lie at a typical point at any edge of two- and four-wheel controlling frameworks.

It might be quite hard to integrate or break down this component for 2 and 4 WS framework over the full scope of (2 and 4 WS) frameworks' points. Because of such sort of troubles, presently day's planner does not utilize Ackermann (2 and 4 WS) framework in four-wheelers, since this component overlooks thorough parameter like dynamic and consistent impacts. In resentment in this course of action, the component

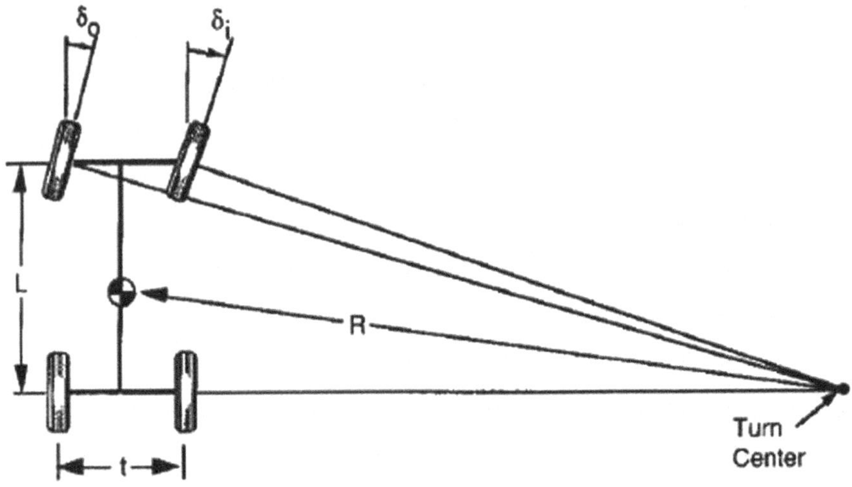

Fig. 48.3 Ackerman two- and four-wheel steering system mechanism

is thorough for low-speed developments, and the privilege and left wheels do not turn by a similar point, in any encompassing paces.

It implies a very hard for vehicles with self-sufficient (2 and 4 WS) framework. The wheels can not be positively given the exact Ackerman turning points. It would legitimately trouble the dynamic treatment of the vehicle, making it hard to control appropriately. Four-wheel steering system is employed in vehicles to achieve better manoeuvrability at high speeds, reducing the turning circle radius of the car and to reduce the driver's steering effort [10].

48.5 Analysis

Here, the main objective is to reduce turning radius. Turning radius is the difference of angles from initial position to the final position of front wheels. A vehicle with an incorrect turning radius will drag the wheels around a corner causing very fast tire wear. In this paper, turning radius will reduce the wear of the vehicle using 4 WS system for easy to turn and safety of the driver. According to the analysis of this system, a kinematic link will lie between the front and rear axles. Such a mechanism is learned by the kinematic analysis of the steering system's geometry. Now, he will calculate the turning radius of a vehicle with the help of geometry of the mechanism. Figures 48.4 and 48.5 used from his (V. Arvind) paper for comparative study between our prototype and ALTO 800 and FORD FIGO [1].

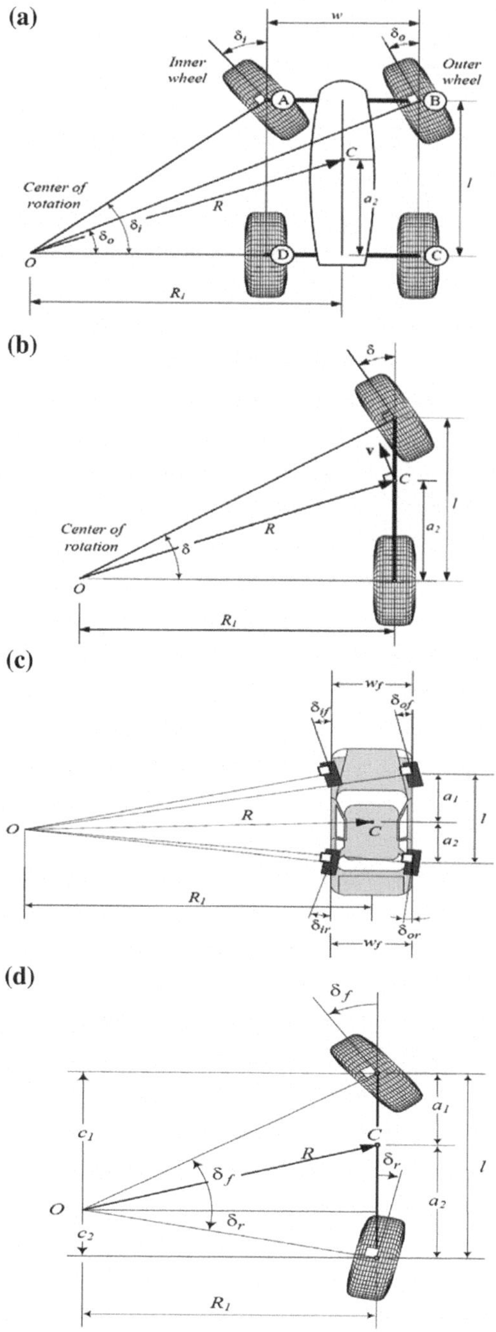

Fig. 48.4 **a** Turning measurement of vehicle [1]. **b** Bicycle model of measuring turning radius [1]. **c** Four-wheel steering system [1]. **d** Bicycle model for four-wheel steering system [1]

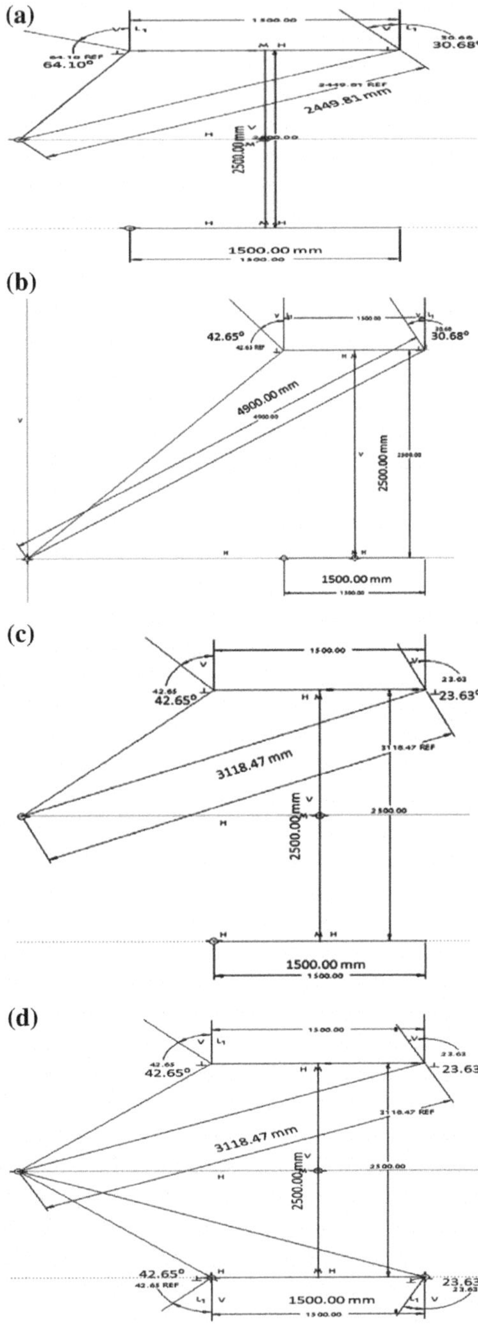

Fig. 48.5 **a** Steered front-wheel angles measurement (FORD FIGO) [1]. **b** Modified steering wheels angles 1 (FORD FIGO) [1]. **c** Modified steering wheels angles 2 (FORD FIGO) [1]. **d** Optimum symmetric four-wheel steering (FORD FIGO) [1]

48.5.1 Turning Radius

In steering system, when vehicle takes a turn from initial position, then the difference in the angles of the front wheels to in a turn is known as turning radius. Due to incorrect turning radius, it will strain (drag) the wheels around a junction (corners). Wearing of tyres occurs due to this reason.

48.5.1.1 Turning Radius of Two-Wheel Steering System (Prototype Based)

The turning radius (R) of the vehicle is usually measured by using the formula as shown in Eqs. (48.1) to (48.3).

$$\cot\,\delta_o - \cot\,\delta_i = w/l \tag{48.1}$$

$$\cot\,\delta = (\cot\,\delta_o + \cot\,\delta_i)2 \tag{48.2}$$

$$R = \sqrt{\left(a_2^2 + l^2 \cot^2 \delta\right)} \tag{48.3}$$

$$\delta_i = 42^\circ$$

$$l = 32$$

$$\begin{aligned} W/l &= 0.846 \\ \cot\,\delta_o &= 0.846 + \cot\,\delta_i \\ \delta_o &= 27.06^\circ \end{aligned}$$

$$\begin{aligned} \cot\,\delta &= (\cot\,\delta_o + \cot\,\delta_i)/2 \\ &= (\cot\,27.06^\circ + \cot\,42)^\circ/2 \\ &= 1534 \end{aligned}$$

$$\delta = 33.09$$

$$a_2 = 17.3\,\text{cm}$$

$$\begin{aligned} R &= \sqrt{(17.3^2 + 32^2 \times 1.532^2)} \\ &= 52.04\,\text{cm} = 520\,\text{mm} \end{aligned}$$

48.5.1.2 Turning Radius of Four-Wheel Steering System (Prototype Based)

$$W = 27.1\,\text{cm}$$

$$\cot\,\delta_{\text{of}} = \cot\,\delta_{\text{if}} = w_f/l - \frac{w_r}{l}.(\cot\,\delta_{\text{of}} - \cot\,\delta_{\text{if}})/(\cot\,\delta_{\text{or}} - \cot\,\delta_{\text{ir}})$$

$$\delta_{\text{of}} = 35^\circ$$
$$\delta_{\text{if}} = 39^\circ$$
$$\delta_{\text{ir}} = 40^\circ$$

$$\cot\,35^\circ - \cot\,39^\circ = 27.1/32 - (\cot\,35^\circ - \cot\,39^\circ)/(\cot\,\delta_{\text{or}} - \cot\,40^\circ)$$
$$0.193 = 0.8468 - 0.8468 \times \{0.193/(\cot\,\delta_{\text{or}} - 1.19)\}$$
$$0.8468 \times 0.193/(\cot\,\delta_{\text{or}} - 1.19) = 0.6538$$

$$\delta_{or} = 34.77^\circ$$

$$C_1 = 16.5\,\text{cm}$$

$$C_2 = 16\,\text{cm}$$

$$a_1 = 14.7\,\text{cm}$$

$$a_2 = 17.3\,\text{cm}$$

$$R = \sqrt{(a_2 - C_2)^2} + C_1^2\cot\,\delta_f$$

$$\cot\,\delta_f = (\cot\,\delta_{if} + \cot\,\delta_{of})/2$$

$$\delta_f = 36.9^\circ$$

$$R = \sqrt{(17.3 - 16)^2} + 16.5 \times 1.33^2$$
$$R = 21.98\,\text{cm} = 219.8\,\text{mm}$$

48.5.1.3 Percentage Reduction in Turning Radius

$$\begin{aligned}\%\,\text{reduction in turning radius} &= \frac{52.04 - 21.98}{52.04} \times 100 \\ &= 57.7\%\end{aligned}$$

48.6 Kinematic Analysis [Using Bicycle Model]

Spentzas et al. [11] said in his paper that "A general kinematical analysis of 4 WS vehicles by considering either the simplified 2 WS (bicycle) or complete 4 WS automobile system.

48.6.1 Four-Wheel Steering System

$$a = GA_f = 14.7\,\text{cm}$$

$$b = GA_r = 17.3\,\text{cm}$$

$$e = GI_m = 2\,\text{cm}$$

$$R = 21.98\,\text{cm}$$

$$\begin{aligned}\beta &= 3.5^\circ \\ \delta_f &= \tan^{-1}\left[\frac{a}{R\cos\beta} + \tan\beta\right] \\ &= \tan^{-1}\left[\frac{14.7}{21.98\cos 3.5} + \tan 3.5\right] \\ &= 36.17\end{aligned}$$

$$\begin{aligned}\delta_r &= \tan^{-1}\left[\frac{b}{R\cos\beta} - \tan\beta\right] \\ &= \tan^{-1}\left[\frac{17.3}{21.98\cos 3.5} - \tan 3.5\right] \\ &= 36.03\end{aligned}$$

48.6.2 Two-Wheel Steering System

$$l = 32\,\text{cm}$$

$$R = 52.04\,\text{cm}$$

$$\tan\,\delta_f = l/R\,\cos\,\beta$$

$$\begin{aligned}\delta_f &= \tan^{-1}\left[\frac{l}{r\,\cos\,\beta}\right]\\ &= \tan^{-1}\left[\frac{32}{52.04\,\cos\,3.5}\right]\\ \delta_f &= 31.63^\circ\end{aligned}$$

48.6.3 Motion Analysis

Number of turns = 5
Steering wheel angle = 1800°
Wheel angle = 70°

$$\begin{aligned}\text{Steering ratio} &= \text{Steering wheel angle/wheel angle}\\ &= 1800/70 = 25.71^\circ\end{aligned}$$

48.7 Comparative Study of Some Four-Wheel System and Our Prototype of Four Wheels

We are going to analyze that how the use of four-wheeler steering system will help to decrease the turning radius of commercial vehicles like FORD FIGO by keeping the practical limitation in mind and compare their results with the prototype that we design.

48.7.1 *Case Study 2—FORD FIGO*

Now, we will take another four-wheeler FORD FIGO and find out the steered angle of this vehicle. We will apply the same procedure as mentioned in the above case study of ALTO 800 [1]. The data of the vehicle considered are as follows:

Wheelbase: 2500 mm (approx.)
Wheel track: 1500 mm (approx.)
Turning radius: 4.9 m
Weight (GVW): 1500 kg (approx.).

As a matter of first importance, front two wheels at the given turning sweep with the assistance of Ford Figo vehicle leaflet's information. During estimation, accept one edge steady and applying four-wheel symmetric directing and dissect the other wheel point with the assistance of geometry. At that point, we discovered the resultant turning span of the CAR utilizing four wheels evenness controlling framework as given underneath.

Now, keeping the other angle constant, we get the resultant turning radius using this four-wheel symmetry steering system.

After this analysis, it is clear that the second method provides the prime solution of turning radius without moderation of the limitation. Here, we get, FORD FIGO car's inner wheel turning radius is 64.10^0 at extreme conditions.

Figure 48.5d shows the optimum solution of the angle of inner wheel that helps to reduce the turning radius of the four-wheeler car. This analysis gives a 36% reduction of turning radius of this four-wheeler.

48.8 Conclusion

From the above analysis of two-wheel and different type of four-wheel model including our prototype model shows in Table 48.1. In which this analysis is converting a steering mode from 2 WS to 4 WS the turning radius get reduces which help automobile to take the sharp turn and prove to be very helpful while driving the car or at the time of changing the lane. This concept may be applied to different cars of

Table 48.1 Comparison between ALTO 800, FORD FIGO and PROTOTYPE

Model	Turning radius in two-wheel mode (in mm)	Turning radius in four-wheel mode (in mm)	% reduction in turning radius
ALTO 800	4600	2877.48	37
FORD FIGO	4900	3118.47	36
PROTOTYPE	520	219.8	57.7

different speeds and weights. By the conservation of angular momentum, if weight and speed of vehicle increase, the then turning radius decreases.

The percentage reduction in turning radius is 37% in ALTO 800 car when switching a steering mode from two wheels to four wheels, while this reduction will be 36% in case of FORD FIGO car. However, in our prototype four-wheel steering, the turning radius has reduced to 57.7%. Thus, our prototype car can turn in small space. Manufacturing cost is reduced due to small size of turning radius.

References

1. Arvind, V.: Optimizing the turning radius of a vehicle using symmetric four wheel steering system. Int. J. Sci. Eng. Res. **4**(12), 2177–2184 (2013)
2. Nithyananth, S., Jagatheesh, S., Madan, K., Nirmalkumar, B.: Convertible four wheels steering with three mode operation. Int. J. Res. Aeronaut. Mech. Eng. **2**(3), 81–89 (2014)
3. Choudhari, D.S.: Four wheel steering system for future. Int. J. Mech. Eng. Robot. Res. **3**(4) (2014)
4. Singh, K.: Automobile Engineering, 12th edn., pp 207–229. Standard Publication Distributors (2017)
5. Abe, M.: Vehicle dynamics and control for improving handling and active safety from four-wheel-steering to direct yaw moment control. J. Multibody Dyn. **213**(4) (2008)
6. Mistry, M.K., et al.: Developing quadra-steering system for multi-purpose of vehicle and give better comfort for driving. Int. J. Adv. Res. Automobile Eng. (2014)
7. Civera, D.N.: Description and design of the steering system for a formula student car (2009–2010)
8. Bhupendra, et al.: Advanced four wheel steering system. Int. J. Res. Eng. Adv. Technol. **3**(2) (2015)
9. Lohith, K., Shankapal, S.R., Monish Gowda, M.H.: Development of four wheel steering system for a car. SASTECH J. **12**(1) (2013)
10. Bhishikar, S., Gudhka, V., Dalal, N., Mehta, P., Bhil, S., Mehta, A.C.: Design and simulation of 4 wheel steering system certified. Int. J. Eng. Innov. Technol. **3**(12) (2014)
11. Spentzas, K.N., Alkhazali, I., Demic, M.: Kinematics of four wheel-steering vehicles. Forsch Ingenieurwes **66**(6), 211–216 (2001). Springer link

Chapter 49
Analysis on Milling of Nanoclay-Doped Epoxy/Carbon Laminates Using Taguchi Approach

Karunesh Kumar, K. K. Singh and R. K. Thakur

Abstract Many components made up of carbon fiber reinforced composite are used by various industries in making complex machines like aircraft, automobile, and aerospace to minimize the weight of the final product and increase fuel economy. So the machining process plays an important role to achieve the final product. Milling process produces good quality and well-defined surface. However, there are some challenges like high delamination factor and surface roughness which are associated with the property of material and cutting parameter of machining. Current work concluded that feed rate was the most significant factor for surface roughness while spindle speed was the most significant factor for the delamination factor. Increase in weight % of nanoclay and spindle speed decreases the surface roughness while increase in spindle speed increases the delamination factor. Decrease in feed rate provides better surface finish and less delamination damage.

Keywords CFRP · Nanoclay · Milling · Delamination · Surface roughness

49.1 Introduction

Application of carbon fiber reinforced composite (CFRP) has increased in recent years in automobile, medical application, aerospace, and aircraft due to excellent properties of CFRP like low weight, high-temperature tolerance, high-specific stiffness, low thermal expansion, high fatigue resistance, corrosion resistance, and specific strength [1]. The properties associated with CFRP are not usual and difficult to obtain in alloys [2]. Epoxy resin has excellent adhesive and mechanical properties, and due to these properties, epoxy resin shows a significant role in polymeric composite materials. In recent decades, polymeric nanocomposites have attracted

---

K. Kumar · K. K. Singh · R. K. Thakur (✉)
Department of Mechanical Engineering, IIT (ISM), Dhanbad 826004, India

S. Yadav et al. (eds.), *Proceedings of International Conference in Mechanical and Energy Technology*, Smart Innovation, Systems and Technologies 174,
https://doi.org/10.1007/978-981-15-2647-3_49

researchers throughout the world in hope of accomplishing different superior properties like thermal, mechanical, fire retardancy, chemical, water absorption, tribology, corrosion resistance, toughness in addition of a small quantity of nanofiller with thermosetting or thermoplastic resin [3–7]. Anisotropy, low heat dissipation, heterogeneity, and inhomogeneity are the property of CFRP. Because of above-mentioned properties, the issues such as high surface roughness, excessive cutting force, and damages in machining process such as fiber pull out, matrix burning and delamination are found, and these should be minimized to get accurate and finish machining.

Zinati et al. [8] performed an experiment on polyamide 6/multiwalled carbon nanotube composite with 3.5 weight percentage using the end milling process. They concluded that MWCNT-doped polyamide gives better surface finish than the neat polyamide. Chen et al. [9] conducted an experimental research on cutting tool wear in the milling process of CFRP laminate. They suggested that cutting edge coating wear dominated initially and later on, rounded cutting edge radius slowly enlarged during milling process and the tool wear slowly expanded to the flank face. Hosakawa et al. [10] performed an experiment on CFRP using diamond-like carbon (DLC)-coated carbide end mills having dissimilar helix angle. They concluded that the effect of the geometry of the end mill was more significant than that of DLC coating. They also observed that the normal force and tangential force reduced significantly on increased helix angle of end mill.

In this paper, carbide end mill cutter of 6 mm diameter with 4 flutes is used for milling process on CFRP laminate. Process parameters like spindle speed (SS), feed rate (FR), and weight % of nanoclay are varied to minimize the delamination factor (F_d) and surface roughness (SR).

49.2 Preparation of Laminate

Nanoclay-embedded epoxy/carbon fabric nanocomposite was selected as the sample material in this research. Ten ply of carbon fiber were used for the preparation of laminate at a stacking order of $(0/90)_s$. A calculated quantity of nanoclay of 0.0 and 2.0 weight % was applied in the matrix to improve the mechanical properties of laminates for the preparation of nanocomposite. Since mechanical and thermal properties of laminates increase up to 3 weight % of nanoclay in epoxy due to strong bonding of the matrix with filler [4, 11]. To prevent from premature curing, hardener (Lapox K-6 AH-31) was introduced and blended with the help of mechanical stirrer at 800 rpm for 5 min. The combination was finally applied by hand layup method to the carbon fiber sheet, and extra resin is squeezed using vacuum bagging technique. The schematic of vacuum bagging is presented in Fig. 49.1.

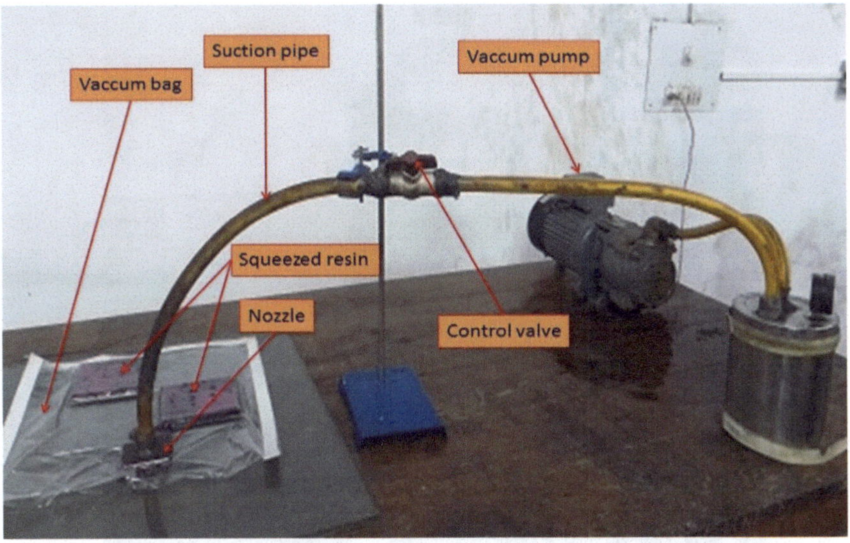

Fig. 49.1 Vacuum bagging setup

49.3 Experimental Setup

The experiment for current work was performed on a vertical end milling machine with maximum SS of 2000 rpm and FR of 160 mm/min as shown in Fig. 49.2. Cemented carbide tool of 6 mm diameter having 4 flutes was used for milling process. To avoid vibration and any type of misalignment, laminate was held rigidly in machine vice on the table of milling machine. Gear mechanism of milling machine was adjusted to change the SS and FR.

49.3.1 Calculation of Surface Roughness

Surface roughness decides the surface texture of the workpiece. It is actually the deviation of surface in perpendicular direction of real surface from its ideal form. Surftest SJ-210 (Mitutoyo America corp.) was used to measure the SR of machined laminate as shown in Fig. 49.3.

49.3.2 Calculation of Delamination Factor

In the present study, the delaminated part is measured by taking a picture using high-resolution camera and then processed into ImageJ software. ImageJ software

Fig. 49.2 Experimental setup of milling machine

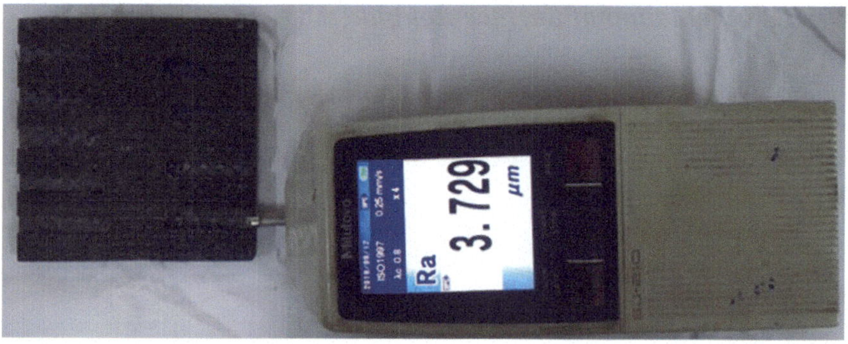

Fig. 49.3 Surface roughness measurement setup

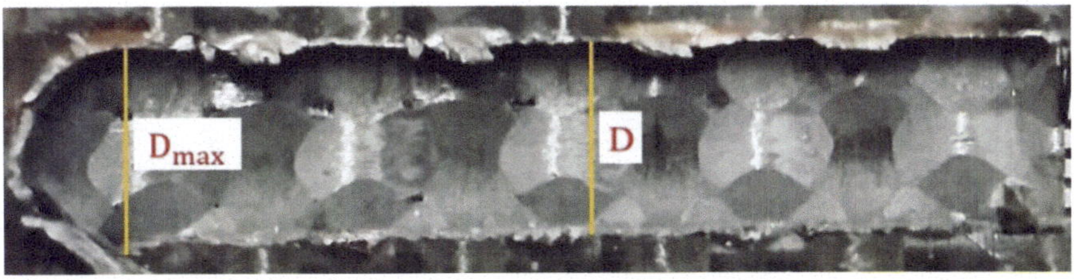

Fig. 49.4 Measurement of delamination in the machined slot

Table 49.1 Factors and their value of levels

Factors	Unit	Level 1	Level 2	Level 3
Spindle speed (SS)	rpm	355	500	710
Feed rate (FR)	mm/min	25	63	100
Nanoclay weight %	–	0.0	2.0	–

is a public domain software. F_d is the ratio of maximum width of cut (D_{max}) to the diameter of the tool (D) as shown in Fig. 49.4.

$$\text{Delamination factor}\,(F_d) = D\text{max}/D \tag{49.1}$$

where

D_{max} Maximum width of machined surface
D Diameter of tool.

49.4 Experimental Setup

The experimental arrangement for this study is based on the technique of Taguchi design of experiment. This technique makes it possible to use a suitable orthogonal array (OA) to study about all experiment variables with a minimum number of experiments. Three important machinability factors are used in end milling to select a suitable orthogonal array as shown in Table 49.1.

49.5 Experimental Setup

Based on the factor and levels, the experimental parameter was set in L18 orthogonal array. SR and F_d were measured at this experimental condition as shown in Table 49.2.

Table 49.2 Calculated value for surface roughness and delamination factor

Exp. No.	Nanoclay weight %	FR	SS	F_d	SR
1	0	25	355	1.101	3.337
2	0	63	355	1.136	3.597
3	0	100	355	1.142	4.623
4	0	25	500	1.124	3.214
5	0	63	500	1.129	3.573
6	0	100	500	1.135	3.658
7	0	25	710	1.139	3.172
8	0	63	710	1.145	3.477
9	0	100	710	1.150	3.514
10	2	25	355	1.100	3.184
11	2	63	355	1.113	3.534
12	2	100	355	1.118	4.171
13	2	25	500	1.110	2.999
14	2	63	500	1.122	3.337
15	2	100	500	1.128	3.582
16	2	25	710	1.134	3.104
17	2	63	710	1.142	3.236
18	2	100	710	1.148	3.498

49.5.1 Analysis of Surface Roughness

SR is analyzed in Minitab 17 by using Taguchi design of experiment. Figure 49.5 obtained after analysis shows that SR decreases on increasing the doping % of nanoclay in CFRP, and also with increasing SS but SR increases with the increase in FR.

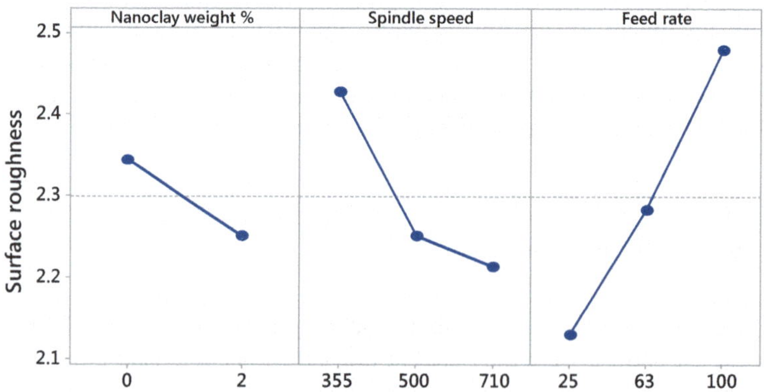

Fig. 49.5 Effect of nanoclay weight %, spindle speed and feed rate on surface roughness

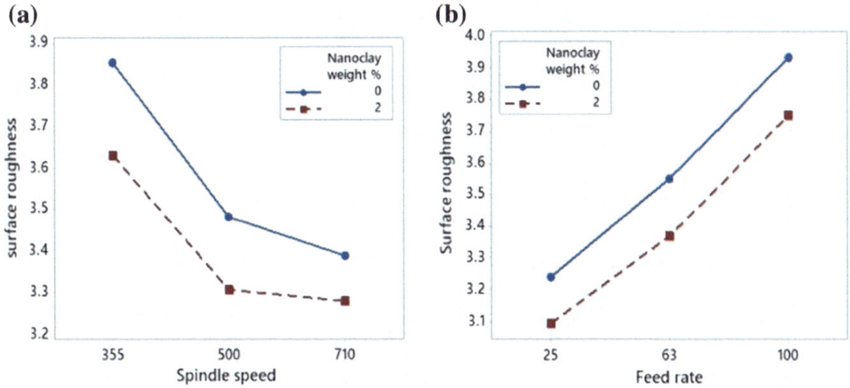

Fig. 49.6 Surface roughness of various amount of nanoclay on variation of **a** spindle speed, **b** feed rate

Table 49.3 Response table for surface roughness

Level	Nanoclay weight %	Spindle speed	Feed rate
1	3.574	3.741	3.168
2	3.394	3.394	3.459
3		3.334	3.841
Delta	0.169	0.407	0.673
Rank	3	2	1

Figure 49.6a shows that SR decreases with increase in SS and doping nanoclay in CFRP. Nanoclay-doped CFRP provides less SR compared to neat CFRP at the same SS. Since increase in SS reduces friction and increase in weight % of nanoclay in CFRP works as lubricant in the machining of laminate. Figure 49.6b shows that SR increases with FR and decreases with doping nanoclay in CFRP. Since increase in FR leads to more vibration in the machining process, nanoclay-doped CFRP provides less SR compared to neat CFRP at the same FR. Table 49.3 shows the response of parameter on SR. From this table, it can be observed that FR is the dominating parameter followed by SS and weight percentage of nanoclay.

49.5.2 Analysis of Delamination Factor

Figure 49.7 shows that Delamination factor decreases on increasing the doping percentage of nanoclay in CFRP but increases with the increase in SS and FR.

Figure 49.8a shows that F_d increases with increase in SS, since plastic deformation of laminate increases on increasing SS that leads to increase in F_d [12]. F_d decreases with doping nanoclay in CFRP since nanoclay works as lubricant in the machining of laminate [6]. Nanoclay-doped CFRP provides less F_d compared to neat CFRP

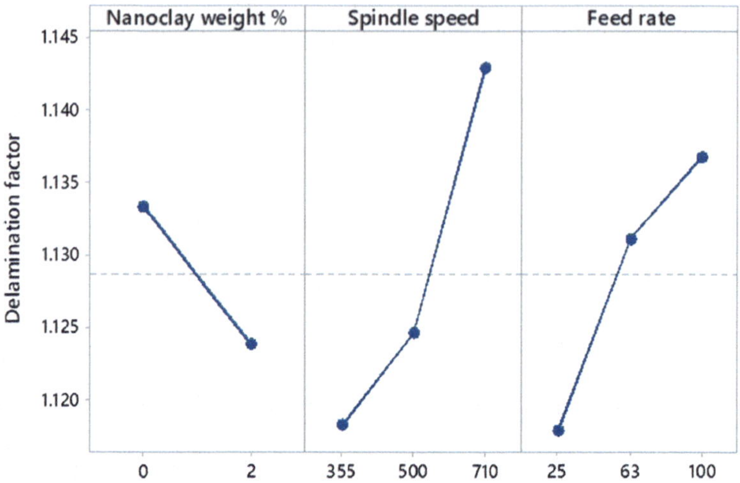

Fig. 49.7 Effect of nanoclay weight %, spindle speed and feed rate on delamination factor

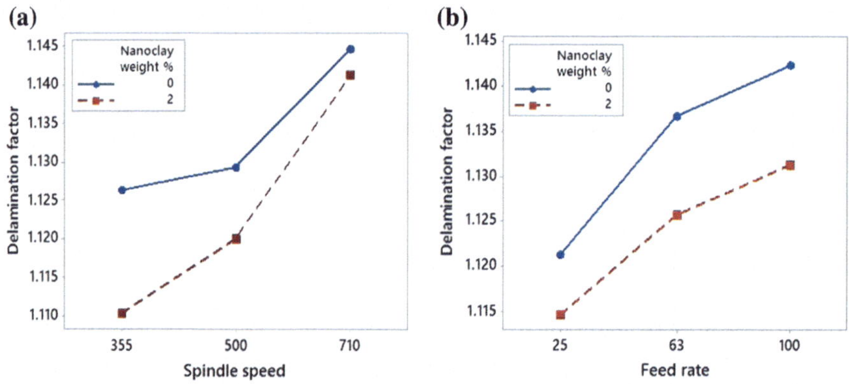

Fig. 49.8 Delamination factor of various amount of nanoclay on variation of **a** spindle speed, **b** feed rate

at the same SS. Figure 49.8b shows that F_d increases with FR and decreases with doping nanoclay in CFRP. Since chip section area and interlaminar shear increase with FR which results in higher delamination [12], nanoclay-doped CFRP provides less F_d compared to neat CFRP at the same FR. Table 49.4 shows the response of parameter on F_d. From this table, it can be observed that SS is the dominating parameter followed by FR and weight percentage of nanoclay.

Table 49.4 Response table for mean of delamination factor

Level	Nanoclay weight %	Spindle speed	Feed rate
1	1.133	1.118	1.118
2	1.124	1.125	1.131
3	–	1.143	1.137
Delta	0.010	0.025	0.019
Rank	3	1	2

49.6 Conclusion

In the present work, end milling is performed on neat and nanoclay-doped CFRP laminate using carbide end mill. Following points can be concluded from the experimental analysis

- Feed rate (FR) significantly affects the surface roughness (SR) followed by spindle speed (SS) and weight % of nanoclay.
- SS significantly affects the delamination factor (F_d) followed by FR and weight % of nanoclay.
- SR decreases with increase in SS and doped % of nanoclay while F_d increases with increase in SS.
- SR and F_d increases with FR.

References

1. Amin, M., Yuan, S., Israr, A., Zhen, L., Qi, W.: Development of cutting force prediction model for vibration-assisted slot milling of carbon fiber reinforced polymers. Int. J. Adv. Manuf. Technol. **94**(9–12), 3863–3874 (2017)
2. Khairusshima, M.K.N., Hassan, C.H.C., Jaharah, A.G., Amin, A.K.M.N.: Tool wear and surface roughness on milling carbon fiber-reinforced plastic using chilled air. J. Asian Sci. Res. **2**(11), 593–598 (2012)
3. Mohan, T.P., Kanny, K.: Influence of nanoclay on rheological and mechanical properties of short glass fiber-reinforced polypropylene composites. J. Reinf. Plast. Compos. **30**(2), 152–160 (2010)
4. Kumar, M.S.S., Raju, N.M.S., Sampath, P.S., Selvam, M.C.P.: Influence of nanoclay on mechanical and thermal properties of glass fiber reinforced polymer nanocomposite. Polym. Nanocomposites **39**(6), 1861–1868 (2016)
5. Xu, Y., Hoa, S.V.: Mechanical properties of carbon fiber reinforced epoxy/clay nanocomposites. Compos. Sci. Technol. **68**(3–4), 854–861 (2008)
6. Thakur, R.K., Sharma, D., Singh, K.K.: Optimization of surface roughness and delamination factor in end milling of graphene modified GFRP using response surface methodology. Materials Today: Proceedings, **19**, 133–139 (2019)
7. Thakur, R.K., Singh, K.K., Sharma, D.: Modeling and optimization of surface roughness in end milling of graphene/epoxy nanocomposite. Materials Today: Proceedings, **19**, 302–306 (2019)
8. Zinati, R.F., Razfar, M.R., Nazockdast, H.: Surface integrity investigation for milling PA 6/MWCNT. Mater. Manuf. Processes **30**(8), 1035–1041 (2015)

9. Chen, T., Gao, F., Li, S., Liu, X.: Experimental study on cutting tool wear in milling carbon fiber composites with spiral staggered diamond-coated milling cutter. Int. J. Adv. Manuf. Technol. **98**(1–4), 413–419 (2018)
10. Hosokawa, A., Hirose, N., Ueda, T., Furumoto, T.: High-quality machining of CFRP with high helix end mill. CIRP Ann-Manuf. Techn. 63, 89–92 (2014)
11. Prabhu, P., Jawahar, P., Balasubramanian, M., Mohan, T.P.: Machinability study of hybrid nanoclay-glass fibre reinforced polyester composites. Int. J. Polym. Sci., 1–11 (2013)
12. Erkan, Ö., Birhan, I., Çiçek, A., Kara, F.: Prediction of damage factor in end milling of glass fibre reinforced plastic composites using artificial neural network. Appl. Compos. Mater. **20**(4), 517–536 (2013)

Chapter 50
Experimental Evaluation of Thermal and Combustion Performance of a DI Diesel Engine Using Waste Cooking Oil Methyl Ester and Diesel Fuel Blends

Prabhakar Sharma and Avdhesh Kr. Sharma

Abstract Experimental evaluation of combustion and thermal performance of dedicated, direct injection, single cylinder, diesel engine (5.2 kW) using waste cooking oil methyl ester (WCOME) and diesel fuel is investigated. Various blends of WCOME with diesel fuel (i.e., B0, B10 and B20) were prepared to evaluate combustion and thermal performance. Kistler make in-vehicle combustion analyzer was interfaced with test engine for high resolution and precise measurements eliminating the experimental uncertainty. Results are encouraging and useful toward better understanding of vital combustion phenomenon. The trends of combustion and thermal performance parameters including in-cylinder combustion peak pressures and peak of rate of change of pressures *up to B*20 (*WCOME* + *Diesel fuel*) are quite similar to neat diesel operation. Thus, neglecting any possibility of harmful long-term effects on dedicated engine hardware such as engine wear and knocking especially when dedicated engine hardware is using WCOME biodiesel blend having different thermo-physico-chemical properties.

Keywords In-cylinder pressure · Waste cooking oil · Biodiesel · Dual fuel · Combustion · Efficiency · Performance

50.1 Introduction

The present scenario of using IC engines for power generation and transportation is not only exploiting the fossil fuel reserves, but also, degrading the environment due to alarming exhaust emissions. Thus, there is an urgent need to shift the focus on high potential alternatives, environment-friendly and cheaper options for power generation. Presently, there are various potential alternatives under investigation (for instance, alcohols, vegetable oils, biogas, producer gas, liquefied petroleum gas and compressed natural gas, etc.). Vegetable oil-based biodiesel is one of the viable

P. Sharma (✉) · A. Kr. Sharma
Department of Mechanical Engineering, D.C.R. University of Science & Technology, Sonepat 131039, Haryana, India

S. Yadav et al. (eds.), *Proceedings of International Conference in Mechanical and Energy Technology*, Smart Innovation, Systems and Technologies 174,
https://doi.org/10.1007/978-981-15-2647-3_50

options (which may offer biodegradable, non-toxic, better inherent lubricity, zero-sulfur, environment-friendly and carbon neutral option [1, 2]). Biodiesel contains almost similar fuel properties to diesel and its use in diesel engine does not demand any hardware retrofitting or modification [3]. Chemically, biodiesel can be described as mono-alkyl esters of long chain fatty acid derived from vegetable oil or animal fat. The chemical process for conversion of vegetable oil or animal fat to biodiesel is known as transesterification [4, 5]. Vegetable oils used for producing biodiesels may be edible (i.e., palm oil, sunflower oil, rapeseed oil, soybean oil, etc.) or non-edible (i.e., jatropha oil, neem oil, karanja oil, rubber seed oil, etc.). The use of virgin edible oils in IC engine as a fuel may not be a feasible option, while non-edible oils have supply chain management issues [6]. Repeated heating of cooking oil can alter the fatty acid composition leading to adverse effects on human health and thus become non-edible (i.e., waste). The same may be procured conveniently from food processing industry and eateries even at low price. It can be converted to biodiesel fuel by transesterification process [7] as use of raw vegetable oils leads to higher maintenance issues (i.e., engine deposits, sticky piston rings, nozzle chocking) due to high viscosity and low volatility of raw vegetable oil.

Several studies were reported on utilization of WCOME in the diesel engine using biodiesel from different feedstock and under different operating conditions. Muralitharan and Vasudevan [8] investigated on sunflower waste cooking oil in a conventional diesel engine under different load conditions and on different compression ratios. Soybean oil-based WCOME was also reported as suitable substitute of diesel engine fuel and reported a decrease in unburnt hydrocarbon, carbon monoxide emissions, while nitrogen oxide was reported on higher side. In study [9], two different waste cooking oils are used, a minor increase in brake specific fuel consumption, and marginal reduction in brake thermal efficiency were reported, while on the emissions front, the biodiesel addition results in the increase in nitrogen oxide emission, while unburnt hydrocarbon and smoke were decreased at all load conditions. Further, precise instrumentations are required to eliminate the experimental uncertainty (i.e., to avoid contradictory measured outcomes/findings). Furthermore, for low blending ratio, i.e., B20 (20% biodiesel + 80% diesel), no engine modifications are required [3, 10–12]. The use of B20 in diesel engine provides reasonable fuel economy, environmental benign and provides less wear and tear due to high lubricity of biodiesel [12]. At this end, the limited studies on combustion analysis followed by more experimental uncertainty vis-à-vis issues related to engine vibration, wear and knock motivated for the present experimental investigations. Therefore, for the present experimental study, the combustion and thermal performance of single cylinder, water cooled, DI diesel engine (without any modifications) has been conducted for two blends (on volumetric basis), namely B10 (10% WCOME biodiesel + 90% diesel), B20 (20% WCOME biodiesel + 80% diesel). Investigations were performed on test rig well instrumented with high precision *in-cylinder combustion analyzer* in order to eliminate/minimize the experimental uncertainty.

Table 50.1 Physico-chemical properties of WCOME and diesel fuel

S. No.	Properties	Units	ASTM D6751	WCOME	Diesel
1	Density	Kg/m^3 at 15 °C	860–900	868	850
2	Kinematic viscosity	cSt at 40 °C	1.9–6.0	3.86	2.87
3	Flash point	°C	>52	158	76
4	Pour point	°C	−15 to 16	−10	3.1
5	Calorific value	kJ/kg	>35,000	38,000	44,000

50.2 Experimental Setup and Experimental Procedure

50.2.1 Test Fuel and Engine

For the preparation of WCO biodiesel, the used cooking/frying oil (which is essentially a mixture of different vegetable oil feedstock such as palm, sunflower, canola, soybean and rice bran) was collected from multiple sources (i.e., local eateries and restaurants) of North Delhi. It was preprocessed (i.e., filtration and heating) for removal of contained impurities in the form of food particles and water content. The free fatty acid of waste cooking oil was obtained using standard titration method and found to be 2.1%, which was treated by using two-stage transesterification process as the free fatty acid content of waste cooking oil was more than 0.5% [13]. In the first stage, acidic catalyst sulfuric acid 10% (by mass) was used keeping molar ratio of methanol: oil at 20:1 at 55 °C for 2 h. In the second stage, alkaline catalyst KOH 1% (by mass) was used at methanol: oil molar ratio of 6:1. Thereafter, this mixture was allowed to settle in separating flasks for 12 h for phase separation. The settled glycerol at the bottom of separating flask was removed leaving behind the *biodiesel*. The physio-chemical properties of WCOME biodiesel were tested at Enkay Test House, New Delhi, and are listed in Table 50.1. The details of engine specifications and associated instrumentations are given in Table 50.2. The block diagram of basic engine test rig is shown in Fig. 50.1.

Installation of in-cylinder combustion analyzer. Kistler Instrument LG make "*Ki-Box To Go*" set—type 2893AK1 combustion analyzer—was installed on diesel engine for recording pressure corresponding to crank angle. The photographic view of test setup well instrumented including in-cylinder combustion analyzer is shown in Fig. 2.

50.2.2 Experimental Procedure

Initially, the engine was run on pure diesel mode for 30 min to stabilize the cooling water and lubricating oil temperatures. Once the engine is warmed up and stabilized, then the test data was recorded in pure diesel mode at various load conditions via

Table 50.2 Engine specifications

S. No.	Part name	Specifications
1	Engine	Make Kirloskar, Model TV1, single cylinder, 4-stroke diesel, water cooled, 5.2 kW, compression ratio 17.5:1, 661 cc
2	Dynamometer	SAJ make, AG-20 model, type eddy current, water cooled, maximum 20 kW@ 2450 rpm
3	Air box	M S fabricated with orifice meter and manometer
4	Temperature sensor	Type RTD, PT100 and thermocouple, Type K
5	Calorimeter	Type pipe in pipe
6	Load indicator	Digital, range 0–50 kg, supply 230 V A/C
7	Load sensor	Load cell type strain gauge, range 0–50 kg
8	Fuel injection timing	23° CA bTDC

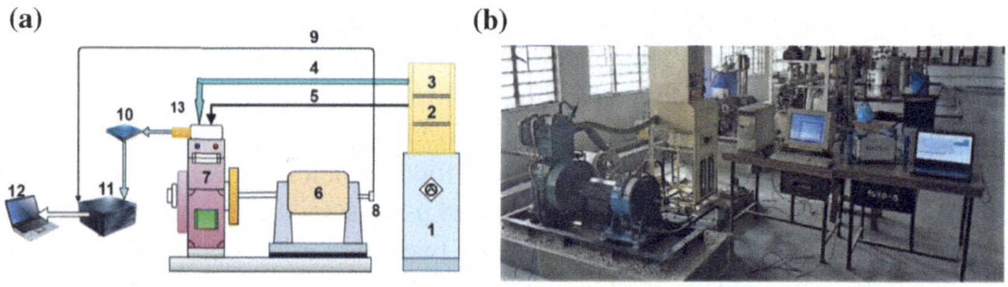

1. Computerised test panel 2. Fuel tank 3. Air box
4. Eddy current dynamometer 5. Fuel line 6. Eddy current dynamometer
7. Diesel engine 8. Crank angle encoder 9. Crank angle signal feed line
10. Signal converter 11. Ki Box 12. Computer with Ki Box cockpit softawre
13. Pressure sensor

Fig. 50.1 **a** Block diagram of computerized test engine setup; **b** Test setup well instrumented with Ki-box in-vehicle combustion analyzer

in-cylinder combustion analyzer. The WCOME biodiesel and diesel fuel blends of B0 (i.e., neat diesel), B10 and B20 on volumetric basis are used. Fuel injection timing and injection pressure were maintained at 23° bTDC and 200 bar, respectively. Measurements were recorded for evaluation of thermal performance and combustion performance of engine.

50.3 Results and Discussion

50.3.1 Combustion Performance

In-cylinder pressure with respect to crank angle. The pressure and crank angle relationship is a good indicator of engine combustion performance. This pressure and

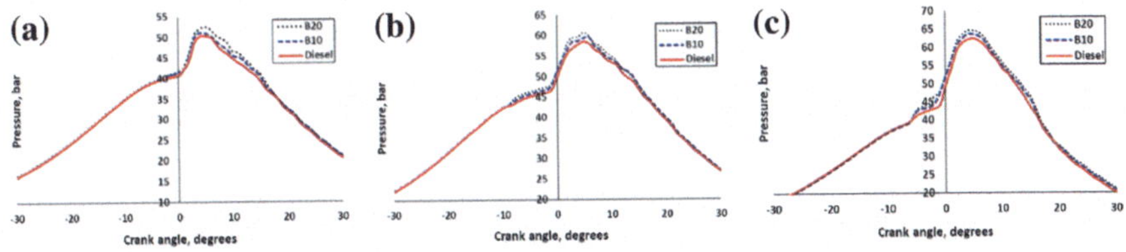

Fig. 50.2 Effect of biodiesel blends on combustion pressure at **a** 50% load, **b** 80% load and **c** full-load conditions, 1500 ± 50 rpm

crank angle relationship for three load conditions (i.e., 50%, 80% and 100% load) for B0, B10 and B20 blends of WCOME biodiesel is presented in Fig. 50.2a–c. It can be observed that the pressure trends increase with increasing load conditions. For all load conditions, a variation in peak pressures for blends B10 and B20 are observed to be within the limit of 1.7%, and 3.4% with respect to neat diesel fuel (B0). Further, the location of peak combustion pressure varies marginally in the range of 4–6° CA for all blends and for all load conditions encountered. At full-load conditions, the in-cylinder pressure trends are observed to be on higher side, which can be attributed due to more fuel consumption at higher loads. Similar results are observed in the literature [9].

Rate of change of in-cylinder pressure with respect to crank angle. The rate of change of pressure is an indicator of how rapidly the rate of change of pressure takes place in the cylinder. It helps identifying any possibility of combustion abnormality or potential damage to engine life. Elevated rate of pressure rise reflects the burning of a large proportion of fuel in the premixed combustion phase [14]. The trends of derivative of change of in-cylinder pressure on crank angle scale (i.e., dP/dθ-θ diagram) for 50%, 80% and 100% load conditions are plotted in Fig. 50.3(a-c). Further, at 50% load condition, the peak rate of pressure rise for B0 (neat diesel), B10 and B20 are observed to be 2.35, 2.37 and 2.55 bar/°, respectively. While, at full-load condition, the peak of rate change of pressure is observed to be 3.55, 3.67 and 3.76 bar/°.

Rate of heat release. Rate of heat release rate is derived from in-cylinder pressure by using thermodynamic calculations followed by geometry of cylinder including crank and connecting rod via inbuilt software. The rate of heat release for B0 (neat

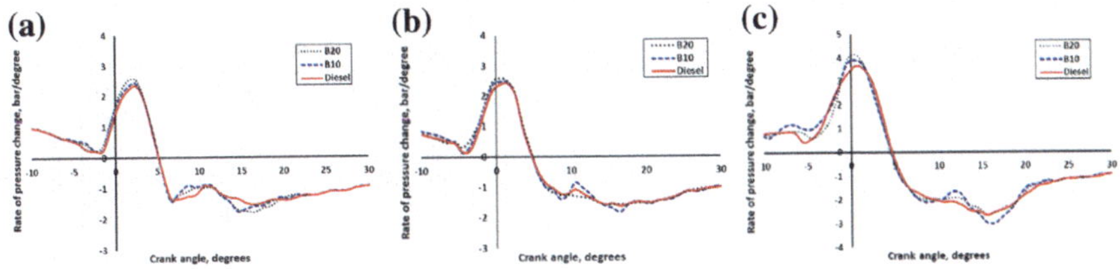

Fig. 50.3 Effect of blends on rate of pressure rise at **a** 50% load, **b** 80% load and **c** full-load conditions, 1500 ± 50 rpm

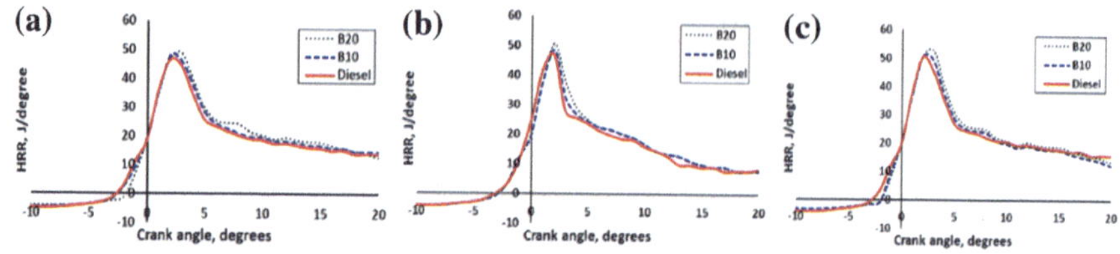

Fig. 50.4 Effect of blends on HRR **a** 50% load, **b** 80% load and **c** full-load conditions

diesel) and B10 and B20 blends at no-load, 50% load, 80% load and full-load conditions at 1500 rpm is plotted in Fig. 50.4a–c. Initially, the rate of heat release was observed to be negative for all test blends and for all load conditions (due to vaporization of diesel fuel and heat loss through engine cylinder during ignition delay period). Thereafter, it shows sharp heat release rate trends (after initialization of premixed combustion) and starts decreasing after peaking at nearly 2–3° after TDC for all blends and load conditions encountered, which followed by mixed-controlled combustion phase. The net heat release rate for pure diesel fuel is marginally lower than B10 or B20 blends for above load conditions. At full load, for both blends, i.e., B10 and B20, the typical increase in rate of heat release is not significant and is within 2% with respect to neat diesel fuel (i.e., B0). Similar findings were reported in the literature [15].

Ignition delay. Pure diesel fuel exhibits poor atomization, increased penetration and slower mixing on higher viscosity in contrast to vegetable and/or animal fat-based biodiesels [16]. The ignition delay is a strong function of cetane number. It influences initiation of combustion, peak pressure and rate of heat release [17]. At P-θ diagram, it can be defined as the crank angle between start of fuel injection and initialization of combustion process. For various fuel blends and engine load conditions, the variations in ignition delay are obtained as shown in Fig. 50.5. On increasing the engine load and blending ratio, the ignition delay becomes shorter. Further, the differences in ignition delay between neat diesel fuel and B20 biodiesel blend for all load conditions are not significant.

Fig. 50.5 Comparison of ignition delay period for diesel only, B10 and B20 blends with 50%, 80% and full-load conditions

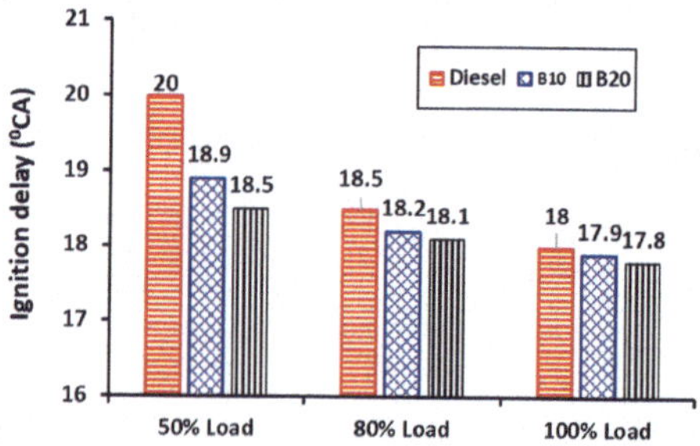

50.3.2 Thermal Performance

Brake specific fuel consumption. Brake specific fuel consumption (BSFC) is defined as the ratio of mass of fuel consumed to brake power output-hour. The trends of BSFC for WCOME biodiesel blends (B0, B10 and B20) were obtained against brake power output as shown in Fig. 50.6a. The BSFC decreases more sharply with initial increase in engine load. The increase in BSFC with blending ratio (i.e., B10 and B20) can be expected due to higher density and low calorific value of WCOME biodiesel than conventional diesel fuel. (Refer Table 50.1).

Brake thermal efficiency. Brake thermal efficiency (BTE) is defined by the fraction brake power output to input energy supplied by the fuel. The trends of BTE against brake power output for WCOME biodiesel blends (i.e., B0, B10 and B20) are shown in Fig. 50.6b. For all these three blends, it can be observed that BTE increases with initial load conditions and then shows decreasing trends after peaking nearly at 80% engine load condition. Increase in BTE with near 80% engine load condition is attributed due to improved combustion. The peak value of BTE for B0, B10 and B20 are observed to be 31.7%, 30.1% and 28.7%, respectively.

Exhaust gas temperature. The engine exhaust gas temperature exhibits the effective conversion of the chemical energy of fuel to motive power output; further delayed combustion due to longer ignition delay results in higher exhaust gas temperature [18]. Herein, the effect on temperature of exhaust gas for biodiesel blends (i.e.,

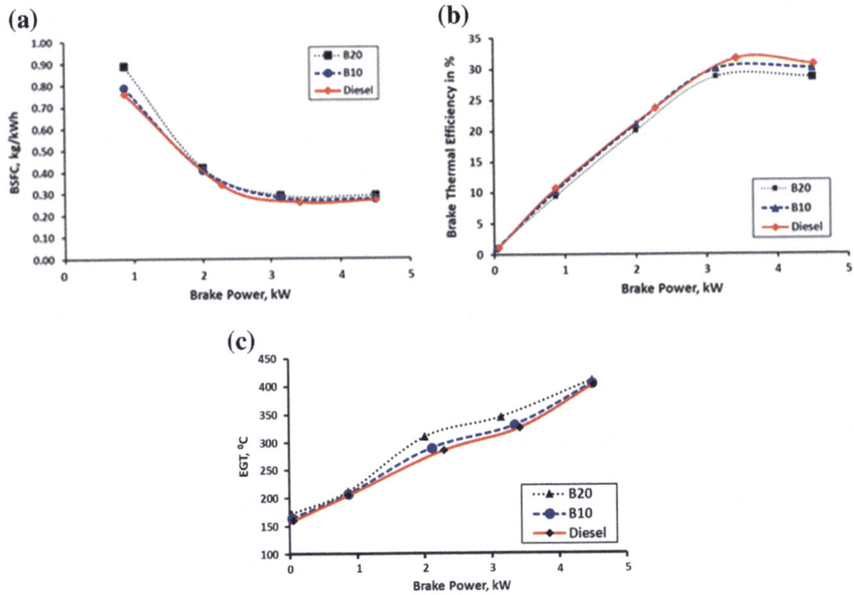

Fig. 50.6 Effect of brake power on **a** brake specific fuel consumption **b** brake thermal efficiency **c** exhaust gas temperature (EGT), 1500 ± 50 rpm

B0, B10 and B20) against engine load is presented in Fig. 50.6c. The exhaust gas temperature found to be increased with engine load and with higher biodiesel blending. However, the influence of blending ratio at low load and full-load conditions is relatively small.

50.4 Conclusions

The combustion and thermal performance of single cylinder water cooled diesel engine using WCOME biodiesel blends with conventional diesel fuel (i.e., B0, B10 and B20) were analyzed carefully via high resolution and precise instrumentation. Following main conclusions are derived out of the present experimental study:

Combustion performance

With blending ratio (B10 and B20), the trends of in-cylinder combustion pressure and rate of change of pressure are observed to be very close to diesel fuel and smooth (without excessive noise and vibrations) during entire operating conditions encountered. Thus, do not show any indication of knocking. At full load, for both blends, i.e., B10 and B20, a marginal incremental difference in rate of heat release has been observed and is within 2% with respect to neat diesel fuel (i.e., B0). Ignition delay becomes shorter with engine load and blending ratio allowing easier auto-ignition. The differences are not significant with respect to conventional diesel fuel.

Thermal performance

A marginally rise in brake specific fuel consumption, while marginal decrement in fuel economy (i.e., brake thermal efficiency), has been observed with blends B10 and B20 with respect to diesel fuel (i.e., B0). Exhaust gas temperature increases with engine load and blending ratio. Although, the influence of blending ratio at lower load and full-load conditions is relatively small.

References

1. Ellabban, O., Abu-Rub, H., Blaabjerg, F.: Renewable energy resources: current status, future prospects and their enabling technology. Renew. Sustain. Energy Rev. **39**(C), 748–764 (2014)
2. Di, Wang, Song, C., Yang, Y., Liu, Y.: Research on the redesign of precision tools and their manufacturing process based on selective laser melting (SLM). Rapid Prototyp. J. **22**(1), 104–114 (2016)
3. Kishore, V.V.N.: Renewable Energy Engineering and Technology, Principles and Practice (2008)
4. Hoekman, S.K., Broch, A., Robbins, C., Ceniceros, E., Natarajan, M.: Review of biodiesel composition, properties, and specifications. Renew. Sustain. Energy Rev. **16**(1), 143–169 (2012)

5. Issariyakul, T., Dalai, A.K.: Biodiesel from vegetable oils. Renew. Sustain. Energy Rev. **31**, 446–471 (2014)
6. Araujo, V.K.W.S., Hamacher, S., Scavarda, L.F.: Economic assessment of biodiesel production from waste frying oils. Bioresour. Technol. **101**(12), 4415–4422 (2010)
7. Atabani, A.E., Silitonga, A.S., Badruddin, I.A., Mahlia, T.M.I., Masjuki, H.H., Mekhilef, S.: A comprehensive review on biodiesel as an alternative energy resource and its characteristics. Renew. Sustain. Energy Rev. **16**(4), 2070–2093 (2012)
8. Muralidharan, K., Vasudevan, D.: Performance, emission and combustion characteristics of a variable compression ratio engine using methyl esters of waste cooking oil and diesel blends. Appl. Energy **88**(11), 3959–3968 (2011)
9. Can, Ö.: Combustion characteristics, performance and exhaust emissions of a diesel engine fueled with a waste cooking oil biodiesel mixture. Energy Convers. Manag. **87**, 676–686 (2014)
10. Bari, S.: Performance, combustion and emission tests of a metro-bus running on biodiesel-ULSD blended (B20) fuel. Appl. Energy **124**, 35–43 (2014)
11. Abedin, M.J., Masjuki, H.H., Kalam, M.A., Sanjid, A., Rahman, S.M.A.: Performance, emissions, and heat losses of palm and *Jatropha* biodiesel blends in diesel engine. Ind. Crops Prod. **59**, 96–104 (2014)
12. Jain, S., Sharma, M.P.: Prospects of biodiesel from *Jatropha* in India: a review. Renew. Sustain. Energy Rev. **14**, 763–771 (2010)
13. Wang, Y., Liu, S.O.P., Zhang, Z.: Preparation of biodiesel from waste cooking oil via two-step catalyzed process. Energy Convers. Manag. **48**(1), 184–188 (2007)
14. Selim, M.Y.E., Radwan, M.S., Elfeky, S.M.S.: Combustion of jojoba methyl ester in an indirect injection diesel engine. Renew. Energy **28**(9), 1401–1420 (2003)
15. Jaikumar, S., Bhatti S.K., Srinivas, V.: Experimental investigations on performance, combustion, and emission characteristics of Niger (*Guizotia abyssinica*) seed oil methyl ester blends with diesel at different compression ratios. Arab. J. Sci. Eng (2018). https://doi.org/10.1007/s13369-018-3538-y
16. Sahoo, P.K., Das, L.M.: Combustion analysis of *Jatropha*, Karanja and Polanga based biodiesel as fuel in a diesel engine. Fuel **88**(6), 994–999 (2009)
17. Ozsezen, A.N., Canakci, M.: Determination of performance and combustion characteristics of a diesel engine fueled with canola and waste palm oil methyl esters. Energy Convers. Manag. **52**(1), 108–116 (2011)

Chapter 51
Effect of Ply Stacking and Fiber Volume Fraction on ILSS of Woven GFRP Laminates

Md. Touhid Alam Ansari, K. K. Singh and Md. Sikandar Azam

Abstract This paper evaluates the effect of the ply stacking sequence and fiber volume fraction on the interlaminar shear strength of glass fiber reinforced polymer (GFRP) laminates. Eight- and 6-ply laminates had two different ply stacking sequences, making them asymmetric and symmetric laminates, while fiber content was higher in case of 8-ply laminates. GFRP laminates were fabricated by hand layup technique followed by press molding. Fiber volume fractions were calculated 56% and 45% for 8- and 6-ply laminates, respectively. Samples were prepared according to the ASTM-D-2344 standard. Tests were performed on the computer-controlled universal testing machine (UTM) with 50 KN load cell at a crosshead speed of 1 mm/min. Results show that the interlaminar shear stress is higher for symmetric laminates compared with asymmetric laminates. It was also found out that an increase in fiber volume fraction decreases ILSS, thus depicting it is a matrix-dominated property. Weibull distribution analysis of the samples was performed in order to examine the static stability of FRP laminates. Laminates with high ILSS had less scattering from the mean value.

Keywords ILSS · Fiber volume fraction · Stacking sequence · Weibull distribution

51.1 Introduction

Owing to the ease of availability of glass fibers and economic processing techniques employed to manufacture components, these are widely used in the field of aerospace, marine, automobile, wind turbine, and sports [1, 2]. Desired properties and functional features of fiber reinforced polymer (FRP) composites can be made comparable to steel by using suitable orientation of fibers and composition. Moreover, these composites have a higher stiffness than aluminum, and the specific gravity is 25% of the steel [3]. Rather than distributed short fibers, utilization of bidirectional fibers has the extra benefit of diminishing the fiber agglomeration and obtaining similar properties of composites in both directions. The laminate properties depend upon

---

Md. T. A. Ansari (✉) · K. K. Singh · Md. S. Azam
Department Mechanical Engineering, Indian Institute of Technology (Indian School of Mines), Dhanbad 826004, India

S. Yadav et al. (eds.), *Proceedings of International Conference in Mechanical and Energy Technology*, Smart Innovation, Systems and Technologies 174,
https://doi.org/10.1007/978-981-15-2647-3_51

two major characteristics, i.e., material properties and geometry [4]. When the material properties, as well as geometry (fiber orientations), are symmetrical about the middle plane, this type of laminate is known as 'symmetrical laminate.' However, lack of symmetry about the mid-surface makes the material non-symmetrical or asymmetrical laminate. Singh et al. [5] studied the stiffness properties of two different stacking sequence FRP laminates and showed that both laminates have almost identical stiffness matrices. Interlaminar shear strength (ILSS) is one of the most significant parameters for calculating the ability of a laminate to resist delamination damage. Generally, short-beam-shear (SBS) experimental technique is used for finding the ILSS of composites. During SBS, the deformation increases proportionally with load up to a peak load. Instantly after the peak load has appeared if the one-third or more load dropped, then it is supposed that the sample failed in lamina shear and the ILSS is further determined using the peak load. In short-beam-shear, test samples are comparatively easy to fabricate and the test itself involves a slight fixturing and hence easy to perform the test. However, the SBS calculates a precise measure of ILSS value only if perfect interlaminar shear failure takes place. Okoli et al. [6] studied the effect of strain rate on the failure of the FRP laminates by performing tensile and three-point bend tests. They found that the fiber/matrix has poor adhesion due to the random orientation of fiber in laminates. Various thermomechanical characteristics of FRP composites are affected due to the strength distribution of fibers. Almeida et al. studied the effect of ply arrangement on interlaminar and in-plane shear properties of GFRP composite and showed that the random fiber-oriented laminates have higher in-plane shear strength [7]. Another important and widely used statistical tool, Weibull analysis, is used for explaining the strength behavior of specimen, based on the hypothesis that failure at the most dangerous defect leads to the overall failure of the samples [8, 9]. Bullock [10] used a two-parameter Weibull model for correlating 4-point flexure and a tensile sample of the unidirectional graphite–epoxy composite. The two-parameter Weibull distribution is mostly valid to the strength distribution of fiber/epoxy composite. [11].

In the present study, ILSS tests were performed to evaluate the interfacial properties considering glass fiber reinforced polymer composite having symmetric and asymmetric 8- and 6-layer layup laminates. The Weibull analysis was also studied in order to investigate the effect of stacking sequence on strength distribution.

51.2 Materials and Fabrication Process

In this work, 2D plain woven E-glass fiber with 610 GSM of surface weight and epoxy compatible sizing procured from S & S polymers, Bengaluru. Bisphenol-A thermosetting epoxy (Araldite LY 556 CS) and hardener (Aradur HY 951) were mixed in a ratio of 10:1 and used as matrix material. Gel time for above-mentioned ratio of epoxy and hardener mixture is 20 min as instructed by the manufacturer.

Laminates were prepared by using the E-glass woven fiber of three different fiber orientations (0°/90°), (±45°) and (±60°). Plies according to the fiber orientation

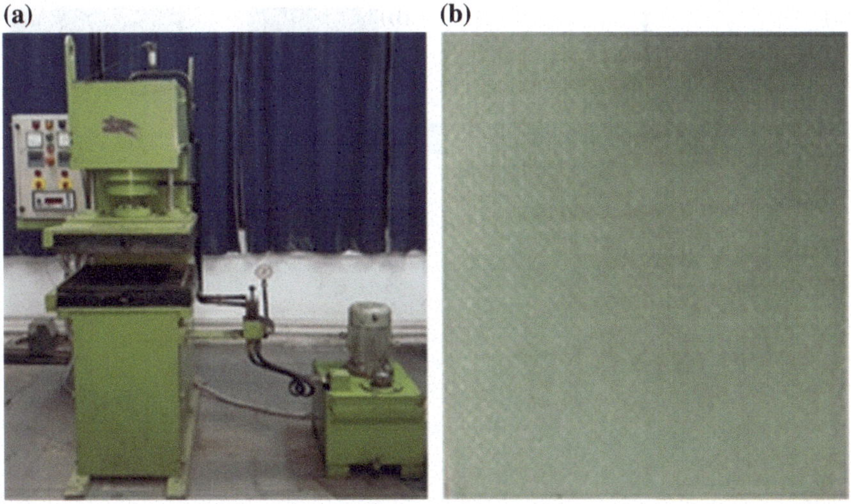

Fig. 51.1 **a** Hydraulic press machine; **b** prepared laminate

were cut from the parent fabric with dimension 28 × 28 cm^2. Symmetrical and asymmetrical laminates were prepared as per the stacking sequence. Laminates were prepared by placing the layers one by one followed by liquid resin as per the proposed ply stacking sequence. Subsequently, excess resin squeezed out by rolling mild steel roller over each layer's surfaces. Eight and six-layered woven GFRP laminates were fabricated by hand layup technique followed by press molding. Hydraulic press machine and prepared laminate are shown in Fig. 51.1. Fiber volume fraction of 8 and 6 plies was found as approximately 56 and 45% as per ASTM D3171-15. Stacking sequences for laminates are:

1. $[(0°/90°)/(\pm45°)/(\pm45°)/(0°/90°)]_2$ and $[(0°/90°)/(\pm60°)/(0°/90°)]_2$ for symmetric laminate.
2. $[(0°/90°)/(\pm45°)/(\pm45°)/(0°/90°)//(\pm45°)/(0°/90°)/(\pm45°)/(0°/90°)]$ and $[(0°/90°)/(\pm60°)/(0°/90°)//(\pm60°)/(0°/90°)/(0°/90°)]$ for asymmetric laminates.

51.3 Testing Procedure

Four specimens of each proposed stacking sequence laminate were prepared as per the ASTM standard. 3-Point bend test of both symmetric and asymmetric type samples was performed on computer-controlled universal testing machine (UTM) Hounsfield H50KS. Tests were performed at a crosshead speed of 1 mm/min. Fixture setup for the tests is shown in Fig. 51.2. ILSS of the composite can be calculated as

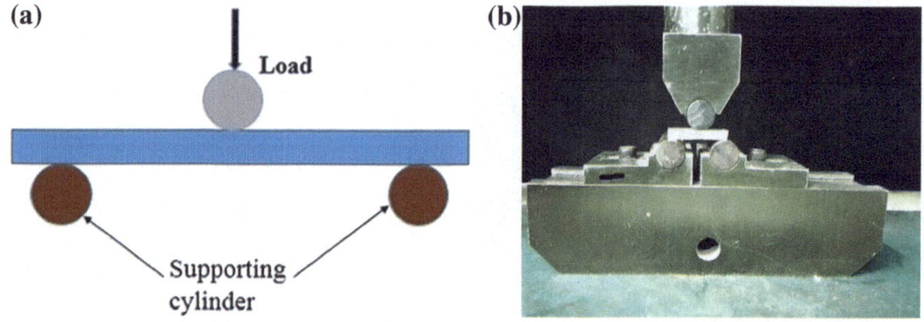

Fig. 51.2 **a** Schematic diagram; **b** fixture setup for interlaminar shear stress

$$F_{sbs} = \frac{P_{\max}}{b \times h} \tag{51.1}$$

where

F_{sbs} = Short-beam strength of samples in MPa
$P_{\max}$ = Peak load observed in N
$b \times h$ = (width) × (thickness) of SBS samples (mm^2).

51.4 Result and Discussion

The loads on the samples were transformed into ILSS by Eq. (51.1) and the normalized deformation by dividing displacement to thickness for representing the deformation of each type of samples. Interlaminar shear stress–strain response of the tested samples is shown in Fig. 51.3, while their corresponding mean value with SD (standard deviation) and coefficient of variance are presented in Table 51.1. The highest average value of ILSS was obtained for symmetric 6-ply material of 25.99 MPa followed by asymmetric 6 plies, symmetric 8 plies, and asymmetric 8 plies having values 24.81 MPa, 24.25 MPa, and 23.17 MPa, respectively. ILSS of a FRP composite is a matrix-dominated property. When the fiber content in the laminates is increased matrix material available for forming adhesion in between the adjacent plies becomes insufficient, thus resulting in poor fiber/matrix bonding which leads to delamination and hence reduced the interlaminar shear strength of the laminate. Also, the stiffness of fiber in laminate increases to a certain value for up to a particular fiber volume fraction beyond which poor transfer of load takes place between the fibers and matrix thus reducing the load-carrying capacity of the laminate.

Fiber stacking sequences and ply angle are combinedly affected the load–displacement response [12]. Crack propagation in a particular ply is not only affected by its own orientation but also influenced by the orientation of adjacent ply [13]. The mismatch angle of the cracked plies has considerable effects on the initiation of crack propagation in the adjacent plies. If ply mismatch angle (90°–θ°) between

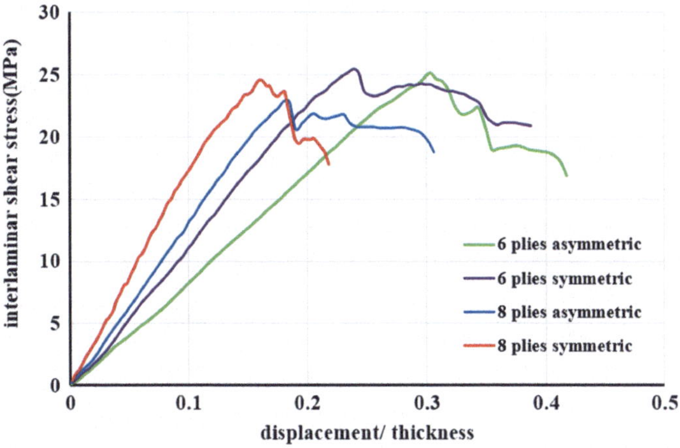

Fig. 51.3 Interlaminar shear stress–displacement/thickness curves

Table 51.1 ILSS of different stacking sequences

Specimen No.	8-Ply symmetric	8-Ply asymmetric	6-Ply symmetric	6-Ply asymmetric
1	27.65	27.25	28.98	28.11
2	25.2	19.1	22.38	21.41
3	23.82	21.78	25.25	23.28
4	20.31	24.54	27.34	26.44
Mean ($\overline{x}$)	24.25	23.17	25.99	24.81
S_{n-1}	2.66	3.04	2.47	2.62
C V(%)	12.64	15.16	10.96	12.19

the adjacent plies is more, then the crack propagation rate is more although similarly oriented adjacent ply resist the propagation of a crack in the through-thickness direction. [14]

For example, two 90° plies favor the condition for propagation in through-thickness, which in turn stops intraply crack and form interply crack to propagate further. Ply with a smaller mismatch angle and the less thickness leads to the more proneness to crack development in the adjacent plies [15]. In symmetric laminates, some ply was become thicker due to the combination of similar adjacent fiber orientation as compared to same single ply which was used in asymmetric laminates. Hence, the crack formation was more in later one and resulting in ILSS value of asymmetric laminates which were decreased.

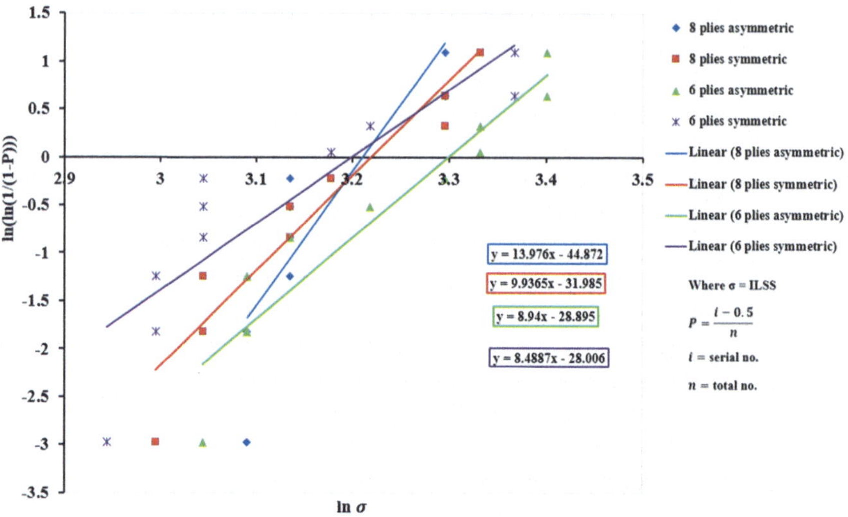

Fig. 51.4 Weibull plot of ILSS data for different laminates

Table 51.2 Weibull distribution parameter values

Laminate type	Shape parameter (β)	Scale parameter (α)
8-Ply asymmetric	13.98	24.79
8-Ply symmetric	9.34	25.01
6-Ply asymmetric	8.94	25.34
6-Ply symmetric	8.48	27.16

The ILSS of all four stacking sequence laminates analyzed in Weibull distribution terms, to evaluate the state of statistical shear damage in the symmetric and asymmetric composite. A plot of four different laminates in Weibull coordinates is shown in Fig. 51.4. Scale parameter (α) and shape parameter (β) was determined from the straight-line equations of plot. These values are listed in Table 51.2. Weibull shape parameter (β) scattering is associated with the result, while scale parameter (α) defines the nominal strength of the material. From the results, it could be ascertained that 6-ply laminates with symmetric arrangement depict the largest nominal strength and less scattering in the result. One could be observed from the figure that the laminate with higher ILSS value had less scattering from mean value.

51.5 Conclusion

Based on the current study, the following conclusion is drawn:

1. The results showed that ILSS of the laminates were depended on the stacking sequence and volume fraction of the laminates. Six-ply symmetric laminates had highest ILSS of 25.99 MPa followed by 6-ply asymmetric, 8-ply symmetric, and 8-ply asymmetric having 24.81 MPa, 24.25 MPa, and 23.17 MPa, respectively.
2. Laminates having higher fiber volume fraction showed lower ILSS value and vice versa.
3. Symmetric laminate showed higher ILSS values as compared to asymmetric laminates.
4. Weibull analysis yielded that the laminates having higher ILSS value, i.e., 6-ply symmetric, showed minimum scattering from the mean value having shape factor (β) 8.48 followed by 6-ply asymmetric, 8-ply symmetric, and 8-ply asymmetric laminates that had the shape factor of approximately 8.94, 9.94, and 13.98, respectively.

References

1. Ansari, M.T.A., Singh, K.K., Azam, M.S.: Fatigue damage analysis of fiber reinforced polymer composites—a review. J. Reinf. Plast. Compos. **37**(9), 636–654 (2018)
2. Gaurav, A., Singh, K.K.: Fatigue behaviour of FRP composites and CNT-Embedded FRP composites: a review. Polym. Compos. **39**(6), 1785–1808 (2018)
3. Awan, G.H., Ali, L., Ghauri, K.M., et al.: Effect of various forms of glass fiber reinforcements on tensile properties of polyester matrix composite. J. Fac. Eng. Technol. **16**(1), 33–39 (2009)
4. Jones, R.M.: Mechanics of Composite Materials, 3rd edn. CRC Press Taylor & Francis, New York (1999)
5. Singh, K.K., Singh, N.K., Jha, R.: Analysis of symmetric and asymmetric glass fiber reinforced plastic laminates subjected to low-velocity impact. J. Compos. Mater. **50**(14), 1853–1863 (2016)
6. Okoli, O.I., Latif, A.A.: An attempt at predicting failure in a random glass/epoxy composite laminate. J. Reinf. Plast. Compos. **21**(11), 1003–1012 (2002)
7. Almeida, J.H.S., Angrizani, C.C., Botelho, E.C., Amico, S.C.: Effect of fiber orientation on the shear behaviour of glass fiber/epoxy composites. Mater. Des. **65**(1), 789–795 (2015)
8. Weibull, W.: A statistical distribution function of wide applicability. J. Appl. Mech. **18**, 293–297 (1951)
9. Thomason, J.L.: On the application of Weibull analysis to experimentally determined single fiber strength distributions. Compos. Sci. Technol. **77**(1), 74–80 (2013)
10. Bullock, R.E.: Strength ratio of composite material in flexural and in tension. J. Compos. Mater. **8**(2), 200–206 (1974)
11. Wang, Z., Xia, Y.: Experimental evaluation of the strength distribution of fibers under high strain rates by bimodal Weibull distribution. Compos. Sci. Technol. **57**(12), 1599–1607 (1998)
12. Naghipour, P., Bartsch, M., Chernova, L., Hausmann, J., Voggenreiter, H.: Effect of fiber angle orientation and stacking sequence on mixed-mode fracture toughness of carbon fiber reinforced plastics: numerical and experimental investigations. Mater. Sci. Eng. A **527**(3), 509–517 (2010)
13. Rehan, M.S.M., Rousseau, J., Gong, X.J., Guillaumat, L., Ali, J.S.M.: Effects of fiber orientation of adjacent plies on the mode I crack propagation in a carbon-epoxy laminate. Procedia Eng. **10**, 3179–3184 (2011)

14. Yokosuka, T., Aoki, T., Ogasawara, T., Ishikawa, T.: Effects of layup angle and ply thickness on matrix crack interaction in contiguous plies of composite laminates. Compos. Part A **36**(9), 1229–1235 (2005)
15. Ansari M.T.A., Singh K.K., Azam M.S.: Effect of stacking sequence and fiber volume fraction on the static mechanical properties of woven GFRP Composite. In: Singh I., Bajpai P., Panwar K. (eds) Trends in Materials Engineering. Lecture Notes on Multidisciplinary Industrial Engineering, pp. 51–58. Springer, Singapore (2019)

Chapter 52
Heat Exchange and Pressure Drop Enhancement Technique with Numerous Inserts in a Circular Tube Using ANSYS

Mohan Gupta, Kamal Sharma and Kuwar Mausam

Abstract The capability of a conventional heat exchanger in transferring heat requires improvement for conveying a considerable proportion of energy at cheaper rate and amount. For augmenting the heat transfer coefficient, dissimilar means have been employed. However, the use of inserts has become an assured method in enhancing heat transfer through endurable escalation of frictional losses. The objective of the study is the examination of a circular tube fitted with multiple inserts with regard to its characteristics related to heat transfer and fluid flow; these inserts are organized in clockwise and anticlockwise orientations.

Keywords Nu · Re · F · Twisted tape insert

52.1 Introduction

Heat exchanger (HE) is a crucial component of refrigeration cycles that provide assistance in the transference of energy by taking into consideration the variation in temperature. Energy transmission takes place from one medium to another. Each fluid passing through the HE shows variance in temperature; therefore, the walls separating each fluid also show the difference in temperature. The thermal functioning of the system depends on the quality of the HE to transmit heat from high-temperature to low-temperature fluid with the intention that the desired amount of heat energy will rapidly get transferred. The designers are enthusiastic to build effective and compressed heat exchangers at the minimum expenditure and working cost.

M. Gupta (✉) · K. Sharma · K. Mausam
GLA University Mathura, Mathura, UP, India

S. Yadav et al. (eds.), *Proceedings of International Conference in Mechanical and Energy Technology*, Smart Innovation, Systems and Technologies 174,
https://doi.org/10.1007/978-981-15-2647-3_52

52.2 Properties of Energy Transfer of Numerous Inserts

52.2.1 Twisted Tape Inserts

The rate of transference of heat of the heat exchangers is fundamentally dependent on the flow pattern of the convective layers and the fluid characteristics related to its thermo-physical nature. The process of heat transfer is encumbered because the sluggish layer of the border attached to the wall provided convective resistance. To augment the transference of heat, any technique for promoting commotion in the vicinity of the heated wall would be beneficial probably for assisting in controlling the simultaneous increase in frictional losses for keeping them under endurable limit. Dissimilar tube inserts have been examined with a focus on achieving improved THP of heat exchangers. Eiamsa-ard et al. [1] carried out experiments for estimating the transference of heat and nature of friction in a circular tube that is attached to complete length TTs with TRs of 6.0 and 8.0 and the space ratios of 1.0, 2.0, and 3.0. The study observed that using the TT inserts in the inner tube of a double pipe HE can considerably escalate the rate of transference of heat. Nevertheless, the friction factor of the tube concurrently increases with the use of TT inserts. Heat transfer (HT) and friction are amplified because of the swirling motion resulting from the secondary flows of the fluid. Eiamsa-ard et al. [2] in their another conducted experiment for analyzing the transference of energy and flow fluid data of the heat exchanging device that used a helical screw tape that was made of stainless steel with 17 mm width and the consent of tube surface $(D - W)/2.0 = 4.0$ mm. The tapes were used with no or with core rod inserts in the double pipe. In the HE, outer diameter is 50 mm and inner tube diameter is 25 mm through which low-temperature and high-temperature water flow as they are used as test fluids, and these fluids remain in shell and tube side, simultaneously. The Re ranged from 2000 to 12,000. The result of the experiment depicted that the transference of heat was better for the loose-fit with no core rod than a helical screw tape along with the core rod. The heat exchange rates were approximately 340% higher upon using the helical screw tape with no core rod insert in place of smooth tube. Though friction reduced by 50% for using a helical screw tape along with a core rod, the heat exchange rates were 25–60% more for the tape with no core rod. Moreover, for the helical screw tape with no core rod the efficiency became double.

Thianpong et al. [3] for examining the compound effect on the transference of heat and friction conducted an experiment with a dimpled tube, where a TT was used as insert, and they used water as working fluid. The outcomes related to the mean HT and the pressure drop was examined using varied permutation of the TR and the pitch ratio for the completely developed flow. They carried out the study using two dimpled tubes with dissimilar pitch ratios ($PR = 0.7$ and 1) and three TT inserts with three dissimilar TRs ($y/w = 3$, 5, and 7). They also used smooth tube for the experiment, which highlighted the advantages of dimpled tube [4]. From the results obtained in this study, it was observed that the friction factor and convective coefficient are higher for the TT inserted dimpled tube than dimple tube with no insert and a plain tube. Furthermore, upon diminishing the TR and pitch ratio values, the heat transfer coefficient (h) and F increased for both the configurations. Eiamsa-ard et al. [5] in another study explored the impact of using delta-winglet tapes as inserts in the circular tubes. For conducting the study, an oblique delta-winglet TT and a straight delta-winglet TT were employed. The delta-winglet TTs used for the experiment had three depths of wing cut ratios 0.110, 0.210, and 0.310 and TRs 3, 4, and 5. The Re ranged from 3000 to 27,000. From the obtained results, it was noticed that with the reduction in TR and increase in the depth of wing cut ratio, there is an increase in the heat transfer coefficient and F.

Twisted tape with cuts

Chang et al. [6] did an experiment using a tube fitted with twisted serrated tape for understanding the flow friction and heat transfer (HT) properties. For enhancing turbulence, modification was made in the typical TT geometry by fixing repeated ribs over the tape surfaces. Owing to large turbulence eddy diffusions, the transference of heat is more through the serrated TT than the simple tube with no tape inserts. In another study, Eiamsa-ard and Promvonge [7] in their study employed the serrated edge TTs as insert for exploring their role in the functioning of heat exchanger. For this study, they considered dissimilar serration depth and width ratio for Re ranging from 4000 to 20,000.

52.3 Contrivance of Heat Transfer and Fluid Flow

The existing literature emphasizes using the TT as insert in a tube, and the insertion helps in mixing of fluid and transmission of heat. With the change in the pattern of fluid flow along the surface of the tube, there is amplification of the transference of heat. Therefore, for understanding heat transfer mechanism, it is crucial to converse about the disruption in fluid flow.

52.3.1 Out-Turn of Ordinarily Twisted Tape

The flow configurations were discussed by Vashistha et al. [8] in their study analyzed the flow patterns for understanding the method involved in fluid flow and transference of heat in a circular tube that was attached with various inserts. The use of TT as inserts initiates swirl motion in the fluid at the core that causes turbulence close to the tube wall. The convective heat transfer enhances because of the generation of centrifugal force as the swirling motion evades the heated wall. Along with the transference of heat, the surface and fluid friction are also affected owing to higher turbulence intensity and broken boundary layer.

52.3.2 Out-Turn of Systematically Spaced Twisted Tape

Eiamsa-ard et al. [1] examined the impact of the regularly spaced TT and found that the existence of huge space between the consecutive nodes results in increasing the free space ratio, which in turn causes faster decomposition of the swirling flow. Hence, for increasing the rate of heat transfer and reducing loss of pressure than plain tube, it is deduced that the value of unity must be more than the optimum value of the free space ratio.

52.4 Experimental Methodology

The investigations are conducted on roundabout cylinder fitted with normal and different turned tapes embed over a wide scope of Re (3500–13,500) with working fluid water to gather the trial information, relating to the warmth move rate and weight drop. Test is likewise done on normal pipe to check the exhibition of numerous turned tape embedded cylinder. The water flow from the tank is controlled by the glove valve, while the stream rate is estimated by the rotameter. Meanwhile, the radiator gives a uniform warmth motion over the cylinder surface. Subsequent to setting the warmth motion and liquid stream rate, the temperatures of the tube surface and at channel and exit are watched. At first, the cylinder surface and leave liquid temperature readings are flimsy and way to deal with higher qualities with the time. It has been seen that roughly the following half-hour, the temperature at all the areas winds up free of the time which affirms that the framework is arrived at the relentless state condition. Not long after arriving at the relentless state condition, the cylinder surface temperature at eight points, delta, and leave liquid temperatures are measured. The distinction of weight head over the test cylinder is estimated by the guide of miniaturized scale manometer. Distinctive arrangement of curved tape supplements is tried over the whole scope of framework and working data by following the above strategy.

52.5 Analysis Made by ANSYS

The models of tabular tube are shown in the figures given below. Figure 52.1, after applying the boundary conditions and geometrical properties and analyzed with ANSYS software, so as per given inputs for heat transfer rate, we are getting higher Nusselt number as shown in Fig. 52.4 the maximum heat transfer rate obtained at the end of the tube and from Fig. 52.5 the given column shows at the circumference of the tube. As the heat transfer rate increases, the Nusselt number also increases, and in modeling and simulation, the maximum heat transfer rate from the ANSYS shows at the end of the tube (Figs. 52.2 and 52.3).

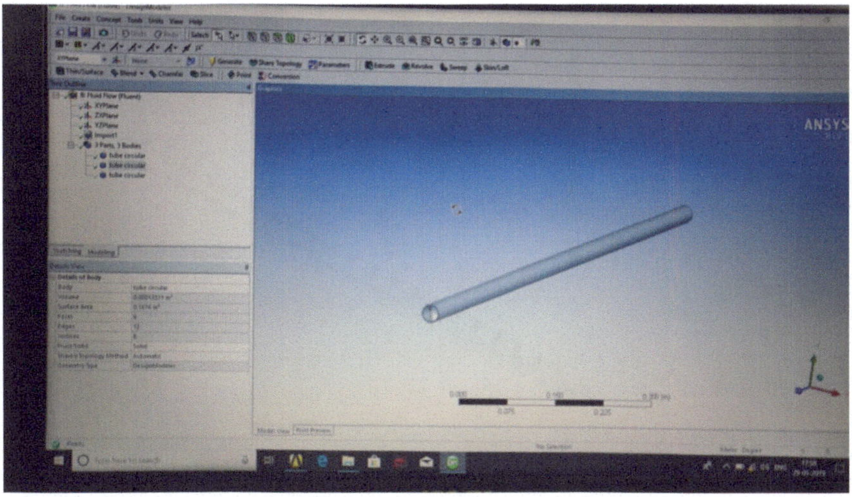

Fig. 52.1 Geometry of plain tabular tube

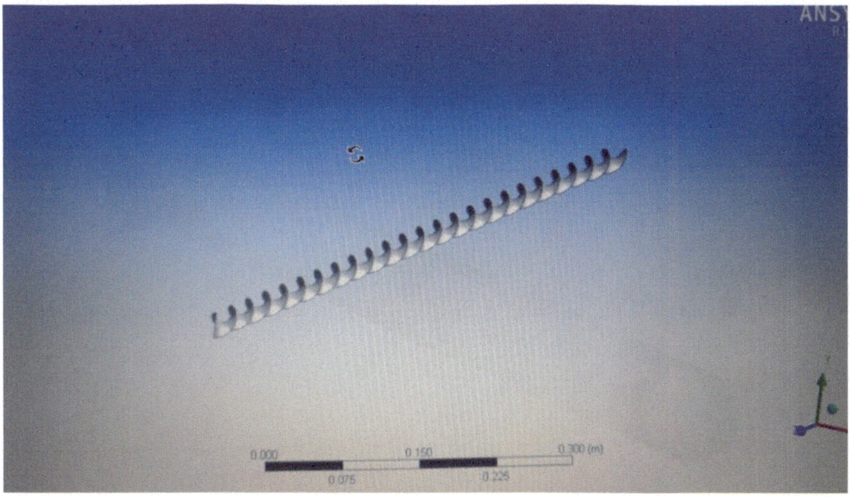

Fig. 52.2 Geometry of inserts

52.6 Conclusions

The conclusion of the above concerning techniques of heat transfer and friction in a circular tube with single, twin and four TT inserts has been demonstrated with the use of water as a fluid for the experiments conducted in dissimilar studies. We examined the differences of HT (heat transfer) and frictional losses for the varied arrangement of TTs. The following conclusion has been derived from this study:

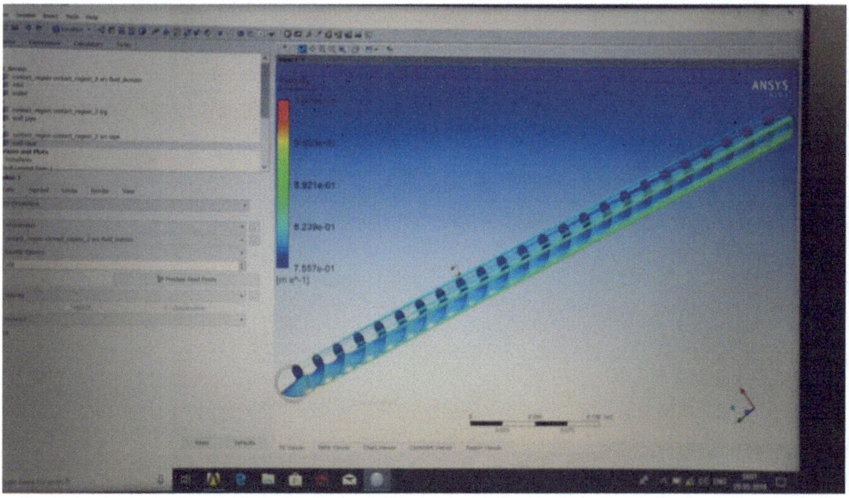

Fig. 52.3 Geometry of inserts inside the tube

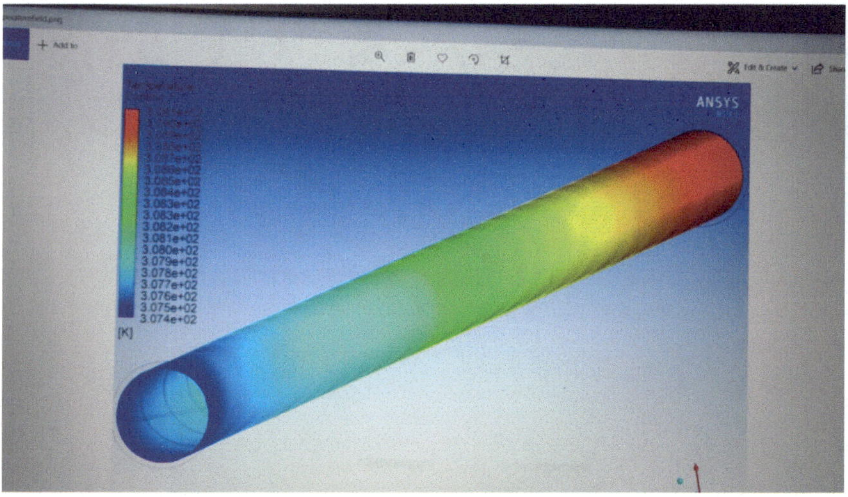

Fig. 52.4 Maximum heat transfer rate

- For the dissimilar types of TTs, the increases in Re cause increase in the Nu and decrease of *F*. Both the Nu and *F* are affected by the Re owing to an increase in the number of TT inserts.
- For all cases, the increases in Re cause decrease in the augmentation ratios of Nu and *F*.
- With a decrease in the TR (*y*) of the TTs, there is an increase in the Nu and *F* values regardless of the variation in Re.

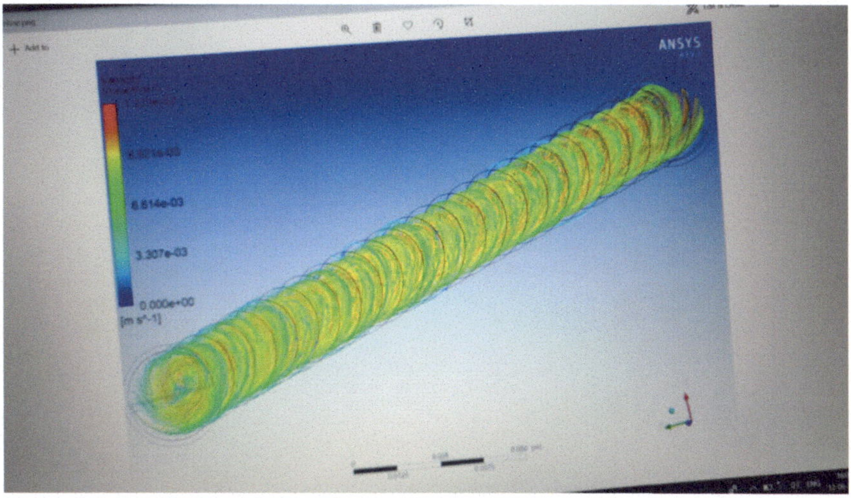

Fig. 52.5 Simulation of maximum heat transfer rate from the ANSYS

- The TT inserts can cause anticlockwise motion and clockwise motion. However, the heat transfer and friction is higher for the former than the latter. The THP factor reaches the highest value with the application of four anticlockwise TTs.

References

1. Eiamsa-ard, S., Thianpong, C., Promvonge, P.: Experimental investigation of heat transfer and flow friction in a circular tube fitted with regularly spaced twisted tape elements. Int. Commun. Heat Mass Transfer **33**(10), 1225–1233 (2006)
2. Eiamsa-ard, S., Promvonge, P.: Heat transfer characteristics in a tube fitted with helical screw-tape with/without core-rod inserts. Int. Commun. Heat Mass Transfer **34**(2), 176–185 (2007)
3. Thianpong, C., et al.: Compound heat transfer enhancement of a dimpled tube with a twisted tape swirl generator. Int. Commun. Heat Mass Transfer **36**(7), 698–704 (2009)
4. Maddah, Heydar, Aghayari, Reza: Factorial experimental design for the thermal performance of a double pipe heat exchanger using Al_2O_3-TiO_2 hybridnanofluid. Int. Commun. Heat Mass Trans. **97**, 92–107 (2018)
5. Eiamsa-ard, S., et al.: Heat transfer enhancement in a tube using delta-winglet twisted tape inserts. Appl. Therm. Eng. **30**(4), 310–318 (2010)
6. Rahimi, M., Shabanian, S.R., Alsairafi, A.A.: Experimental and CFD studies on heat transfer and friction factor characteristics of a tube equipped with modified twisted tape inserts. Chem. Eng. Process. **48**(3), 762–770 (2009)
7. Murugesan, P., et al.: Heat transfer and pressure drop characteristics in a circular tube fitted with and without V-cut twisted tape insert. Int. Commun. Heat Mass Transfer **38**(3), 329–334 (2011)
8. Vashistha, C., Patil, A.K., Kumar, M.: Experimental investigation of heat transfer and pressure drop in a circular tube with multiple inserts. Appl. Therm. Eng. **96**, 117–129 (2016)

Chapter 53
Numerical Estimation of Interfacial Heat Flux Using Inverse Heat Conduction Method

Muhammad Muneef Sadiq and Mohammad Asif

Abstract Heat dissipation at the interface of mold and casting plays an important role to reduce the casting defects and for the production of better castings. Therefore, the study of interfacial heat transfer during solidification of castings is of utmost significance. This paper focuses on the numerical estimation of heat flux at the inner surface of a hollow cylinder by using the 'inverse heat conduction' method. The 'inverse heat conduction' method involves the estimation of boundary conditions from the knowledge of transient temperature distribution inside a heat-conducting body. In the present work, one-dimensional heat transfer in the radial direction of the cylinder is considered assuming insulation at the top and bottom. Conjugate gradient method with adjoint problem has been employed to solve the inverse problem of transient heat flux estimation. Inputs to the inverse problem are one-dimensional transient simulated temperature data within the cylinder. Simulated data are generated by solving the direct problem for four test functions of heat flux, viz., cosine, triangular, polynomial and step function. Random error terms having zero mean and two different standard deviation values are used to form simulated data. The estimated results have been compared with the exact values used to generate the simulated data. The estimated results are found to be in good match with the exact one for all the test functions with some deviations at the sharp corners. This proves the worth of the inverse method for the estimation of transient interfacial heat flux during metal casting.

Keywords Metal casting · Inverse heat transfer method · Transient temperature data · Interfacial heat transfer

M. M. Sadiq (✉) · M. Asif
Mechanical Engineering Department, Aligarh Muslim University, Aligarh, India
e-mail: mmsadiq@myamu.ac.in

M. Asif
e-mail: masif@zhcet.ac.in

S. Yadav et al. (eds.), *Proceedings of International Conference in Mechanical and Energy Technology*, Smart Innovation, Systems and Technologies 174,
https://doi.org/10.1007/978-981-15-2647-3_53

53.1 Introduction

Heat flux is an important parameter in modeling and designing systems involving heat transfer. It has applications in estimating unknown heat flux from a furnace or heater. Moreover, it is also useful in the study of heat transfer in IC engines, die-casting arrangements, etc. Estimation of transient heat flux is an essential parameter for the thermal analysis of a heat transfer system in the design process. For instance, in the case of die casting, interfacial heat transfer is an important parameter in designing the die-casting system and the selection of other parameters as well. Thermal contact conductance can also be calculated using interfacial heat flux and temperature gradient at the interface. The estimation of transient heat flux would also help in applying metallic coatings on cylindrical surfaces in order to control heat transfer. Estimation of transient heat flux at the wall surface which is subjected to heat is difficult using direct conventional methods. 'Inverse heat transfer' method is very popular in heat transfer analysis where direct property estimation is not possible or very difficult to achieve since the sensors at the surface would disturb the phenomena. Therefore, the 'inverse heat transfer' method is utilized to estimate transient heat flux at the surface of the wall providing the transient temperatures data near the wall as an input [1].

Huang et al. [2] estimated thermal contact conductance numerically by using the conjugate gradient method during metal casting. The direct problem involving the solidification of metal has been solved by the enthalpy method. Singh et al. [3] inspected a robust inverse method for estimation of transient wall heat flux using temperature data of single thermocouple, instead of using data of a row of thermocouples fixed inside the wall. Wang et al. [4] presented the work on the squeeze casting of an aluminum alloy A356 into a plate-shaped product in H13 steel die. Utilizing the measured transient temperature data inside the die, the interfacial heat transfer coefficient at the interface of die and casting was determined by applying an inverse approach. Zhang et al. [5] established an inverse conduction model to determine the interfacial heat flux and heat transfer coefficient between cylindrical sand mold and metal. The casting was done by the lost foam process. Kim et al. [6] examined the interfacial heat transfer coefficient between die and casting with respect to the interface temperature. Die used is cylindrical in shape made up of copper metal. Aluminum metal is used for casting. Zhu et al. [7] estimated heat flux on a triangular wall using the 'inverse heat conduction problem (IHCP).' Huang and Wu [8] estimated unknown boundary heat flux utilizing boundary temperature measurements. Conjugate gradient method (CGM) was used to solve a hyperbolic heat conduction problem. Chen and Yang [9] calculated space-dependent heat flux at the interface of roller and workpiece during the rolling process. An inverse method based on the conjugate gradient method was used.

After an extensive literature review, we were unable to find an 'easy to understand' and efficient method to estimate one-dimensional transient heat flux at the surface of a hollow cylindrical body. Moreover, the efficiency of the used method, i.e., 'conjugate gradient method' in cylindrical coordinates for strict test cases involving a triangular and a step function of heat flux was a research void present in the archival literature. In the present work, heat flux has been estimated at the inner surface of the hollow cylinder which is assumed to be insulated at the top and bottom. Thus, the heat flux is considered to be one-dimensional in the radial direction. Transient temperature variation data inside the cylinder wall, thermo-physical properties of cylinder material and boundary condition at the outer surface are taken as input to the inverse problem. The boundary condition at the inner surface, i.e., transient heat flux is the unknown function in the inverse problem. Heat conduction equations are discretized using the 'central space forward time' method and programmed in MATLAB.

53.2 Problem Formulation

The technique of the 'conjugate gradient method (CGM)' of minimization with 'adjoint problem' for function estimation has been utilized. Prior information about the functional form of the unknown heat flux is not required in this method. Steps involved in the solution of the inverse problem by this approach are described below.

53.2.1 Direct Problem

The direct problem involves the calculation of transient temperatures within the cylinder wall when the thermal and physical properties of the material and the boundary conditions for the cylindrical specimen are known. We have assumed that the heat transfer is taking place in the radial direction only. Specimen considered in the problem and grid points are shown in Fig. 53.1.

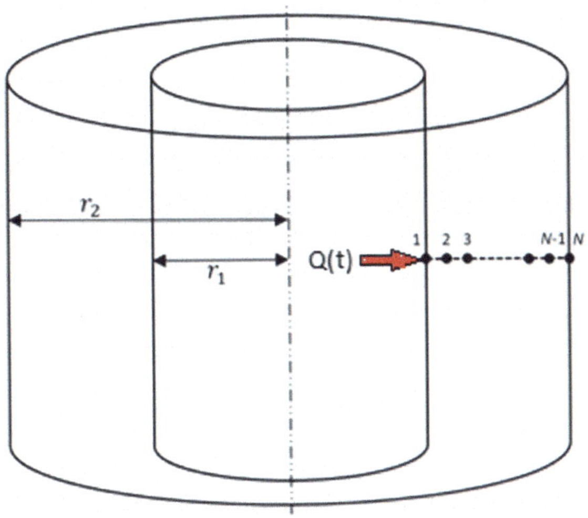

Fig. 53.1 Specimen geometry showing heat flux and grid points

Initially, the cylinder is at the temperature T_o and the boundary conditions for the cylinder are uniform heat flux $Q(t)$ at the inner surface and convection at the outer surface with heat transfer coefficient h. Equations for the direct problem are given as:

$$\frac{\partial^2 T}{\partial r^2} + \frac{1}{r}\frac{\partial T}{\partial r} = \frac{1}{\alpha}\frac{\partial T}{\partial t} \quad \text{at}\, r_1 < r < r_2, t > 0 \tag{53.1a}$$

$$Q(t) = -k\frac{\partial T}{\partial r} \quad \text{at}\, r = r_1, t > 0 \tag{53.1b}$$

$$-k\frac{\partial T}{\partial r} = h(T_N - T_\infty) \quad \text{at}\, r = r_2, t > 0 \tag{53.1c}$$

$$T = T_o \quad \text{at}\, r_1 \geq r \leq r_2, t = 0 \tag{53.1d}$$

53.2.2 *Inverse Problem*

The inverse problem is formed by considering the heat flux, $Q(t)$ as unknown but everything else which is required to solve the direct problem as known. As shown in Fig. 53.1, we have divided the cylindrical wall into N grid points and thermocouples are located from grid points 2 to $N - 1$. All the thermocouples are located at equal distances of Δr. Temperature data are taken with these thermocouples up to the time $t = t_f$ and denoted by; $Y(t) \equiv Y_i$, $i = 1,2 \ldots N$.

Therefore, the inverse problem is formed by utilizing the abovementioned measured temperature readings Y_i and known thermo-physical properties to estimate the unknown heat flux $Q(t)$ over the total time period t_f. Following function is to be minimized in order to obtain the solution of the inverse problem:

$$S[Q(t)] = \int_0^{t_f} \left[\sum_{i=2}^{N_1 - 1} (T_i - Y_i)^2 \right] \mathrm{d}t \tag{53.2}$$

where T_i and Y_i are the estimated and measured temperatures, respectively.

53.2.3 *Sensitivity Problem*

The sensitivity problem is obtained from the direct problem (Eqs. 53.1a, 53.1b, 53.1c and 53.1d). At first, we replace $Q(t)$ by '$Q(t) + \Delta Q(t)$' and $T(r,t)$ by '$T(r,t) + \Delta T(r,t)$' in the direct problem and obtain new expressions. Then the original direct problem is subtracted from the obtained expressions. Now after neglecting the second-order terms, the sensitivity problem for the sensitivity function, $\Delta T(r, t)$ is obtained. The sensitivity function is required for the computation of the search step size β^k (Eq. 53.7), required in iterative solution [1].

53.2.4 Adjoint Problem

In the present inverse problem, the estimated temperatures should satisfy a constraint, which is the heat conduction problem for the cylindrical wall. Therefore, a Lagrange multiplier, $\lambda(r, t)$, comes into picture which is actually needed for the computation of the gradient equation (Eq. 53.3). The *Adjoint problem* has to be solved for the determination of the Lagrange multiplier $\lambda(r, t)$ [1].

53.2.5 Gradient Equation

The following expression has been obtained for the gradient $\nabla S[Q(t)]$ of the functional $S[Q(t)]$:

$$\nabla S[Q(t)] = \left[\frac{\lambda(r_1, t)}{K}\right] \tag{53.3}$$

53.2.6 Iterative Procedure

The iterative procedure for the estimation of heat flux $Q(t)$ is as follows:

$$Q^{k+1}(t) = Q^k(t) - \beta^k d^k(t) \tag{53.4}$$

where β^k is the *search step size* in going from iteration k to iteration $k+1$ and $d^k(t)$ is the *Direction of Descent* given by:

$$d^k(t) = \nabla S\big[Q^k(t)\big] + \gamma^k d^{k-1}(t) \tag{53.5}$$

where γ^k is the *Conjugation Coefficient* and is determined from the expression as:

$$\gamma^k = \frac{\int_0^{t_f}\left\{\nabla S\big[Q^k(t)\big]\right\}^2 dt}{\int_0^{t_f}\left\{\nabla S\big[Q^{k-1}(t)\big]\right\}^2 dt} \quad \text{for } k = 1, 2\ldots \text{ with } \gamma^k = 0 \text{ for } k = 0 \tag{53.6}$$

The search step size β^k is determined by minimizing the functional $S[Q(t)]$ given by Eq. (53.2), which leads the following expression for β^k:

$$\beta^k = \frac{\int_0^{t_f}\left\{\sum_{i=2}^{(N-1)}(T_i - Y_i)\Delta T_i\right\}\mathrm{d}t}{\int_0^{t_f}\left\{\sum_{i=2}^{(N-1)}[\Delta T_i]^2\right\}\mathrm{d}t} \tag{53.7}$$

where T_i are the solutions of the direct problem (53.1a, 53.1b, 53.1c and 53.1d), obtained by using the current estimate for $Q(t)$ while the sensitivity functions ΔT_i are the solutions of the sensitivity problem, obtained by setting $\Delta Q(t) = d^k(t)$.

53.2.7 Stopping Criterion

The stopping criterion is given by

$$S[Q(t)] < \varepsilon \tag{53.8}$$

where $S[Q(t)]$ is computed with Eq. (53.2). The tolerance ε is chosen such that smooth solutions are obtained with measured temperatures containing random errors. It is assumed that the solution is sufficiently accurate when

$$|Y(t) - T[x_{\text{meas}}, t;\ Q(t)]| \approx \sigma \tag{53.9}$$

where'σ' is the standard deviation of measurement errors. Thus 'ε' is obtained from Eq. (53.2) using Eq. (53.9) as:

$$\varepsilon = nt_f\sigma^2 \tag{53.10}$$

where n is the total number of thermocouples.

53.2.8 Computational Algorithm

Generate the simulated data $Y(r, t)$ using different test functions as described below. Suppose an initial guess $Q^0(t)$ is available for the function $Q^k(t)$. Set $k = 0$ and then:

1. Solve the direct problem (Eqs. 53.1a, 53.1b, 53.1c and 53.1d) using the initial guess $Q^k(t)$, in order to compute $T(r, t)$.
2. Using $Y(r,t)$ and $T(r,t)$, check the stopping criterion (Eq. 53.8). Continue if not satisfied.
3. Now solve the adjoint problem to compute $\lambda(r_1, t)$.

4. Knowing $\lambda(r_1, t)$, compute each component of gradient vector $\nabla S[Q(t)]$ from Eq. (53.3).
5. Knowing the gradient $\nabla S[Q(t)]$, compute γ^k from Eq. (53.6) and then the direction of descent $d^k(t)$ from Eq. (53.5).
6. By setting $Q^k(t) = d^k(t)$, solve the sensitivity problem to obtain $\Delta T(r, t)$.
7. Knowing $\Delta T(r, t)$, compute the search step size β^k from Eq. (53.7).
8. Knowing the search step size β^k and the direction of descent $d^k(t)$, compute the new estimate $Q^{k+1}(t)$ from Eq. (53.4) and return to step 1.

53.3 Numerical Implementation

To solve the above inverse problem, measured transient temperature distribution data inside the cylinder wall are required. In present numerical estimation, simulated measurement data are provided instead of experimentally measured data.

53.3.1 Simulated Measurements

Simulated temperature measurements at fixed radial thermocouple locations of the wall have been obtained from the solution of the direct problem (53.1a, 53.1b, 53.1c and 53.1d) at the measurement locations by using various test functions of $Q(t)$. The measurements containing random errors are simulated by adding an error term as:

$$Y(r_{\text{meas}}, t) = Y_{\text{ex}}(r_{\text{meas}}, t) + \omega\sigma \tag{53.11}$$

where $Y(r_{\text{meas}}, t)$: Simulated measurements containing random errors
$Y_{\text{ex}}(r_{\text{meas}}, t)$: Exact (errorless) simulated measurements
ω is random variable with normal distribution, zero mean and unitary standard deviation and σ is the standard deviation of the measurement error.

Thereafter, the inverse problem has been solved using the conjugate gradient method as explained above. The stability of the inverse problem solution has been examined for various levels of measurement errors. Both exact and inexact simulated temperature measurements have been considered but all the other parameters used in the analysis are supposed to be errorless. The physical problem analyzed is as follows:

Consider a hollow cylinder with an inner and outer radius as: $r_1 = 75$ mm & $r_2 = 100$ mm respectively. Cylinder is made up of copper having thermal conductivity $k = 385$ w/mk and diffusivity $\alpha = 1.11 \times 10^{-4}$. The wall is divided into six grid points ($N = 6$) with space steps, $\Delta r = 5 \times 10^{-3}$ m and the time step Δt is chosen as 1 s. Discretization has been done by the *Implicit form* of finite difference method as it is *unconditionally stable* and *fast convergent*. To test the adopted inverse methodology, different test functions on the basis of their applications have been employed which are as follows:

Case 1: *Cosine function of* $Q(t)$. It is a cyclic function that denotes the oscillating heat flux which is the case in many thermal systems.

$$Q(t) = Q_o \mathrm{Cos}\left(\frac{\pi t}{18}\right);\ \text{where } Q_o = 10^5 w/m^2 \tag{53.12}$$

Case 2: *Step function in* $Q(t)$. This function tests the algorithm and method of the solution as it contains discontinuities and sharp corners which are considered as difficult to be recovered by an inverse analysis [1]. This is the step-up heat transfer function which is generally found in IC engines.

$$Q(t) = \begin{cases} 1 \times 10^5, & 0 \le t \le 45 \\ 5 \times 10^3, & 45 \le t \le 100 \end{cases} \tag{53.13}$$

Case 3: *Triangular function in* $Q(t)$. The triangular function is also chosen as it contains sharp corners, which are considered to be difficult to achieve by inverse methodology.

$$Q(t) = \begin{cases} 2 \times 10^3 t, & 0 \le t \le 44 \\ (200 - 2t) \times 10^3, & 44 \le t \le 100 \end{cases} \tag{53.14}$$

Case 4: *Polynomial function in* $Q(t)$. In the metal casting, the heat flux at the interface of casting and mold generally varies as a polynomial function in time during the solidification of the metal. Therefore to simulate the case of solidification of metal in die casting, the heat flux is assumed to be a cubic polynomial in time as follows:

$$Q(t) = C_1 + C_2 t + C_3 t^2 + C_4 t^3 \tag{53.15}$$

where the constants $C_1 = 10^5$, $C_2 = -900$, $C_3 = 100$ and $C_4 = -1$.

To check the accuracy of results using the present inverse methodology, percentage errors have also been calculated in each case by the following expression.

$$\%\text{Error} = \frac{\sum_{i=1}^{i=N} \left| \frac{Q_{\text{exact}} - Q_{\text{est}}}{Q_{\text{exact}}} \right|}{N} \times 100 \tag{53.16}$$

53.4 Results and Discussion

Four test cases on the basis of their significance have been selected to analyze the reliability of the inverse method in predicting the heat flux with time. Based on recommendations, a 6% run advanced to starting time and 6% overrun in final time has been done to get the reduction in the errors for both errorless measurements and measurements with random errors due to the effects of the initial guess on the solution and because of the null gradient at the final time [1]. For Figs. (53.2, 53.3, 53.4, 53.5), Q_{exact} refers to the prescribed value of heat flux, used to estimate the simulated temperature measurements and Q_{est} refers to estimated heat flux resulted by solving the inverse problem for different values of the standard deviation of measurement errors.

Case 1: Cosine function of $Q(t)$. In the first case, cosine function of 'heat flux' is used for generation of simulated data with the standard deviations of measurements as, $\sigma \approx 0$ (no measurement error), $\sigma = 0.2$ and $\sigma = 0.5$. Using this data as input for the inverse problem, estimated heat flux is obtained and compared to the exact

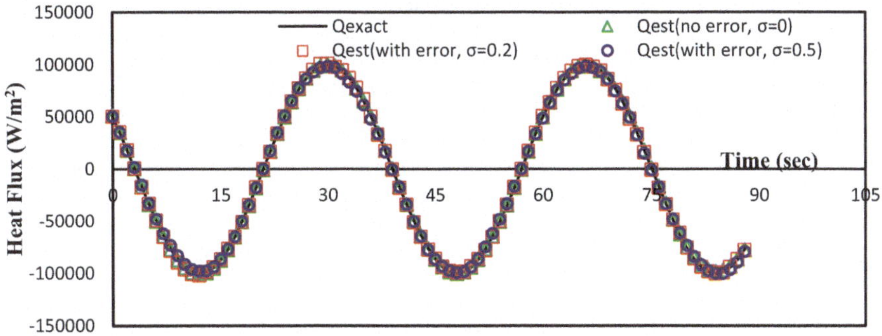

Fig. 53.2 Variation of estimated heat flux for cosine function with different random errors

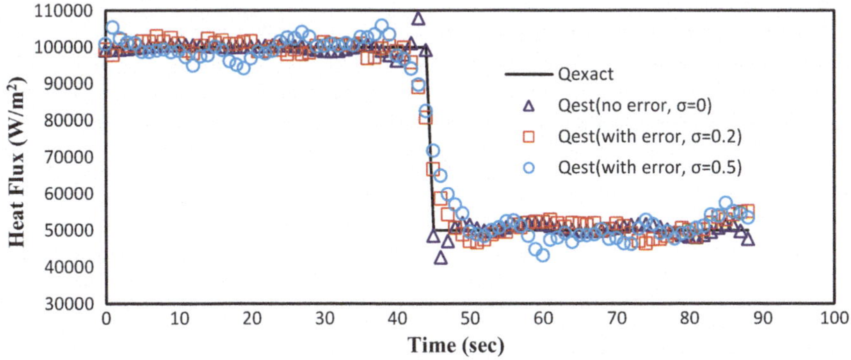

Fig. 53.3 Variation of estimated heat flux for step function with different random errors

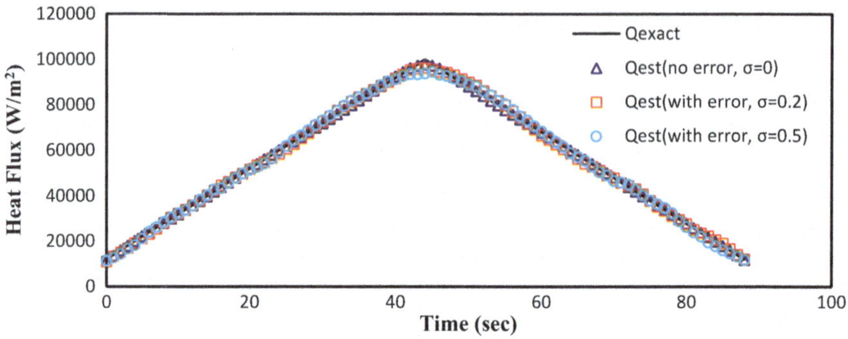

Fig. 53.4 Variation of estimated heat flux for triangular function with different random errors

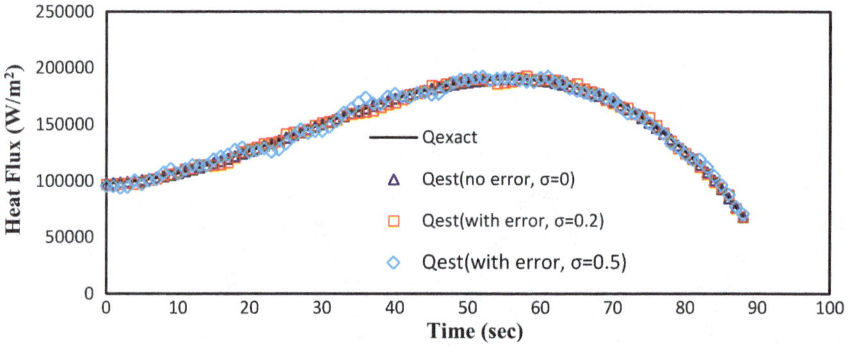

Fig. 53.5 Variation of estimated heat flux for polynomial function with different random errors

function as shown in Fig. 53.2. From the graphs, it is observed that the estimated heat flux with no measurement error is very close to the exact value with a percentage error of 0.59%. But when the measurement errors are included estimated, heat flux differs from the exact value mostly at peaks and valleys of the curve, this is inherent with the inverse solution. Further, the results for $\sigma = 0.2$ are better than for $\sigma = 0.5$ which is also evident from the % error values which are 1.45% $\sigma = 0.2$ and 1.86% for $\sigma = 0.5$.

Case 2: **Step function of** $\boldsymbol{Q(t)}$. Estimated results of heat flux with time for step function have been shown against the exact function in Fig. 53.3. From Fig. 53.3, it is apparent that the results of estimated heat flux agree well with the exact function but deviates at the sharp corners. Further, deviations are found to increase when adding measurement errors to the data. The results are also supported by the corresponding percentage error values for no measurement errors, measurement errors with $\sigma = 0.2$ and $\sigma = 0.5$ are 1.6%, 3.4%, and 4.7%, respectively. It can be noted that error values are maximum for step function. However, the scheme is able to recover the step function with reasonable accuracy.

Case 3: **Triangular function of** $Q(t)$. The results of transient heat flux for the triangular function have been shown in Fig. 53.4. From Fig. 53.4, it is clear that the estimated results of the heat flux function match very well with the exact one with deviations only at the vertex. Further, the percentage error values are 0.29%, 1.81% and 2.55% for no measurement errors, measurement errors with $\sigma = 0.2$ and $\sigma = 0.5$, respectively. The present results demonstrate the suitability of the scheme for triangular function with good accuracy.

Case 4: **Polynomial function of** $Q(t)$. Figure 53.5 presents the results of the heat flux for a cubic polynomial test function. The estimated results show good accuracy with the exact function. However, accuracy reduces while adding random errors. It is also confirmed from the percentage error values, calculated as 0.12%, 1.06% and 2.02% for no measurement errors, measurement errors with $\sigma = 0.2$ and $\sigma = 0.5$, respectively. The results of the polynomial test function demonstrate the suitability and applicability of the present inverse analysis for estimating the interfacial heat flux in the metal casting applications.

53.5 Conclusions

The function estimation inverse technique utilizing the conjugate gradient method with the adjoint problem has been employed to estimate the unknown function, i.e., heat flux with time at the inner surface of the hollow cylinder. Assuming one-dimensional heat conduction in radial direction, simulated temperature data have been generated with no error and with random errors. To test the technique, the simulated temperature data have been obtained for four types of functions of heat flux $Q(t)$, on the basis of their applicability in the heat transfer systems, namely cosine function, step function, triangular function and polynomial function. It is concluded that for all the test functions, estimated heat flux agrees very well with the exact one with deviations at the sharp corners, which is obvious with the inverse analysis. However, the deviations between the estimated heat flux and exact one increase using the simulated measurement with random errors, for all the test functions of heat flux $Q(t)$. The results confirmed that the present inverse technique employing the conjugate gradient method is an efficient and powerful technique. Further, the results demonstrate the suitability and significance of the present inverse methodology for the study of interfacial heat flux in metal casting applications, specifically during the solidification of the metal.

References

1. Ozisik, M.N., Orlande, H.R.B.: Inverse Heat Transfer: fundamentals and Applications. Taylor & Francis, New York (2000)
2. Huang, C.H., Ozisik, M.N., Sawaf, B.: Conjugate gradient method for determining unknown contact conductance during metal casting. Int. J. Heat Mass Transf. **35**, 1779–1786(1992)
3. Singh, S.K., Yadav, M.K., Sonawane, R., Khandekar, S., Muralidhar, K.: Estimation of time-dependent wall heat flux from single thermocouple data. Int. J. Therm. Sci. **115**, 1–15 (2017)
4. Wang, F., Ma, Q., Meng, W., Han, Z.: Experimental study on the heat transfer behaviour and contact pressure at the casting-mold interface in squeeze casting of aluminum alloy. Int. J. Heat Mass Transf. **112**, 1032–1043 (2017)
5. Zhang, L., Tan, W., Hu, H.: Determination of the heat transfer coefficient at the metal-sand mold interface of lost foam casting process. Heat Mass Transf. **52**, 1131–1138 (2016)
6. Kim, H., Cho, I., Shin, J., Lee, S., Moon, B.: Solidification parameters dependent on interfacial heat transfer coefficient between aluminum casting and copper mold. ISIJ Int. **45**, 192–198 (2005)
7. Zhu, Y., Liu, B., Jiang, P., Fu, T., Lei, Y.: Inverse heat conduction problem for estimating heat flux on a triangular wall. J Thermophys. Heat Transf. **31**, 1–6 (2016)
8. Huang, C.H., Wu, H.H.: Aninverse hyperbolic heat conduction problem in estimating surface heat flux by the conjugate gradient method. J. Phys. D Appl. Phys. **39**, 4087–4096 (2006)
9. Chen, W.L., Yang, Y.C.: Inverse problem of estimating the heat flux at the roller/workpiece interface during a rolling process. Appl. Therm. Eng. **30**, 1247–1254 (2010)

Chapter 54
Modelling and Analysis of Skew Hybrid Sandwich Plate

S. M. Shaukat Rafi, M. Naushad Alam and Najeeb ur Rahman

Abstract Sandwich plates provide a lightweight and stiff construction for structures where weight saving is an important consideration. Skew sandwich plates can be found in structures such as the swept wings of airplanes which can be idealised as oblique plates. This work presents structural dynamic analysis of a skew hybrid sandwich plate. The sandwich plate is made up of four laminate faces and a soft core (90/0/C/0/90), with cantilever boundary conditions. The skew plate is a hybrid composite with piezoelectric bottom and top layers bonded to their elastic substrate. A moderately thick skew sandwich plate is modelled with ANSYS using an eight-node shell element for the cross-ply laminate faces and a 20-node solid for modelling the soft core of the skew plate. The free flexural vibration behaviour of the skew hybrid sandwich plate has been investigated for different skew angles. The results of the investigation will help in acquiring a better insight about the behaviour of laminated sandwich plates with bonded piezoelectric layers. The results presented here were compared with those found in literature. The motive of incorporating piezoelectrics is that the piezoelectric layers may be used as sensors and actuators for shape control and vibration suppression applications in structural members.

Keywords Hybrid sandwich plate · Skew angle · Free vibrations · Finite element method

54.1 Introduction

Sandwich plates have recently been widely used in structures that require a lightweight and stiff construction. Skew sandwich plate finds application in aircraft structures, the sub-structure of the swept wing of an airplane can be idealised as an oblique plate. Oblique or skew plates also find application in civil engineering structures such as bridges and buildings.

Many investigators have studied the structural behaviour of isotropic and orthotropic skew sandwich plates. Dynamic response of a truss-core sandwich panel

S. M. Shaukat Rafi (✉) · M. Naushad Alam · N. Rahman
Department of Mechanical Engineering, Zakir Husain College of Engineering and Technology, Aligarh Muslim University, Aligarh 202002, India

S. Yadav et al. (eds.), *Proceedings of International Conference in Mechanical and Energy Technology*, Smart Innovation, Systems and Technologies 174,
https://doi.org/10.1007/978-981-15-2647-3_54

with homogeneous orthotropic materials for a thick plate was studied by Lok and Cheng [1]. Closed-form solution for a clamped plate was obtained for both 2-D and 3-D finite elements. Wang et al. [2] investigated sandwich plates made up of orthotropic core and laminate faces; the free vibration analysis of the plates was presented. The analysis was performed for different boundary conditions with varying aspect ratios. Kapuria and Kulkarni [3] carried the dynamic analysis of smart plate with segmented piezoelectrics using a four-node quadrilateral element. The finite element was developed using the coupled improved zigzag theory. Ashour [4] studied the free vibration behaviour of thin laminated skew plates with clamped edges. The natural frequencies of the plate were determined and results were obtained for various skew angles for different composite laminates. Yasin et al. [5] studied vibration control of smart structures with patched piezoelectrics. Their study involved solving eigenvalue problem for obtaining undamped natural frequencies and mode shapes of the plate considering two cases; one without electromechanical coupling and the other with electromechanical coupling. Kanasogi and Ray [6] developed a finite element model for analysing the active constrained layer damping of smart skew cross-ply laminated composites and smart skew angle-ply composite plates. Singh and Chakrabarti [7] developed a C_0 finite element model for studying the vibration and buckling characteristics of laminated composite and skew sandwich plates. The finite element was based on the higher-order zigzag theory. Sayyad and Ghugal [8] reviewed the literature available for analysing the free vibration behaviour of multilayered laminated composites and sandwich plates. Asemi et al. [9] analysed the static and dynamic behaviour of functionally graded rectangular and skew plates, considered the effects of material gradient index and skew angles, for different boundary conditions. Joseph and Mohanty [10] studied skew sandwich plates with a viscoelastic core and the faces made up of different materials; one face has a layer of functionally graded material and the other face consisting of a layer of elastic material. The free vibration and buckling behaviour of the plate was analysed considering different skew angles and aspect ratios. Mandal et al. [11] studied laminated composite skew plates with cut-outs and carried free vibration analysis of the plate both experimentally and numerically. Her and Chen [12] investigated the stresses developed in a sandwich composite plate by the use of embedded piezoelectric layers in the plate. Jin et al. [13] presented a thermo-mechanical analysis of multilayered composite beams based on mixed-form global–local higher-order theory, including transverse normal thermal deformations. Gupta and Ghosh [14] analysed the static and transient behaviour of sandwich plates by performing an isogeometric analysis based on the first-order shear deformation theory (FSDT).

After an exhaustive literature survey, it can be concluded that although lot of work has already been done, still there exists scope for detailed analysis of skew hybrid plates. The motivation behind this work is the possibility of using piezoelectric layers with sandwich plates for shape control and vibration suppression applications. A nonrectangular skew hybrid sandwich plate with varying skew angles has been considered for the analysis.

54.2 Modelling and Analysis of Hybrid Skew Plate

In order to assess free vibration response of skew hybrid sandwich plate with cantilever support, various skew angles have been considered. The configuration of the plate is a multilayered sandwich with (0/90/core/90//0) orientation. The core of the plate is made up of a softer material 2 having a thickness of 0.64 h, where h is the total thickness of the sandwich plate. The soft core is bounded on each side by two cross-ply (0/90) graphite–epoxy face layers of material 1, having a thickness of 0.04 h for each layer. The sandwich plate is further bounded by two lead zirconate titanate (PZT-5A) layers, each of thickness 0.1 h bonded to the top and bottom faces of the plate. The skew hybrid plate is modelled in ANSYS environment. The model uses eight-nodded shell-99 element to model the composite laminates and solid-95 element for modelling the soft core. The geometry of the skew sandwich plate with cantilever support is shown in Fig. 54.1.

Following are the material properties for the hybrid sandwich plate:

Material-1(Faces)

$E_1 = 131.1$ GPa, $E_2 = E_3 = 6.9$ GPa, $G_{12} = G_{13} = 3.588$ GPa, $G_{23} = 2.332$ GPa, $\upsilon_{12} = \upsilon_{13} = 0.32$, $\upsilon_{23} = 0.49$, $\rho = 1000$ kg/m^3

Material-2 (Soft Core)

$E_1 = 0.0002208$ GPa, $E_2 = 0.0002001$ GPa, $E_3 = 2.76$ GPa, $G_{12} = 0.01656$ GPa, $G_{13} = 0.5451$ GPa,

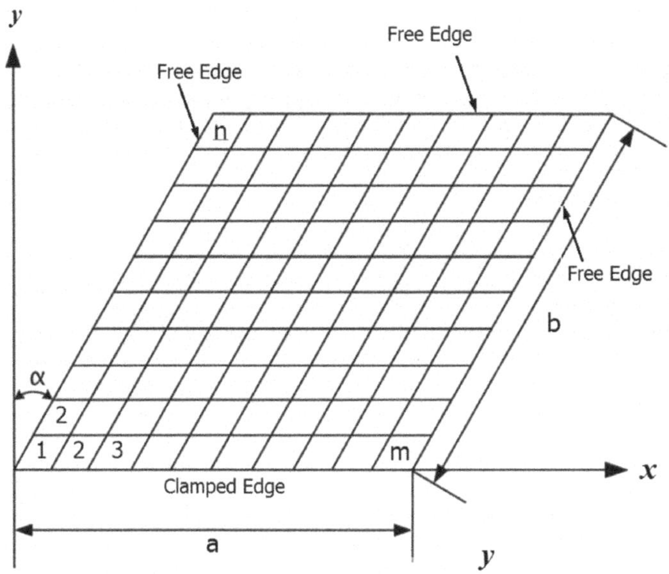

Fig. 54.1 Finite element mesh for the skew cantilever plate

$G_{23} = 0.4554$ GPa, $\upsilon_{12} = 0.99$, $\upsilon_{13} = \upsilon_{23} = 3 \times 10^{-5}$; $\rho = 70$ kg/m^3

PZT-5A (Piezoelectric layers)

$E_1 = E_2 = 61.0$ GPa, $E_3 = 53.2$ GPa, $G_{12} = 22.6$ GPa, $G_{13} = G_{23} = 21.1$ GPa, $\upsilon_{12} = 0.35$, $\upsilon_{13} = \upsilon_{23} = 0.38$, $d_{31} = d_{32} = -171$, $d_{33} = 374$, $d_{15} = d_{24} = 584$ pm/V; $\varepsilon_{11} = \varepsilon_{22} = 15.3$; $\varepsilon_{33} = 15.0$ μF/m; $\rho = 7600$ kg/m^3.

54.3 Results and Discussion

The first six natural frequencies ($\bar{\omega}_n$) of the skew hybrid sandwich plate have been obtained for a skew angle of 30° using a 24 × 24 mesh size, with $S = a/h = 10$ for b/a = 1. The circular frequencies are non-dimensionalised as $\bar{\omega}_n = \omega_n a S(\rho_0/Y_0)^{1/2}$ with $Y_0 = 6.9$ GPa and $\rho_0 = 1000$ kg/m^3. The results for the first six natural frequencies of the plate have been compared with the values given in the Ref. [3] and found to be in good agreement, as can be seen in Table 54.1.

The non-dimensionalised frequencies for the moderately thick skew hybrid cantilever plate ($S = 10$) with different skew angles i.e. 15°, 30° and 45° are given in Table 54.2. The natural frequencies of the skew hybrid plate were found to increase with an increase in the skew angle. This is due to the fact that increase in skew angle has a direct effect on the plate's stiffness; stiffness increases with an increase in the skew angle.

The non-dimensionalised frequencies versus mode numbers are plotted and the trend of the graphs is found to be similar for all skew angles, as shown in Fig. 54.2. From the graph, it is clear that the non-dimensionalised frequencies of skew plates are approximately the same at mode number 4 for all the skew angles.

Further, it was observed that the mode shapes 3 and 4 get interchanged for skew angles 30° and 45°. The first four flexural mode shapes for the skew hybrid plates are plotted in Figs. 54.3, 54.4, 54.5, 54.6, 54.7 and 54.8 for different skew angles, i.e. $\alpha = 15°$, 30° and 45°.

Table 54.1 Non-dimensionalised natural frequencies $\bar{\omega}_n$ with skew angle 30° and $S = 10$

Skew angle α	$\bar{\omega}_n$	Present	Reference [3]
30°	$\bar{\omega}_1$	1.768	1.827
	$\bar{\omega}_2$	3.479	3.575
	$\bar{\omega}_3$	7.092	7.537
	$\bar{\omega}_4$	8.078	8.116
	$\bar{\omega}_5$	8.668	9.011
	$\bar{\omega}_6$	11.757	12.151

Table 54.2 Non-dimensionalised natural frequencies of skew hybrid cantilever plate with $S = 10$

Skew angle α	Non-dimensionalised frequencies	
15°	$\bar{\omega}_1$	1.719
	$\bar{\omega}_2$	3.055
	$\bar{\omega}_3$	6.720
	$\bar{\omega}_4$	7.975
	$\bar{\omega}_5$	8.499
	$\bar{\omega}_6$	10.553
30°	$\bar{\omega}_1$	1.768
	$\bar{\omega}_2$	3.479
	$\bar{\omega}_3$	7.092
	$\bar{\omega}_4$	8.078
	$\bar{\omega}_5$	8.668
	$\bar{\omega}_6$	11.757
45°	$\bar{\omega}_1$	1.872
	$\bar{\omega}_2$	4.258
	$\bar{\omega}_3$	7.670
	$\bar{\omega}_4$	7.904
	$\bar{\omega}_5$	9.330
	$\bar{\omega}_6$	13.062

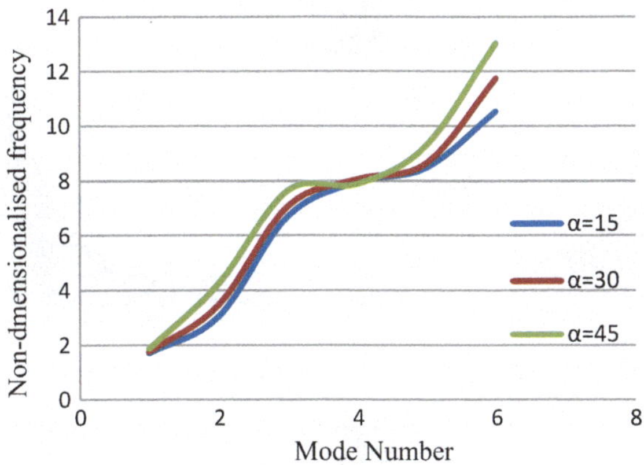

Fig. 54.2 Effect of skew angle α on non-dimensionalised fundamental frequencies

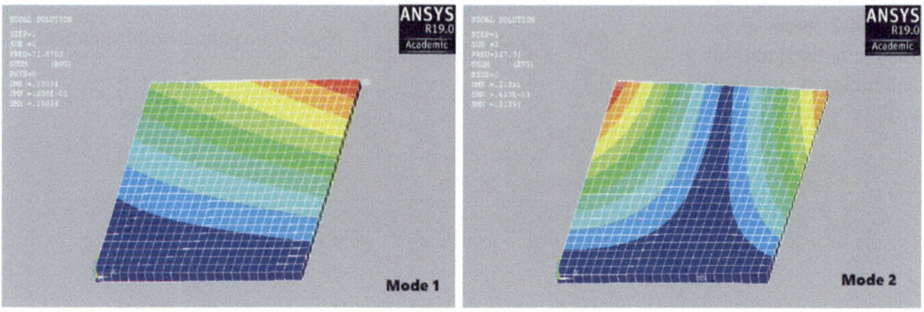

Fig. 54.3 Mode shapes (modes 1 and 2) of a skew hybrid cantilever plate with $\alpha = 15°$

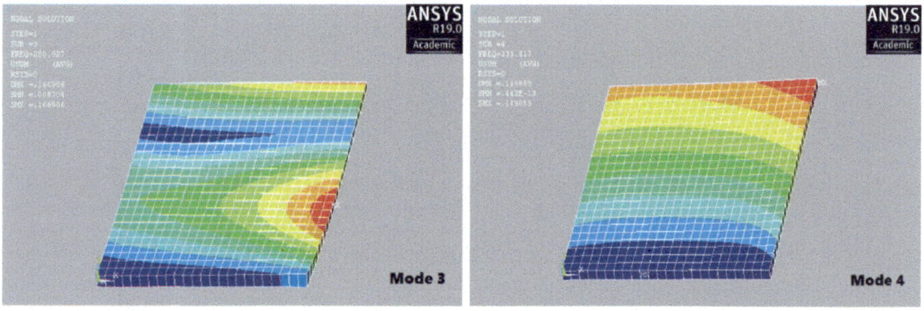

Fig. 54.4 Mode shapes (modes 3 and 4) of a skew hybrid cantilever plate with $\alpha = 15°$

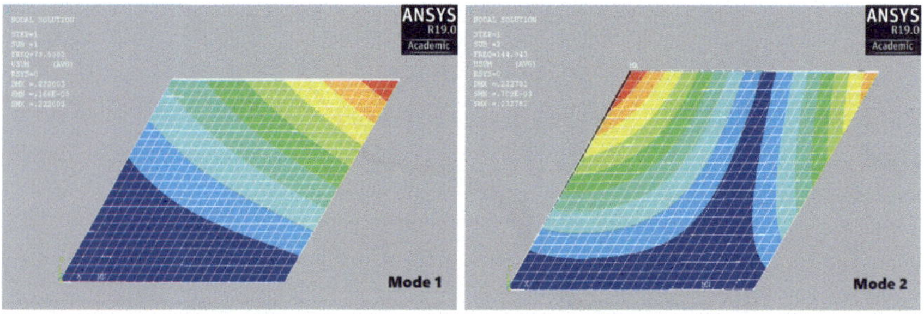

Fig. 54.5 Mode shapes (modes 1 and 2) of a skew hybrid cantilever plate with $\alpha = 30°$

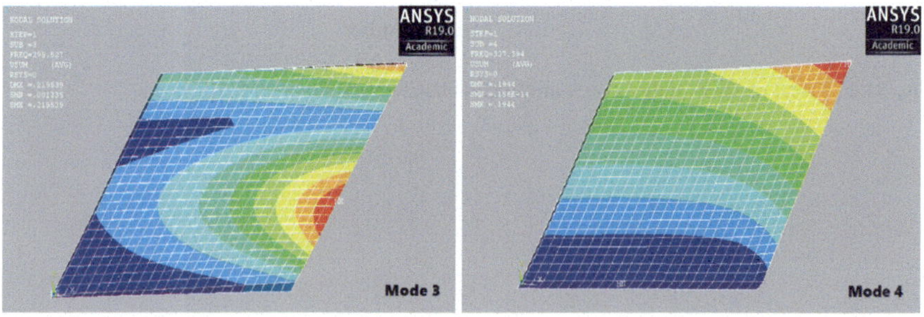

Fig. 54.6 Mode shapes (modes 3 and 4) of a skew hybrid cantilever plate with $\alpha = 30°$

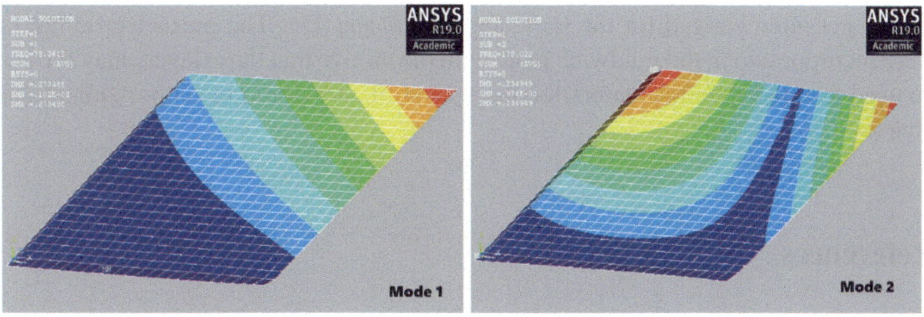

Fig. 54.7 Mode shapes (modes 1 and 2) of a skew hybrid cantilever plate with $\alpha = 45°$

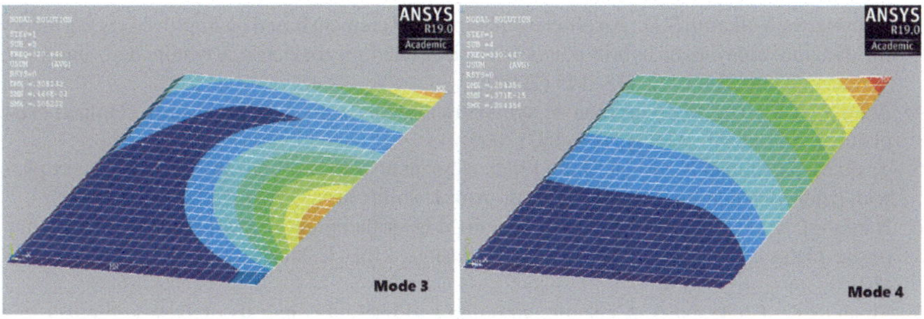

Fig. 54.8 Mode shapes (modes 3 and 4) of a skew hybrid cantilever plate with $\alpha = 45°$

54.4 Conclusions

The dynamic characteristics of a moderately thick skew hybrid sandwich cantilever plate with varying skew angles have been analysed. The hybrid plate is a five-layered sandwich laminate of orientation (0/90/core/90//0) with two cross-ply graphite–epoxy faces (0/90) of material 1, and a core of softer material 2. The top and bottom faces of the sandwich plate are bounded by a layer of lead zirconate titanate (PZT-5A), thus forming a skew hybrid plate. The thickness of each face sheet is 0.04 h, the thickness of piezoelectric layers is 0.1 h and that of the core is 0.64 h. The plate is modelled in ANSYS environment with a 24 × 24 mesh size using eight-nodded shell-99 element for the laminate faces and a solid-95 element for modelling the soft core. The study shows that the skew angle significantly affects the sandwich plate's dynamic character. The first six natural frequencies of the hybrid sandwich plate were obtained for different skew angles. It was observed that an increase in the skew angle of the plate increases the natural frequency; this is due to the fact that an increase in the skew angle increases the stiffness of the plate. The natural frequency is greater for 45° as compared to 15° and 30° skew angles. It has further been observed that the non-dimensionalised natural frequencies are approximately same in the fourth flexural mode for all skew angles and the third and fourth flexural modes get interchanged for the skew angles 30° and 45°. The piezoelectric layers were incorporated in the sandwich plate with the motive that these layers may further be used as sensors and actuators for shape control and vibration suppression of the plate.

References

1. Lok, T.S., Cheng, Q.H.: Free vibration of clamped orthotropic sandwich panel. J. Sound Vib. **229**(2), 311–327 (2000)
2. Wang, C.M., Ang, K.K., Yang, L., Watanabe, E.: Free vibration of skew sandwich plates with laminated facings. J. Sound and Vib. **235**(2), 317–340 (2000)
3. Kapuria, S., Kulkarni, S.D.: An efficient quadrilateral element based on improved zigzag theory for dynamic analysis of hybrid plates with electroded piezoelectric actuators and sensors. J. Sound and Vib. **315**(1–2), 118–145 (2008)
4. Ashour, A.S.: The free vibration of symmetrically angle-ply laminated fully clamped skew plates. J. Sound Vib. **323**(1), 444–450 (2009)
5. Yasin M.Y., Ahmad N., Alam M.N.: Finite element analysis of actively controlled smart plate with patched actuators and sensors. Latin Am. J. Solids Struct. **7**(3), 227–247 (2010)
6. Kanasogi, R.M., Ray, M.C.: Active constrained layer damping of smart skew laminated composite plates using 1–3 piezoelectric composites. Int. J. Mech. Maters. Des. **8**(3), 1–17 (2013)
7. Singh, S.K., Chakrabarti, A.: Static, vibration and buckling analysis of skew composite and sandwich plates under thermo mechanical loading. Int. J. Appl. Mech. Eng. **18**(3), 887–898 (2013)
8. Sayyad, S.S., Ghugal, Y.M.: On the free vibration analysis of laminated composite and sandwich plates, A review of recent literature with some numerical results. Compos. Struct. **129**, 177–201 (2015)
9. Asemi, K., Salami, S.J., Salehi, M., Sadighi, M.: Dynamic and static analysis of FGM skew plates with 3D elasticity based graded finite element modeling. Lat. Am. J. Solids Struct. **11**(3), 504–533 (2014)
10. Joseph, S.V., Mohanty, S.C.: Buckling and free vibration analysis of skew sandwich plates with viscoelastic core and functionally graded material constraining layer. In: Proc. IMechE. J. Aerosp. Eng. **233**(1), 369–389 (2017)
11. Mandal, A., Ray, C., Haldar, S.: Free vibration analysis of laminated composite skew plates with cut-out. Arch. Appl. Mech. **87**(9), 1511–1523 (2017)
12. Her, S.-C., Chen, H.-Y.: Stress analysis of sandwich composite beam induced by piezoelectric layer. J. Appl. Biomater. Funct. Mater. **16**(1S), 132–139 (2018)
13. Jin, Q., Mao, Z., Hu, X., Yao, W.: Thermo-mechanical analysis of multilayered composite beams based on a new mixed global-local model. J. Compos. Mater. **153**(28–30), 3963–3978 (2019)
14. Gupta, A., Ghosh, A.: Static and transient analysis of sandwich composite plates using isogeometric analysis. Mech. Adv. Mater. Struct. **26**(1), 81–87 (2019)

Chapter 55
Exergy Analysis of an Air Conditioning System Using Air-Cooled Condenser at Different Ambient Conditions with Different Volume Flow Rates of Air

Taliv Hussain, Adnan Hafiz and Akramuddin

Abstract In this paper, an analysis of exergy of an air conditioner having condenser of air-cooled type is presented. The prime objectives of this paper are to analyse all the four main components (condenser, expansion valve, compressor and evaporator) of vapour compression refrigeration system individually and then to quantify and identify the sites having the largest exergy losses. The exergy efficiency is also calculated at different ambient conditions, i.e. (30, 32, 34, 36 and 38 °C) with different volume flow rates of air, i.e. (0.08, 0.1, 0.12, 0.14 and 0.16 m^3s).The results showed that exergy loss is highest in the compressor. The order of exergy loss is (Exergy Destruction)$_{compressor}$ > (Exergy Destruction)$_{condenser}$ > (ExergyDestruction)$_{evaporator}$ > (ExergyDestruction)$_{expansionvalve.}$ In addition, the exergy efficiency of the system varies from 31.10 to 34.52%.

Keywords Exergy loss · Exergy efficiency and vapour compression refrigeration system

Nomenclature

Wc	Compressor work
Qc	Refrigerating effect
Mr	Refrigerant mass flow rate
h1	Refrigerantenthalpy at exit of evaporator or inlet of compressor in kJ/kg
h2	Refrigerant enthalpy at exit of compressor or inlet of condenser in kJ/kg
h3	Refrigerant enthalpy at exit of the condenser or inlet of expansion valve in kJ/kg
h4	Refrigerantenthalpy at exit of expansion valve or entry of evaporator in kJ/kg
S1	Refrigerantentropy at the exit of evaporator or inlet of compressor in kJ/kg-K

T. Hussain (✉) · A. Hafiz · Akramuddin
Department of Mechanical Engineering, Aligarh Muslim University, Aligarh, India 202002

S. Yadav et al. (eds.), *Proceedings of International Conference in Mechanical and Energy Technology*, Smart Innovation, Systems and Technologies 174,
https://doi.org/10.1007/978-981-15-2647-3_55

S2	Refrigerantentropy at exit of compressor or inlet of condenser in kJ/kg-K
S3	Refrigerant entropy at exit of condenser or inlet of expansion valve in kJ/kg-K
S4	Refrigerantentropy at exit of expansion valve or entry of evaporator in kJ/kg-K
Te	Evaporator temperature
Tc	Condenser temperature
To	Ambient temperature
ηexergy	Exergy efficiency

55.1 Introduction

The atmospheric temperature has increased so much because of the adverse effects of global warming. Thus, even in the rural areas, air conditioners and refrigerators have become an integral part of human life [1–3]. Human comfort is much important to the extent for the industrial and domestic areas are concerned and due to which air conditioning bears a heavy cost. To make a such system more efficient and environment-friendly, main difficulty is to utilise few energy and to have the minimum losses [4, 5]. To predict all the thermodynamic cycles and determination of irreversibility attached to all processes is very important [6, 7]. Thus, irreversibility associated with a process is qualitatively determined by exergy analysis. It helps the engineers and scientists to focus on the areas where the improvement can be done and the effectiveness of vapour compression refrigeration system can also be further improved [8]. The exergy losses can be determined by the second law-based analysis [9, 10]. Padilla et al. [11] performed exergy analysis of domestic VCRS with R413a and R12. They investigated that the performance in terms of irreversibility, power consumption and exergy efficiency of R413a is finer as that of R12, so R12 can be changed with R413a in domestic VCRS. The way to determine different losses as well as COP and exergetic efficiency of the cycle has been described by proper example. Yumrutas et al. [12] performed exergy analysis-based exploration of the effect of evaporator and condenser temperature on VCRS cycle in terms of COP, pressure losses, exergy losses and second law efficiency. He concluded that disparity in temperature of condenser has very less effect on exergy losses of expansion valve and compressor. Also, exergy efficiency and first law efficiency increase but total exergy losses of system decrease on increasing the operating temperature of condenser and evaporator. Park et al. [13] described through EEVs mass flow features of R410a and R22. They showed that with the increase in sub-cooling, EEV opening and inlet pressure, the mass flow rates passing through EEVs also increase. Nikolaidis and Probert [9] studied two-stage vapour compression refrigeration plant and determined the effect of change in temperatures of evaporator and condenser analytically. Venkataramanamurthy et al. [14] conducted an experimental test for vapour compression refrigeration system and analysed the energy flow and comparisons of the second law efficiency of R22 and its replacement R-436b.

In this paper, the exergy analysis of all the four main components (condenser, expansion valve, compressor and evaporator) of VCRS is individually analysed. The exergy loss and efficiency are calculated at different ambient conditions with different volume rates of air.

55.2 Experimental Set-up

Figure 55.1 represents the experimental set-up of an air conditioner. In this experimental set-up, R-410a is used as refrigerant. The chief components of an air conditioning system are compressor, condenser, throttling device and evaporator.

Formulae Used:

Exergy balance for the control volume can be described as:

$$\begin{aligned} EX &= \sum (mex)in - \sum (mex)\ out \\ &\quad + \left(\sum \left(Q(1 - T0/T)in - \sum (Q(1 - To/T)\ out\ \right] \pm \sum W \end{aligned}$$

Fig. 55.1 Air conditioning system

The component-wise exergy balance equation is as follows:

(1) Compressor:

$$EX_D\,comp = Ex1 + Wc - Ex2 = mr(To\,(s2 - s1))$$

(2) Condenser:

$$\begin{aligned}(EX_D)cond &= Ex2 - Qc(1 - T0/Tc) - Ex3\\ &= mr(h2 - To\,s2) - Qc(1 - T0/Tc) - mr(h3 - To\,s3)\end{aligned}$$

(3) Expansion device:

$$(EX_D)\,exp = Ex3 - Ex4 = mr(To(s2 - s1))$$

(4) Evaporator:

$$\begin{aligned}(EX_D)evap &= Ex4 + Qe(1 - To/Tr) - Ex1\\ &= mr(h4 - To\,s4) + Qe(1 - To/Tr) - mr(h1 - To\,s1)\end{aligned}$$

The total exergy destruction in the VCRS is the sum of exergy destruction in the four main components of the VCRS and is expressed as follows:

(5) Total exergy destruction:

$$(EXD)\,total = (EXD)\,comp + (EXD)\,cond + (EXD)\,exp + (EXD)\,evap$$

(6) Exergy effiency:

$$\boxed{\eta_{exergy} = \frac{Qe}{Wc}\left(1 - \frac{To}{Te}\right)}$$

55.3 Results

Tables (55.1, 55.2, 55.3, 55.4 and 55.5) show the exergy losses at different ambient temperatures with different volume flow rates of air.

Table 55.1 Exergy loss at constant ambient temp of 30 °C with different volume flow

Vol flow rate of air (m^3/s)	(EX_D) Comp KJ/kg	(EX_D) Cond KJ/Kg	(EX_D) evap KJ/kg	(EX_D) exp valve KJ/kg	(EX_D) tot (KJ/kg)	ηexegy (%)
0.08	189.4	164.2	120.7	81.7	556.18	31.10
0.1	158.3	135.7	123.1	83.4	500.76	32.05
0.12	151.7	132.1	127.4	86.9	499.30	32.94
0.14	151.7	132.3	127.4	86.9	498.48	33.06
0.16	147.4	129.9	129.2	86.9	493.53	33.43

Table 55.2 Exergy loss at constant ambient temp of 32 °C with different volume flow

Vol flow rate of air (m^3/s)	(EX_D) Comp KJ/kg	(EX_D) Cond KJ/Kg	(EX_D) evap KJ/kg	(EX_D) exp valve KJ/kg	(EX_D) tot (KJ/kg)	ηexegy (%)
0.08	165.0	140.4	125.4	82.2	513.16	31.47
0.1	161.7	140.7	130.4	87.5	520.24	32.57
0.12	158.1	139.4	132.9	89.33	520.24	32.43
0.14	154.4	137.8	135.5	91.13	518.97	33.43
0.16	153.2	139.7	139.7	91.13	527.11	33.79

Table 55.3 Exergy loss at constant ambient temp of 34 °C with different volume flow

Vol flow rate of air (m^3/s)	(EX_D) Comp KJ/kg	(EX_D) Cond KJ/Kg	(EX_D) evap KJ/kg	(EX_D) exp valve KJ/kg	(EX_D) tot (KJ/kg)	ηexegy (%)
0.08	169.3	138.7	104.14	67.26	479.51	32.31
0.1	169.3	138.7	105.86	68.98	482.94	32.32
0.12	171.2	142.5	104.46	67.2	485.42	33.81
0.14	170.6	143.3	110.15	72.43	496.51	33.90
0.16	167.3	141.2	109.19	70.7	488.45	34.11

Table 55.4 Exergy loss at constant ambient temp of 36 °C with different volume flow

Vol flow rate of air (m^3/s)	(EX_D) Comp KJ/kg	(EX_D) Cond KJ/Kg	(EX_D) evap KJ/kg	(EX_D) exp valve KJ/kg	(EX_D) tot (KJ/kg)	ηexegy (%)
0.08	128.6	104.2	108.0	64.21	405.22	32.33
0.1	125.5	110.1	113	64.14	412.71	33.18
0.12	125.9	114.8	120.7	69.4	430.91	34.16
0.14	125.3	109.2	121.7	74.8	431.00	34.28
0.16	123.3	109.7	126.8	80.4	440.21	34.52

Table 55.5 Exergy loss at constant ambient temp of 38 °C with different volume flow

Vol flow rate of air (m^3/s)	(EX_D) Comp KJ/kg	(EX_D) Cond KJ/Kg	(EX_D) evap KJ/kg	(EX_D) exp valve KJ/kg	(EX_D) tot (KJ/kg)	ηexegy (%)
0.08	129.4	112.8	135.6	91.2	469.13	32.32
0.1	124.1	110.4	141.74	94.8	471.11	32.70
0.12	123.5	111.1	139.2	91.9	464.91	33.67
0.14	122.9	110.7	136.7	87.4	457.71	33.55
0.16	121.7	111.1	135.1	83.8	451.61	34.16

55.4 Graphs and Discussion

From Figs. 55.2 and 55.3, it can be seen that exergy loss of compressor is maximum among all other components but it decreases with increase in volume flow rate while increases with increase in ambient temperature. But among all the parts of air conditioning system, exergy loss is minimum in expansion valve.

From Fig. 55.4, it can be found that total exergy loss increases continuously with increment in ambient temperature. As with increase in the ambient temperature, COP of system decreases.

From Fig. 55.5, it conveys that total exergy loss increases with volume flow rate of refrigerant but it obtains an optimum value of nearly 0.1 m^3/s.

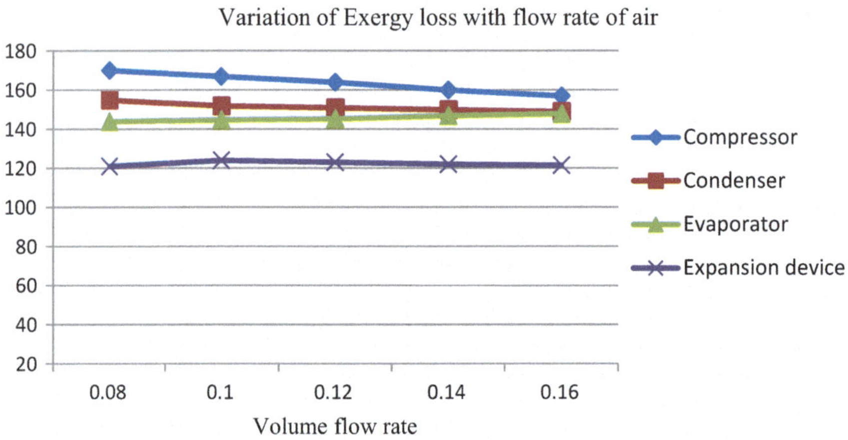

Fig. 55.2 Shows variation of exergy loss with flow rate of air

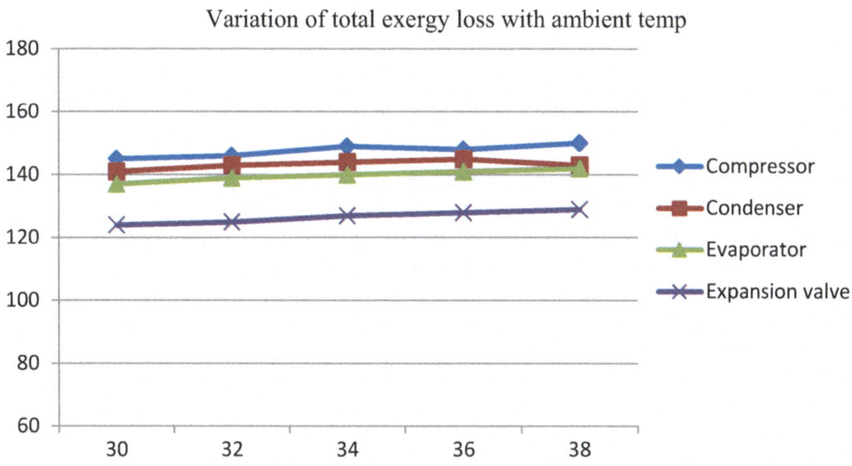

Fig. 55.3 Shows variation of exergy loss at different temperature

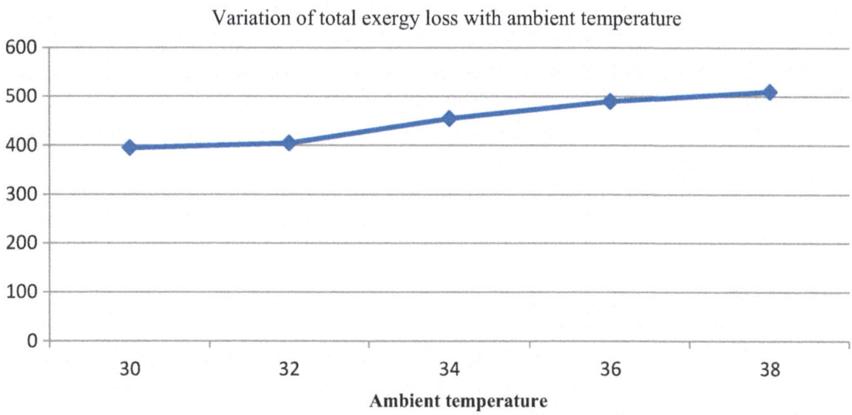

Fig. 55.4 Shows variation of total exergy loss with ambient temperature

55.5 Conclusions

Exergy analysis is a method to describe the process and this further helps in reducing the thermodynamic losses happening in these processes. It is a salient feature in explaining the different energy flows in the process and in the final run helps to reduce losses happening in the main system. The following results are drawn:

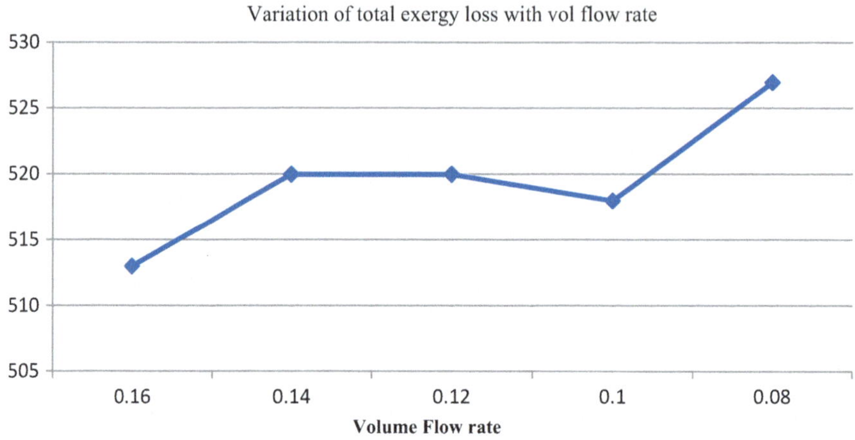

Fig. 55.5 Variation of total exergy loss with volume flow rate

- However, the volume flow rate and ambient temperature do affect the exergy losses for all components, but highest losses occur in compressor for all the cases. The order of exergy loss is $(EX_D)_{compressor} > (EX_D)_{condenser} > (EX_D)evaporator > (EX_D)expansion$ valve.
- The exergy efficiency of the VCRS varies from 31.10 to 34.52% at different ambient conditions for different volume rates of air.

References

1. Ansari, A.A., Goyal, V., Yahya, S.M., Hussain, T.: Experimental investigation for performance enhancement of a vapor compression refrigeration system by employing several types of water-cooled condenser. Sci. Technol. Built Environ. **24**(7), 793–802 (2018)
2. Hughes, B.R., Chaudhry, H.N., Ghani, S.A.: A review of sustainable cooling technologies in buildings. Renew. Sustain. Energy Rev. **15**(6), 3112–3120 (2011)
3. Duan, Z., Zhan, C., Zhang, X., Mustafa, M., Zhao, X., Alimohammadisagvand, B., Hasan, A.: Indirect evaporative cooling: past, present and future potentials. Renew. Sustain. Energy Rev. **16**(9), 6823–6850 (2012)
4. Caliskan, H., Dincer, I., Hepbasli, A.: Exergoeconomic, enviroeconomic and sustainability analyses of a novel air cooler. Energy Build. **55**, 747–756 (2012)
5. Costelloe, B., Finn, D.: Thermal effectiveness characteristics of low approach indirect evaporative cooling systems in buildings. Energy Build. **39**(12), 1235–1243 (2007)
6. Maheshwari, G.P., Al-Ragom, F., Suri, R.K.: Energy-saving potential of an indirect evaporative cooler. Appl. Energy **69**(1), 69–76 (2001)
7. Rosen, M.A., Dincer, I., Kanoglu, M.: Role of exergy in increasing efficiency and sustainability and reducing environmental impact. Energy policy **36**(1), 128–137 (2008)
8. Arora, A., Kaushik, S.C.: Theoretical analysis of a vapour compression refrigeration system with R502, R404A and R507A. Int. J. Refrig **31**(6), 998–1005 (2008)
9. Nikolaidis, C., Probert, D.: Exergy-method analysis of a two-stage vapour-compression refrigeration-plants performance. Appl. Energy **60**(4), 241–256 (1998)

10. Mairaj, M., Siddiqui, S.A., Hafiz, A.: Energetic and exergetic analysis of some models of vapour absorption chillers using lithium bromide and water. J. Basic Appl. Eng. Res. **2**(4), 326–329 (2015)
11. Padilla, M., Revellin, R., Bonjour, J.: Exergy analysis of R413A as replacement of R12 in a domestic refrigeration system. Energy Convers. Manag. **51**(11), 2195–2201 (2010)
12. Yumrutaş, R., Kunduz, M., Kanoğlu, M.: Exergy analysis of vapor compression refrigeration systems exergy. Int J **2**(4), 266–272 (2002)
13. Park, C., Cho, H., Lee, Y., Kim, Y.: Mass flow characteristics and empirical modeling of R22 and R410A flowing through electronic expansion valves. Int. J. Refrig. **30**(8) 1401–1407 (2007)
14. Venkataramanamurthy, V.P., Kumar Senthil, P.: Experimental Comparative energy, exergy flow and second law efficiency analysis of R22, R436b vapour compression refrigeration cycles. Int. J. Sci. Technol. **2**(5), 1399–1412 (2010)

Chapter 56
Modelling and Simulation of MEMS Graphene Pressure Sensor for HealthCare Devices

Meetu Nag, Monica Lamba, Kulwant Singh and Ajay Kumar

Abstract In this paper, sensitive piezoresistive graphene pressure sensor is designed and analysed for applications in field of biomedical for healthcare devices. The relationship between applied pressure and output voltage was studied by finite element modelling and simulation method in COMSOL. The results show that sensitivity of 3.89 mV/psi is achieved on application of 0–100 psi pressure. The effect of frequency variation on sensitivity is also investigated for this pressure sensor. Nonlinearity is calculated as a function of applied pressure and obtained output voltage. Linear behaviour is noticed between 40 and 90 psi.

Keywords Graphene · Biosensor · Piezoresistive · Healthcare devices · Modelling and simulation · COMSOL

56.1 Introduction

Graphene is said to be one of the most compatible material for various applications such as measuring blood pressure, sensing respiratory flow and tactile sensing [1]. Graphene due to its unique properties such as strength, flexibility, transparency and biocompatibility makes it a promising material for fabrication of pressure sensor for healthcare devices [2]. Pressure sensing technology also plays a vital role in improving the performance of the pressure sensor in terms of sensitivity and linearity. Different types of sensing technology are utilized, such as piezoresistive, piezoelectric and capacitive. Piezoresistive technology shows good results for various dynamic systems [3]. In biomedical field, dynamic movement of the human body such as wrist pulse movement, tactile movements (finger bending, knee bending), sense of respiratory flow can be measured by graphene pressure sensor [4]. Chun et al. designed a graphene pressure sensor for tactile movements and obtained sensitivity of 0.24 KPa^{-1} at low pressure and 0.039 KPa^{-1} at high pressure [5]. Wang

M. Nag (✉) · A. Kumar
Department of Mechatronics Engineering, Manipal University Jaipur, Jaipur, India

M. Lamba · K. Singh
Department of Electronics and Communication Engineering, Manipal University Jaipur, Jaipur, India

S. Yadav et al. (eds.), *Proceedings of International Conference in Mechanical and Energy Technology*, Smart Innovation, Systems and Technologies 174,
https://doi.org/10.1007/978-981-15-2647-3_56

et al. monitored eye blinking and pulse measurement in real-time system by using graphene and ultrathin medical tape [6]. From the literature review, it can be revealed that graphene is used as a suitable material for pressure sensor in healthcare devices. In this paper, piezoresistive sensing technology is used for the measurement of the pressure. Graphene is used as a sensing material in wheatstone bridge configuration which converts physical quantity into electrical signal [3]. On the application of applied pressure, the sensing element shows the change in resistance which is converted into output voltage with the help of bridge configuration. The dimension of the sensing element and diaphragm is selected by simulating the model on different parametric sweeps. The optimum combination of piezoresistors such as length, breadth and thickness and diaphragm thickness is taken for maximum sensitivity.

56.2 Modelling and Finite Element Analysis

Major functional part of the pressure sensor is the diaphragm. The dimensions of the diaphragm play an important role in defining the correct stress generated area [7]. COMSOL Multiphysics is used to simulate the model of the graphene pressure sensor. Potential of 3.3 V is taken at one terminal of the bridge configuration and other terminal is grounded. Proper insulation is provided between the different work planes during designing the geometry of the pressure sensor. The dimension of the device is taken as 4 mm $\times$4 mm, silicon is taken as a substrate material, and combination of gold and tin is used for making contact points for measuring the output voltage. For modelling, different property values are taken as shown in Table 56.1[3].

The sensitivity of the pressure sensor is the ratio of obtained output potential due to change in resistance of the piezoresistors. Figure 56.1 shows the top and bottom view of the graphene pressure sensor. Due to the application of applied pressure, deformation in diaphragm is noticed which generates the stress on the piezoresistors. The placement of each piezoresistor is taken such that it covers the tensile and compressed stress regions. Therefore, two piezoresistors are placed at the edges of the fixed diaphragm and other two are placed nearer to the centre which is having less stress region [8]. The resistors are connected in wheatstone bridge configuration to measure the change in resistance in the form of output voltage. The expression of output voltage is given by Eq. 56.1 where $R1$, $R2$, $R3$, $R4$ are the resistance of the piezoresistors.

$$V_{\text{out}} = \left(\frac{R2}{R1 + R2} - \frac{R4}{R3 + R4} \right) V_{\text{in}} \tag{56.1}$$

On the application of applied pressure of range between 0 and 100 psi, the variation in generated stress is obtained. Figure 56.2 shows the generated stress on the sensor. From the figure, it can be shown that major variation is at the edges of the diaphragm. The amount of displacement is based on the natural frequency of the system. In this paper, eigen frequency is also plotted on the application of applied pressure.

Table 56.1 Material properties of silicon and gold as used in COMSOL

Material	Properties	Value	Unit
Graphene (piezoresistor)	Young's modulus	1.0	T Pa
	Poisson's ratio	0.17	Unit less
	Electrical conductivity	102	S/m
	Thermal conductivity	5000	Wm^{-1} K^{-1}
	Density	2000	kg m^{-3}
	Relative permittivity	2.14	Unit less
Silicon (substrate material)	Young's modulus	170×10^9	Pa
	Poisson's ratio	0.28	Unit less
	Heat capacity at constant pressure	700	J $Kg^{-1}.K^{-1}$
	Thermal conductivity	130	Wm^{-1} K^{-1}
	Density	2329	kg m^{-3}
	Coefficient of thermal expansion	2.6×10^{-6}	K^{-1}
Gold/tin (contact pads)	Relative permittivity	1	Unit less
	Electrical conductivity (σ)	4.5×10^6	S/m

(a) (b)

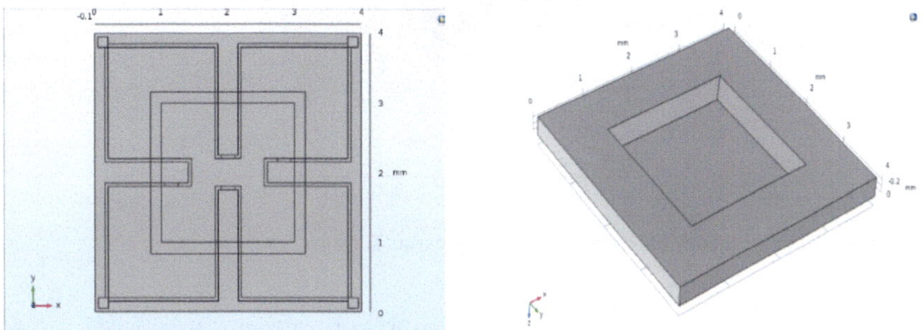

Fig. 56.1 Model of graphene pressure sensor. **a** Top view of the MEMS graphene pressure sensor. **b** Bottom part of the MEMS graphene pressure sensor

Eigen frequency is also called as natural frequency of the system [9]. Here, using graphene as a sensing material increases the natural frequency of the system which in turn increases the deformation in the system. Deformation in the diaphragm on small amount of applied pressure increases the sensitivity of the pressure sensor. Figure 56.3 shows the eigen frequency diagram of the graphene pressure sensor.

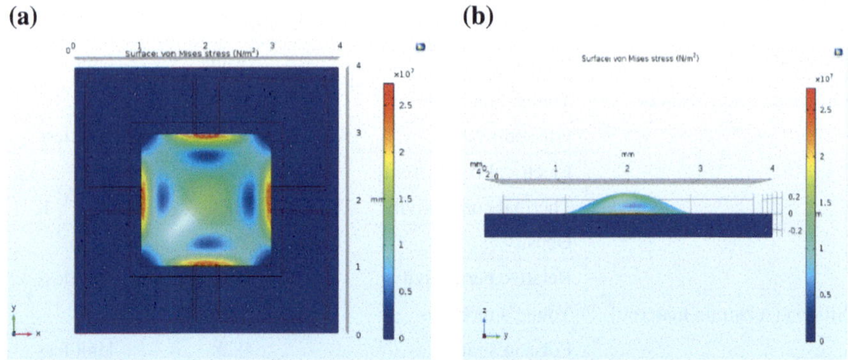

Fig. 56.2 Von Mises stress diagram of the MEMS graphene pressure sensor

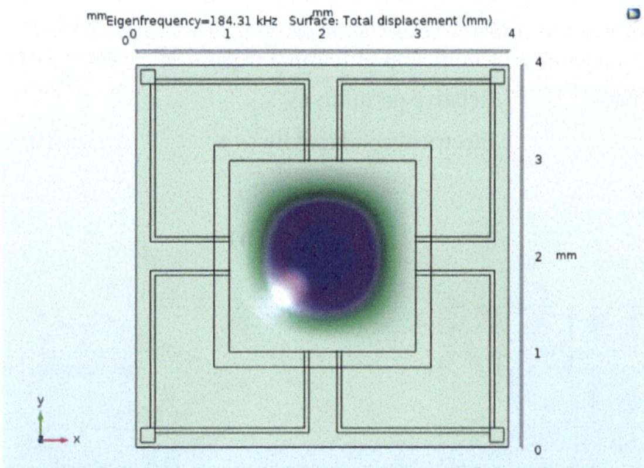

Fig. 56.3 Eigen frequency diagram of graphene pressure sensor

The value of eigen frequency varies in different modes. Ideally, six eigen frequency are calculated by the simulation which gives the idea of maximum displacement [10]. Figure 56.4 shows the six eigen frequency having different values ranges from 184.31 to 675 kHz.

56.3 Result and Discussion

Graph between output voltage and applied pressure is plotted in Fig. 56.5. Sensitivity of 3.89 mV/psi is obtained from graphene pressure sensor. It is noticed that on increasing the applied pressure the value of output voltage increases. Figure 56.6 shows graph between nonlinearity versus applied pressure. Linearity is noticed from 40 to 90 psi. From this paper, the designed pressure sensor shows linear variation at higher

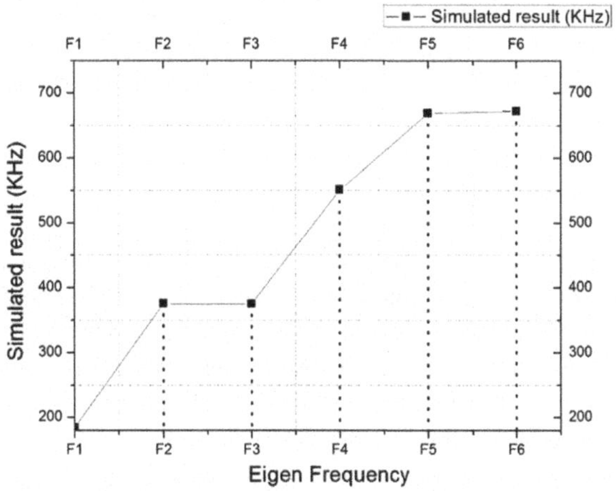

Fig. 56.4 Eigen frequency analysis of the designed model

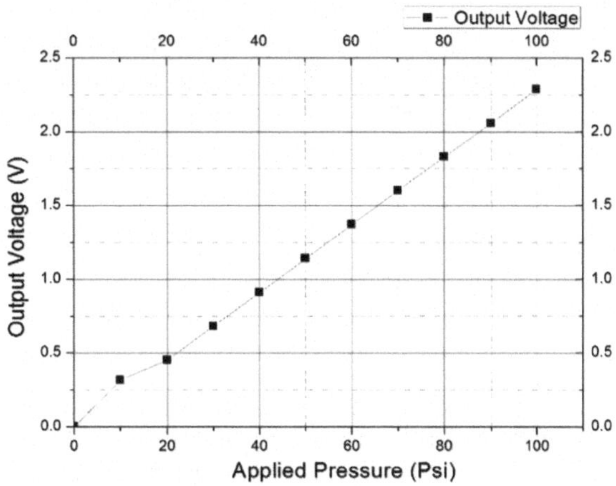

Fig. 56.5 Relation between output voltage and applied pressure

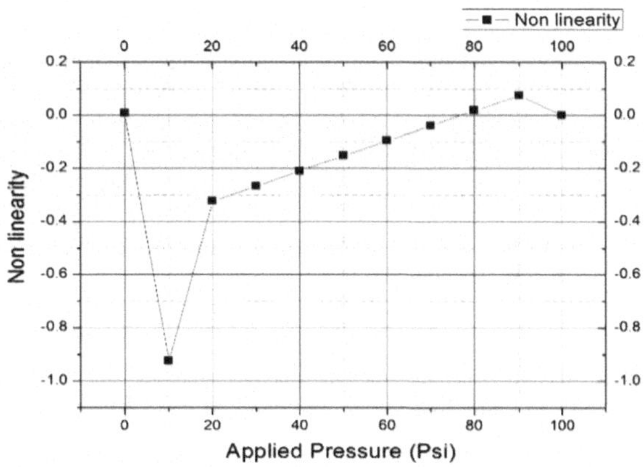

Fig. 56.6 Relation between applied pressure and nonlinearity

temperature that shows the temperature compatibility of graphene pressure sensor at higher temperature with maximum sensitivity.

56.4 Conclusion

From the simulation results, it is noticed that there is a linear relation between applied pressure and generated output voltage due to generated strain. Maximum sensitivity of 3.89 mV/psi is obtained for graphene pressure sensor which is a desirable factor for utilizing graphene pressure sensor in healthcare devices. Placement of piezo resistor in tensile and compressive stress regions is also one of the important factors in increasing the sensitivity of the graphene pressure sensor. Dimensions of the diaphragm are also chosen on the basic of low nonlinearity in the output. However, considering the fabrication of the graphene pressure sensor, synthesis of graphene also improves the characteristic of the device. By taking CVD monolayer graphene and proper transfer of graphene also effects the sensitivity of the pressure sensor. Even the packaging also induces nonlinearity in the system. If each factor is analysed thoroughly, then it helps to reduce the nonlinearity in the output of the pressure sensor.

References

1. Amoli, V., Kim, J.S., Kim, S.Y., Koo, J., Chung, Y.S., Choi, H., Kim, D.H.: Ionic tactile sensors for emerging human-interactive technologies: a review of recent progress. Adv. Funct. Mater (2019)
2. He, J., et al.: A universal high accuracy wearable pulse monitoring system via high sensitivity and large linearity graphene pressure sensor. Nano Energy **59**, 422–433 (2019)
3. Nag, M., Singh, J., Kumar, A., Kulwant, P.A.A.: Sensitivity enhancement and temperature compatibility of graphene piezoresistive MEMS pressure sensor. Microsyst. Technol. **5**, 2019 (2011)
4. Yang, H., Xue, T., Li, F., Liu, W., Song, Y.: Graphene: diversified flexible 2D material for wearable vital signs monitoring. Adv. Mater. Technol. **4**(2), 1–20 (2019)
5. Chun, S., Kim, Y., Oh, H.S., Bae, G., Park, W.: A highly sensitive pressure sensor using a double-layered graphene structure for tactile sensing. Nanoscale **7**(27), 11652–11659 (2015)
6. Wang, L., Loh, K.J., Chiang, W.H., Manna, K.: Micro-patterned graphene-based sensing skin for human physiological monitoring, 1–21 (2018)
7. Suja, K.J., Kumar, G.S., Nisanth, A., Komaragiri, R.: Dimension and doping concentration based noise and performance optimization of a piezoresistive MEMS pressure sensor. Microsyst. Technol. **21**(4), 831–839 (2015)
8. Singh, K., Joyce, R., Varghese, S., Akhtar, J.: Fabrication of electron beam physical vapor deposited polysilicon piezoresistive MEMS pressure sensor. Sens. Actuators A Phys (2015)
9. Yusof, N., Bais, B., Majlis, B.Y., Soin, N.: Mechanical analysis of MEMS diaphragm for bladder pressure monitoring. In: Proc. 2017 IEEE Reg. Symp. Micro Nanoelectron, RSM 2017, pp. 22–25 (2017)
10. Ralib, A.A.M., Zamri, N.N.B.N.M., Hazadi, A.H.F., Rahim, R.A., Za'bah, N.F., Saidin, N.: Investigation on the mass sensitivity of quartz crystal microbalance gas sensor using finite element simulation. Bull. Electr. Eng. Inf. **8**(2), 460–469 (2019)

Chapter 57
Preparation and Properties of Novel Sulfonated Pentablock Copolymer (sPBC) Membrane for PEM Fuel Cell

M. Yusuf Ansari, S. J. A. Rizvi and Inamuddin

Abstract Fuel cell is considered as a promising alternative energy device to convert chemical energy into electrical energy. Proton exchange membrane fuel cell (PEMFC) has special attention, in all types of fuel cell, due to its high power density, low temperature start-up, portable and cleanliness. Nafion (perfluorosulfonic acid) is successfully used as membrane for PEMFC but because of its limitations other alternative materials are also explored. In aromatic hydrocarbon, especially block copolymer, has been very attractive for membrane material due to its implicit properties in the field of ionic transportation, water retention and repulsion. In this work, sulfonated pentablock copolymer (sPBC, Nexar^{TM}) has been worked out as an alternative to Nafion for PEMFC membrane. The sPBC membrane was developed by casting method. Material properties were analyzed with the help of FTIR, XRD and TGA. Membrane characterization test for fuel cell such as water uptake, ion exchange capacity (IEC) and proton conductivity was carried out and then compared with Nafion. Proton conductivity at different temperatures was also evaluated, and it was found that Nexar has higher conductivity than Nafion. Conductivity–temperature relationship is used to calculate activation energy, and it was found that Nexar has lower activation energy than Nafion. Transport model for proton was also predicted using activation energy.

Keywords PEM fuel cell · Nexar · Nafion

57.1 Introduction

Fuel cell is an electrochemical device used to convert chemical energy into electrical energy without any mechanical route. There are various categories of fuel cell depending upon the nature of fuel, type of ion movement, type of electrolyte. Among them, proton exchange membrane fuel cell (PEMFC) has special attention due to its

M. Y. Ansari · S. J. A. Rizvi (✉)
Department of Petroleum Studies, ZHCET, AMU, Aligarh, India

Inamuddin
Department of Applied Chemistry, ZHCET, AMU, Aligarh, India

S. Yadav et al. (eds.), *Proceedings of International Conference in Mechanical and Energy Technology*, Smart Innovation, Systems and Technologies 174,
https://doi.org/10.1007/978-981-15-2647-3_57

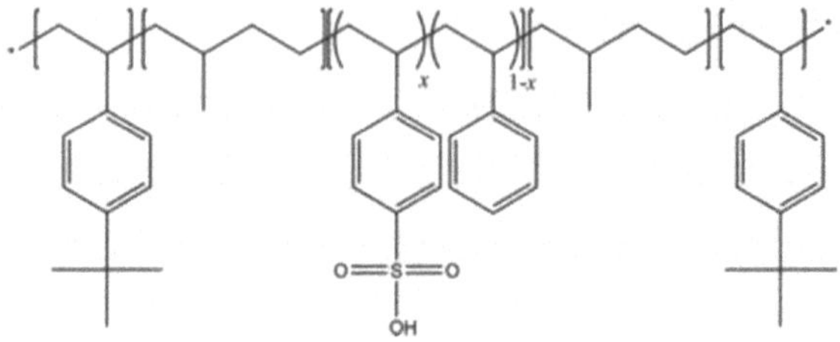

Fig. 57.1 Structure of Nexar

high power density, easy start-up, portable and cleanliness. Membrane used in PEM fuel cell is considered as the heart of cell and performance depends upon it [1–4].

The most widely used material for membrane is Nafion because of its high conductivity, excellent chemical and mechanical stability [5–8]. But due to its high cost and low performance at high temperature and methanol crossover in direct methanol fuel cell (DMFC), other alternative materials are also explored [6–10]. Aromatic hydrocarbon, having low cast and showing strong chemical and mechanical stability, is considered as good alternative for membrane material in fuel cell. In aromatic hydrocarbon, particularly block copolymer has been very attractive for membrane material due to its implicit properties [11–15]. Block polymers show interesting morphologies because of its self-assembled nanostructures. Due to these merits, there is continuous growth in using block copolymers for proton exchange membrane over the last decades [11, 16–30].

The modified and advanced copolymer similar to styrene–ethylene–butylene–styrene (SEBS) block copolymer was commercially developed by Kraton Polymer LLC, USA, with trade name of Nexar™. It is a sulfonated styrene pentablock copolymer as shown in Fig. 57.1. The polymer possesses good mechanical strength and flexibility because it consists of tert-butyl styrene (tBS) end blocks and ethylene-co-propylene (EP) blocks. The middle styrene block of the pentablock copolymer can be sulfonated with a desired ion exchange capacity (IEC) [31]. Membrane from Nexar™ polymers is used in many applications successfully, such as in alcohol pervaporation [31, 32], in desalination [33], for IPMC actuator [34–36], in oil–water separation [37] and for water purification [38]. In this work, Nexar™ MD9200 grade is used to develop membrane for proton exchange membrane fuel cell.

57.2 Experimental

57.2.1 Chemicals

Nexar™ MD9200 solution, a propriety chemical, was obtained by Kraton Polymer LLC, Houston, Texas, USA; film of Nafion 117, 0.180 mm thick supplied by Alfa Aesar; tetrahydrofuran (THF) was supplied by CDH. All the chemicals were used as received.

57.2.2 *Membrane Preparation*

The Nexar™ solution was diluted with a fixed amount of THF at room temperature with continuous stirring. A homogenous solution is obtained after two hours. This solution is cast in petri dish at room temperature. Slow evaporation rate is required to avoid cracks in membrane.

57.3 Characterization of Membrane

57.3.1 *Characterization of Martial*

Characterization of Nexar sample was carried out to explore its feasibility for proton exchange membrane. FTIR spectrum, XRD analysis and TGA were carried out.

57.3.2 *Water Uptake and Swelling Ratio*

The known weight of dry membrane sample was kept in deionised water for overnight for water absorption; then, this membrane was taken out, soaked with blotting papers and then weighted. The percentage of water uptake is calculated as follows:

$$\text{Water uptake (WU, \%)} = \frac{W_{\text{Wet}} - W_{\text{Dry}}}{W_{\text{Dry}}} \times 100 \tag{57.1}$$

where W_{Wet} = weight of wet membrane; W_{Dry} = weight of dry membrane

Swelling ratio of membrane is the percentage change in the length of wet membrane to dry membrane. It can be calculated as follows;

$$\text{Swelling Ratio, \%} = \frac{L_W - L_d}{L_d} \times 100 \tag{57.2}$$

where L_w and L_d were the lengths of wet and dry membrane, respectively.

57.3.3 *Ion Exchange Capacity (IEC)*

Titration method is used to determine IEC of membrane. The sample was dipped into NaCl (1 M) solution for 24 h and then titrated with 0.1 M NaOH using phenolphthalein as indicator. The IEC value was obtained using following formula:

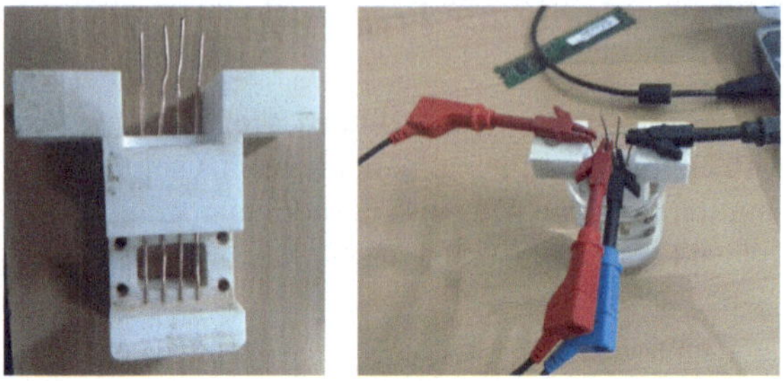

Fig. 57.2 Probe conductivity cell

$$\mathrm{IEC}\left(\frac{\mathrm{meq}}{g}\right) = \frac{V_{\mathrm{NaOH}} \times M_{\mathrm{NaOH}}}{W_{\mathrm{Dry}}} \quad (57.3)$$

where

$V_{\mathbf{NaOH}}$ is NaOH volume consumed in titration, ml; $M_{\mathbf{NaOH}}$ molarity of NaOH.

$W_{\mathbf{Dry}}$ is the dry weight of membrane sample being dipped in NaCl solution in gm.

57.3.4 Ionic Conductivity

Ionic conductivity is measured by Autolab potentiostat PGSTAT30. A four-probe ionic conductivity measurement system was indigenously built (Fig. 57.2). Membrane with dimension of 30 × 10 mm was used to measure proton conductivity under fully hydrated condition. FRA analysis was carried out in 1–100 kHz range. The ionic conductivity, σ(S/cm), is calculated by the equation

$$\text{Conductivity}, \sigma \ (\mathrm{s/cm}) = l/(R \times A) \quad (57.4)$$

where

l is the distance between electrodes (cm); **A** is the cross-sectional area of membrane (cm^2); **R** is the membrane resistance value from the AC impedance data (Ω).

57.4 Results and Discussion

57.4.1 Characterization of Material

XRD is a very useful tool to examine the crystallinity of the membrane and hence their symmetry and regularity. The XRD graph for Nexar^{TM} membrane sample is shown in Fig. 57.3. Because of very high noise-to-signal ratio, the diffracted peaks were observed which proposed an amorphous membrane sample.

The FTIR spectra of the Nexar™ are shown in Fig. 57.4. The band at 3414 cm^{-1} shows O–H stretching. Bands at around ~1411 and ~1223 cm^{-1} are due to the presence of S = O stretching vibration which confirms the existence of block styrene sulfonate unit of Nexar polymer. The sharp bands around 1600 and 1465 cm^{-1} can be described due to the C = C stretching vibration of Nexar pentablock copolymer.

The TGA curve for Nexar membrane is shown in Fig. 57.5. The TGA curve of the Nexar showed weight loss up to 350 °C is about 22%, which represents the presence of moisture. A sharp weight loss from 22 to 85% at around 400 °C may be due to the thermal degradation of sulfonic acid groups. At 480 °C onward, almost smooth horizontal section is shown, which represents the complete formation of the oxide form of the membranes.

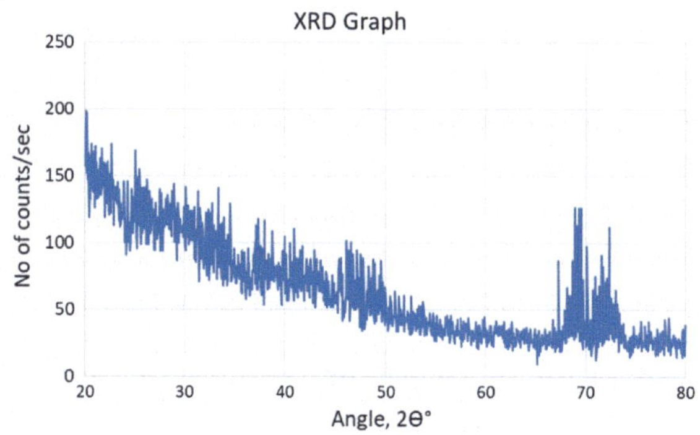

Fig. 57.3 XRD of Nexar

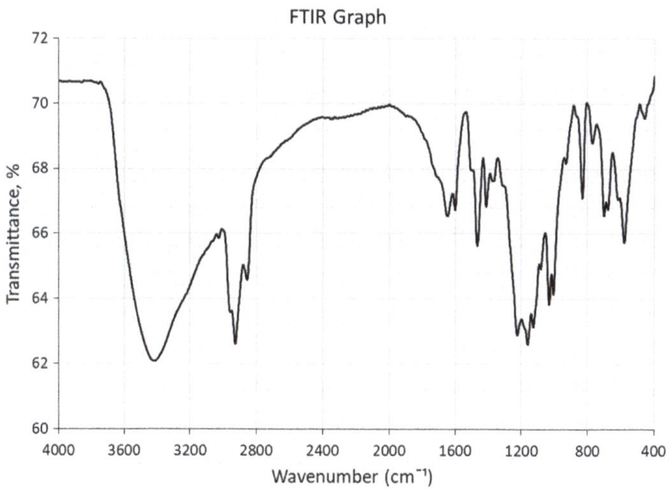

Fig. 57.4 FTIR of Nexar

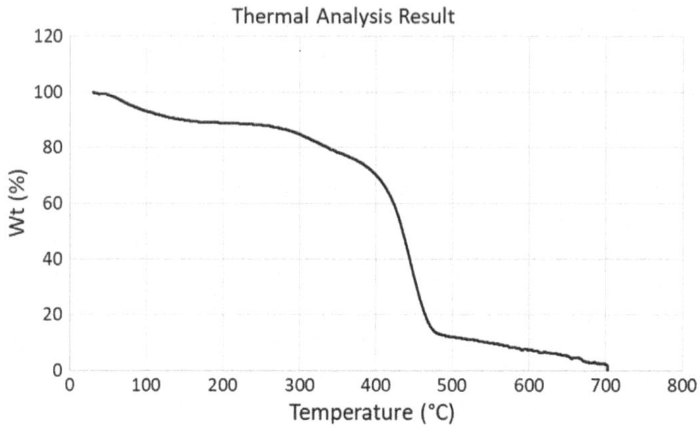

Fig. 57.5 TGA of Nexar

57.4.2 *Water Uptake and Swelling Ratio*

Water uptake is an indication of water-holding power of membrane. It helps to transport ions across membrane. Higher value of water is required but too high value of water uptake could cause mechanically weaker membrane and that is not desirable. The value of water uptake for Nexar is found as 125%. This value is higher as compared to Nafion, 15.5%. This high value is due to more number of sulfonic acid groups and higher value of IEC [39].

Swelling ratio of Nexar was observed as 26.6% and Nafion was 10.0%. The swelling ratio of Nexar is not as high as water uptake in comparison with Nafion, and this is due to hydrophobic block present in Nexar.

57.4.3 *Ion Exchange Capacity (IEC)*

IEC indicates the moles of $SO3^-$ present per gm of dry membrane. Higher value of IEC is always recommended. The IEC value of Nexar is found as 2.01 meq/gm, whereas Nafion is showing 0.91 meq/gm. Nafion is having almost half the value of IEC Nexar.

57.4.4 *Conductivity and Activation Energy*

Proton conductivity is the measure of membrane performance. Electricity generated by fuel cell is directly related to proton conductivity. The proton conductivity of Nexar and Nafion membranes is shown in Fig. 57.6 for different temperature under

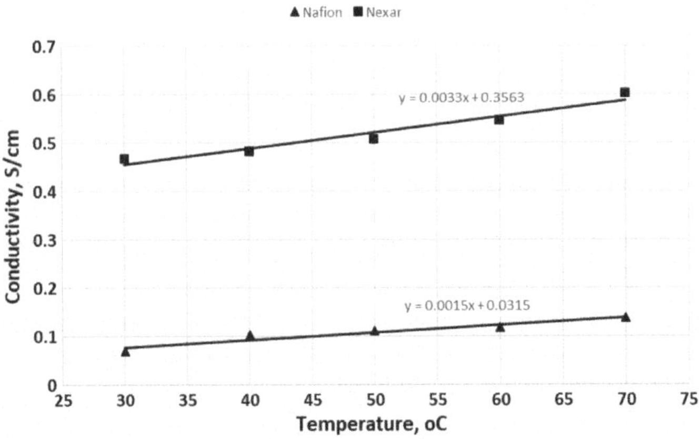

Fig. 57.6 Conductivity comparison of Nexar and Nafion

100% humidity. Conductivity of Nexar is greater than Nafion as shown in Fig. 57.6. The increase in conductivity of Nexar is due to more number of sulfonic groups or high IEC as compared to Nafion 117. From Fig. 57.6, it is clear that conductivity of Nexar is more sensitive than Nafion with the change in temperature.

Conductivity–temperature relationship can be used to calculate activation energy which is minimum energy required for proton transport. It can be calculated using following equation:

$$\sigma = \sigma_o e^{-\left(\frac{E_a}{k_B T}\right)} \tag{57.3}$$

where σ_o = pre-exponential factor; k_B = Boltzmann constant; T = temperature, K.

The Arrhenius plot is used to calculate activation energy, Ea, value. Nexar requires lower activation energy than Nafion, which means that Nexar requires less effort to transport proton than Nafion. Transportation of proton in the membrane can be explained in two ways, vehicle (diffusion) mechanism and Grotthuss (hopping) mechanism. The obtained results for Nexar and Nafion indicate that Grotthuss mechanism supports the transportation of proton, and it is more dominated in Nexar.

57.5 Conclusion

The comparative analysis between Nexar and Nafion for different parameters was analyzed. High value of IEC leading to higher value of conductivity for Nexar shows promising alternative material for PEM fuel cell, but high value of water uptake puts negative impression that causes more swelling and weaker in terms of mechanical strength. However, activation energy supports Nexar for having lower energy requirement than Nafion.

References

1. Vernekar, S.P.: Polymer electrolyte for fuel cells-an overview what is a fuel cell ?
2. Zaidi, S.M.J.: Research trends in polymer electrolyte membranes for PEMFC. In: Javaid, S.M., Matsuura, T. (eds.) Polymer Membranes for Fuel Cells, Z. Springer, Boston, MA (2009)
3. Peighambardoust, S.J., Rowshanzamir, S., Amjadi, M.: Review of the proton exchange membranes for fuel cell applications. Int. J. Hydrogen Energy **35**(17), 9349–9384 (2010)
4. Wang, Y., Chen, K.S., Mishler, J., Chan, S.C., Cordobes, X.C.: A review of polymer electrolyte membrane fuel cells: technology, applications, and needs on fundamental research. Appl. Energy **88**(4), 981–1007 (2011)
5. Chien, H.C., Tsai, L.D., Huang, C.P., Kang, C.Y., Lin, J.N., Chang, F.C.: Sulfonated graphene oxide/Nafion composite membranes for high-performance direct methanol fuel cells. Int. J. Hydrogen Energy **38**(31), 13792–13801 (2013)
6. Yang, H.N., Lee, W.H., Choi, B.S., Kim, W.J.: Preparation of Nafion/Pt-containing TiO_2/graphene oxide composite membranes for self-humidifying proton exchange membrane fuel cell. J. Membr. Sci. **504**, 20–28 (2016)
7. Hasani-sadrabadi, M.M., Mokarram, N., Azami, M., Dashtimoghadam, E., Majedi, F.S., Jacob, K.I.: Preparation and characterization of nanocomposite polyelectrolyte membranes based on Nafion ionomer and nanocrystalline hydroxyapatite. Polymer (Guildf) **52**(5), 1286–1296 (2011)
8. Zhang, B., Cao, Y., Jiang, S., Li, Z., He, G., Wu, H.: Enhanced proton conductivity of Nafion nanohybrid membrane incorporated with phosphonic acid functionalized graphene oxide at elevated temperature and low humidity. J. Membr. Sci. **518**, 243–253 (2016)
9. Shao, Z., Joghee, P., Hsing, I.: Preparation and characterization of hybrid Nafion–silica membrane doped with phosphotungstic acid for high temperature operation of proton exchange membrane fuel cells. J. Membr. Sci. **229**, 43–51 (2004)
10. Lin, C.W., Lu, Y.S.: Highly ordered graphene oxide paper laminated with a Nafion membrane for direct methanol fuel cells. J. Power Sour **237**, 187–194 (2013)
11. Hickner, M.A., Ghassemi, H., Kim, Y.S., Einsla, B.R., McGrath, J.E.: Alternative polymer systems for proton exchange membranes (PEMs). Chem. Rev. **104**(10), 4587–4612 (2004)
12. Li, N., Guiver, M.D.: Ion transport by nanochannels in ion-containing aromatic copolymers. Macromolecules **47**(7), 2175–2198 (2014)
13. Sankir, M.: Proton exchange membrane fuel cell systems based on aromatic hydrocarbon and partially fluorinated disulfonated poly (arylene ether) copolymers. Virginia Polytechnic Institute and State University (2005)
14. Smitha, B., Sridhar, S., Khan, A.A.: Solid polymer electrolyte membranes for fuel cell applications—a review. J. Membr. Sci. **259**, 10–26 (2005)
15. Higashihara, T., Matsumoto, K., Ueda, M.: Sulfonated aromatic hydrocarbon polymers as proton exchange membranes for fuel cells. Polymer (Guildf) **50**(23), 5341–5357 (2009)
16. Lim, Y., et al.: Synthesis and characterization of sulfonated poly (arylene ether ketone sulfone) block copolymers containing multi-phenyl for PEMFC. Int. J. Hydrogen Energy **38**(1), 631–639 (2013)
17. Seo, D.W., et al.: Preparation and characterization of block copolymers containing multi-sulfonated unit for proton exchange membrane fuel cell. Electrochim. Acta **86**, 352–359 (2012)
18. Piboonsatsanasakul, P., Wootthikanokkhan, J., Thanawan, S.: Preparation and characterizations of direct methanol fuel cell membrane from sulfonated polystyrene/poly (vinylidene fluoride) blend compatibilized with poly (styrene)-b-Poly (methyl methacrytlate) block copolymer. J. Appl. Polym. Sci. 1325–1336 (2007)
19. Sannigrahi, A., Takamuku, S., Jannasch, P.: Block copolymers combining semi-fluorinated poly (arylene ether) and sulfonated poly (arylene ether sulfone) segments for proton exchange membranes. Int. J. Hydrogen Energy **39**(28), 15718–15727 (2014)
20. Bae, J.: Control of microdomain orientation in block copolymer thin films by electric field for proton exchange membrane. Adv. Chem. Eng. Sci. **04**(02), 95–102 (2014)

21. Zhang, X., Liu, S., Yin, J.: Synthesis and characterization of a new block copolymer for proton exchange membrane. J. Membr. Sci. **258**(1–2), 78–84 (2005)
22. Kim, J., Kim, B., Jung, B.: Proton conductivities and methanol permeabilities of membranes made from partially sulfonated polystyrene-block-poly (ethylene-ran-butylene)-block-polystyrene copolymers. J. Memb. Sci. **207**(1), 129–137 (2002)
23. Bae, B., Yoda, T., Miyatake, K., Uchida, M., Uchida, H., Watanabe, M.: Sulfonated poly (arylene ether sulfone ketone) multiblock copolymers with highly sulfonated block. fuel cell performance. J. Phys. Chem. B **114**(32), 10481–10487 (2010)
24. Lee, H.S., Roy, A., Lane, O., Dunn, S., McGrath, J.E.: Hydrophilic-hydrophobic multiblock copolymers based on poly (arylene ether sulfone) via low-temperature coupling reactions for proton exchange membrane fuel cells. Polymer (Guildf) **49**(3), 715–723 (2008)
25. Chen, L., Hallinan, D.T., Elabd, Y.A., Hillmyer, M.A.: Highly selective polymer electrolyte membranes from reactive block polymers. Macromolecules **42**(16), 6075–6085 (2009)
26. Saito, T., Moore, H.D., Hickner, M.A.: Synthesis of midblock-sulfonated triblock copolymers. Macromolecules **43**(2), 599–601 (2010)
27. Xu, K., Li, K., Khanchaitit, P., Wang, Q.: Synthesis and characterization of self-Assembled sulfonated poly(styrene-b-vinylidene fluoride-b-styrene) triblock copolymers for proton conductive membranes. Chem. Mater. **19**(24), 5937–5945 (2007)
28. Park, M.J., et al.: Increased water retention in polymer electrolyte membranes at elevated temperatures assisted by capillary condensation. Nano Lett. **7**(11), 3547–3552 (2007)
29. Shi, Z., Holdcroft, S.: Synthesis and proton conductivity of partially sulfonated poly ([vinylidene difluoride-co-hexafluoropropylene]-b-styrene) block copolymers. Macromolecules **38**(10), 4193–4201 (2005)
30. Elabd, Y.A., Napadensky, E., Sloan, J.M., Crawford, D.M., Walker, C.W.: Triblock copolymer ionomer membranes: part I. Methanol and proton transport. J. Membr. Sci. **217**(1–2), 227–242 (2003)
31. Shi, G.M., Zuo, J., Tang, S.H., Wei, S., Chung, T.S.: Layer-by-layer (LbL) polyelectrolyte membrane with NexarTM polymer as a polyanion for pervaporation dehydration of ethanol. Sep. Purif. Technol. **140**, 13–22 (2015)
32. Zuo, J., Shi, G.M., Wei, S., Chung, T.S.: The development of novel Nexar block copolymer/ultem composite membranes for C2-C4 alcohols dehydration via pervaporation. ACS Appl. Mater. Interfaces **6**(16), 13874–13883 (2014)
33. Geise, G.M., Freeman, B.D., Paul, D.R.: Characterization of a sulfonated pentablock copolymer for desalination applications. Polymer (Guildf) **51**(24), 5815–5822 (2010)
34. Inamuddin, Khan, A., Luqman, M., Dutta, A.: Kraton based ionic polymer metal composite (IPMC) actuator. Sens Actuators A Phys. **216**, 295–300 (2014)
35. Gao, R., Wang, D., Heflin, J.R., Long, T.E.: Imidazolium sulfonate-containing pentablock copolymer–ionic liquid membranes for electroactive actuators. J. Mater. Chem. **22**(27), 13473 (2012)
36. Huang, F., Largier, T.D., Zheng, W., Cornelius, C.J.: Pentablock copolymer morphology dependent transport and its impact upon film swelling, proton conductivity, hydrogen fuel cell operation, vanadium flow battery function, and electroactive actuator performance. J. Membr. Sci. **545**, 1–10 (2018)
37. Duong, P.H.H., Chung, T.-S., Wei, S., Irish, L.: Highly permeable double-skinned forward osmosis membranes for anti-fouling in the emulsified oil-water separation process. Environ. Sci. Technol. **48**(8), 4537–4545 (2014)
38. Geise, G.M., Lee, H.-S., Miller, D.J., Freeman, B.D., McGrath, J.E., Paul, D.R.: Water purification by membranes: the role of polymer science. J. Polym. Sci. Part B Polym. Phys. **48**(15), 1685–1718 (2010)
39. Lee, J.K., Kerres, J.: Synthesis and characterization of sulfonated poly (arylene thioether) s and their blends with polybenzimidazole for proton exchange membranes. J. Membr. Sci. **294**, 75–83 (2007)

Chapter 58
Experimental Validation for Electrical, Thermal and Overall Energy Generation from Open Low Concentration Ratio-Based Photovoltaic Hybrid Collector

Rohit Tripathi, Deepak Sharma, Nitin K. Gupta, G. N. Tiwari, T. S. Bhatti and V. K. Dwivedi

Abstract In the present communication, an experimental validation has been taken to analyze how are experimental results close to the theoretical results for 100% covered low concentration-based photovoltaic hybrid (PV hybrid) collector. The analysis has been attempted in composite climate in New Delhi, India, where all types of weather conditions exist. This validation has been attempted for PV cell temperature, electrical efficiency of module, outlet fluid temperature, electrical gain, overall thermal energy and exergy for a clear day, January month, New Delhi, India. Here, the area of collector of proposed system is 0.605 m^2 which is equal to PV module area also. The low ratio concentration technology is used as compound parabolic concentrator (CPC). The aperture area is 1.21 m^2. The aperture to receiver area ratio is 2. The mathematical expression for electrical efficiency of proposed system has also been obtained. The mass flow rate of fluid, H_2O, is 0.08 kg/s with open-loop system process. It is observed that the results of theoretical and experimental are found to be fair agreement with each other.

Keywords PVT · CPC · Glass to glass PV · Electrical efficiency · Overall energy

---

R. Tripathi (✉) · D. Sharma · N. K. Gupta
School of Electrical, Electronics and Communication Engineering, Galgotias University, G. Noida, U.P, India

G. N. Tiwari
Centre Bag Energy Research Society (BERS), Mahamana Nager, Varanasi, U.P 221005, India

R. Tripathi · T. S. Bhatti
Centre for Energy Studies, Indian Institute of Technology Delhi, New Delhi 110016, India

V. K. Dwivedi
Galgotias College of Engineering and Technology, G. Noida, U.P, India

S. Yadav et al. (eds.), *Proceedings of International Conference in Mechanical and Energy Technology*, Smart Innovation, Systems and Technologies 174,
https://doi.org/10.1007/978-981-15-2647-3_58

58.1 Introduction

The green energy highly demanded due to shortage of fossil fuels in all over the world. This green energy also dominates in favor of reduction of carbon emission to fossil fuels. Green energy has ability to fulfill human's demand of energy in significant proportion. The most usable solar energy receiver is photovoltaic system which can produce heat as well as power or energy simultaneously by solar radiation. The very first design of PV thermal was developed by Kern and Russell [1]. The PVT has four components as follows: solar cells or module, toughened glass top, black absorber plate integrated with copper tubes and one aluminum channel box. Some area of receiver portion is covered by photovoltaic cells or modules, and rest is enclosed by glass. Just below the receiver area, the black tube in absorber plate is placed. This type of arrangement is kept in aluminum casing. A theoretical and experimental study of PVT using techniques of tube in plate [2]. The evolution of performance of dual PV collector with different configurations has been carried out with different fluids and characteristics [3, 4]. The life cycle cost and production of hybrid PV collector which connected in series have been studied [5].

To enhance the overall energy gain from the convectional PVT, the concentration technology has been integrated with PVT which deals higher possibility of more overall solar system performance. A compound parabolic concentrator (CPC) having less than 4 CR. A CPC has an additive benefit that increases the input solar radiation on receiver and maximum incident solar rays come on receiver area after reflection from the reflectors [6]. The expression for useful energy/exergy of number of identical photovoltaic module PVT water collector has been studied [7].

Comparative study of PVT with air and water has been discussed [8, 9]. Mathematical modeling for N-PVT-CPC has been presented [10]. Further, four PVT family collectors have been studied and compared with each other on the annual gain performance basis. Overall gain and power generation from hybrid collector have been presented for two different systems. Recent studies have been presented for ther/elect performance of series connected PV/T [11–19].

In the present study, an attempt has been carried for validation of outlet fluid temperature, overall energy and exergy for theoretical and experimental results by fully covered concentrated PVT, January month, New Delhi.

58.2 Working Principle and System Setup

The concentrated system has been roofed fully with glass to glass PV module. The area of PV module and receiver area are equal due to fully covered by PV. The effective receiver area is 0.605 m^2. The absorber plate has been considered in collector which is placed just below to PV module (Fig. 58.1).

The concentrator and module area of concentrated PVT has been treated as $A_{a:}A_r$ = 2:1, respectively. The sun irradiance falls on aperture area of concentrated PVT.

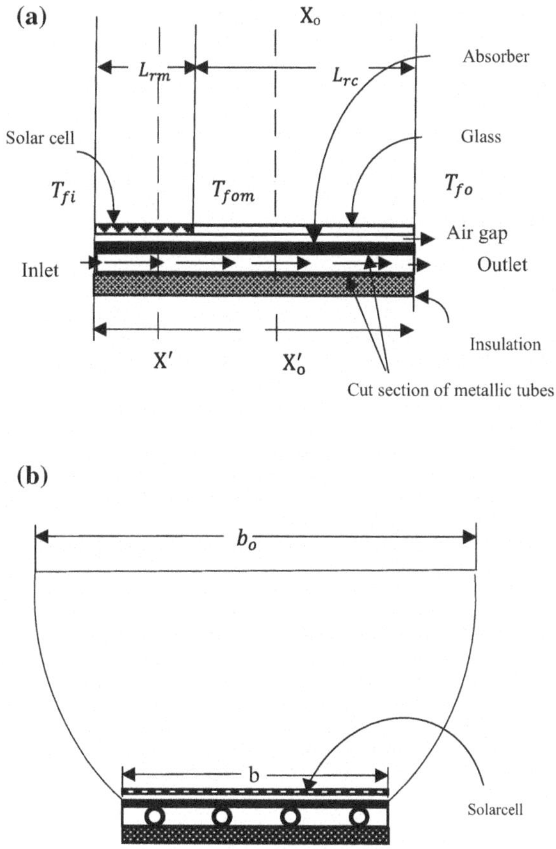

Fig. 58.1 **a, b** Internal view of concentrated photovoltaic thermal collector [10]

At the same time, some part of sun irradiance reaches on PV cells and heat is generated on PV cells which decrease electrical generation. To overcome this issue, PVT collector is used where heat from PV is indirectly moved to the heat absorber plate. So temperature of solar cells decreases by this function which increases electrical energy/power and the temperature of blackened copper plate rises. The fluid or water extracts this excess heat from absorber tube in plate. This increases the temperature of the fluid flowing under tubes below the absorber. The flow rate of H_2O has been fixed throughout the operation at 0.008 kg/s.

58.3 Modeling

Some certain presuppositions have been taken during writing the energy balance equations for the present system [5]

58.3.1 Concentrated PVT Water Collector

- For Photovoltaic module:

$$\rho\alpha_c\tau_g\beta_c I_b A_{am} = \left[U_{tc,a}(T_c - T_a) + U_{tc,p}(T_c - T_p)\right]A_{rm} + \rho\eta_m I_b A_{am} \quad (58.1)$$

- For black copper plate:

$$\rho\alpha_c\tau_g^2(1 - \beta_c)I_b A_{am} + U_{tc,p}(T_c - T_p)A_{rm} = F' h_{pf}(T_p - T_f)A_{rm} + U_{tp,a}(T_p - T_f) \quad (58.2)$$

- For flowing water under tube below plate:

$$\dot{m}_f c_f \frac{\mathrm{d}T_f}{\mathrm{d}x}\mathrm{d}x = F' h_{pf}(T_p - T_f)A_{rm} b\,\mathrm{d}x \quad (58.3)$$

And the outlet temperature of hybrid CPV/T with fluid has been evaluated $T_f = T_{fo1}$ at $x = \mathrm{Lr}$.

$$T_{fo} = \left[\frac{PF_2(\alpha\tau)_{m,\mathrm{eff}} I_b}{U_{L,m}} + T_a\right]\left[1 - \exp\left\{\frac{-A_r F' U_{L,m}}{\dot{m}_f c_f}\right\}\right] + T_{fi}\exp\left[\frac{-A_r F' U_{L,m}}{\dot{m}_f c_f}\right] \quad (58.4)$$

Through Eq. (58.4), the rate of available thermal energy by CPV/T has been calculated by below equation [6]

$$\dot{Q}_{\mathrm{uthe}} = \dot{m}_f c_f (T_{fo} - T_{fi}) \quad (58.5)$$

The rate of usable heat exergy has been attained for system by Eq. (58.6) as following [6]

$$\dot{E}_{\mathrm{xthe}} = \dot{m}_f c_f (T_{fo} - T_{fi}) - \dot{m}_f c_f (T_a + 273)\ln\left(\frac{T_{fo} + 273}{T_{fi} + 273}\right) \quad (58.6)$$

By solving Eqs. (58.1–58.3), the average copper plate temperature ($\overline{T}_p$) and PV cell temperature ($\overline{T}_c$) can be evaluated. The electrical efficiency of photovoltaic cells of CPVT collector has been obtained by analytically following expression

$$\eta_c = \eta_0\left[1 - \beta_0(\overline{T}_c - T_0)\right] \quad (58.7)$$

The above expression can further be rewritten as following

$$\eta_c = \frac{\eta_0\left[1-\beta_0\left[\frac{1}{U_{tc,p}+U_{tc,a}}\left[\begin{array}{c} I_b\left[\begin{array}{c}\rho\tau_g\alpha_c\beta_c\left(\frac{A_a}{A_r}\right)+\frac{U_{tc,p}}{U_{L2}+F'h_{pf}}\left\{(\alpha\tau)_{2,\text{eff}}+PF_1\rho\alpha_c\tau_g\beta_c\left(\frac{A_a}{A_r}\right)\right\} \\ +\frac{U_{tc,p}F'h_{pf}PF_2}{U_{l,m}(U_{L2}+F'h_{pf})}\left[(\alpha\tau)_{2,\text{eff}}+\left\{PF_1\rho\alpha_c\tau_g\beta_c\left(\frac{A_a}{A_r}\right)\right\}\right]\left(1-\frac{A_rF_{rm}}{F'}\right)\end{array}\right] \\ +T_a\left[U_{tc,a}+\frac{U_{tc,p}U_{L2}}{U_{L2}+F'h_{pf}}+\frac{U_{tc,p}F'h_{pf}}{(U_{L2}+F'h_{pf})}\left\{\left(1-\frac{A_rF_{rm}}{F'}\right)\right\}\right] \\ +T_{fi}\left[\frac{U_{tc,p}F'h_{pf}A_rF_{rm}}{(U_{L2}+F'h_{pf})}\right]\end{array}\right]-T_0\right]\right]}{\left[1-\frac{\eta_0\beta_0I_b}{(U_{tc,a}+U_{tc,p})}\left[\rho\tau_g\beta_c\left(\frac{A_a}{A_r}\right)+\left\{\frac{U_{tc,p}}{(U_{L2}+F'h_{pf})}PF_1\rho\tau_g\beta_c\left(\frac{A_a}{A_r}\right)\right\}+\frac{U_{tc,p}F'h_{pf}}{U_{L,m}(U_{L2}+F'h_{pf})}PF_2\rho\tau_g\beta_c\left(1-\frac{A_rF_{Rm}}{F'}\right)\right]\right]} \tag{58.8}$$

The efficiency of electrical gain from photovoltaic module of proposed system is given as

[Module efficiency] = [transitivity of glass of module] × [packing factor of module] × [PV cell efficiency]

$$\eta_m = \tau_g\beta_c\eta_c \tag{58.9}$$

The rate of usable electrical energy gain from concentrated PVT collector can be obtained by below expression

$$\dot{Q}_{x\text{ele}} = \sigma A_{am}I_b\eta_m \tag{58.10}$$

The rate of overall heat energy of system can be found by the following communication.

$$\dot{Q}_{\text{overall,the}} = \dot{Q}_{u\text{the}} + \frac{\dot{Q}_{x\text{ele}}}{0.38} \tag{58.11}$$

The rate of overall heat exergy from concentrated hybrid PV/T collector will be evaluated as:

Through Eqs. (58.6) and (58.10), an overall or complete exergy has been evaluated as follows (Fig. 58.2)

$$\dot{E}_{xu} = \dot{E}_{x\text{th}} + \dot{Q}_{x\text{ele}} \tag{58.12}$$

58.4 Statistical Analysis

See Table 58.1.

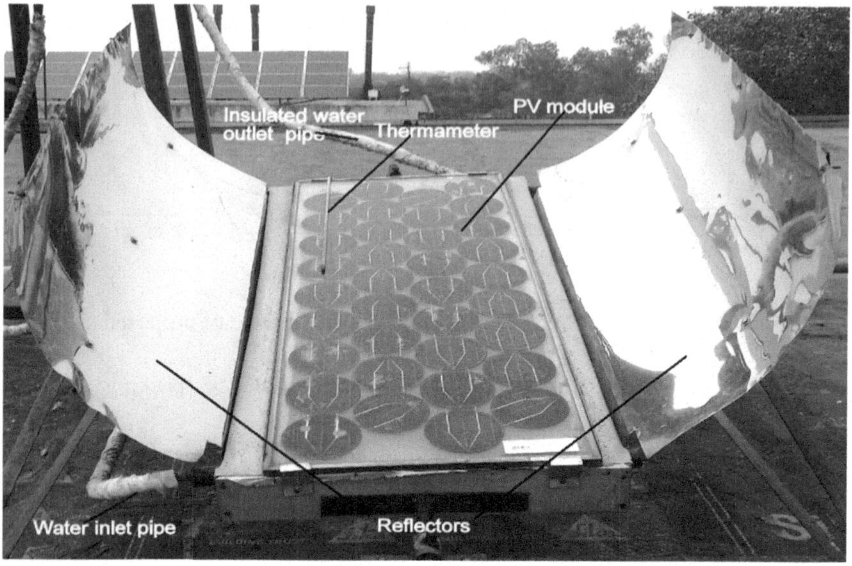

Fig. 58.2 Actual picture for concentrated hybrid photovoltaic thermal (PV/T) receiver

Table 58.1 Statistical analysis parameters

S. No.	Terms	Equations	Description
1	Correlation coefficient (r)	$r = \frac{n\sum X_i Y_i - \sum X_i \sum Y_i}{\sqrt{n\sum X_i^2 - (\sum X_i)^2}\sqrt{n\sum Y_i^2 - (\sum Y_i)^2}}$	X_i = theoretical value. Y_i = exp. Value. ith = No. of readings. n = total no. of counted readings
2	RMS percent deviation (e)	$e = \sqrt{\frac{\sum (e_i)^2}{n}}$	$e_i = \left(\frac{X_i - Y_i}{X_i}\right) \times 100$

58.5 Results and Discussions

The pattern of I_B, beam radiation and T_a, temperature on hourly basis for theoretical and experimental results has been given in Figs. 58.3 and 58.4. Figure 58.5 shows cell temperature and electrical efficiency of module of proposed system on hourly basis. It is noted that the behavior of PV cell and its electrical efficiency are inversely proportional to each other with effect of temperature. According to the values of 'r' and 'e,' it has good agreement between these results.

Outlet temperature of proposed system on hourly basis has been shown in Fig. 58.6. Here, it can be seen that hypothetical and experimental results are very

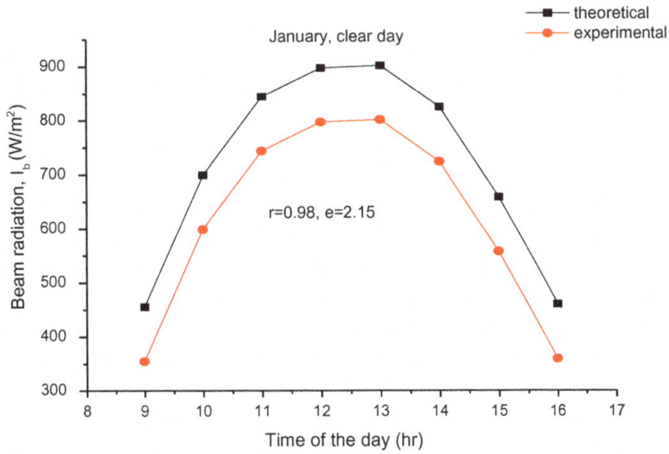

Fig. 58.3 Available beam radiation on CPVT collector on hourly basis in a clear day, January, New Delhi

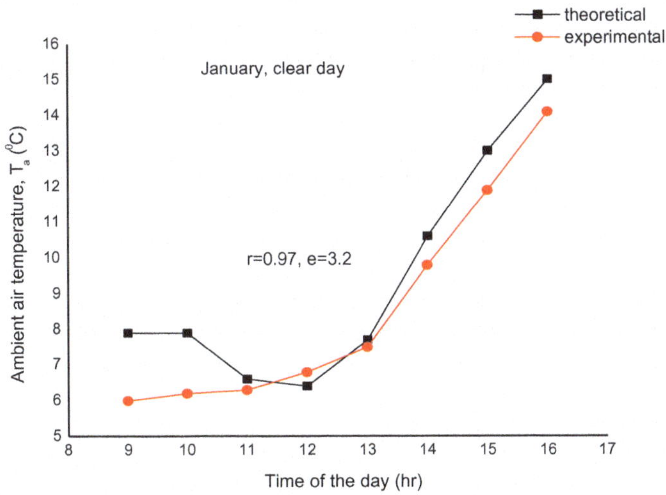

Fig. 58.4 Ambient temperature on hourly basis

close to each other, and the acceptable values of '*r*' and '*e*' have also been obtained.

Hourly variation of overall thermal energy and exergy of fully covered concentrated PVT water system has been shown in Fig. 58.7. Here, it is observed that the theoretical and experimental results for overall thermal energy and exergy have been

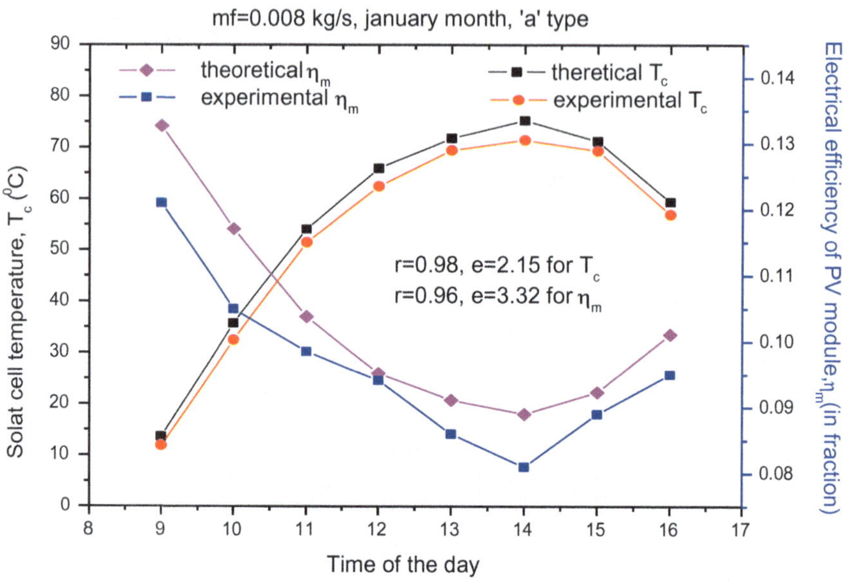

Fig. 58.5 PV cell temp. and elect. efficiency of module on hourly basis for theoretical and experimental results in same climate condition

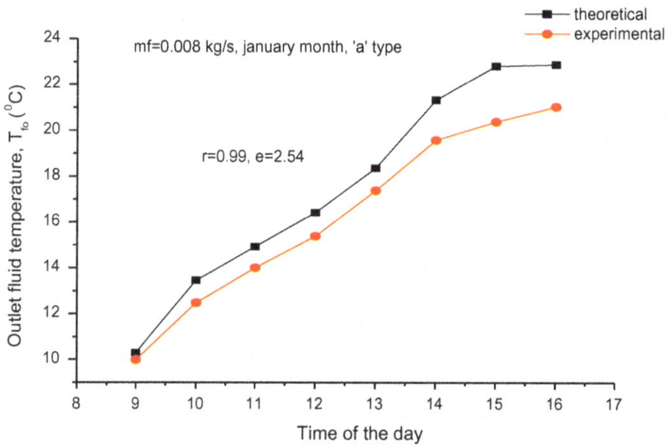

Fig. 58.6 Outlet fluid temperature of system for theoretical and experimental results for a clear day

shown through correlation coefficient and root mean SPD value. The values of '*r*' and '*e*' are found to be permissible and fair for the obtained results. The experimental daily maximum overall heat energy and exergy have been attained as 2.52 kWh and 0.75 kWh, respectively.

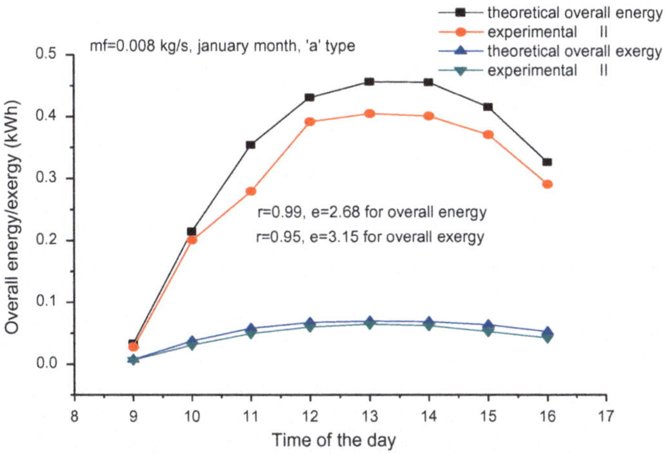

Fig. 58.7 Overall heat energy/exergy of system on hourly basis in same climate condition

58.6 Conclusions

On the above-obtained results from computation, some conclusions have been taken place as below:

- The validation for PV cell, ambient temperature, efficiency of module and outlet temperature of theoretical and experimental results indicate fair agreement.
- The max. water temperature at outlet has obtained as 20 °C with $m_f = 0.008$ kg/s in the month of January.
- It is also noted that such system is self-sustainable which can run the motor or water pump and circulate fluid from system to storage tank and then desired place.

Appendix

Following certain terms have been followed in modeling and numerical computation for the present study:

$$U_{tc,a} = \left[\frac{1}{h_o} + \frac{L_g}{K_g}\right]^{-1}; v = 1 \text{ m/s}; v' = 1 \text{ m/s};$$

$$h_o = 5.7 + 3.8v;$$

$$U_{tc,p} = \left[\frac{1}{h_i} + \frac{L_g}{K_g}\right]^{-1};$$

$$h_i' = 2.8 + 3v'; h_i = 5.7;$$

$$U_{L2} = U_{L1} + U_{tp,a};$$
$$U_{L1} = \frac{U_{tc,p}U_{tc,a}}{U_{tc,p} + U_{tc,p}};\ U_{L,m} = \frac{h_{pf}U_{L2}}{F'h_{pf} + U_{L2}};$$
$$U_{tp,a} = \left[\frac{1}{U_{tc,a}} + \frac{1}{U_{tc,p}}\right]^{-1} + \left[\frac{1}{h_i'} + \frac{1}{h_{pf}} + \frac{L_i}{K_i}\right]^{-1}$$

References

1. Kern, E.C., Russell, M.C.: Combined photovoltaic and thermal hybrid collector systems.In: Proceedings of the 13th IEEE Photovoltaic Specialists, Washington, DC, pp. 1153–1157 (1978)
2. Chow, T.T.: Performance analysis of photovoltaic–thermal collector by explicit dynamic model. Sol. Energy **75**, 143–152 (2003)
3. Bhargava, A.K., Garg, H.P., Agarwal, R.K.: Study of a hybrid solar system-solar air heater combined with solar cells. Energy Convers. Manag. **31**(5), 471–479 (1991)
4. Hendrie, S.D.: Evaluation of combined photovoltaic/thermal collectors. In: Proceedings of International Conference ISES, vol. 3, Atlanta, Georgia, pp. 1865–1869
5. Dubey, S., Tiwari, G.N.: Analysis of PV/T flat plate water collectors connected in series. Sol. Energy **83**, 1485–1498 (2009)
6. Atheaya, D., Tiwari, A., Tiwari, G.N., Al-Helal, I.M.: Analytical characteristic equation for partially covered photovoltaic thermal (PVT) compound parabolic concentrator (CPC). Sol. Energy **111**, 176–185 (2015)
7. Tiwari, G.N., Mishra, R.K., Solanki, S.C.: Photovoltaic modules and their applications: a review on thermal modeling. Appl. Energy **88**, 2287–2304 (2011)
8. Tiwari, A., Sodha, M.S.: Performance evaluation of hybrid PV/thermal water/air heating system: a parametric study. Renew. Energy **31**(15) 2460–2474 (2006)
9. Dubey, S., Sandhu, G.S., Tiwari, G.N.: Analytical express for electrical efficiency of PV/T hybrid air collector. Appl. Energy **86**, 697–705 (2009)
10. Tripathi, R., Tiwari, G.N., Al-Helal, I.M.: Thermal modelling of N partially covered photovoltaic thermal (PVT)–Compound parabolic concentrator (CPC) collectors connected in series. Sol. Energy **123**, 174–184 (2016)
11. Tripathi, R., Tiwari, G.N., Bhatti, T.S., Dwivedi, V.K.: 2-E (Energy-Exergy) for partially covered concentrated photovoltaic thermal (PVT) collector. Energy Proc. **142**, 616–623 (2017)
12. Saini, V., Tripathi, R., Tiwari, G.N., Al-Helal, I.M.: Electrical and thermal assesment of N partially covered PVT-compound parabolic concentrator collector connected in series, for different solar cell materials. Appl. Therm. Eng. **128**, 1611–1623 (2018)
13. Tripathi, R., Tiwari, G.N., Dwivedi, V.K.: Overall energy, exergy and carbon credit analysis of N partially covered photovoltaic thermal (PVT) concentrating collector connected in series. Sol. Energy **136**, 260–267 (2016)
14. Tripathi, R., Tiwari, G.N.: Energetic and exergetic analysis of N partially covered photovoltaic thermal-compound parabolic concentrator (PVT-CPC) collectors connected in series. Sol. Energy **137**, 441–451 (2016)
15. Tripathi, R., Tiwari, G.N., Dwivedi, V.K.: Overall energy and exergy performance of partially covered N-photovoltaic thermal (PVT)-compound parabolic concentrator (CPC) collectors connected in series. In: IEEE International Conference on Power Electronics, Intelligent Control and Energy Systems (ICPEICES), pp. 12–17 (2016)
16. Tripathi, R., Tiwari, G.N.: Annual energy, exergy and environmental benefits of N half covered concentrated photovoltaic thermal (CPVT) air collector. Adv. Smart Grid Renew. Energy-Lect. Notes Electric. Eng **435**, 113–127 (2018)

17. Tripathi, R., Tiwari, G.N.: Annual performance evaluation (energy and exergy) of fully covered concentrated photovoltaic thermal (PVT) water collector: an experimental validation. Sol. Energy **146**, 180–190 (2017)
18. Tripathi, R., Tiwari, G.N., Dwivedi, V.K.: Energy matrices evaluation and exergoeconomic analysis of series connected N partially covered (glass to glass PV module) concentrated photovoltaic thermal collector: at constant flow rate mode. Energy Convers. Manag. **145**, 357–370 (2017)
19. Tripathi, R., Tiwari, G.N.: Energy matrices, life cycle cost, carbon mitigation and credits of open-loop N concentrated photovoltaic thermal (CPVT) collector at cold climate in India: a comparative study. Sol. Energy **186**, 347–359 (2019)

Chapter 59
Analysis of Performance Parameters of Active Solar Still

Etkaf Hasan Khan, Eram Neha, Mohd Atif Wahid, Islam Nawaz and G. N. Tiwari

Abstract Water is the need of every living entity on earth. The rapid growth of population and technology is leaving very less space for the trees and the greenery which in turn is affecting the rain cycles and thus narrowing the lifeline rivers of our cities and villages. So we are moving toward the groundwater which is also reducing in level with every passing day. Researchers and scientists are coming with new ideas of rainwater harvesting and many more, but we can also reutilize the water by making it clean with the help of solar energy. Solar stills are the devices used for distilling the impure and salinated water in potable one. This technology has been used for very first time around 400 years in BC by Aristotle. With years passing by, the technology of still has got advanced. Now, the more efficient active hybrid still is being used and analyzed for further possibility of the efficiency enhancement.

Keywords Active still · Collector · Hybrid PV/T · Solar still · Water

59.1 Introduction

Earth's surface is surrounded by water similar to the human body. However, out of all the water, only less than 1% is freshwater. The major availability of freshwater is in lakes, rivers, and underground. About one-third of the freshwater potentially available can only be used by humans. Around the world, approximately 7.7 billion people

---

E. H. Khan · I. Nawaz
Department of Mechanical Engineering, Jamia Millia Islamia, New Delhi, India

E. Neha (✉)
Galgotias College of Engineering and Technology, Greater Noida, Uttar Pradesh, India
e-mail: eramneha@gmail.com

M. A. Wahid
Delhi Technical Campus, Greater Noida, Uttar Pradesh, India

G. N. Tiwari
Centre for Energy Studies, IIT Delhi, New Delhi, India

S. Yadav et al. (eds.), *Proceedings of International Conference in Mechanical and Energy Technology*, Smart Innovation, Systems and Technologies 174,
https://doi.org/10.1007/978-981-15-2647-3_59

are facing great difficulty in accessing clean drinking water. Around the globe, more than 28 countries are not having sufficient water to carry their agricultural needs and do economic development. Four-fifths of the deserts around the world are suffering from cyclic droughts.

Currently, India is facing nastiest water crisis as bestowed by a media report of July 9, 2018. It is being predicted that in coming 2 years, up to 21 Indian cities will face a groundwater crisis, including Delhi, Chennai, Bengaluru, and some other major cities. NITI Aayog data have conferred that there is a huge need for "urgent and improved" management of water resources. There is high-to-extreme water stress being faced by over nearly 600 million Indians. Annually, due to inadequate access to safe water, about 200,000 people are dying. The situation is likely to get worse by 2050 as the demand will cross the supply way far, as projected by a report [1].

To solve this freshwater unavailability, desalination plants based on renewable energy are the perfect choice without causing any harm to the environment. In solar distillation, the greenhouse effect is utilized to fetch clean water from impure water for making it potable and human use. Solar distillation comes out to be one of the most economical and practical methods for large-scale production of clean distilled water from saline [2]. The major drawback of various distillation processes is the high energy cost of evaporation. Solar desalination plants have the advantage of abundant solar energy available at no cost. Another advantage is that these plants can be installed at any remote or far location having no infrastructure. Solar distillation is way above the other evaporation-type distillation methods since it utilizes the free source of energy [3]. The basic drinking water requirements of man can be fulfilled by solar still. Indian land is having numerous amount of solar energy, being a tropical country. In different regions of the continent, the average daily solar radiation varies between 4 and 7 kWh/m^2. In India, about 5000 trillion kWh of solar energy is received every year [4]. Solar stills are sufficient to provide potable water to rural houses in India since this technology is quite feasible and has low operational costs. India is suffering from a water crisis with variable intensity throughout the year, and the growth of population is fueling it [5].

There are various important parameters upon which the efficiency of active still depends directly or indirectly. This research is being done to compile and analyze all the parameters responsible for the variation in the performance of the active solar still.

59.2 History

Solar still first came into existence in the year 1872. Classification can be done as follows.

59.2.1 *Passive Still*

These types have a working in which the water of the basin is heated by direct solar energy. Researchers have published lots of work in this field. It has been reported that passive stills are slow-working devices, and they produce a daily total output of 2.25 kg/m^2 day in summer. Detailed studies have shown that among all positions of the angle of condensing cover, the maximum yearly freshwater output and efficiency are achieved at a 15-degree angle [6].

59.2.2 *Active Still*

In an active still, heat is supplied directly through sun and indirectly through waste hot water from industries or solar collectors. A thermal model is developed, adopted with integrated active solar still for different solar collectors [7]. This model was validated with data obtained from experiments performed in December 2005 at water depth of 0.05 mt. The result shows that active still having integration with evacuated-type tube collector gives an output of 4.24 kg/m^2-day [8].

59.3 Experimental Setup

The side view of self-sustaining active solar still connected with hybrid (PV/T) is shown in Fig. 59.1. Following are system parts:

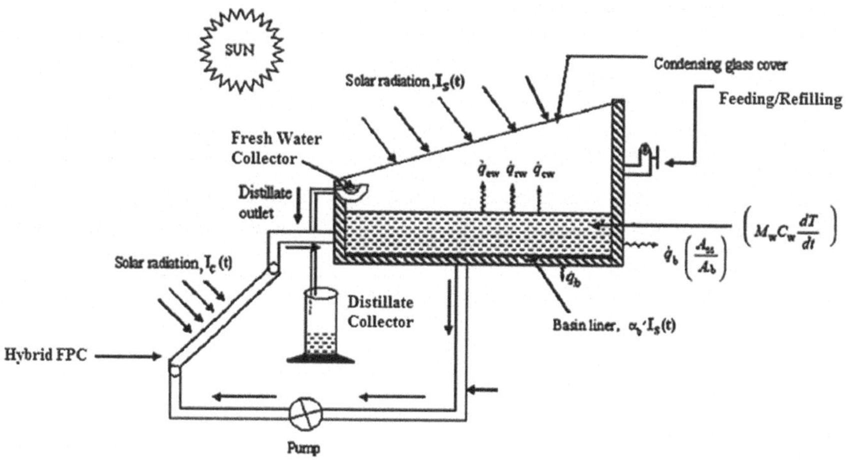

Fig. 59.1 Schematics of the experimental setup of active still with a flat-plate collector [17]

(i) Solar still,
(ii) FPC integrated with PV plate,
(iii) Motor pump running on DC current.

A flat-plate collector is connected in series to the basin of solar still connected through insulated pipes. The active still is attached with two flat-plate collectors to the basin of the solar still. It is connected in series to the basin to heat the water in the still. A PV module of dimension 0.55 m × 1.20 m of power rating 75 W is coupled to a collector at the bottom, where the low-temperature water is entering. The PV module runs a motor of DC supply.

59.3.1 Hybrid PV/T Collectors

Hybrid PV/T collector absorbs solar energy and converts it into heat and electricity simultaneously. A typical PV/T collector has a collector and a PV module. It has peak efficiency of the order of 5–20%. It is also attached with an absorber plate on the back of the PV module, which acts as a heat-removing device. The absorber plate keeps the temperature of the PV module in the range to give optimum efficiency and simultaneously providing the absorb heat to the basin of the still [9].

59.3.2 Solar Still

Solar still is the device working on the greenhouse effect of solar energy, and that intends to vaporize the brackish or saline water which gets collected after condensing on the top cover of the still [10].

Stills can be classified as passive and active. There are so many parameters that affect the still's performance, e.g., water level in the basin, basin material, velocity of air, solar radiation, surrounding temperature, and angle of inclination. The efficiency of all type of still is governed by the difference between the temperature of water and glass cover [11]. Passive solar still has the only source for heating the basin water, i.e., direct solar energy, which results in lower productivity. Another type of still was developed, which has other heating and energy sources for elevating the water temperature, known as active solar stills.

59.3.3 DC Motor Pump

The pump helps in circulating the water from the collector to the still and then again from the still to the collector, to the temperature of the water inside the still. The motor runs by the electricity produced PV module of the hybrid collector [12]. S. Kumar

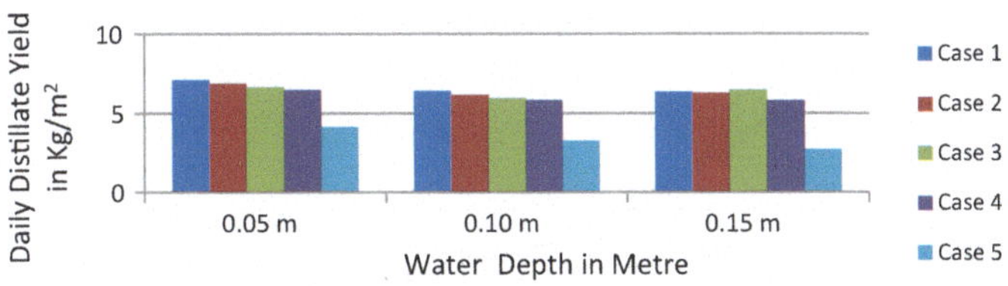

Fig. 59.2 Effect of pump working duration on the daily output of freshwater in the hybrid still for various depths, in March 2007 [13]

and A. Tiwari experimentally analyze the pump running time effect for different time durations at different water depths and taking five cases like a pump running for 9 h, 7 h, 5 h, 3 h, and no pump running. The results show that there is no significant change in the yield when the pump running is decreased by 4 h from case 1 to case 3, but there is a huge drop in the output if the pump running time is further decreased by 2 h. The results are shown below in Fig. 59.2.

59.4 Performance of Active Solar Distillation System

Solar radiation intensity, surrounding temperature, and air velocity are some of the non-controllable parameters that cannot be changed; these parameters are controlled by the climate and geographical location [14]. The parameters that can be controlled are water depth, glass cover inclination, material of still, water temperature, and insulation thickness. These controllable parameters are the ones responsible for the variable productivity of the different types of modified stills [15]. This is the case of an active still. Following are the parameters that have to be discussed; they affect the performance of the still in one way or the other.

59.4.1 Depth of Water

Output of the still is inversely proportional to the depth of the water during daytime. It has been experimentally proven by many researchers [25]. Experimental results showed that the daily yield for PV/T hybrid active still is 7.22 kg, 4.94 kg, and 5.28 kg for the water levels of 0.05 m, 0.10 m, and 0.15 m, respectively (see Fig. 59.3). The thermal efficiencies of the hybrid active solar still are found to be as 19.4, 15.4, and 15.2%, for water depths of 0.05 m, 0.10 m, and 0.15 m, respectively. Finally, hybrid active solar still is evaluated for overall thermal efficiency, and it comes out to be as 48.3, 44.3, and 44.1 percent for respective water depths of 0.05, 0.10, and 0.15 m, and the results are almost 20% higher than the passive solar still in the month of April 2006 [16].

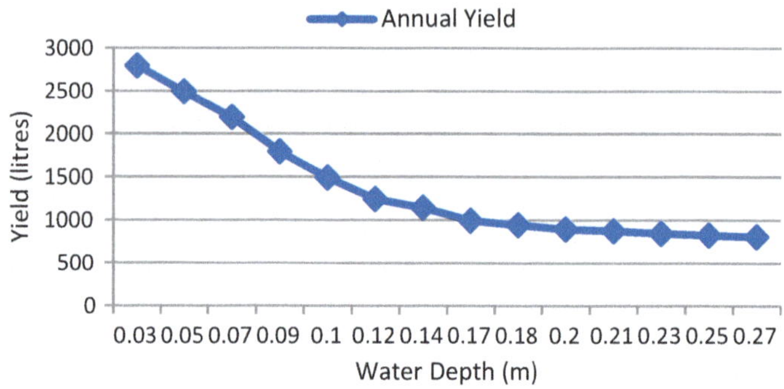

Fig. 59.3 Annual yield of an active solar still with different water depths [17]

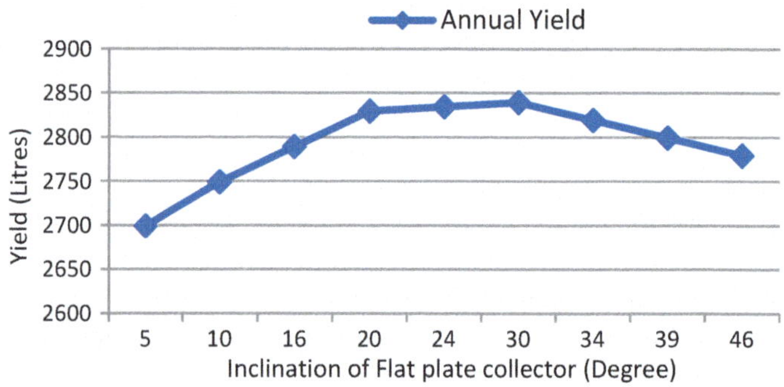

Fig. 59.4 Output (annual) of an active solar still with different collector tilts [17]

59.4.2 Glass Cover Angle, Thickness, and Material

To receive the sun rays falling normally over all the year, the cover inclination angle is kept equal to latitude angle (see Fig. 59.4) [17]. Decreasing the thickness and increasing the thermal conductivity increase the still productivity [27].

59.4.3 Fabrication Material

The body of the still is to be perfectly insulated. Earlier, wood, cast iron, and steel were being used with some insulation materials on the outside body. Later, fiber-reinforced plastics are introduced. Now glass-reinforced plastics are being used. GRP are more than two times long-lasting than FRP [33]. Mat of glass wool (40%) is coated with unsaturated polymer resin (60%) to make the body. Vimal Dimri et al. [34] analyzed experimentally a solar still with flat plate using different collector of various cover materials. The finding shows that yield is directly related to thermal conductivity of

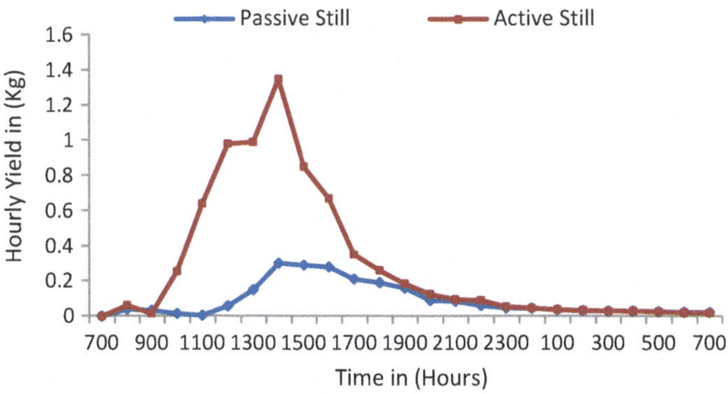

Fig. 59.5 Hourly output comparison of passive and active solar stills at 0.05 m water depth [14]

condensing cover material. Due to higher conductivity, copper provides better results compared to glass/plastic.

59.4.4 *Temperature of Water*

The rate of evaporation of impure water is controlled by the difference in temperature of water in the still and the temperature of the glass cover. The water temperature in passive still increases at a very slow pace. In active stills, thermal energy from waste heat or other external sources is utilized to raise the temperature of water in the basin. External sources could be either industrial or other low energy waste, energy fed from flat-plate and hybrid PV/T collectors [18].

S. Kumar and A. Tiwari in their experimental analysis of PV/T hybrid still find out that the temperature of the basin in the still is much higher than that of the passive still. And due to the higher temperatures, the yield is 5.5 times higher in winter and 3.5 times higher in summer than that of the passive still (see Fig. 59.5) [14].

Below in the Table 59.1 has been discussed the output efficiency of various types of modified stills.

59.5 Conclusion

Following are the points that can be inferred from the above analysis:

(1) Solar still is best suited for the region where abundant solar energy and ample open space are available.
(2) The performance of a solar still mainly depends on the parameter of water depth, cover inclination, body of still, and temperature of water.

Table 59.1 Various types of solar stills and their efficiency

Solar still type	Output efficiency (%)	Year/Reference
Solar still (simple basin type)	30	1981, [19]
Solar still (single slope type)	23–31	2008, [20]
Solar still (double slope type)	25–34	2008, [21]
Solar still (multiple wick type)	34	1981, [22]
Solar still (low-cost thermoformed type)	39	2009, [23]
Solar still (double-effect multi-wick type)	50–60	1992, [24]
Capillary film distiller (one stage)	50–55	1998, [25]
Solar still (tilted wick type)	53	2011, [26]
Solar still with FPC (forced circulation mode)	66	1990, [27]
Solar still with FPC (natural circulation mode)	40	2005, [28]
Solar still (active regenerative type)	25	1993, [29]
Vertical solar still with FPC	50	1987, [30]
Solar still parabolic concentrator	38	2004, [31]
Regenerative still with parabolic concentrator	70	1996, [32]

(3) The difference in temperature between water and the cover of the still accelerates the productivity and hence net output.
(4) Solar stills are the best alternative for potable water in remote and arid areas.
(5) The quality of water obtained is similar to that obtained by other solar stills.

59.6 Future Scope

Since a lot of researches have also been conducted already in this field, there is also scope of research. Experimental analysis of full PV module and separate flat-plate collector can be done.

References

1. Science & Technology section of first post webpage: https://www.firstpost.com/tech/science/indias-water-crisis-bengaluru-delhi-chennai-hyderabad-among-21-cities-to-run-out-of-groundwater-by-2020-4590221.html. Accessed 02 Aug 2019
2. Wazwaz, J., Salmi, H., Hallak, R.: Solar thermal performance of a nickel-pigmented aluminium oxide selective absorber. Renew. Energy **27**, 277–292 (2002)
3. Orel, Z.C., Gunde, M.K., Hutchins, M.G.: Spectrally selective solar absorbers in different non-black colours. In: Proceedings of WREC VII (CD–ROM), Cologne, Germany (2002)
4. Konttinen, P., Lund, P.D., Kilpi, R.J.: Mechanically manufactured selective solar absorber surfaces. Sol. Energ. Mat. Sol. **C.79**, 273–283 (2003)

5. Francia, G.: A new collector of solar radiant energy. In: Proceedings of the United Nations Conference on New Sources of Energy, p. 572, paper No. F6; Rome, Italy (1961)
6. Slaman, M., Griessen, R.: Solar collector overheating protection. Sol. Energy **83**, 982–987 (2009)
7. Kumar, K.R., Reddy, K.S.: Thermal analysis of solar parabolic trough with porous disc receiver. Appl. Energy **86**, 1804–1812 (2009)
8. Ackermann, J.A., Ong, L.E., Lau, S.C.: Conjugate heat transfer in solar collector panels with internal longitudinal corrugated fins—part I: overall results. Forsch. Ingenieurwes. **61**, 84–92 (1995)
9. Sopian, K., Alghoula, M.A., Alfegib, E.M., Sulaimana, M.Y., Musab, E.A.: Evaluation of thermal efficiency of double-pass solar collector with porous–nonporous media. Renew. Energy **34**, 640–645 (2009)
10. Martinopoulos, G., Missirlis, D., Tsilingiridis, G., Yakinthos, K., Kyriakis, N.: CFD modeling of a polymer solar collector. Renew. Energy **35**, 1499–1508 (2010)
11. Tian, Y., Zhao, C.Y.: A numerical investigation of heat transfer in phase change materials (PCMs) embedded in porous metals. Energy **36**, 5539–5546 (2011)
12. Tian, Y., Zhao, C.Y.: Heat transfer analysis for phase change materials (PCMs). In: Proceedings of the 11th International Conference on Energy Storage Stockholm, Sweden (2009)
13. Saha, S.K., Mahanta, D.K.: Thermodynamic optimization of solar flat-plate collector. Renew. Energy **23**, 181–193 (2001)
14. Kumar, S., Tiwari, A.: Int J Energy Res **32**, 847–858 (2008)
15. Selmi, M., Al-Khawaja, M.J., Marafia, A.: Validation of CFD simulation for flat plate solar energy collector. Renew. Energy **33**, 383–387 (2008)
16. Kumar, S., Tiwari, A.: Design, fabrication and performance of a hybrid photovoltaic/thermal (PV/T) active solar still. Energy Convers Manag **51**, 1219–1229 (2010)
17. Singh, H.N., Tiwari, G.N.: Monthly performance of passive and active solar stills for different Indian climatic condition. Desalination **16**, 145–150 (2004)
18. Bergene, T., Lovvik, O.M.: Model calculations on a flat-plate solar heat collector with integrated solar cells. Sol. Energy **55**, 453–462 (1995)
19. Tonui, J.K., Tripanagnostopoulos, Y.: Improved PV/T solar collectors with heat extraction by forced or natural air circulation. Renew. Energy **32**, 623–637 (2007)
20. Hegazy, A.A.: Comparative study of the performances of four photovoltaic/thermal solar air collectors. Energ. Convers. Manage. **41**, 861–881 (2000)
21. Brogren, M.: Low-concentrating photovoltaic systems with parabolic reflectors. Licentiate Thesis, Division of Solid State Physics, Uppsala University (2001)
22. Fujisawa, T., Tani, T.: Annual exergy evaluation on photovoltaic–thermal hybrid collector. Sol. Energ. Mat. Sol. C. **47**, 135–148 (1997)
23. Huang, B.J., Lin, T.H., Hung, W.C., Sun, F.S.: Performance evaluation of solar photovoltaic/thermal systems. Sol. Energy **70**, 443–448 (2001)
24. Rai, S.N., Dutt, D.K., Tiwari, G.N.: Some experimental studies of single basin solar still. Energy Convers. Manag. **30**(2), 149–153 (1990)
25. Yadav, Y.P.: Analytical performance of a solar still integrated with a flat plate solar collector: thermosiphon mode. Energy Convers. Manag. **31**(3), 255–263 (1991)
26. Badran, O.O., Al-Tahaineh, H.A.: The effect of coupling a flat plate collector on the solar still productivity. Desalination **183**, 137–142 (2005)
27. Kumar, S., Tiwari, G.N.: Performance evaluation of an active solar distillation system. Energy **21**(9), 805–808 (1996)
28. Tiwari, G.N., Sinha, S.: Parametric studies of active regenerative solar still. Energy Convers. Manag. **34**(3), 209–218 (1993)
29. Yadav, Y.P., Prasad, A.S.: Performance analysis of a high temperature solar distillation system. Energy Convers. Manag. **36**(5), 365–374 (1995)
30. Garcia Rodriguez, L., Gomez Camacho, C.: Design parameter selection for a distillation system coupled to a solar parabolic through collector. Desalination **122**, 195–204 (1999)

31. Bechir, C., Adel, Z., Slimane, G.: Desalination of brackish water by means of a parabolic solar concentrator. Desalination **217**, 118–126 (2007)
32. Kumar, S., Sinha, S.: Transient model and comparative study of concentrator coupled regenerative solar still in forced circulation mode. Energy Convers. Manag. **37**(5), 629–636 (1996)
33. Wang, R.Z., Zhai, X.Q.: Development of solar thermal technologies in China. Energy **35**, 4407–4416 (2012)
34. Tiwari, G.N., Vimal, D., Arvind, C.: Parametric study of an active and passive solar distillation system: energy and exergy analysis. Desalination **242**, 1–18 (2009)

Chapter 60
Eutectics Usage to Enhance the Efficiency of the Solar Photovoltaic Modules

Gopal Nandan, Nafees A. Wani, Ravi Kant, P. K. Rohatgi and Naveen Kumar

Abstract With the increasing demand for renewable energy sources and the huge gap between the demand and supply conditions, measures have been implemented to enhance the productivity of the existing renewable energy sources. The sun supplies an ample amount of light and heat, which can be converted into abundant and safer solar energy. Photovoltaic (PV) cells can be employed for the conversion of abundant solar energy to useful electricity. But the issue of heating up of the PV cells drastically reduces the conversion efficiency of the system. Hence, there is a need for some additional arrangement to this system that can arrest the temperature rise of PV. This paper is a review of such systems that employ eutectic for the thermal regulation of the PV panel. It is a passive cooling technique which can be utilized very easily and cheaply .

Keywords Solar photovoltaics · Thermal efficiency · Electrical efficiency · Phase change materials · Cut-off voltage

60.1 Introduction

The concepts of photovoltaics and photoelectric effect have been going around for more than a century. Still, it is not able to utilize it in a large and fully profitable scale, the reason being its decreasing nature of electrical efficiency with the increase in surface temperature above the ambient value. It can convert both dispersed and concentrated (visible and ultraviolet) solar incident radiations into electricity [1–4]. The infrared solar radiation is converted into heat, due to which the temperature of panel may increase [5]. The temperature rise of the PV panel, output power, fill factor and conversion efficiency decreases, and thermal lattice vibrations increase [6]. In the long term, there is an effect on the performance of the panel, if operated at high temperature. The electrical efficiency, open-circuit voltage, short-circuit current and

G. Nandan · N. A. Wani · P. K. Rohatgi
ME Department, Amity University, Uttar Pradesh, Noida, India

R. Kant (✉) · N. Kumar
Galgotias College of Engineering & Technology, Greater Noida, India
e-mail: ravikantknl@gmail.com

S. Yadav et al. (eds.), *Proceedings of International Conference in Mechanical and Energy Technology*, Smart Innovation, Systems and Technologies 174,
https://doi.org/10.1007/978-981-15-2647-3_60

fill factor of the panel are decreased every year [7]. It is observed that the PV panels' performance decreases above the ambient working temperature. There is a drop of about 0.45–0.5% for every 1 °C gain in the cell temperature. The efficiency is dependent on the cell material and its technology. This varies from 10% for organic PV to almost 45% for multi-junction cells [8]. A low value of efficiency reduces the usefulness and advantages of the PV panel. The scientists and researchers attempted various techniques for cooling the system to application of various mechanical attachments to the system. Sun trackers are used to getting maximum insolation throughout the day; periodic cleaning of the dust accumulated over the panel either physically or using robotic vacuum cleaners or robots, for example, NO-water Mechanical Automated Dusting Device, anti-reflective coatings to eliminate the wastage of reflected radiations, etc., are some of the techniques apart from the cooling techniques to enhance the efficiency of the PV panel [9]. Several cooling techniques like natural or forced convection (working fluids like air, water, nanofluids, etc.), heat pipe cooling, thermoelectric cooling and application of various phase change materials (PCMs) have been reported in the literature. The PCMs have high heat capacity and can store and release 5–14 times of heat per unit volume as compared to sensible heat storage [10]. Certain desirable thermodynamic, kinetic and chemical properties are must for the deployment of this PCM in the application. Moreover, cost considerations and the easy availability of these materials are also important. The preferred operating temperature of the PV module is 25 °C. Hence, the PCM should have a suitable phase transition temperature, high latent heat of transition and good heat transfer. The charging and discharging of heat are governed by the thermal conductivity of PCM.

The implementation of these technologies requires initial costs and maintenance cast. Natural and forced convection using natural air circulation will be at low cost. But, it will require more space to maintain proper mass flow rate, which is disadvantageous. There will be low value of heat transfer coefficient. The heat capacity of air is also very low. This scenario arises in systems like the building-integrated PV systems. On the other hand, active cooling using air is coolant in the air circulation and will be more effective due to more convective heat transfer coefficient but has the disadvantage of consumption of additional energy, the liability of maintenance, etc. The use of PCM proves to be a promising thermal management option for a PV system [11]. This paper focuses on the application of eutectics for thermal management to improve electrical conversion efficiency extensively.

60.2 Overview of Temperature Regulation of Solar Photovoltaic Panels

The electrical power conversion efficiency of the module is highly influenced by the temperature of the module [12]. Researchers have reported several cooling arrangements to operate at lower temperature of the PV module. It is summarized in Fig. 60.1.

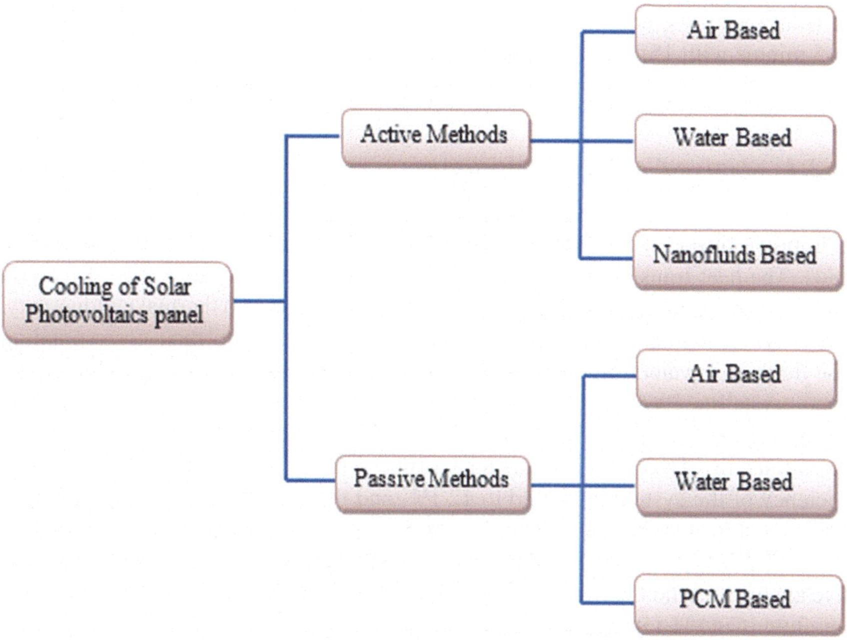

Fig. 60.1 Overview of cooling techniques applied in solar photovoltaic panel

These techniques can be of sensible heat storage type, thermo-chemical storage type or latent heat storage type. These techniques have been tried in either inactive way or passive way. Among them, the latent heat energy storage system proves to be the most useful and promising option owing to its compactness and stability of temperature throughout the phase transition stage. The sufficient reduction in the panel temperature has been reported by researchers worldwide.

This technique requires additional power to maintain the fluid flow above/beneath the PV panel. There is an involvement of externally powered sources like fans or pumps which facilitate the heat extraction as part of the cooling action apart from the natural phenomenon of conduction, convection, etc., in the active technique. The heat transfer is increased by the flow of fluids. These techniques are preferable in situations where higher and stable solar irradiation is available so that the added efficiency of the solar panel after active cooling is more than the energy requirement of the cooling system. Air-based, water-based and nanofluid-based active techniques have been reported extensively in this decade. On the back surface of modules, a parallel array of the duct was provision for maintaining the airflow resulted in an increase in the electrical efficiency to 12.5% from 8.6% without cooling [13]. In the forced cooling air system, the system power output increased by 8% [14]. Due to good heat transport property, water is a more suitable candidate as compared to air for active cooling technique. The water flow has been maintained either over the top face or under the rear portion of the PV panel using pumps. By water flow over the front panel, the cooling rate of 2 °C/min when the experiment was performed for 5 minute with 15-minute pauses under a fixed flow of 0.48 kg/s [15].

60.2.1 Active Cooling Techniques

To improve further heat transfer, nanofluids (oxide-based) are used instead of base fluids. The heat transfer characteristics of nanofluid are better than that of its base fluid. They possess relatively higher heat capacities, improved thermal conductivity, convective heat transfer coefficient, viscosity, etc., relative to their corresponding base fluids like water, oil, etc [16–25]. The use of Al_2O_3–water nanofluid to cool a mono-crystalline PV/T panel in the weather conditions of Malaysia, showed a decline in panel temperature from 79.1 to 42.2 °C for a 0.2 L/s flow rate and a nanofluid concentration ratio of 0.3% for 24 min under a 1000 W/m^2 irradiation [26]. The electrical efficiency of the system increased to 12.1%, and thermal efficiency reached 34.4% at the end of the experiment. Michael and Iniyan [27] used a copper sheet lamination over a 37 W PV panel, area 0.24 m^2, to reduce the thermal resistance and a comparative analysis method between 0.05% volume concentrations of copper oxide–water nanofluid (CuO/H_2O), and water was employed for the performance enhancement. The experiment was conducted under a constant laminar flow of 0.01 kg/s and was found that the thermal efficiency was improved to 45.76%. Maximum electrical efficiency of 6.18% and 7.62% was obtained using nanofluid with and without glazing, respectively. It was also concluded from the experiment that the electrical efficiency using nanofluid cooling, under identical conditions, is lesser than that using normal water cooling, but the thermal efficiency is greater for nanofluid cooled mechanism. Advanced materials like the carbon nanotubes (CNTs), spectrally tailorable multiparticle nanofluid filters, etc., are also employed in cooling techniques. Hjerrild et al. [28] used suspended 0.026 wt% core–shell silver–silica ($Ag–SiO_2$) nano-discs and carbon nanotubes (CNTs) in water. Maximum electrical efficiency of 6.6 + 0.5% and the greatest merit function were observed with this concentration of $Ag–SiO_2$ in water and electricity: thermal energy worth factor of 3 was obtained. Among the more diluted $Ag–SiO_2$ solutions tested, a higher merit function was obtained from $Ag–SiO_2$ particles suspended in CNT base fluid. The heat transfer is enhanced by flow of fluids [29–42].

60.2.2 Passive Cooling Techniques

Those cooling techniques do not require the aid of an external power source to extract heat from the PV panel. The heat is exacted from the panel and dissipated in the surrounding. Metals having a high value of thermal conductivity like copper and aluminium with or without application of fins are referred to enhance heat transfer to the surroundings. Several passive cooling arrangements used by the scholars are discussed below very briefly.

60.2.2.1 Air-Based Passive Cooling

Thermal grease applied to PV cell with aluminium fins was compared with a PV cell without any heat sink using a passive cooling technique [43]. A 9% relative gain in electric efficiency was achieved. Gains of 8, 27, 46 and 65% in output power were observed when the incident input insolation was 200, 400, 600 and 800 W/m^2, respectively. A maximum cooling of 25.2% was achieved for an insolation of 800 W/m^2, in turn, increasing the power output by around 20% and improving the performance. The main drawbacks of this system are that the fluctuation in PV cell temperature is very high and during peak periods the panel temperatures also become very high. An air-based cooling system for a solar PV system has also been presented by Du et al. [5].

60.2.2.2 Water-Based Passive Cooling

The water was used to cool down the back surface of a mono-crystalline PV module by employing a capillary effect achieved using spirally wrapped cotton wick structures immersed in water. The module temperature was successfully reduced by around 20 °C which means around 30% temperature reduction during the experiment time [44]. Due to the low thermal properties of air and water, it is efficient to transfer heat from the panel in hot regions. The heat transfer will be more using water as compared to air. The places where the water is a limiting factor, it is not advised to use. The water-based cooling system has been summarized in the ref. number [45].

60.2.2.3 Phase Changing Material-Based Passive Cooling

During a phase change, heat loss or gain happens at a constant temperature. The PCM is attached beneath the panel. Initially, the temperature panel as well as PCM is lower. The temperature of the combined system is lower than the phase change temperature of PCM. Hence, when the temperature of panel rises, the temperature of PCM also rises sensibly and heat is gained from the panel. Once the temperature of PCM reaches the melting point, then it initiates melting and absorbs heat gradually. This process stops after complete melting of PCM; later on, the temperature of PCM increases sensibly. The period of phase change depends on the volume, thermal property and chemical structure of PCM [46].

The thickness of the PCM layer has a significant effect on cooling. A very high or very low thickness of the layer reduces the heat transfer efficiency of the PV–PCM system. The optimum thickness of PCM behind the panel should be from 3.9 to 5.3 cm to maintain the PV at a lower temperature [47]. The low-temperature solidification of the PCM, especially at night, is also affected by the density of the PCM used. Experiments have shown that the top layers, in case of high-density PCMs, consume the heat after melting in the form of sensible heat, leaving the bottom layers either partially melted or not melted, during the day times [48]. These bottom layers lead to additional thermal resistance and in turn affect the solidification of the top layers at night. Hence, an optimized size and thickness of the PCM should be modelled before its application into the PV panels. Such a system was found to be inefficient where a 0.01-m-thick layer of organic paraffin-based RT-22 and a 0.015-m-thick layer of petroleum-based ceresine were attached separately to different PV panels [49].

60.3 Eutectic

Eutectics are a mixture of more than one substance taken in a fixed proportion, which melts and freezes at a lower temperature relative to their respective individual melting or freezing points. Bauer et al. have presented state-of-the-art mixtures of salts and recent developments [50]. The eutectic having melting point less than 30 °C is best suited for the application in the SPV cooling. Each of the substances changes phase propitiously and forms a mixture of the component crystals during crystallization without segregation [10]. The eutectics can be classified into three categories: (a) organic–organic mixture, (b) inorganic–inorganic mixture and (c) organic–inorganic mixture. Sharma et al. [10] have covered numbers of suitable eutectics of all three categories. Commonly used PCM with its properties has been summarized in Table 60.1.

60.4 Result and Discussion

Hasan et al. [52] used four different heat sinks and attached with PV panel to investigate the performance of different types of PCM. Electrical and thermal efficiencies were evaluated in two different climatic conditions. The system is not economical in Ireland but economical in Pakistan due to climatic condition. They experimented eutectic mixture of capric acid–palmitic acid. The mixture contained 75.2% by weight of 98% pure capric acid and 24.8% by weight of 98% pure palmitic acid. The peak temperature was reduced by 12 °C. Economic analysis was also performed.

60.5 Conclusions

Eutectic-based photovoltaic system proves to be an efficient and promising alternative solution for cooling. Other cooling techniques are also discussed briefly. The cooling techniques were broadly classified into active and passive cooling techniques which

Table 60.1 List of eutectics suitable for cooling of solar PV panel

PCM	Composition (weight %)	Melting/freezing point (°C)	Latent heat (kJ/kg)	Thermal conductivity (W/mK)
Capric–lauric acids [51]	-	20.6	179.0^{L} 188.0^{S}	0.139 0.143
Capric–palmitic acids [51, 52]	-	22.4	195.0	0.143
$C_{14}H_{28}O_2$ + $C_{10}H_{20}O_2$ [10]	34 + 66	24	147.7	–
$CaCl_2$ + $MgCl_2 \cdot 6H_2O$ [10]	50 + 50	25	95	–
CH_3CONH_2 + NH_2CONH_2 [10]	50 + 50	27	163	–
$Ca(NO_3) \cdot 4H_2O$ + $Mg(NO_3)_3 \cdot 6H_2O$ [10]	47 + 53	30	136	–
$CH_3COONa \cdot 3H_2O$ + NH_2CONH_2 [10]	40 + 60	30	200.5	–
Decanoic acid [12]	-	27–32 °C	152.7	0.372^{S} 0.153^{L}

L = liquid and S = solid

were further sub-categorized into air-based, water-based and nanofluid-based cooling techniques for simplicity, and the effort was to make an overview of research works that utilized various prominent and existing cooling techniques before specializing into the PCM cooling techniques. The main focus of this paper was to cover the usage of eutectics in detail in the temperature regulation of the PV panels. The conclusion based on the analysis of various research works is given below. The inorganic PCMs have higher thermal conductivity than their organic counterparts and are non-flammable. They also have high latent heat per unit volume and are cost-effective in application. But they have problems with their stability and may decompose, dehydrate or undergo phase segregation during working. Apart from that, they can be corrosive and usually require nucleating and thickening agents to prevent sub-cooling and phase segregation. On an economic front, this set-up is yet to be proven advantageous.

References

1. Philipps, S.P., Bett, D.A.W., Horowitz, K., Kurtz, S.: Current status of concentrator photovoltaic (CPV) technology. Tech. Rep. (National Renewal Energy Laboratory, Colorado, USA) (2015)
2. Suresh, A.K., Khurana, S., Nandan, G., Dwivedi, G., Kumar, S.: Role on nanofluids in cooling solar photovoltaic cell to enhance overall efficiency. Mater. Today Proc. **5**(9), 20614–20620 (2018)
3. Sharma, R., Gupta, A., Nandan, G., Dwivedi, G., Kumar, S.: Life span and overall performance enhancement of Solar Photovoltaic cell using water as coolant: a recent review. Mater. Today Proc. **5**(9), 18202–18210 (2018)
4. Rathour, R.S., Chauhan, V., Agarwal, K., Sharma, S., Nandan, G.: Cooling of solar photovoltaic cell: using novel technique. In: Lecture Notes in Mechanical Engineering, pp. 521–529. Springer Singapore (2019)
5. Luque, A., Hegedus, S. (eds.): Handbook of Photovoltaic Science and Engineering. Wiley (2011)
6. Radziemska, E.: The effect of temperature on the power drop in crystalline silicon solar cells. Renew. Energy **28**(1), 1–12 (2003)
7. da Cunha, J.P., Eames, P.: Thermal energy storage for low and medium temperature applications using phase change materials—a review. Appl. Energy **177**, 227–238 (2016)
8. Machniewicz, A., Knera, D., Heim, D.: Effect of Transition Temperature on Efficiency of PV/PCM Panels. Energy Procedia **78**, 1684–1689 (2015)
9. Nazar, R.: Improvement of efficiency of solar panel using different methods. Int. J. Electr. Electron. Eng. (Jan 2015)
10. Sharma, A., Tyagi, V.V., Chen, C.R., Buddhi, D.: Review on thermal energy storage with phase change materials and applications. Renew. Sustain. Energy Rev. **13**(2), 318–345 (2009)
11. Islam, M.M., Pandey, A.K., Hasanuzzaman, M., Rahim, N.A.: Recent progresses and achievements in photovoltaic-phase change material technology: a review with special treatment on photovoltaic thermal-phase change material systems. Energy Convers. Manag. **126**, 177–204 (2016)
12. Kawtharani, F., Kawtharani, A., Hammoud, M., abd Ali Shaito, A.H., Assi, A., Assi, I.: Cooling PV modules using phase change material. In: 29th International Conference on Microelectronics (ICM) (2017)
13. Teo, H.G., Lee, P.S., Hawlader, M.N.A.: An active cooling system for photovoltaic modules. Appl. Energy **90**(1), 309–315 (2012)
14. Krauter, S., Araújo, R.G., Schroer, S., Hanitsch, R., Salhi, M.J., Triebel, C., Lemoine, R.: Combined photovoltaic and solar thermal systems for facade integration and building insulation. Sol. Energy **67**(4–6), 239–248 (1999)
15. Moharram, K.A., Abd-Elhady, M.S., Kandil, H.A., El-Sherif, H.: Enhancing the performance of photovoltaic panels by water cooling. Ain Shams Eng. J. **4**(4), 869–877 (2013)
16. Sekhar, T.V.R., Nandan, G., Prakash, R., Muthuraman, M.: Investigations on viscosity and thermal conductivity of cobalt oxide- water nano fluid. Mater. Today Proc. **5**(2), 6176–6182 (2018)
17. Sekhar, T.V.R., Prakash, R., Nandan, G., Muthuraman, M.: Preparation of Co_3O_4–H_2O nanofluid and application to CR-60 Concentrating Solar Collector. Prog. Ind. Ecol. Int. J. **11**(3), 227–246 (2017)
18. Sekhar, T.V.R., Prakash, R., Nandan, G., Muthuraman, M.: Performance enhancement of a renewable thermal energy collector using metallic oxide nanofluids. Micro Nano Lett. **13**(2), 248–251 (2018)
19. Sekhar, T.V.R., Prakash, R., Nandan, G., Muthuraman, M.: Pressure drop characteristics and Efficiency enhancement by using TiO_2–H_2O nanofluid in a sustainable solar thermal energy collector. Int. J. Environ. Sustain. Dev. (2018)
20. Sekhar, T.V.R., Nandan, G., Prakash, R., Muthuraman, M.: Modeling a renewable energy collector and prediction in different flow regimes using CFD. Mater. Today Proc. **5**(2), 4563–4574 (2018)

21. Shukla, A.K., Sharma, A., Sharma, M., Nandan, G.: Thermodynamic investigation of solar energy-based triple combined power cycle. Energy Sources, Part A: Recover. Util. Environ. Eff. **41**(10), 1161–1179 (2018)
22. Krishn, S., Goyal, M., Nandan, G., Kumar, S., Kumar, P., Shukla, A.K.: Pool boiling using nanofluids: a review. In: Lecture Notes in Mechanical Engineering, pp. 325–336. Springer Singapore (2019)
23. Agrawal, T., Ajitkumar, R., Prakash, R., Nandan, G.: Sodium silicide as a hydrogen source for portable energy devices: a review. Mater. Today Proc. **5**(2), 3563–3570 (2018)
24. Vadhera, J., Sura, A., Nandan, G., Dwivedi, G.: Study of phase change materials and its domestic application. Mater. Today Proc. **5**(2), 3411–3417 (2018)
25. Sekhar, T.V.R., Nandan, G., Prakash, R.: Parabolic trough solar collectors. Eur. J. Sci. (EJS) pp. 43–53 (2018)
26. Yu, W., Xie, H.: A review on nanofluids: preparation, stability mechanisms, and applications. J. Nanomater. **2012**, 1–17 (2012)
27. Michael, J.J., Iniyan, S.: Performance analysis of a copper sheet laminated photovoltaic thermal collector using copper oxide—water nanofluid. Sol. Energy **119**, 439–451 (2015)
28. Hjerrild, N.E., Mesgari, S., Crisostomo, F., Scott, J.A., Amal, R., Taylor, R.A.: Hybrid PV/T enhancement using selectively absorbing Ag–SiO_2/carbon nanofluids. Sol. Energy Mater. Sol. Cells **147**, 281–287 (2016)
29. Rawat, K.S., Thakur, H.C., Nandan, G.: CFD analysis of a Pentagonal Rib over absorber plate of a solar air heater. In: 3rd International Conference on Manufacturing Excellence, pp. 191–196. Amity University Uttar Pradesh, Noida, India (2016)
30. Kumar, S., Nandan, G., Singh, G.K.: Numerical study of natural convection and radiative heat transfer of heated inner cylinder placed inside a isothermally cooled circular enclosure. In: Third International Conference on Manufacturing Excellence, pp. 179–183. Amity University Uttar Pradesh, Noida, India (2016)
31. Anamika, Thakur, H.C., Nandan, G.: A review on solar air heater performance using different artificial roughened Rib. In: International Conference Emerging Development In Engineering and Technology (ICEDET-2016), pp. 347–349. Shivalik College of Engineering, Dehradun, India (2016)
32. Sekhar, T.V.R., Nandan, G., Prakash, R.: Parabolic trough solar collectors. Eur. J. Sci. (EJS) **1**(1), 43–53 (2018)
33. Nandan, G., Sahoo, P.K., Kumar, R., Chatterjee, B., Mukhopadhyay, D., Lele, H.G.: Experimental investigation of sagging of a completely voided pressure tube of Indian PHWR under heatup condition. Nucl. Eng. Des. **240**(10), 3504–3512 (2010)
34. Nandan, G., Majumdar, P., Sahoo, P.K., Kumar, R., Chatterjee, B., Mukhopadhyay, D., Lele, H.G.: Study of ballooning of a completely voided pressure tube of Indian PHWR under heat up condition. Nucl. Eng. Des. **243**, 301–310 (2012)
35. Nandan, G., Lele, H.G., Sahoo, P.K., Chatterjee, B., Kumar, R., Mukhopadhyay, D.: Experimental investigation of heat transfer during LOCA with failure of emergency cooling system. In: 5th International Conference on Heat Transfer, Fluid Mechanics and Thermodynamics (2007)
36. Nandan, G., Sahoo, P.K., Kumar, R., Chatterjee, B., Mukhopadhyay, D., Lele, H.G.: Experimental investigation of sagging and ballooning for a completely voided pressure tube of Indian PHWR under heatup condition. World Acad. Sci. Eng. Technol. **61**, 197–204 (2010)
37. Nandan, G., Sahoo, P.K., Kumar, R., Chatterjee, B., Mukhopadhyay, D., Lele, H.G.: Experimental investigation of sagging and ballooning in LOCA for Indian PHWR. Eng. Lett. **18**(3) (2010)
38. Nandan, G., Sahoo, P.K., Kumar, R., Chatterjee, B., Mukhopadhyay, D., Lele, H.G.: Thermomechanical behavior of pressure tube of Indian PHWR at 20 bar pressure. World Acad. Sci., Eng. Technol. **61**, 205–213 (2010)
39. Majumdar, P., Chatterjee, B., Nandan, G., Mukhopadhyay, D., Lele, H.G.: Assessment of the code "PTCREEP" for IPHWR pressure tube ballooning study. J. Press. Vessel. Technol. Trans. ASME **133**(1) (2011)

40. Nandan, G., Sahoo, P.K., Kumar, R., Chatterjee, B., Mukhopadhyay, D., Lele, H.G.: Thermo-mechanical response of pressure tube during degraded cooling condition event in a pressurised heavy water reactor. In: 19th National and 8th ISHMT-ASME Heat and Mass Transfer Conference. JNTU Hyderabad, India (2008)
41. Nandan, G., Sahoo, P.K., Kumar, R., Chatterjee, B., Mukhopadhyay, D., Lele, H.G.: Thermo-mechanical behavior of pressure tube of Indian PHWR at 20 bar Pressure. In: International Conference on Mechanical, Industrial, and Manufacturing Engineering, Cape Town, South Africa (2010)
42. Kumar, R., Nandan, G., Sahoo, P.K., Chatterjee, B., Mukhopadhyay, D., Lele, H.G.: Ballooning of pressure tube under LOCA in an Indian pressurised heavy water reactor. In: 14th International Heat Transfer Conference, pp. 319–323. American Society of Mechanical Engineers Digital Collection, Washington, DC, USA (2010)
43. Cuce, E., Bali, T., Sekucoglu, S.A.: Effects of passive cooling on performance of silicon photovoltaic cells. Int. J. Low-Carbon Technol. **6**(4), 299–308 (2011)
44. Chandrasekar, M., Suresh, S., Senthilkumar, T., Karthikeyan, M.G.: Passive cooling of standalone flat PV module with cotton wick structures. Energy Convers. Manag. **71**, 43–50 (2013)
45. Sharma, R., Gupta, A., Nandan, G., Dwivedi, G., Kumar, S.: Life span and overall performance enhancement of solar photovoltaic cell using water as coolant: a recent review. In: 8th International conference on Materials Processing and Characterization (ICMPC 2018). GRIET, Hyderabad, India (2018)
46. Cabeza, L.F., Castell, A., Barreneche, C., de Gracia, A., Fernández, A.I.: Materials used as PCM in thermal energy storage in buildings: a review. Renew. Sustain. Energy Rev. **15**(3), 1675–1695 (2011)
47. Khanna, S., Reddy, K.S., Mallick, T.K.: Optimization of solar photovoltaic system integrated with phase change material. Sol. Energy **163**, 591–599 (2018)
48. Chandel, S.S., Agarwal, T.: Review of cooling techniques using phase change materials for enhancing efficiency of photovoltaic power systems. Renew. Sustain. Energy Rev. **73**, 1342–1351 (2017)
49. Klugmann-Radziemska, E., Wciso, P., Ryms, M., Denda, H.: The possibility of phase change materials (PCM) usage to increase efficiency of the photovoltaic modules. In: MATEC Web of Conferences 18, 03008 (2014)
50. Bauer, T., Laing, D., Tamme, R.: Recent progress in alkali nitrate/nitrite developments for solar thermal power applications. In: Molten Salts Chemistry and Technology, pp. 543–553. Wiley (2014)
51. Hasan, A., McCormack, S.J., Huang, M.J., Norton, B.: Characterization of phase change materials for thermal control of photovoltaics using differential scanning calorimetry and temperature history method. Energy Convers. Manag. **81**, 322–329 (2014)
52. Hasan, A., McCormack, S., Huang, M., Norton, B.: Energy and cost saving of a photovoltaic-phase change materials (PV-PCM) system through temperature regulation and performance enhancement of photovoltaics. Energies **7**(3), 1318–1331 (2014)

Chapter 61
Empirical Modelling of Kerf Characteristics in Laser Profile Cutting of Ni-Superalloy

P. Joshi, A. Sharma and V. Singh

Abstract Laser cutting is the most acceptable method to create complex geometries where it is difficult to cut sheet metal with close tolerances and high precision. In this study, empirical models for top kerf deviation and kef taper have been developed using response surface methodology (RSM). Box–Behnken Design (BBD)-based experiments on Ni-superalloy sheet have been conducted in laser cutting process. Lamp current, duration of pulse, frequency of pulses and speed of cutting are taken as process variables. The prime application of RSM is to build up a functional correlation between input and output parameters of the process. This relationship is analysed using analysis of variance (ANOVA) for the adequacy of models, and subsequently, adequate models have been used to predict the process behaviour.

Keywords Laser cutting · Superalloy sheet · Response surface methodology · Analysis of variance

61.1 Introduction

Advanced fabrication industries like aircrafts, automobiles, shipbuilding and nuclear industries demand superalloys due to their lightweight, high strength, chemical stability and resistance to wear at superior temperatures [1–3]. Superalloys may be classified based on the main constituent as iron, cobalt and nickel. The distinctive applications and exceptional mechanical properties of the superalloys restrict the unconventional sheet cutting processes to produce complicated profiles with close tolerances. Laser cutting is the suitable technique to cut sheet metals in required dimensions with proper accuracy and precision.

In laser cutting process, converging beam of laser light interacts with the sheet metal surface which causes heating. Subsequently, the focussed region of sheet metal converts into molten state after that molten material can be ejected from the cutting

P. Joshi · A. Sharma
Jaypee University of Engineering and Technology, A-B Road, Raghogarh, Guna, M.P., India

V. Singh (✉)
Galgotias College of Engineering and Technology, Greater Noida, U.P., India
e-mail: vivek.singh@galgotiacollege.edu

S. Yadav et al. (eds.), *Proceedings of International Conference in Mechanical and Energy Technology*, Smart Innovation, Systems and Technologies 174,
https://doi.org/10.1007/978-981-15-2647-3_61

zone by using the assist gas jet. The harmful by-products are restricted up to a well-defined minimal region, and in this way the process may be termed as environment-friendly [4]. Laser cutting has various advantages over other sheet cutting processes such as high cutting speed, capable to cut intricate shapes with close tolerances and no physical interaction [5]. The process can be automized easily so that operating cost is reduced. In modern fabrication industries, the prime concern is to obtain the highly accurate and precisely cut sheet metals for further processing. Such results can be achieved by using laser cutting process where accuracy of cut profile has been quantified in terms of deviation and taper angle. Due to the convergent and divergent nature of laser beam profile, the cutting of sheet metals is remained a challenging research area from the beginning to get the remarkably good quality cut [6–8]. The geometrical features of laser cut kerf (i.e. width, deviation and taper of the kerf) have been analysed by researchers to optimize the cutting parameters. Usually, parameter optimization is required for the cost-effective and proficient cutting of hard-to-cut materials.

Thawari et al. [9] analysed the effects of overlapping of laser spot in Nd-YAG laser cutting of Hastelloy sheet to observe the kerf width. They operated laser in pulsed wave mode and concluded that the width of kerf increases with the increment of spotoverlap, however surface roughness decreases. Cutting of austenitic stainless steel using Nd-YAG laser was reported by Ghany and Newishy [10]. They observed that the nature of variation of kerf width depends on some laser cutting variables significantly such as power, pulse repetition rate and gas pressure.

Dubey and Yadava [11] employed Taguchi methodology (TM) and response surface methodology (RSM) together to maximize the removal rate and minimize the kerf width simultaneously in laser cutting of steel sheet. Further, they employed principle component analysis coupled with TM to optimize the laser cutting of superalloy sheet [3]. The influences of gas pressure, pulse duration and frequency were examined by Rao and Yadava [12] in profile cutting of SUPERNI 718 sheet. They optimized laser cut kerf geometry using grey relational analysis (GRA). Sharma et al. [13] optimized laser cut characteristics using TM. They conducted experiments based on L_{27} orthogonal array (OA) for profile cutting. They found the value of taper angle is more for curved profile than straight profile. Sharma and Yadava [14, 15] used combination of TM and RSM to build up the model for average kerf taper in profile cutting of Al-alloy sheet of 0.7 mm thickness.

In this paper, empirical modelling for deviation at top side (TKD) and taper angle (KT) of kerf has been carried out using RSM in Nd-YAG laser cutting of Inconel 718 sheet. Lamp current (LC), duration of pulse (PW), frequency of pulses (PF) and speed of cutting (CS) have been taken as process variables for the study. Further, the experiments have been performed using BBD and their results are used to develop the response surface models. Subsequently, the predicted results obtained from the developed models have been analysed.

61.2 Experimentation

In the present study, 250W Nd-YAG laser cutting system has been used to conduct the experiments on Ni-based superalloy (Inconel-718) sheet. Compressed air is used for ejecting the molten sheet metal from the localized area after interaction with laser. Laser beam and compressed air are passed through the same nozzle which is having a diameter of 1.5 mm. Hence, it can be stated that the flow of air is co-axial with laser beam. During experimentation, the distance between focussing lens and sheet metal (35 mm), nozzle tip distance (2 mm), thickness of the workpiece (0.7 mm), nozzle diameter and air pressure (10 kg/cm^2) were kept constant.

Lamp current (LC) for the generation of laser, duration of pulse (PW), frequency of pulses (PF) and speed of cutting (CS) are considered as laser cutting parameters. The range of cutting parameters to obtain the through cut were chosen on the basis of initial phase of experiments. After deciding the upper and lower limit of laser cutting parameters, experiments were planned based on three levels and four factors (laser cutting parameters) using BBD. Values of these variables with respect to each level are listed in Table 61.1. For developing and regulating the quadratic models, BBD is commonly used. It is rotatable and more useful for small number of parameters than central composite design (CCD) [16]. Researchers have used BBD in optimization of laser cut characteristics [17, 18]. Farthest conditions are not considered in this design because it does not contain such parameter settings where highest and lowest levels of parameters are used together [19]. Design is more capable than full factorial design (FFD) because with the increase in number of parameter, the FFD becomes costly design [20].

The curved profile is considered as a one-fourth sector of circle, and the length of this sector (arc) is 28.65 mm. Here, geometry of laser cut has been investigated by considering TKD and KT in the laser cut kerf. Measured values of top and bottom kerf widths are used for finding TKD and KT. The observations are taken at five different locations along the path of laser cut. The observations have been taken at different orientations (i.e. 0°, 22.5°, 45°, 67.5° and 90°) from the base along the cut.

TKD is the difference of maximum and minimum values of top kerf width, and it can be calculated as follows:

$$\text{TKD} = \text{Maximum TKW} - \text{Minimum TKW} \tag{61.1}$$

Table 61.1 Process variables and their levels

Cutting variables	Levels		
	L1 (−1)	L2 (0)	L3 (+1)
LC (A)	200	220	240
PW (ms)	2	3	4
PF (Hz)	4	8	12
CS (mm/min)	20	40	60

The value of KT in degree has been calculated using the following formula

$$KT = \left| \frac{(TKW - BKW) \times 180}{2\Pi \times t} \right| \quad (61.2)$$

where thickness of workpiece (t) is used in mm. The calculated values of responses corresponding to each trial are given in Table 61.2.

Table 61.2 Experimental results based on BBD

Exp. No.	LC	PW	PF	CS	TKD (mm)	KT (°)
1	+1	0	−1	0	0.629	2.17400
2	0	0	0	0	0.369	2.11900
3	−1	0	0	+1	0.087	1.96600
4	+1	0	+1	0	0.523	5.88900
5	−1	+1	0	0	0.22	4.30100
6	0	0	+1	+1	0.337	2.27000
7	0	+1	0	−1	0.499	5.28100
8	+1	+1	0	0	0.517	9.08819
9	0	−1	0	−1	0.36	6.16463
10	0	+1	−1	0	0.375	5.56933
11	0	0	+1	−1	0.472	7.69917
12	0	−1	+1	0	0.218	6.15581
13	0	−1	0	+1	0.158	5.36649
14	−1	0	−1	0	0.193	3.29838
15	−1	0	0	−1	0.298	3.62028
16	0	0	−1	−1	0.438	3.09200
17	0	0	−1	+1	0.127	6.21800
18	+1	0	0	−1	0.777	5.16600
19	0	+1	+1	0	0.305	6.56590
20	0	+1	0	+1	0.201	6.75551
21	+1	−1	0	0	0.357	4.20235
22	0	0	0	0	0.316	3.25400
23	0	−1	−1	0	0.171	3.93337
24	−1	0	+1	0	0.192	0.66600
25	0	0	0	0	0.417	2.25800
26	−1	−1	0	0	0.217	5.26948
27	+1	0	0	+1	0.53	7.93900

61.3 Response Surface Methodology (RSM)

RSM comprises of statistical techniques which are useful for developing the functional relationship between various process parameters and responses. RSM finds applications in certain conditions where a large number of variables influence some performance measures significantly. RSM makes use of regression analysis (RA) on experimental data obtained from designed experiments. RA can further be used to develop response surface models for finding the approximate minima or maxima. These developed models show the mathematical relation between input and output parameters [21]. A generic response surface model can be expressed as follows:

$$y = \beta_0 + \sum \beta_i x_i + \sum \beta_{ii} x_{ii}^2 + \sum \beta_{ij} x_i x_j \quad (61.3)$$

where β_0, β_i, β_{ii} and β_{ij} all are the coefficients of regression and x_i, $x_i x_j$ are the process parameters which influence the process performance. Regression coefficients are determined by using the regression analysis where data for analysis have been collected from the experimentation. In the present study, experiments have been performed using BBD. This rotatable design has been used for small number of input parameters. In BBD, response surface models can be developed with less number of experiments [19]. The design is more efficient than central composite design (CCD) and full factorial design. The analysis of variance (ANOVA) is the required part of this method which has been carried out at last to check the competency of the models.

61.4 Results and Discussion

The results of TKD and KT obtained from the experimentation are given corresponding to each setting of experiments in Table 61.2. Further, these results have been used for analysis in MINITAB statistical software. Subsequently, mathematical models for TKD and KT have been developed which are expressed as follows:

$$\begin{aligned} y_{\mathrm{TKD}} = {} & 3.19625 - 0.03984 \times \mathrm{LC} - 0.18150 \times \mathrm{PW} + 0.10237 \times \mathrm{PF} - 0.00372 \\ & \times \mathrm{CS} + 0.00011\mathrm{LC}^2 - 0.07562 \times \mathrm{PW}^2 - 0.00174 \times \mathrm{PF}^2 + 0.00003 \times \mathrm{CS}^2 \\ & + 0.00196 \times \mathrm{LC} \times \mathrm{PW} - 0.00033 \times \mathrm{LC} \times \mathrm{PF} - 0.00002 \times \mathrm{LC} \times \mathrm{CS} \\ & - 0.00731 \times \mathrm{PW} \times \mathrm{PF} - 0.00120 \times \mathrm{PW} \times \mathrm{CS} + 0.00055 \times \mathrm{PF} \times \mathrm{CS} \end{aligned} \quad (61.4)$$

$$\begin{aligned} y_{\mathrm{KT}} = {} & 174.105 - 0.955 \times \mathrm{LC} - 30.280 \times \mathrm{PW} - 3.437 \times \mathrm{PF} - 0.778 \times \mathrm{CS} \\ & + 0.001\mathrm{LC}^2 + 2.366 \times \mathrm{PW}^2 + 0.030 \times \mathrm{PF}^2 + 0.004 \times \mathrm{CS}^2 \\ & + 0.073 \times \mathrm{LC} \times \mathrm{PW} + 0.020 \times \mathrm{LC} \times \mathrm{PF} + 0.003 \times \mathrm{LC} \times \mathrm{CS} \\ & - 0.077 \times \mathrm{PW} \times \mathrm{PF} + 0.028 \times \mathrm{PW} \times \mathrm{CS} - 0.027 \times \mathrm{PF} \times \mathrm{CS} \end{aligned} \quad (61.5)$$

Table 61.3 ANOVA results

Source	TKD		KT	
	F-value	*p*-value	*F*-value	*p*-value
Regression	10.17	0.000	8.80	0.000
Linear	0.88	0.505	9.33	0.001
Square	3.41	0.044	10.43	0.001
Interaction	0.81	0.579	8.60	0.001
Lack of fit	1.99	0.381	2.44	0.326
	$S = 0.06820$, R-sq = 92.2%, R sq (adj) = 83.2%		$S = 0.9181$, R-sq = 91.1%, R-sq (adj) = 80.8%	

After developing the response surface model, the results of models have been analysed. The S value for deviation and taper angle are found as 0.06820 and 0.9181, respectively, in the regression analysis, and these are smaller. The R-Sq and R-Sq (adj) values for TKD are 92.2% and 83.2% and for KT are obtained as 91.1% and 80.8%, respectively. The values of R-square and R-square (adjusted) are reasonably high for both models. Hence, the developed models are good enough for fitting the data in cutting of superalloy sheet. Subsequently, F-value and p-value have been obtained in the analysis to test the capability of models and these results are given in Table 61.3. From the results obtained, it can be observed that the source of regression and square effects has been found significant in the analysis because their p-values are lower than 0.05 for each response. F-value for lack of fit in the analysis has been computed for TKD and KT as 1.99 and 2.44, respectively. F-distribution value from the standard table (at 95% confidence level) is found as 19.43 which is higher than the computed values. Therefore, it can be stated that the developed response surface models for the TKD and KT are reasonable at 95% confidence level [21].

Residuals for each experimental run have been determined by using the developed response surface models. The difference of observed values and forecasted values can be termed as residuals, and the probabilities of these residuals in case of deviation and taper may be observed from Fig. 61.1. It may be observed from the figure that the data are randomly distributed along the straight line which indicates that the data are normally distributed. The maximum variations between observed and predicted values for TKD are found as 0.3 and 0.5 and for KT as 4.0 and 4.0 from Fig. 61.2. These plots are not showing any certain pattern. Hence, the fitted models are adequate. The nature of variation of residuals with respect to observation number is shown in Fig. 61.3. The figure shows that the distribution of residuals is equal in positive and negative directions. So it can be stated that the data are not dependent.

Developed response surface models have been used for the prediction of TKD and KT values, and these predicted values have been compared with the original values obtained from BBD-based experimentation. These results are given in Table 61.4 against each trial. The maximum percentage error in the prediction of TKD and KT values has been found as 14.06% and 14.18%, respectively, by using the response surface models.

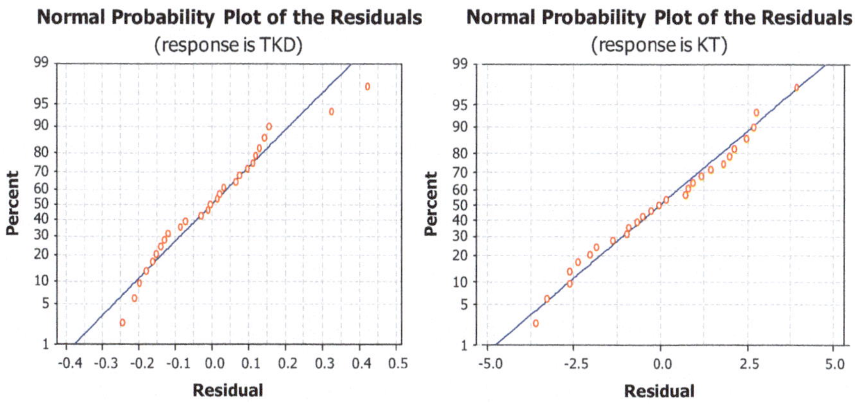

Fig. 61.1 Normal probability plot of TKD and KT

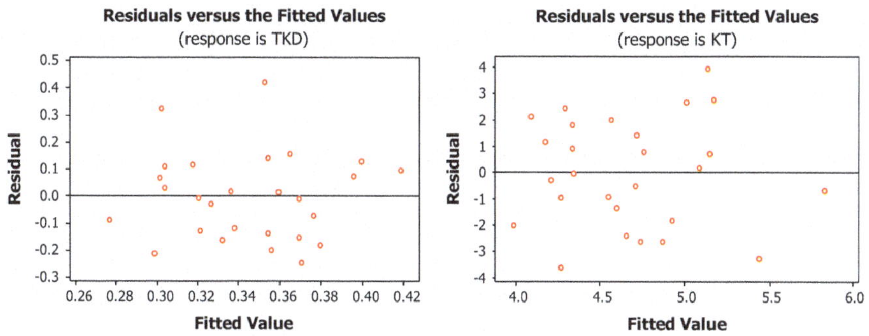

Fig. 61.2 Plot of residuals versus the fits for TKD and KT

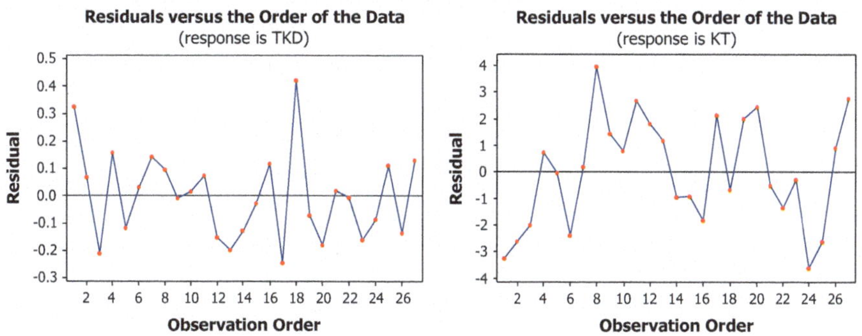

Fig. 61.3 Plot of residuals versus observation order for TKD and KT

Table 61.4 Predicted results of TKD and KT using RSM along with experimental results

Exp. No.	TKD (mm)		KT (°)	
	Experimental	Predicted	Experimental	Predicted
1	0.629	0.5755	2.17400	2.77913
2	0.369	0.36733	2.11900	2.54367
3	0.087	0.13442	1.96600	2.07487
4	0.523	0.542	5.88900	6.77962
5	0.22	0.17042	4.30100	3.18889
6	0.337	0.28608	2.27000	2.72832
7	0.499	0.49583	5.28100	6.39956
8	0.517	0.60325	9.08819	8.67228
9	0.36	0.34183	6.16463	6.45775
10	0.375	0.33658	5.56933	5.81756
11	0.472	0.43208	7.69917	7.09059
12	0.218	0.24958	6.15581	5.56623
13	0.158	0.15583	5.36649	5.23675
14	0.193	0.16867	3.29838	3.39658
15	0.298	0.35042	3.62028	4.37319
16	0.438	0.50108	3.09200	1.9862
17	0.127	0.17908	6.21800	6.17911
18	0.777	0.72275	5.16600	4.71578
19	0.305	0.29708	6.56590	6.03143
20	0.201	0.21383	6.75551	7.45121
21	0.357	0.41875	4.20235	4.66699
22	0.316	0.36733	3.25400	2.54367
23	0.171	0.17208	3.93337	4.12649
24	0.192	0.24017	0.66600	1.04969
25	0.417	0.36733	2.25800	2.54367
26	0.217	0.14292	5.26948	5.03792
27	0.53	0.47075	7.93900	6.84474

61.5 Conclusions

In this paper, empirical modelling for kerf deviation and taper has been carried out using response surface methodology during laser cutting of Inconel 718 sheet. The following conclusions have been drawn from this study:

1. The empirical model for kerf deviation and taper angle in laser cutting of Inconel 718 sheet has been found adequate.

2. Linear, quadratic and interaction influences of process variables are found considerable for TKD model.
3. The average prediction errors for TKD and KT models have been found as 14.06% and 14.18%, respectively.

References

1. Kalpakjian, S., Schmid, S.R.: Manufacturing Engineering and Technology. Addison Wesley Private Limited, India (2000)
2. Sharma, A., Yadava, V.: Modeling and optimization of cut quality characteristics for pulsed Nd-YAG laser curved profile cutting of Ni-based superalloy thin sheet. Lasers Eng. **31**(5/6), 351–382 (2015)
3. Dubey, A.K., Yadava, V.: Multi-objective optimization of Nd:YAG laser cutting of nickel-based super alloy sheet using orthogonal array with principle component analysis. Opt. Lasers Eng. **46**(2), 124–132 (2008)
4. Sharma, A., Yadava, V.: Experimental analysis of Nd-YAG laser cutting of sheet materials—a review. Opt. Laser Technol. **98**, 264–280 (2018)
5. Madic, M., Anttucheviciene, J., Radovanovic, M., et al.: Determination of laser cutting process conditions using the preference selection index method. Opt. Laser Technol. **89**, 214–220 (2017)
6. Steen, W.M.: Laser Material Processing. Springer, New York (1991)
7. Chryssolouris, G.: Laser Machining—Theory and Practice (Mechanical Engineering Series). Springer, New York (1991)
8. Dubey, A.K., Yadava, V.: Laser beam machining—a review. Int. J. Mach. Tools Manuf. **48**, 609–628 (2008)
9. Thawari, G., SarinSundar, J.K., Sundararajan, G., et al.: Influence of process parameters during pulsed Nd: YAG laser cutting of nickel-base superalloys. J. Mater. Process. Technol. **170**, 222–239 (2005)
10. Ghany, K.A., Newishy, M.: Cutting of 1.2 mm thick austenitic stainless steel sheet using pulsed and CW Nd: YAG laser. J. Mater. Process. Technol. **168**, 438–447 (2005)
11. Dubey, A.K., Yadava, V.: Multi-objective optimization of laser beam cutting process. Opt. Laser Technol. **40**(3), 562–570 (2008)
12. Rao, R., Yadava, V.: Multi-objective optimization of Nd-YAG laser cutting of thin super alloy sheet using grey relational analysis with entropy measurement. Opt. Laser Technol. **41**(8), 922–930 (2009)
13. Sharma, A., Yadava, V., Rao, R.: Optimization of kerf quality characteristics during Nd:YAG laser cutting of nickel based superalloy sheet for straight and curved cut profiles. Opt. Lasers Eng. **48**, 915–925 (2010)
14. Sharma, A., Yadava, V.: Modelling and optimization of cut quality during pulsed Nd-YAG laser cutting of thin Al-alloy sheet for straight profile. Opt. Laser Technol. **44**, 159–168 (2012)
15. Sharma, A., Yadava, V.: Modelling and optimization of cut quality characteristics during Nd: YAG laser cutting of al-alloy for curved profile. Opt. Lasers Eng. **51**, 77–88 (2013)
16. Bhemuni, V.P., Chalamalasetti, S.R.: A review on hard turning by using of experiments. J. Manuf. Sci. Prod. **13**(3), 209–219 (2013)
17. Eltawahni, H.A., Olabi, A.G., Benyounis, K.Y.: Investigating the CO_2 laser cutting parameters of MDF wood composite material. Opt. Laser Technol. **43**(3), 648–659 (2011)
18. Eltawahni, H.A., Hagino, M., Benyounis, K.Y., Inoue, T., Olabi, A.G.: Effect of CO_2 laser cutting process parameters on edge quality and operating cost of AISI316L. Opt. Laser Technol. **44**(4), 1068–1082 (2012)

19. Ferreira, S.L.C., Bruns, R.E., Ferreira, H.S., Matos, G.D., David, J.M., Brandao, G.C., da Silva, E.G.P., Portugal, L.A., dos Reis, P.S., Souza, A.S., dos Santos, W.N.L.: Box-Behnken design an alternative for the optimization of analytical methods. Anal. Chim. Acta. **597**(2), 179–186 (2007)
20. Pandey, A.K., Dubey, A.K.: Modeling and optimization of kerf taper in pulsed laser cutting of duralumin sheet. In: International Manufacturing Science & Engineering Conference (MSEC), pp. 1–8 (2012)
21. Montgomery, D.C.: Design and analysis of experiments. Willy India, New York (2004)

Chapter 62
Thermal Modeling and Analysis of Double-Slope Solar Still

A. K. Sethi, V. K. Dwivedi, Sudhanshu Sharma and A. K. Sharma

Abstract Freshwater is an important prerequisite for sustenance of human being as well as the quality of life. It is required for human settlement, industry, agriculture, forest, recreation, etc. It is very precious and limited, needs to be treated with utmost care. In the twenty-first century, the freshwater demand increases due to rise in population, water wastage and water pollution due to industrialization as these factors adversely affecting its availability, accessibility and quality. Hence, there is an urgent need for fulfillment of water and sustainable water management through non-convention energy resources. In this study, solar still in Delhi NCR region has been made and thermal model on the basis of energy balance equation has been developed. This model enabled to calculate the variation of the water and condensing cover temperature (inner) and hourly distillate yield and validate the results.

Keywords Thermal model · Solar still · Statistical analysis · Distillate yield

Nomenclature

C	Specific heat capacity ($\text{Jkg}^{-1}\,°\text{C}^{-1}$)
Gr.Pr	Product of Grash of number and Prandtl number
I_t	Solar radiation (W/m^2)
I_C	Hourly incident solar radiation on collector plate
L	Latent heat of vaporization of water (J/kg)
m_w	Hourly distillate
M_w	Daily yield (kg)
n	Constant
Nu	Nusselt number
P_{ci}	Vapor pressure at T_{ci} (N/m)

A. K. Sethi (✉) · V. K. Dwivedi · S. Sharma
Galgotias College of Engineering & Technology, Greater Noida, India
e-mail: anilkrsethi@galgotiacollege.edu

A. K. Sharma
Institute of Engineering & Technology, Lucknow, India

S. Yadav et al. (eds.), *Proceedings of International Conference in Mechanical and Energy Technology*, Smart Innovation, Systems and Technologies 174,
https://doi.org/10.1007/978-981-15-2647-3_62

P_w	Vapor pressure at T_w (N/m^2)
q_{ew}	Evaporative heat transfer rate (W/m^2)
$\dot{Q}_{ew}$	Energy for evaporation of water vapors (Wm^{-2})
t	Time (s)
Δt	Small interval of time (s)
T_a	Ambient air temperature
T_{ci}	Temperature of condensing cover
T_w	Water temperature
α'_b	Solar flux absorbed by baseliner
α'_g	Fraction of solar energy absorbed by condensing cover
α'_w	Fraction of solar energy absorbed by water mass
α	Absorptivity of collector
τ	Transitivity of collector
ε	Emissivity
ΔT	Temperature difference (°C)
σ	Stefan–Boltzman's constant (Wm^{-2} K^{-4})

Subscript

E	East side
W	West side
i	Inner side
g	Glass
w	Water

62.1 Introduction

The empirical relations of heat transfer coefficient for a conventional solar still have been analyzed and presented by Dunkle [1]. Morse and Read [2] included the thermal capacity of the system and accordingly carried out a transient analysis. They have expressed the heat fluxes as the functions of the condensing cover temperature. Thus, the glass temperature has been obtained by a graphical solution. Nawayseh et al. [3] obtained the heat and mass transfer coefficients in solar desalination on the basis of humidification process. Shukla and Sorayan [4] developed thermal model for solar still and determined the heat transfer coefficient by using glass cover temperatures and instantaneous efficiency and distillate yield for double-slope distillation systems were derived by Shukla and Rai [5]. Dev and Tiwari [6] have developed characteristic equations for passive solar still for solar intensity, water and ambient temperature as the parameter and found that lower water depth and condensing cover inclination of

45° gives better performance. Kumar et al. [7] worked on passive and hybrid solar still to calculate the convective heat transfer coefficient for a water depth of 5 cm. Thermal analysis of solar still based on energy balance equations was carried out by Morad et al. [8] and revealed that active solar still produces more freshwater as compared with passive solar still. Feilizadeh et al. [9] studied the effects of distance between water surface and condensing cover distance in the solar still and revealed that distillate yield can increase by varying the distance. Gnanaraj et al. [10] studied single and double basin solar still and found that double basin still enhanced the distillate yield by 58%. Layek [11] studied the effects of different absorbing materials in solar still and found that distillate increased by adding absorbing material in the solar still. In this research, the relationship between water, glass temperature and hourly distillate has been derived analytically and compared with experimental data.

62.2 Experimental Setup

The line diagram of solar still under study has been shown in Fig. 62.1. Solar still of basin area (A_1) 2 m^2 was made by using low thermal conductivity material and painted black to increase absorption of solar radiation integrated with flat plate collector area

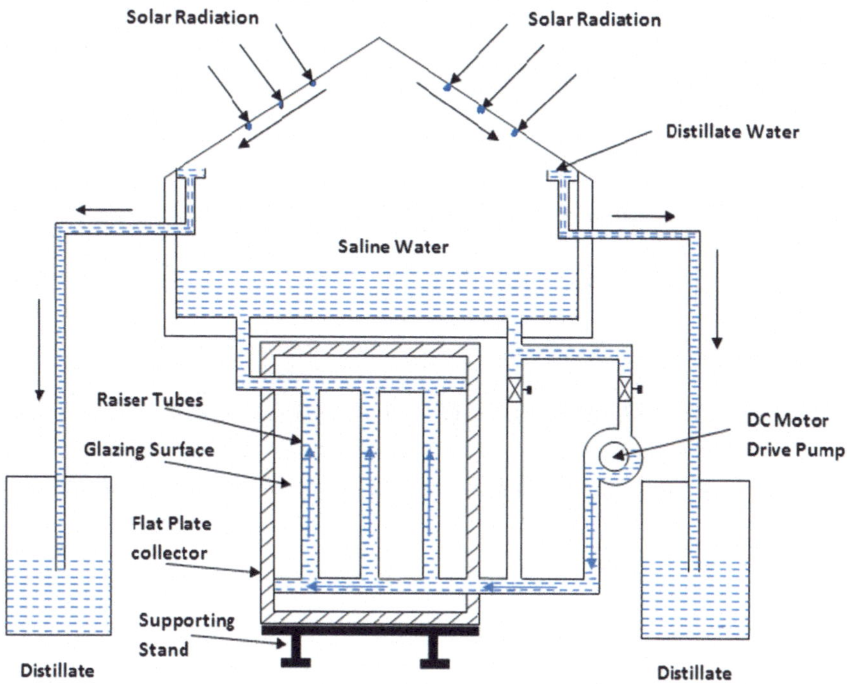

Fig. 62.1 Line diagram of experimental setup

(A_c) of 2 m^2. The plain transparent glass of 3 mm thickness and inclined at 25° is used for condensing cover.

The distillate yield, ambient temperature, water temperature, inner glass temperature and solar intensity were recorded during sunshine hour as shown in Table 62.1.

62.3 Thermal Modeling Solar Still

Solar radiation passes through transparent condensing cover and is absorbed by the water in the solar still and heated the water and form vapor that rises and condensed when comes in contact with glass cover. The different modes of heat transfer are shown in Fig. 62.2.

Thermal model on the basis of energy balance equation has been developed with these assumptions.

- Hourly intensity of solar radiation and temperature are constant;
- Temperature gradient is negligible for the water in the basing and condensing cover of solar still;
- Basin is airtight and no heat loss through leakage;
- Coefficients of heat transfer convective (h_{cw}) and evaporative (h_{ew}) are temperature dependent.

The energy balance equations are written as follows:

$$\left\{\alpha'_g I_{tE} + h_{1\,wE}(T_w - T_{ciE})\right\} A_2 = h_{2g}(T_{ciE} - T_a) A_2 \quad (62.1)$$

where $h_{1\,wE} = h_{cwE} + h_{ewE} + h_{rwE}$

$$\left\{\alpha'_g I_{tW} + h_{1wW}(T_w - T_{ciW})\right\} A_2 = h_{2g}(T_{ciW} - T_a) A_2 \quad (62.2)$$

where, $h_{1\,wW} = h_{cwW} + h_{ewW} + h_{rwW}$

$$\alpha'_b A_1 (I_{tE} + I_{tW}) = \{h_3(T_b - T_w) - h_b(T_b - T_a)\} A_1 \quad (62.3)$$

$$\left\{\alpha'_w (I_{tE} + I_{tW}) + h_3(T_b - T_w)\right\} A_1 - \{h_{1\,wE}(T_w - T_{ciE}) - h_{1\,wW}(T_w - T_{ciW})\} A_2 + \dot{Q}_u = \left((\mathrm{MC})_w \frac{dT_w}{dt} A_1\right) \quad (62.4)$$

where

$$\dot{Q}_u = A_c F'[(\alpha\zeta) I_C - h_4(T_w - T_a)] \quad (62.5)$$

Table 62.1 Hourly variation of solar intensity, various temperatures and yield for double-slope active solar still at 0.03 m water depth on June 22, 2018

Time (h)	T_a (°C)	T_w (°C)	I_c (W/m^2)	T_{ciE} (°C)	I_{tE} (W/m^2)	m_{wE} (kg)	T_{ciW} (°C)	I_{tW} (W/m^2)	m_{wW} (kg)
7–8 AM	28	29	160	28	140	0.02	30	120	0.01
8–9 AM	30	35	280	34	260	0.07	35	220	0.06
9–10 AM	31	40	380	36	360	0.09	35	280	0.06
10–11 AM	33	44	420	40	420	0.07	39	380	0.05
11–12 N	34	48	540	42	500	0.17	41	440	0.14
12–1 PM	35	52	600	45	540	0.24	49	520	0.21
1–2 PM	36	54	660	50	620	0.31	55	700	0.33
2–3 PM	36	55	700	54	720	0.38	60	740	0.39
3–4 PM	36	51	540	58	460	0.43	61	560	0.43
4–5 PM	36	49	440	53	400	0.36	56	460	0.35
5–6 PM	35	49	240	48	220	0.31	52	300	0.33

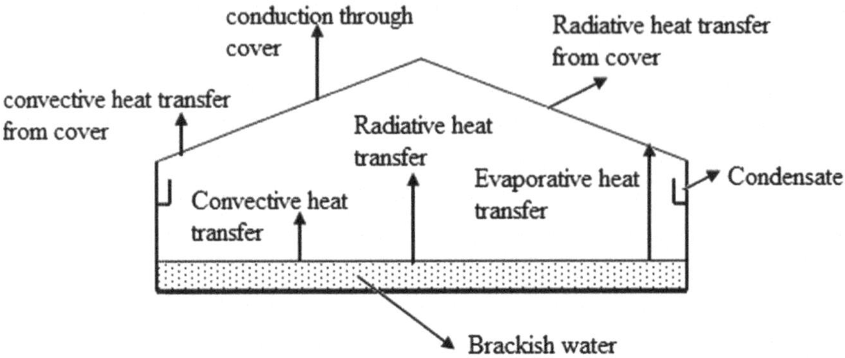

Fig. 62.2 Modes of heat transfer

Equations (62.1–62.4) represent for east-facing glass cover, west-facing glass cover, baseliner and water mass, respectively.

The expressions for T_{ciE}, T_{ciW}, and T_{b} were obtained from Eqs. (62.1–62.3) are as follows:

$$T_{\text{ciE}} = \frac{\alpha'_{\text{g}} I_{\text{tE}} + h_{1\,\text{wE}} T_{\text{w}} + h_{2\text{g}} T_{\text{a}}}{h_{2\text{g}} + h_{1\,\text{wE}}} \tag{62.6}$$

$$T_{\text{ciW}} = \frac{\alpha'_{\text{g}} I_{\text{tW}} + h_{1\,\text{wW}} T_{\text{w}} + h_{2\text{g}} T_{\text{a}}}{h_{2\text{g}} + h_{1\,\text{wW}}} \tag{62.7}$$

$$T_{\text{b}} = \frac{\alpha'_{\text{b}} (I_{\text{tE}} + I_{\text{tW}}) + h_3 T_{\text{w}} - h_{\text{b}} T_{\text{a}}}{h_3 - h_{\text{b}}} \tag{62.8}$$

The values of T_{ciE} and T_{ciW} and T_{b} were substituted into Eqs. (62.4) and (62.4) reduces into the following forms:

$$\frac{\text{d}T_{\text{w}}}{\text{d}t} + aT_{\text{w}} = f(t) \tag{62.9}$$

where $a = \frac{1}{(MC)_{\text{w}}}\left[\frac{2h_3 h_{\text{b}}}{h_3 + h_{\text{b}}} + \frac{h_{2\text{g}} h_{1\,\text{wE}}}{h_{2\text{g}} + h_{1\,\text{wE}}} + \frac{h_{2\text{g}} h_{1\,\text{wW}}}{h_{2\text{g}} + h_{1\,\text{wW}}} + A_{\text{c}} F' h_4\right]$and,

$$f(t) = \frac{1}{(MC)_{\text{w}}}\left[\left(\alpha'_{\text{w}} + \frac{\alpha'_{\text{b}} h_3}{h_3 + h_{\text{b}}}\right)(I_{\text{tE}} + I_{\text{tW}}) + A_{\text{c}} F' \{(\alpha\zeta) I_{\text{C}} + h_4 T_{\text{a}}\} + \frac{h_{\text{tE}}\left(\alpha'_{\text{g}} I_{\text{tE}} + h_{2\text{g}} T_{\text{a}}\right) + h_{\text{tW}}\left(\alpha'_{\text{g}} I_{\text{tW}} + h_{2\text{g}} T_{\text{a}}\right)}{h_{2\text{g}} + h_{1\,\text{wE}}} + \frac{h_3 h_{\text{b}}}{h_3 + h_{\text{b}}} T_{\text{a}}\right]$$

In order to solve Eq. (62.9), time interval was considered small and $f(t)$ and heat transfer coefficients were also constant during this time interval.

The solution of Eq. (62.9) with boundary condition becomes

$$T_W = \frac{\overline{f(t)}}{a}\left[1 - \exp(-a\Delta t)\right] + T_{W0}\exp(-a\Delta t) \tag{62.10}$$

The obtained values of T_{ciE}, T_{ciW}, T_W from Eqs. (62.6), (62.7) and (62.10) become initial conditions for calculation of the next set of value. Similarly, other sets of value for different time intervals have been calculated by this procedure.

The heat transfer coefficient h_{cw} [1], h_{ew} [12] and h_{rw} [13] were determined by the following equations.

$$h_{cw} = 0.884\left[(T_W - T_{ci}) + \frac{(P_W - P_{ci})(T_W + 273.15)}{268.9 \times 10^3 - P_W}\right]^{1/3} \tag{62.11}$$

$$h_{ew} = 16.273 \times 10^{-3} h_{cw}\frac{(P_W - P_{ci})}{(T_W - T_{ci})} \tag{62.12}$$

$$h_{rw} = \frac{\sigma_c\left(T_W^2 + T_C^2\right)(T_W + T_C)}{\frac{1}{\epsilon_w} + \left(\frac{1}{\epsilon_c} - 1\right)\frac{A_1}{A_c}} \tag{62.13}$$

To evaluate mass rate for condensing covers [14], the following expression has been used:

$$m_w = \frac{0.01623}{L}.\frac{\lambda}{L_v}A_1.t.(P_w - P_{ci}).C(\mathrm{Gr.Pr})^n \tag{62.14}$$

when predicted values are validated with the experimental value, Chapra and Canale [15] have given the following expression for coefficient of correlation (R_c) and root mean square of percent deviation (R_e);

$$R_c = \frac{N\sum X_i Y_i - \left(\sum X_i\right)\left(\sum Y_i\right)}{\sqrt{N\sum X_i^2 - \left(\sum X_i\right)^2}\sqrt{N\sum Y_i^2 - \left(\sum Y_i\right)^2}} \tag{62.15}$$

and,

$$R_e = \sqrt{\frac{\sum (e_i)^2}{N}} \tag{62.16}$$

where

$$R_{ei} = \left[\frac{X_i - Y_i}{X_i}\right] \times 100$$

where the number of observations (N), predicted value (X_i) and experimental value (Y_i).

62.4 Results and Discussion

Thermal model of the above experimental setup has been developed and ambient, water, condensing cover temperatures and distillate yield were taken for discussion. The recorded ambient temperature and solar intensity are shown in Figs. 62.3 and 62.4, respectively.

From Fig. 62.4, it is observed that solar intensities falling on east and west condensing surfaces are different during same time period. This is the reason, for a given value of water temperature, the values of glass temperatures for east and west condensing surfaces are different.

The graph between predicted values of water temperature calculated by Eq. (62.10), condensing cover temperature in east and west side by Eqs. (62.6 and 62.7) and distillate yield by Eq. (62.14) versus experimental value are shown in Figs. 62.5, 62.6 and 62.7, respectively. The values of R_c and R_e are shown in Table 62.2.

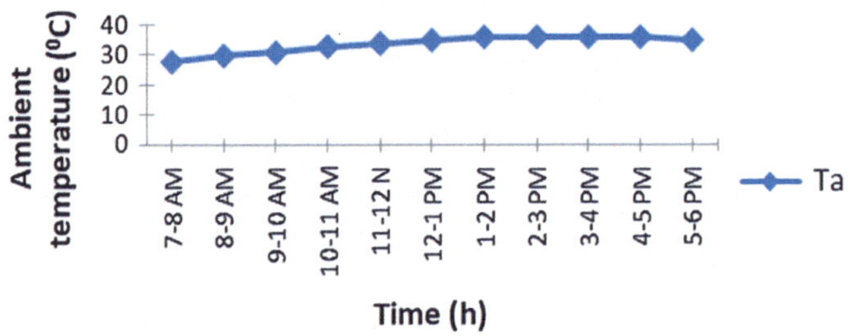

Fig. 62.3 Recorded ambient temperature

Fig. 62.4 Solar intensity on condensing cover and collector plate

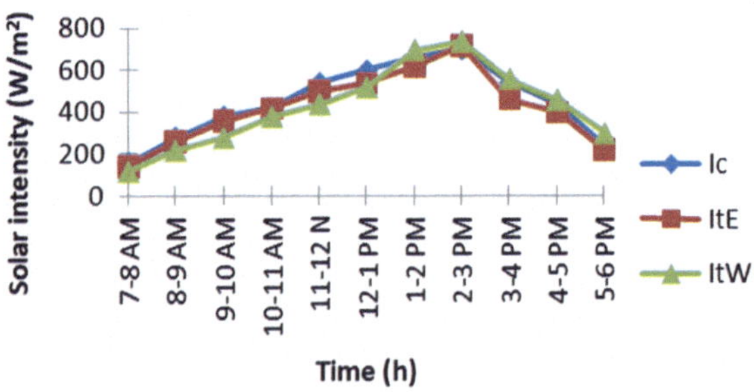

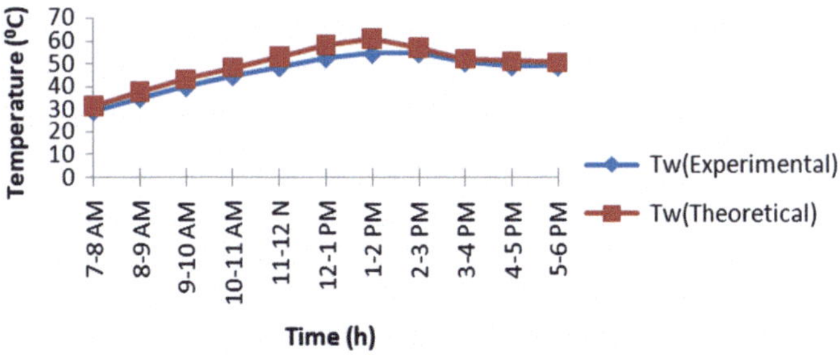

Fig. 62.5 Predicted and experimental water temperature

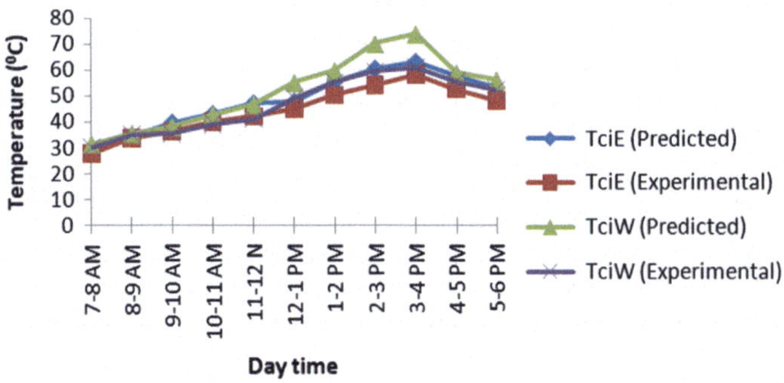

Fig. 62.6 Variation in condensing cover temperature

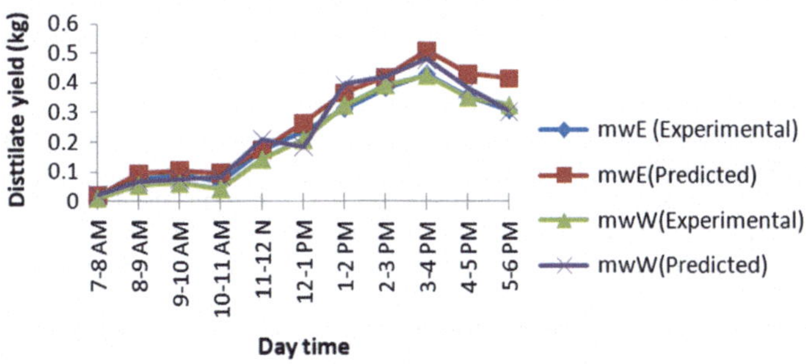

Fig. 62.7 Predicted and experimental distillate yield

Table 62.2 Value of coefficient of correlation and root mean square of percent deviation

Parameter	R_c	R_e
T_W	0.985	7.09
T_{ciE}	0.996	7.99
T_{ciW}	0.984	9.96
m_{wE}	0.990	17.42
m_{wW}	0.982	19.56

62.5 Conclusions

The close proximity between predicted value and experimental value validates the thermal model for this experimental setup.

References

1. Dunkle, R.V.: Solar water distillation: the roof type still and a multiple effect diffusion still. In: Proceedings International Heat Transfer Conference, vol. 5, p. 895, University of Colorado, USA (1961)
2. Morse, R.N., Read, W.R.W.: A rational basis for the engineering development of a solar still. Sol. Energy **12**(1), 5–17 (1968)
3. Nawayseh, N.K., Farid, M.M., Al-Hallaj, S., Al-Timimi, A.R.: Solar desalination based on humidification process—I. Evaluating the heat and mass transfer coefficients. Energy Convers. Manag. **40**(13), 1423–1439(1999)
4. Shukla, S.K., Sorayan, V.P.S.: Thermal modeling of solar stills: an experimental validation. Renew. Energy **30**(5), 683–699 (2005)
5. Shukla, S.K., Rai, A.K.: Analytical thermal modeling of double slope solar still by using inner glass cover temperature. Therm. Sci. **12**(3), 139–152 (2008)
6. Dev, R., Tiwari, G.N.: Characteristic equation of a passive solar still. Desalination **245**(1–3), 246–265 (2009)
7. Kumar, S., Tiwari, G.N., Gaur, M.K.: Development of empirical relation to evaluate the heat transfer coefficients and fractional energy in basin type hybrid (PV/T) active solar still. Desalination **250**(1), 214–221 (2010)
8. Morad, M.M., El-Maghawry, H.A., Wasfy, K.I.: Improving the double slope solar still performance by using flat-plate solar collector and cooling glass cover. Desalination **373**, 1–9 (2015)
9. Feilizadeh, M., Estahbanati, M.K., Ahsan, A., Jafarpur, K., Mersaghian, A.: Effects of water and basin depths in single basin solar stills: an experimental and theoretical study. Energy Convers. Manag. **122**, 174–181 (2016)
10. Gnanaraj, S.J.P., Ramachandran, S., Christopher, D.S.: Enhancing the design to optimize the performance of double basin solar still. Desalination **411**, 112–123 (2017)
11. Layek, A.: Exergetic analysis of basin type solar still. Eng. Sci. Technol. Int. J. **21**(1), 99–106 (2018)
12. Cooper, P.I.: Digital simulation of experimental solar still data. Sol. Energy **14**(4), 451–468 (1973)
13. Chauk, S.S., Fan, L.S.: Heat Transfer in Packed and Fluidized Beds. Handbook of heat transfer (1998)
14. Joshi, P., Tiwari, G.N.: Effect of cooling condensing cover on the performance of N-identical photovoltaic thermal-compound parabolic concentrator active solar still: a comparative study. Int. J. Energy Environ. Eng. **9**(4), 473–498 (2018)
15. Chapra, S.C., Canale, R.P.: Numerical Methods for Engineers, 1st edn. McGraw-Hill, New York (1989)

Chapter 63
HVAC Systems and Environmental Controls in Hospital Operation Theatres

Taliv Hussain, Pragati Agarwal and Adnan Hafiz

Abstract Operation theatres in a hospital require effective ventilation, heating and air-conditioning systems and design for the patients in order to have desired comfort zone within the space. This paper provides a proper methodology so as to discover the effects of varying inner as well as outer thermal conditions of an operation theatre on the comfort zone of the patients. The main objective is to develop the various heating, ventilation and air-conditioning systems and environmental controls like availability of proper space, temperature and humidity control, for effective ventilation use of air changes per hour method, for effective cooling and environmental controls use of new efficient techniques and for proper air cleaning use of high-efficiency particulate air (HEPA) filters, etc. that all can provide the required comfort zone in the hospital operation theatres. This paper also summarizes the work of some researchers and scientists that was carried out in order to achieve comfort zone in the different hospital operation theatres with different inner and outer thermal conditions.

Keywords Heating · Ventilation and air-conditioning systems · Environmental controls · Operation theatre · Inner and outer thermal conditions · Comfort zone

63.1 Introduction

The thermal conditions of an operation theatre like temperature, pressure, humidity, etc. greatly affect the working environment as well as well-being of the medical staff and the patients. Nowadays, with the high rate development in the medical sector has led to the introduction of hi-tech medical equipment but these machines have specific temperature requirement in order to work properly and give accurate results. This advancement in medical science has also led to the usage of highly effective medicines, and these also require very specific temperature range to show their full effect and cure the patient. Surgeons also have their own temperature requirements in order to work comfortably in their heavy surgical suits with high success rates. The health of the patient is also affected by the temperature of the operation theatre. Therefore, we can say that the hospital's operation theatres are the most demanding

T. Hussain (✉) · P. Agarwal · A. Hafiz
Department of Mechanical Engineering, Aligarh Muslim University, Aligarh 202002, India

S. Yadav et al. (eds.), *Proceedings of International Conference in Mechanical and Energy Technology*, Smart Innovation, Systems and Technologies 174,
https://doi.org/10.1007/978-981-15-2647-3_63

independent indoor areas. Various equipments and machines in the operation theatres also contribute to the temperature rise, for example, the surgery lamps alone produce a lot of heat, thus creating an unfit environment for the surgeons as well as the patient. Therefore, the rise in temperature due to various internal loads inside the operation theatres must also be taken into account, apart from the temperature requirement of the staff and patient, while designing an air-conditioning system for the operation theatre. A comfortable climate is defined as one having temperature range from 18 to 25 °C, a relative humidity between 30 and 50% and an air velocity of 0.1–0.2 m per second [1]. Therefore, indoor temperature of the operation room should be within the allowable range of the required working conditions, and also, special attention must be given to the heat addition due to various internal loads for the proper designing of an air-conditioning system.

This paper also summarizes the work of various scientists and researchers that was done in order to obtain the comfort level in the various hospital operation theatres with different thermal conditions.

63.1.1 General Design Parameters

A number of design parameters have to be taken into consideration while designing an air-conditioning system for an operation theatre. The designing also depends on the size and number of the operation theatres in the hospital. Atypical layout of an operation theatre involves the main operation room, anaesthesia room, reception patient area, sterile material rooms, recovery room, scrub room and auxiliary spaces (e.g. storage rooms, staff locker rooms, etc.). The corridor separates aseptic and septic areas. A proper space area must prevent direct communication of operation room with additional use spaces (e.g. storage locker rooms, offices, etc.) so as to secure the aseptic condition and reduce risk of bacterial infection [2]. A properly designed layout of an operation room not only reduces the risk of functional problems but also enhances the performance of the employees. Various design parameters that should be taken into consideration while designing an air-conditioning system for operation theatre include.

63.1.1.1 Availability of Space

A major problem faced by hospitals is the availability of space. A usual practice in today's hospitals is that for the purpose of fulfilling the need to increase the capacity of an operation theatre, they turn to renovation which only increases the number of operation rooms while decreasing the size. Even today's operation theatres face problems of availability of space in small dimensions. This creates an unfavourable environment for the medical staff as well as the patient. Another problem faced is the spaces with improper functions, for example, the sterilization rooms, which could be shifted outside the operation rooms as they unnecessarily occupy space and also lead to the contamination of the inside air due to the chemicals and bacteria released by them. So places with these kinds of functions should be located at some other part of the hospital so that the free space can be used to satisfy the space requirement of the medical staff and patient. Also some space is required in the operation room to store the medical supplies as well as the equipments in order to ensure proper functioning of the facility. So proper space should be provided for their storage in the operation room as we cannot use separate lockers for them because they will unnecessarily reduce free space and also cause health and safety risk and also contribute to the pollution source especially when they are not given proper care and attention or they become old. Generally, pathogenic organisms do not adhere to ceilings or walls readily unless the surface becomes moist, sticky or damaged [3]. Therefore, wall coverings of an operation room should be water resistant especially at places where there is contact with body fluids, in order to minimize risk of infections. Ideally, the walls and ceilings should have a smooth surface which could be cleaned easily with minimum chance of the bacteria to adhere.

According to latest researches, it has been found out that, in general, the finished ceiling height should be minimum of 3 metres above the floor and the walls should avoid corners [4, 5].

Ahmed [6] in his research found out that based on the type of operation, policies and procedures of infection control, management and quality assurance, the most effective design which satisfies the minimum area condition and complies with infection control is 'single-corridor system', where there is only one corridor to have sterilized materials, medical staff and patient taking into account the separation between sterilized and non-sterilized materials.

63.1.1.2 Temperature

Temperature of an operation room should be maintained within allowable range to ensure the acceptable working conditions. Hospital's operation rooms are the most demanding independent indoor areas as a lot of factors lead to the temperature variation. One must give special attention to the heat generation due to various internal loads, for example, surgical lamps by proper designing of the air-conditioning system and exhaust and the temperature of the supply air (Table 63.1).

a Use of higher or lower temperatures is acceptable according to the patient's medical condition.
b Reference supply airflow rate of 0.67 m^3/s.

Based upon the above table, the desirable indoor temperature is usually 22–24 °C, but it can be changed according to the requirement of the operations or the type of operation carried out in the operation room, for example, we require a lower temperature for cardiac surgeries where as a high temperature is required for paediatric surgeries because the babies are more sensitive to lower temperatures. However, one cannot satisfy the temperature requirements of all the occupants in the operation room because the surgeons have their own temperature requirements as they have to

Table 63.1 Recommended indoor conditions for an operation room [2]

Temperature (°C)	Relative humidity (%)	Ventilation	References
20–23	30–60	Positive pressurization. Minimum 15 ACH (air changes per hour), with at least 3 ACH (air changes per hour) of outer air. All the recirculated and newly air should be filtered with filters having an efficiency of at least 90%. Air entrance should be at ceiling and the exhaust should be at floor, in the rooms that are not designed for horizontal laminar flow	American Institute of Architects [7]
17–27	45–55	Positive pressurization. Minimum 25 ACH (air changes per hour), with at least 5 ACH (air changes per hour), of the outdoor air	ASHRAE [8]
20–24	30–60	A minimum positive pressurization of 2.5 Pa. Primary Supply Diffusers and non-aspirating Minimum 20 ACH (air changes per hour), with at least 4 ACH (air changes per hour), of outdoor air	ASHRAE [9].
22–26	50–60	Positive pressurization. If the height of the room is 3 m, the ACH (air changes per hour) should be 60 $m^3/m^2/h$, or else it should be 20	German Institute for Standardization [10]
18–24	50–60	Positive pressurization with 15% excess air. Airflow rates should be 70–85 m^3/h. However, during off-use hours, airflow can be reduced lower to 30% of the full load conditions	Technical Chamber of Greece [11, 12] Hellenic Health Ministry [13]

work in their heavy surgical clothing in order to avoid infection, whereas anaesthetist requires a slightly lower temperature. Therefore, in general, surgeons are likely to feel slightly warm to warm [14], whereas the anaesthetist and the nurses in contrast feel slightly cool to cool thermal conditions.

Also, the asymmetric radiations from the surgical lights also create thermal discomfort for the medical staff. One must also consider the allowable temperature ranges for the medical equipments and medicines in order to design a proper air-conditioning and ventilation system as the medicines also have their own temperature requirements in order to work properly and the medical equipments have to be maintained under certain temperature limits otherwise it can lead to the corrosion or damage which in turn lead to lower success rates of the operations and higher chances of postoperative infections.

63.1.1.3 Humidity

Humidity of an operation room should also be kept taken within the allowable ranges as it affects the thermal condition as well as the hygiene. Using the thumb rule, for example, 54 L/(s kW), or temperature only design approach often leads to the system which is not able to fulfil strict inner requirements [15]. Humid environment often leads to bacterial or microbial growth as humid environment favours the growth of microorganisms which will lead to the postoperative infections on the wound of the patient. Also, low humidity leads to coagulation of blood which is unfavourable during an operation that is why humidity of an operation room should be maintained according to the requirement of the operation so that to minimize the chances of infection and carry out the operation with high success rates. Due to this, humidification and dehumidification processes are required. Thus, in order to fulfil these humidity requirements one should keep in mind these psychrometric processes while designing an HVAC system. According to Table 63.1, it can be seen that the recommended level of indoor relative humidity should be between 30 and 60%. Researchers have established that for operation theatres that use inflammable materials or volatile liquids, should maintain a relative humidity of about 60%.

63.1.1.4 Ventilation

In the eighteenth century, the provision of separate rooms for operations came into picture mainly for teaching and demonstration of how surgery is carried out [16]. However, it was only after the Second World War when the official documents suggested the positive pressure ventilation to minimize airborne contamination of wound. Paestur, in the mid-nineteenth century, was the first to recognize contamination by air [17]. He soaked wounds and hands in dilute carbolic acid solution and subsequently disinfected air with spray. The death rate from sepsis following amputation fell from 46 to 15%. Carbolic spray was toxic and therefore was replaced by aseptic surgical techniques afterwards. Today, proper operation room ventilation plays an effective role in ensuring health and occupant safety including airborne contamination [18] and infection control [19, 20] even though the infection rates are still unacceptably high [21].

Howard summarizes a reduction in the number of colony-forming units (cfu) which leads to a corresponding reduction in the incidence of wound sepsis [22].

Increased chances of postoperative infections will be there if the ventilation system, installed to provide high-quality air to the operation theatres, do not meet the required level of quality [23].

In addition, proper ventilation also reduces the chances of postoperative infection in patients going through deep hole surgery. The main function of an operation theatre ventilation system is to stop airborne contaminants from going into the patient's wounds and cause infection. The bacteria can get inside the wounds either by falling directly on the wound or by falling on the exposed surface of the surgical instruments which are used to carry out the surgery.

Areas with various levels of microbiological control are used for surgical operations. There are mainly two types:

1. Conventionally ventilated operations suits
2. Ultraclean ventilated operation theatres.

Ultraclean ventilated operation suites: The evidence confirming the relation of ultraclean ventilation and decreased postoperative infection is obtained from the studies of prosthetic joint surgery conducted by MRC [24]. The use of prophylaxis antibiotics and the influence of ultraclean ventilation in reducing the chances of post-operative infections are inseparable. The MRC did not prevent the use of prophylaxis antibiotics, but the later publications based on the analysis of data from these trials revealed that combination of ultraclean air and prophylaxis antibiotics gives extra benefit. Ultraclean air is formed by using a laminar flow which is designed to advance particle-free air or ultraclean air over operation theatre at a uniform velocity. This air is filtered by the high-efficiency particulate air (HEPA) filter which takes out the particles of greater than 2.3 μm in diameter with an efficiency of 99.97%. It is approximated that the microorganisms are also washed as particles. The rooms that are close to operation theatre do not need this type of treatment because there is no chance of infection as wound has already been bandaged and bacteria cannot contaminate it via air [25, 26]. Therefore, the use of ultraclean air is only needed in operation theatre. For this, the filter should be placed inside the operation room. It can be either located on the wall (for horizontal) or on the ceiling (for vertical flow). The air is first treated by the use of two additional lower-grade filters [25, 27].

However, this system, that is, HEPA filter in combination with laminar airflow, is mainly needed for operations like orthopaedic implant surgery where the risk of surgical site infections is high [24–26, 28]. It would require a large number of patients, undergoing the same operation, operated by same surgeons under same conditions differ only in bacterial counts to prove that the ultraclean ventilation has an independent effect [28–30]. Ultraclean ventilation is mostly designed to provide a particle-free area around the patient while carrying out the operation in order to minimize the risk of postoperative infection. In a multicenter study, inspecting over 8000 knee and hip replacements, it was showed that the ultraclean air minimizes risk of surgical site infection (SSI). This study also compared the effects of ultraclean air with and without microbial prophylaxis.

Conventionally ventilated operation suites: Most of the modern hospitals have operation theatres with conventional plenum ventilation in which the air is filtered using filters of efficiency 82.95% to remove airborne particles of size greater than or equal to 5 μm [31]. There are anecdotal reports that describe the effect of enhanced ventilation facilities or natural ventilation (i.e. absence of artificial ventilation) on the postoperative infection risks. It was noticed that the rise in postoperative infections for the period when the ventilation system of the theatre was substandard and the temperatures were high. Subsequently, when a new ventilation system was installed, there was no infection during the hot months of the following year and incidence for both hot and cold months was alike, thus implying that uncomfortable conditions in

Table 63.2 Surgical site infection (SSI) with respect to the combined use of ultraclean air and prophylactic antibiotics [28]

	With antibiotics cases			Without antibiotics cases		
Operation room conditions	Total	With SSI	SSI (%)	Total	With SSI	SSI (%)
Ultraclean ventilation	1279	9	0.7	516	8	1.6
Conventional ventilation	2968	24	0.8	1161	39	3.4

theatre may have described the rise in infection. Ayliffe has reported that for a conventionally ventilated operation theatre, the infection rates are higher compared with the previous year when there was no artificial ventilation in the theatre 12.6% versus 8.8% [32]. The details of operation performed were not provided, and patients were not matched during two observation periods, but air counts were lowered by 50% when the theatre was ventilated. Table 63.2 shows surgical site infection (SSI) for ultraclean ventilation and conventional ventilation.

63.1.1.5 Hygiene

Surgical site infections are a serious problem in today's medical science. There are a number of factors that can lead to surgical site infection, e.g.: contamination of the indoor environment, type of operation performed, immune state of the patient, insertion of foreign particles or implants, surgeon's skill, adequacy and timing of antimicrobial prophylaxis and appropriateness of surgical preparation. One needs to have the knowledge of the source and transport of the foreign infection-causing bacteria in order to control surgical site infections. Most of the surgical site infection pathogens originate from the skin of the staff or the patient [33–35]. These infections can either be superficial or deep. Superficial ones can be cured without much complication, but the deep ones can become serious and may lead to re-operation. Therefore, special care must be taken of hygiene in order to reduce surgical site infections. It is important to prevent the contamination of the instruments used in the theatres whether they are laid down in a preparation room or in the theatre itself. Free zones should be sufficiently large in order to accommodate the need of instruments so that there would be a better assurance of prevention of contamination. The choice of theatre dress, education of personnel, patient preparation and draping and design of the ventilation system of the operation theatres also play important role in reducing the bacterial count. These infections and other sources of contamination in the operation theatre should be minimized in order to reduce high rate of health care and morbidity costs associated with these infections.

Types of OT (operation theatres): As per new guidelines for air conditioning in OTs by NABH, operation theatres are divided into two groups:
Super specialty OT: It includes OTs for orthopaedics, cardiothoracic, neurosciences and transplant surgery.

Table 63.3 Latest guidelines given by the NABH, for a general OT

Design parameter	Value/range
Air changes per hour (ACH)	Minimum 20 ACH with at least 4 ACH of fresh air
Air velocity	The airflow should be unidirectional and downwards on the OT table with an air face velocity of 25–35 FPM (feet per minute)
Positive pressure	Positive pressurization of at least 2.5 Pa (0.01 in. of water)
Temperature and relative humidity	It should be about 210 ± 3 °C (except for joints replacement where it should be kept 180 ± 2 °C) with a corresponding relative humidity of about 20–60%, though the ideal RH is considered to be 55%

General OT: It includes OTs for ophthalmology, FRU OT, district hospital OTs and all other basic surgical disciplines.

Table 63.3 shows latest guidelines given by the NABH, for a general OT. Also, according to the Australasian Health Facility Guidelines, for an operating room having an area of 60 m^2, 24 h of operation and an occupancy of one patient and 5–12 staff members should have a positive pressure, proper exhaust system and air filtration using HEPA filters.

63.2 Conclusions

From the study performed, we found out that while designing an air-conditioning system for an operation theatre, one should take into consideration the thermal conditions and comfort levels required by the surgeons, medical staff, patients, the recommended working environment for the medicines and medical equipment inside the operation theatre. However, one cannot satisfy the temperature requirements of all the occupants in the operation theatres simultaneously because the surgeon requires a somewhat lower temperature to work in the heavy surgical dress, whereas the anaesthetist requires a comparatively high temperature. But from the study, we found out that in general, the desirable temperature is usually 22–24 °C. Humidity, hygiene and space availability also play important role in determining the air-conditioning system. A lower humidity may result in the coagulation of blood, whereas higher humidity can cause growth of bacteria. As mentioned above, the relative humidity of the operation theatre should be between 20 and 60%. Hygiene of an operation theatre helps to ensure high success rates of the operation performed and lower the risk of infections. Availability of space is also an important factor as it in turn helps to increase the comfort level of the occupants in the operation theatre. Therefore, proper space and hygiene should also be maintained inside the operation theatre. The ventilation system plays an important role especially in maintaining the hygiene and reduces the infection rates. The type of ventilation used also depends on the type of the operation that is being performed in the operation theatre because as discussed above, ultraclean ventilated operations suits are required for orthopaedic implant operations where the chances of infections are higher. Therefore, various factors are to be considered while designing an air-conditioning system for an operation theatre and they all are interrelated so if any one of the factories remains unsatisfied, it may lead to postoperative infections which may cause a risk to health of the patient and also the health of the medical of the occupants in the operation theatre and also adds up to the healthcare costs.

References

1. Mangram, A.J., Horan, T.C., Pearson, M.L., Silver, L.C., Jarvis, W.R.: Hospital infection control practices advisory committee. guideline for prevention of surgical site infection. Infect. Control Hosp. Epidemiol. **20**(4), 247–280 (1999)
2. Balaras, C.A., Dascalaki, E., Gaglia, A.: HVAC and indoor thermal conditions in hospital operation rooms. Energy Build. **39**(4), 454–470 (2007)
3. Noskin, G.A., Peterson, L.R.: Engineering infection control through facility design. Emerg. Infect. Dis. **7**(2), 354 (2001)
4. American Institute of Architects (AIA): Guidelines for Design and Construction for Health Care Facilities. AIA Press, Washington (2006)
5. Chefurka, T., Nesdoly, F., Christie, J.: Concepts in flexibility in healthcare facility planning, design, and construction. Acad. J. (2005)
6. Ahmed, K.S.: Medical planning: operating theatre design and its impact on cost, area and workflow. In: International Conference on Bioinformatics and Biomedical Engineering. pp. 325–332, Springer, Cham (2017)
7. Ninomura, P., Rousseau, C., Bartley, J.: Updated guidelines for design and construction of hospital and health care facilities. ASHRAE J. **48**(6), H33 (2006)
8. Handbook, A.S.H.R.A.E.: HVAC systems and equipment. In: American Society of Heating, Refrigerating, and Air Conditioning Engineers. pp 1–10, Atlanta, GA (1996)
9. Perez, P., Holloway, J., Ehrenfeld, L., Cohen, S., Cunningham, L., Miley, G.B., Hollenbeck, B.L.: Door openings in the operation room are associated with increased environmental contamination. Am. J. Infect. Control **46**(8), 954–956 (2018)
10. DIN 1946: Ventilation and Air-Conditioning. Part 4. Ventilation in Hospitals, vol. 4, Deutsches Institut für Normung e.V., Beuth-Verlag, Berlin (1999)
11. Technical Directive 2423/86: Building Installations, Air-Conditioning of Building Spaces, Technical Chamber of Greece, Athens (1987)
12. Technical Directive 2425/86: Building Installations, Elements of HVAC Load Calculations for Building Spaces, Technical Chamber of Greece, Athens (1987)
13. Hoffman, P.N., Williams, J., Stacey, A., Bennett, A.M., Ridgway, G.L., Dobson, C., Humphreys, H.: Microbiological commissioning and monitoring of operation theatre suites. J. Hosp. Infect. **52**(1), 1–28 (2002)
14. Mora, R., English, M.J., Athienitis, A.K.: Assessment of thermal comfort during surgical operations/discussion. ASHRAE Trans. **107**, 52 (2001)
15. Murphy, J.: Temperature and humidity control in surgery rooms. ASHRAE J. **48**(6), H18 (2006)
16. Essex-Lopresti, M.: Operation theatre design. Lancet **353**(9157), 1007–1010
17. Bannister, G.: (v) Prevention of infection in joint replacement. Curr. Orthop. **16**(6), 426–433 (2002)
18. Persson, M., Van der Linden, J.: Wound ventilation with carbon dioxide: a simple method to prevent direct airborne contamination during cardiac surgery? J. Hosp. Infect. **56**(2), 131–136 (2004)

19. Fox, N.J.: Space, sterility and surgery: circuits of hygiene in the operation theatre. Soc. Sci. Med. **45**(5), 649–657 (1997)
20. Humphreys, H., Taylor, E.W.: Operation theatre ventilation standards and the risk of postoperative infection. J. Hosp. Infect. **50**(2), 85–90 (2002)
21. Blom, A.W., Gozzard, C., Heal, J., Bowker, K., Estela, C.M.: Bacterial strike-through of re-usable surgical drapes: the effect of different wetting agents. J. Hosp. Infect. **52**(1), 52–55 (2002)
22. Howorth, F.H.: Prevention of airborne infection during surgery. Lancet **325**(8425), 386–388 (1985)
23. Estates, N.H.S.: Health technical memorandum 2025. Ventilation in healthcare premises. Part 3. Validation and verification (1994)
24. Lidwell, O.M., Lowbury, E.J., Whyte, W., Blowers, R., Stanley, S.J., Lowe, D.: Effect of ultraclean air in operation rooms on deep sepsis in the joint after total hip or knee replacement: a randomised study. Br. Med. J. (Clin. Res. Ed.) **285**(6334), 10–14 (1982)
25. Dharan, S., Pittet, D.: Environmental controls in operation theatres. J. Hosp. Infect. **51**, 79–84 (2002)
26. Humphreys, H., Stacey, A.R., Taylor, E.W.: Survey of operation theatres in Great Britain and Ireland. J. Hosp. Infect. **30**, 245–252 (1995)
27. Scherrer, M.: Hygiene and room climate in the operation room. Minim. Invasive Ther. Allied Technol. **12**(6), 293–299 (2003)
28. Lidwell, O.M., Elson, R.A., Lowbury, E.J., Whyte, W., Blowers, R., Stanley, S.J., Lowe, D.: Ultraclean air and antibiotics for prevention of postoperative infection: a multicenter study of 8,052 joint replacement operations. Acta Orthop. Scand. **58**(1), 4–13 (1987)
29. Hanssen, A.D., Osmon, D.R., Nelson, C.L.: Instructional course lectures, the American Academy of Orthopaedic Surgeons-prevention of deep periprosthetic joint infection. JBJS **78**(3), 458–471 (1996)
30. Pittet, D., Ducel, G.: Infectious risk factors related to operation rooms. Infect. Control Hosp. Epidemiol. **15**(7), 456–462 (1994)
31. Standard, M.: Filter units, protective clothing, gas-mask components and related products: performance test methods. Tech. Rep. MIL-STD-282, US Government Printing Office, Washington, DC, USA (1956)
32. Ayliffe, G.A.J.: Role of the environment of the operation suite in surgical wound infection. Rev. Infect. Dis. **13**(Supplement_10), S800–S804 (1991)
33. Bitkover, C.Y., Marcusson, E., Ransjö, U.: Spread of coagulase-negative staphylococci during cardiac operations in a modern operation room. Ann. Thorac. Surg. **69**(4), 1110–1115 (2000)
34. Whyte, W., Hambraeus, A., Laurell, G., Hoborn, J.: The relative importance of routes and sources of wound contamination during general surgery I. Non-airborne. J. Hosp. Infect. **18**(2), 93–107 (1991)
35. Whyte, W., Hambraeus, A., Laurell, G., Hoborn, J.: The relative importance of the routes and sources of wound contamination during general surgery II. Airborne. J. Hosp. Infect. **22**(1), 41–54 (1992)

Chapter 64
Optimization of Turning Parameter for AISI 4140 by Vegetable Oil as Cutting Fluid

Nandini M. Jadhav and Dayanand A. Ghatge

Abstract The aim of this research is to optimize the turning parameters for AISI 4140 by using vegetable oil as cutting fluid with help of coated carbide insert of TNMG 160408. The output parameters selected for this research are MRR, power consumption. The analysis of output parameters is done experimentally with input parameters of speed, feed, depth of cut and cutting oil. Three different types of vegetable oil are selected for machining that are castor oil, soybean oil, palm oil. The work experimentation carried out with L_{27} orthogonal array design with four cutting parameters, i.e., speed, feed, depth of cut and oil. Optimal conditions for power consumption and MRR were determined using signal-to-noise (S/N) ratio.

Keywords CNC turning · AISI 4140 · CVD-coated insert · Vegetable oils · Minitab17 · ANOVA

64.1 Introduction

Material removing processes like turning, milling and drilling are explained above. For the experimentation, we are using turning process. In turning operation, unwanted material removes from workpiece by using cutting tool. With the use of turning process such as rotational, typically axisymmetric parts are produced that have different features, i.e., taper, holes, groove. Turning is suitable to remove external material from workpiece; the same process is carried out for boring but difference is internal material removed from workpiece to male hole. Whether with a boring tool or turning is called "facing." Turning operation carried out manually by traditional lathe machine under supervision. But, nowadays, automated machines are highly used to reduce human effort and to obtain higher accuracy. With the use of automated machines, productivity get increases. Nowadays, the most common type of automation is computer numerical control known as CNC. (CNC is also commonly used with many other types of machining besides turning.) Cutting fluids are used in machining process to minimize friction, cool the workpiece. Cutting fluid helps to

N. M. Jadhav · D. A. Ghatge (✉)
K.B.P College of Engineering, Satara, India
e-mail: dayanand.ghatge@kbpcoes.edu.in

S. Yadav et al. (eds.), *Proceedings of International Conference in Mechanical and Energy Technology*, Smart Innovation, Systems and Technologies 174,
https://doi.org/10.1007/978-981-15-2647-3_64

and improves surface quality and reduces tool wear. Machine surface can be protected from corrosion by using cutting fluids. Cutting fluids also reduce the cutting forces. Cutting fluid helps to protect tool from damage and gives protection to workpiece from corrosion. It absorbs heat which is produced during machining.

64.2 Literature Survey

Adegbuyi et al. used mild steel as workpiece for experimentation. They analyze the effect of cutting fluids on the mechanical properties of mild steel in a turning operation. Turning was done under dry condition with help of three coolants. They found that palm kernel oil performed very well as cutting fluid which gives better surface finish, good chip formation and also helps in reduction of heat generated [1].

Hamdan et al. used three different cutting fluids while machining operation. The three cutting fluids were chosen that are soluble oil, neat oil and semi-synthetic cutting fluids. It observed that soluble oil does not drastically change with the feed rates and cutting velocity. With proper selection of machining parameter, water-mixed cutting fluids (soluble oil and semi-synthetic) give better performance on surface roughness results [2].

Avila and Abrao used AISI 4340 for tuning operation with the use of three cutting oils. The turning operation carried out with help of mixed alumina insert. The output parameters are tool life, surface finish and chip form. They found that at high cutting speeds, surface finish is obtained with use of cutting fluid. They conclude that to reduce the scatter in the surface roughness value, cutting fluids are responsible [3].

Isik worked on An Experimental Investigation on Effect of Cutting Fluids in Turning with Coated Carbides Tool. The cutting tool used in this research is CVD-coated carbide TiC + Al_2O_3 TiN insert (ISO P25). The type of inserts which are used in machining is DNMG 150608. The results are compared with wet-cooled and dry turning. He concluded that CVD-coated carbide TiC + Al_2O_3 + TiN cutting tool performed better during wet machining mode [4].

Deepak et al. did experimentation on optimization of turning parameters of OHNS steel with help of cutting fluid. The results were compared with dry and wet turning operations. They performed turning operations on lathe machine by using OHNS steel as workpiece material. They concluded that the improvement of the cutting performance and surface finish obtained with help of minimum cutting fluid [5].

Babur et al. did experimentation on influence of vegetable-based cutting fluids on the wear in drilling. They worked on semi-synthetic commercial cutting fluid, sunflower and canola oils. Canola-based cutting fluid gives best performance with respect to other cutting fluids at the constant cutting conditions because it has higher lubricant properties [6].

64.3 Experimental Setup and Cutting Conditions

Experimentation conducted on CNC lathe machine with wet condition by using three different cutting fluids such as castor oil, soybean oil, palm oil. The experimentation conducted on AISI 4140 alloy steel rod in operation of turning is cutting speed, feed, depth of cut, cutting fluid. The workpiece material was cut into 27 pieces of dimension 36 mm in diameter and 25 mm in length (Table 64.1).

64.3.1 Machine Setup for Turning

Turning of AISI 4140 stainless steel is carried out on CNC lathe machine. The specifications of machine used are given as follows (Fig. 64.1; Table 64.2).

Table 64.1 Machining parameters

Workpiece	AISI 4140
Cutting speed (mm/min)	50, 100, 150
Feed	0.12, 0.24, 0.36
Doc	0.5, 0.75, 1
Cutting fluid	Castor oil, soybean oil, palm oil
Insert	TNMG 160408 (CVD coated)

Fig. 64.1 CNC lathe machine

Table 64.2 Specification of CNC lathe machine

Maximum turning diameter	250 mm
Spindle speed	45–4500 rpm
Spindle motor power	7.5 kW

64.3.2 Cutting Tool

For machining operation, cutting tool plays main role. Selection of cutting tool depends on machining operation. For this experimentation, we selected CVD-coated TNMG 160408 insert as cutting tool.

64.4 Result and Discussion

64.4.1 Material Removal Rate

The material removal rate, MRR, can be defined as the volume of material removed divided by the machining time that is how much amount of material is removed during machining time. Another way to calculate material removal rate is initial weight of workpiece before machining and final weight of workpiece after machining divided by time taken for operation (Table 64.3).

$$\text{MRR} = \frac{\text{initial W.T} - \text{final W.T}}{\text{Time}}$$

64.4.2 Power Consumption

Power consumption is found with power capacity analyzer. While performing experiment, power capacity analyzer is attached to the CNC machine of main supply to measure the power consumption. During each turning experiment, power consumption noted down. It is summarized in the table given below (Fig. 64.2; Table 64.4).

$$\cos \emptyset * \sqrt{3} * V * I$$

$\cos \emptyset = 0.37 =$ lead factor.

64.4.3 Input and Output Parameters

See Table 64.5.

Taguchi Design of Experiments

The Taguchi method defines two types of factors: control factors and noise factors. Control factor is process design parameter that can be controlled, e.g., equipment

Table 64.3 Experimental result for MRR

Expt. No.	Material removal rate (g/min)
1	19.28
2	27.85
3	34
4	34.28
5	55.38
6	60
7	30
8	80
9	96
10	26.67
11	42.35
12	53.34
13	60
14	90
15	127.5
16	85.71
17	120
18	180
19	66
20	80
21	90
22	110
23	120
24	*190*
25	108
26	156
27	180

setting, material used for production. Noise factor is difficult to control, e.g., temperature or humidity. Taguchi is new engineering design used for optimization that improves quality of products and processes. Taguchi method makes products or processes robust. Taguchi method helps in evaluating and implementing improvements in products, processes, materials, equipment and facilities.

64.4.4 Analysis of Variance for Material Removal Rate

See Fig. 64.3.

Fig. 64.2 Power consumption instrument

ANOVA Main Plots For Material Removal Rate

Graph shows cutting parameters versus material removal rate. For material removal rate speed and depth of cut, both are highly affected parameters. For maximum cutting speed and depth of cut, we got maximum material removal rate. Soybean oil gives better performance for material removal rate. ANOVA for material removal rate larger value is better. When feed rate increases, material removal rate also increases. For maximum value of depth of cut, maximum material removal rate is obtained it means that when depth of cut increases, material removal rate also increases. Soybean oil is better fluid for maximum material removal rate. Maximum material removal rate is 190 g/min which obtains at maximum cutting speed and maximum depth of cut for soybean oil. Next better fluid for material removal rate is palm oil.

Table 64.4 Experimental result for power consumption

Expt. No.	Power consumption (kW)
1	*515.95*
2	648.93
3	757.97
4	587.76
5	675.52
6	792.55
7	598.40
8	813.82
9	837.76
10	808.50
11	960.10
12	1077.12
13	1042.54
14	957.74
15	1236.69
16	1082.44
17	1194.14
18	1361.69
19	1114.35
20	1340.42
21	1773.92
22	1324.46
23	1359.03
24	1856.37
25	1343.07
26	1606.37
27	1906.90

64.4.5 ANOVA for Power Consumption

Main Effect Plot For Power Consumption

See Fig. 64.4.

Graph shows cutting parameters versus power consumption. During machining process energy is used to drive components such as spindle, feed and tool. ANOVA for power consumption smaller is better. For minimum speed and depth of cut minimum, power consumption is obtained. Power consumption increases with increase in speed. For power consumption, castor oil gives better output as compared with the other two oils. Better output obtained for power consumption with minimum cutting speed,

Table 64.5 Inputs and output parameter

S. No.	Cutting speed (RPM)	Feed (mm/rev)	Doc (mm)	Fluids	MRR (g/min)	P.C. (kW)
1	443	0.12	0.5	C	19.28	515.95
2	443	0.12	0.75	P	27.85	648.93
3	443	0.12	1.00	S	34.00	757.97
4	443	0.24	0.5	C	34.28	587.76
5	443	0.24	0.75	P	55.38	675.52
6	443	0.24	1.00	S	60.00	792.55
7	443	0.36	0.5	C	30.00	598.4
8	443	0.36	0.75	P	80.00	813.82
9	443	0.36	1.00	S	96.00	837.76
10	885	0.12	0.5	C	26.67	808.5
11	885	0.12	0.75	P	42.35	960.1
12	885	0.12	1.00	S	53.34	1077.12
13	885	0.24	0.5	C	60.00	1042.54
14	885	0.24	0.75	P	90.00	957.74
15	885	0.24	1.00	S	127.50	1236.69
16	885	0.36	0.5	C	85.71	1082.44
17	885	0.36	0.75	P	120.00	1194.14
18	885	0.36	1.00	S	180.00	1361.69
19	1327	0.12	0.5	C	66.00	1114.35
20	1327	0.12	0.75	P	80.00	1340.42
21	1327	0.12	1.00	S	90.00	1772.92
22	1327	0.24	0.5	C	110.00	1324.46
23	1327	0.24	0.75	P	120.00	1359.03
24	1327	0.24	1.00	S	190.00	1856.37
25	1327	0.36	0.5	C	108.00	1343.07
26	1327	0.36	0.75	P	156.00	1606.37
27	1327	0.36	1.00	S	180.00	1906.9

feed and depth of cut. For feed rate, power consumption is gradually increased. For power consumption, castor oil gives better performance as compare to other oils used.

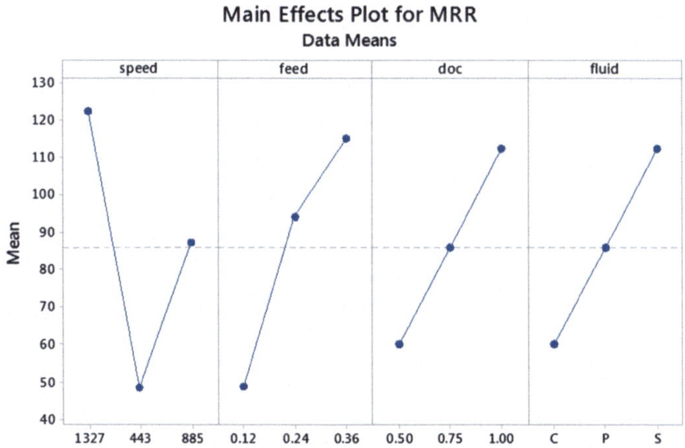

Fig. 64.3 Main effect plot for MRR

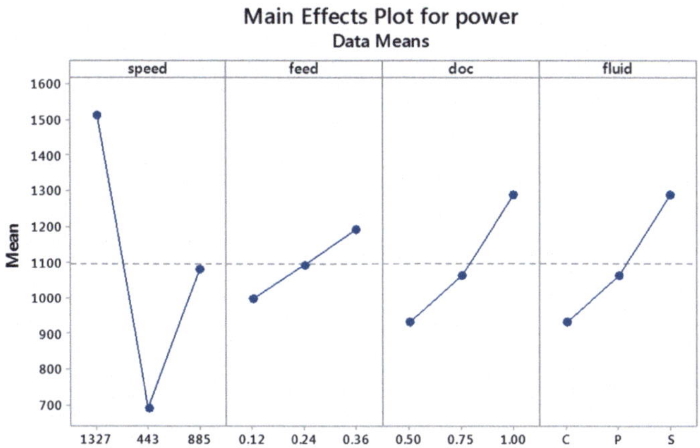

Fig. 64.4 Main effect plot for power consumption

64.5 Conclusion

For material removal rate speed and depth of cut, both are highly affected parameters. We got good results with use of vegetable oil as cutting fluid for material removal rate and power consumption. For power consumption, we got 515.65 KW better value by using castor oil. For material removal rate, we got maximum value of 190 g/min with use of soybean oil. It is concluded that for power consumption, castor oil is better as compared to the other two oils.

References

1. Adegbuyi, P.A.O., Lawal, G., Oluseye, O., Odunaiya, G.: Analyzing the effect of cutting fluids on the mechanical properties of mild steel in a turning operation. https://doi.org/10.5251/ajsir.2011.2.1.1.10
2. Hamdan, A., Fadzil, M., Abou-El-Hossein, K.A., Hamdi, M.: Performance evaluation of different types of cutting fluid in the machining of AISI 01 hardened steel using pulsed jet MQL system
3. Avila, R.F., Abrao, A.M.: The effect of cutting fluids on the machining of hardened AISI 4340 steel. J. Mater. Process. Technol. **119**, 21–26 (2001)
4. Isik, Y.: An experimental investigation on effect of cutting fluids in turning with coated carbides tool. J. Mech. Eng. **56**(3), 195–201 (2010)
5. Johnson, R.D.J., Wins, K.L.D., Raj, A., Beatrice, B.A.: Optimization of cutting parameters and fluid application parameters during turning of OHNS steel. In: 12th Global Congress on Manufacturing and Management, GCMM (2014)
6. Ozcelika, B., Kuram, E., Demirbas, E., Sik, E.: Effects of vegetable-based cutting fluids on the wear in drilling. Sadhana **38**(4), 687–706 (2013). Indian Academy of Sciences

Chapter 65
Optimization of Process Parameters of Galvanizing Steel in Resistance Seam Welding Using RSM

Umesh Kumar Vates, B. P. Sharma, Nand Jee Kanu, Naveen Anand Daniel, Sivaraos Subramanian and Priyanshu Sharma

Abstract In present research, response surface methodology (RSM) is used to investigate the effects of control parameters on critical responses in resistance seam welding (RSW) process. Four critical controllable input variables, weld pressure, weld current, overlapping and material thickness were selected to analyse the effects on the responses weld time (T_w), surface roughness (R_a) and hardness (H) in RSW. Central composite design (CCD) array is selected to conduct the experiments and to estimate the model adequacy test in RSW. Correlation coefficient (R^2) values were observed for surface roughness, weld time and hardness, respectively. Responses were optimized using critical variables weld pressure, weld current, material thickness and overlapping, respectively, with influence of RSM. The optimal process parameters are introduced and tested in industrial environment.

Keywords Resistance seam welding · Response surface methodology · Central composite design

65.1 Introduction

Welding, the process used to adhere two or more pieces of metal together by applying pressure or thermal energy, is a reliable, precise and cost-effective method for joining materials. This technique is widely accustomed by manufacturers to join alloys and metals [1, 2]. Resistance seam welding (RSW) has two revolving wheel electrodes which are shown in Fig. 65.1. Spot welds are made along the lap joint in overlapping [11,12]. The process produces the airtight joints which involve producing the gasoline tanks, automobile mufflers and various others fabricated sheet-metal vessel

U. K. Vates (✉) · B. P. Sharma · N. A. Daniel · P. Sharma
Amity University Uttar Pradesh, Noida, India
e-mail: ukvates@amity.edu

N. J. Kanu
SVNIT, Surat, India

S. Subramanian
Faculty of Manufacturing Engineering, Universiti Teknikal Malaysia Melaka, Melaka, Malaysia

S. Yadav et al. (eds.), *Proceedings of International Conference in Mechanical and Energy Technology*, Smart Innovation, Systems and Technologies 174,
https://doi.org/10.1007/978-981-15-2647-3_65

Fig. 65.1 Copper electrode wheel

[3, 14]. The gap between the weld nuggets in resistance seam welding on based on the motion of the electrode wheels relative to the application of the weld current. Traditional method like continuous motion welding was being adapted to join such sheet metal with constant wheel velocity [13]. The current is turned on at certain time intervals consistent with the desperate spacing between spot welds along the repeated seam to overlapping the weld spots generated. If the frequency of current is decreased sufficiently, gap between the weld spots and electrode is converted into roll spot welding [2, 4, 5].

It is verdict that welding process parameters like weld time, welding current, thickness of metal plate, weld pressure and electrode force were critically considered into the account to establish the range of adequate nugget formation and surface roughness. There are many more critical influencing parameters may also be considered for response optimization in seam welding.

65.2 Experimental Set-up

Various experimentations were conducted to analyse the effects of critical machining parameters on seam welding process. Present studies emphasized on investigation for weld pressure, weld current, overlapping and material thickness on weld time, surface roughness (R_a) and hardness in resistance seam welding. The galvanizing

Fig. 65.2 Machine set-up

steel is being selected as workpiece material due to its emergent range of applications in the field of corrosion resistance in various industries for present research work. Experiments were conducted on resistance seam welding (KSW-100C) machine. A copper wheel is used as an electrode having dimension of (Ø300 × 20) and metal workpiece thickness of 18 gauge, 20 gauge and 24 gauge are selected for experimentation. The composition and properties of such materials are given in Tables 65.1 and 65.2, respectively. Workpiece material, set-up electrode are given also in Figs. 65.3 and 65.2, respectively. Response surface methodology (RSM) experimental design was selected to conduct the possible experiments having four factors on three different level, which given in Table 65.3.

Workpiece: Galvanize steel is one of the broadly used materials applied for shield metal arc welding, spot welding, resistance seam welding and to prevent the corrosion/rusting. In recent days, many car manufacturers are started to adapt the galvanized sheets in the automotive industries [6, 14]. Two different type coatings were generally applied in the automotive industry for manufacture the steel sheet components [7]. It is clear that galvanized coating is made through pure zinc coating and gal annealed is made by iron alloy coating [8]. The workpieces observed as silver colour but actually it is zinc. Galvanizing coating a metal provides it anti-corrosion behaviours. The use of protecting zinc coating prevents the possibly oxidation corrosion. Galvanized steel is a cost operative alternative materials and as austenitic stainless steel in order to avoid the corrosion rate. Galvanic corrosion happens when two metals of different electrochemical properties are placed into contact with one another in the presence of electrolyte. It has been observed that anode corrodes quickly than cathode. The zinc is used for galvanizing harmony towards being the

Table 65.1 Chemical composition of work material

Steel	Element	C	Si	P	S	Mn	Ti	Ni	Cr	Cu	Al	Zn	Fe
Galvanized	By weight	0.005	0.004	0.011	0.005	0.096	0.046	0.008	0.013	0.018	0.036	0.003	99.7

Table 65.2 Mechanical properties of work material

Steel	Tensile strength (MPa)	Yield stress (MPa)	Elongation (%)
Galvanizing	295.99	156	47.87

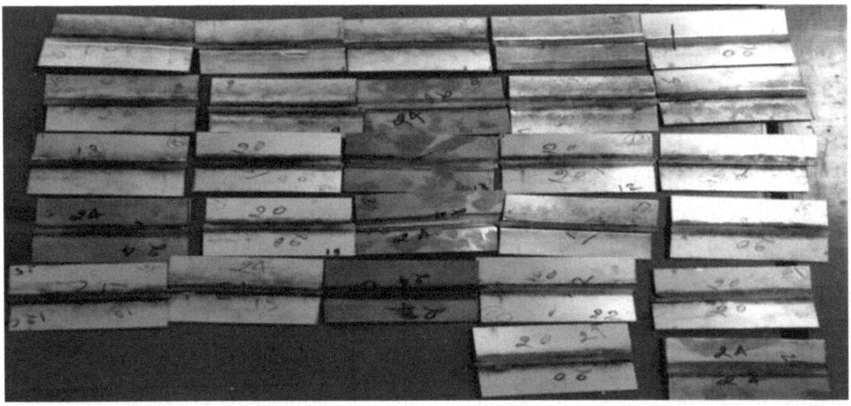

Fig. 65.3 Workpiece (after machining)

Table 65.3 Factors and levels

S. No	Levels	Weld pressure (Kg/cm^2)	Weld current (A)	Overlapping (mm)	Material thickness (gauge)
1	Maximum (+1)	5.5	140	1.2	24
2	Moderate (0)	5.0	130	1	20
3	Minimum (−1)	4.5	120	0.8	18

anode when in contact with many different types of metals. Galvanize zinc coating ranging 0.001 to 0.004 in. is used depending on the various industrial product and its applications [9].

Surface Roughness Measurements

Centre line average value is the simplest parameter to represent the R_a. In present research, surface roughness was measured using contact type Mitutoyo Surface Tester which being available with Amity University, Noida, Uttar Pradesh. CLA value (R_a) has been measured using the different orientation and taking the mean. It was measured three times also for similar to each reading to record the average value. The parameters that affect weld time, hardness and surface roughness are weld pressure, weld current, overlapping and material thickness. Optimal parametric combination yields the technological quality controllable aspect of a product, which directly influence on the manufacturing cost of the product. R_a is demarcated as the reckoning value of the profile of surface from the centreline along the working span. The average value of reading up to two decimal place of microns will be obtained the least count 1 nm (manometer) as Figs. 65.4 and 65.5.

Hardness test: The capability of materials to repel penetration, scratching, abrasion and cutting are known as hardness. It is the assets by which material resists also from permanent deformation. Various hardness tests are available. Knoop hardness test is generally done to measure microhardness [10]. Hardness test is carried out on, Vickers hardness tester (Model: VM-50) within Balmer Lawrie & Co. Ltd. (Fig. 65.6).

The second-order model is being used to analyse and the mathematical model developed between the process variable and responses. The second-order model in Eq. 65.1 explains the behaviour of the system.

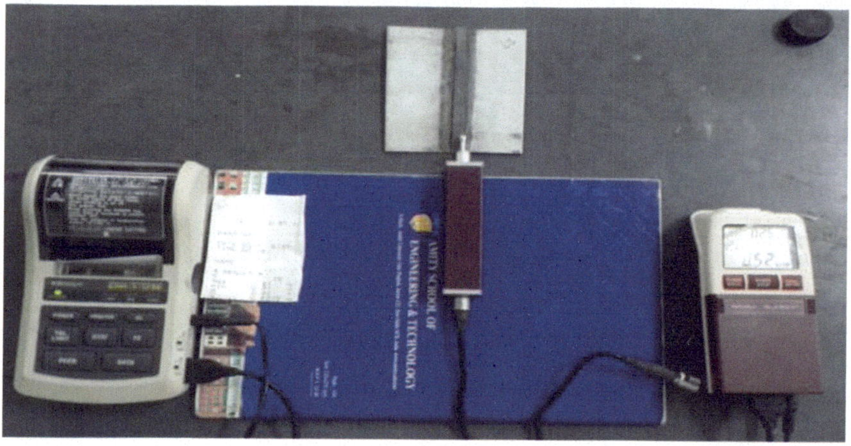

Fig. 65.4 Surface roughness measurement-TR 200 (calibrated with least count: 0.02 μm)

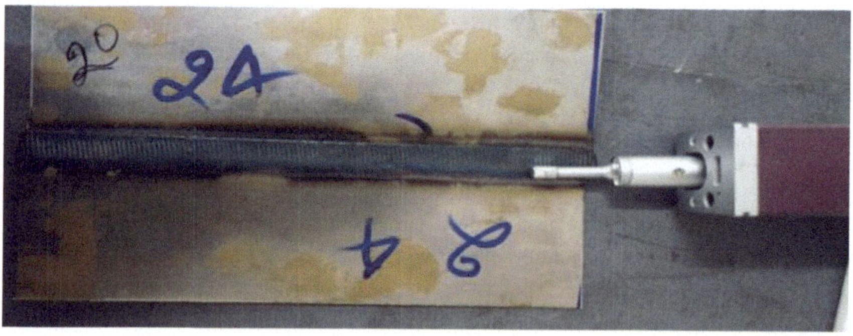

Fig. 65.5 Surface roughness measurement with probe

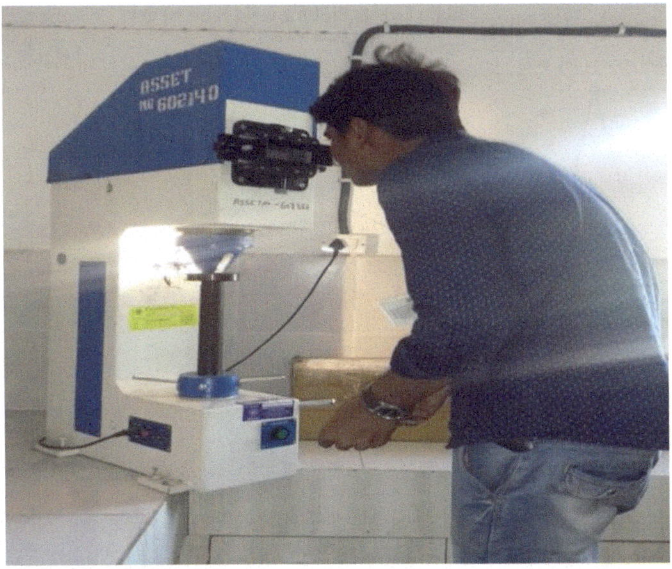

Fig. 65.6 Hardness measurement

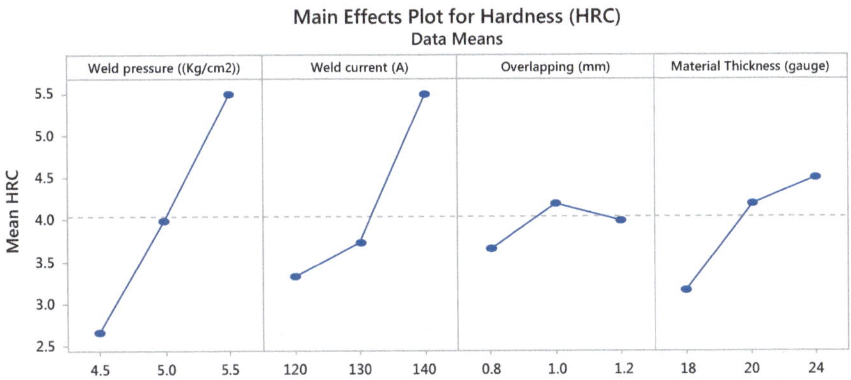

Fig. 65.7 Main effect plot

$$Y = \beta_o + \sum_{i=1}^{k} \beta_i X_i + \sum_{i=1}^{k} \beta_i i X_i^2 + \sum_{i,j=1, i \neq j}^{k} \beta_{ij} X_i X_j + \in \tag{65.1}$$

where Y is the corresponding response, X_i is the input variables, X_i^2 and X_i, X_j are the squares and interaction terms, respectively, of these input variables.

Estimated regression coefficients for weld time (s)

$$R\text{-Sq} = 76.02\% \quad R\text{-Sq(pred)} = 0.00\% \quad R\text{-Sq(adj)} = 48.05\%$$

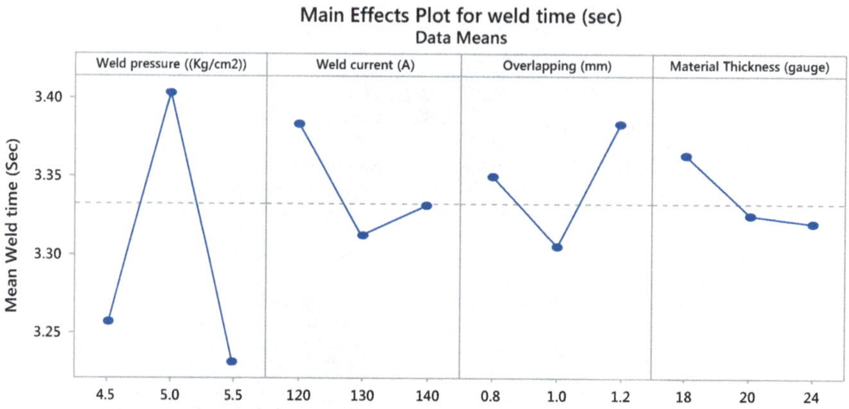

Fig. 65.8 Main effect plot for weld time

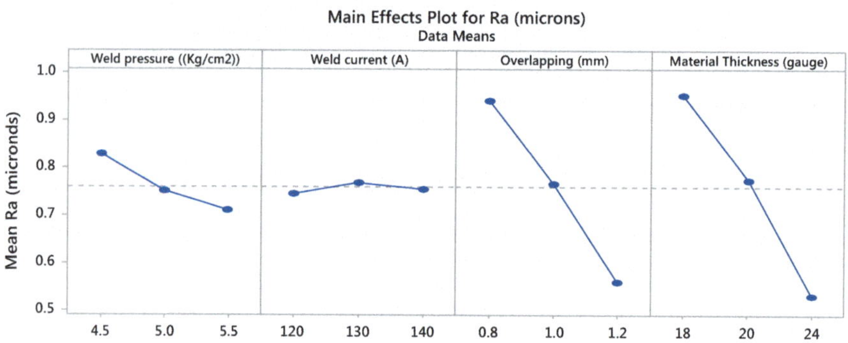

Fig. 65.9 Main effect plot for R_a

Estimated regression coefficients for hardness

$$R\text{-Sq} = 84.35\% \quad R\text{-Sq(pred)} = 32.73\% \quad R\text{-Sq(adj)} = 66.08\%$$

Estimated regression coefficients for R_a

$$R\text{-Sq} = 96.28\% \quad R\text{-Sq(pred)} = 0.00\% \quad R\text{-Sq(adj)} = 57.64\%.$$

65.3 Result and Discussion

In the present investigation, intelligence approach (MORO) is being used to combine optimization of R_a, weld time and hardness. Optimal values of influencing parametric combinations are investigated like weld pressure, weld current, material thickness and overlapping as given in Fig. 65.10.

Figure 65.7 shows the effect of weld pressure, weld current overlapping and material thickness on hardness; here, it observed hardness is continuously increase with increase in weld pressure in other hand when weld current is increase then hardness is increase slowly but after some time it has been increased rapidly, the effect of overlapping on the hardness is also predicted in this figure show that when work material overlapping is increase, hardness is also increase but beyond the moderate value it goes to slowly decrease. Effect of the material thickness on the hardness, it is observed that hardness is rapidly increase with material thickness but hardness is slowly increase when the material thickness goes to beyond the moderate value.

The effect of weld pressure, weld current overlapping and material thickness on weld time shown in Fig. 65.8, it represents that initially weld time increase with increase in weld pressure but weld pressure goes to 5 PSI the response weld time decrease. Effect of weld current on weld time is also estimated in this figure shows that weld time decrease with increase weld current after reached the value of weld current 130 it (weld time) tends to increase. Effect of overlapping on weld time shown in this figure presented that initially the value of overlapping is increase with decrease in weld time. After reaching the value of overlapping 1 mm. weld time initially increases. The effect of material thickness on weld time is also forecast in this figure. Here, it shows that initially weld time decrease with increase material thickness. After reaching the value of material thickness 20 gauge, weld time ignite decrease.

Figure 65.8 shows the effect of weld pressure, weld current, overlapping and material thickness on surface roughness. It indicates that when weld pressure increase with decrease in surface roughness after reach the moderate value of weld pressure

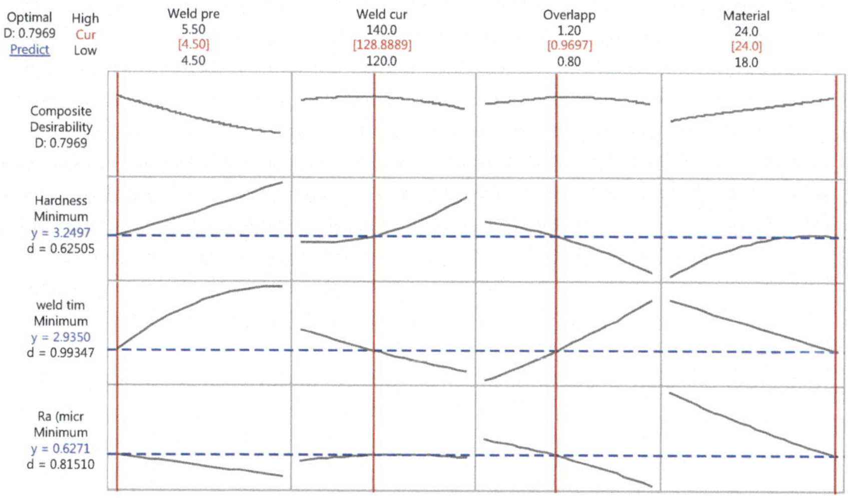

Fig. 65.10 Optimization plot for the responses

response tends to slowly decrease. Effect of weld current on R_a is also presented here, initially, weld current is increased with very slowly increase in weld time but weld time decrease very slowly when weld current value goes to beyond 130. Effect of overlapping R_a represents that value of R_a decrease at any point of increasing overlapping of work material. In this Fig. 65.9, it also estimated the effect of independent critical parameters on R_a, here represent that value of surface roughness decreased with increasing of material thickness.

In this study, four significant variables are under consideration to obtain the polynomial regression modelling for simplicity, a quadratic model of surface roughness, hardness and weld time are considered. The unknown coefficients are determined from the experimental data as present in Table 65.4. T values are calculated for 95% level of confidence and the factor having more than 0.5 are considered insignificant. The final response equation for R_a, weld time and hardness are nonlinear in nature, it is clear that a linear polynomial will not be able to predict the responses accurately, therefore, the second-ordered model (quadratic model) is found to be an adequate model for resistance seam welding process [7]. The ANOVA for the curtailed quadratic model depicts the value of coefficient of performance (R^2) for R_a, weld time and hardness as 96.28, 76.02 and 84.35% which signifies the variable in the response the higher of R^2 indicates the better fitting.

65.4 Conclusion

In the present study, it is verdict that resistance seam welding process parameters significantly affect the surface roughness, weld time and hardness. Second-order response model of these parameters was developed and fond that weld pressure, weld current, material thickness and overlapping with other parameters significantly affected the R, weld time and hardness (Fig. 65.10).

Experiment conducted on "KSW-100C resistance seam welding having copper-zinc electrode on galvanizing steel workpiece. RSM techniques were also implemented to predict the R, weld time and hardness. Correlation coefficient (R^2) values were observed 96.28%, 76.02% and 84.35% for surface roughness weld time and hardness, respectively. Responses (weld pressure, weld current, material thickness and overlapping) were also optimized as 0.6271, 2.9350 and 3.2497, respectively, using critical values of variables weld pressure, weld current, material thickness and overlapping as 4.50, 120.0, 18.0 and 0.80, respectively, with the influence of response surface methodology techniques. The research finding of the present study based on RSM models can be used effectively in welding of galvanizing steel in order to obtained best possible weldability, hardness and minimum surface roughness. RSM predicted result was also verified with almost 5.32% deviation by measuring the responses directly through instruments. This research can also help researchers and industries for developing a robust, reliable knowledge-based easily prediction of R_a, weld time and hardness without experiment in resistance seam welding (RSW) process for galvanizing steel.

Table 65.4 Experimental observations

S. No.	Weld pressure (Kg/cm^2)	Weld current (A)	Overlapping (mm)	Material thickness (gauge)	R_a (μm)	Weld time (s)	Hardness increased (HRC)
1	5	120	1.2	20	0.51	3.47	3
2	5	140	1	18	1.286	3.69	5
3	4.5	130	0.8	20	1.69	2.93	4
4	5	140	0.8	20	0.57	3.29	5
5	5	130	0.8	18	1.29	3.54	1
6	5	140	1.2	20	0.476	3.29	6
7	4.5	140	1	20	0.673	3.21	4
8	5	140	1	24	0.826	3.36	6
9	5.5	130	0.8	20	1.066	3.50	5
10	5	130	1	20	1.063	3.41	5
11	5	120	0.8	20	0.65	3.56	3
12	5.5	130	1.2	20	0.79	3.11	6
13	5.5	130	1	24	0.45	3.49	6
14	4.5	130	1.2	20	0.633	3.69	2
15	5	120	1	18	1.256	3.27	4
16	4.5	120	1	20	0.80	3.40	2
17	4.5	130	1	18	0.673	3.30	1
18	5	130	1.2	24	0.386	3.34	4
19	5.5	120	1	20	0.62	3.16	4
20	5	120	1	24	0.64	3.44	4
21	5	130	1	20	0.766	3.31	3
22	5.5	140	1	20	0.7	3.15	7
23	4.5	130	1	24	0.503	3.01	3
24	5	130	1.2	18	0.57	3.40	3
25	5.5	130	1	18	0.643	2.98	5
26	5	130	0.8	24	0.393	3.28	4
27	5	130	1	20	0.606	3.40	4

References

1. Khosravi, A., Halvaee, A., Hasannia, M.H.: Weldability of electro galvanized versus galvanized interstitial free steel sheets by resistance seam welding. Mater. Des. **44**, 90–98 (2013)
2. Benypounis, K.Y., Olabi, A.G., Hashmi, M.S.J.: Effect of Laser Welding Parameters on the Heat Input and Weld-Bead Profile. School of Mechanical and Manufacturing Engineering, Dublin City University Press (2018)
3. Ambroziak, A., Korzeniowski, M.: Using resistance spot welding for joining aluminium elements in automotive industry. Arch. Civ. Mech. Eng. **10**(1), 5–13 (2010)

4. Bian, J., Zhu, Y., Liu, X.H., Wang, G.D.: Development of hot dip galvanized steel strip and its application in automobile industry. J. Iron. Steel Res. Int. **13**, 47–50 (2006)
5. Danial, K., Mostafaei, A., Ali, A.A.: Resistance spot welding joints of AISI 316L austenitic stainless steel sheets. Hase Transform. Mech. Prop. Microstruct. **14**, 44–57 (2017)
6. Volek, C.W., Begeman, M.L.: Effect of zinc coating on resistance seam welding of sheet steel. Resist Weld. Manuf. Assoc. Bull. **30**, 243–253 (2017)
7. Kou, S.: Welding Metallurgy. NewYork (1987)
8. Minitab16, Minitab User Manual Release 14 MINITAB Inc, State College, PA, USA, (2003)
9. Song, Q., Zhang, W., Bay, N.: An experimental study determines the electrical contact resistance in resistance welding. Weld. J. **84**(5), 79–89 (2005)
10. Abdel-Aleem, H., Kato, M., Nishio, K., Yamaguchi, T., Furukawa, K.: Joining of 1050/5052 by resistance seam welding and materials evaluation of joints. Weld. Int. Sel. J. Light Metal Weld. Constr. **19**, 199–204 (2015)
11. Inoue, T., Suzuki, M., Okabe, T., Matsui, Y.: Development of advanced electric resistance welding (ERW) line pipe. In: Mighty Seam TM with High Quality Weld Seam Suitable for Extra-Low Temperature Services, JFE Technical Report No. 18, pp. 18–22 (2013)
12. Raza, M.A., Kashyap, S.K..: The effect of welding on mechanical and microstructural properties of materials-a critical review (2018)
13. Emre, H.E., Kaçar, R.: Development of weld lobe for resistance spot-welded TRIP800 steel and evaluation of fracture mode of its weldment. Int. J. Adv. Manuf. Technol. **83**, 1737–1747 (2016)
14. Chen, J., Yuan, X., Hu, Z., Sun, C., Zhang, Y., Zhang, Y.: Microstructure and mechanical properties of resistance-spot-welded joints for A5052 aluminum alloy and DP 600 steel. Materials Characterization **120**, 45–52 (2016)

Chapter 66
An Analytic Hierarchy Process (AHP)-Based Multi-criteria Evaluation and Priority Analysis for Best FWH Substitution of Solar Aided Thermal Power Plant

Shailendra Pratap Singh, Subrata Kumar Ghosh and Vijay Kumar Dwivedi

Abstract Utilization of solar energy in lieu of traditional feed water heating process in thermal power plants could be a milestone when the world is impending on to analyze critically the present scenario of energy consumption and liberation of greenhouse gases to the atmosphere. For it to occur, a theoretical cycle is thus proposed, which utilizes a feed water heat exchanger (FWHE) based on solar assisted system (parabolic trough), which would work in tandem to the heaters of a 210 MWe thermal power plant. Out of the listed six cases (LPH-2 to HPH-7), a congruent alternative has been proposed in this work, which relies on multi-criteria decision approach, based on analytical hierarchy process (AHP). AHP has been applied on the listed criteria, i.e., improvement in energy efficiency of plant, improvement in exergy efficiency of plant, Reduction in unit heat rate, improvement in exergy efficiency of solar field, solar contribution, and reduction in fuel consumption. After a thorough analysis, HPH-6 has been found as the most suitable alternative and the least one is LPH-2. The outcomes relied on its weightage and the relative significance/importance that have been assigned to them. Outcomes could differ with a change in the significance of various criteria, based on the necessities of the situation.

Keywords Analytic hierarchy process · Priority analysis · Solar-aided thermal power plant

S. P. Singh (✉) · S. K. Ghosh
Indian Institute of Technology (ISM), Dhanbad 826004, India

V. K. Dwivedi
Galgotias College of Engineering & Technology, Greater Noida 201306, India

S. Yadav et al. (eds.), *Proceedings of International Conference in Mechanical and Energy Technology*, Smart Innovation, Systems and Technologies 174,
https://doi.org/10.1007/978-981-15-2647-3_66

66.1 Introduction

An amalgamation of solar power with the traditional thermal power stations has an enormous potential that drives the very motive to attain a green-based power generation. With this process of combining the two, we could meet the targets of renewable energy, which in turn reduces the plant emissions and cost of fuel consumption. In this work, the motive could be achieved by using a solar-based FWHE in tandem to the traditional fossil fuel-based regenerative feed water heaters, this would in turn help us to reduce the fossil fuel consumption and its harmful emissions.

The analytic network process (ANP) and analytic hierarchy process (AHP) are multiple criteria decision-making (MCDM) methods. MCDM is a set of ideas, strategies, and procedures created to help decision-makers to settle on intricate decisions in an orderly and organized manner. A category of MCDM is multiple criteria discrete alternative problems. There are a few proposed strategies for taking care of discrete alternative problems, some of these are based on multiple attribute utility theory (MAUT) used by Keeney et al. [1], AHP/ANP by Saaty [2], other methods can be TOPSIS by Hwang et al. [3], PROMETHEE by Brans et al. [4] and ELECTRE by Roy [5]. A progressive point by point depiction of these techniques can be found in Pomerol-Barba et al. [6] and Belton Stewart et al. [7, 8] made an exhaustive bibliometric investigation on the improvement of MCDM strategies for the period 1992–2007 in both practical and theoretical situations. It is not simple to develop a mathematical model based on MCDM. Bouyssou et al. [9] said that 'there is no best model available for MCDM systems.' With its pros and cons, its application varies with the change in context. Diaz-Balteiro et al. [10], Kumar et al. [11], Shen et al. [12], and Dos Santos et al. [13] used MCDM techniques for decision making in sustainable development. Ho et al. [14] claimed that AHP is the outstanding and frequently used technique for supporting decision making among all the accessible methods, while Zhang et al. [15] mentioned that the AHP is an MCDM strategy that utilizes pair-wise examination, in the light of a numerical scale.

An AHP-TOPSIS-based framework was proposed by Choudhary et al. [16] for selecting the most efficient and suitable location for TPP. Stein EW [17] used AHP technique for ranking power generation technologies (renewable and non-renewable) on the basis of technical, financial, socio-economic, and environmental criteria. Cook et al. [18] analyzed the data of 40 power units to investigate relative operating efficiencies by using two-stage hierarchy model. AHP technique was used by Chatzimouratidis et al. [19–22] to examine ten types of renewable and non-renewable power plants. Initially, the impact of power generation by different sources on expectation for everyday comforts of nearby networks was assessed. Further, sensitivity analysis is being conducted for overall investigation of power units after considering technical, financial, and sustainability aspects. Another study is conducted on the same units for subjective and objective assessments. Pablo et al. [23] applied an AHP/ANP for selection of solar-aided TPP investment projects. Another analyst Yagmur Levent [24] had utilized another framework for by using AHP to evaluate and prioritize the equipment in a thermal power plant. Ren et al. [25] developed the MCDM framework

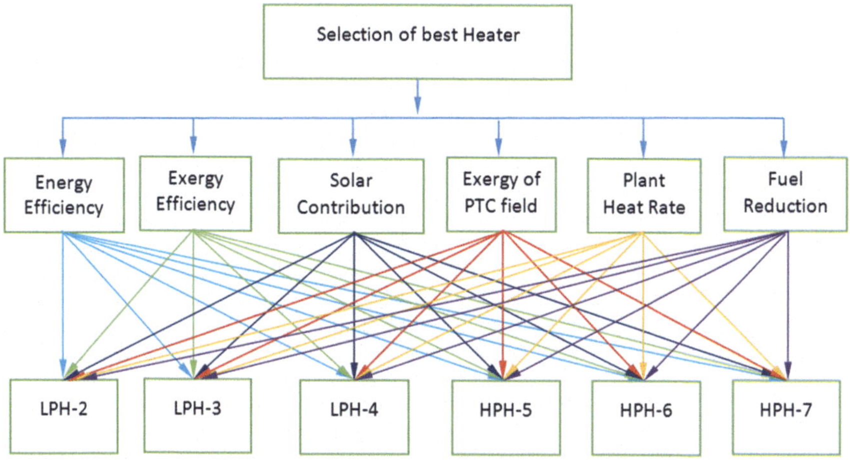

Fig. 66.1 Framework structure. *Source* Based on Saaty [2]

(nine metrics in four dimensions) for power production sustainability and security assessment, and a total of nine metrics in four dimensions including availability and security of supply, affordability and reliability, energy and economic efficiency, and environmental stewardship.

The motive behind using an ANP/AHP approach is to (i) enable decision-maker to examine complex problem by breaking a fundamental issue into less difficult and moderate sub-issues, (ii) ANP ought to be utilized, in case of interdependencies between criteria and alternatives, (iii) reliable decision making needs the elaborate study of interdependencies/priorities between criteria and alternatives. Out of numerous MCDM techniques available, AHP is used for the heater selection purpose and a framework is developed considering some specific criteria. The criteria for comparison were the main features of different heaters as shown in Fig. 66.1. Selection of best heater depends upon many factors like area of solar field, energy efficiency and exergy efficiency of the plant, solar contribution, exergy efficiency of solar field, unit heat rate (UHR), reduction in fuel consumption, ash content, CO_2 reduction, and various other factors. So the parameter selection was an important task to identify the most suitable substitution. AHP is used here with the help of Super Decision Software (open source) to rank six available cases.

66.2 Descriptions of Reference Plant

A coal-fired thermal power station with generating a capacity of 210 MWe based on Russian Technology has been selected as a reference unit for the study purpose This unit operates on sub-critical pressure, based on the Rankine cycle with reheating and regeneration. Figure 66.2 illustrates the integration of the reference unit with solar

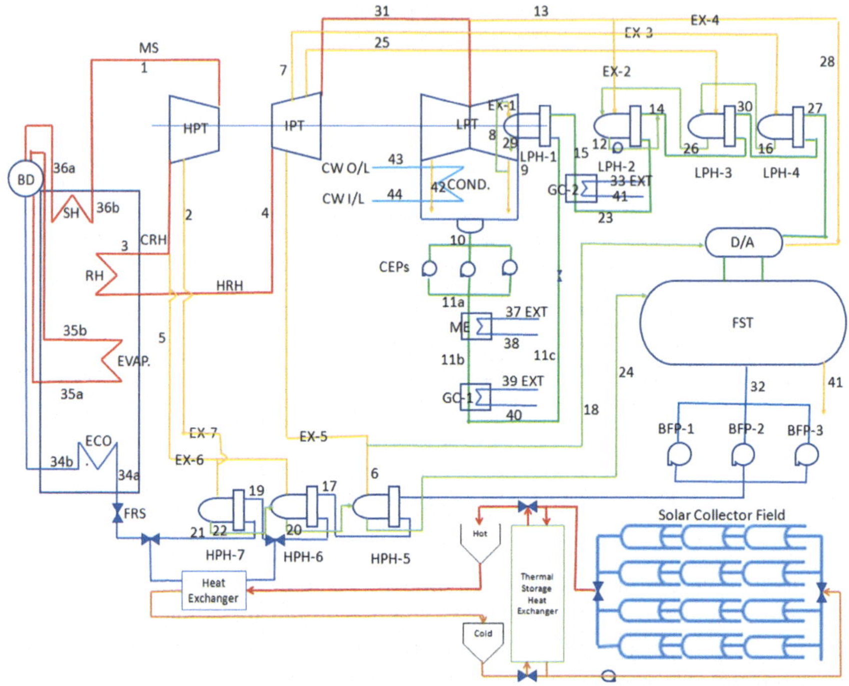

Fig. 66.2 Schematic diagram of solar integrated thermal power plant. *Source* The authors & NTPC, India

in place of high-pressure heater (HPH)-7.

All feed water heaters operate on different parameters. The data related to feedwater heaters for reference power plant has been shown in Table 66.1.

Table 66.1 Feed water heater parameters of reference unit plant

Parameter	Unit	LPH-2	LPH-3	LPH-4	HPH-5	HPH-6	HPH-7
Inlet temperature	°C	68.2	99.5	122.9	165.2	180	218.4
Outlet temperature	°C	99.5	122.9	152.1	180	218.4	239
Inlet flow	kg/s	145	145	145	190	190	190
Inlet enthalpy	kJ/kg	284.4	416.2	517.42	697	771.7	941.42
Outlet enthalpy	kJ/kg	416.2	517.42	642.30	771.7	941.42	1035.15
Solar power required	MW	19.11	14.68	18.11	14.19	32.25	17.81

Source The authors

66.3 Selection of Best Heater for Substitution

It consists of the following steps to use this tool.

66.3.1 Formulate the Model and Identify the Problem

The initial phase for utilization of the software is to build up the structure through which one can recognize the connection between various components and finalize the problem. The framework structure as demonstrated in Fig. 66.3 is created, and the issue is to select and rank the heaters from various access technologies.

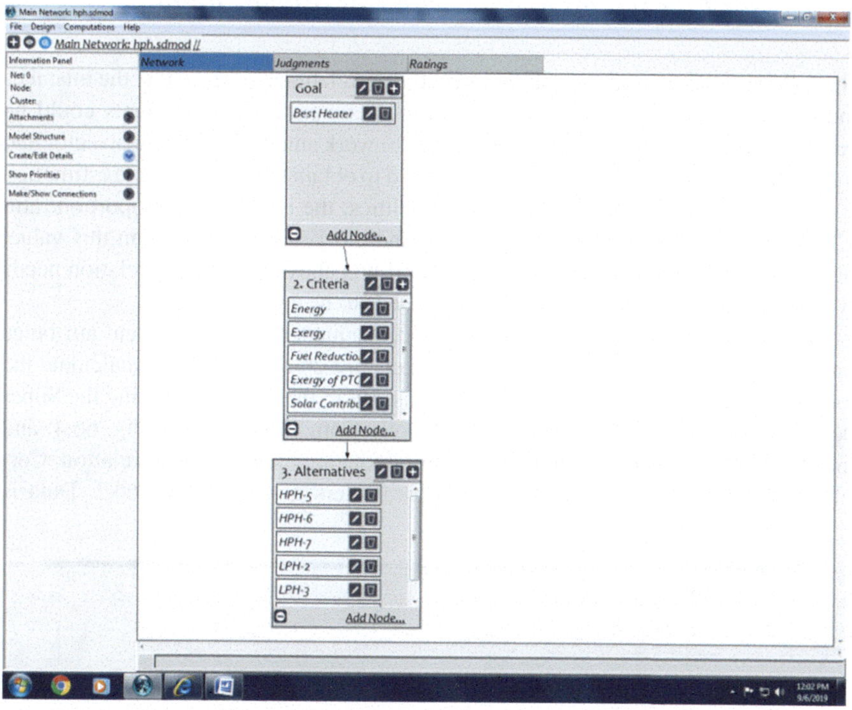

Fig. 66.3 Framework structure on software. *Source* The authors

66.3.2 Make the Framework/Matrix for the Pair-Wise Correlation Among Various Components

The subsequent step is to formulate the grid for analysis between every component/traits on which the choice depends. It helps in assigning significance to various characteristics, so that one can choose the required parameters later effortlessly for correlations. One can utilize the scale from 1 to 9, where 1 implies slightest significance and 9 implies most significant. While comparing two parameters and utilize reciprocals of these scales for reverse examinations.

66.3.3 Standardize the Matrix/Grid and Calculate Weight for Eigenvector and the with Consistency Ratio

The following stage was to institutionalize the correlations result so that the mistakes and disarray between significances of various comparative components could be expelled effortlessly for the entire procedure to work smoothly. The eigenvector and eigenvalues for every component are computed to obtain the most extreme estimation of the eigenvector. After computing the qualities, the consistency proportion/ratio (CR) is being checked, which ought to be under 0.1. If it is lesser than this value, then it means the comparisons are good and if it is above then the correlation needs to be rechecked, and the consistency ratio needs to be enhanced.

The CR value helps to maintain the logical sequence among different attributes while correlating them. The eigenvalue and eigenvector are needed to calculate the consistency ratio if using manual calculations, but in the event of utilizing the Super Decision Software the CR can be seen directly from software as in Fig. 66.4, and correlation ratio can be altered by enhancing/reducing the parameters relation. Correlation is created after considering the parameters given in Table 66.2. Data is

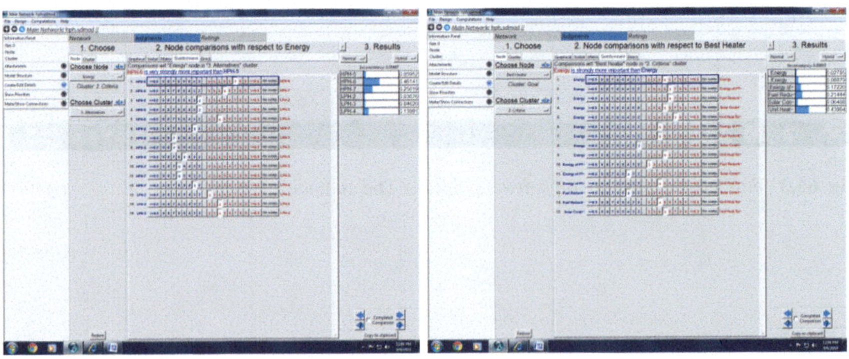

Fig. 66.4 Relative significance between criteria and alternatives with CR comparison. *Source* The authors

Table 66.2 Analysed values of criteria for AHP investigation

Parameter	Unit	HPH-7	HPH-6	HPH-5	LPH-4	LPH-3	LPH-2
Improvement in energy efficiency of plant	%	4.98	6.69	2.28	3.27	2.04	1.82
Improvement in exergy efficiency of plant	%	7.67	9.43	4.90	5.92	4.65	4.43
Reduction in unit heat rate	%	8.17	11.25	3.11	5.00	2.64	2.21
Improvement in exergy efficiency of solar field	%	9.48	9.00	8.57	7.65	6.78	5.62
Solar contribution	%	3.06	5.50	2.40	3.06	2.47	3.18
Reduction in fuel consumption	%	4.08	5.62	1.55	2.50	1.31	1.10

Source The authors

generated after a brief calculation.

66.3.4 *Super-Matrix Making, Its Analysis and Ranking the Heaters*

The correlation is being done using the software depending upon all the criteria and then super-matrix is formed as shown in Table 66.3. Afterward, the weight of various components is investigated with valid consistency ratio. Therefore, the relative significance among each of the components is generated as shown in Table 66.4, which makes the examination substantially simpler.

66.4 Results and Discussion

All the conceivable options/alternatives were assessed based on the weight of different attributes and the relative ranking among them is generated. The outcome reveals how much one heater is more advantageous in contrast with others based on the weight assigned.

As a result of the above comparison as shown in Table 66.4, HPH-6 ranked first and LPH-2 ranked last. However, the results relied on the weight criteria and the relative significance/importance assigned to them (decision-maker dependent). Different results could be acquired by changing the significance of various criteria depending upon the necessities of the situation, the ranking and selection of heater are a laborious process, and it required an in-depth knowledge of all the parameters, which impact the heater selection; however, the usage of any software tool makes the selection process substantially easier.

Table 66.3 Weighted super matrix

Clusters	RPEn	RPEx	Ex(PTC)	FR	SC	UHC	HPH-5	HPH-6	HPH-7	LPH-2	LPH-3	LPH-4	Best heater
RPE_n	0.00000	0.00000	0.00000	0.00000	0.00000	0.00000	0.00000	0.00000	0.00000	0.00000	0.00000	0.00000	0.02785
RPE_x	0.00000	0.00000	0.00000	0.00000	0.00000	0.00000	0.00000	0.00000	0.00000	0.00000	0.00000	0.00000	0.08079
E_x(PTC)	0.00000	0.00000	0.00000	0.00000	0.00000	0.00000	0.00000	0.00000	0.00000	0.00000	0.00000	0.00000	0.17220
FR	0.00000	0.00000	0.00000	0.00000	0.00000	0.00000	0.00000	0.00000	0.00000	0.00000	0.00000	0.00000	0.21444
SC	0.00000	0.00000	0.00000	0.00000	0.00000	0.00000	0.00000	0.00000	0.00000	0.00000	0.00000	0.00000	0.06488
UHR	0.00000	0.00000	0.00000	0.00000	0.00000	0.00000	0.00000	0.00000	0.00000	0.00000	0.00000	0.00000	0.43984
HPH-5	0.05952	0.08057	0.17798	0.05588	0.05345	0.05441	0.00000	0.00000	0.00000	0.00000	0.00000	0.00000	0.00000
HPH-6	0.46141	0.42923	0.26421	0.48675	0.50721	0.48900	0.00000	0.00000	0.00000	0.00000	0.00000	0.00000	0.00000
HPH-7	0.25619	0.23165	0.33383	0.25623	0.10814	0.25872	0.00000	0.00000	0.00000	0.00000	0.00000	0.00000	0.00000
LPH-2	0.03676	0.05565	0.03516	0.03066	0.13658	0.03000	0.00000	0.00000	0.00000	0.00000	0.00000	0.00000	0.00000
LPH-3	0.04620	0.05948	0.07179	0.03802	0.05894	0.04058	0.00000	0.00000	0.00000	0.00000	0.00000	0.00000	0.00000
LPH-4	0.13991	0.14341	0.11703	0.13245	0.13569	0.12730	0.00000	0.00000	0.00000	0.00000	0.00000	0.00000	0.00000
Best heater	0.00000	0.00000	0.00000	0.00000	0.00000	0.00000	0.00000	0.00000	0.00000	0.00000	0.00000	0.00000	0.00000

Source The authors

Table 66.4 Alternative ranking

Graphic	Alternatives	Total	Normal	Ideal	Rank
	HPH-5	0.0391	0.0782	0.1756	4
	HPH-6	0.2227	0.4454	1.0000	1
	HPH-7	0.1295	0.2591	0.5817	2
	LPH-2	0.0201	0.0402	0.0903	6
	LPH-3	0.0241	0.0483	0.1084	5
	LPH-4	0.0644	0.1288	0.2893	3

Source The authors

66.5 Conclusion

This study intends to assess a priority investigation of feed water heaters used in a TPP after solar substitution. The technique proposed in this work contributes theoretically and methodologically to better comprehend the complex decision-making process. Conclusion, hereby, sees a lot varying range of solar power requirement by heaters from 14 to 33 MW for the pre-existing thermal power plants, mandating an aperture area of 3–7 ha. (Source: the authors). With such a huge variation, it is not possible to select the best possible regenerative feed water heater along with the solar integration. AHP framework proposed by Saaty [2] is being used to prioritized the substitution by using six criteria (improvement in energy efficiency of plant, improvement in exergy efficiency of plant, reduction in unit heat rate, improvement in exergy efficiency of solar field solar contribution, and reduction in fuel consumption) and six alternatives (all six heaters) Super Decision Software is attributed toward defining the correlation and finding the respective weightages between criteria's and alternatives. Finally, the relative significance is generated, which is lowest for LPH-2 (0.0201) and highest for HPH-6 (0.2227). After normalizing, it is found that there is the difference of 41.83% in relative significance between first and second ranking alternatives, whereas 90.97% in first and last alternative. HPH-6 has been selected as the first alternative along with solar-based FWHE to satisfy the technical parameters and challenges for a very motive toward green power cogeneration.

Acknowledgements The authors are greatly acknowledging the valuable support provided by NTPC Ltd., Badarpur, Delhi, India and DME, IIT-Dhanbad for the necessary facilities to work.

References

1. Keeney, R.L., Raiffa, H.: Decisions with multiple objectives: preferences and value trade-offs. IEEE Trans. Syst. Man. Cybern. **9**, 403 (1979). https://doi.org/10.1109/tsmc.1979.4310245
2. Saaty, T.L., Vargas, L.G.: Decision Making With the Analytic Process Network Process. Springer Science and Business Media, New York, NY (2008)

3. Hwang, C.L., Yoon, K.: Multiple Attribute Decision Making Methods and Applications: A State-of-the-Art Survey. Springer-Verlag, Berlin Heidelb, New York (1981)
4. Brans, J.P., Vincke, P., Mareschal, B.: How to select and how to rank projects : the PROMETHEE method. Eur. J. Oper. Res. **24**, 228–238 (1986). https://doi.org/10.1016/0377-2217(86)90044-5
5. Roy, B.: The outranking approach and the foundations of ELECTRE methods. Theory Decis. **31**, 49–73 (1991). https://doi.org/10.1007/978-3-642-75935-2_8
6. Pomerol, J., Barba-Romero, S.: Multicriterion Decision in Management: Principles and Practice. Kluwer Academic Publishers, New York (2000)
7. Belton, V., Stewart, T.: Multiple Criteria Desicion Analysis: An Integrated Approach. Kluwer Academic Publishers, London (2002)
8. Wallenius, J., Dyer, J.S., Fishburn, P.C., Steuer, R.E., Zionts, S., Deb, K.: Multiple criteria decision making, multiattribute utility theory: recent accomplishments and what lies ahead. Manage. Sci. **54**, 1336–1349 (2008). https://doi.org/10.1287/mnsc.1070.0838
9. Bouyssou, D., Marchant, T., Pirlot, M., Perny, P., Tsoukias, A., Vincke, P.: Evaluation and Decision Models: A Critical Perspective. Kluwer Academic Publishers, London, Dordrecht (2000)
10. Diaz-Balteiro, L., González-Pachón, J., Romero, C.: Measuring systems sustainability with multi-criteria methods: a critical review. Eur. J. Oper. Res. **258**, 607–616 (2017). https://doi.org/10.1016/j.ejor.2016.08.075
11. Kumar, D., Garg, C.P., Evaluating sustainable supply chain indicators using fuzzy AHP: case of Indian automotive industry. Benchmarking **24**, 1742–1766 (2017). https://doi.org/10.1108/bij-11-2015-0111
12. Shen, K.Y., Tzeng, G.H.: Advances in multiple criteria decision making for sustainability: modeling and applications. Sustainability **10**, 1–7 (2018). https://doi.org/10.3390/su10051600
13. Dos Santos, P.H., Neves, S.M., Sant'Anna, D.O., Oliveira, C.H. de, Carvalho, H.D.: The analytic hierarchy process supporting decision making for sustainable development: an overview of applications. J. Clean. Prod. **212**, 119–138 (2019). https://doi.org/10.1016/j.jclepro.2018.11.270
14. Ho, W., Ma, X.: The state-of-the-art integrations and applications of the analytic hierarchy process. Eur. J. Oper. Res. **267**, 399–414 (2018). https://doi.org/10.1016/j.ejor.2017.09.007
15. Zhang, Y., Mohsen, J.P.: A Project-based sustainability rating tool for pavement maintenance. Engineering **4**, 200–208 (2018). https://doi.org/10.1016/j.eng.2018.03.001
16. Choudhary, D., Shankar, R.: An STEEP-fuzzy AHP-TOPSIS framework for evaluation and selection of thermal power plant location: a case study from India. Energy **42**, 510–521 (2012). https://doi.org/10.1016/j.energy.2012.03.010
17. Stein, E.W.: A comprehensive multi-criteria model to rank electric energy production technologies. Renew. Sustain. Energy Rev. **22**, 640–654 (2013). https://doi.org/10.1016/j.rser.2013.02.001
18. Cook, W.D., Green, R.H.: Evaluating power plant efficiency: a hierarchical model. Comput. Oper. Res. **32**, 813–823 (2005). https://doi.org/10.1016/j.cor.2003.08.019
19. Chatzimouratidis, A.I., Pilavachi, P.A.: Multicriteria evaluation of power plants impact on the living standard using the analytic hierarchy process. Energy Policy **36**, 1074–1089 (2008). https://doi.org/10.1016/j.enpol.2007.11.028
20. Chatzimouratidis, A.I., Pilavachi, P.A.: Sensitivity analysis of technological, economic and sustainability evaluation of power plants using the analytic hierarchy process. Energy Policy **37**, 788–798 (2009). https://doi.org/10.1016/j.enpol.2008.10.009
21. Chatzimouratidis, A.I., Pilavachi, P.A.: Sensitivity analysis of the evaluation of power plants impact on the living standard using the analytic hierarchy process. Energy Convers. Manag. **49**, 3599–3611 (2008). https://doi.org/10.1016/j.enconman. 2008.07.009
22. Chatzimouratidis, A.I., Pilavachi, P.A.: Technological, economic and sustainability evaluation of power plants using the analytic hierarchy process. Energy Policy **37**, 778–787 (2009) https://doi.org/10.1016/j.enpol.2008.10.009
23. Aragonés-Beltrán, P., Chaparro-González, F., Pastor-Ferrando, J.-P., Pla-Rubio, A.: An AHP (analytic hierarchy process)/ANP (analytic network process)-based multi-criteria decision

approach for the selection of solar-thermal power plant investment projects. Energy **66**, 222–238 (2014). https://doi.org/10.1016/j.energy.2013.12.016
24. Yagmur, L.: Multi-criteria evaluation and priority analysis for localization equipment in a thermal power plant using the AHP. Energy **94**, 476–482 (2016). https://doi.org/10.1016/j.energy.2015.11.011
25. Ren, J., Dong, L.: Evaluation of electricity supply sustainability and security: Multi-criteria decision analysis approach. J. Clean. Prod. **172**, 438–453 (2018). https://doi.org/10.1016/j.jclepro.2017.10.167

Chapter 67
Ignition and Combustion Characteristics of Impinging Diesel and Biodiesel Blended Sprays Under Diesel Engine-Like Operating Conditions

Sanaur Rehman, Firoj Alam and Mohammad Adil

Abstract In present work, ignition and combustion characteristics of impinging diesel and biodiesel blended diesel sprays are examined experimentally in constant volume chamber at diesel engine-like experimental conditions. Comparative study of ignition delay (ID) and duration of combustion (DOC) for two selected fuels are done using various transducers or sensors. Commercially available diesel and biodiesel blended diesel (50% diesel and 50% biodiesel by v/v) fuels are investigated. Biodiesel produced from waste vegetable oil is chosen for blending. Operating conditions include hot surface temperature (HST) range (350–450 °C) and cylinder air pressure (CP) range (2–4 MPa) at 300 bar-fuel injection pressure (IP). It is observed that ID and DOC of each fuel reduce significantly with increase in HST and CP. It is investigated that ID and DOC of blended sprays are substantially lesser than diesel sprays at present experimental conditions. Lower magnitudes of ID and DOC of blended sprays are attributed to distinct differences in properties of selected fuels. Highest fall in ID of biodiesel blended sprays than diesel sprays is found at lower HST of 623 K for all CP. Variation of ID of both fuels is found to be nonlinear with HST. It is also deduced that differences in DOC of two selected fuels mitigate at higher HST (673 K) for all CP and at higher CP (40 bar) for all HST values. Higher biodiesel content blended fuels seem to be an attractive choice in DI diesel engines having fuel impingement.

Keywords Ignition and combustion characteristics · Biodiesel blend · Constant volume chamber · Impinging sprays

67.1 Introduction

The search of alternative fuels for diesel engine has increased in recent years due to increasing environmental concerns, stringent emissions regulations, depleting fossil fuel reserves and rising petroleum oil prices. Among various alternative fuels,

S. Rehman (✉) · F. Alam · M. Adil
Department of Mechanical Engineering, AMU, Aligarh, India
e-mail: sanaur.rehman@zhcet.ac.in

S. Yadav et al. (eds.), *Proceedings of International Conference in Mechanical and Energy Technology*, Smart Innovation, Systems and Technologies 174,
https://doi.org/10.1007/978-981-15-2647-3_67

biodiesel fuel is focussed for use in transportation area due to its available production technology and compatibility of biodiesel fuel usage in existing engines infrastructure [1]. In recent years, biodiesel has been emerged as a possible alternative fuel for diesel engine due to insignificant changes in engine hardware [2]. Biodiesel blended diesel fuels have a great influence on engine performance and emissions due to significantly different physical and chemical properties [3–5]. Further research is required to know the detailed effects of impinging biodiesel blended diesel fuels on performance, combustion and emissions in existing engines without any major modifications. Recent developments in smaller engine sizes having common-rail injection direct-injection (DI) systems (for high efficiency) leads to cause significant spray-wall/piston interactions and results in excessive fuel impingement on piston surfaces [6, 7]. A study by Rao [8] investigated that in high-speed DI diesel engine about 75% of injected fuel reached piston surface under higher loads. In small-bore high injection pressure DI diesel engines, spray-wall impingement occurs and affects significantly combustion, thermal efficiency and emissions formation [9–11]. Nevertheless, exhaust emissions are significantly affected by piston surface temperature [12]. Hot surface ignition appreciably affects ignition and combustion characteristics of fuel sprays [13–18].

67.2 Experimental Section

Experimental setup and its block diagram used in the present study are shown in Fig. 67.1. Commercial diesel fuel (DF) and blended diesel (50% diesel + 50% biodiesel by v/v) are used for comparative study at various operating conditions. DF properties are shown in study [14]. Biodiesel used in the present work is prepared from waste vegetable oil through transesterification process. The diagrams and specifications of components employed in experimental setup are given in study [14].

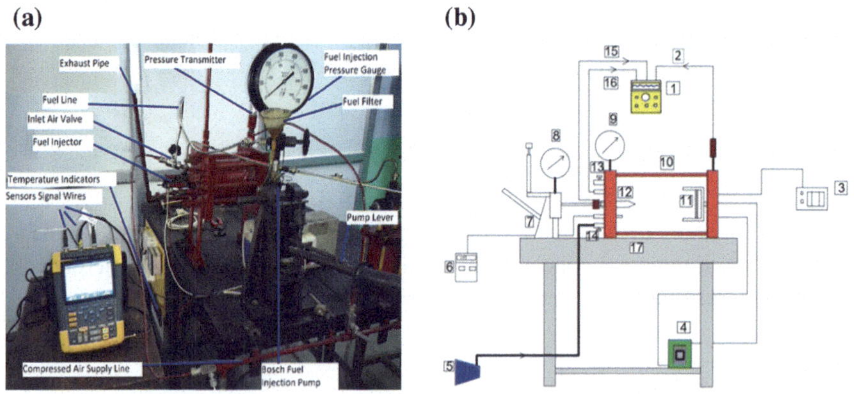

Fig. 67.1 **a** Experimental setup, **b** block diagram of experimental setup

Cylindrical combustion chamber has mechanical fuel injector and hot surface plate and each are mounted on opposite side of the chamber. Heating element (electric wire) placed behind hot plate heated it. Fuel injector has pintle nozzle for producing hollow conical sprays, which impinge on hot surface/plate. Bosch fuel injection pump is mounted on mechanical system and connected with injector via fuel line. The plunger of injection pump is operated by lever. Lever is actuated by hand manually. Fuel IP is obtained with pressure gauge placed on fuel pump. Cylinder air pressure (CP) or pre-firing pressure is obtained with pressure gauge placed on combustion chamber. Piezoresistive pressure transducer mounted on combustion chamber senses pressure rise due to combustion. K-type thermocouples indicate hot surface temperature (HST) and cylinder air temperature on indicator. Inlet valve and exhaust valve are mounted on same end of chamber. Reciprocating air compressor produced compressed air at various pressures. Various components of experimental setup are given in Table 67.1.

67.2.1 Experimental Methodology

In starting, purging of the combustion chamber is done by compressed fresh air. Temperature of hot surface (HST) is maintained at value of 623 K (350 °C) and compressed air (from reciprocating compressor) is supplied into the combustion chamber through inlet valve is supplied at pressures of 20 bar (2 MPa), 30 bar (3 MPa) and 40 bar (4 MPa). At each value of cylinder pressure (CP), DF and biodiesel blended diesel (BD50) fuel are injected into the CVCC by Bosch injector (through pintle nozzle) at fixed IP (300 bar). At a particular experimental condition, injection, ignition and combustion processes of impinging sprays of DF and BD50 are recorded and saved on the screen of scopemeter. ID is measured in millisecond (ms) as shown in Fig. 67.2. It is to be noted that ID measured in this study is optical ID because it is measured with the help of flame luminosity using photo sensor.

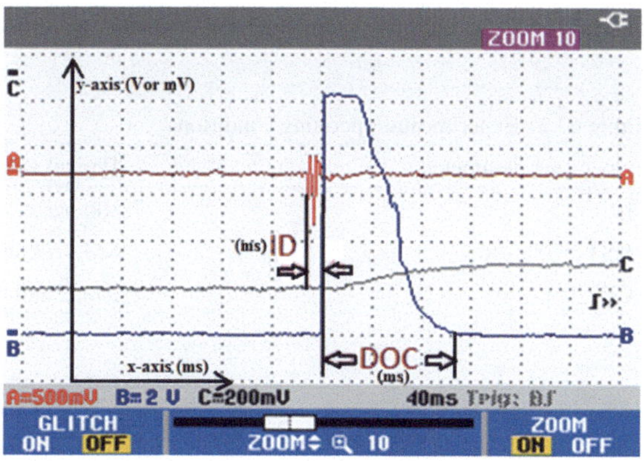

Fig. 67.2 Measurement of ID and DOC of spray with the help of zoomed image of scopemeter at any particular operating condition

Table 67.1 Various components of experimental setup

S. No.	Components	S. No	Components
1	Four channel scopemeter	10	Control volume combustion chamber
2	Signal from piezoresistive pressure transmitter	11	Hot surface plate for ignition
3	Temperature controller and indicator	12	Single hole pintle nozzle
4	Transformer	13	Exhaust valve
5	Reciprocating air compressor	14	Inlet valve
6	Air temperature indicator	15	Signal from needle lift sensor
7	Fuel pump assembly	16	Signal from photo sensor
8	Pressure gauge for IP	17	Table
9	Pressure gauge for CP		

Duration of combustion/burn (DOC) is measured in ms as difference between point of start of combustion (SOC) and end of combustion (EOC) as shown in Fig. 67.2. Similarly, at HST of 673 K and CP are changed and two fuels are injected to obtain different sets of readings. Lastly, at HST of 673 K, again CP are varied and readings are taken with the help of scopemeter. After noting each reading, exhaust gases are expelled out from combustion chamber by fresh compressed air. At each condition, four readings are taken and the average value of four readings is considered at that condition to ensure accurateness of obtained reading.

67.2.2 Experimental Conditions

Combustion of DF and BD50 impinging sprays is studied under various operating conditions similar to diesel engine-like conditions. Experimental conditions used in present work are presented in Table 67.2. Four readings are taken at each set of condition (e.g., 300 bar-fuel IP, 40 bar-CP and 723 K-HST) and average value is calculated to reduce the error and uncertainty in determination of ID and DOC.

Table 67.2 Experimental/operating conditions

Operating parameters	Typical values
IP	300 bar
HST	623, 673 and 723 K
CP	20, 30 and 40 bar

67.3 Results and Discussion

Ignition delay (ID) is an essential ignition characteristic of DF and similar fuels [19]. ID in any injection-combustion system is generally described as the time difference between point of start of fuel injection and point of start of combustion/ignition process [6] in ms or crank angle (CA). In ID period, fuel which is already injected into combustion chamber atomized, vaporized and mixed with heated surrounding air and residual gases and fuel vapor-air mixtures prepares for auto-ignition phenomenon leading to premixed combustion phase. ID greatly influences engine design and performance [3, 4, 19]. Figure 67.3 shows the effect of HST and CP on ID of pure diesel (DF) and biodiesel blended diesel fuel (BD50) sprays at different operating conditions. Figure 67.3a shows the effect of varying HST from 623 to 723 K on ID at constant CP for both fuels. It can be observed in Fig. 67.3a that with increase in HST values, ID significantly reduces for impinging sprays of both DF and BD50 at all operating conditions. Increase in HST causes reduction in ID for impinging sprays [6, 9, 11]. The cause for reduced ID with the high HST is due to the quicker fuel evaporation of impinging DF and BD50 sprays near and onto the hot surface at high temperature. Quicker evaporation process of impinging fuel on hot surface promotes quick mixing process of fuel and air near impingement hot surface. Faster mixing process of fuel and air results in smaller ID of both impinging fuel sprays. Moreover, increase in HST causes increase in cylinder air temperature substantially and subsequently reduces ID [6]. Increase in HST leads to decrease in ID gradually as investigated [20, 21]. It is important to observe that HST effect on ID mitigates as HST increase to higher values for both DF and BD50 sprays. This effect is also reported by other studies [9, 18, 20]. Figure 67.3b indicates the effect of CP variation (20–40 bar) on ID at constant HST for both fuels. It can be noted from Fig. 67.3b that rise in CP from 20 bar (2 MPa) to 40 bar (4 MPa), ID of both fuels significantly reduces at all present experimental conditions. Fall in ID with increase in CP may be attributed to the higher cylinder pressure of air, which enhances cylinder air density. Also, high density of cylinder air has decreases auto-ignition temperature (AIT) of injected fuel and therefore the ID reduces with the rise in CP [19]. Moreover, high air density produces more air entrainment into the impinging fuel sprays which enhance fuel-air mixing process significantly.

67.3.1 *Comparison of ID of DF and BD50 Sprays*

It is found in Fig. 67.3 that BD50 fuel has remarkably smaller ID than DF at all operating conditions of the present study. The smaller ID period of BD50 blend than DF is attributed to high cetane number (CN) of biodiesel as compared to DF [1, 3, 19]. Due to higher CN of biodiesel blend (BD50), it combust earlier than DF and results in shorter ID period. Shorter ID of impinging BD50 sprays are also due to lower compressibility and higher viscosity of biodiesel than DF [1, 4, 19]. Similar

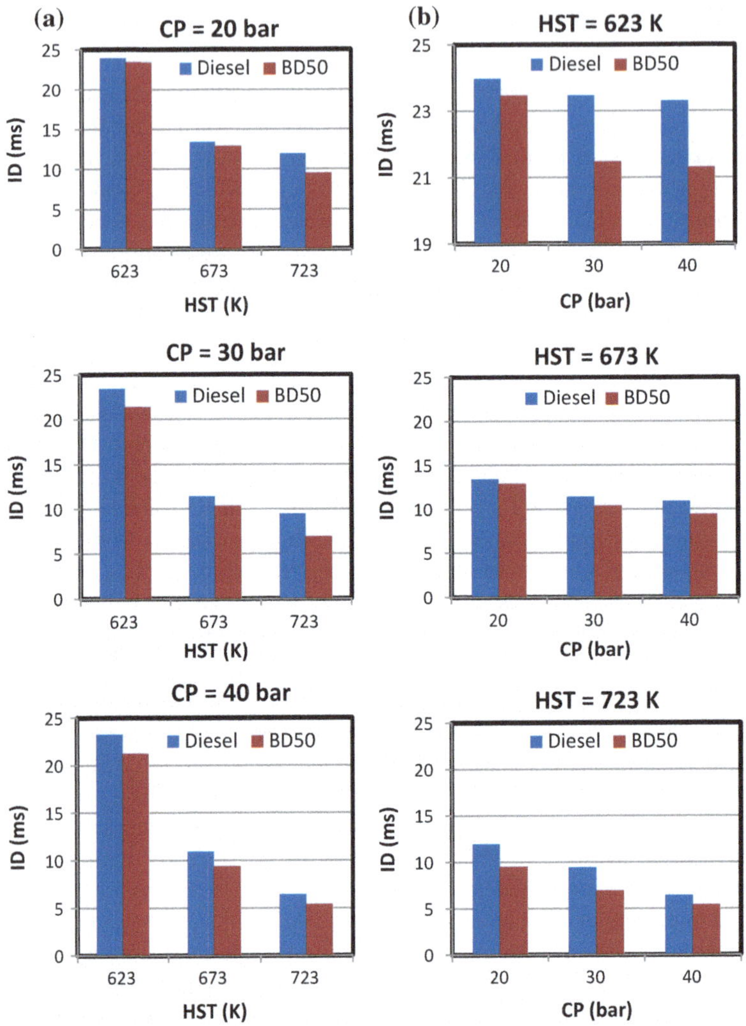

Fig. 67.3 **a** Effect of HST on ID of DF and BD50 sprays at different CP values, **b** effect of CP on ID of DF and BD50 sprays at different HST values for 300 bar IP

results of biodiesel blend (BD20) against DF were obtained by Agarwal [2]. A study on single-cylinder and four-stroke diesel engine also found short ID of various biodiesel blends of different feedstock [20]. Shorter ID period of BD50 sprays will also results in reduced premixed combustion phase and subsequently smaller peak heat release rate (PHRR). Also, it is observed in Fig. 67.3 that highest reduction in ID of impinging sprays of BD50 than DF sprays is found at lower HST of 623 K for all CP values.

67.3.2 Combustion Characteristic

Duration of combustion (DOC) is also called as the duration of burn (DOB) and it is basically complete combustion duration of impinging sprays. In other words, it is the time difference between point of start of combustion (start of flame formation) and point of end of combustion (flame extinguishes) in ms. Figure 67.4 shows that DOC of impinging sprays of DF and BD50 significantly affected by the change in HST and CP values. Figure 67.4a indicates the variation of DOC of both fuels sprays with increase in HST at constant CP values. It is observed from Fig. 67.4a that when HST is increased from 623 to 723 K, it causes reduction in DOC substantially at all CP values (2, 3 and 4 MPa). Reduction in DOC with increase in HST at constant CP indicates that combustion process of impinging DF and BD50 sprays becomes faster at higher HST values. The reason for reduction in DOC with increase in CP and HST is the faster fuel-air mixing rate due to high air entrainment into fuel spray at higher CP and faster evaporation of impinging fuel sprays at higher HST. The reduction in DOC with increase in HST and CP are also reported in the studies [9, 14].

67.3.3 Comparison of DOC of DF and BD50 Sprays

Combustion characteristic of fuel heavily depends on its properties such as CN, oxygen content, heat of combustion and bulk modulus [1–3]. Significant differences in combustion characteristic of DF and biodiesel blended fuels (BD50) are due to the distinct variation in properties of these fuels. Higher amount of oxygen in biodiesel blended (BD50) sprays causes increase in fuel-air mixture homogeneity and hence results in quick combustion or reduced combustion duration (DOC) of BD50 fuel. Therefore, it is observed that DOC of impinging BD50 sprays is noticeably smaller than impinging DF sprays at nearly all present operating conditions as presented in Fig. 67.4. Moreover, biodiesel fuel-air mixture achieves nearly stoichiometric conditions and therefore combustion of biodiesel or biodiesel blend is more complete and quicker than DF [4, 5, 21]. It is interesting to note that differences in DOC of DF and BD50 sprays at higher HST of 723 K for all CP and at higher CP of 40 bar for all HST values are mitigating.

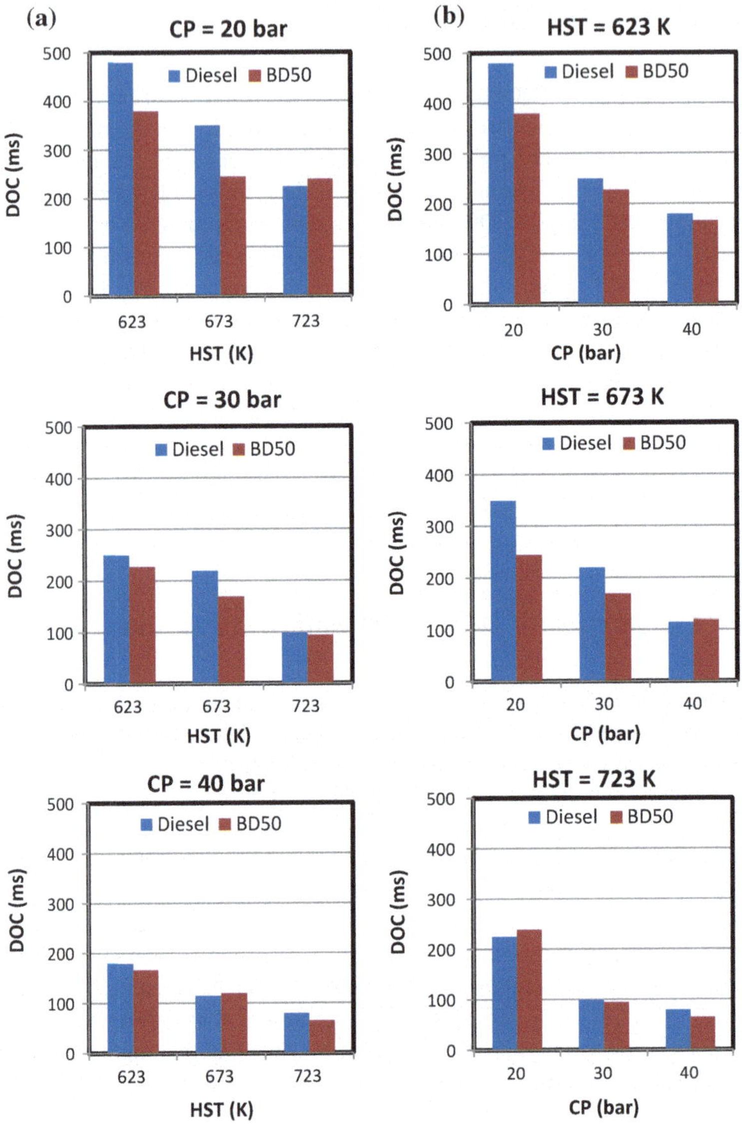

Fig. 67.4 **a** Effect of HST on DOC of DF and BD50 sprays at different CP, **b** effect of CP on DOC of DF and BD50 sprays at different HST values for 300 bar IP

67.4 Conclusion

It is found that ID and DOC of impinging sprays of each fuel reduce significantly with increase in HST and CP. It is also investigated that ID and DOC of impinging BD50 sprays are substantially shorter as compared to impinging DF sprays at all present experimental conditions. Lower magnitudes of ID of BD50 sprays are due to distinct differences in properties of biodiesel such as high CN, lower compressibility and higher viscosity than DF fuel. Moreover, it is found that hot surface ignition effect on ID of DF and BD50 sprays diminishes at higher HST above 673 K. Variation of ID with HST of impinging sprays of both fuels is observed to be nonlinear. Highest reduction in ID of impinging BD50 sprays than DF sprays is found at lower HST of 623 K for all CP. Lower DOC of BD50 sprays attributed to the higher oxygen amount in biodiesel blend than DF. It is deduced that differences in DOC of two selected fuels mitigate at higher HST (673 K) for all CP and at higher CP (40 bar) for all HST values. Higher biodiesel content blended fuels seem to be an attractive choice in DI engines having substantial fuel impingement due to improved combustion.

References

1. Ng, J.H., Ng, H.K., Gan, S.: Advances in biodiesel fuel for application in compression ignition engines. Clean. Techn. Environ. Policy **12**, 459–493 (2010)
2. Agarwal, A.K., Dhar, A., Gupta, J.G., Kim, W., Choi, K., Lee, C.S., Park, S.: Effect of fuel injection pressure and injection timing of Karanja biodiesel blends on fuel spray, engine performance, emissions and combustion characteristics. Energy Convers. Manag. **91**, 302–314 (2015)
3. Aldhaidhaw, M., Miron, L., Chiriac, R., Badescu, V.: Autoignition process in compression ignition engine fueled by diesel fuel and biodiesel with 20% rapeseed biofuel in diesel fuel. J. Energy Eng. **144**(5), 04018049 (2018)
4. Kale, P., Ragit, S.S.: An overview of ignition delay and combustion of biodiesel fuelled in CI engine. In: International Conference on Emanations in Modern Technology and Engineering ((ICEMTE-2017), vol. 5 (2017)
5. Aldhaidhaw, M., Chiriac, R., Badescu, V.: Ignition delay, combustion and emission characteristics of diesel engine fueled with rapeseed biodiesel—a literature review. Renew. Sustain. Energy Rev. **73**, 178–186 (2017)
6. Heywood, J.B.: Internal Combustion Engines Fundamentals, 2nd edn. McGraw Hill International, New York (2011)
7. Ladommatos, N., Xiao, Z., Zhao, H.: The effect of piston bowl temperature on diesel exhaust emissions. Proc. Inst. Mech. Eng. Part D J. Automob. Eng. **219**(3), 371–388 (2005)
8. Rao, K.K., Winterbone, D.E., Clough, E.: Combustion and emission studies in a high speed DI diesel engine. In: IMechE International Conference, C448/070, pp. 117–129 (1992)
9. Feng, L., Chen, B., Liu, H., Yao, M., Geng, C.: Combustion characteristics of wall-impinging diesel fuel spray under different wall temperatures. In: SAE Technical Paper 2017-01-2251 (2017)
10. Liu, H., Chen, B., Feng, L., Wang, Y.: Study on fuel distribution of wall-impinging diesel spray under different wall temperatures by laser-induced exciplex fluorescence (LIEF). Energies **11**, 1–14 (2018)
11. Du, W., Zhang, Q., Zhang, Z., Lou, J., Bao, W.: Effects of injection pressure on ignition and combustion characteristics of impinging diesel spray. Appl. Energy **226**, 1163–1168 (2018)
12. Rehman, S., Zaidi, K.: Experimental study of hot surface temperature effect on diesel and blended diesel exhaust emissions. In: International Conference on energy and Environment, Organized by IST and JNTUHCEH, JNTUH, Hyderabad, 15–17 Dec 2014
13. Rehman, S.: Sensors based measurement techniques of fuel injection and ignition characteristics of diesel sprays in DI combustion system. Alexandria Eng. J. **55**(3), 2391–2403 (2016)

14. Rehman, S.: Hot surface ignition and combustion characteristics of sprays in constant volume combustion chamber using various sensors. Cogent Eng. **5**, 1464879 (2018)
15. Shahabuddin, M., Liaquat, A.M., Masjuki, H.H., Kalam, M.A., Mofijur, M.: Ignition delay, combustion and emission characteristics of diesel engine fueled with biodiesel. Renew. Sustain. Energy Rev. **21**, 623–632 (2013)
16. Hiroyasu, H., Kadota, T., Arai, M.: Supplementary comments: fuel spray characterization in diesel engines. In: Mattavinad, J.N., Amann, C.A. (eds.), pp. 369–408. Plenum Press (1980)
17. Rehman, S., Zaidi, K.: Study of ignition delay of certain blended petroleum fuel sprays in a cylindrical combustion chamber. Int. J. Sci. Technol. ST Rev. **1**(2), 65–76 (2012). ISSN: 2231-5160
18. Varsheny, A., Singh, P., Varsheny, M., Alam, S.S.: Ignition delay control in constant volume combustion systems. In: AIAA Propulsion and Energy Forum, pp. 1–9 (2019)
19. Ganesan, V.: Internal Combustion Engines. McGraw Hill Education Private Limited, New Delhi (2014)
20. Sahoo, P.K., Das, L.M.: Combustion analysis of Jatropha, Karanja and Polanga based biodiesel as fuel in a diesel engine. Fuel **88**, 994–999 (2009)
21. Rakopoulos, C.D., Antonopoulos, K.A., Rakopoulos, D.C., Hountalas, D.T., Giakoumis, E.G.: Comparative performance and emissions study of a direct injection diesel engine using blends of diesel fuel with vegetable oils or bio-diesels of various origins. Energy Convers. Manag. **47**, 3272–3287 (2006)

Chapter 68
Design Optimization of Double-Stage Thermoelectric Refrigerator

Sudhanshu Sharma, V. K. Dwivedi and A. K. Sethi

Abstract In the present work, a two-stage thermoelectric refrigerator that consists of 50 elements has been analyzed. A non-dimensional parameter 'x' has been considered that represents the ratio of number of elements on hot side and that on cold side. Coefficient of performance (COP) has been calculated to evaluate the operating performance of the double-stage thermoelectric refrigerator with all possible values of 'x' keeping the sum of elements constant in two stages. Coefficient of performance is the most desirable performance measuring parameter of a refrigerating device. This work obtains the values of current input as 8 amp and 'x' as 1.63 at which COP has the maximum value, i.e., 0.5046, the best performance of the device with this configuration.

Keywords Thermoelectric refrigerator · Laws of thermodynamics · COP · Rate of refrigeration

68.1 Introduction

A thermoelectric generator (TEG) can be utilized as heat engine (using Seebeck effect) and a thermoelectric refrigerator (TER), or heat pump (using Peltier effect). They have colossal potential in comparison to other options. As there is no moving component, thus it increases durability, reduces maintenance, and improves the span of life of the system. With alteration in direction of current, the device can be used as refrigerator or heat pump. The non-existence of refrigerant removes the hazards of spillages to the environment which are genuine issues in conventional vapor compression refrigeration systems. As there is no compressor used, it eliminates the noise in the system. TER's are more suitable in areas such as small electronic circuits, laser diodes, communication gadgets, medical, pharmaceutical, and defense operations where meticulous control of temperature is more important than other parameters.

S. Sharma (✉) · V. K. Dwivedi · A. K. Sethi
Department of Mechanical Engineering, Galgotias College of Engineering and Technology, Greater Noida, Uttar Pradesh 201306, India

S. Yadav et al. (eds.), *Proceedings of International Conference in Mechanical and Energy Technology*, Smart Innovation, Systems and Technologies 174,
https://doi.org/10.1007/978-981-15-2647-3_68

TERs are better choice for the green environment as the ozone depleting working substances do not exist in the system. Due to these advantages, the thermoelectric are being used in wide zones. A new setup of a double-stage TER has been studied in which the two stages are supplied power by two different thermoelectric generators. A three-dimensional thermoelectric demonstrate was considered to compare the performance of the modern and original designs for various thermocouple number ratios and working conditions [1]. A density-based topology optimization approach was utilized to optimize the cooling control and adequacy of thermoelectric coolers by spatially disseminating two distinctive thermoelectric materials in a two-dimensional plan space [2]. A test was conducted to assess the cooling performance of two-stage thermoelectric coolers (TECs) under current pulse operation so as to realize a greater temperature drop [3]. Cooling characteristics of 2-jointed, 4-jointed, and 6-jointed thermoelectric modules are tentatively assessed and examined beneath the same extent of input electrical control [4]. A study that clarifies the rising intrigued of thermoelectric innovation and applications have been reported. This depicts the standards of thermoelectricity and presents a clarification of current and up and coming materials [5]. A three-dimensional numerical demonstrate of a compact TEC, which took under consideration the temperature-dependent TE material properties were examined [6]. An exo-reversible thermodynamic model of the annular thermoelectric cooler considering Thomson effect in conjunction with Peltier, Joule and Fourier heat conduction has been investigated using exergy analysis [7]. A three-dimensional numerical model has been analyzed for the optimization design of a compact TEC utilizing finite element method [8]. Modern applications of the thermoelectric generator, counting the structure optimization which essentially influences the thermoelectric generator have been looked into [9]. Impact of Thomson coefficient on the execution of a thermoelectric model has been analyzed. [10]. An optimization model based on the entropy generation minimization method is analyzed for a thermoelectric cooler [11]. A thermoelectric cooler was attached into the system to upgrade the method of both humidification and dehumidification and its execution was tried for different working conditions [12].

68.2 Two-Stage Thermoelectric Refrigerator

A thermoelectric refrigerator works on the principle of Peltier effect. According to Peltier effect, when electrical current is supplied through a pair of p- and n-type semi-conductor material elements, one end of the pair gets cold and another end becomes hot. Therefore, such a combination can be used for cooling or heating purposes or as a refrigerator and heat pump according to the requirement.

Number of semi-conductor elements which form a thermoelectric refrigerator can be placed in single stage or in multi-stages. Comparison of single-stage and double-stage TER with equal number of thermoelectric elements has been done [13]. A two-stage TER can be formed by connecting two single-stage TER's. The total number of thermoelectric elements can be divided into two stages equally or

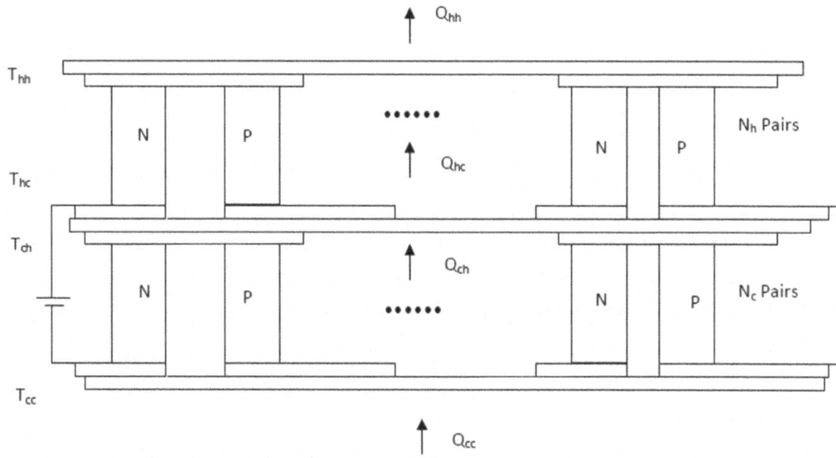

Fig. 68.1 Schematic diagram of double-stage TER

unequally. A two-stage TER is shown in Fig. 68.1. Equations involving heat transfers at all sides of this model can be written as.

$$Q_{cc} = \left[\alpha I_c T_{cc} - \frac{I_c^2 R}{2} - k(T_{ch} - T_{cc})\right] N_c \tag{68.1}$$

$$Q_{ch} = \left[\alpha I_c T_{ch} + \frac{I_c^2 R}{2} - k(T_{ch} - T_{cc})\right] N_c \tag{68.2}$$

$$Q_{hc} = \left[\alpha I_h T_{hc} - \frac{I_h^2 R}{2} - k(T_{hh} - T_{hc})\right] N_h \tag{68.3}$$

$$Q_{hh} = \left[\alpha I_h T_{hh} + \frac{I_h^2 R}{2} - k(T_{hh} - T_{hc})\right] N_h \tag{68.4}$$

where Q_{cc} is the cooling capacity at the cold side of the colder stage; Q_{ch} is the rate of discharge of heat at the hot side of colder stage; Q_{hc} is the cooling capacity at the colder side of the hotter stage; and Q_{hh} is the rate of discharge of heat at the hot side of the hotter stage. I_c is the current flowing in colder side and I_h is the current flowing in hotter side of TER. In this analysis, the two stages are electrically is series, so the current $I_c = I_h$. N_h and N_c are number of semi-conductor elements in colder and hotter side of TER, respectively. For this analysis, $N_h + N_c = 50$ has been considered. T_{cc} and T_{ch} denote the temperatures of the cold side and hot side of colder stage and T_{hc} and T_{hh} denote the cold side and hot side temperature of hotter side. It can be caught on that, there could be an intersection temperature T_m which can be calculated by putting $Q_{ch} = Q_{hc}$. Since the two stages are thermally in series arrangement so that the heat rejected from colder stage will be totally provided to hotter stage.

Hence, $Q_{\text{ch}} = Q_{\text{hc}}$ gives

$$T_{\text{m}} = \frac{k(T_{\text{cc}} + xT_{\text{hh}}) + \left(xj^2 + 1\right)\frac{1}{2}I_{\text{c}}^2 R}{k(x+1) + I_{\text{c}}\alpha(xj - 1)} \tag{68.5}$$

where $x = N_{\text{h}}/N_{\text{c}}$ and $j = I_{\text{h}}/I_{\text{c}}$.

Eventually, the heat balance equations can be written as:

$$Q_{cc} = \left[\alpha I_c T_{cc} - \frac{I_c^2 R}{2} - k(T_m - T_{cc})\right] N_c \tag{68.6}$$

$$Q_{\text{ch}} = \left[\alpha I_{\text{c}} T_{\text{m}} + \frac{I_{\text{c}}^2 R}{2} - k(T_{\text{m}} - T_{\text{cc}})\right] N_{\text{c}} \tag{68.7}$$

$$Q_{\text{hc}} = \left[\alpha I_{\text{h}} T_{\text{m}} - \frac{I_{\text{h}}^2 R}{2} - k(T_{\text{hh}} - T_{\text{m}})\right] N_{\text{h}} \tag{68.8}$$

$$Q_{\text{hh}} = \left[\alpha I_{\text{h}} T_{\text{hh}} + \frac{I_{\text{h}}^2 R}{2} - k(T_{\text{hh}} - T_{\text{m}})\right] N_{\text{h}} \tag{68.9}$$

COP of two-stage TER:

$$(\text{COP})_{\text{act}} = \frac{Q_{\text{cc}}}{W_{\text{hh}} + W_{\text{cc}}} = \frac{Q_{\text{cc}}}{(Q_{\text{hh}} - Q_{\text{hc}}) + (Q_{\text{ch}} - Q_{\text{cc}})} = \frac{Q_{\text{cc}}}{(Q_{\text{hh}} - Q_{\text{cc}})} \tag{68.10}$$

W_{hh} and W_{cc} are the work inputs in hotter and colder sides individually. Since a two-stage TER is a combination of two single-stage TER's associated thermally in series, so the aggregate work input would be equal to the sum of the work input to individual stages. But Q_{cc} would be the heat absorbed at the cold side plate only.

In this trial, N_{h} and N_{c} are number of semi-conductor elements in first stage (cold side) and second stage (hot side), respectively, and $N_{\text{h}} + N_{\text{c}} = 50$. There may be three possibilities to select N_{h} and N_{c} such that $N_{\text{h}} > N_{\text{c}}$, $N_{\text{h}} < N_{\text{c}}$ and $N_{\text{h}} = N_{\text{c}}$. A parameter x has been considered which is the ratio of elements in hot side and cold side in a two-stage TER. To begin with, $x = 1$ has been considered to determine the COP of the considered model. $x = 1$ represents the case when the number of elements is equal, i.e., 25–25 elements in both the stages.

It can be observed from Fig. 68.2 and Table 68.1 that the highest value of COP is obtained at 8 amp. The highest value of COP obtained at this current is 0.441. This value encourages for making some attempts to enhance the value of COP in double-stage thermoelectric cooler. Double-stage TER has flexibility to change number of semi-conductor elements in the two stages. In this model, all combinations of these elements have been tried to enhance the value of COP keeping total number of elements as constant, i.e., 50. These various combinations may be formed by keeping 1 element at colder side and 49 elements at hotter side, i.e., $x = 49$. Now adding one element at colder side and subtracting one element from hotter side the model will

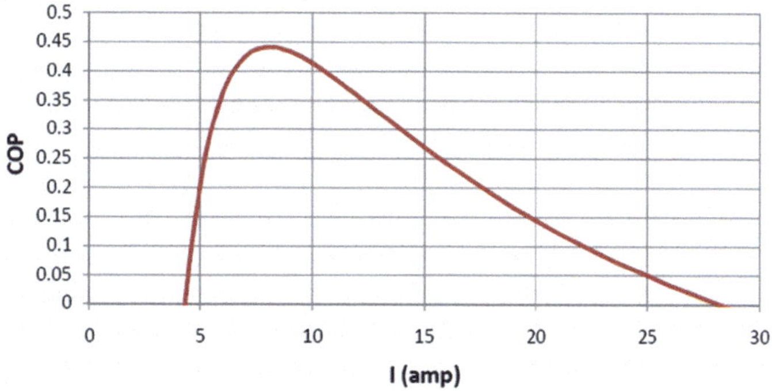

Fig. 68.2 COP of double-stage TER for $x = 1$

Table 68.1 COP of double-stage TER for $x = 1$

Current I (amp)	COP for $x = 1$
8	0.441

eventually have 49 elements at colder side and 1 element at hotter side. Therefore, the parameter x may vary from 49 to 0.0204.

The model was run for the different values of x with current as variable and it was found that the highest value of COP is obtained at $x = 1.63$, shown in Fig. 68.3. x would be 1.63 when 31 thermoelectric elements are placed in hot side and 19 thermoelectric elements are placed in cold side. It can be observed that the highest value of COP obtained is 0.5046. Which is 12.6% higher that the COP obtained at $x = 1$. Only the neighboring values of highest value of COP has been reported here. The values have been compiled in Table 68.2.

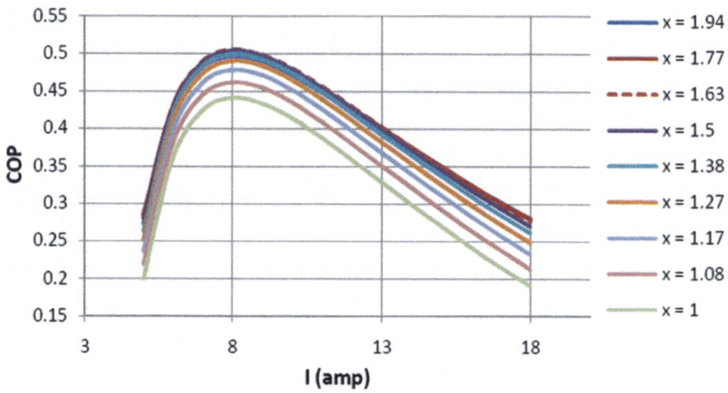

Fig. 68.3 COP of double-stage TER for variable x

Table 68.2 COP of double-stage TER at $I = 8$ amp and for different 'x'

x	1.94	1.77	1.63	1.5	1.38	1.27	1.17	1.08	1
COP	0.4957	0.5019	0.5046	0.5036	0.4988	0.04,903	0.4778	0.4614	0.4411

68.3 Conclusion

A two-stage TER with same and diverse number of thermoelectric elements on hotter side and colder side has been examined. COP has been calculated to measure the performance. As COP is the most desirable performance measuring parameter of a refrigerating device. This work shows the values of current input and 'x' at which COP has the maximum value. The results may be useful for a user to obtain the best performance according to the application and desired effect. TERs are more useful in the applications where precise control of the temperature is needed. So this work may be extended to design a TER for such applications.

References

1. Lin, L., Zhang, Y.F., Liu, H.B., Meng, J.H., Chen, W.H., Wang, X.D.: A new configuration design of thermoelectric cooler driven by thermoelectric generator. Appl. Therm. Eng. **160**, 114087 (2019)
2. Lundgaard, C., Sigmund, O.: Design of segmented thermoelectric Peltier coolers by topology optimization. Appl. Energy **239**, 1003–1013 (2019)
3. Li, W.K., Chang, J.H., Amani, M., Yang, T.F., Yan, W.M.: Experimental study on transient supercooling of two-stage thermoelectric cooler. Case Stud. Therm. Eng. **14**, 100509 (2019)
4. Dizaji, H.S., Jafarmadar, S., Khalilarya, S.: Novel experiments on COP improvement of thermoelectric air coolers. Energy Convers. Manag. **187**, 328–338 (2019)
5. Pourkiaei, S.M., Ahmadi, M.H., Sadeghzadeh, M., Moosavi, S., Pourfayaz, F., Chen, L., Yazdi, A.P., Kumar, R.: Thermoelectric cooler and thermoelectric generator devices: a review of present and potential applications, modeling and materials. Energy (2019)
6. Gong, T., Wu, Y., Gao, L., Zhang, L., Li, J., Ming, T.: Thermo-mechanical analysis on a compact thermoelectric cooler. Energy **172**, 1211–1224 (2019)
7. Manikandan, S., Kaushik, S.C.: Energy and exergy analysis of an annular thermoelectric cooler. Energy Convers. Manag. **106**, 804–814 (2015)
8. Gong, T., Gao, L., Wu, Y., Zhang, L., Yin, S., Li, J., Ming, T.: Numerical simulation on a compact thermoelectric cooler for the optimized design. Appl. Therm. Eng. **146**, 815–825 (2019)
9. He, W., Zhang, G., Zhang, X., Ji, J., Li, G., Zhao, X.: Recent development and application of thermoelectric generator and cooler. Appl. Energy **143**, 1–25 (2015)
10. Lee, H.: The Thomson effect and the ideal equation on thermoelectric coolers. Energy **56**, 61–69 (2013)
11. Zhu, L., Yu, J.: Optimization of heat sink of thermoelectric cooler using entropy generation analysis. Int. J. Therm. Sci. **118**, 168–175 (2017)
12. Yıldırım, C., Soylu, S.K., Atmaca, İ., Solmuş, İ.: Experimental investigation of a portable desalination unit configured by a thermoelectric cooler. Energy Convers. Manag. **85**, 140–145 (2014)
13. Sharma, S., Dwivedi, V.K., Pandit, S.N.: Exergy analysis of single stage and multi stage thermoelectric cooler. Int. J. Energy Res. **38**(2), 213–222 (2014)

Chapter 69
State of the Art of Working of Indian Small Manufacturing Sectors: A Case Study

Brijesh Singh, Manoj Kumar Lohumi and Manohar Singh

Abstract Small sectors help in the country's social and economic development with their effective, efficient, flexible and innovative entrepreneurial spirit. To survive and prosper in this agile world, these need to improve performance and competitiveness. In this research work, MSMEs are studied for their working practices. The investigations, based on surveyed MSMEs of India, cited that least orders are received through Internet. Produced quality does not suit to global marketplaces. Implementation of suitable type of advance manufacturing technologies with information technology tools can help these sectors to improve their performance.

Keywords Micro, small and medium enterprises · MSMEs · Small and medium enterprises · SMEs · Performance of SMEs

69.1 Introduction

Small sectors are contributing appreciably in employment generation (at less capital investment with respect to large enterprises), nation's GDP, exports and in industrialization of the rural and backward areas. Small and medium enterprises (SMEs) are significant driver of economic development [1], which is supported by Tahir et al. [2] who found that Malaysian SMEs comprise about 99.2% of total business establishments and contributing 32% of GDP. According to China Banking News, China's SMEs accounted for about 90% of all entities, about 80% of employment, more than 70% of patents, over 60% of GDP and more than 50% of tax revenues [3]. Further, in UK, according to research from the Department for Business, Energy and Industrial Strategy, at the start of 2018, a massive 99.9% of the 5.7 million businesses in the UK are SMEs. These accounted for 60% of all private sector jobs in the UK, a total of 16.3 million. Hence, SMEs are crucial to the UK's economy, and their contribution is increasing every year [4]. In India, there is about 32% contribution of micro, small and medium enterprises (MSMEs) in nation's gross value added. During 2015–16,

B. Singh (✉) · M. K. Lohumi · M. Singh
Department of Mechanical Engineering, Galgotias College of Engineering and Technology, Greater Noida 201306, India
e-mail: brijesh.singh@galgotiacollege.edu

S. Yadav et al. (eds.), *Proceedings of International Conference in Mechanical and Energy Technology*, Smart Innovation, Systems and Technologies 174,
https://doi.org/10.1007/978-981-15-2647-3_69

there are 633.8 lakh unincorporated non-agriculture MSMEs in the country engaged in different economic activities providing employment to 11.10 crore workers [5]. With this in mind, a study, aiming to know about the performance, adoption of information technology (IT)/electronic commerce (EC)/software for various purposes, status of rejection, quality of produced products, threats and factors influencing the growth of MSMEs, was planned as a part of this case study.

69.2 Review of Literature

Manufacturing systems are well recognized for improving living standard of people and for wealth generation [6]. Global competition, technological change and demanding customers are creating a more knowledge intensive, turbulent, complex and uncertain environment [7]. With these, manufacturers are facing unprecedented levels of pressure resulting from availability of foreign products, globalization of business, new product introductions by competitors, rapid technological innovation, shorter product life, unanticipated customer shift and advances in manufacturing and IT [8]. Chuang et al. [9] quoted that worldwide, manufacturers are adopting advance manufacturing technologies (AMTs), just to address changing needs of the customer without delay in an effective manner and to maintain competitiveness.

The manufacturing enterprises are either large-scale enterprises or SMEs. The World Bank states that formal SMEs contribute nearly 60% of total employment and approximately 40% of national income in emerging economies [10]. So, SMEs are drivers of economic growth, job creation and for poverty alleviation [11]. Majama and Magang [12] added that SMEs contribute immensely to the economy of a country and are characterized by low performance and high failure rate which is often blamed on lack of resources such as funds, land and skilled labour. The research of [13] showed there are ten major barriers faced by SMEs that include competition barriers, financial access, price of energy, technology, inefficient production cost, economic factors, management skill, process, limitation of sales and raw material. In addition to these, the barriers that affect the manufacturing system significantly include poor availability of skilled/expert workforce, lack of training to management and employees, poor infrastructure, poor acceptability of employees for changing conditions and salary discrepancies. Workers in modern manufacturing environments not only need training in depth (level of proficiency in a skill) but also breadth (different skills) as well [14]. According to [15], performance of enterprise is severely affected by the way of organization and management of production. In addition to this, skills, capabilities and participation of workforce also affect the enterprise performance [16]. Further, IT has remarkable effect on productivity of manufacturing enterprises [17]. In today's open market, customers can reach manufacturers across the world to choose their preferred ones, owing to advancements in reliable and affordable ICT [18]. Rahman [19] added that social networking, computerized record and online marketing are the main variables that found to be significant to SME survival.

69.3 Methodology

For the purpose of case study, the expo organized by Indian Industries Association was visited and a survey through questionnaire was conducted. Questions of questionnaire were framed on five-point Likert scale. Findings are elaborated using descriptive statistics (mean, SD and frequency distribution). Cronbach's alpha has been calculated for validity and reliability of collected responses. Further, one-way ANOVA, to ensure whether these influence significantly or not, has been computed. To elaborate further, figures and tables are introduced at suitable places.

69.4 Findings of the Survey

69.4.1 Performance of MSMEs

The parameters considered for measuring the performance of enterprises included annual turn-over, sales growth (past and projected for three years), export growth (past and projected for three years) and investment growth (past and projected for three years). The outcomes are presented in Table 69.1.

Referring Table 69.1, it can be observed that the past sales growth was nearly equally distributed ranging from 10 to 20% (about 33.3% respondents), 21 to 50% (about 30% respondents) and 51 to 75% (about 36.7% respondents), while projected sales growth was expected to be high (about 40% respondents targeted for 51–75% growth). There was no change in export growth (in past and projected, about 53.3% respondents quoted for export growth up to 10%), while invested growth was slightly increased from past to projected. Annual turn-over for maximum of participating enterprises was up to Rs. 50 lakhs (about 46.7% respondents). Regarding the export growth, these enterprises might be doing indirect export and it can be inferred here that many of the MSMEs are looking for increase in sales growth and will plan for investment.

69.4.2 Adoption of IT/EC/Software

Here, the usage/adoption of IT/EC/software in the organization was referred as mentioned in Table 69.2. For these responses, Cronbach's alpha is 0.669 ($N = 8$).

From Table 69.2, it can be observed that the difference in mean among various MSMEs is significant ($p < 0.05$ at 95% confidence level) for online advertisement, order receiving and sales. Hence, this could not be considered as potential use/adoption of IT/EC/software in all MSMEs. Further, referring the highest mean, it can be inferred that all the MSMEs were using IT/EC/software for payroll preparation, HR affairs, billing and collection purposes, documentation, reporting and

Table 69.1 Performance of participating MSMEs

Measure of performance	Number of respondents (%)					Total respondents, %
	Up to 10%	10–20%	21–50%	51–100%	>100%	
Sales growth (Past, over 3 years)	0	10 (33.3)	9 (30.0)	11 (36.7)	0	30, 100
Sales growth (Projected for 3 years)	0	3 (10)	15 (50.0)	12 (40)	0	30, 100
Export growth (Past, over 3 years)	16 (53.3)	7 (23.3)	6 (20.0)	1 (3.3)	0	30, 100
Export growth (Projected for 3 years)	16 (53.3)	7 (23.3)	6 (20.0)	1 (3.3)	0	30, 100
Investment growth (Past, over 3 years)	2 (6.7)	20 (66.7)	7 (23.3)	1 (3.3)	0	30, 100
Investment growth (Projected for 3 years)	3 (10.0)	17 (56.7)	8 (26.7)	2 (6.7)	0	30, 100
Annual turn-over	<50 L	50–100 L	101–200 L	201–500 L	>500 L	30, 100
	14 (46.7)	2 (6.7)	4 (13.3)	5 (16.7)	5 (16.7)	

Table 69.2 State of adoption of IT tools in MSMEs

Parameters for adoption of IT/EC/software	Mean	SD	*F*-value (significance)
Documentation/reporting and monitoring	4.27	0.91	0.838 (0.444)
Payroll/HR/billing/collection	4.27	1.05	2.633 (0.090)
Online advertisement/order receiving/sale	4.13	1.22	9.239 (0.001)
Communication with customers/suppliers/dealers, etc.	3.43	1.17	0.205 (0.816)
Customers' and suppliers' information	3.27	1.14	0.777 (0.470)
Customer service/feedback/complaints	3.13	1.07	0.705 (0.503)
Research and development	1.7	1.09	0.380 (0.688)
Production/distribution and logistics	1.57	0.77	0.567 (0.574)

monitoring purposes. IT is generally used for communication with customers, suppliers, dealers, etc., for gathering information about customers and suppliers and for customers' service, feedback and complaints. IT/EC/software finds its least or negligible adoption/use in R&D and in production, distribution and logistics. Abdullah and Zain [20] quoted that ICT helps SMEs in expanding their business globally. In the same line, [21] inferred that ICTs influence growth of economies and businesses' efficiency and facilitate innovation through diffusion processes, usage practices and commercial success.

69.4.3 Rejection Rate and Quality of Produced Products

To analyse the state of production system, the question related to rejection rate, availability of resources, machines, etc., and quality of produced products was asked. Almost all the enterprises (about 88%) replied that total rejection rate during production of products was up to 5% (while about 20% enterprises replied for 6–10% rejection). The amount of scrap/wastage produced from the raw material during the entire production was also up to 5% for about 90% enterprises (while about 10% enterprises replied for 6–10% wastage). Cheraghi et al. [22] stated that to be more competitive in today's business environment, enterprises are striving to improve their manufacturing processes by increasing workers' productivity and eliminating waste from operations. Marri et al. [23] quoted that to remain competitive, SMEs must deliver products to customer at minimum possible cost, the best possible quality and the minimum lead time. Regarding the quality of produced products, the outcomes are also presented in Table 69.3.

Table 69.3 Quality of produced products

Quality parameter	Mean	SD	Frequency (number and % of enterprises)				
			Nil	Up to 10%	11–20%	21–50%	More than 50%
Excellent quality (fit for export)	3.93	0.98	1 (3.3)	2 (6.7)	3 (10.0)	16 (53.3)	8 (26.7)
Good quality (fit for national market)	4.50	0.63	2 (6.7)	11 (36.7)	17 (56.7)	Nil	Nil
Acceptable quality (fit for local market)	1.93	1.20	15 (50.0)	8 (26.7)	2 (6.7)	4 (13.3)	1 (3.3)

Referring Table 69.3, it can be stated that most of the MSMEs were producing their products for national/regional markets only when quality was 'good' (mean = 4.50 and SD = 0.63). With this, it can be observed that quality of produced products cannot meet global standards. Hence, there exists a need to upgrade MSMEs technologically, so that these enterprises can compete in today's competitive and agile global market. By employing automated machines for production, enterprises can exploit the benefits of customized production with improved quality and shorter lead times with less rejection. Ding and Karunamuni [24] suggested that for manufacturers, achieving economy, efficiency, productivity and quality in production is an important objective for business success.

69.4.4 Threats

In this section, various threats to MSMEs were analysed, as mentioned in Table 69.4. The Cronbach's alpha has value of 0.834 ($N = 8$). Refer Table 69.4 for the outcomes.

From Table 69.4, it can be observed that there is no significant difference of mean ($p < 0.05$ at 95% confidence level) for any of the parameters related to considered threats. On the basis of highest mean, the most severe threat to MSMEs was cost of power (electricity) and fuel like diesel, LPG, CNG, etc. To run the enterprise, power at cost-effective rates is essentially required, and nowadays, its availability/supply and tariff are unfavourable, which is slowly ruining their survival. This threat was followed by cost of raw material needed for production. Fluctuations in the cost of raw materials are mainly due to volatile markets and are also offering difficulty in operating business smoothly and thus in achieving targeted growth. The finding of [25] suggested that SME sectors of Indonesia are still faced with the challenges, such as limited access to capital, raw material, information technology, qualified human resources, marketing the product or services, lacked of guidance from government, exchange rate and high interest rate.

Table 69.4 Threats to MSMEs of India

Threats to MSMEs	Mean	SD	F-value (significance)
Electricity cost/fuel as diesel/LPG/CNG	4.37	0.81	0.036 (0.965)
Cost of raw material needed for production	3.87	1.14	3.298 (0.052)
Due to presence of high capital-intensive units in same type of product/band	3.80	0.66	2.658 (0.088)
Government policies on SEZ/tax benefits	3.80	0.71	0.306 (0.739)
Quality of product made by rivals	3.80	1.06	0.242 (0.787)
Loan interest rates offered by banks	3.47	1.11	0.137 (0.873)
Investment on advertisement	3.10	0.96	0.073 (0.929)
Cost of product made by competitors	3.00	1.11	0.073 (0.929)

Table 69.5 Factors for growth of MSMEs of India

Factors influencing the growth of MSMEs	Mean	SD	*F*-value (significance)
Improvement in the product quality	4.83	0.38	0.466 (0.633)
Use of technological upgradation/implementation of AMT	4.57	0.57	2.529 (0.098)
Training the employees	4.27	0.69	0.087 (0.917)
Welfare of employees/motivation, etc.	4.07	0.69	0.060 (0.942)
Reduction in the inventory	4.07	1.31	0.917 (0.412)
Use of ergonomic conditions for worker	3.97	0.81	0.197 (0.823)
Use of software/IT solutions	3.80	0.76	0.292 (0.749)
Reduction in cost of production	3.80	1.16	0.355 (0.704)
Use of technical consultant/champion	3.57	0.97	0.340 (0.715)
Reduction in rejection rate	3.53	1.07	1.616 (0.217)
Optimization in purchase	3.50	0.94	0.194 (0.825)

69.4.5 Factors that Can Influence Growth of MSMEs

Many factors that may affirm the growth of MSMEs are tried to be analysed, as outlined in Table 69.5. The Cronbach's alpha is 0.618 ($N = 11$). Refer Table 69.5.

From Table 69.5, it can be observed that there is no significant difference of mean ($p < 0.05$ at 95% confidence level) for any of the parameters related to factors needed for growth of MSMEs. On the basis of highest mean, the top five factors or measures to be considered for accelerated growth included improvement in product quality, adoption of AMT or technological upgradation, training of employees or workforce, welfare and motivation of employees or workforce and reduction in the inventory.

69.5 Discussion

In the light of the available literature and finding of the survey, it can be said that SMEs can grow and prosper if they use newer technologies and ICT. This will also help in growth of country's economy, export and employment generation. In this respect, identified barriers that included infrastructure, technology and skilled labour must be overcome. Majority of Indian small sectors have poor capabilities to manufacture globally acceptable quality products. Significant rejection needs to be reduced. Increased orders may help MSMEs to prosper, for which larger customer base is required. Larger customer base can be acquired through offering reliable, cost-effective and quality products for sale at more and more locations. Manufactured products' quantity, quality and cost are affected by non-affordable and discontinuous power supply. Further, prevailing fluctuating markets (cost of raw materials) are added threats. With this, MSMEs are not able to compete in global market places.

The solution is using of suitable type of AMT with ICT. In addition to this, workforce state needs to be improved through training/retraining, motivating them and having favourable and ergonomic working conditions.

69.6 Conclusions

The outcome of the case study outlined that MSMEs of India are generally using IT/EC/software for office works that include payroll preparation, HR affairs, billing and collection purposes, documentation, reporting and monitoring purposes. MSMEs are critically suffered by quality of produced products. The manufactured products are not meeting global standards, and significant rejection is generated. Severe threats are observed owing to cost of power (electricity), fuel like diesel, LPG, CNG, etc., and cost of raw material needed for production. Most of the respondents suggested that for improvement in product quality, adoption of AMT or technological upgradation, training of employees, welfare and motivation of employees and reduction in the inventory are required. By adopting suitable AMT, an enterprise can explore improvement in quality, increased sales, lowering production cost and reduction in labour cost. The risks that can debar successful adoption of suitable AMT include opposition by workforce, followed by inadequate workers trainings. The workforce of Indian MSMEs is not eager to accept the adoption of suitable AMT in their enterprises as they used to oppose and provoke others to oppose. But, the prominent employees' culture includes reducing rejection rate, quality improvement and reducing pollution. Further, MSMEs are using IT for order receiving and dispatch, majorly.

References

1. Obi, J., Ibidunni, A.S., Tolulope, A., Olokundun, M.A., Amaihian, A.S., Borishade, T.T., Fred, P.: Contribution of small and medium enterprises to economic development: evidence from a transiting economy. Data Brief **18**, 835–839 (2018)
2. Tahir, H.M., Razak, N.A., Rentab, F.: The contributions of small and medium enterprises (SME's) on Malaysian economic growth: a sectorial analysis. In: Proceedings of the 7th International Conference on Kansei Engineering and Emotion Research, pp. 704–711 (2018)
3. Web 1: China Banking News. Available at http://www.chinabankingnews.com/2018/06/22/smes-account-60-chinas-gdp-beijing-mulls-inclusion-loans-mlf-collateral/ (downloaded on Oct 03, 2019) (2018)
4. Web 2: SMEs in UK. Available at https://www.merchantsavvy.co.uk/uk-sme-data-stats-charts/ (downloaded on Oct 03, 2019) (2018)
5. Web 3: Economic Survey of India 2017–18. Available at https://mofapp.nic.in/economicsurvey/economicsurvey/pdf/120-150_Chapter_08_Economic_Survey_2017-18.pdf (downloaded on Oct 03, 2019) (2018)
6. Mital, A., Pennathur, A., Huston, R.L., Thompson, D., Pittman, M., Kaber, D.B., Crumpton, L., Bishu, R.R., Rajurkar, K.P., Rajan, V., McMulkin, J.E.M., Deivanayagam, S.,

Ray, P.S., Sule, D.: The need for worker training in advanced manufacturing technology (AMT) environments: a white paper. Int. J. Ind. Ergonomics **24**, 173–184 (1999)
7. Ahuja, I.S.: An evaluation of impact of technology upgradations on manufacturing performance. Int. J. Indian Cult. Bus. Manag. **9**(2), 229–247 (2014)
8. Karim, M.A., Smith, A.J.R., Halgamuge, S.K., Islam, M.M.: A comparative study of manufacturing practices and performance variables. Int. J. Prod. Econ. **112**, 841–859 (2008)
9. Chuang, M., Yan, Y.S., Lin, C.T.: Production technology selection: deploying market requirements, competitive and operational strategies, and manufacturing attributes. Int. J. Comput. Integr. Manuf. **22**(4), 345–355 (2009)
10. Ndiaye, N., Razak, L.A., Nagayev, R., Ng, A.: Demystifying small and medium enterprises' (SMEs) performance in emerging and developing economies. Borsa Istanbul Rev. **18**(4), 269–281 (2018)
11. Wang, Y.: What are the biggest obstacles to growth of SMEs in developing countries?—an empirical evidence from an enterprise survey. Borsa Istanbul Rev. **16**(3), 167–176 (2016)
12. Majama, N.S., Magang, T.I.T.: Strategic planning in small and medium enterprises (SMEs): a case study of Botswana SMEs. J. Manag. Strategy **8**(1), 74–103 (2017)
13. Irjayanti, M., Azis, A.M.: Barrier factors and potential solutions for Indonesian SMEs. Procedia Econ. Financ. **4**, 3–12 (2012)
14. Singh, H., Khamba, J.S.: Evaluating the barriers for enhancing the utilization level of advanced manufacturing technologies (AMTs) in Indian manufacturing industry. Available at www.iiis.org/CDs2008/CD2009SCI/IMETI2009/PapersPdf/F565XC.pdf (2008)
15. Chodur, M., Palka, P.: Impact of AMT investments on effectiveness and competitiveness of manufacturing systems. J. Acad. Res. Econ. **1**, 63–79 (2009)
16. Demeter, K., Kolos, K.: Marketing, manufacturing and logistics: an empirical examination of their joint effect on company performance. Int. J. Manuf. Technol. Manag. **16**(3), 215–233 (2009)
17. Abri, A.G., Mahmoudzadeh, M.: Impact of information technology on productivity and efficiency in Iranian manufacturing industries. J. Ind. Eng. Int. (2014). https://doi.org/10.1007/s40092-014-0095-1 (published online on Dec 17, 2014)
18. Nagalingam, S.V., Lin, G.C.I.: CIM-still the solution for manufacturing industry. Robot. Comput. Integr. Manuf. **24**, 332–344 (2008)
19. Rahman, N.A., Yaacob, Z., Radzi, R.M.: An overview of technological innovation on SME survival: a conceptual paper. Procedia Soc. Behav. Sci. **224**, 508–515 (2016)
20. Abdullah, N., Zain, S.N.: The internationalization theory and Malaysian small and medium enterprises (SMEs). Int. J. Trade Econ. Financ. **2**(4), 318–322 (2011)
21. Cuevas-Vargas, H., Estrada, S., Larios-Gomez, E.: The effects of ICTs as innovation facilitators for a greater business performance. Evidence from Mexico. Inf. Technol. Quant. Manag. (ITQM 2016) **91**, 47–56 (2016)
22. Cheraghi, S.H., Dadashzadeh, M., Soppin, M.: Comparative analysis of production system through simulation. J. Bus. Econ. Res. **6**(5), 87–104 (2008)
23. Marri, H.B., Grieve, R.J., Gunasekaran, A., Kobu, B.: Government-industry university collaboration on the successful implementation of CIM in SMEs: an empirical analysis. Logistics Inf. Manag. **15**(2), 105–114 (2002)
24. Ding, J.T., Karunamuni, R.J.: Modeling information sharing to improve just in time purchasing vendor evaluation. Int. J. Manuf. Technol. Manag. **13**(1), 30–54 (2008)
25. Eravia, D., Handayani, T., Julina: The opportunities and threats of small and medium enterprises in Pekanbaru: comparison between SMEs in food and restaurant industries. Procedia Soc. Behav. Sci. **169**, 88–97 (2015)

Chapter 70
Microstructural, Pyroelectric and Energy Storage Performance of Rare Earth-Doped $Ba_{0.15}R_{0.04}Ca_{0.15}Zr_{0.1}Ti_{0.9}O_3$ Ceramics

Prateek Mishra, K.S. Srikanth, Shubham Agarwal, Sajal Agrawal and Harshit Kumar Gautam

Abstract In this article, the microstructural, pyroelectric and energy storage performance of rare earth-doped lead-free ceramics, namely ($BaTiO_3$ based) $Ba_{0.15}R_{0.04}Ca_{0.15}Zr_{0.1}Ti_{0.9}O_3$ (BCZTO-R; R=Ce, Nd and Ho) is investigated. The Ce, Ho and Nd were taken as the rare earth dopant in pristine BCZTO ceramics. Structural analysis indicated that the samples were in a single perovskite phase as there were no secondary peaks observed in X-ray diffraction analysis. Microstructural observation indicates the well densification with bimodal grain size distribution suggesting that the larger grains grew at the expense of smaller grains. The pyroelectric performance of the pure ceramic and rare earth-doped BCZTO was evaluated based on experimental observations. The pyroelectric coefficient improved from 10×10^{-4} to 14×10^{-4} C/cm^2 at their phase transition temperature. However, the pyroelectric coefficient increased from 3 to 4.6 at room temperature for Nd-doped BCZT. Also, on exposure to the transient temperature gradient, the Nd-doped sample had a higher open-circuit voltage of 685 mV when compared with 465 mV in base ceramic under consideration. The figure of merits (FOMs) for voltage responsivity (F_v), current responsivity (F_i), detectivity (F_d) and energy harvesting (F_e and F_e*) was evaluated for samples. Further, the energy storage performance also increased for Nd-doped specimen over pure ceramic. All these improvements could possibly replace its lead-based counterparts for pyroelectric applications. We have also pre-processed our laboratory data for removing noises and to get more accurate results also a computer program is proposed that can reproduce our experimental work and can be used to find out how much voltage the ceramic can produce when kept under a particular temperature gradient. On Validating our programmatic data with the experimental data we observed that the results are in close proximity to the experimental data.

P. Mishra (✉)
School of Computer Science Engineering, Galgotias University, Greater Noida, UP 201310, India

K.S. Srikanth · S. Agarwal · S. Agrawal · H. K. Gautam
School of Mechanical Engineering, Galgotias University, Greater Noida, UP 201310, India

S. Yadav et al. (eds.), *Proceedings of International Conference in Mechanical and Energy Technology*, Smart Innovation, Systems and Technologies 174,
https://doi.org/10.1007/978-981-15-2647-3_70

Keywords Pyroelectric · Relative permittivity · Lead-free · Ferroelectric · Ceramics

70.1 Introduction

Ceramic materials are very well known in all the engineering domains because of its properties and its wide range of applications make it more popular among researchers and industrialists. Ceramic materials are not new to us, though it is being used for several decades, and because of this ceramics have a wide range of market for detectors, thermal imaging applications and sensors [1–3]. Because of its wide use in varieties of products, it has always been a topic of research to explore cost-effective materials as well as new applications for sufficient usage of ceramics. The pyroelectric property of ceramics makes them even more unique as they possess thermo-electric conversion ability. This unique ability is credited to the fact that these materials show high sensitivity toward change in temperature because of the change in net polarization. For improving the efficiency of pyroelectric substances, an ideal balance between specific heat, dielectric constant, pyroelectric coefficient [4–6] and dielectric loss is a must. The chemical and physical compositional modification method is such a method that has adapted in previous time to chemically tune their pyroelectric performance [6–8]. In this field, permeable ceramics increase scientist's interest since they are suitable materials for various pyroelectric applications. From Pb-free ceramics, $BaTiO_3$ is the oldest ceramics recognized and is widely investigated from past time because of its excellent ferroelectric, dielectric and piezoelectric properties [9–11]. These findings encouraged global research of $BaTiO_3$-based substances for the replacement of lead-based ceramic with new lead-free substances for different dielectric purposes. Therefore, the article deals with the pyroelectric analysis of rare earth-doped BCZTO (R=Ce, Ho and Nd)-based substances. Energy storage capacity is also present in ferroelectric materials. Hence, in this article, we also discussed energy storage characteristics for all understudy compositions. We have also analyzed our experimental results, and a program was written which calculates maximum voltage produced by samples in a certain temperature range. The programmatically generated values can also be used to calculate voltage produced by the samples in various temperature ranges.

70.2 Experimental Procedure

The ceramics with compositions $Ba_{0.15}R_{0.04}Ca_{0.15}Zr_{0.1}Ti_{0.9}O_3$ are formed using a solid-state reaction channel. For this purpose, pure reagents like TiO_2 (99% pure), $BaCO_3$ (99.99% pure), CaO (99% pure), ZrO_2 (99% pure) and oxides or carbonates of precious earth materials Nd, Ce and Ho were used as base elements. These powders were scaled and blended in a ball mill as per their stoichiometric proportion.

Further, the mix was calcined at 13,000 °C for around 5 h to excite the solid-state reaction. After that, the powders were pressed and made into circular pills of 1 mm thick and 12 mm diameter heavy after crushing and then using 2 %wt of polyvinyl alcohol (PVA) which acts as a binding ointment. Further, sintering was performed at 15,000 °C for 4 h. Following, the density of the sample was calculated using the Archimedes principle. The pureness of all prepared specimens was characterized by X-ray diffraction (XRD) operating CuK α-radiation with a scan step of 2° per minute. Scanning electron microscopy (SEM) (FEI SEM NOVA Nanosem 450, Hillsboro, OR) was done to study the morphology of the cover of the specimen. Before SEM, both the surfaces of the specimen were polished. A silver electrode was deposited on both specimens for doing the electrical analyses. The polarization-electric field (P-E) hysteresis loops were also measured at the different electric field and different temperatures using a remodeled sawyer tower circuit (Marine India, New Delhi, India). For studying dielectric properties, dielectric constants and dielectric loss were calculated at room temperature using an impedance analyzer (Agilent E4990A, Agilent Technologies Inc., Santa Clara, CA). The pyroelectric properties of all the ceramics were measured by subjecting the material to the transient temperature gradient by keeping it in front of hot and cold air.

70.3 Results and Discussion

70.3.1 Microstructural Characteristics

Figure 70.1 shows the XRD patterns of pure and doped BCZTO ceramics. It is clear from the figure that solid-state reaction has occurred in the specified heating

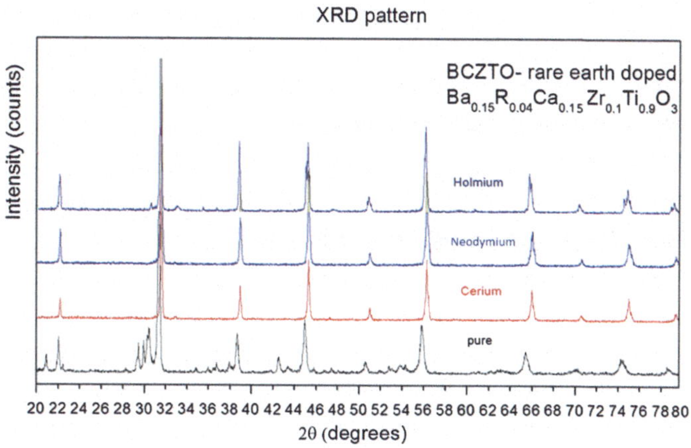

Fig. 70.1 XRD pattern of all samples at room temperature

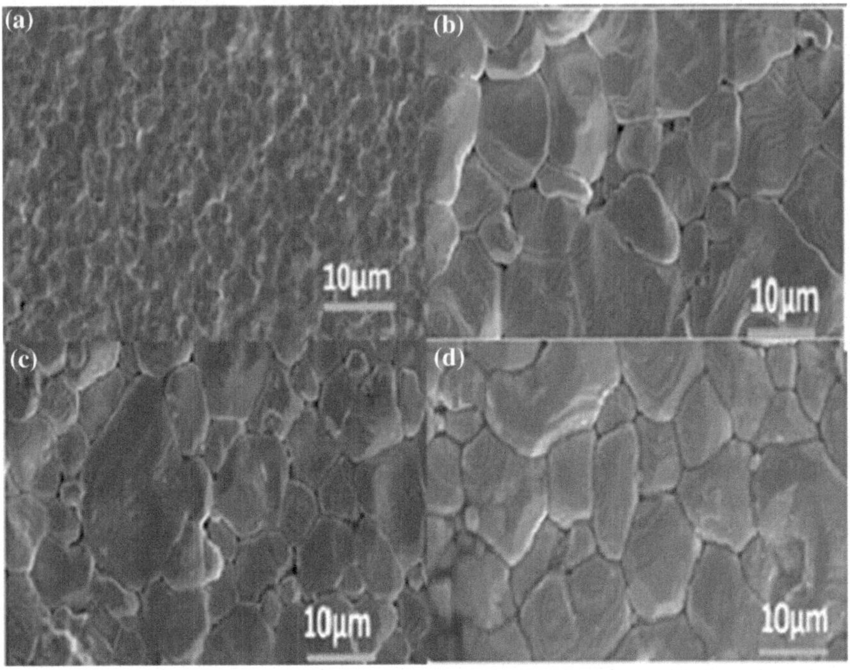

Fig. 70.2 SEM image of **a** pure BCZTO, **b** Ce-BCZTO, **c** Nd-BCZTO, **d** Ho-BCZTO samples

condition, and the sample has a single phase. The lack of any secondary peaks suggests that the prepared sample has a single perovskite phase. The SEM measures all the sintered ceramics which is presented in Fig. 70.2. It shows that all the ceramics have well densification with interconnected grains with a bimodal grain distribution. That means the larger grains are increased on the account of grains which are of smaller size. Further, the Archimedes principle was used to calculate the density of all the sintered ceramics use, and the density of pure BCZTO, Ce added BCZTO, Nd added BCZTO and Ho added BCZTO was 5.42, 5.45, 5.43 and 5.48 gm/cc which were nearly 95% of all theoretical density.

70.3.2 Polarization-Electric Field Ferroelectric Characteristics

In order to assess the ferroelectric performance of the investigated ceramics, we have taken the polarization-electric field hysteresis loop in different running temperature and the corresponding electric field as shown in Fig. 70.3. In this direction, Fig. 70.3a shows the P-E loops of pure BCZTO. This gives a saturated hysteresis loop with the highest polarization of about 19.48 $\mu C/cm^2$ with the remnant polarization of

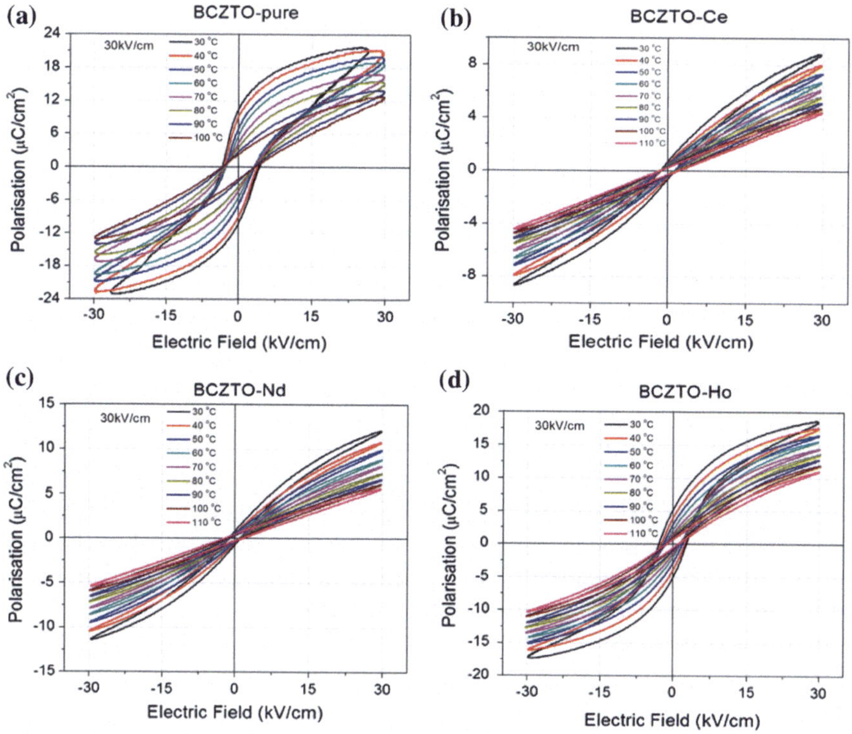

Fig. 70.3 Polarization-electric field loops of **a** pure BCZTO, **b** Ce, **c** Ho and **d** Nd-doped BCZTO at various temperatures

5.45 μC/cm^2 at 303 K and at 30 kV/cm electric field. Similarly, Fig. 70.3b–d shows the hysteresis loop of rare earth-doped BCZTO ceramics. All the loops are well saturated. Upon heating, the loops shrank, and this happened because on increasing the temperature it losses the dipoles order arrangement BCZTO at a higher temperature (>353 K) and becomes straight which is a signature of ferroelectric to paraelectric transformation. This phase change temperature was found to be decreasing (toward room temperature) with the addition of rare earth elements.

70.3.3 Pyroelectric Performance

For evaluating pyroelectric performance of the investigated composition, the P-E loops had been taken for all the samples at 30 kV/cm electric field and at various temperatures from 303 K for a different of 10 K from P-E loop, the parameters that represent it like remnant polarization (P_r) was extracted, and pyroelectric coefficients were calculated using a static method which is shown by Eq. (70.1)

Table 70.1 Pyroelectric coefficient of all the samples

Material	Operating temperature (K)	Pyroelectric coefficient ($\times 10^{-4}$ C/cm^2)
BCZTO	303	3.0
Ce-BCZTO	303	3.2
Nd-BCZTO	303	4.6
Ho-BCZTO	303	4.0

$$p = \mathrm{d}P_r/\mathrm{d}t \tag{70.1}$$

where t stands for temperatures, and P_r stands for remnant polarization. The values of the pyroelectric coefficient at room temperature (303 K) and a phase transition temperature are depicted in Table 70.1.

The static methods which are used cannot be considered as live signals. For evaluating the live signals out of these pyroelectric materials, the materials were susceptible to transient temperature gradient while being exposed to cold and hot air. Figure 70.4a shows schematically the representation while being exposed to cold as well as hot air and material generating electric signals.

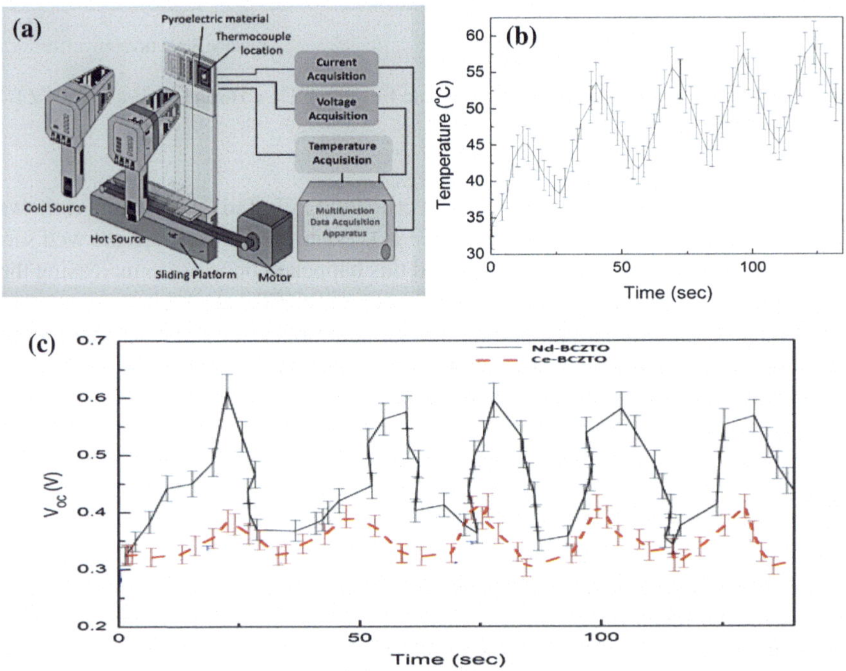

Fig. 70.4 **a** Setup to measure temperature profile and voltage, **b** temperature profile on the material surface and **c** the open-circuit voltage of the Nd-doped BCZTO sample

Table 70.2 Pyroelectric figure of merits of all the samples

Material	Operating	F_e* (pm^3/J)	F_v (m^2/C)	F_i (pm/V)	F_d (10^{-5} Pa$^{-0.5}$)	F_e (J/m^3 K^2)
BCZTO	303	2.3	0.01	185	20	8.1
Ce-BCZTO	303	2.4	0.012	200	23	8.2
Nd-BCZTO	303	3.2	0.024	294	32	8.9
Ho-BCZTO	303	2.5	0.014	223	24	8.3

Now, Fig. 70.4b depicts the profile of temperature for a ceramic as the function of time. The variation of temperature is shown by the graph, where it varied up to phase transition temperature. The temperature gradient is essential for generating electric voltage V as shown in Fig. 70.4c. The peak voltage of pure sample for the open circuit was 465 mV which was raised for rare earth-doped BCZTO and was at peak value for Nd-doped BCZTO, i.e., 685 mV. The signals cannot be contemplated as pure primary as most of the samples got varied warming alongside the width of the material. Now, splitting up the secondary and primary responses becomes crucial to gauge the red signals. We have also evaluated figures of merits (FOMs) of pyroelectric materials for all the samples at room temperature. The FOMs are parameters to gauge the performance of pyroelectric material. The FOMs include voltage responsivity (F_v), detectivity (F_d), energy harvesting (F_e and F_e*) and current responsivity (F_i) which were calculated for all the specimens. The current responsivity (F_i) can be mathematically written as $F_i = p/C_v$, where C_v denotes specific heat capacity of the material. Fi was found to be 185 pm/V for pure BCZTO which is increased to 294 pm/V for Nd-doped BCZTO. The voltage responsivity is expressed as (F_v) $= p/C_v\varepsilon\varepsilon_0$, where ε and ε_0 are dielectric constant and permittivity of free space, respectively. The energy harvesting FOMs (F_e) $= p^2/\varepsilon\varepsilon_0$ and (F_e*) $= p^2/\varepsilon\varepsilon_0 C_v^2$ have been used in the present study. The FOM that have high detectivity can be written as (F_d) $= p/(C_v\sqrt{\varepsilon\varepsilon_0}\tan\delta)$. Similar to F_i, the other figure of merits also showed a tremendous improvement over the perovskite material under study. These FOMs are a cumulative effect of dielectric constant, dielectric loss, pyroelectric coefficient and specific heat. For larger FOMs, low dielectric constant and the high pyroelectric coefficient are acceptable. The values of all FOMs at room temperature are shown in Table 70.2.

70.3.4 Energy Storage Performance

For comparing the electric energy storage performance of BCZTO ceramic with a different set of rare earth-doped BCZTO ceramics, the ESD or energy storage density is determined first with the help of the data which was generated from P-E loops.

In Fig. 70.5, a comparison of a sample's electric ESD is also shown. Usually, the ESD in any ferroelectric substance can be calculated by integrating it under the area

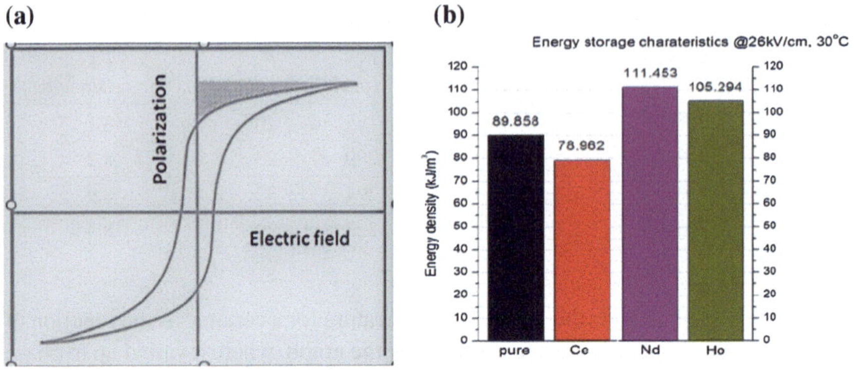

Fig. 70.5 **a** Procedure to calculate energy storage density and **b** energy storage density as a function of dopant

inside the P-E loop of the polarization and the discharged curve, in the period of P_r and P_s as shown in Fig. 70.5a. Mathematical expression for calculating the ESD is shown in Eq. 70.2.

$$W = \int_{p_r}^{p_s} E \cdot \mathrm{d}P \quad (70.2)$$

where W is for stored ESD in the material, E is for applied electric field, and P_s and P_r indicate saturation and remnant polarization, respectively. The ESD is shown with the help of a bar graph in Fig. 70.5b, and it comes out to be around 89.849 kJ/m^3 for BCZTO which increased to 111.453 kJ/m^3 after adding Nd into it. However, Eq. (70.2) shows only retrievable energy density when we integrate it w.r.t the P-E discharging curve. A known fact is that the total amount of energy for charging the ferroelectric capacitor equals to the sum of recoverable energy (W) and dielectric loss (W_{loss}). Every loss included is already existing in the form of heat/noise present inside the substance. The region circumscribed under the charging and discharging curve constitutes unrecoverable energy. The equation number 3 shows the total amount of energy required for charging the capacitor.

$$W_{\text{total}} = W + W_{\text{loss}} \quad (70.3)$$

On observing equation numbers 70.2 and 70.3, we can calculate the energy storage efficiency of the material using Eq. 70.4.

$$\eta = W/W_{\text{total}} \quad (70.4)$$

The equation above shows that we can obtain high performance by decreasing the dielectric loss of every observation which is recommended strongly in support of the Nd-doped BCZTO sample as it is among the favorite materials for improving the electrical energy storage.

70.3.5 The Programmatic Approach

With the experimental data which we collected during a series of experiments, a program was written in a Python programming language which helps us to calculate voltage performance of a pyroelectric substance in a particular temperature range. The program aims to calculate voltage which can be produced by samples. The program takes two temperature values in °C as input and considers it as one cycle. One cycle here means the substance under observation will be having some minimum temperature (first input temperature) whose temperature will be increased at some level, i.e., peak temperature (second input temperature), and then we decrease the temperature to its earlier temperature (base temperature). The highest temperature of the substance is referred to as the peak temperature of the substance (here). The program works only when the temperature range is less than 75 °C as above the Curie temperature(>75 °C) the material becomes paraelectric.

For writing the program, we divided one cycle into two parts. In the first part, the temperature of the substance is increased by 1 °C, and in the second part, the temperature is decreased by 1 °C. We calculated the rate at which the voltage is produced in both the cases and observed that while increasing the temperature of the pure substance by 1 °C there is a constant growth of 0.058 V and with the decrease in temperature by 1 °C the voltage growth is of 0.076 V. Similarly, in the case of Ce-BCZTO sample, the rate at which the voltage produced while increasing the temperature of the material by 1 °C is a constant growth of 0.027 V and with the decrease in temperature by 1 °C the voltage growth is of 0.036 V.

Program for calculating voltage produced

```
def up_starting_voltage(temp,peak_up,base_volt,base_temp):
        up_voltage = (temp - base_temp) * peak_up + base_volt
    return up_voltage
def down_starting_voltage(max,temp,base_up,peak_down):
        down_voltage = base_up - (max - temp) * peak_down
        return down_voltage
```

The above two programming functions are the important part of the program in which the function up_starting_voltage calculates the voltage produced by the substance when the temperature is increased. And the second function down_starting_voltage calculates the voltage produced when the temperature is decreased. Numerical values which are used to call these functions are in Fig. 70.6.

On plotting the voltages produced by the program, we can observe the consistency in the voltage by the samples and thus can verify for the correctness of the experiment from the scattered plot obtained (Fig. 70.7a, b).

```
Enter the Base Temprature :: 32
Enter the Peak Temprature :: 40
```

Voltage (100% Pure)	Voltage (Ce-BCZTO)		Temperature
0.157	0.044	-	32
0.215	0.071	-	33
0.273	0.098	-	34
0.331	0.125	-	35
0.389	0.152	-	36
0.447	0.179	-	37
0.505	0.206	-	38
0.563	0.233	-	39
0.487	0.197	-	40
0.41	0.161	-	39
0.334	0.124	-	38
0.257	0.088	-	37
0.181	0.052	-	36
0.105	0.016	-	35
0.028	-0.02	-	34

Fig. 70.6 Programmatic values of voltage for particular temperature range

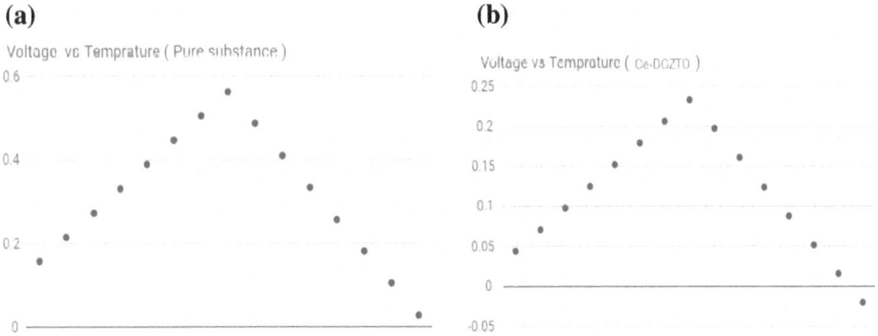

Fig. 70.7 **a** Scatter plot of voltage produced by pure substance in a particular temperature range **b** Scatter plot of voltage produced by Ce-BCZTO in a particular temperature range

70.4 Conclusion

(1) Rare earth-doped $Ba_{0.15}R_{0.04}Ca_{0.15}Zr_{0.1}Ti_{0.9}O_3$ (R=Ce, Nd and Ho) perovskite ceramics were fabricated by conventional solid-state reaction route, and the influence of dopant Ce, Ho and Nd on the microstructural pyroelectric and energy storage performance of BCZTO ceramics was investigated in detail.

(2) Microstructural observation indicates well densification with a bimodal grain size distribution for all samples.

(3) The pyroelectric coefficient of Nd-doped BCZTO ceramics was found to be a maximum of 4.6 $\mu C/cm^2$ at room temperature. The open voltage (V_{oc}) was found to be 685 mV for an anti-doped sample when compared with 465 mV for a pure sample.

(4) All the FOMs show tremendous improvement over pure samples under study.

(5) The energy storage density of the Nd-doped structure was found to be 111.453 kJ/m^3 which was higher than the pure sample.
(6) So, all these improvements could possibly replace lead-based counterparts for pyroelectric applications.
(7) The values generated by the program in Sect. 70.3.5 are similar to the experimental values collected within the same temperature range and thus verify that it can be used to reproduce our experiments.

References

1. Zaho, T., Jiang, W., Niu, D., et al.: Flexible pyroelectric device for scavenging thermal energy from chemical process and as self-powered temperature monitor. Appl. Energy **195**, 754–760 (2017)
2. Lau, S., Cheng, C., Choy, S., et al.: Lead-free ceramics for pyroelectric applications. J. Appl. Phys. **103**, 104105 (2008)
3. Chen, Y., Zhang, Y., et al.: A flexible PMN-PT ribbon-based piezoelectric-pyroelectric hybrid generator for human-activity energy harvesting and monitoring. Adv. Electron. Mater. **3**, 1600540 (2017)
4. Sassi, U., Parret, R., Nanot, S., Bruna, M., Borini, S., et al.: Graphene-based mid-infrared room-temperature pyroelectric bolometers with ultrahigh temperature coefficient of resistance. Nat. Commun. **8**, 14311 (2017)
5. Zhuo, F., Li, Q., Gao, J., Yan, Q., Zhang, Y., Xi, X., Chu, X.: Phase transformations, anisotropic pyroelectric energy harvesting and electrocaloric properties of (Pb, La)(Zr, Sn, Ti)O_3 single crystals. Phys. Chem. Chem. Phys. **19**, 13534–13546 (2017)
6. Vaish, M., Sharma, M., Vaish, R., Chauhan, V.: Harvesting thermal energy (via radiation) using pyroelectric materials (PZT-5H): an experimental study. Ferroelectr. Lett. **44**, 35–41 (2017)
7. Mangalam, R., Agar, J., Damodaran, A., Karthik, J., Martin, L.: Improved pyroelectric figures of merit in compositionally graded $PbZr_{1-x}Ti_xO_3$ thin films. ACS Appl. Mater. Interfaces **5**, 13235–13241 (2013)
8. Lang, S., Das-Gupta, D.: Handbook of advanced electronic and photonic materials and devices. In: Nalwa, H.S. (ed.) Ferroelectric and Dielectrics, Chap. 1. Pyroelectricity: Fundamentals and Applications, vol. 4, pp. 22–50 (2001)
9. Balakt, A., Shaw, C., Zhang, Q.: Enhancement of pyroelectric properties of lead-free $0.94Na_{0.5}Bi_{0.5}TiO_3$-$0.06BaTiO_3$ ceramics by La doping. J. Eur. Ceram. Soc. **37**, 1459–1466 (2017)
10. Srikanth, K., Mishra, U., Modi, Y., Srivastava, V., Gupta, R.: Microstructural, pyroelectric and energy storage performance of rare earth doped $Ba_{0.15}R_{0.70}Ca_{0.15}Zr_{0.1}Ti_{0.9}O_3$ (R=Ce, La & Gd) lead-free ceramics. In: AIP Conference Proceedings, vol. 2142, p. 070036 (2019)
11. Srikanth, K., Patel, S., Vaish, R.: Enhanced electrocaloric effect in glass-added $0.94Bi_{0.5}Na_{0.5}TiO_3$-$0.06BaTiO_3$ ceramics. J. Aust. Ceram. Soc. **53**, 523 (2017)

Chapter 71
Additive Manufacturing in Wind Energy Systems: A Review

Ummed Singh, M. K. Lohumi and Harish Kumar

Abstract Additive manufacturing is the process of converting suitable format of CAD design file into final product using layer-by-layer deposition of material. Generally, wind energy system components are of large size. So, transportation of these components becomes a problem which can be solved by onsite production of the parts. The wind turbine blades should be designed aerodynamically which results in complex geometry, and this issue can be solved using additive manufacturing. Not only manufacturing, repairing of the parts is also possible with the use of additive manufacturing. Additive manufacturing is a better choice for producing low volume products. Initially, additive manufacturing was used to build prototype of wind turbine parts for testing purpose. But, nowadays, final end-use products are being produced by additive manufacturing. Big area additive manufacturing (BAAM) process is developed by department of Energy's Oak Ridge National Laboratory (ORNL) working with TPI Composites, which is able to print parts with the capacity of 1000 cubic inches per hour. Equipment developer companies are working to develop more economical and large capacity additive manufacturing systems to make printing of wind energy systems more easier and affordable.

Keywords Additive manufacturing · Rapid prototyping · Wind turbine blade · Big area additive manufacturing (BAAM) · Wind energy

71.1 Introduction

71.1.1 Introduction to Additive Manufacturing

Additive manufacturing is the process of converting the suitable format of sliced 3D CAD design file into final product using layer-by-layer deposition of material.

U. Singh (✉) · H. Kumar
National Institute of Technology Delhi, New Delhi, Delhi 110040, India

M. K. Lohumi
Galgotias College of Engineering and Technology, G B Nagar 201306, India

S. Yadav et al. (eds.), *Proceedings of International Conference in Mechanical and Energy Technology*, Smart Innovation, Systems and Technologies 174,
https://doi.org/10.1007/978-981-15-2647-3_71

Initially, additive manufacturing was developed as stereolithography process by 3D systems and commercialized in 1989 [1, 2], and now, AM is widely being used in almost every field of manufacturing and with various type of material. Due to its sustainability, rapidity, ability to print complex parts, compactness, and prototype making, AM is growing in large scale [3, 4]. AM development took place in a significant way from prototype making to printing of finished product and repairing capabilities [5]. Extensive research is being done in the field of affordable cost, optimum speed, more variety of material printability, and quality assurance about the final product [6]. Additive manufacturing is more attractive because it is environmentally friendly, lesser wastage of material, lesser carbon emission, and requires relatively less workforce.

Additive manufacturing includes the following steps [7]:

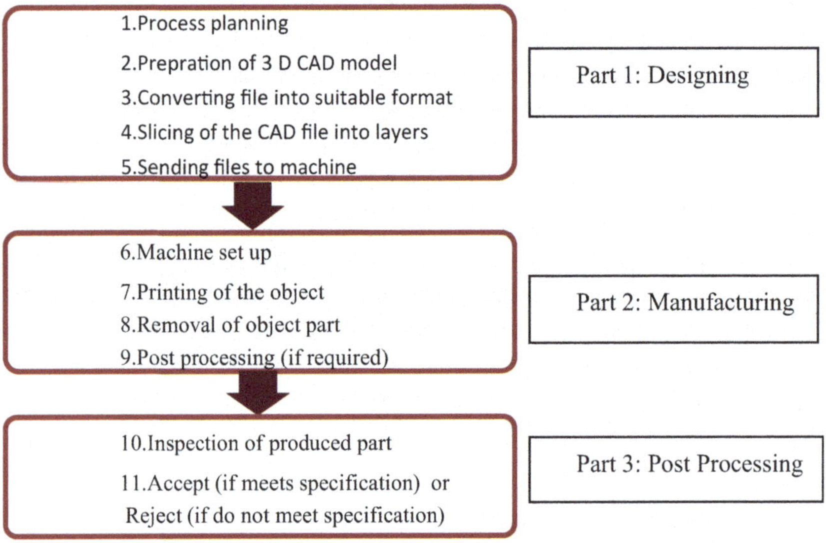

According to the form of printing material additive manufacturing is classified into four categories:

1. Liquid state
2. Filament/paste form
3. Powder form
4. Solid sheet.

Further details are discussed in Fig. 71.1.

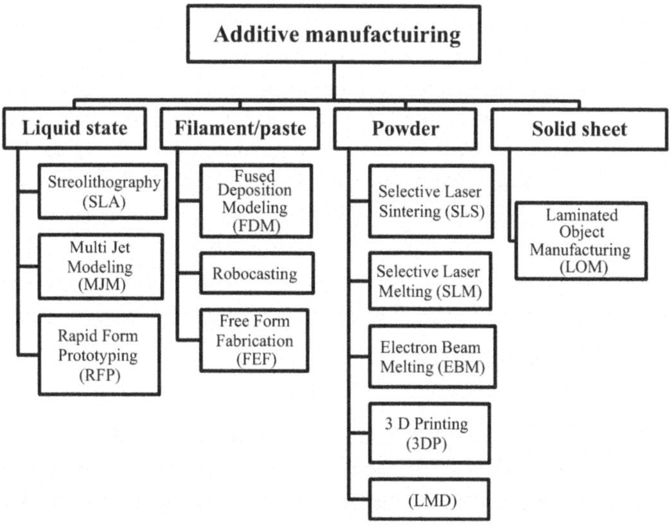

Fig. 71.1 Classification of AM according to the form of printing material

71.2 Applications of Additive Manufacturing

71.2.1 Aerospace Industry

Due to small batch size and lightweight, it is the most suitable industry for AM. Component in aerospace industry is made up of special alloy materials which are time taking and hard to produce such as alloys of titanium, nickel, ceramic, and steel; these can be printed easily using AM [8]. Concept laser manufactured turbine blade using Inconel 718 material and selective laser melting (SLM) [9]. Optomec developed satellite and jet engine components using laser engineered net shaping (LENS) [10]. Thales Alenia Space Agency developed Koreasat 5A and 7 satellites using direct metal laser melting (DMLM) in 2015 [11]. Xue and Islam presented the way of repairing tips and seal of destroyed Inconel 738 blade using LENS process [12]. Structure of fuel injector and leading edges of hypersonic vehicles, nose cones of missiles, and nozzle throats of spacecraft propulsion engines can be manufactured using AM technique [12].

71.2.2 BioMedical

Extensive research is being performed in the field of biocompatible material in additive manufacturing. Facility of mass customization attracts additive manufacturing in biomedical field due to different designs and shapes of product according to customers like dental prosthesis, hip joint, and hearing aids [13]. Biocompatible material such as Ti–6Al–4V, polymers, ceramics, and many more are being used to print various implants [14]. Arcam developed implants of shoulder, hips, and acetabular cup using EBM technique [15]. Cell maneuvering techniques are developed like tissue building using syringe-based cell deposition [16] and cell printing using inkjet printers [17, 18]. Recently, researchers at Tel Aviv University printed a personalized heart using bio-ink printer [19].

71.2.3 Automotive Industry

Production by conventional method takes a lot of initial investment, but AM prints the product for less cost and gives freedom of customization. Volkswagen and Kinazo Design manufactured ENDURO e1 electric bicycle of aluminum metal using direct metal laser sintering (DMLS) method in 2017 [20, 21].

In year 2014, first Strati electric car is presented at the International Manufacturing Technology Show in Chicago printed using BAAM facility of ORNL [22].

71.2.4 Defense Application

Additive manufacturing has huge potential in defense applications because it can reduce the weight of weapons and helps in onsite manufacturing and repairing of the machines. Additive manufacturing technology has been adopted by defense sectors in 1983 [23]. The US' Defense Advanced Research Projects Agency (DARPA) has started the 'Open Manufacturing' scheme for the research of AM technology in defense sectors [24]. The US Army is planning to use MELD additive manufacturing technology to repair military vehicles [25].

71.2.5 Energy Generation System

Bourell et al. developed bipolar plate for fuel cell made of graphite composite by using SLS-based method [26–29].

71.3 Importance of Wind Energy

Renewable energy is the clean and sustainable source of energy that helps to reduce carbon emission and many more pollution-causing factors; due to this, the world is

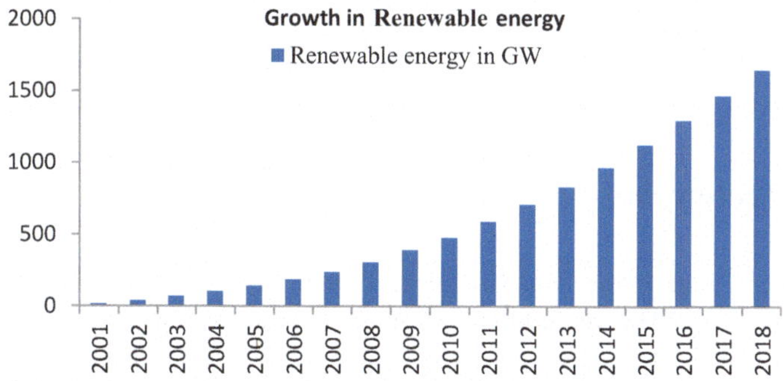

Fig. 71.2 IEA report renewable 2018 (International Energy Agency) [32]

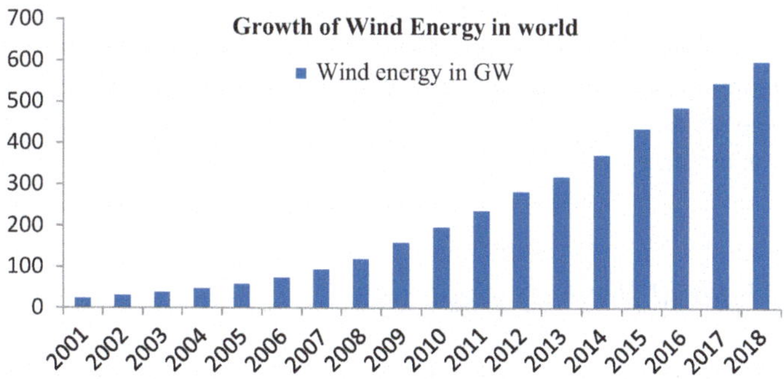

Fig. 71.3 WWEA report (World Wind Energy Association) [35]

shifting toward renewable energy sources to fulfill the energy demand [30]. Wind power contribution in world reached 597 GW in 2018 [31]. A report indicates that contribution of renewable energy will be around 30% in the total demand of energy of the world by 2023 [32]. In 2016, the total contribution of renewable energy was 16,893.54 TWh out of which the share of wind energy was 959.53 TWh, and it is growing rapidly [33].

The growth rate of wind energy source was 10% in recent years [33, 34] (Figs. 71.2 and 71.3).

71.4 Additive Manufacturing in Wind Energy System

Wind energy systems convert energy of wind into the electricity or mechanical power by means of aerodynamically designed airfoils or wind turbine blade [36]. Wind energy systems are most suitable for additive manufacturing application because of small batch size, ability to print complex geometries, significant relief in transportation problem of big parts, comparatively less initial investment, advantage of human safety, prototyping facility, and capability to accommodate frequent design variations.

71.4.1 Prototyping

In year 2015, Sean Poole and Russell Phillips printed various prototypes of turbine blades using FDM method and PLA material to find the optimum design of mini HAWT [37]. They designed turbine blade in LabVIEW‘ program using many parameters (output of chord, output of pitch, radius, etc.) then the 3D CAD model which is prepared using Rhinoceros 3D software and converted into 3D printer suitable file using Repetier-Host software and finally printed self-developed 3D printer based on a Prusa which works on Marlin open-source software. Material used was PLA due to low thermal contraction than ABS. These turbine blades are tested at Kestrel's outdoor wind tunnel at average wind speed around 9 m/s and the turbine rotational speeds around 500 rpm. They found it time saving, cost effective, and sufficiently accurate.

71.4.2 Material Property Analysis

In the year 2019, A. Ermakova, A. Mehmanparasta, and S. Ganguly reviewed Wire + Arc Additive Manufacture (WAAM) method (type of direct energy deposition method) for offshore wind energy systems and obtained the following results [38] (Table 71.1).

Table 71.1 Property analysis for wind system material

S. No.	Property	Material	Value compared to wrought material
1.	Yield and ultimate tensile strength	Steel specimen	Similar
2.	Charpy impact test	Steel specimen	Higher
3.	Fracture toughness	Ti–6Al–4V specimen	Slightly high
4.	Number of cycles to failure on S-N curve	304L stainless steel	More

71.4.3 End-Use Product

Sandia National Laboratories (SNL) developed a 100-m all-glass baseline blade for a 13.2 MW capacity HAWT with E-LT-5500 fiberglass with EP-3 epoxy glass material [39]. After the analysis, they found that the strain in the spar cap and trailing edge is under allowable limit. Fatigue life was estimated 1290 years using slope parameter according to the recommendation of GL. Report suggested the possibility of further improvement in aerofoil design, weight reduction, and addition of joints to accommodate manufacturing of blade in parts and better flutter prediction technique.

71.4.4 Wind Turbine Molds

ORNL worked with TPI Composites to use the BAAM system to manufacture a wind turbine blade mold [40]. The blade was 13 m long with 16 sections containing heating channels. They prepared CAD model; then, they printed it and added fiberglass on it and machined it followed by installed heating channel and frame and assembled it [41]. BAAM system is able to print around 1000 cubic inches per hour and material cost lowered to ($3–$5/lb) with high velocity and less required energy intensity. They are developing the system (WHAM) to print the final end-use turbine blade directly by additive manufacturing.

71.4.5 Supply Chain of Turbine Blade Parts

Conventionally, molds are transported to another region by ships or other ways; now, all these costs can be saved by simply transferring the CAD design file [42]. The damaged part of blade can be replaced at site by 3D printed part instead of replacing entire blade. Additively manufactured mold was more economical than conventionally manufactured part for any quantity, and it will further reduce with maturity of AM. Production using AM was time and cost saving with compared to conventional methods.

71.4.6 Repairing of Wind Turbine Parts

Generally, repairing of turbine is done by glue a patch to wounded place, micro-architected patches particle reinforced, multielement patches assembled during the repair, fast adhesive curing, nano-modification of the coating [43].

Niklas Wahlström and Oscar Gabrielsson selected fluid rotary joint for repairing application because it had more failure frequency [42]. They achieved 44% weight reduction, increased life expectancy, and ease of assembly. The cost of storage and transportation reduced significantly, and they recommended the use of AM for spare parts.

71.5 Discussion

Additive manufacturing technology is growing rapidly in every field of application and material. It is attractive due to sustainability, eco-friendly nature, mass customization, rapid manufacturing, reduced transportation and handling cost, ease of data storage, and many reasons. A lot of research work is being done on the optimum speed for best results and less lead time, quality assurance due to difference in each product, capacity to accommodate the variety of designs, affordable cost that anyone can use 3D printed products. Additive manufacturing is applied in prototyping, end-use product manufacturing, and onsite repairing of wind energy systems, and it proved to be acceptable in each application with a confidence that in upcoming time manufacturing time and the cost will reduce. ORNL and Ingersoll Machine Tool Company are developing wide high additive manufacturing (WHAM) system with a build volume of around 25 ft width, 20 ft height, and 100 ft length.

References

1. Bourell, D.L., Beaman Jr., J.J., Leu, M.C., Rosen, D.W.: A brief history of additive manufacturing and the X. roadmap for additive manufacturing (2014)
2. Frazier, W.E.: Metal additive manufacturing: a review. J. Mater. Eng. Perform. **23**, 1917–1928 (2014)
3. Shamsaei, N., Yadollahi, A., Bian, L., Thompson, S.: An overview of direct laser deposition for additive manufacturing; Part II: mechanical behavior, process parameter optimization and control. Add. Manuf. **8**, 12–35 (2015)
4. Levy, G., Schindel, R., Kruth, J.: Rapid manufacturing and rapid tooling with layer manufacturing (LM) technologies. State of the art and future perspectives. CIRP Ann. **52**(2), 589–609 (2003)
5. Eyers, D.R., Potter, A.T.: Industrial additive manufacturing: a manufacturing systems perspective. Comput. Ind. **92–93**, 208–218 (2017)
6. Cozmei, C., Caloian, F.: Additive manufacturing flickering at the beginning of existence. Procedia Econ. Financ. **3**, 457–462 (2012)
7. González, D.S., González Álvarez, A.: Additive manufacturing feasibility study & technology demonstration. EDA Additive Manufacturing State of the Art & Strategic Report (2018)
8. Guo, N., et al.: Additive manufacturing: technology, applications and research needs. Front. Mech. Eng. **8**(3), 215–243 (2013)
9. Compact laser GMBH. http://www.concept-laser.de/ (visited on 30 Sept 2019)
10. Optomec. http://www.optomec.com/ (visited on 30 Sept 2019)
11. http://www.3ders.org/articles/20190322-thales-alenia-space-takes-3d-printing-into-series-production-for-satellites-and-spacecraft.html (visited on 30 Sept 2019)
12. Xue, L., Islam, M.U.: Laser consolidation—a novel one-step manufacturing process for making net-shape functional components. In: Cost Effective Manufacturing via Net-Shape Processing, pp. 15-1–15-4. Neuilly-sur-Seine, France (2006)

13. Ahuja, B., Karg, M., Schmidt, M.: Additive manufacturing in production—challenges and opportunities. In: Proceedings of SPIE, vol. 9353, p. 935304
14. Pandey, K., et al.: Some bio compatible material used in medical practices—a review. Int. J. Pharm. Sci. Res. **7**(7), 2748–2755 (2016)
15. Arcam AB: http://www.arcam.com. Accessed on 24 Sept 2019
16. Chang, R., Emami, K., Wu, H., Sun, W.: Bio fabrication of a three dimensional liver micro-organ as an in vitro drug metabolism model. Biofabrication **2**(4), 045004 (2010)
17. Cui, X., Boland, T.: Human microvasculature fabrication using thermal inkjet printing technology. Biomaterials **30**(31), 6221–6227 (2009)
18. Boland, T., Xu, T., Damon, B., Cui, X.: Application of inkjet printing to tissue engineering. Biotechnol. J. **1**(9), 910–917 (2006)
19. Noor, N., Shapira, A., Edri, R., Gal, I., Wertheim, L., Dvir, T.: 3D printing of personalized thick and perfusable cardiac patches and hearts. Adv. Sci. 1900344
20. www.kinazo.com/e1/ (2019)
21. https://kinazo-design.com/e-bike/ (2019)
22. Love, L.J.: Utility of big area additive manufacturing (BAAM) for the rapid manufacture of customized electric vehicles. ORNL/TM-2014/607 CRADA/NFE-14-04988
23. Coykendall, J., et al.: 3D opportunity in aerospace and defense (2014)
24. Boosting Confidence in New Manufacturing Technologies. DARPA, 29 May 2015, available at http://www.darpa.mil/news-events/2015-05-29 (2019)
25. https://3dprintingindustry.com/news/meld-additive-manufacturing-selected-by-u-s-army-for-vehicle-repair-137392/ (2019)
26. Chen, S., Bourell, D.L., Wood, K.L.: Fabrication of PEM fuel cell bipolar plates by indirect SLS. In: Proceedings of the Solid Freeform Fabrication Symposium, Austin, TX, pp. 244–256 (2004)
27. Chen, S., Murphy, J., Herlehy, J., Bourell, D.L., Wood, K.L.: Development of SLS fuel cell current collectors. Rapid Prototyping J. **12**(5), 275–282 (2006)
28. Alayavalli, K., Bourell, D.L.: Fabrication of modified graphite bipolar plates by indirect selective laser sintering (SLS) for direct methanol fuel cells. Rapid Prototyping J. **16**(4), 268–274 (2010)
29. Alayavalli, K., Bourell, D.L.: Fabrication of electrically conductive, fluid impermeable direct methanol fuel cell (DMFC) graphite bipolar plates by indirect selective laser sintering (SLS). In: Proceedings of the International Solid Freeform Fabrication Symposium, Austin, TX, pp. 186–193 (2008)
30. Gielena, D., Boshella, F., Sayginb, D., Bazilianc, M.D., Wagnera, N., Gorinia, R.: The role of renewable energy in the global energy transformation. Energy Strategy Rev. **24**, 38–50 (2019)
31. https://wwindea.org/blog/2019/02/25/wind-power-capacity-worldwide-reaches-600-gw-539-gw-added-in-2018/ (2018)
32. Renewables 2018: market analysis and forecast 2018 to https://www.iea.org/renewables2018/ (2019)
33. Ritchie, H., Roser, M.: Global renewable energy consumption over the long-term. https://ourworldindata.org/renewable-energy (2019)
34. Global Renewable Generation Continues its Strong Growth, New IRENA Capacity Data. https://www.irena.org/newsroom/pressreleases/2018/Apr/Global-Renewable-Generation-Continues-its-Strong-Growth-New-IRENA-Capacity-Data-Shows (2019)
35. https://library.wwindea.org/global-statitistics-1980/ (2019)
36. Panwar, N.L., Kaushik, S.C., Kothari, S.: Role of renewable energy sources in environmental protection: a review. Renew. Sustain. Energy Rev. **15**, 1513–1524 (2011)
37. Poole, S., Phillips, R.: Rapid prototyping of small wind turbine blades using additive manufacturing. In: Pattern Recognition Association of South Africa and Robotics and Mechatronics. International Conference (PRASA-RobMech), Port Elizabeth, South Africa, 26–27 Nov (2015)
38. Ermakova, A., Mehmanparast, A., Ganguly, S.: A review of present status and challenges of using additive manufacturing technology for offshore wind applications. Procedia Struct. Integrity **17**, 29–36 (2019)

39. Griffith, D.T., Ashwill, T.D.: The Sandia 100-meter all-glass baseline wind turbine blade: SNL100-00, SAND2011-3779. Unlimited Release Printed June 2011
40. Post, B.K., et al.: Big area additive manufacturing application in wind turbine molds. In: 28th Annual International Solid Freeform Fabrication Symposium, August, Austin, Texas, US (2017)
41. https://www.ornl.gov
42. Wahlström, N., Gabrielsson, O.: Additive manufacturing applications for wind turbines. Master of Science Thesis MMK 2017:98 MKN 191 (2017)
43. Kraynyk, B.C., Janzen, D.W.: Innovating the supply chain of wind energy through the application of additive manufacturing. MS Thesis (2017)

Chapter 72
Identification of Safe Machining Range for Laser Drilling of Basalt–Glass Hybrid Composite Using Artificial Neural Network

Akshay Jain, Bhagat Singh and Yogesh Shrivastava

Abstract Prediction of optimal process parameters concerning high hole circularity with least heat-affected zone (HAZ) has been presented in this paper. In this work, the first stage is to fabricate a new basalt–glass-based hybrid composite. In the second stage, laser-based drilling has been conducted on fabricated composite. In the third stage, artificial neural network has been used to obtain safe machining zone. To validate the determined safe machining zone, validation experiments have been performed.

Keywords Basalt–glass composite · HAZ · Safe machining range · ANN

72.1 Introduction

Composite materials are widely used nowadays. Fabrication of composite depends on the desired properties. Research has been conducted since the last decade on the applicability of basalt and glass fibers [1–3]. Basalt can be found over the earth surface. It is a natural volcanic rock. Basalt is an eco-friendly, resistance to high temperature and highly chemical stable material [4–9]. Basalt fiber can be used for the production of marine equipments, fabrication of circuit board and the application where high resistance to acid required. Basalt fibers have the low tensile mechanical strength which can be increased by hybridizing it with E glass fiber. Hybrid composite, i.e., basalt with glass fiber shows the better mechanical properties. From the past literature study, it has been found that basalt–glass hybrid composite has better applicability.

Moreover, the processing of hybrid composite is a difficult task. Delamination, burr formation and fiber pullout occur when composite is processed by conventional machining [10, 11]. These limitations will result in poor surface quality. Several

---

A. Jain · B. Singh
Mechanical Engineering Department, Jaypee University of Engineering and Technology, Guna, India

Y. Shrivastava (✉)
Mechanical Engineering Department, Galgotias College of Engineering & Technology, Greater Noida, UP, India

S. Yadav et al. (eds.), *Proceedings of International Conference in Mechanical and Energy Technology*, Smart Innovation, Systems and Technologies 174,
https://doi.org/10.1007/978-981-15-2647-3_72

studies have been conducted by the researchers on the processing of composite materials using unconventional machining and other conventional machining techniques [12–22]. From the literature study, it has been observed that for the machining of composite, laser beam machining is the best alternative. However, it has certain limitations like heat-affected zone. For better quality of drilled hole by laser drilling, the selection of optimal process parameters is required. In this work, the hole circularity and heat-affected zone (HAZ) are considered as drilling quality parameters.

72.2 Experimental Details

Laser beam drilling on fabricated composite has been performed by Nd: YAG laser system, which is operated in pulsed mode. This laser system is available at Raja Rammana Center of Advanced Technology (RRCAT), Indore, India. The input variables and levels have been shown in Table 72.1.

72.2.1 Hole Circularity

$$\text{Hole circularity} = \frac{\text{minimum diameter}}{\text{maximum diameter}} \tag{72.1}$$

Hole circularity can be measured by using Eq. 72.1. For the better quality of drilled hole, hole circularity should be maximum. Stereo optical microscope of 160× magnification has been used for the measurement of hole circularity and heat-affected zone. Figure 72.1 shows the representation of desired measurement.

72.2.2 Heat-Affected Zone (HAZ)

Design of experiment and measured value of desired responses has been shown in Table 72.2.

Table 72.1 Input variables and their levels

Process parameters	L_1	L_2	L_3
A	160	180	200
B	20	25	30
C	8	9	10

Here, *A* is lamp current (amp), *B* is pulse frequency (Hz) and *C* is Air–gas pressure (bar)

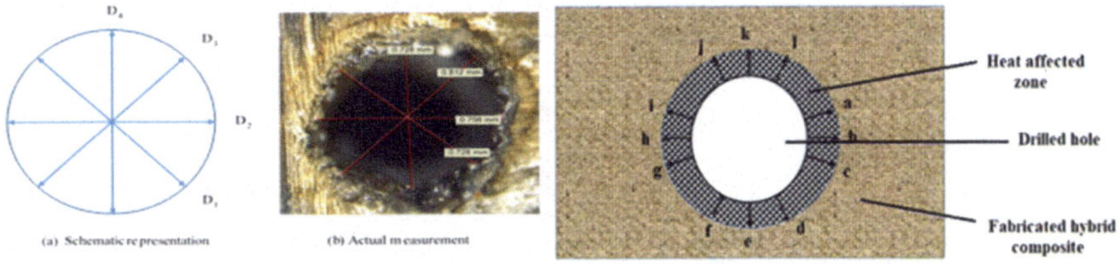

Fig. 72.1 Schematic representation of measurement

Table 72.2 Design of experiment and measured value of desired responses

S. No.	Lamp current (*A*)	Pulse frequency (*B*)	Air pressure (*C*)	Hole circularity	HAZ
1	160	20	8	0.903	351.26
2	160	20	9	0.728	167.174
3	160	20	10	0.782	253.96
4	160	25	8	0.882	273.402
5	160	25	9	0.701	125.37
6	160	25	10	0.838	281.037
7	160	30	8	0.858	301.45
8	160	30	9	0.797	155.053
9	160	30	10	0.856	325.12
10	180	20	8	0.960	308.726
11	180	20	9	0.702	112.38
12	180	20	10	0.817	148.062
13	180	25	8	0.846	223.23
14	180	25	9	0.665	72.771
15	180	25	10	0.718	169.54
16	180	30	8	0.888	278.452
17	180	30	9	0.729	142.56
18	180	30	10	0.808	266.411
19	200	20	8	0.918	334.769
20	200	20	9	0.848	253.004
21	200	20	10	0.797	275.62
22	200	25	8	0.920	302.663
23	200	25	9	0.793	187.366
24	200	25	10	0.825	276.937
25	200	30	8	0.869	298.336
26	200	30	9	0.837	288.251
27	200	30	10	0.912	356.368

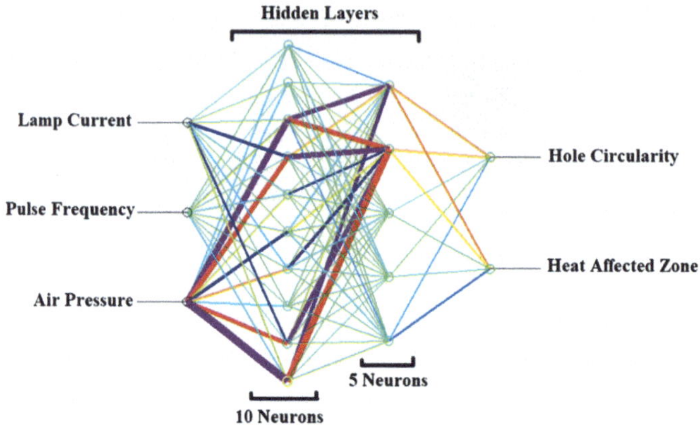

Fig. 72.2 ANN architecture

72.3 Artificial Neural Network (ANN)

In the ANN, input parameters are input layer and responses are considered as output layers. Architecture of ANN is shown in Fig. 72.2. After modeling by ANN, predicted values have been calculated and the deviation between experimental and predicted values is shown in Table 72.3.

72.4 The Dependency of Quality Parameters

To ascertain the effects of input variables on responses, viz. hole circularity and HAZ, contour plots have been drawn using the developed models. While drawing the contour plots, two parameters at a time and three parameters at a time have been considered separately. The conversion for coded values into actual has been shown in Table 72.4.

In Figs. 72.3, 72.4 and 72.5, contour plot shows that hole circularity is high with low HAZ when air pressure is kept at higher value with lamp current and pulse frequency kept at middle value. Flushing efficiency is one of the reasons for this trend. When material is melted through the laser beam the molten material is supposed to be vaporized and access material is supposed to be flushed out using the assisted air pressure. If the rate of material melting is higher than the required flushing pressure is more. However, the air pressure range selected in the present work is limited so that it does not affect the heat flow. Therefore, for the available air pressure, the molten material should be limited. The amount of material melted depends on the lamp current and pulse frequency. Higher the lamp current and pulse frequency more will be heat available to melt the material. To obtain significant values of input parameters concerning acceptable hole circularity and minimum

Table 72.3 Measured experimental values versus predicted values

A	*B*	*C*	Experimental hole circularity	Predicted hole circularity	Percentage deviation in hole circularity	Experimental heat-affected zone	Predicted heat-affected zone	Percentage deviation in heat-affected zone
160	20	9	0.728	0.7434736	1.9854047	167.174	153.366333	8.26
200	20	9	0.848	0.856626	1.017217	253.004	246.4603882	2.5864
160	30	9	0.797	0.7424985	6.8383312	155.053	157.246109	1.4144
200	30	9	0.924	0.897232	2.8969697	288.251	287.5664368	0.2375
160	25	8	0.882	0.8928545	1.2306689	273.402	264.5616455	3.2335
200	25	8	0.92	0.9106077	1.0209022	302.663	288.2413635	4.7649
160	25	10	0.838	0.8406363	0.3145943	281.037	244.7944336	12.896
200	25	10	0.825	0.8136901	1.370897	276.937	236.2915802	14.6768
180	20	8	0.96	0.9263083	3.5095521	308.726	314.43927	1.8506
180	30	8	0.888	0.9297377	4.7001914	301.273	322.2781372	6.9721
180	20	10	0.817	0.7966371	2.492399	148.062	199.8400879	34.9705
180	30	10	0.808	0.8650762	7.0638861	266.411	293.7349243	10.2563
180	25	9	0.665	0.6701936	0.7809925	72.771	75.7704163	4.1217
180	25	9	0.665	0.6701936	0.7809925	72.695	75.7704163	4.2306
180	25	9	0.663	0.6701936	1.0850075	72.397	75.7704163	4.6596
180	25	9	0.664	0.6701936	0.9327711	73.015	75.7704163	3.7738
180	25	9	0.664	0.6701936	0.9327711	72.916	75.7704163	3.9147
Average % deviation					2.291%			7.23%

Table 72.4 Conversion for coded values into actual values

Code	Lamp current (*A*)	Pulse frequency (*B*)	Air pressure (*C*)	Hole circularity	Heat-affected zone (HAZ)
0	160	20	8	Minimum (<0.7)	Minimum (<120)
0.5	180	25	9	0.7–0.8	120–200
1	200	30	10	Maximum (>0.8)	Maximum (>200)

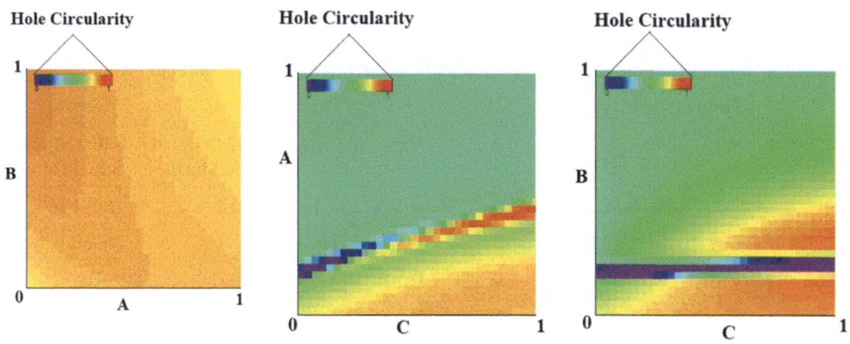

Fig. 72.3 Variation in hole circularity with two parameters at a time

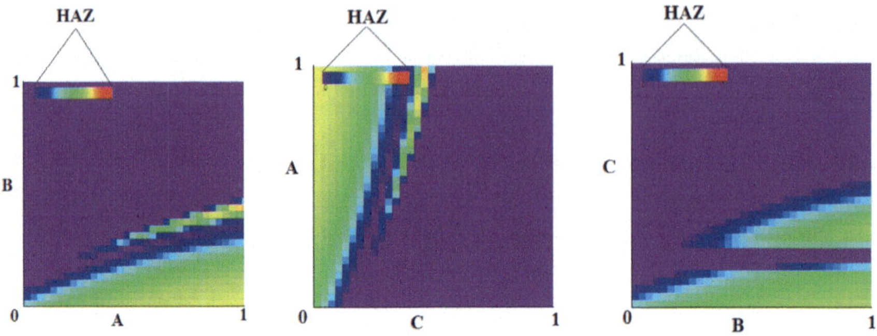

Fig. 72.4 Variation in heat-affected zone with two parameters at a time

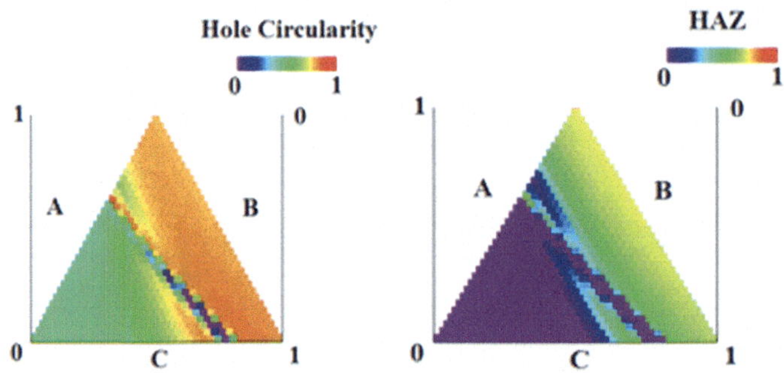

Fig. 72.5 Variation of responses while taken three variables at a time

heat-affected intersection of contour plots have been taken and the resulting range has been listed in Table 72.5. The obtained safe machining range has been verified by validation experiments. The optimal setting of input parameters for validation experiments has been listed in Table 72.6. The obtained results are represented in Fig. 72.6 and Table 72.6. It is evident from the results that the adopted methodology is significant.

Table 72.5 Intersection range in coded form for hole circularity and heat-affected zone

	Hole circularity			HAZ				
Variables	A_N	B_N	C_N	A_N	B_N	C_N	Safe range in coded form	Safe range in actual form
A_N	–	0–0.5	0–0.4	–	0–0.5	0–1	0–0.4	160–176
B_N	0.5–1	–	0–1	0.4–1	–	0.6–1	0.6–1	26–30
C_N	0–1	0.3–1	–	0.5–1	0.5–1	–	0.5–1	9–10

where A_N, B_N and C_N are the coded values of input parameters A, B and C, respectively

Table 72.6 Validation of predicted safe range

S. No.	*A*	*B*	*C*	Hole circularity	HAZ (μm)
1	160	28	9	0.783	127.41
2	170	30	10	0.817	131.52
3	160	27	10	0.802	153.74

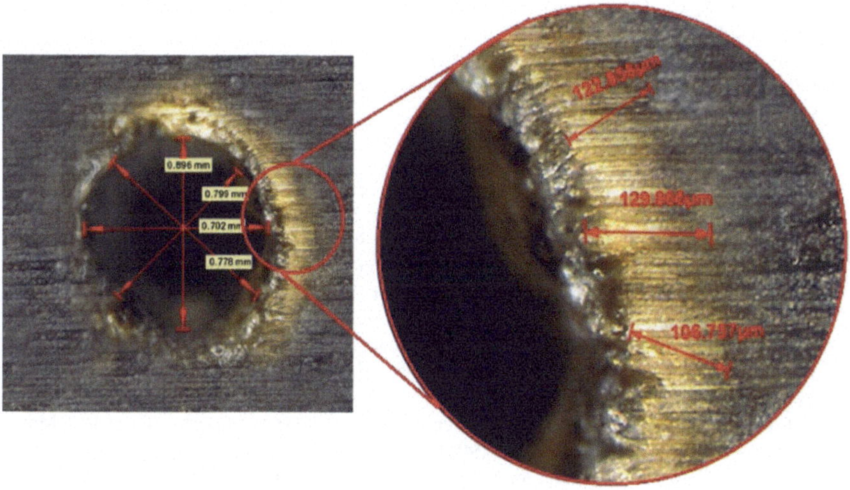

Fig. 72.6 Obtained values of hole circularity and heat-affected zone at $A = 160$ A, $B = 30$ Hz and $C = 9$ bar

72.5 Conclusion

In this paper, basalt–glass-based a new composite (hybrid) has been fabricated and drilled by laser beam machine. The input parameters selected during the drill operation have been explored to identify the dependency of outputs, viz. hole circularity and HAZ on inputs. For exploring the dependency, mathematical models of the responses have been developed using artificial neural network. The deviation between the experimental and predicted values of hole circularity and HAZ was only 2.29 and 7.23% which indicates that the developed models are appropriate. Moreover, a safe machining range of input parameters has been determined. From the validation results, it evident that the obtained safe machining range is true and if the fabricated composite is drilled by keeping the input parameters in the determined range the HAZ will be minimum with favorable hole circularity. The proposed approach seems helpful for the researchers to determine a safe machining range for any work material and thus to achieve better quality of drilled hole with higher accuracy and precision.

References

1. Pavlovski, D., Mislavsky, B., Antonov, A.: CNG cylinder manufacturers test basalt fibre. Reinf. Plast. **51**(4), 36–39 (2007)
2. Fiore, V., Di Bella, G., Valenza, A.: Glass–basalt/epoxy hybrid composites for marine applications. Mater. Des. **32**(4), 2091–2099 (2011)
3. Fiore, V., Scalici, T., Di Bella, G., Valenza, A.: A review on basalt fibre and its composites. Compos. B Eng. **74**, 74–94 (2015)
4. Sim, J., Park, C.: Characteristics of basalt fiber as a strengthening material for concrete structures. Compos. B Eng. **36**(6–7), 504–512 (2005)
5. Černý, M., Glogar, P., Sucharda, Z., Chlup, Z., Kotek, J.: Partially pyrolyzed composites with basalt fibres–mechanical properties at laboratory and elevated temperatures. Compos. A Appl. Sci. Manuf. **40**(10), 1650–1659 (2009)
6. Wei, B., Cao, H., Song, S.: Tensile behavior contrast of basalt and glass fibers after chemical treatment. Mater. Des. **31**(9), 4244–4250 (2010)
7. Quagliarini, E., Monni, F., Lenci, S., Bondioli, F.: Tensile characterization of basalt fiber rods and ropes: a first contribution. Constr. Build. Mater. **34**, 372–380 (2012)
8. Petrucci, R., Nisini, E., Ghelli, D., Santulli, C., Puglia, D., Sarasini, F., Minak, G., Kenny, J.: Mechanical and impact characterisation of hybrid composite laminates based on flax, hemp, basalt and glass fibers produced by vacuum infusion. In: ECCM15, 15th European Conference on Composite Materials (2012)
9. Gutnikov, S., Manylov, M., Lipatov, Y.V., Lazoryak, B., Pokholok, K.: Effect of the reduction treatment on the basalt continuous fiber crystallization properties. J. Non-Cryst. Solids **368**, 45–50 (2013)
10. Lau, W., Yue, T., Lee, T., Lee, W.: Un-conventional machining of composite materials. J. Mater. Process. Technol. **48**(1–4), 199–205 (1995)
11. Hejjaji, A., Singh, D., Kubher, S., Kalyanasundaram, D., Gururaja, S.: Machining damage in FRPs: laser versus conventional drilling. Compos. A Appl. Sci. Manuf. **82**, 42–52 (2016). https://doi.org/10.1016/j.compositesa.2015.11.036
12. Herzog, D., Jaeschke, P., Meier, O., Haferkamp, H.: Investigations on the thermal effect caused by laser cutting with respect to static strength of CFRP. Int. J. Mach. Tools Manuf **48**(12), 1464–1473 (2008). https://doi.org/10.1016/j.ijmachtools.2008.04.007
13. El Aoud, B., Boujelbene, M., Bayraktar, E., Salem, S.B.: Optimization of kerf quality during CO2 laser cutting of titanium alloy sheet Ti-6Al-4V and pure Titanium Ti. In: Mechanics of Composite, Hybrid and Multifunctional Materials, vol. 5, pp. 213–219. Springer (2019)
14. Zhang, Y., Li, J., Yang, R., Liu, T., Yan, Y.: Analysis of kerf quality on ultrafast laser cutting of anode material for lithium-ion battery. Opt. Lasers Eng. **118**, 14–21 (2019)
15. Mohamed, S.B., Ab Rashid, R., Muhamad, M., Ismail, J.: Composite materials and types of machining. In: Down Milling Trimming Process Optimization for Carbon Fiber-Reinforced Plastic, pp. 1–14. Springer (2019)
16. Negarestani, R., Li, L., Sezer, H., Whitehead, D., Methven, J.: Nano-second pulsed DPSS Nd: YAG laser cutting of CFRP composites with mixed reactive and inert gases. Int. J. Adv. Manuf. Technol. **49**(5–8), 553–566 (2010)
17. Mathew, J., Goswami, G., Ramakrishnan, N., Naik, N.: Parametric studies on pulsed Nd: YAG laser cutting of carbon fibre reinforced plastic composites. J. Mater. Process. Technol. **89**, 198–203 (1999)
18. Davim, J.P., Barricas, N., Conceicao, M., Oliveira, C.: Some experimental studies on CO2 laser cutting quality of polymeric materials. J. Mater. Process. Technol. **198**(1–3), 99–104 (2008)
19. Raza, M.S., Datta, S., Bule, B., Saha, P.: Parametric study of laser cutting of carbon fibre reinforced polymer (CFRP) and the effect of fibre orientation on cutting quality. Adv. Mater. Process. Technol. 1–11 (2019)
20. Li, M., Li, S., Yang, X., Zhang, Y., Liang, Z.: Effect of lay-up configuration and processing parameters on surface quality during fiber laser cutting of CFRP laminates. Int. J. Adv. Manuf. Technol. **100**(1–4), 623–635 (2019)

21. Oh, S., Lee, I., Park, Y.-B., Ki, H.: Investigation of cut quality in fiber laser cutting of CFRP. Opt. Laser Technol. **113**, 129–140 (2019)
22. Rahimi, M.H., Shayganmanesh, M., Noorossana, R., Pazhuheian, F.: Modelling and optimization of laser engraving qualitative characteristics of Al-SiC composite using response surface methodology and artificial neural networks. Opt. Laser Technol. **112**, 65–76 (2019)

Chapter 73
3D Printer Nozzle Design and Its Parameters: A Systematic Review

Prashant Kaduba Kedare, S. A. Khan and Harish Kumar

Abstract Additive manufacturing is a process in which layer by layer injection or introduction of material is used to build objects or products. There are various types of 3D printing methods of which, in fused deposition modeling (FDM), the nozzle is the component which extrudes the material. That makes nozzle as a crucial part of the FDM process. The change in any aspect of the nozzle design affects the final outcome of 3D printing. Nozzle affects the printing time, geometrical errors, surface finish, pressure changes and temperature variations of the material. Surface finish and texture are affected as nozzle is the last thing that is in contact with the print material. Integration of proper design and process planning for design of nozzle impacts the development time and cost. Different nozzles for different or particular materials are developed as the feeding technique, temperature coefficient and speed of printing affect the final print. As the feasibility of 3D printing is increasing, the use of colored materials with different or/and multiple materials is the focus. Visualization of prototypes is improved due to color implementation. Multiple nozzles or multinozzel 3D printers drop the printing time considerably. Nozzle diameter is an important parameter of the nozzle design. Pressure and temperature in the feeding tube and heating block are managed by optimum nozzle diameter and size. In this paper, we will focus on the different aspects and challenges of designing 3D printing nozzles.

Keywords Fused deposition modeling · Rapid prototyping · Heating block · Finite element analysis · Road width

---

P. K. Kedare (✉) · H. Kumar
National Institute of Technology Delhi, New Delhi, India

S. A. Khan
Galgotias College of Engineering and Technology, G B Nagar, India

S. Yadav et al. (eds.), *Proceedings of International Conference in Mechanical and Energy Technology*, Smart Innovation, Systems and Technologies 174,
https://doi.org/10.1007/978-981-15-2647-3_73

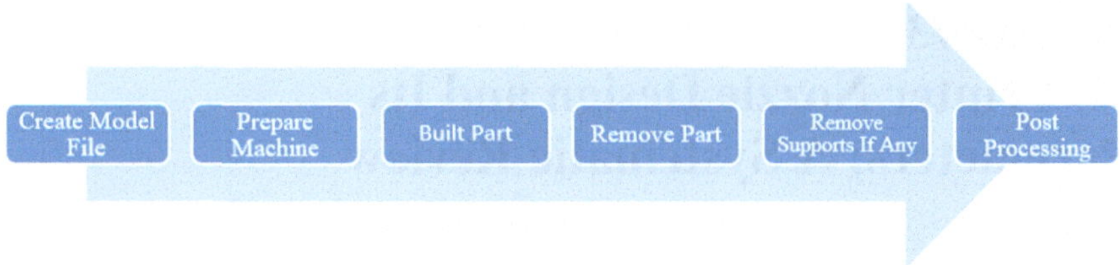

Fig. 73.1 Process details for 3D printing

73.1 Introduction

There are four main types of rapid prototyping or additive manufacturing techniques namely selective laser sintering, fused deposition melting, stereolithography and laminated object manufacturing. In all of the above, only in fused deposition modeling nozzle is used in order to deposit the material layer by layer with the help of nozzle and the effects of the nozzle can be seen.

In FDM process, *x*- and *y*-axis movement for 3D printing is controlled by the arm which deposits the material, whereas the table on which the printing is to be done moves in the *z*-axis direction [1]. The material used is wound on a spool which feeds the material for heating to a certain temperature in order to get the material to fuse with the previously deposited layer. The heating and controlling of the outflow are done with the help of the nozzle [2].

The process of FDM is first the model of the object which is to be produced is formed in a suitable CAD software which can give the detailed structure of the object. 3D scanners can also be used on an already existing object or the prototype which is made beforehand. This file is then converted into an .stl format by the use of software which is commonly called as a slicer [3]. This software basically cuts the model design into layers which are to be printed. The printer follows guided path of the design in order to print the design accordingly (Fig. 73.1).

As shown in the above figure, the process of 3D printing begins with creating the model file and ends with post-processing of the finished product [4].

This paper is focused mainly on the nozzle of the FDM 3D printer, the effects and analysis of the nozzle along with the changes caused by the changes done in the nozzle design.

73.2 Multi-colors Through Nozzle

Most of the FDM printers are limited in extend to print only monochrome or single type and single color materials. But for trending use multicolor objects are more in character and for that Shanling Han et al. designed a nozzle for color mixing. Shanling Han et al. used unigraphics software for designing the model. The Y shape of the nozzle design has an angle of 30° which is not only good for the flow but also

Table 73.1 General properties of ABS material [5]

Item	Quantity
Density (kg/m^3)	1020
Elastic modulus (Gpa)	2
Poisson's ratio	0.394
Specific heat capacity $(J/(kg \cdot K))$	1386
Thermal conductivity $(W/(m \cdot K))$	0.2256
Viscosity $(Pa \cdot s)$	84

for heat dissipation. The design of the nozzle device contains fixed block, nozzle, heating block and throat pipes. The heating block has holes for thermistor, throat pipe and the nozzle which can be attached by screw threads. ICEM CFD software is used to mesh the design model which is designed in UG [5]. In the meshing process, the finite element model is then moved to fluent software in order to get fluid analysis. In this analysis method, ABS is used as the printing material and mixing of these materials is studied (Table 73.1).

General properties of ABS material and its effects:

Shanling et al. did multiple experiments with the mixing of ABS material in order to get the best color effect, which can only be obtained by properly mixing of the completely melted material. The focus here is that the proper heating and mixing of the different color ABS material and its effect on the nozzle. After that the material is moved to deposit layer by layer and optimizing the deposition also gives an edge for better surface finish and tensile strength [6]. While the mixing is taking place the change in temperature is also affected by the different filament speeds. In this paper, different filament speeds are treated in order to get the best mixing and speed of printing. The behavior of the ABS material depends on its properties [7].

The inlet speed of the ABS material affects the temperature at the throat. If the temperature of the ABS is too high the printing process will show strings between ends of each cut or stop of the print. Also if the temperature is more than recommended there will be leakages in the print [8]. If the temperature is less than desired, the mixing of two colors or ABS materials coming would not be uniform [9]. Due to the Y structure, there is a chance of blockage and removing the blockage is difficult that is why change in design is necessary. The change in the angle of inlet of two materials might reduce the chance of blockage as well as the change in diameter of the flow channel can give better heat dissipation performance. The simulation is done for a time period of 60 s so as to get the flow analysis of computational fluid dynamics. The experiments are done with speeds of both inlets to be same at 0.2 mm/s, 0.5 mm/s 1 mm/s and also with different speeds of 0.2 mm/s and 0.5 mm/s, respectively. The proper mixing of two color material depends on the intersection temperature reached. The wall temperature is specified by experimentation in order to melt the ABS material but flow of the material is continuously absorbing the heat from the walls causing a large temperature gradient. As the mixing depends on the temperature at intersection, which needs to be 494 K so as the material melted and able to mix properly. This shows that the mixing of materials does not only depend on the heating temperature but also on the inlet speeds. By mixing the two materials in the head, in order to change the color, the rapid change is not possible as the mixed material existing in the nozzle is not able to shift from mixed to single color instantaneously. Because of this, there is a visible color shifting from mixed to single color in the print.

73.3 Multi-nozzle Printing

Tim Reiner et al. showed that by simply applying small geometric offset while printing using two nozzles, with two different colors tone can be varied which produces patterned look to the print object. Printing an object in full color is more desirable but is not that affordable for small printers [10]. Tim Reiner et al. used the Makerbot Replicator 2X desktop printer in order to carry out the experiment with the material having black and white color material using two separate nozzles [11]. Most of the researchers in the field focus more on the geometric form of the object, optimization of the print process and reducing print time, but in this method the focus is on the appearance of the printed shape along with geometric form. There are certain multicolor 3D printers available which are able to print the object with more than two colors but the cost of these printers is too high. The work done by author is more focused on contrast generated with two colors by placing the print of black and white layer in the successive print layer and also giving an angle to each direction of print to the previous causing a checkered pattern from outside. Multiple colors can be fed into single nozzle as well as we have seen above for different colors but the rapid color changes are difficult to achieve that can be overcome by giving each layer with separate color which is given by separate nozzle. But as this requires changing of the nozzle repeatedly, it increases the print time. In this method, the height of the print increases there starts to appear certain misalignment between layers which is visible through naked eye. Any offset in the print layer causes the black and white colors to mix together giving colored spots in the print. To avoid this, the sine pattern is given to the outline of the print during slicing. By phase shifting while printing by π, the first print is given in black and white layer is given in the second layer [12]. This experiment is easier with simple patterns as we go for complicated patterns the sin frequency gets complicated to manage. For arbitrary or complicated patterns, sin frequency needs to be adjusted constantly or allow for error in the print [13].

The printed object with giving a phase shift in layers causes the nozzles to leak at unexpected places due to continuous starting and stopping print material. Also, as the nozzles glide over the printed material it causes it to smear already printed material. To avoid this, the distance between the nozzles can be changed or some simple shape is printed alongside the print in order to keep the nozzles primed. The shift in layer gives an in between tone of two colors when looked from a distance also it provides grip to the object. This is a good method for small printers with at least two print heads or nozzles, if there are more than two nozzles in the printer more tone can be achieved with more color variations. With a single color, there can be given the texture which might not be visible easily but will be good for feel and touch. The deposition of the material also affects the tensile strength of the produced material as shown by Wang et al. The part deposited in *Z*-axis direction gives higher tensile strength as compared to other axes [14]. Also, the surface roughness depends more on the layer thickness than road width and speed of deposition [15].

73.4 Diameter of Nozzle

Ramanath et al. showed by finite element analysis (FEA) where the flow of polycaprolactone (PCL) is observed by changing the nozzle diameter to 0.4 mm instead of 0.2 mm. This resulted in reducing the pressure drop across the liquefier or heating block. The accuracy of the finished product depends on the consistency and proper liquefying of the material is seen. But to have a proper consistency of the material pressure drop along the liquefier is needed to be kept as minimum as possible [16]. Increasing the nozzle diameter though increases the surface roughness as the material leaves gap in between two layers. For this optimization of the material, diameter is needed to be done. In order to get optimized nozzle diameter for better quality and precise product, Nor Aiman Sukindar et al. did research in his paper. The force created by the pulleys while pushing the material into the heating chamber is constant throughout. The change in the nozzle diameter affects the pressure change along the heating block [17, 18]. But to optimize the nozzle diameter not just the pressure drop but we have to focus on extrusion time as well. In this paper, the material used is PLA for testing. The process used for experimentation was as follows: (Fig. 73.2).

In this method, it is assumed that the flow within the heating block is steady state but in reality the flow is changing as the temperature while heating is fluctuating continuously.

The boundary conditions regarding the heating block or liquefier are very important which will conduct the FEA simulation. Rapid prototyping through 3D printing

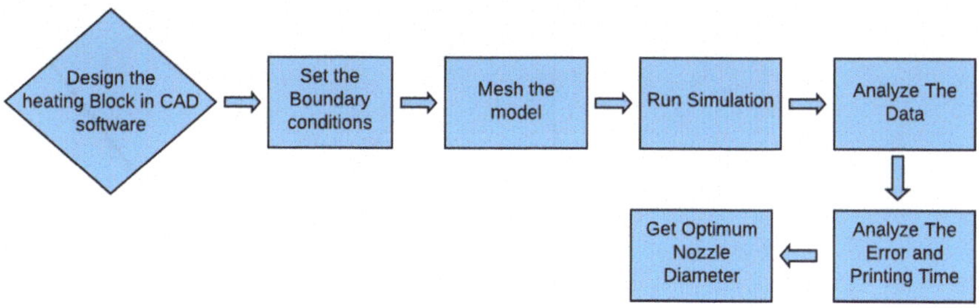

Fig. 73.2 Simulation and experimentation process

is generally an unsteady-state process, but in order to simplify the model, it is considered as steady state with change in flow over time is considered as negligible [19]. Some other assumptions are also considered which are as follows:

1. Flow is considered laminar.
2. The PLA material flow in the heating block along its walls is considered as zero.
3. The temperature of the heating block is considered as constant and it is considered to be perfectly insulated.

The main change to be seen while changing the diameter of the nozzle is the pressure drop across the heating block. But a nozzle is comprised of different sections as shown. At every part of the nozzle according to its geometry the pressure drop is calculated, which is achieved by Malenka who has comprised the following equations [11] (Fig. 73.3).

1. $\Delta P = \Delta P_A + \Delta P_B + \Delta P_C$
2. $\Delta P_A = 2L_\alpha \left(\frac{\vartheta}{\emptyset}\right)^{1/m} \left(\frac{m+3}{(D/2)^{m+1}}\right)^{1/m} \exp\left[\alpha\left(\frac{1}{T} - \frac{1}{T_\alpha}\right)\right]$
3. $\Delta P_B = \left(\frac{2m}{3\tan(\theta/2)}\right)\left(\frac{1}{d^{3/m}} - \frac{1}{D^{3/m}}\right) \times \left(\frac{D^2}{2}(m+3)2^{m+3}\right)^{1/m} \times \exp\left[\alpha\left(\frac{1}{T} - \frac{1}{T_\alpha}\right)\right]$
4. $\Delta P_C = 2L_b \left(\frac{\vartheta}{\emptyset}\right)^{1/m} \left(\frac{(m+3)(D_1/2)^2}{(D_2/2)^{m+3}}\right)^{1/m} \exp\left[\alpha\left(\frac{1}{T} - \frac{1}{T_\alpha}\right)\right]$

As shown above, the total pressure drop (ΔP) is calculated by adding all the pressure drops from all the regions shown those regions are *A*, *B* and *C*. In the above equations, Φ and m are power-law fit parameters and θ gives the nozzle angle. The pressure drop can be seen as the analysis is done for 0.2 mm to 0.25 mm, 0.3 mm,

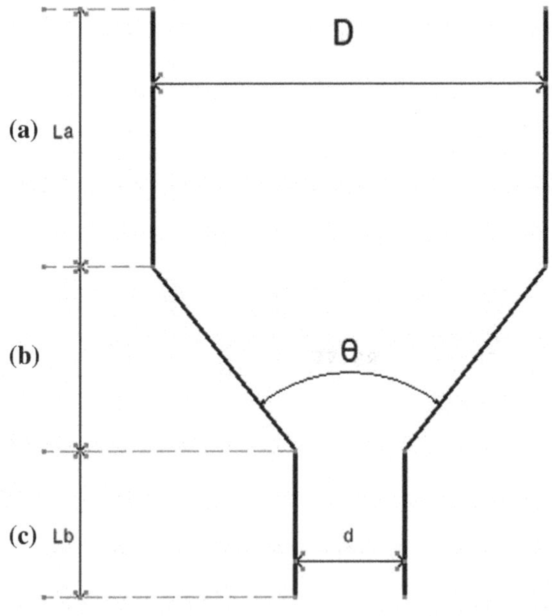

Fig. 73.3 General diagram for nozzle tip

0.35 mm and 0.4 mm, and it shows that as the diameter is reducing the pressure drop increases. The same thing is seen in the Ramanath's research as well as [16], who observed the polycaprolactone in fused deposition modeling. It is observed that while using the 0.2 mm diameter nozzle produces, three times the pressure drop that of when 0.4 mm is used.

Some researchers like Liang and Ness proposed for dealing with change in *L*/*D* ratio where *L* is the length of the conical edge and *D* is the outlet diameter. When this ratio is increased that is when diameter is reduced the pressure drop increases makes it important to select an optimum nozzle diameter in order to have consistency in extruding the material [20].

The diameter change in nozzle also affects the road width. Road width means the larger diameter of the extruded layer and layer thickness is the smaller diameter. From this, it is observed that road width affects with change in diameter of the nozzle, as the diameter decreases the road width becomes inconsistent that affects the surface finish of the product. But as we go higher the road width becomes more consistent but the gap between the layers increases, the strength increases.

73.5 Effect of Temperature of Nozzle

Quality of printing depends on the used raw material as well as the processing parameters. One of those processing parameters is the temperature of the nozzle head. The mechanical properties of the printed material are greatly affected by the printing temperature and the temperature of the nozzle tip. Head temperature affects the processing variables like density of the product, orientation, scaling, etc. [21]. In the experiments carried out by the author, the test objects were printed with ISO 527 on an open-source printer. Material used in printing is PLA of 1.75 mm which is extruded with single nozzle of 1.5 mm diameter and the cooling fan speed is kept at 30% for all tests. Tensile test is done on the printed product to test the printed product. To do the above tests, the samples are printed at different temperatures of 180, 200, 220 and 240 °C [22]. The experiment shows that with increase of temperature the tensile strength increases along with the Young's modulus alas it has a negligible effect on elongation of the single extrude. With the increase of nozzle head temperature, the strain at break value increases from 34 to 56 MPa. The inconsistency of temperature along the print shape causes a lack of adheration between the layers. This shows that the mechanical properties of the printed material are dependent on the nozzle temperature but they can be getting in control by post curing of the product or prototype.

Due to maintained temperature of the nozzle tip, the error in the finished product is reduced. Bonding between layers is observed to be better causing the tensile strength of the product to be maximum. Surface finish of the product is improved as the material settles properly and air gaps between layers are reduced.

73.6 Conclusion

While designing 3D printer nozzles certain parameters like its diameter, shape, temperature, material used and number of nozzles used should be considered properly. 3D printers have a major future scope, but in order to get there, the accuracy of printing, reducing time of print and cost is essential which can be achieved with proper management and design of nozzle and its affecting parameters. Multi-nozzle printers can be used to produce texture in the product by changing the phase angle while printing along with that they can be used to produce multiple products at the same time. Optimum diameter of the nozzle is necessary to be used in order to have lower pressure drop across the heating block so as to get consistent flow of material. The temperature of the nozzle affects the mechanical properties of the material which is to be considered while designing a nozzle. All in all combining each of the parameters and using them in order to perfect 3D printing is the goal.

References

1. Gokhare, V.G., Raut, D.N., Shinde, D.K.: A Review Paper on 3D-Printing Aspects and Various Processes Used in the 3D-Printing, vol. 6(06) (2017)
2. Galantucci, L.M., Bodi, I., Kacani, J., Lavecchia, F.: Analysis of dimensional performance for a 3D open-source printer based on fused deposition modeling technique. Procedia CIRP **28**, 82–87 (2015)
3. Rosen, D.W.: Research supporting principles for design for additive manufacturing. Virtual Phys. Prototyping **9**(4), 225–232 (2014)
4. Tran, N.-H., Nguyen, V.-C., Nguyen, V.-N.: Study on design and manufacture of 3D printer based on fused deposition modeling technique. Int. J. Eng. Adv. Technol. (IJEAT) **6**(6) (2017). ISSN: 2249 – 8958
5. Han, S., Xiao, Y., Qi, T., Li, Z., Zeng, Q.: Design and analysis of fused deposition modeling 3d printer nozzle for color mixing. **2017**, Article ID 2095137
6. Thrimurthulu, K., Pandey, P.M., Venkata Reddy, N.: Optimum part deposition orientation in fused deposition modeling. Int. J. Mach. Tools Manuf. **44**(6), 585–594 (2004)
7. Dawoud, M., Taha, I., Ebeid, S.J.: Mechanical behaviour of ABS: an experimental study using FDM and induction moulding techniques. J. Manuf. Processes **21**, 39–45 (2016) (Elsevier)
8. Haghsefat, K., TingTing, L., Haghsefat, S.: Structure design of FDM full color 3D printer extrusion device. **3**(4) (2018)
9. Dudek, P.: FDM 3D printing technology in manufacturing composite elements. Arch. Metall. Mater. **58**(4), 1415–1418 (2013)
10. Reiner, T., Carr, N., Mech, R., Štava, O., Dachsbacher, C., Miller, G.: Dual-Color Mixing for Fused Deposition Modeling Printers, vol. 33(2) (2014)
11. Melenka, G.W., Schofield, J.S., Dawson, M.R., Carey, J.P.: Evaluation of dimensional accuracy and material properties of the MakerBot 3D desktop printer. Rapid Prototyping J. **21**(5), 618–627 (2015)
12. Hasan, M., Fuchs, M., Matusik, W., Pfister, H., Rusinkiewicz, S.: Physical reproduction of materials with specified subsurface scattering. ACM Trans. Graph. (Proc. SIGGRAPH) **29**(4), 61:1–61:10 (2010)
13. Vagdevi, G., Umapathy, Srikanth, M.: Dual extruder indigenous 3D printer. Int. J. Electron. Electr. Comput. Syst. **7**(3) (2018)

14. Wang, CC., Lin, T.-W., Hu, S.-S.: Optimizing the rapid prototyping process by integrating the Taguchi method with the Gray relational analysis. Rapid Prototyping J. **13**(5), 304–315 (2007)
15. Anita, R., Arunachalam, S., RadhaKrishnan, P.: Critical parameters influencing the quality of prototypes in fused deposition modelling. J. Mater. Process. Technol. **118**, 385–388 (2001)
16. Ramanath, H.S., Chua, C.K., Leong, K.F., Shah, K.D.: Melt flow behaviour of Poly-E-Caprolactone in fused deposition modelling. J. Mater. Sci.: Mater. Med. **19**(7), 2541–2550 (2008)
17. Sukindara, N.A., Ariffinb, M.K.A., Hang Tuah Baharudinb, B.T., Jaafarb, C.N.A, Ismailb, M.I.S.: Analyzing the effect of nozzle diameter in fused deposition modeling for extruding polylactic acid using open source 3d printing. J. Teknologi (2016)
18. Wu, W., Ye, W., Wu, Z., Geng, P., Wang, Y., Zhao, J.: Influence of layer thickness, raster angle, deformation temperature and recovery temperature on the shape-memory effect of 3D-printed polylactic acid samples. Materials **10**, 970 (2017)
19. Novakova-Marcincinova, L., Novac-Marcincin, J.: Verification of mechanical properties of ABS materials used in FDM rapid prototypng technology. Proc. Manuf. Syst. **8**(2), 87–92 (2013)
20. Turner, B.N., Strong, R., Gold, S.A.: A review of melt extrusion additive manufacturing processes: I. Process design and modeling. Rapid Prototyping J. **20**(3), 192–204 (2014)
21. Boparai, K.S., Singh, R., Singh, H.: Development of rapid tooling using fused deposition modeling: a review. Rapid Prototyping J. **22** (2016)
22. Behzadnasab, M., Yousefi, A.A.: Effects of 3D printer nozzle head temperature on the physical and mechanical properties of PLA based product. In: 12th International Seminar on Polymer Science and Technology, 2–5 Nov 2016

Chapter 74
Modeling and Analysis of Composite Helical Compression Spring

Akhileshwar Nirala, A. K. Richhariya, N. Kumar, V. K. Dwivedi and Manohar Singh

Abstract The challenges for the automobile manufacturers are to improve the quality along with the performance of the components and reduction in weight and manufacturing cost. Therefore, the investigation is going for alternative materials to design automobile parts and improves the strength to weight ratio, which is fuel-efficient. As alternative materials, composites are used widely. The use of composite material produces in mass level which reduces the manufacturing cost and higher strength. The present work is keen to study the feasibility of composite helical compression spring used in suspension systems. The comparative study was done with conventional steel IS4544 Grade 3 (Tata Indica Vista) helical compression spring to the composites helical spring in the present study on a load of 4542.3 N. EGlass/Epoxy and IS4544 Grade 3 are used as the composite constituent, in which EGlass/Epoxy is used inner or outer layer and IS4544 Grade 3 vice versa. Composite helical spring has found better mechanical properties than conventional. Solid works 2013 is used for the modeling of helical compression spring, and commercially available ANSYS 14.5 software is used for the static and modal analysis. It is observed from the study that stresses and deformation generated in the composite material helical compression spring are reduced, and strain energies stored by the composite material helical compression spring are increased.

Keywords Spring · Composite materials · EGlass/Epoxy · Mechanical properties · FEM

74.1 Introduction

To absorb the energy due to the impacts, helical compression springs are commonly used to form a flexible link. The link deflects when loading has done and restored the objects after that the disturbing forces are removed. That is why these springs

A. Nirala (✉) · A. K. Richhariya · N. Kumar · V. K. Dwivedi · M. Singh
Department of Mechanical Engineering, Galgotias College of Engineering and Technology, Greater Noida 201310, India

S. Yadav et al. (eds.), *Proceedings of International Conference in Mechanical and Energy Technology*, Smart Innovation, Systems and Technologies 174,
https://doi.org/10.1007/978-981-15-2647-3_74

are used in the suspension system. S. S. Bhatia et al. designed helical compression spring which was fabricated by conventional steel material (IS 4454 Grade 3) and then analyzed comparative study at the two loading conditions of 1500 N (normal loads) and 3000 N (overload). They have done modeling and analysis of the helical spring using CREO 2.0 and ANSYS 14.5 to predict the stresses, deflections and strain energies. They have found that weight of conventional steel helical compression spring (IS4454 Grade 3) was 1.433 kg; whereas, EGlass/Epoxy and Carbon/Epoxy composite material helical compression springs weigh only 0.233 kg and 0.02912 kg, respectively, and weight saving of 83.74% and 97.96%, respectively, for composites as compared to conventional steel (IS4454 Grade3) helical compression spring. Authors also found that stresses induced in the composite helical compression spring were much lower than conventional steel (IS4454 Grade 3) spring and higher deflection than the conventional steel (IS4454 Grade 3) helical compression spring. Therefore, the suspension system seems to be less rigid and ensures the smooth riding of the vehicle with fewer jerks. They have observed that the deformation in the composite materials springs at a load of 3000 N (severe overloading conditions) was greater than the maximum allowable deflection; whereas, it was within the safe limit at the normal loading conditions of 1500 N. Hence, the use of composite materials may not be recommended for severe overloading conditions [1]. M. Bakhshesh et al. have compared experimentally to FEM analysis for uniform loading in the light vehicle suspension system and found good agreement. They have replaced conventional spring materials to these composite materials such as EGlass/Epoxy, Carbon/Epoxy and Kevlar/Epoxy. It has been investigated that the orientation of fiber in composite influenced the stress value in compression spring [2]. Some other authors have analyzed the shock-absorbing capacity of bikes for varying the wire diameter of spring. They have done modeling and analysis of shock absorber for 160 cc bike in proE and ANSYS software, and it has been designed for a modified shock absorber by reducing the wire diameter [3]. D. A. Budan and T. S. Manjunatha established the practicability of composite coil spring at the place of metal coil spring. They have fabricated three different composite springs as carbon fiber, glass fiber and combination of carbon and glass fiber. Composite compression spring has a lower weight and higher corrosion resistance than conventional materials used for spring, and it is fuel-efficient for any automobile. Composite spring can be used at a higher temperature and higher load condition, but design and manufacturing are difficult, that is why it is not popular [4]. T. M. Mulla et al. analyzed the mechanical properties of helical compression spring for a three-wheeled vehicle which is used for public transport and carrying freight as well. To design this kind of spring, they have considered elastic characteristics and fatigue strength. To maintain elastic characteristics along with weight reduction is a big task for researchers, and they have done in a well manner. They have considered structural reliability and done finite element analysis for stress distribution in spring [5]. Niranjan Singh has been widely applied in the suspension system of flexible leaf springs fixed in the four corners in a horse-drawn carriage and relate with complex control algorithms of the recent automobile. The design objective of the suspension system is to isolate the vehicle mass to the uneven road and keep up contact with the wheel and road surface. The author has used numerical as well as an experimental method to analyze and calculate fatigue strength and shear stress by the finite element method [6]. Suresh. G et al. have designed and experimented analysis of fiber-reinforced polymer and epoxy resin reinforced with nano-clay composite helical spring. Epoxy resin reinforced with nano-clay composite helical spring showed improved mechanical and tribological properties and high corrosion resistance and has a high factor of safety at low weight; therefore, it is economical. Authors have studied comparative study between steel helical spring and composite helical spring and found better properties such as load-carrying capacity, stiffness and weight savings in composite helical spring. The study was done by authors to consider conventional steel spring dimensions and fabricate the same composite spring in their study [7]. B. Praveen et al. have investigated to reduce the spontaneous failure and enhance working life by design optimization with the variation of wire diameter of helical compression spring. Their research indicates that by increasing the wire diameter of spring von Mises stresses decreases. Therefore, the spring rate increases by reducing deflection, and hence, premature failure reduces [8]. Authors studied to take four different laminate composites such as unidirectional laminates (UL), rubber core (RC) UL, UL with the braided outer layer (BOL) and RC UL with BOL. And they incorporated the mechanical properties of UL, RCUL, UL with BOL and RC UL with BOL [9]. S. Patter et al. have also analyzed disparity in deformation value as well as maximum shear stress value by analytical and FEM with the varying load on that [10].

The present study is showing better mechanical properties than reported research data which is helpful for further research work to design and fabricate new composite compression spring for the automobile sector.

74.2 Materials Selection and Design Equation

This chapter incorporates the governing equation used in the design of helical compression spring. Moreover, material selection along with is also discussed in the latter part of the chapter. Strain energy (U) considered first as

$$U = \left(\sigma^2\right)/(2 \times \rho \times E) \tag{74.1}$$

Here, under the same stress (σ) condition, specific strain energy will be higher when Young's modulus (E) and density (ρ) will be lower. If we consider a helical compression spring with the axial load, then shear stress can be determined as follows:

$$\tau = (K \times 8 \times F \times D) / (\pi \times d^3) \tag{74.2}$$

where τ = shear stress, F = load on the spring, D = inside diameter of helical compression spring, and d = wire diameter of helical compression spring.

Equation 74.2 shows that shear stress increases by increasing the Wahl's factor (K), and we can calculate K by using Eq. 74.3 as follows:

$$K = (4C - 1)/(4C - 4) + 0.615/C \quad (74.3)$$

Here, C is the spring index which is given by ($C = Dd$), and the deflection in the spring is estimated by

$$\delta = \left(8 \times F \times n \times D^3\right)/\left(G \times d^4\right) \quad (74.4)$$

where δ = deflection, and n = number of active coil.

Steel and its variants are the conventional choices for the manufacturing of spring used for the automobile industry as well as other industrial machinery. The steel offers a better combination of mechanical properties such as strength, ductility, hardness and toughness. Presently, steel graded as IS4454 is used for making automotive helical compression spring. Tables 74.1 and 74.2 show the mechanical property of IS4454 Grade 3 steel and EGlass/Epoxy, respectively.

For the boundary condition and load calculation, a helical compression spring of Tata Indica V2 has been considered which has some specification weight of the vehicle 1145 kg. The present study is considered for five persons in car/vehicle of weight (80 × 5 = 400) 400 kg, and the total weight of the vehicle is = 155 kg. Total load of the vehicle when car is fully filled =545 × 9.8 = 15,141 N. In these type of cars, 60% load is carried by front wheels so the load on a single wheel is = 15,41 × 60,100 × 12 = 4542.3 N after using Eqs. (74.1)–(74.4) we have got as the value of

Table 74.1 Mechanical properties of IS4454 Grade 3

S. No.	Specification	Value
1.	Density	0.00785 kg/cm^3
2.	Young's modulus	2.1 × 10^5 MPa
3.	Poisson's ratio	0.3
4.	Ultimate tensile strength (UTS)	1500 MPa
5.	Yield tensile strength	1100 MPa
6.	Coefficient of thermal expansion (/°C)	12 × 10^{-6}

Table 74.2 Material properties of EGlass/Epoxy

S. No.	Specification of EGlass/Epoxy	Value (MPa)
1.	Tensile modulus (along *X*-axis)	4.3 × 10^4
2.	Tensile modulus (along *Y*-axis)	6.5 × 10^3
3.	Tensile modulus (along *Z*-axis)	6.5 × 10^3
4.	Tensile strength	900
5.	Compressive strength	450

C is 10.17 and K is 1.14, shear stress on the helical compression spring 931.75 MPa and deflection of the spring 188.22 mm.

74.3 Modeling and Analysis

Figure 74.1a is the reference design of the helical spring of Tata Indica V2 2008, and the dimensions of helical compression spring are measured by vernier calipers. The first model of Tata Indica was introduced in 1999 by Tata Motors. In Tata Indica V2, MacPherson strut suspension system has been used to absorb the shock and to provide better driver comfort while the vehicle is moving.

This is used in the front wheel of the vehicle. The helical compression spring of conventional steel IS4544 Grade 3 is designed on the Solid Works 2013. The coil is designed according to Tables 74.3 and 74.4, while for the composite helical compression spring the assembly is made in the Solid Works 2013. The composite spring is made with steel and EGlass/Epoxy, steel is taken inside which has 8 mm diameter, and composite is taken outside which has 4 mm diameter Fig. 74.1 (a) Helical compression spring of Tata Indica (b) shows the CAD model of helical compression spring made of conventional steel of IS4454 Grade 3 and (c) shows the CAD model of same helical compression spring made of composite materials. In the composite helical compression spring, the spring is made of steel and EGlass/Epoxy.

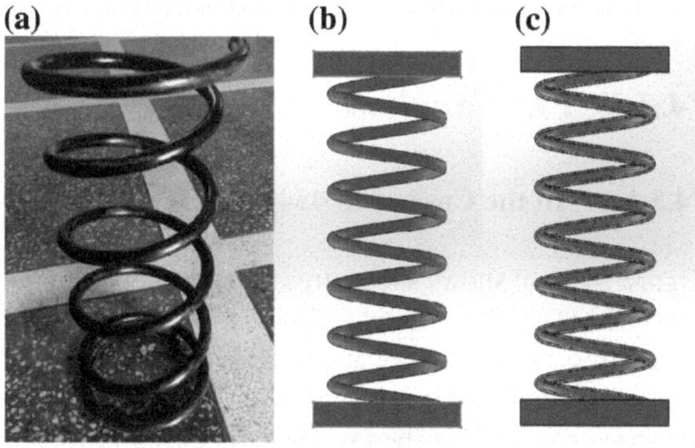

Fig. 74.1 **a** Helical compression spring of Tata Indica, **b** CAD model of helical compression spring of conventional steel of IS4454 Grade 3 and **c** CAD model of same helical compression spring of composite materials

Table 74.3 Dimension of helical coil spring

S. No.	Dimensions	Size in mm
1.	Free length (Helical coil spring)	340
2.	Diameter (Spring wire)	12
3.	Diameter (Mean coil)	122
4.	Number of actual turns	8
5.	Number of active turns	6.5

Table 74.4 Detail specification of helical compression spring under analysis

Parameter numerical	Value
Total weight of vehicle	1545 kg
Total load on vehicle	15,141 N
Load on the single coil spring	4542.3 N
Type of spring	Helical compression
Spring wire diameter	12 mm
Spring free length	340 mm
Spring turns	6.5

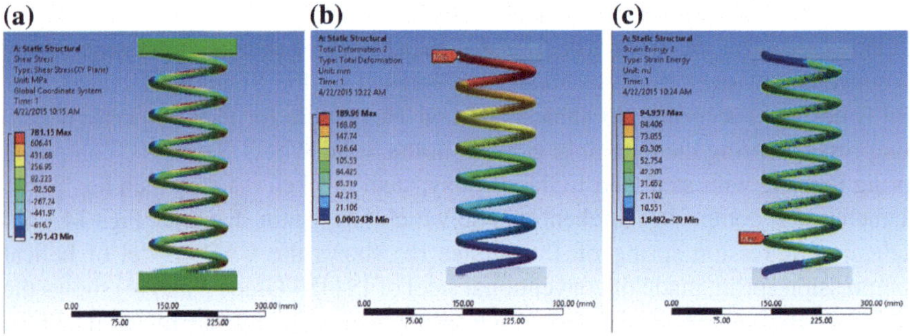

Fig. 74.2 **a** Shear stress on the IS 4544 Grade 3 material helical compression spring, **b** Total deformation on the IS 4544 Grade 3 material helical compression spring and **c** Strain energy in the IS 4544 Grade 3 material helical compression spring

74.3.1 FEM Analysis

74.3.1.1 In the Case of IS 4544 Grade 3 Helical Compression Spring

Figure 74.2a–c shows shear stress, strain energy and maximum deformation in FEM Analysis of IS 4544 Grade 3 helical compression spring. It shows the maximum shear stress, strain energy and maximum deformation. When the load is 4542.3, the maximum shear stress is 781.15 MPa, the maximum deformation is 189.96 mm, and the strain energy absorbed by the coil spring is 94.95 mJ.

74.3.1.2 In the Case of IS 4544 Grade 3 and EGlass/Epoxy When EGlass/Epoxy Is Inner Material and IS4544 Grade 3 Steel Is Outside Material (6E-12S)

Figure 74.3a–c shows shear stress, deformation and strain energy in FEM of hybrid helical compression spring. This case study consists of first 6 mm EGlass/Epoxy inner material and an outer 6 mm overlay IS 4544 Grade 3 steel. Hybrid helical compression

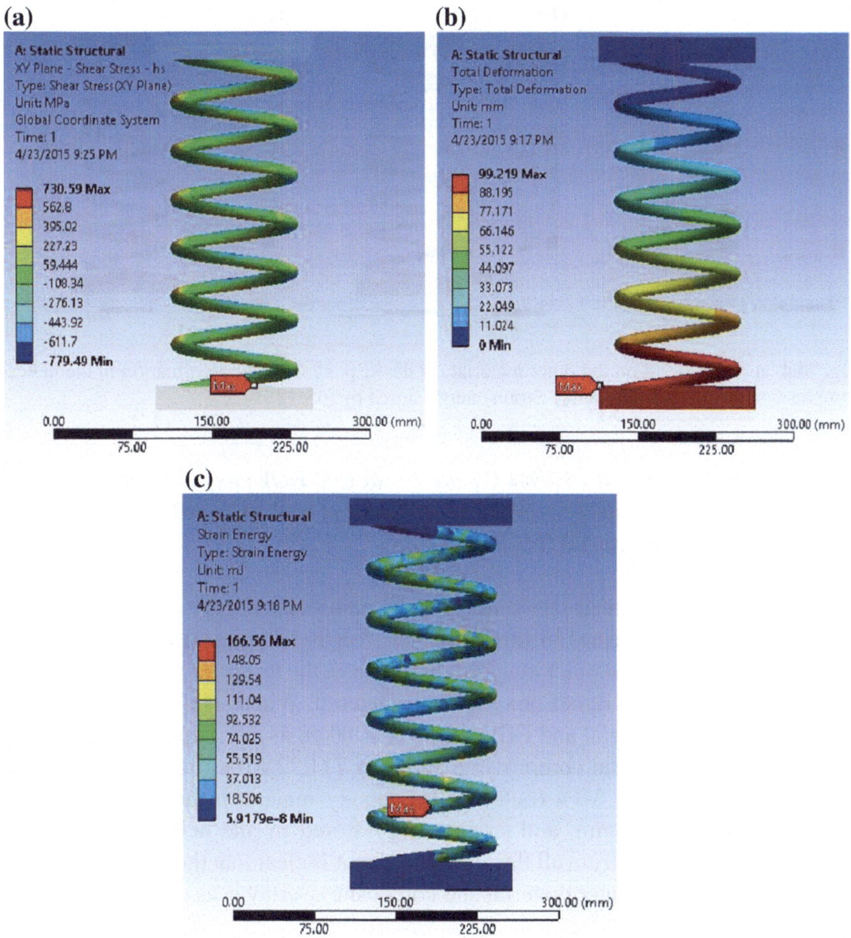

Fig. 74.3 **a** Shear stress on the inner and outer material is approximate similar, **b** Total deformation on composite helical spring and **c** Strain energy on the composite material

spring is subjected to the same loading and boundary condition as that of IS4544 Grade 3 conventional steel. The maximum shear stress in the helical compression spring is 650–730.59 MPa on the inner composite material, the maximum shear stress on the outer steel material is 600–695.8 MPa, the maximum total deformation is 99.219 mm, the strain energy is 166.56 mJ, and the results are shown in Fig. 74.3.

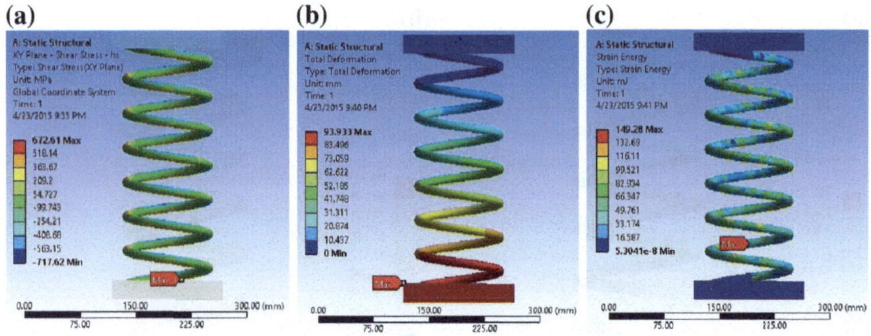

Fig. 74.4 **a** Shear stress on the outer material (EGlass/Epoxy) **b** Total deformation in the helical compression spring (6S-12E) and **c** Strain energy stored by (6S-12S)

74.3.1.3 In the Case of IS 4544 Grade 3 and EGlass/Epoxy When IS4544 Grade 3 Steel Is Inner Material and EGlass/Epoxy Is Outside Material (6S-12E)

Figure 74.4a–c shows shear stress, deformation and strain energy in FEM of hybrid helical compression spring. In this case, the steel is taken as inner material, and EGlass/Epoxy is taken as outer material. In this case, the results are better when steel is outer material and EGlass/Epoxy is inner material. When steel of IS4544 Grade 3 is taken as inner material and EGlass/Epoxy is taken as outer material maximum, shear stress on the helical compression spring is 711.22 MPa (inner IS4544 grade 3 material) and 672.61 MPa (outer EGlass/Epoxy material), the maximum total deformation is 93.933 mm, and strain energy stored by the helical compression spring is 149.28 mJ. So from all the above results, it is clear that the results are better when steel is taken as inner material and composite overlay is as outside material.

74.4 Conclusion

In the case of IS 4544 Grade 3 helical compression spring at load 4542.3 N, the maximum shear stress is 781.15 MPa, the maximum deformation is 189.96 mm, and the strain energy absorbed by the coil spring is 94.95 mJ. In the second case as hybrid helical compression spring study which consists of first 6 mm EGlass/Epoxy inner material and an outer 6 mm overlay IS 4544 Grade 3 steel, the maximum shear stress in the helical compression spring is 650–730.59 MPa on the inner composite material, the maximum shear stress on the outer steel material is 600–695.8 MPa, the maximum total deformation is 99.219 mm, and the strain energy is 166.56 mJ. Further when steel is taken as inner material and EGlass/Epoxy is taken as outer material, then maximum shear stress on the helical compression spring is 711.22 MPa (inner IS4544 grade 3 material) and 672.61 MPa (outer EGlass/Epoxy material), the maximum total deformation is 93.933 mm, and strain energy stored by the helical compression spring is 149.28 mJ. So from all the above results, it is clear that the strain energy is better when steel is taken as outer material in the composite. Total deformation reduces 50%, and strain energy enhanced by 43% compared to conventional compression spring.

References

1. Bhatia, S.S., Bergaley, A.: Analysis of the design of helical compression spring to study the behavior of steel and composite used as spring materials. IJESRT, 576–583 (2014). ISSN: 2277-9655
2. Bakhshesh, M., Bakshesh, M.: Optimization of steel helical spring by composite spring. Int. J. Multidisciplinary Sci. Eng. **3**(6), 47–51 (2012)
3. Prince Jerome Chirstopher, J., Pavendhan, R.: Design and analysis of two wheeler shock absorber coil spring. IJMER, 134–140 (2014)
4. Abdul Budan, D., Manjunatha, T.S.: Investigation on the Feasibility of Composite Coil Spring for Automotive Applications. WASET, vol. 4, pp. 577–581 (2010)
5. Mulla, T.M., Kadam, S.J., Vaibhav, S.: Finite element analysis of helical coil compression spring for three wheeler automotive front suspension. IJMIE, vol. 2, pp. 74–77 (2012). ISSN No. 2231–6477
6. Singh, N.: General review of mechanical spring used in automobile suspension system. IJAERS, pp. 115–122 (2014). E-ISSN2249–8974
7. Suresh, G., Vignesh, R., Aravinth, B., Padmanabhan, K., Thiagarajan, A.: Fabrication and analysis of nano composite cylindrical helical spring. IJIRSET **3**, 1208–1213 (2014). ISSN: 2319-8753
8. Praveen Kumar, B., Sampath Rao, P.: Investigation on reduction in premature failure of locomotive coil springs. IJERT **3**, 224–229 (2014). ISSN: 2278-0181
9. Chiu, C.-H., Hwan, C.-L., Tsai, H.-S., Lee, W.-P.: An experimental investigation into the mechanical behaviors of helical composite springs. SCIENCE DIRECT, Compos. Struct. **77**, 331–340 (2007)
10. Pattar, S., Sanjay, S.J., Math, V.B.: Static analysis of helical compression spring. IJRET, 835–838 (2014). eISSN: 2319-1163, pISSN: 2321-7308

Chapter 75
Various Tuning and Optimization Techniques Employed in PID Controller: A Review

M. Prabhat Dev, Sidharth Jain, Harish Kumar, B. N. Tripathi and S. A. Khan

Abstract PID controllers are implemented in more than 90% of the control system applications. In this review paper, various tuning methods have been presented and comparison of established algorithmic tuning methods has been discussed on system response. There have been many approaches used in the past for tuning and obtaining optimized gain factors such as Ziegler–Nichols method, genetic algorithm (GA), particle swarm optimization (PSO) method, and artificial neural network (ANN). The primary goal of this paper is to establish a proper understanding about different tuning and optimization methods and their effect on process efficiency and stability. The secondary goal is to provide a pathway for future development of a tuning algorithm for a high-temperature research grade furnace controller, based on machine learning (ML). This leads to higher controller efficiency over a predefined finite set of ramp–hold cycles, ensuring lesser rise and settling time, reduced or no overshoot, minimized mean squared error, and maximum stability. Critical manufacturing processes like investment casting, metal injection molding, and other thermal cycling processes like physical vapor deposition/chemical vapor deposition, e-waste processing, which require precise control of temperature are expected to be benefited by ML-integrated PID parameter auto-tuning and control.

Keywords PID · Tuning algorithm · Z-N tuning · ANN · GA · PSO · Control system · Machine learning

---

M. Prabhat Dev · H. Kumar
National Institute of Technology Delhi, New Delhi, India

S. Jain (✉) · B. N. Tripathi
Meerut Institute of Engineering & Technology, Meerut, Uttar Pradesh, India
e-mail: siddhartha.jain@miet.ac.in

S. A. Khan
Galgotias College of Engineering and Technology, G. B. Nagar, Uttar Pradesh, India

S. Yadav et al. (eds.), *Proceedings of International Conference in Mechanical and Energy Technology*, Smart Innovation, Systems and Technologies 174,
https://doi.org/10.1007/978-981-15-2647-3_75

75.1 Introduction

PID controllers can overcome any fluctuations that occur in the process due to various internal and external factors. They have a simple and robust control structure [1] and can overcome disturbances [2]. Single or multiple degree-of-freedom systems can be controlled. It allows the users to tune the controller to control the process path. Some of the most common places where these controllers are used include flight systems, temperature controllers, autonomous drones, etc., compared to other controllers; PID controllers have a simple control structure which can incorporate any optimization algorithms. The development of microcontrollers integrated with the control system allows the user to implement various tuning algorithms and to collect data points from the process and improve the process parameters further based on the data points [3] (Fig. 75.1).

$$u(t) = k_p e(t) + k_i \int_o^t e(t)\,\mathrm{d}t + k_d \frac{\mathrm{d}e(t)}{\mathrm{d}t}$$

75.1.1 Significance of Gain Factors

Path which the process follows depends on the gain factors. This way one can fine-tune the process path by varying the gain factors accordingly. k_p, the proportional factor works based on the present condition of the system. k_i, the integral factor works based on the past error data and eliminates steady-state error [4], and k_d, the derivative factor predicts the future of the process by keeping track of rate of change of error. This way, a perfect balance is maintained so that the system can smoothly reach the set point from the initial condition without any fluctuations. The goal is to tune the controller so that the plant has a quick response, lesser rise and settling time, maximum stability, and minimum or no overshoot.

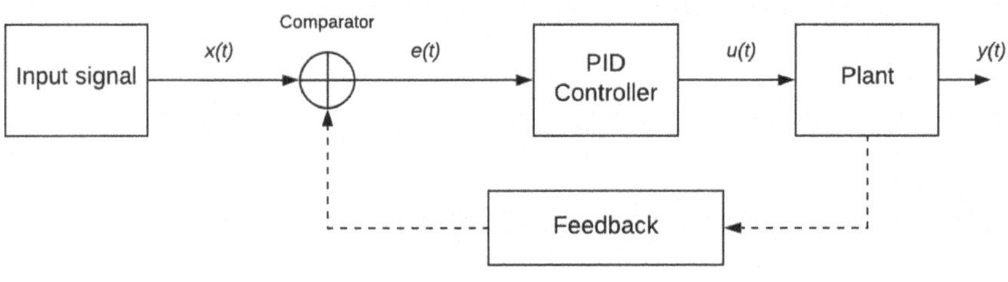

Fig. 75.1 Basic layout of a PID controlled plant

75.1.2 Problems Faced with Integral and Derivative Parts

If the plant is saturated (reaches its physical limit) and the error still exists, then the integral increases the process value but resulting in no change in the plant output because of the saturation. And once the load is off the plant, the controller rapidly increases the output resulting in a high overshoot. Similarly in the derivative part of the controller, the high-frequency signals are amplified. These types of signals are usually from the sensor and are barely noticeable. The reason this phenomenon has high impact on high-frequency signals is because of the higher slope of the wave. Filtering and set point weighting techniques can be used to filter out the noise during the operation [5].

There are various methods by which these problems can be neglected. One of the effective methods is to clamp the integrator. This implies turning off the integrator once saturation is reached. The problem with the derivative part can be solved by implementing a low-pass first-order filter which filters out signals which are over a certain frequency limit. The lower frequency noise even when amplified through the derivative part does not affect the output of the plant.

75.2 Stability of the System

In order to find if the system is stable, unstable, oscillating, etc., we need to find the step response of the plant. This process is a crucial step in the design as the controller will be modeled based on the nature of the system, and hence, stability analysis is performed with the utmost care. The stability of the system will depend on how well the controller is tuned. Most of the controllers are not properly tuned [3], and hence, tuning the controller according to the process requirement is a crucial part of control engineering. Disturbances tend to reduce the output efficiency of the system, and using a closed-loop transfer function helps in minimizing these disturbances [6].

75.3 Analyzing the Transfer Function

Consider a transfer function $T(s)$

$$C(s) = R(s) \cdot T(s)$$

Where $R(s)$ is the response function. Various values of $R(s)$ for different tests. $R(s) = 1/s$ for step response, 1 for impulse response and $1/s^2$ for Ramp response.

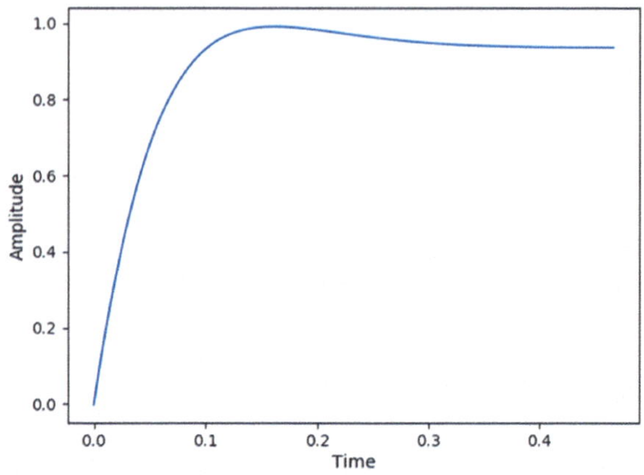

Fig. 75.2 Step response output

75.3.1 *Step Response of a Transfer Function*

For obtaining the step response, we need the output in terms of $C(t)$. Inverse Laplace transform is used for this conversion.

$$L^{-1}[C(s)] = \frac{1}{s}L^{-1}[T(s)]$$

After this step, we get the equation in terms of C(t). {Since $L^{-1}[C(s)] = C(t)$}

Once we obtain the equations in terms of time-dependent parameters, we can find out the time step response of the transfer function. From the step response graph, we can determine the nature of the system by looking at the system's progression with time. The following graph shows the step response of a transfer function (Fig. 75.2).

75.4 Tuning Methods

75.4.1 *Z-N Tuning Approach*

There are various methods that have been used previously for obtaining the gain parameters. Among them, Zeigler–Nichols is one of the most popular methods used for tuning. In [7], lag and plant reaction time is introduced as quantitative measures of controllability. This method is only applicable for plant function which has S-shaped step response (monotonic) [8]. The Z-N tuning involves setting the integral and derivative gains to zero and then slowly increasing the proportional gain starting from zero until the step response of the plant output reaches a state where it has a stable and constant oscillation cycle. Based on the time period of a cycle T_U and the respective proportional gain value K_U, the other gain parameters are obtained.

Other than conventional Z-N tuning approach, modified Z-N methods have also been used [9] which showed better performance than the conventional method (minimum rise time, minimum overshoot, and minimum settling time).

75.5 Modern Tuning Methods

The integration of microprocessors in the controller led many researchers to tweak and tune the control algorithm to obtain the desired output. This led to an increase in the efficiency of the controller and development of more efficient and state-of-the-art tuning methods [10]. This also allows us to make use of the processing capacity and hence to integrate advanced algorithms incorporating modern techniques such as machine learning, neural network, genetic algorithm, deep learning, and various other optimization methods.

75.5.1 Artificial Neural Network (ANN)-Based Tuning

Artificial neural network is inspired by the working of neurons in the brain. This method can be used for tuning and obtaining optimized gain parameters for the controller. There are certain process steps involved in this method. Initially, the network is trained by using input data. Each input value is influenced by weight functions. Genetic algorithm can be used for optimizing these weight functions. During this process, the network recognizes the system and learns about the response of the system to the input. Selecting the right network is the crucial part of modeling. A number of parameters should be minimum as possible to minimize the complexity of the network [11]. The error value obtained during the process can be used to backpropagate and modify the weight functions of the network [12]. This leads to a more efficient tuning of the controller. Deep learning has also been used as a tool for designing the controller [13] in which the deep leaning network was integrated. Artificial neural networks can also be used for system identification process [14]. This is the process by which the network can automatically derive the system model. This way we can train a neural network to generate a mathematical model and use this model for obtaining tuning parameters.

75.5.2 Genetic Algorithm Approach

Genetic algorithm approach has also been used for tuning the controller and obtains the optimized gain parameters. In genetic algorithm, the fittest variables are picked from a data set. It is an iteration process in which a batch of fittest parameters is found at the end. GA approach yielded better results and faster response, better rise, and

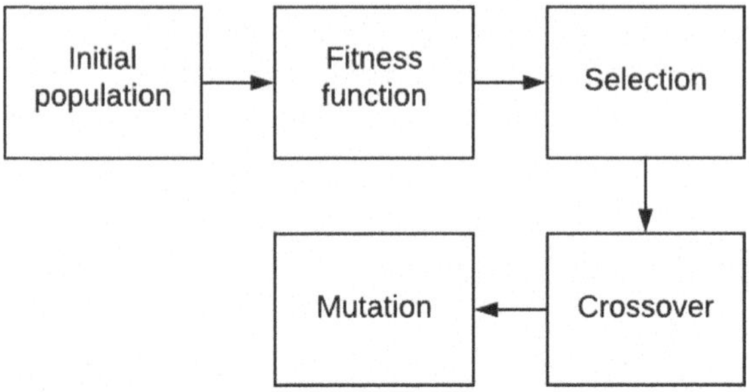

Fig. 75.3 Genetic algorithm structure

settling time compared to conventional tuning methods [15]. GA usually involves five processes (Fig. 75.3).

First-order lag plus time delay (FOLPD) is a commonly occurring problem. In [16], GA has been used to overcome this problem. Modified genetic algorithm can improve the dynamic response of the system [17].

75.5.3 Particle Swarm Optimization Method

The method is inspired by swarm intelligence which is a behavior shown by certain birds and fishes. This is an iterative process by which the algorithm tries to find an optimal solution based on predefined quality measures. It is a process of computational intelligence involving evolutionary computing. In [18], the author used internal model control (IMC)-based approach in which the PID controller is tuned with a single tuning parameter. This parameter is called IMC filter coefficient. This technique yields a better solution for the problem with larger time delays [18]. The comparison between conventional Z-N tuning and PSO algorithm was shown by the author in [19] in which PSO-based tuning provided better step response performance.

75.6 Future Scope

Based on the study of various tuning and optimization methods, further research will be continued to develop a PID controller for a high-temperature research grade furnace. This controller will be employed with an algorithm which will be developed based on conventional tuning methods integrated with machine learning. This way the controller will be trained to overcome any discontinuities in the system and provide precise heat cycle control. This furnace will be employed to burn out additively printed 3D investments in SLA and PLA polymers, wherein the surface finish of aluminum alloy castings will be correlated with the precision of investment burn-out

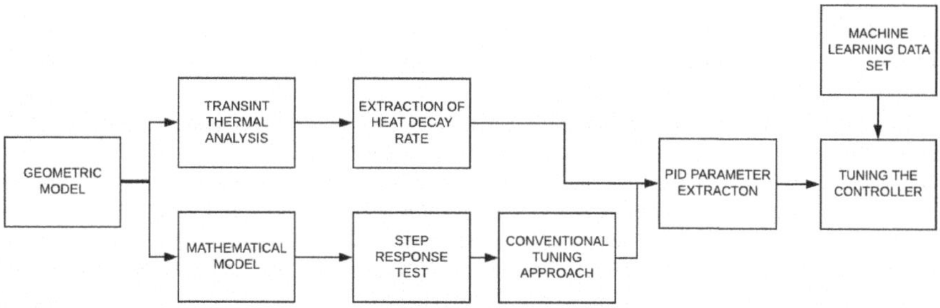

Fig. 75.4 Process flow for tuning the controller

cycle control. This testing will only be possible with the development of a furnace controller precise enough to control the heating cycles, capable of advanced tuning and process variable control over continuous multiple ramp–hold steps.

75.6.1 Tuning Approach

The algorithm will be based on basic inputs obtained from conventional tuning methods and from computational fluid dynamics (CFD) simulation of the furnace to extract heat decay rate. It will then be integrated with machine learning to make the controller more effective and to withstand if some discontinuities arise in the system (Fig. 75.4).

The CFD and conventional tuning approach will provide a basis for developing the tuning algorithm. The second step would be to obtain a process transfer function through system identification. This transfer function will be compared to the theoretically obtained function and make some changes if necessary. Machine learning will then be employed to the modified tuning algorithm obtained from the previous process. This way the controller should be able to auto-tune the gain parameters if necessary in order to produce a stable and precisely controlled plant output.

75.7 Conclusion

Various tuning algorithms and their control behavior were studied and analyzed in this research paper. Discontinuities are a commonly occurring phenomenon in any system. These discontinuities tend to drive the system toward instability which may take some time to settle. These instabilities affect the outcome of any process. In order to avoid these instabilities, the controller requires a robust algorithm which should be able to handle these discontinuities. Machine learning is a powerful method for analyzing these discontinuities, learn from past data and control the process within specified limits with minimal or no human intervention. This results in reducing the mean squared error and increasing the control efficiency.

Acknowledgements This review paper has been prepared to develop an academic understanding of basic PID tuning and optimization methods for developing an AI-Based Controlled Environment Tubular Furnace of Maximum Working Temperature 1200 °C: Project funded under Collaborative Research and Innovation Program (CRIP) through TEQUP(III) of Dr. A. P. J. Abdul Kalam Technical University (AKTU), Lucknow, Uttar Pradesh (India). Dr. Sidharth Jain (P.I.) and Dr. B. N. Tripathi (Co-P.I.) are both faculty at the Mechanical Engineering Department, Meerut Institute of Engineering and Technology, Meerut, Uttar Pradesh (India).

References

1. Aarabi, A., Shahbazian, M., Hadian, M.: Improved closed loop performance and control signal using evolutionary algorithms based PID controller. In: 2015 16th International Carpathian Control Conference (ICCC)
2. Johnson, M.A., Moradi, M.H.: PID Control New Identification and Design MethodsISBN-10: 1-85233-702-8, ISBN-13: 978-1-85233-702-5 Springer Science+Business Media springeronline.com (2005)
3. O'Dwyer, A.: Handbook of PI and PID Controller Tuning Rules, 3rd edn. Published by Imperial College Press 57 Shelton Street Covent Garden London WC2H 9HE (2009)
4. Åström, K.J., Hägglund, T.: Advanced PID control. Department of automatic control, Lund Institute of Technology, Lund University, Copyrights by Instrument Society of America, ISBN 1-55617-516-7 (1998)
5. Astrom, K.J.: PID Control. Control System Design. Department of automatic control, Lund University (2002)
6. Ahmad, A.A.A, Hussein, E.M.: Effect of disturbance on closed-loop control system. Int. J. Innov. Res. Sci. Eng. Technol. **3**(8) (2014)
7. Ziegler, J.G., Nichols, N.B., Rochester, N.Y.: Optimum settings for automatic controllers, J. G. Ziegler and N. B. Nichols: optimum settings for automatic controllers. Trans. ASME **64**, 759–768 (1942)
8. Basilio, J.C., Matos, S.R.: Design of PI and PID controllers with transient performance specification. IEEE Trans. Educ. **45**(4) (2002)
9. Kushwah, M., Patra, A.: PID controller tuning using Ziegler-Nichols method for speed control of DC motor. Int. J. Sci. Eng. Technol. Res. **03**(13), 2924–2929 (2014)
10. Åström, K.J., Hägglund, T.: PID Controllers: Theory, Design and Tuning. Instrument Society of America, USA (1995)
11. Malekabadi, M., Haghparast, M., Nasiri, F.: Air condition's PID controller fine-tuning using artificial neural networks and genetic algorithms. Computers **7**, 32 (2018). https://doi.org/10.3390/computers7020032 (Published: 21 May 2018)
12. Kumar, R., Vardhan, H., Bharadwaj, S.: Temperature control system using artificial neural network. Res. Appl. (IJERA) **3**(4), 672–675 (2013). ISSN: 2248-9622, www.ijera.com
13. Cheon, K., Kim, J., Hamadache, M., Lee, D.: On replacing PID controller with deep learning controller for DC motor system. J. Autom. Control Eng. **3**(6) (2015)
14. Prasad, V., Bequette, B.W.: Nonlinear system identification and model reduction using artificial neural networks. Comput. Chem. Eng. (2003)
15. Mantri, G., Kulkarni, N.R.: Design and optimization of PID controller using genetic algorithm. Int. J. Res. Eng. Technol. **02**(06) (2013)
16. Mirzal, A., Yoshii, S., Furukawa, F.: PID parameters optimization by using genetic algorithm. Graduate School of Information Science and Technology Hokkaido University Sapporo, Japan

17. Bagis, A.: Determination of the PID controller parameters by modified genetic algorithm for improved performance. J. Inf. Sci. Eng. **23**, 1469–1480 (2007)
18. Vincent, A.K., Nersisson, R.: Particle swarm optimization based PID controller tuning for level control of two tank system. IOP Conf. Ser.: Mater. Sci. Eng. **263**, 052001 (2017). https://doi.org/10.1088/1757-899x/263/5/052001, 14th ICSET-2017
19. Solihin, M.I., Tack, L.F., Kean, M.L.: Tuning of PID controller using particle swarm optimization (PSO). In: Proceeding of the International Conference on Advanced Science, Engineering and Information Technology (2011)

Chapter 76
Exergoeconomic and Enviroeconomic Analyses of Double Slope Solar Desalination Unit Loaded With/Without Nanofluid: A Comparative Study

Desh Bandhu Singh, Gagan Bansal, J. K. Yadav, Navneet Kumar and Manohar Singh

Abstract The analysis of basin-type double slope desalination unit based on solar energy is the demand of time as there is an acute shortage of drinking water throughout the globe particularly developing and underdeveloped countries. Only less than 1% of water is within the reach of human beings and this small amount of water is also getting polluted day by day due to huge population growth, industrial revolution, and many other anthropogenic activities. The technology of purifying water is known to human beings from sixteenth century as per the history; however, this technology has not been popular till date due to low output, high initial investment, and non-availability of technician at local level. The exergy analysis locates the losses taking place at different components of the system and hence, it provides a way to minimize the exergy loss occurring at different points of the system. Exergoeconomic analysis relates the exergy with uniform end-of-year annual cost and it provides a way to improve the performance of the system. On the other hand, enviroeconomic analysis controls the pollution of the environment by giving motivation for reducing pollutants and encourages the use of renewable energy technology which does not create pollution. This paper deals with the review of basin-type double slope solar still loaded with/without nanofluid on the basis of exergoeconomics and enviroeconomic parameters. It has been observed that the use of water-based nanofluid enhances the performance of double slope solar desalination unit on the basis of exergoeconomics and enviroeconomic parameters. At last, various types of double slope passive and active systems have been compared and discussion has been provided followed by the recommendation.

Keywords Exergoeconomic parameter · Enviroeconomic parameter · Double slope solar desalination unit · Comparative study

---

D. B. Singh · G. Bansal
Mechanical Engineering Department, Graphic Era (Deemed to be University), Dehradun, Uttarakhand, India

J. K. Yadav (✉) · N. Kumar · M. Singh
Galgotias College of Engineering and Technology, Greater Noida, UP 201306, India

S. Yadav et al. (eds.), *Proceedings of International Conference in Mechanical and Energy Technology*, Smart Innovation, Systems and Technologies 174,
https://doi.org/10.1007/978-981-15-2647-3_76

76.1 Introduction

Solar distiller unit is based on solar energy technology with an objective of purifying polluted water and the fresh water so obtained can be used for varieties of application including drinking. The solar distiller unit consists of a box-type structure made up of wood/concrete/glass reinforced plastic/fiber reinforced plastic/other locally available materials, and covered by condensing surface generally made up of glass. Glass is used for condensing surface because of its unique characteristic of allowing short wavelength of radiation to pass through it and prevent the longer wavelength radiation. This characteristic helps in the development of green house effect which is working principle of solar distiller unit. The condensing surface is inclined slightly to facilitate the condensed water for trickling down. If inclination is provided in two directions, the system is called double slope solar distiller unit. It can be either passive type or active type. Passive solar distiller unit takes heat directly from sun, whereas active solar distiller unit takes heat directly from sun falling on the condensing surface as well as from external source like collector. The present review paper summarizes the performance of double slope solar distiller unit on the basis of exergoeconomic and enviroeconomic parameters. The exergoeconomic parameter can be defined as either exergy gain per unit uniform end-of-year annual cost with an aim to maximize or exergy loss per unit uniform end-of-year annual cost with an aim of minimization.

76.2 Passive Double Slope Solar Still (DSSS) Loaded With/Without Nanofluid

The double slope solar desalination unit works on green house effect and it takes heat directly from the sun. No external source is provided to add heat to the basin. The depth of water in the basin is generally low as working temperature is low due to comparatively lower heat gain from solar radiation. The hourly production of fresh water decreases with the increase in the value of water depth in the basin. Different researchers have tried to increase the output of passive type solar desalination unit by proposing various designs. Some researchers have even used nanoparticles and enhancement in the yield was obtained. However, result is not too much appreciable. The output of this type of distiller is still low and needs to be improved for making this type of system for purifying dirty water popular as well as competitive. Singh et al. [1] have made a drastic evolution in the field of characterization of solar energy. They also gave the comparative cost analysis of passive double slope solar desalination system. Also the exergoeconomic and enviroecnmic analyses based on energy as well exergy were made. The energy cost based on exergoeconomic parameters were found as discussed in Table 76.2, and also revealed the outputs of various other parameters. Time dependent analysis also played the effective role in parametric study of the passive DSSS. The monthly energy/exergy gain, hourly deviation of thermal heat transfer, the thermal efficiency, thermal energy/exergy, variation of EPT, LCCE, disparity of exergoeconomic parameter, and annual productivity, etc., for DSSS were graphed for systematic analysis. In 2016, Sohata and Tiwari [2] showed that the Al_2O_3, H_2O-based nanofluid provide relatively recovering productivity of the passive DSSS. Some of the conclusions made related to experimental exergoeconomics and enviroeconomic parametric study say that total thermal energy and the productivity of hybrid set up system without heat exchangers (HE) (Type A) deliver better output characteristics than system with helically coiled HE (Type B) and also the conventional HE (Type C). [2]. The annual performance for the CuO–water-based type gives superior efficiency when used in Type A and Type B set up, whereas Al_2O_3–H_2O-based nanaofluid gives superior results with Type C. The obtained data also reflected that the EPBT, the EPF, and the LCCE matrices are improved by incorporating H_2O-based nanofluids in the hybrid systems. Though the review regarding the solar collectors with various similar characterizations was already discussed by various researchers but the attempt is made to restructure the updated trends. Further, Type A and Type B give higher values of CO_2 justification and environmental cost yearly by uniting the water-based nanofluids, whereas Type C gives higher values with ($Al_2O_{3)}$ water-based nanofluids for the DSSS. Sahota and Tiwari [3] highlighted the effect of nanofluids to evaluate the performance of DSSS using various descriptive characteristic curves. Numerous characteristics equations were derived, analyzed, and described for the passive DSSS. It is interesting to note that higher thermal energy efficiency was obtained for nanofluids (i.e., aluminum oxide—50.34%; Titanium oxide—46.10%; and copper oxide—43.81%) in contrast to basefluid (37.78%). By taking several similar assumptions, the overall thermal exergy was 4.92% for basefluid but it is obtained to be 14.1%, 12.38%, and 9.75% for Al_2O_3, $TiO_{2,}$ and CuO, respectively.

The energy matrices, the exergoeconomic analysis, and the enviroeconomic analyses for passive DSSS (using the set up as shown in Fig. 76.1) were also conducted by researchers Sohata et al., with water-based nanofluids [4]. The set up used metallic nanofluids for the characterization. The system incorporates aluminum oxide, titanium oxide, and copper oxide water-based nanofluids that enhance the annual productivity by 19.1%, 10.38%, and 5.25%, respectively; energy by 26.76%, 19.36%, and 12.96%, respectively, and exergy by 37.77%, 25.55%, and 11.99%, respectively, of the passive DSSS. The metallic oxides used as nanoparticles enhance the characteristics of the solar still. For the same constituents of water-based nanofluids in DSSS, the exergoeconomic and enviroeconomic parameters are analyzed for unlike interest rates, i.e., $I = 4\%$, 8%, and 10% and life span (maximum 50 years). The final results of various efficiency parameters in DSSS depend on the ambient temperature and the solar intensity, i.e., the constant environmental conditions. The optimum range of variations for DSSS is discussed and published by the researchers and academicians [5]. The schematic representation of passive DSSS is shown in Fig. 76.1. The figure shows the descriptive view of the DSSS loaded with various metallic NPs.

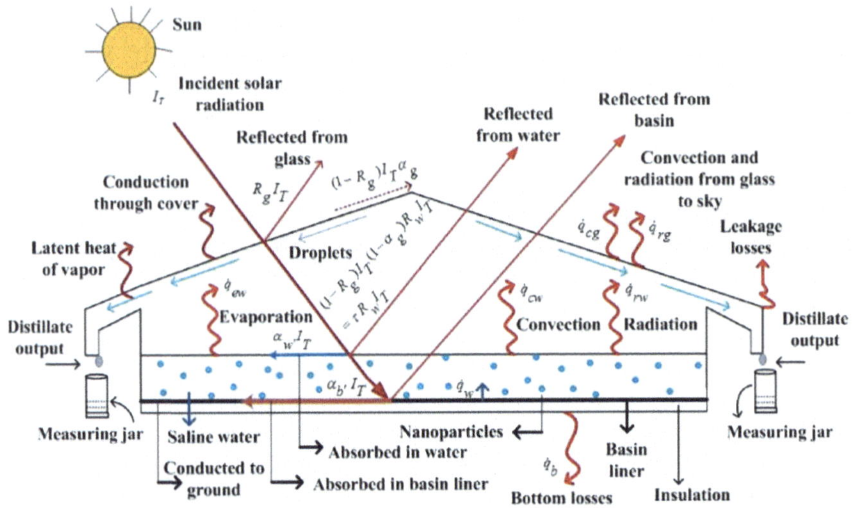

Fig. 76.1 Descriptive image of the passive DSSS loaded with metallic nanoparticles (NPs) [5]

76.3 Active Double Slope Solar Still Loaded With/Without Nanofluid

Ranjan and Kaushik [5] reviewed the active solar distiller installations and parametric study. The amalgamation of various research attempts related to solar distillation system before 2013 is done. The aim behind the review attempt was to identify the research gap in the current field and to optimize the productivity. The applications like steam turbine, gas turbines, boilers, and the cogeneration systems were related to the optimization path. At the efficient sunshine areas, comparing the integrated and conventional solar desalination systems, the technological applications for integrated system will be better choice [5]. Singh [6] systematically developed both the passive DSSS and the active DSSS set ups and installed various modified systems in order to characterize the optimality of passive as well as active DSSS with N-identical PV thermal panel for in-depth exergoeconomic and enviroeconomic analyses. The simulations were also identified for the developed set up. The change in constants, climatic conditions, time durations, etc., was also analyzed. The sum-up statements say that the exergoeconomic and enviroeconomic parameters show superior results for N-PVT-FPC-DS followed by N-PVT-CPC-DS and CDS. However, the productivity characteristic follows the reverse trend. Singh and Tiwari [7, 8] reported the performance of basin-type solar desalination unit coupled with concentrators.

The DSSS is used along with nanofluid for exergoeconomic and enviroeconomic characterizations using advance computational as well as experimental set ups. Sahota and Tiwari [9] attention on supporting and optimizing the nanoparticles and the basin fluid mass for DSSS with/without heat exchanger. The intention behind developing the variations of set ups was to get more and more viable results. Exergy of the given set up, overall thermal energy at different locations, cost analysis of the overall set up, and the productivity of the water-based nanofluids installations were optimized. The arrangements were further optimized using CO_2–H_2O-based nanofluids. Later the similar comparisons between the CSS and hybrid system are also highlighted. Sahota et al. in 2016 derived the analytical characteristics equations for passive DSSS with helical-coiled HE-loaded with several types of nanofluids [10]. Here, two different systems were considered for analysis. Type I—Active DSSS coupled with N-identical PVT Flat plate collectors that are connected parallel and in series (N-PVT-FPC) and operating without helical HE; and Type II—for active DSSS coupled with series connected partially covered (N-PVT-FPC) and operating with helical HE. Specific quantity of nanofluids, basin fluid, with 4 collectors and 0.03 kg/s of massflow rate was used for analysis. The instantaneous gain or we can say the loss in thermal energy efficiency, the productivity change, thermal energy and exergy, etc., parameters were analyzed for both Type I and Type II for 0.25% concentration of copper oxide, aluminum oxide, and titanium oxide metallic NPs separately [10]. The magnitudes of rise and fall of parameters are effectively published in the article for the active DSSS [10] Final conclusions were made for both Type I and Type II that says instantaneous gain and loss thermal energy efficiency, maximum change in temperature; HTCs (both convective as well as evaporative); thermal energy and exergy; thermal exergy efficiency; and productivity are found to be higher using CuO–water-based nanofluid follows the descending order as CuO (highest), followed by Al_2O_3, then TiO_2 and finally water. Overall performance for Type I is considered better in active DSSS [6]. Figure 76.2 shows the descriptive view and schematics representation of the active DSSS loaded with various nanoparticles. The diagram gives the clear cut image of the difference between active and passive DSSS set ups. Sharshir et al. [11] also enlighten the current research by providing

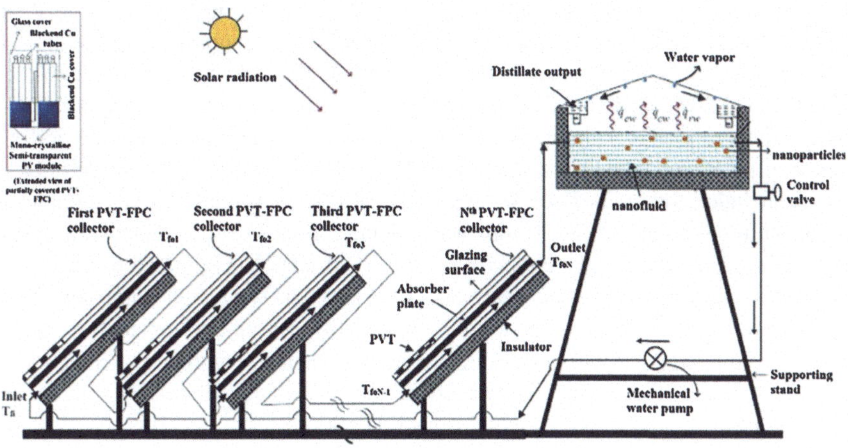

Fig. 76.2 Descriptive view of the active DSSS set up loaded with nanoparticles [9]

Table 76.1 Diurnal efficiency of various compositions [11]

Objective	Components	Percent
Diurnal efficiency in DSSS for various components	Graphite	4.32%
	CuO nanofluids	3.78%
	CSS	2.63%
Diurnal efficiency of CSS components	Basin	11.89%
	Glazier	17.66%
	Water	63.61%
Diurnal exergy efficiencies for components of the still with CuO nanofluid	Basin	13.40%
	Glazing cover	23.18%
	Saline water	80.83%
Diurnal exergy efficiencies for components of the still with graphite nanofluid	Basin liner	32.00%
	Glazing cover	23.76%
	Saline water	93.35%
Diurnal energy efficiencies for the still	Graphite	41.18%
	CuO nanofluids	38.61%
	CSS	29.17%
Cost of water production	MSS + CuO	0.20 RMB/L
	MSS + Graphite	0.21 RMB/L
		00.2 RMB/L

the comprehensive review of modified solar still (MSS) and conventional solar still (CSS) (Table 76.1).

Identification of different properties and applications in the field of exergoeconomic and enviroeconomic parametric analyses leads to development of large solar applications. Ibrahim et al. [12] also performed the exergoeconomic analysis for cost optimization. Outstanding outcomes were obtained related to cost optimizations.

Fathy et al. [13] used oil finned-serpentine loop heat exchanger coupled parabolic through collector (PTC) along with DSSS for analyzing and optimizing the performance of the set up using experimental approach for three different configurations. They concluded that solar collector with PTC has better productivity as compared to conventional solar still at 20 mm water depth. Also 28.1% higher overall productivity is detected in solar collector with tracked PTC as compared to fixed and 142.3% as compared to conventional solar collector [13]. The results outlay an effective research gap for the modification of the existing set ups. Many such research works are characterized with the similar set up but changed parameters. One noticeable point in the research by Fathy et al. is that the solar still performs better in summer for all kinds of experimental set up. Daily efficiency for Type A, Type B, and Type C set ups are 36.87%, 23.26, and 29.81%, respectively. Finally, based on the review of various research papers on double slope solar still for exergoeconomic and enviroeconomic analyses reveals that the properties of solar collector and overall efficiency are modified with the use of nanofluids, various photovoltaic plates/tubes, heat exchangers, collectors shape and dimensions, load capacity, etc. Also the improved results are observed with minor manipulations in the existing set ups. Recently, Singh [14] evaluated tubular solar collector with known specification integrated with double slope solar still to characterize the productivity, the exergoeconomic and the enviroeconomic parameters for four different set up cases keeping certain constants as discussed in cases 1–4. Four different setups cases with N-identical evaluated tubular collector integrated with DSSS are made for comparative evaluation. Case 1: N-(PVT)-(FPCs) with DS, Case 2: PVT-(CPCs) incorporated without DS, Case 3: Case 2 incorporated with DS and Case 4: conventional DSSS. Different cases being very different in cost estimation and annual productivity, therefore, Table 76.2 illustrates the comparative study of annual productivity and cost of production. Use of photovoltaic panel is in the trend for power development through solar still. The evaluation of 4Es, i.e., enviroeconomic, exergoeconomic, energy, and exergy is done on the new incorporation of solar still (SS) along with PV panels [15]. The best part in the research by Elbar et al. is that they installed the PV panel behind the double solar still which reduces the requirement of the extra area. That is the area required was the area which is engaged by the solar still. This innovative set up helps the PV module to work like the reflector. Various different set ups were examined, analyzed, and compared. Some of them were: the conventional solar still (CSS), the CSS with PV panel (CSS + PV), and the solar cell with PV and black still wood fiber inside basin, (CSS + PV + BSWF) and SS with PV as a reflector only (CSSS + PV as a reflector only). The manipulation in experimental set up increased the energy efficiency as identified and listed [1]. Various conclusions were made by Elbar et al. as the applications of PV with solar compared to CSS was found to give negative payback time. The enviroeconomic and exergoeconomic characterizations were satisfactory. But the use of PV as reflector along with CSS was termed as the best installation. Various concluding remarks were made from the comparison of four different systems of solar still to characterize energy, exergy, enviroeconomic, and exergoeconomic. Comparison of overall accumulative freshwater yields was made and CSS with PV and BSWF system showed almost 31.48% rise [15].

Energy efficiency and exergy efficiency also increase with inclusions of PV, BSWF, and PV as reflector and Elbar et al. revealed that use of PV as reflector only in CSS gives maximum efficiency [15]. EPBT results with energy and exergy parametric approach were also separately diagnosed [15]. The exergoeconomic parameters increase with lifetime of the system. The final and overall analyses revealed that using

Table 76.2 Comparative study of annual productivity and cost of production

	Reduction in cost of portable water (%) as compared to Case 1	Rise in annual productivity as compared to Case 1
Case 2	74.45	50.97
Case 3	136.17	55.24
Case 4	82.98	37.82

only PV as a reflector gives the maximum overall efficiency. The results worked as a research gap analysis for various researchers working in the similar field.

76.4 Conclusions

An effort has been made to review the effect of nanofluid on the performance of basin-type double slope solar desalination unit. From the current review work, the following conclusions have been made:

i. The output of passive double slope solar still loaded with/without nanofluid is not appreciable and need to be improved for making it popular as well as competitive.
ii. The demerits of low outputs of passive type double slope solar desalination unit can be overcome by active double slope solar desalination unit.

Recommendations

The exergoeconomic and enviroeconomic parameters of solar desalination unit loaded with nanofluid can further be improved by optimizing the mass flow rate, number of collectors, and water depth with the help of optimization techniques.

Acknowledgements Authors would like to thank Graphic Era deemed to be University and Galgotias College of Engineering and Technology for providing all sorts of amenities, support, and facilities to carry on the current work effectively and efficiently. Authors also acknowledge the hierarchical mentors for providing assistance whenever needed.

References

1. Singh, D.B., Tiwari, G.N., Al-Helal, I.M., Dwivedi, V.K., Yadav, J.K.: Effect of energy matrices on life cycle cost analysis of passive solar stills. Solar Energy **134**, 9–22 (2016)
2. Sahota, L., Tiwari, G.N.: Effect of Al_2O_3 nanoparticles on the performance of passive double slope solar still. Solar Energy **130**, 260–272 (2016)
3. Sahota, L., Tiwari, G.N.: Effect of nanofluids on the performance of passive double slope solar still: a comparative study using characteristic curve. Desalination **388**, 9–21 (2016)
4. Sahota, L., Shyam, Tiwari, G.N.: Energy matrices, enviroeconomic and exergoeconomic analysis of passive double slope solar still with water based nanofluids. Desalination **409**, 66–79 (2017)
5. Ranjan, K.R., Kaushik, S.C.: Energy, exergy and thermo-economic analysis of solar distillation systems: a review. Renew. Sustain. Energy Rev. **27**, 709–723 (2013)
6. Singh, D.B.: Exergoeconomic and enviroeconomic analyses of N identical photovoltaic thermal integrated double slope solar still. Int. J. Exergy **23**(4), 347–366 (2017)
7. Singh, D.B., Tiwari, G.N.: Exergoeconomic, enviroeconomic and productivity analyses of basin type solar stills by incorporating N identical PVT compound parabolic concentrator collectors: a comparative study. Energy Convers. Manag. **135**, 129–147 (2017)
8. Singh, D.B., Tiwari, G.N.: Performance analysis of basin type solar stills integrated with N identical photovoltaic thermal (PVT) compound parabolic concentrator (CPC) collectors: a comparative study. Sol. Energy **142**, 144–158 (2017)

9. Sahota, L., Tiwari, G.N.: Exergoeconomic and enviroeconomic analyses of hybrid double slope solar still loaded with nanofluids. Energy Convers. Manag. **148**, 413–430 (2017)
10. Sahota, L., Shyam, Tiwari, G.N.: Analytical characteristic equation of nanofluid loaded active double slope solar still coupled with helically coiled heat exchanger. Energy Convers. Manag. **135**, 308–326 (2017)
11. Sharshir, S.W., et al.: Energy and exergy analysis of solar stills with micro/nano particles: a comparative study. Energy Convers. Manag. **177**(July), 363–375 (2018)
12. Ibrahim, A.G.M., Rashad, A.M., Dincer, I.: Exergoeconomic analysis for cost optimization of a solar distillation system. Sol. Energy **151**, 22–32 (2017)
13. Fathy, M., Hassan, H., Salem Ahmed, M.: Experimental study on the effect of coupling parabolic trough collector with double slope solar still on its performance. Sol. Energy **163**(August 2017), 54–61 (2018)
14. Singh, D.B.: Exergo-economic, enviro-economic and productivity analyses of N identical evacuated tubular collectors integrated double slope solar still. Appl. Therm. Eng. **148**, 96–104 (2017)
15. Elbar, A.R.A., Yousef, M.S., Hassan, H.: Energy, exergy, exergoeconomic and enviroeconomic (4E) evaluation of a new integration of solar still with photovoltaic panel. J. Clean. Prod. **233**, 665–680 (2019)

Chapter 77
A Study on Coal Ash Slurry Flow at Higher Solid Concentrations in Pipeline

Navneet Kumar, Kanwar Pal Singh, V. K. Dwivedi, J. K. Yadav, Sudhir Kumar and Navin Kumar

Abstract In India, thermal energy accounts for more than 70% of electricity production and millions of tons of coal are burned in these thermal power plants. Thus, large quantities of coal ash (fly ash and bottom ash) are produced and the current level of production is about 120 million tons per year. Out of this, approximately 20% is bottom ash and the rest 80% is fly ash. The ash produced in India usually has higher specific gravity as Indian coal has a much higher content of non-combustible matter. Also, the majority of the thermal power plant in India disposes of both the materials, namely fly ash and bottom ash to ash ponds using the same pipeline. The knowledge of slurry rheology is very vital for the design of a slurry pipeline particularly for the dense phase conveying system. Since the pilot plant loop tests at these concentrations are tedious, time-consuming, and complex in nature, the slurry pipeline designers have been adopting the empirical approach for slurry pipeline design based on the rheological model of the slurry. From the vast study of literature, an attempt has been made to highlight the various influencing parameters like concentration of solid, rheological properties, and chemical additives that affect the flow of coal ash slurry in the long-distance pipelines.

Keywords Coal ash · Slurry · Rheological behavior · Shear rate · Shear stress

77.1 Introduction

A slurry pipeline system is used for conveying solid particles using fluid which is generally water as a carrier. Slurry pipeline transportation has been one of the progressive technologies for conveying a large number of materials over long distances [1–3]. The mineral ores in mining and process industries, coal ash in thermal power

N. Kumar (✉) · J. K. Yadav
Galgotias College of Engineering & Technology, Greater Noida 201306, Uttar Pradesh, India

K. P. Singh · S. Kumar · N. Kumar
Greater Noida Institute of Technology, Greater Noida 201306, Uttar Pradesh, India

V. K. Dwivedi
Madan Mohan Malaviya University of Technology, Gorakhpur, Uttar Pradesh, India

S. Yadav et al. (eds.), *Proceedings of International Conference in Mechanical and Energy Technology*, Smart Innovation, Systems and Technologies 174,
https://doi.org/10.1007/978-981-15-2647-3_77

plants disposal of waste materials like tailing and waste materials are a few examples that can be mentioned. These materials need to be stored or transported in industries for processing purposes or disposal to nearby locations. Compared to the conventional mode of transport, slurry transportation has been established as an economic and ecological method. The design of the slurry pipeline system depends on many parameters. At present, slurry pipelines are mostly designed on the basis of the laboratory investigations and semi-empirical correlations. The operational and design factors of the slurry pipelines are still the topic of investigation. This is due to the fact that the design and operation of such pipeline depend on large numbers of parameters. Thus, the pipeline designer's ultimate objective is to develop a methodology that will enable him to optimize all the relevant variables required for hydraulic design on the basis of data from bench-scale tests as well as pilot plant loop tests.

The fly ash produced is a very fine powder, whereas bottom ash is coarse. About a decade ago, utilization of power generation waste which is primarily coal ash was one of the major problems of thermal power plants, especially for solid-fuel power plants. At present, it is possible to utilize the coal ash for road embankments, cement bricks, etc. Efforts are being made to utilize all the ash being produced in power plants and it is estimated that in our country about 35% of the production of ash is being utilized at present. The remaining 65% still needs to be transported and stored safely for subsequent consumption. At present, these materials are being transported by slurry pipelines at low concentration (10–20% by weight). This leads to very high power consumption excessive erosion wear due to higher transportation velocity and the existence of a highly skewed concentration profile making the system very uneconomical. Today, many industries are adopting this mode of transport for both short and long distance transportation because it offers many advantages such as reliability, reduced availability, year-round availability, access to a remote location, and automation possibilities. Additional features of these pipelines, which make them more attractive as a means of transportation, include traffic congestion, air, and noise pollution, and accidents. The long-distance transport of solids by pipelines is generally more economical than the conventional mode of transport [4, 5].

77.2 Literature Review

In order to have some insight into the present state of knowledge, an extensive literature survey has been conducted and reported in this chapter. An attempt has also been made to highlight the gaps in the knowledge regarding the slurry pipelines which in turn useful for deciding the scope of the current study.

77.2.1 Slurry Pipeline Flow and Effecting Parameters

Slurry flow is a complex flow of a continuous liquid that transports the dispersed solid in suspension and is transported by the drag forces of the liquid acting on the particles. It depends on the properties of the solid particles in relation to the properties of the liquids. It influences the dependence of the pressure drop as a function of the flow rate and the magnitude of the pressure drop, as well as other performance characteristics. Knowing the different regimes of slurry flow is very important to understand the behavior of slurry. The rheological behavior of the slurry is mainly expressed by (apparent, plastic, zero, or infinite) viscosity and (shear or yield) stress. These behaviors are significantly influenced by various physical and chemical properties of the slurry samples [6]. When solid particles are added to the transport liquid by increasing the solids concentration, the viscosity of the suspension is increased because of physical interaction in the particles during the dispersion of solid particles in the liquid. Rutgers [7] observed that viscosity of the samples increased linearly at low solid concentrations and after a certain value of solid concentration, the viscosity increases marginally at a very high rate. This is because at high concentration, the inter-particle distance becomes smaller and increases the attractive potential and collisions between the particles. Rutgers concluded that the effect of hydrodynamic interactions dominates at low to medium solids concentration. According to Cheng [8], the particle frictional contact dominates for solid concentration ranging from medium to high and at very high solids concentration the particle effect dominates the hydrodynamic effects. The other key parameters affecting the rheological properties of the slurry are particle size and distribution. According to He et al. [9], it is because for slurries with finer particles the surface area per unit mass increases providing greater inter-particle attractions than for the coarser ones at a fixed solids concentration. Working temperature is also influencing parameter which affects the rheological properties of slurry [10]. Studies carried out in the last phase of the previous century with other materials have shown that at higher concentrations slurry follows Bingham model with non-Newtonian behavior [11]. For the case of slurry flow at very high concentration, the flow regime becomes homogeneous and the fluid becomes highly non-Newtonian. Hence, it is possible to operate the pipeline at velocities in the laminar flow regime. High-concentration slurry transportation (50% by weight) reduces the skewness in concentration profile. Reduction in the skewness of concentration profile and velocity is expected to reduce the erosion wear of pipeline and hence increase the life span of the pipeline [12, 13].

77.2.2 Effect of Rheological Parameters on Coal Ash Slurry Flow

The study of the flow of solid particles and their deformation under applied forces is known as rheology and it is measured regularly using a rheometer. Considering the flow characteristics of the suspension dependents upon the shear stress and shear rate relationship, it is very essential to have complete knowledge of the rheological properties of slurry samples particularly for the dense phase slurry conveying system for the design of the slurry pipeline.

77.2.3 Rheological Mathematical Models

The shear stress and the viscosity of the slurry are usually convenient to mathematically express as a function of shear rate. The rheological behavior of the slurry is defined by various models as listed in Table 77.1. Several investigations have studied the rheology of suspensions of different materials covering almost all the rheological aspects. The rheological property of slurry depends on working temperature, solids concentrations, particle size, and shape [14]. Kolar [15] has studied the rheological properties of suspensions to establish the effect of polymer additive (polyacrylamide) on viscosity. He has suggested that the visco-elastic properties of the liquid phase strongly influence the flow of suspensions. Concentration of additive and solid particles, pipe diameter, and material of pipe walls and roughness of their surface are the other factors that were found to influence the pressure drop. Horsely [16] has investigated the rheological properties of a gold slime using a Hakke rotary viscometer and 40 mm diameter test loop. Rheograms have been obtained for 75% concentration, (by weight) with sodium phosphate as an additive. The results show that visco-metric data can be used to predict pressure gradients in slurry pipelines as long as the concentration is kept high enough. Kundu and Peterson [17] have highlighted the use of a new transport medium, which requires at most, half the amount of water. Due to superior particle suspension ability, stability, and visibility of the foam, it has a much higher solid carrying capacity than water. The other advantage that has been highlighted is in the reduction or almost elimination of some of the processes required for the slurry preparation and drying of transported materials.

Table 77.1 Rheological models of fluids

S. No.	Model name	Equation
1	Hershal–Bulkley [18]	$\tau = \tau_h + K_h\dot{\gamma}^p$
2	Sisko [19]	$\tau = \eta_s\dot{\gamma} + K_s\dot{\gamma}^m$
3	Casson [20]	$\tau^{1/2} = \tau_c^{1/2} + \eta_c\dot{\gamma}^{1/2}$
4	Bingham plastic	$\tau = \tau_b + \eta_b\dot{\gamma}$
5	Power-law	$\tau = K_p\dot{\gamma}^n$

77.2.4 Effect of Chemical Additives on Slurry Flow Parameters

Kaji et al. [21] have studied the effect of two different types of dispersants, ionic and non-ionic, on highly concentrated coal-water mixtures, and have thrown light on the mechanisms of the particle dispersion in both mixtures. Walmsley and Duffy [22] have highlighted the use of flocculated fiber at low concentrations to support coarse and fine coal particles in suspension as well as to improve the slurry flow characteristics by reducing the friction toss (by 30%). The flocculated fiber also acts as a drag-reducing agent. Thomas and Sobata [23] have highlighted the use of additives, which could be any other waste being added to the fly ash slurry disposal system, allowing the simultaneous solution to two disposal problems at much lower pressure drop. Seshadri et al. [24] and Verma et al. [25] have determined the rheological characteristics of fly ash slurry by performing experiments over five samples of fly ash slurry of various particle size ranges and concentration with and without additive. They found that deflocculating additives like sodium hexametaphosphate, when introduced in the fly ash slurry, modified the rheological properties of the slurry and thus resulted in a substantial decrease in total pressure loss.

77.3 Conclusion

On the basis of an extensive literature survey, it is seen that there is still a significant gap in knowledge on the selection of optimized parameters for hydraulic design of the slurry pipeline, especially at high concentrations for fine slurries Indian fly ash is one of them. It is well known that coal ash (FA and BA) produced in various thermal power plants across India vary widely in their physical and chemical properties due to the variation in the combustion efficiency and quality of the coal used. Literature reveals that different rheological models proposed are not universal due to the complex dependence of rheological properties on several parameters namely particle size distribution, concentration, and other chemical properties. The existing flow models are not applicable to all slurries at high concentrations. This shortcoming has motivated the author to study the rheology of coal ash slurries at high concentrations. Studies are also undertaken to achieve the reduction in the values of rheological parameters by using additives and to investigate their impact on slurry pipeline design.

Acknowledgements Support under the Collaborative Research and Innovation Program (CRIP) funding through TEQIP-III of Dr. A. P. J. Abdul Kalam Technical University Uttar Pradesh Lucknow (Ref. No. AKTU/Dean-PGSR/2019/CRIP/21) is gratefully acknowledged.

References

1. Abulnaga, B.E.: Slurry systems handbook. McGraw-Hill, New York (2002)
2. Baker, P.J., Barry, E.A.J., Bonnington, S.T.: A Guide to Slurry Pipeline Systems. British Hydromechanics Research Association, Fluid Engineering (1979)
3. Brown, N.P., Heywood, N.I. (eds.): Slurry Handling: Design of Solid-Liquid Systems. Springer, Science & Business Media (1991)
4. Link, J.M., Faddick, R.R., Lavingia, N.J.: Slurry pipeline economics. In: Society of Mining Engineers. AIME Annual Meeting, Dallas, Texas (1974)
5. Seshadri, V., Malhotra, R.C., Anand, S.: Hydraulic transportation of mineral ores, a state of the art report, Internal Report, Fluid Mechanics Laboratory, I.I.T. Delhi (1975)
6. He, M., Wang, Y., Forssberg, E.: Slurry rheology in wet ultrafine grinding of industrial minerals: a review. Powder Technol. **147**(1–3), 94–112 (2004)
7. Rutgers, I.R.: Relative viscosity and concentration. Rheol. Acta **2**(4), 305–348 (1962)
8. Cheng, D.C.H.: Viscosity-concentration equations and flow curves for suspensions. Chem. Ind. **10**, 403–406 (1980)
9. He, M., Wang, Y., Forssberg, E.: Parameter effects on wet ultrafine grinding of limestone through slurry rheology in a stirred media mill. Powder Technol. **161**(1), 10–21 (2006)
10. Shenoy, A.V.: Drag reduction with surfactants at elevated temperatures. Rheol. Acta **15**(11–12), 658–664 (1976)
11. Bunn, T.F., Chambers, A.J.: Experiences with dense phase hydraulic conveying of vales point fly ash. In: 4th International Proceedings on Bulk Materials, Storage, Handling and Transportation, Institution of Engineers, Australia (1992)
12. Seshadri, V.: Basic process design. In: Proceedings of Workshop on Hydraulic Transportation of Mineral Ores, IIT, Delhi (1979)
13. Seshadri, V.: Basic process design for a slurry pipeline. In: Proceedings of the Short Term Course on Design of Pipelines for Transporting Liquid and Solid Materials, IIT, Delhi (1982)
14. Clarke, B.: Rheology of coarse settling suspensions. Trans. Inst. Chem. Eng. **45**, T251–T256 (1967)
15. Kolar, V.: Visco-elastic fluid solid flow in pipes. In: Proceedings of Hydrotransport, vol. 1, pp. 225–242. British Hydromechanics Research Association, UK (1970)
16. Horsley, R.R.: Viscometer and pipe loop tests on gold slime slurry at very high concentrations by weight, with and without additives. In: Proceedings of Hydrotransport, vol. 8 (1982)
17. Kundu, N., Peterson, G.P.: Transmission of solid particulates using a two-phase medium. J. Energy Res. Technol. **109**(1), 35–39 (1987)
18. Herschel, W.H., Bulkley, R.: Measurement of consistency as applied to rubber-benzene solutions. Am. Soc. Test Proc. **26**(2), 621–633 (1926)
19. Sisko, A.W.: The flow of lubricating greases. Ind. Eng. Chem. **50**(12), 1789–1792 (1958)
20. Casson, N.: A flow equation for pigment-oil suspensions of the printing ink type. Rheol. Disperse Syst. (1959)
21. Kaji, R., Muranaka, Y., Otsuka, K., Hishinuma, Y.: Effect of ions on the rheology of CWM. In: Proceedings of 7th International Symposium on Coal Slurry Fuels Preparation and Utilization, pp. 16–23 (1985)
22. Walmsley, M.R.W., Duffy, G.G.: Reducing pipeline pressure gradients of coarse-coal slurries with wood pulp fibres. In: Proceeding of Hydrotransport, vol. 11, pp. 295–308 (1988)
23. Thomas, A.D., Sobota, J.: Influence of additives on energy loss in pipeline flow of fly ash mixtures. In: Proceedings of Hydrotransport, vol. 15, pp. 329–343 (2002)
24. Seshadri, V.: Rheology of fly ash slurries at high concentration and its application to the design of high concentration slurry disposal system (HCSD). In: Proceedings of Fly Ash India, pp. 1–10, New Delhi (2005)
25. Verma, A.K.: Effect of additive on head loss in the high concentration slurry disposal of fly ash. J. Inst. Eng. (India) **89**, 3–10 (2008)

Author Index

S. Yadav et al. (eds.), *Proceedings of International Conference in Mechanical and Energy Technology*, Smart Innovation, Systems and Technologies 174,
https://doi.org/10.1007/978-981-15-2647-3

The manufacturer's authorised representative in the EU is Springer Nature Customer Service Centre GmbH, Europaplatz 3, 69115 Heidelberg, Germany. If you have any concerns regarding our products, please contact ProductSafety@springernature.com

Printed and bound by CPI Group (UK) Ltd, Croydon, CR0 4YY
30/09/2026
02225557-0018